何金海论文选

何金海 等 著

图书在版编目(CIP)数据

何金海论文选 / 何金海等著. — 北京：气象出版社，2020.10
 ISBN 978-7-5029-7281-3

Ⅰ.①何… Ⅱ.①何… Ⅲ.①东亚季风-文集 Ⅳ.①P425.4-53

中国版本图书馆 CIP 数据核字（2020）第 177331 号

何金海论文选
He Jinhai Lunwenxuan

出版发行：气象出版社			
地　　址：北京市海淀区中关村南大街 46 号		邮政编码：100081	
电　　话：010-68407112（总编室）　010-68408042（发行部）			
网　　址：http://www.qxcbs.com		E-mail：qxcbs@cma.gov.cn	
责任编辑：黄海燕		终　　审：吴晓鹏	
责任校对：张硕杰		责任技编：赵相宁	
封面设计：博雅锦			
印　　刷：北京建宏印刷有限公司			
开　　本：787 mm×1092 mm　1/16		印　　张：41.25	
字　　数：1056 千字		彩　　插：8	
版　　次：2020 年 10 月第 1 版		印　　次：2020 年 10 月第 1 次印刷	
定　　价：280.00 元			

本书如存在文字不清、漏印以及缺页、倒页、脱页等，请与本社发行部联系调换。

何金海教授

▲ 2011年，荣获江苏省科学技术奖一等奖

▲ 1996年，庆祝恩师叶笃正先生八十华诞之际，与恩师夫妇及全体同门合影

◀ 20世纪80年代中期，接待来学校访问的恩师叶笃正先生

▶ 2013年11月，参加在南京召开的"陶诗言先生学术思想专题报告会"，纪念陶诗言先生逝世一周年

▲ 1983年10月，访问马里兰大学

▲ 1983年，与夏威夷大学访问学者合影

▲ 1983年10月，访问马里兰大学时在白宫前留影

▲ 1985年，为来访的夏威夷大学 Murakami Takio 教授做翻译

▲ 1985年，在家宴请来校访问的夏威夷大学 Murakami Takio 教授

▲ 20世纪90年代，赴日参加中日季风合作科研工作

▲ 1995年，"东亚季风研究"成果获国家自然科学奖二等奖，项目主要完成人合影

▲ 1998年4月，参加1998年南海季风爆发日期研讨会

▲ 1989年，在西安参加优秀青年气象科技工作者会议

▲ 1993年，参加第五次中美季风学术研讨会

◀ 2004年，参加杭州季风会议，与部分专家合影

▲ 2008年5月，参加在北京召开的东亚副热带季风研讨会，并作主题报告

▲ 2007年7月，作为发起者和副主任委员，参加在上海召开的中国气象学会副热带气象委员会成立大会

▲ 2016年，参加梅雨监测指标专家论证会

▲ 2017年，在郑州参加第34届中国气象学会副热带气象委员会成立10周年大会

▲ 2018年11月，参加在合肥举办的第35届中国气象学会副热带气象委员会年会

▲ 2012年，作为"973"项目课题组长，在课题启动会上作报告

▲ 2018年，作为科学咨询委员会委员，指导广州热带海洋气象研究所工作

◀ 2019年6月，参加南京信息工程大学ICAR会议并与世界大气海洋顶尖科学家Yamagata交流

◀ 2019年，参加海峡两岸民生气象论坛

◀ 1997年7月，培养的第一届博士研究生毕业

▶ 与南京气象学院大气物理系第一届校友及老系主任王鹏飞先生合影

▲ 2019年，六十周年校庆活动启动会上作为学校代表点亮倒计时球

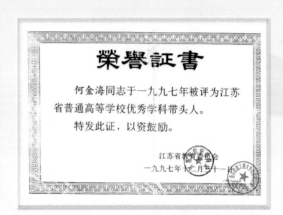

序 一

独具慧眼识天气,高瞻远瞩观季风
——祝贺何金海教授八十寿辰

何金海老师是南京信息工程大学(原南京气象学院)大气科学学院教授。作为该校第一届毕业生,于1965年毕业后至中央气象局气象科学研究所(即现在的中国气象科学研究院)攻读研究生,师从叶笃正先生。1975年8月,何金海老师回到南京信息工程大学工作,至今已在该校工作45年(在职35年)。在过去长达六十年的气象生涯中,他为学校的建设和发展做出了重大贡献。如今他已是桃李满天下,在大气科学界发表的著作享誉国内外,是南京信息工程大学名师中的代表人物。

我与何金海教授的相识和相知主要源于三次季风合作:20世纪80年代的中美季风合作,90年代的中日季风合作和第一次南海季风试验(SCSMEX,1995—2000年)。他为这三个科学计划贡献了重要力量。一个突出的例子是他指导他的学生们在南海季风试验前,专门编写和评述了南海和东亚夏季风指标,这为南海季风试验中各国季风观测和研究确立了共同使用的参考标准。这些具有标准型的指标有些至今仍被沿用。他在20世纪80年代中美合作中,与夏威夷大学的村上多喜雄(Murakami Takio)教授一起研究了季风中的低频振荡,这是我读到的最早论述这个问题的文章之一。他在中日季风合作中精心编撰了一本文集,历史性地记载了两国科学家在东亚季风研究中所做出的重要成果,受到了后来学者的普遍赞誉,其中就包括东亚季风系统的定义、组成和影响。他在季风研究中的重要成果获得了1995年国家自然科学二等奖,是主要获奖人之一。

何金海教授对中国天气学和大气科学也做出了重要贡献。他在中国天气学的研究中最突出的特点是能从复杂的天气学事实中提炼出规律性和本质的东西,使人一看就懂,一听就明白。近年来他多次与我谈到中国东亚季风的季节变化会导致旱涝急转,这是季节气候预测的难点,又是关键点。这个观点和研究成果目前已为广大预报员所接受,并在我国的气候预测中成为一个重要的预报指标。

近年来,他以更广阔的视野致力于大气科学学科的建设,不但带领团队编写了《大气科学概论》一书,而且还翻译了国际享名的《大气科学》(第二版)一书。这是由美国华盛顿大学华莱士(美国科学院院士)和霍布斯教授编写的名著。该书自1972年编写出版后已有近50年历史,几乎是各国院校大气专业课程使用的经典教材。何教授及其同事们以很大的精力完成了这本名著的翻译与出版(科学出版社,2008年),这不但是我们大气科学界的幸事,且作为大气科学的一个范本,也对我国地学界的通识教育有很大帮助。尤其可贵的是他还为学生们开设"大气科学概论"课,并带领团队编写了中国版的《大气科学概论》。他讲课生动,引人入胜,深受学生欢迎。

在我与何金海教授交往的四十多年间，感到他的为人和处事有很强的包容性，凡事力争以团结为上，促进合作，争取把事情共同办好。他对学生热心扶植，真心真意，尽心帮助学生，提携后辈，因而他深受师生们的爱戴。他也曾热情推荐一些优秀学生攻读我的研究生，经过我们的共同执教，现在他们已事业有成，在国内外工作中有出色的表现与成就。

何金海教授是中国天气学的著名专家，是东亚季风研究的领军人物，是热心关怀和扶植后辈的名师，是具有科学合作和包容精神的典范。时值何金海教授八十华诞之际，《何金海论文选》正式出版，这既是他六十年寒耕暑耘的辛勤收获，也将是广大气象科技工作者喜闻乐见的重要文献。

春秋不老，日月长明。我衷心祝愿他生日快乐，健康长寿！

中国工程院院士 丁一汇

2020 年 1 月 4 日

序 二

探索东亚季风的奥秘和培养大气科学人才是何金海教授恒久的追求。

何金海教授是国际著名季风专家,他投身气象事业六十年来,致力于东亚季风的科学研究,在东亚夏季风建立、季节进程、年际和年代际变异的多尺度特征和变化机制等方面取得了系统性创新成果,是国际季风研究领域有广泛影响的学术带头人之一。他的研究成果推进了东亚季风基本理论框架的建立,深化了对东亚热带季风本质与变化规律的认识,为我国季风研究跻身世界大气科学前沿作出了重要贡献。

何金海教授和合作者发现,东亚副热带夏季风的建立独立并早于南海夏季风,由此提出我国汛期应以东亚副热带季风及其雨带的建立为标志,汛期会商时间应由原来的每年4月份提早到3月份;发现了东亚季风雨带推进过程存在"旱涝并存、旱涝急转"的重要现象并揭示了其成因;突破"梅雨天气"的传统认识,创新性地提出了"梅雨季节"这一气候概念,……这些创新性科研成果对季风研究的发展具有重要意义,为我国旱涝预测提供了新的科学依据,由此发展的预测技术和方法能够显著提高预测水平。

何金海教授在我的学术生涯中,不仅是我的忘年交,也是我的知心友。除了在学术上向他学到了不少东西外,我深深感觉到他在培养学生中所倾注的热情和爱。他始终用心去关爱学生,启发学生对知识的渴求,鼓励学生质疑解难。他在大气科学人才培养事业中硕果累累,培养了一百多名硕士、博士研究生,他们工作在世界各地,有很多已经成长为国际知名学者、教学名师、国家或省级首席专家,可谓桃李满天下!这部《何金海论文选》,是其研究成果的荟萃,是其毕生奉献的精华。

祝何金海教授福如东海,寿比南山,身体健康,永远快乐!

中国科学院院士

2020年4月于复旦大学

贺　词

　　庚子新春，偶闻何金海教授八十华诞之事，惊喜！感叹！时光飞逝，岁月留下了何教授多少奋斗的足迹、丰满的收获和同仁们赞美的词句啊！何教授，八十华诞快乐！

　　认识何教授始于 20 世纪 90 年代初，中国科学院大气物理研究所组织的多个国际、国内学术研讨会都能见到他的身影，睿智、诙谐、和气是他的符号。自那段时间之后，见到何教授的次数就更多了。后来，我受命担任中国科学院大气物理研究所所长，与何教授共事很多，受益匪浅。对大气物理研究所和我个人的工作，他始终是积极而认真的，往返京宁之间，甚是频繁，其辛苦可想而知。更重要的是，他总会对一些问题提出独到的见解，贡献独到的智慧和力量。

　　生活之中，何教授经常和年轻一辈人打成一片，是很多学生和年轻学人的良师益友，这缘于他开朗、乐观和积极的人生态度和性格特质。在学术和育人两个方面，他的奋斗和成就令人瞩目。如今何教授"桃李满天下"，更愿此文选"翰墨传四方"。

　　当此之时，我愿送上对何金海教授八十华诞的衷心祝福！庆祝他半个多世纪的辛勤耕耘、光辉业绩，并祝愿他活力永驻，在科坛和讲坛再创辉煌。祝愿何金海教授健康长寿！

中国科学院院士　王会军

2020 年 3 月 4 日于南京

前　言

何金海教授是大气科学界著名专家，是东亚季风研究领域的领军人物。先生在东亚季风变异规律和机理、天气气候预测、大气季节内振荡领域取得了众多具有重要影响力的创新性科研成果，为提升东亚季风的科学认知、提高中国旱涝预测水平做出了卓越的贡献。为表达敬仰和对先生八十寿辰暨从事大气科学研究事业六十周年的庆贺，我们将先生在学术生涯不同时期、针对东亚季风多个侧面研究的代表性论文结集成册，以便后学能更好地领会学习先生在季风研究领域的真知灼见及其不断进取的科学实践，从中受益并传承发扬其学术精神。作为学生，回忆起我与先生的过往，总能被先生博学而精专、求真而务实的科研精神所打动，为先生严于律己、豁达包容、积极向上、坚韧执着的处事风范所震撼，更为先生诲人不倦、爱如己出、奖掖后学、春风化雨的育人之道而动容。

先生献身大气科学事业60年，合作撰写或主编了具有重要学术价值的专著8部、译著1部、教材1部，论文400余篇。60年来，先生一直奋战在科研一线，推动并直接参与了中美季风合作、中日季风合作、南海季风试验等重大科学试验。同时，先生将毕生所学融汇到日常教学之中，一直耕耘在三尺讲台，主讲本科新生专业课"大气科学概论""天气学原理和方法"和研究生课程"大气科学进展""东亚季风"等。60年来，先生培养出了多名大气科学界国际知名科学家和一大批国内外优秀气象业务骨干，其中包括加拿大气象与海洋学会总统奖获得者加拿大环境部林海研究员、日本文部省青年科学家奖获得者罗京佳教授、纽约州立大学奥尔巴尼分校周黎明教授，高等院校及科研院所知名教授、研究员、优秀青年英才以及业务部门国家级首席等。

《何金海论文选》（以下简称《论文选》）于2018年8月开始选编，于2020年4月交付出版社。先生亲自为本书挑选了最能代表其学术观点的49篇论文，大致涵盖了他各个时期的重要学术观点。《论文选》所摘录的论文时间跨度为1983—2016年，以先生为第一作者的为主，有两篇为合作论文，也选刊了部分先生直接指导和参与研究的论文，其中第一作者均为他所指导的研究生。按照内容，全书分为三个部分：第一部分为东亚季风机理研究，第二部分为有关东亚季风气候预测的成果，第三部分为东亚季风的低频变化（季节内振荡）研究。附录部分包括了先生至今发表的主要论文（选）和专著的目录，以便读者查找。由于论文时间跨度近40年，原刊开本、版式各异，许多插图模糊难辨，为了确保图书质量，我们决定重新录入制版，对于年份久远、模糊不清的图片，尽可能进行了修补或重绘。为尽量遵从历史原貌，对原文中使用的名词术语、公式符号、编辑风格等基本不作改动，仅对原文中明显的印刷错误进行了校正，并对编排格式进行了统一。论文出处均在每篇文章首页下端进行了标注。通过层层把关校稿，尽最大努力保证书稿质量。南京信息工程大学《大气科学学报》执行主编智协飞等审阅了全书并提出了宝贵意见和建议。

全书的统稿主要由祁莉、吕江津、朱志伟完成。此外,还应该特别鸣谢南京信息工程大学大气科学学院和科技处的大力支持,以及先生的弟子们在论文照片搜集、书稿校对等方面付出的大量心血,他们高度的责任心和认真态度保证了《论文选》的高质量完成;衷心感谢气象出版社黄海燕等同志在书稿编辑、排版、审校等方面的辛勤劳动,没有她/他们的耐心帮助和细致工作,就没有本书的顺利出版。由于编印时间仍较紧张,疏漏差错在所难免,还恳请各位读者批评指正。

"解落三秋叶,能开二月花。"先生从事气象事业六十年、专注季风研究四十载,先生的人格魅力也像季风一样对一批又一批的学生起着潜移默化而又深刻的影响。先生筚路蓝缕,以启山林,在东亚季风领域取得了一系列开创性的卓越成就,然而他仍有许多预期和设想并没有完全完成。让人欣慰的是,先生的后学仍继续沿着先行者开辟的道路奋勇前行。在向先生崇敬致礼之际,让我们奋发图强、矢志不渝,继续沿着先行者所开辟的宏图大道,共同为中国大气科学事业的不断发展贡献力量。

中国气象局副局长

2020 年 4 月

目 录

序一
序二
贺词
前言

第一部分　季风机理

A Study of Circulation Differences Between East-Asian and Indian Summer Monsoons
　　with Their Interaction ……………………………………………………………… (3)
T_{BB}资料揭示的亚澳季风区季节转换及亚洲夏季风建立的特征 ………………… (15)
北太平洋海温异常与东亚夏季风相互作用的年代际变化 ……………………………… (22)
关于南海夏季风建立的大尺度特征及其机制的讨论 …………………………………… (32)
东亚季风指数及其与大尺度热力环流年际变化关系 …………………………………… (42)
40 a 南海夏季风建立日期的确定 ……………………………………………………… (53)
黑潮地区海温影响南海夏季风爆发日期的数值试验 …………………………………… (58)
中南半岛影响南海夏季风建立和维持的数值研究 ……………………………………… (65)
南海夏季风建立期间副高带断裂和东撤及其可能机制 ………………………………… (78)
东亚副热带季风特征及其指数的建立 …………………………………………………… (89)
有关东亚季风的形成及其变率的研究 …………………………………………………… (101)
Possible Mechanism of the Effect of Convection over Asian-Australian "Land Bridge"
　　on the East Asian Summer Monsoon Onset ………………………………………… (113)
Characteristics of the Onset of the Asian Summer Monsoon and the Importance of
　　Asian-Australian "Land Bridge" ……………………………………………………… (125)
关于东亚副热带季风和热带季风的再认识 ……………………………………………… (143)
Seasonal Transition Features of Large-Scale Moisture Transport in the Asian-Australian
　　Monsoon Region ………………………………………………………………………… (153)
A Review of Recent Advances in Research on Asian Monsoon in China ……………… (173)
Discussion of Some Problems as to the East Asian Subtropical Monsoon …………… (203)
Seasonal Cycle of the Zonal Land-Sea Thermal Contrast and East Asian Subtropical
　　Monsoon Circulation …………………………………………………………………… (222)

东亚副热带季风雨带建立特征及其降水性质分析 (230)
Seasonal Transition of East Asian Subtropical Monsoon and Its Possible
　　Mechanism (240)
东亚与北美东部降水和环流季节演变差异及其可能机理分析 (253)
江南雨季地理区域及起止时间的客观确定 (269)
The East Asian Subtropical Summer Monsoon: Recent Progress (282)
Relationship Between the Seasonal Transition of East Asian Monsoon Circulation and
　　Asian-Pacific Thermal Field and Possible Mechanisms (308)

第二部分　季风气候预测

1979年6月东亚和南亚上空的水汽通量 (329)
江淮入梅的年际变化及其与北大西洋涛动和海温异常的联系 (339)
Vertical Circulation Structure, Interannual Variation Features and Variation
　　Mechanism of Western Pacific Subtropical High (351)
宁夏春季沙尘暴与北极海冰之间的遥相关关系 (365)
西北地区夏季降水异常及其水汽输送和环流特征分析 (376)
华南前汛期不同降水时段的特征分析 (384)
北半球环状模和东北冷涡与我国东北夏季降水关系分析 (393)
东北冷涡的"气候效应"及其对梅雨的影响 (401)
北太平洋中纬度负海温异常对副热带高压影响的数值试验 (411)
长江中下游夏季旱涝并存及其异常年海气特征分析 (421)
江淮梅雨期降水经向非均匀分布及异常年特征分析 (429)
江淮梅雨气候变化研究进展 (438)
江淮梅雨建立的年际变化及其前期强影响信号分析 (447)
1961—2001年青藏高原大气热源的气候特征 (457)
一种新的 El Niño 海气耦合指数 (468)
秋季北极海冰与欧亚冬季气温在年代际和年际尺度上的不同联系 (484)

第三部分　季风低频变化

On the 40-50 day Oscillations During the 1979 Northern Hemisphere Summer
　　Part Ⅰ: Phase Propagation (501)
1979年夏季我国东部各纬带水汽输送周期振荡的初步分析 (536)
The Southern Hemisphere Mid-Latitude Quasi-40-day Periodic Oscillation with Its Effect
　　on the Northern Hemisphere Summer Monsoon Circulation (543)
Meridional Propagation of East Asian Low-Frequency Mode and Midlatitude
　　Low-Frequency Waves (554)
Numerical Experiment with Processes for Effect of Australian Cold Air Activity on

East-Asian Summer Monsoon ……………………………………………………… (565)
Impacts of SST and SST Anomalies on Low-Frequency Oscillation in the Tropical
　　Atmosphere ……………………………………………………………………………… (576)
Another Look at Influences of the Madden-Julian Oscillation on the Wintertime
　　East Asian Weather ……………………………………………………………………… (586)
MJO 对华南前汛期降水的影响及其可能机制 ………………………………………… (605)
Meridional Propagation of the 30- to 60-day Variability of Precipitation in the East
　　Asian Subtropical Summer Monsoon Region: Monitoring and Prediction ……… (614)

附录1 何金海教授主要著(译)作和论文 ……………………………………………… (634)
附录2 何金海教授重要年表 ……………………………………………………………… (645)

何金海论文选

第一部分 壹

季风机理

A Study of Circulation Differences Between East-Asian and Indian Summer Monsoons with Their Interaction[*]

Zhu Qiangen　He Jinhai　Wang Panxing

(Nanjing Institute of Meteorology, Nanjing 210044)

Abstract: Primarily based on the 1979 FGGE data an analysis is made of the circulation differences between the East-Asian and Indian summer monsoons together with their oscillation features and also the interplay between various monsoon systems originating from the fact that the Asian monsoon area is divided into the East-Asian and Indian regions, of which the former is demarcated into the Nanhai (the South China Sea) and the Mainland subregions.

1　Introduction

The Asian monsoon studies have been focused on the Indian system in view of its great intensity and other profound features. In recent years Chinese researchers have indicated that the East-Asian and Indian summer monsoons are separate systems Cheng et al.[1-2] shows that both are separate systems with different properties and major elements of each own. During the period of the intense ITCZ when the Nanhai (the South China Sea) summer monsoon (SM) is active, the Indian SM is often in a break-off state, and vice versa. In 1979, for instance, an evident off-break happened in the Indian SM whereas the East-Asian SM was quite active. Jin et al.[3] demonstrates that the East-Asian SM has considerable influence on the Indian one, but there is only slight effect of the latter on the former as regards the surface pressure and low-level geopotential height fields and that the interface between both systems is around 95°～100 °E. Liang et al.[4] indicates that in the transitional season the Nanhai SM originates upstream from the Indian SM and in mid-summer from a cross-equatorial flow. Yet You[5] shows that the East-Asian SM results from the Indian.

It should be pointed out that the above studies are limited mainly to the Nanhai-western Pacific area without reference to China's mainland. On the other hand, while much research has been carried out into the SM over the mainland by Chinese investigators, no distinction is made between the SM over the mainland and India, especially over the mainland and Nanhai. In fact, as shown in the present work, the Nanhai SM is not just the eastward extension of

---

[*] 本文发表于《Advances in Atmospheric Sciences》,1986 年第 3 卷第 4 期,466-477.

the Indian monsoon and the mainland SM is not the northward extension of the Indian and Nanhai SM's, either. They are independent systems. The aim of this paper is to make analysis of the East-Asian and Indian SM and also a preliminary study of the differences between the subsystems over the Mainland and Nanhai.

In addition, the articles cited above show that, in general, the East-Asian system affects the Indian one as regards the interaction. However, this study indicates that the opposite type of effect can not be ignored.

2 Differences in circulations between the East-Asian and Indian SM systems before and after the establishment

2.1 Differences in the low-level circulations

Through a synoptic statistics study of the daily 700-hPa flow fields of both regions over the period of May-July 1979 major differences are obtained of the low-level circulations over these two SM areas prior and subsequent to the establishment. They are described as follows.

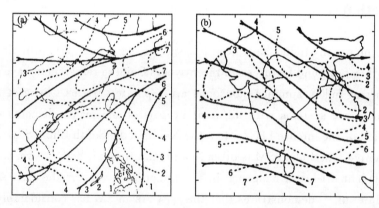

Fig. 1 The mean flow field at 700-hPa for May-July1979. a-the East-Asian region; b-the Indian SM region. Solid lines with arrows represent streamlines and broken isotaches in units of m·s^{-1}

(1) Difference in the mean flow fields

Fig. 1 depicts that a) the WSW and WNW flows are prevalent for the East-Asian and the Indian regions, respectively; b) for the former the mean flow field exhibits an amplitude variation in direction $3/4\pi$ (WNW-SSE) and a greater relative variability[①]of the windspeeds 0.486 while for the latter only $3/8\pi$ (NW-WSW) and 0.343; c) the mean flow field over the East-Asian region displays a marked convergence around 30°N while that over the Indian SM area essentially the parallelism of flows.

---

① The relative variability is defined as the ratio of the average to the mean variance, of the change in speed with position in pace.

The difference of the prevalent current in direction is primarily due to the effect of the mean west flow past the Qinghai-Xizang (Tibetan) Plateau in a roundabout way while the difference in the complexity of the mean flow fields is based upon the fact that the East-Asian SM system consists in essence of the Mainland and Nanhai subsystems and the Indian system is simple in nature.

(2) Difference in characteristics of time sequences in the departure flow fields

By applying the empirical orthogonal function (EOF) method to the vector fields major characteristics vectors (CV) are obtained of the time sequences of the departure flow fields in both the regions for May-July 1979, together with the ratio of these CV fitted with the total variance of the departure fields, ρ_h. A comparison of ρ_h indicates that the first three of the CV are much more significant than the others for the East-Asian region while only the first CV is most important for the Indian SM area. In view of their consequence they are termed major CV of both the regions respectively (Figs. 2a-c and 3a).

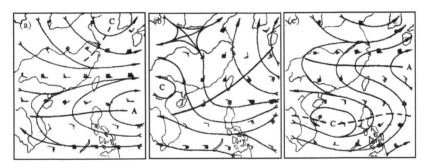

Fig. 2 Major CV of the time sequences in the daily departure flow fields at 700-hPa over the period of May-July 1979 in the East-Asian SM region. a-the first CV ; b-the second ; c-the third. The windspeed is 10^2 times that of the normalized CV in units of $m \cdot s^{-1}$

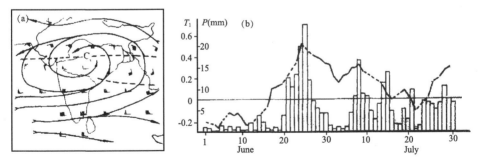

Fig. 3 a. The first CV of the time sequence for the Indian SM region; the others are as in Fig. 2;
b. Relationship between the time coefficients (T_1) of the Indian first CV (curve) and the daily average rainfall (P) of 21 station in the central plain of China (blocks) for June-July 1979

As shown in Fig. 2, the subtropical high as the most important system in East Asia is illustrated in terms of the first three CV with the difference in position and orientation. Analysis indicates that when the time coefficients are positive (negative), the first CV shows a situation typical of the first rainy season over South China (the second period and the wet period in North China); when the coefficients are positive, the third CV demonstrates a typical situation of the Changjiang-Huaihe plum-rain season; the second CV is capable of indicating a situation favorable for vapor transport to all the above rainbands.

The Indian depression is a unique system of importance to this region, which can be shown in terms of the first CV, its center being in the middle and northern parts of the subcontinent, covered by a continental high when the time coefficients are negative.

Based on the analysis of the evolution of time coefficients of the major CV of the East-Asian SM region, it is found that the May-July circulation evolution is featured by the progressive northward movement of the subtropical high; the weather processes of the rainband; the establishment of SM by the northward extension. And similar analysis made for the Indian region indicates that the May-July circulation evolution is marked principally by the replacement of the continental high by the Indian depression; the weather processes of the dry by wet season; and the establishment by an outbreak. In addition, the time coefficients of the Indian first CV reflect a medium-term oscillation of the SM during its prevalent activities, as shown in Fig. 3b by the quasi-synchronous change in the first CV time-coefficient curve for the Indian region and that of the daily averaged rainfall of 21 stations on the central plain of the country over the period of June-July 1979.

2.2 Differences in the upper-level circulations

Fig. 4 delineates the change in the 100-hPa divergence field versus time at 120° (East Asian) and 65° E (between the eastern Arabian Sea and West India), respectively. A comparison of both the cases indicates that the upper-level circulations of the two SM systems are clearly displayed in the divergence fields. The Nanhai (5°~15° N) divergence field is established in the third pentad of May, corresponding to the building-up of SM (12 May) in this region. This divergence belt persists till the fifth pentad of July. Obviously, it is an upper-level offsetting current necessary to the updraft from the low-level SM trough after the Nanhai SM has been established and it is the upper-level reflection of the trough with its precipitation. Hence the movement of this belt indicates that of the trough or SM. On the other hand, for the region between the Arabian Sea and India (around 20°N) the divergence belt is formed in the fourth pentad of June, which corresponds to the time of the Indian SM establishment, about 7 pentads later than the starting of the SM over the Nanhai. To the north of the South China Sea divergence belt there exists an evident convergence belt with a region of descending air, relative to the western Pacific subtropical high at low levels. For the West Indian SM area, south of the upper-level divergence band lies an intense convergence belt, in relation to the equatorial anticyclone at low levels. In addition, another divergence belt shows up over the mainland to the north of the Nanhai convergence band. With respect to the rainband of the first rainy season in South China, this belt is maintained in the neighborhood of 25° N from the third pentad of May to the fourth of June; associated with the plum-rain season over the Changjiang reaches, it moves northward to about 30°N between the fourth pentad of June and the same period of the next month, before its moving further northward and putting an end to the plum-rain period. In the meantime the Nanhai divergence and convergence belts march toward the north; so do the related low-level SM trough and subtropical high. It follows that the East-Asian SM region is concerned with two upper-level divergence belts in the north and south, which correspond to two low-level convergence belts and monsoon rainbands. In view of the fact that both of the divergence bands are established in the third pentad of May, the SM system starts simultaneously over the Nanhai and Mainland. showing a tendency for the wind to move northward synchronously except that the Nanhai SM system travels at a lower speed and over a shorter distance than the Mainland type. Over the Arabian Sea-Indian region, however, only one divergence belt is available and maintained around 20°N, which is related to the low-level Indian SM trough along with the rainfall.

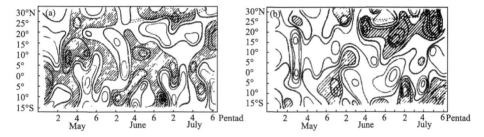

Fig. 4 The time section of the pentad-averaged divergence field at 100-hPa for May-July 1979. a-the section at 120 and b-at 65°E. The shaded area shows intense divergence; the heavy solid line denotes zero-value divergence; the double-broken the ridge line of the South-Asian high

It is worth noting that both the Arabian Sea-India and the Nanhai divergence belts lie within the easterlies south of the South-Asian high, which are homologous to the low-level monsoon troughs, whereas the mainland divergence belt, relative to the low-level convergence band north of the subtropical high, lies within the westerly flow about or north of the ridge line of the South-Asian high. Therefore, the SM system associated with the upper-level divergence belt in the former case is tropical while the Mainland SM system related to the belt in the latter is subtropical in nature.

Based on the analysis of the evolution of the pentad mean u and v at the 100-hPa level (figures omitted) of the Nanhai (15°N, 115°~125°E) and the Indian SM region (15°N, 70°~80°E) for May-July 1979, it is found that a strong northeast wind over the sea after the third pentad of May is evidently the result of the westward by north movement of the South-Asian high center over this region, while for the Indian area a northeast wind begins only after the third pentad of June, i. e. , about a month later than does an east wind in this region (the third-fourth pentad of May), which is closely linked to the eastward movement of the Iranian high. In view of the synchronous establishment of the low-level SM and upper northeast wind the seasonal transition of the upper-level circulation patterns over these two regions has significant differences in the setting-up process as well as in time, as demonstrated by the analyses above.

2.3 Difference in the low-latitude circulations

Two vertical sections of the regions along 7.5°N are selected for investigating the difference between the above SM systems in the circulation patterns at low latitudes. Fig. 5 illustrates the first and second CV of the u-component fields, which make to the square of the norm a contribution 42.4 and 13.0 %, respectively.

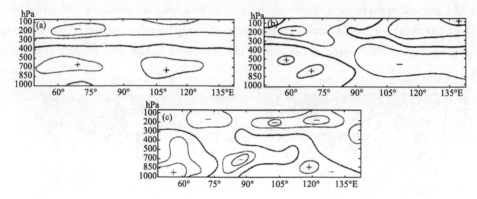

Fig. 5a and 5b. The CV of u-component fields in the 7.5°N vertical sections for May-August 1979. a-the first CV; b-The second CV. The sings + and-denote a west and east wind, respectively. c. The pattern of the first CV in the v-component field in a vertical section along 7.5°N for May-August 1979. The signs+and−denote a south and north wind, respectively

As indicated by the first CV in Fig. 5a, for low latitudes, the interface between the east and west wind is located around the 400-hPa level; the Nanhai and Indian SM regions each have an area of a stronger vertical shear of the zonal wind to the southeast, with the shear being even stronger in the area relative to the Indian region; since the first CV time coefficients in this period remain positive, the vertical disposition of the east and west winds is kept unchanged. However, the second CV indicates an opposite disposition over the low-latitude areas associated with the two SM regions, that is, the disposition over the Indian sector agrees with that as shown by the first CV, and that over the East-Asian region is to the contrary. The second CV time coefficients change to positive from negative in the mid-decade of May, leading to the intensification of the vertical shear of winds over the low-latitude Indian Ocean owing to the superimposition of these two CV and to an appreciable diminution of the shear over the Nanhai and the area to the east, with the total effect that the Walker circulation cell (an E-W vertical pattern) over the low-latitude Indian Ocean becomes reinforced from the mid-decade of May; the change of the second CV time coefficients from positive to negative in sign during the last decade of August suggests that the Walker cell in the western Pacific at low latitudes gets strengthened whereas that over the Indian Ocean is weakened.

As a consequence, there exists significant difference in the low-latitude E-W vertical circulation pattern in relation to the two SM systems.

Fig. 5c delineates a vertical section where the first CV in the v-component field makes a contribution of 20.6% to the square of the norm, much higher than do the others. The fact that the first CV time coefficients of the v-component field are positive for May-August 1979 for the most part results in the general pattern of the north and south winds in the low-latitude section as illustrated in Fig. 5c. This figure shows the difference in the low-latitude meridional circulations between the SM systems. The meridional circulation associated with the East-Asian system consists of a (weak) low-level cross-equatorial northward flow over Indonesia and a (strong) upper southward flow to the southeast of the South-Asian high while the circulation with respect to the Indian system is made up of a (strong) low-level Somalian jet and a (weak) southward flow to the southeast of the Iranian high. The interface between the meridional circulation cells is over the range from 80° to 90°E (at 7.5°N), where the circulation pattern is characterized by regions of a north wind, strong at low and weak at upper levels. The above-mentioned features are also typical of the circulation in the vertical section at the equator. Consequently, the two SM circulation cells on the low-latitude meridional planes associated with the two monsoon systems have significant properties different from each other.

Analysis indicates that the curves of the time coefficients of the three CV mentioned above have a nearly-40-day oscillation. The first CV in the u and v fields illustrates an in-phase oscillation of the vertical circulation cells (with an upper-level northeasterly and low-level southwesterly wind) at low latitudes in relation to both the SM system; the second CV in the u-component field shows an antiphase oscillation of the zonal wind component in the low-latitude vertical sections relative to the SM systems. The latter case agrees quite well with the conclusion of Jin et al.[3].

3 Interaction between the SM systems

As indicated in the introduction, the authors inclined toward varied emphasis on the role of one of the SM systems in the mutual interplay. So far, the complicated problem has not been examined systematically-that is, only a small number of cases have been analyzed, with diagnostic analysis made merely of some of the aspects and dynamic studies just in the infancy. Therefore, it is of value to carry out extensive diagnostics of the problem. In the present work this problem is investigated through wave propagation, vapor transport and energy exchange and better understanding has been gained.

3.1 Wave propagation

Fig. 6 shows the time-dependent evolution of the 850-hPa geopotential height H and the west-wind component u along 15°N. As illustrated in Fig. 6a, between 50° and 120°E H has a

40-day periodic oscillation for both the systems but the amplitude in the East-Asian SM region is smaller and 1-2 pentads later than the one in the Indian area. It follows that the wave motion is eastwards, damping, until it disappears east of 120°E.

Fig. 6b shows that (1) over the Nanhai region an east wind is shifted to a west wind for the third pentad of May (during this period the SM system is established) with a 40-day periodic oscillation (mainly the alternation of the winds) and a smaller speed < 10 m·s^{-1}; (2) in the fourth pentad of June (when the Indian monsoon breaks out) over the Indian area a weak west wind is greatly intensified, attaining a maximum speedy >20 m·s^{-1} in the center of the wind belt, followed by a nearly-40-day oscillation of the wind intensity. Similar to the behavior of the wave motion in the height field, waves in the u field are carried eastwards, damping and vanishing to the east of 135°E.

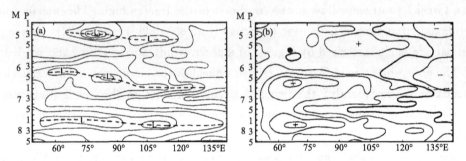

Fig. 6 The time-dependent evolution of the 850-hPa geopotential height H (a) and the west-wind component u (b) along 15°N in units of 10 geopotential meters and m·s^{-1}, respectively. The isoline interval is 2 units. P-Pentad; M-Month

Similar analyses made of the temperature and moisture fields (figures omitted) indicate that they are marked by a significant periodic oscillation. After the SM prevails over the Nanhai, the temperature at 850-hPa drops below 20°C and the moisture increases to above 14 g·kg^{-1}, while in India when the monsoon has broken out (the fourth pentad of June) both the fields at the level show analogous change. Like the alternation of the east and west winds and oscillation of the west wind intensification mentioned above, the two element fields, particularly the moisture one, have a nearly-40-day oscillation, which is propagated eastwards in phase with that of the H and u fields.

It can be seen from the above that after the establishment of SM over the Nanhai and Indian Ocean the associated fields of H, u, q and T on the 850-hPa surface along 15°N undergo considerable variation. The first three, in particular, display a nearly-40-day inphase oscillation which is propagated towards the east.

Analyses of the 850-hPa power-spectrum curve (Fig. 7a) and the time-dependent evolution of the pentad mean height at 100-hPa (Fig. 7b) indicate that the height fields at low and upper levels have a quasi-40-day oscillation which is propagated eastwards and then northwards. The transfer of a high-value (low-value) area from India to a high-value (low-value) area south of the Changjiang River of China takes, on the average, about 2～3 pentads, which is in agreement with the conclusion of He et al[7], that it takes approximately 10~12 days for the vapor fluxes of a maximal vertical integration to be transported eastward through India (75°E) to the Nanhai and then northward across its boundary (18.75°N).

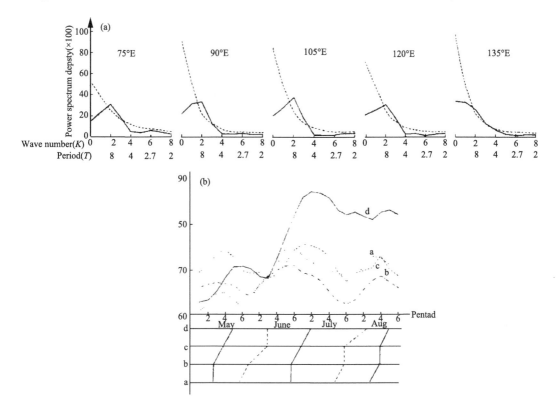

Fig. 7 The 850-hPa power-spectrum curves (7a, top) and the time-dependent evolution of the pentad mean height at 100-hPa (7b, left) for some regions. Fig. 7a shows the spectrum curves at 75°, 90°, 105°, 120°and 135 °E along 15°N for May-August 1979 (solid line) and the confidence $a=0.05$ (broken line); Fig. 7b depicts pentad mean-height curves: a-India (15°~20°N, 70°~80°E); b-Bay of Bengal (10°~15°N, 80°~100°E); c-Nanhai(15°~20°N, 110°~125°E); d-souther side of Changjiang River (25°~30°N, 110°~125°E). The graph at the bottom indicates lines connecting the peak-(solid) and Valley-points (broken line)

3.2 Vapor transport

The band-pass (with a 40-50-day oscillation) filtered data are analyzed from the vertical-integration vapor-flux field for May-September 1979. This frequency band is selected for use just because it is a band having a significant oscillation for a vapor flux (uq) field at low latitudes. The first and second CV fields obtained are illustrated in Fig. 8. The first CV field (where the ratio of a fitting to the total variance is 50%) indicates clearly that a positive-value center is between 10 and 15°N and a negative to the northeast, a zero-value line coinciding with the South-Asian SM trough. This means that when the time coefficients are positive, a vapor-transporting west wind is strong on the southern side of the trough and so is an east wind on the northern side, and vice verse. The second CV field (34%) shows that a positive center is shifted northeastwards with respect to that of the first CV field and a negative region shows up to the southwest. Since the curve of the first CV time coefficients goes by 1/4 wavelength in phase ahead of that of the second (figure omitted), the vapor transport oscillation is propagated northeastwards, implying that a large amount of water is transferred from the Indian into the East-Asian SM system with a low-frequencyoscillation in intensity.

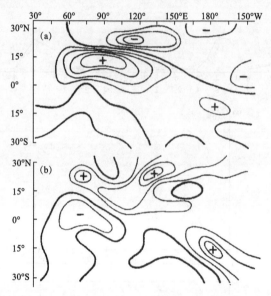

Fig. 8　The normalized CV for a vertical-integration vapor transport (in terms of the bandpass filtered data of a 40-50-day periodic oscillation). a-the first CV; b-the second CV. The heavy solid line denotes a zero-value line. The isoline interval is 0.08

3.3　Energy exchange between both SM systems

The transport of various energies through a vertical section (0°~40°N, 100°E) for August-September 1982 is calculated with the data of 8 upper-air stations (48694, 48568, 48455, 48327, 56739, 56247, 56046 and 52652) in the vicinity of the interface of both the SM systems (about 100°E). For calculation purpose, the sector is divided into 4 regions, i. e., two parts in vertical by the 500-hPa surface and two portions in horizontal by the boundary of the east and west winds at 300-hPa. These regions are labelled as I and II (north of the boundary and above and below the surface, respectively), III and IV (south of the boundary and above and below the surface, respectively). The calculation results are given in Table 1.

From the table the following can be summed up concerning the energy exchange between the two SM systems from the period of its prevalent activities to the retreat.

(1) The calculation of the whole section shows that sensible and latent heat and kinetic energies are carried to the East-Asian from the Indian SM region (including the Plateau in the north) whereas potential energy is transferred from the former to the latter region. The

calculation of the total energy indicates an eastward transfer of energy.

Table 1 Energy-Flux Density through the 100°E Vertical Section* (Units: J · hPa^{-1} · cm^{-1} · s^{-1})

Energy	Region						Whole Section
	I	II	I+II	III	IV	III+IV	
$uC_pT(10^5)$	2.34	0.65	1.68	−1.24	0.89	−0.1	0.48
$uL_q(10^3)$	1.88	2.70	2.20	−0.34	7.61	3.92	3.49
$u\varphi(10^5)$	1.07	0.13	0.706	−1.07	0.07	−0.462	−0.09
$uK(10^2)$	4.04	0.15	2.53	−1.52	0.22	−0.58	0.44
$uE(10^8)$	3.43	0.80	2.41	−2.53	1.03	−0.52	0.43

* C_pT-sensible heat energy; L_q-latent hear energy; φ-potential energy; K-kinetic energy; E-total energy.

(2) Evidently the Nanhai SM region transports the total energy to the Indian and this is accomplished via the westward transfer of sensible heat and potential energy by the upper-level easterlies. In the lower part of the troposphere the total energy and all components of energy are carried eastward. In particular, the eastward-transported latent heat energy (in form of vapor) is by no means neglected, although its amount is less than those of potential and sensible heat energy. A quite large amount of the vapor is turned into SM rainfall for the East-Asian region.

(3) The Qinghai-Xizang Plateau transports various types of energy to the Mainland SM subregion.

It follows that for the energy exchange between the East-Asian and Indian SM regions (the latter includes the Plateau in the north) energies are transferred eastwards and for that between the Nanhai and Indian SM systems the total energy is carried westward. It should be noted that the eastward transfer of latent heat energy is of particular importance to the Mainland SM rainfall.

4 Concluding remarks

This article indicates that the East-Asian and the Indian SM systems are characterized by significant differences and strong interaction. The complexity of the lower-tropospheric flow field and some other aspects of the East-Asian system originates from two subsystems (i. e., the Nanhai tropical and Mainland subtropical SM systems) different in essence as its components while the Indian system is simply tropical in nature. So complicated is the interplay between both the systems in forms of wave propagation, vapor transport and energy exchange. Particular emphasis is on the interaction in a low-frequency oscillation way extremely important to low latitudes. The results show that the effect of the Indian on the East-Asian SM system takes the first place. Some studies, however, emphasize a reverse effect which, as demonstrated by Zhu et al.[8], may be concerned with the interplay in a secondary low-frequency-oscillation way. Such effect can be visualized as low-latitude easterly waves progressing from the Nanhai to the Bay of Bengal.

References

[1] Chen Longxun, Jin Zuhui, et al. Medium-range oscillation in tropical circulation during summer over Asia [J]. Acta Oceanologica Sinica, 1983, 5: 575-586 (in Chinese).

[2] Chen Longxun, Luo Shaohua. An analysis of atmospheric circulation at low latitudes over the western Pacific during the strong and the weak ITCZ periods[M]. Proceedings by Institute of Atmospheric Physics, Academia Sinica, Science Press, 1979, 8: 77-85 (in Chinese).

[3] Jin Zuhui, Chen Longxun. On the medium-range oscillation of the East Asia monsoon circulation system and its relation with the Indian monsoon system[C]//Proceedings of the Symposium on the Summer Monsoon in Southeast Asia. People's Press of Yunnan Province, 1982: 204-217 (in Chinese with English abstract).

[4] Liang Biqi, Liang Mengxuan, Xu Xiaoying. The cross-equatorial current in the lower troposphere and the summer monsoon in the South China Sea[C]//Proceedings of the Symposium on the Summer Monsoon in Southeast Asia. People's Press of Yunnan Province, 1982: 39-48 (In Chinese with English abstract).

[5] You Liyu. The influence of seasonal transit of circulations in the Northern and Southern Hemispheres on the activities of summer monsoon[C]//Proceedings of the Symposium on the Summer Monsoon in Southeast Asia. People's Press of Yunnan Province, 1982: 30-44 (in Chinese with English abstract).

[6] Zhang Jijia, Wang Panxing, et al. Synoptic and statistical analysis of the seasonal transformation of the low-latitude atmospheric circulation and the effect of diabatic heating by the Qinghai-Xizang Plateau[J]. Journal of Nanjing Institute of Meteorology, 1983, 1: 1-13 (in Chinese with English abstract).

[7] He Jinhai, Murakami T, Nakazawa T. Circulation with 40-50 day oscillation and changes in moisture transport over Asian monsoon in 1979 summer[J]. Journal of Nanjing Institute of Meteorology, 1984, 2: 163-175 (in Chinese with English abstract).

[8] Zhu Qiangen, Wang Xin. Energy exchange between the eastern and western monsoon regions in Asia during summer and its Oscillation[J]. Journal of Nanjing Institute of Meteorology, 1985, 3: 266-275 (in Chinese with English abstract).

T_{BB}资料揭示的亚澳季风区季节转换及亚洲夏季风建立的特征[*]

何金海[1]　朱乾根[1]　Murakami M[2]

(1. 南京气象学院,南京 210044; 2. 日本气象研究所)

摘要: 采用日本气象研究所提供的 T_{BB} 资料分析了亚澳季风区的季节转换及亚洲夏季风建立的特征。发现 4 月季节转换就已开始,5—6 月是季节突变时期,6 月亚洲地区夏季风形势完全建立。冬季位于苏门答腊的对流中心由冬入夏沿着海洋大陆"大陆桥"和中印半岛系统地向北偏西移动。由此引起的大尺度环流改变导致了东南亚和印度地区夏季风的建立。而南海—西太平洋地区夏季风的建立则与中印半岛对流的活跃及副热带 T_{BB} 高值带(相应于西太平洋副高)的连续东撤紧密相关。

关键词: T_{BB}资料;亚澳季风区;季节转换;亚洲夏季风的建立

1 引　言

最近十多年来,随着季风研究的深入,人们对季风环流系统的认识取得了重大进展。陶诗言等[1]首先指出,东亚夏季风与南亚(印度)夏季风是两个相对独立的亚洲季风子系统。朱乾根等[2]则提出东亚夏季风又可划分为南海—西太平洋热带夏季风和中国大陆—日本的副热带夏季风。与此同时,南半球盛行印尼—北澳冬季风。这些季风系统既相互独立又相互联系,构成了一个完整的季风体系。北半球冬季,东亚东北风可以越过赤道进入南半球转为印度尼西亚—北澳的夏季风。南北半球冬、夏季风的季节转换构成亚澳季风的季节循环。

夏季风的建立及相应的季节转换特征一直是人们关心的重要问题。大气环流的 6 月和 10 月突变早为我国气象学家(叶笃正等[3])所阐明。然而日益增多的证据表明,亚澳季风区中低纬度与夏季风建立相关的季节转换在 6 月之前就已发生,北半球春季对于确定季风随后的演变以及海—气—陆(地)耦合系统的年际变率是至关重要的。Webster 等[4]发现春季是海—陆—气系统失去其"记忆"最快的季节,在这一季节,赤道环流特别"脆弱",因此其对于边界条件的变化和来自热带地区以外的随机强迫比较敏感。Yasunari[5]则指出,与季风—ENSO 系统有关的降水和环流异常与春末夏初改变异常符号的年循环有一种锁相关系。他提出了一种"季风年"(Monsoon Year)的概念,它开始于一年的 4—5 月,且结束于下一年的相同月份。这个"季风年"的定义表明,北半球春季对于确定冬、夏季风的季节转换有重要意义。本文的目的是试图通过 T_{BB} 资料来分析这一关键时期亚澳地区的季节转换特征,并讨论各个地区夏季风

---

[*] 本文发表于《热带气象学报》,1996 年第 12 卷第 1 期,34-42。其英文版发表于《Journal of Tropical Meteorology》,1997,1:18-26。

建立的特征和差异以及形成这些差异的可能机制。

2 资料和方法说明

本文采用的资料是日本气象研究所提供的 GMS 观测的黑体辐射温度,范围为 80°E～160°W,60°S～60°N;分辨率为 1°×1°经纬度。

有必要指出,在无云和少云区内,T_{BB} 表示的是地面黑体辐射温度,其值较高,因此 T_{BB} 高值区常与高压系统相对应;而在有云区内,T_{BB} 表示的为云顶黑体辐射温度,其值较低,对流越强,云顶越高,T_{BB} 越低,因而 T_{BB} 低值区为多云区,低值中心常表示强对流中心。但在冬季的中高纬度和高原地区,地面温度较低,因此这里即使无云 T_{BB} 也较低,分析时要小心判断。此外,在中高纬地区,纬度越高,地面温度越低,T_{BB} 也越低,故其等值线多呈纬向分布,冬半球这种特征更为显著。当这些等值线向低纬伸展形成低槽时,与周围同纬度相比,槽中 T_{BB} 较低可以判断其为云区。因此我们可以在中高纬地区分析 T_{BB} 的槽线以表示多云带。我们的分析经验表明,T_{BB} 场高低值中心和槽脊线分布与 500 hPa 高度场形势非常相似。

3 结果分析

3.1 多年平均图上季节转换的特征

图 1a,b,c,d 展示了多年平均逐月(3—6 月)T_{BB} 的水平分布。

由图可以看出,3 月 T_{BB} 的分布特征依然与 1 月(图略)的特征相同。亦即,在南半球副热带地区存在一条 T_{BB} 高值带,其脊线位于 25°S 左右,主要中心在澳大利亚中部,与南半球副高压带和澳大利亚高压中心相对应;北半球副热带也存在一条横贯东西的高值带,其脊线位于 15°～20°N,主要中心在西太平洋,同时在南海和孟加拉湾东北也各有一个中心,它们分别与这一地区副热带高压带及相应的中心相对应;另外在赤道地区也存在一较窄的高值带,自东向西伸展,位于 170°E 以东(取 272 线为特征线),与双 ITCZ 之间的相对高压带相对应。

与上述高值带相间分布的低值带是:(1)南北半球中纬度的低值带或低值区,其上有若干小槽(低值槽);(2)赤道南侧的低值带,横贯整个研究区域(80°E～160°W),在其上有 4 个显著的低值中心,分别位于苏门答腊(Sumatra),婆罗洲(Borneo),苏拉威西(Sulawesi)和新几内亚(New Guinea)4 个岛屿上空,表明这些岛屿上空有很强的对流活动。从新几内亚低中心向南伸出两条低槽,位于北澳上空,与澳大利亚北部的夏季风相对应。

但是到了 4 月,上述特征则发生了显著变化(如图 1b 所示),主要表现在:(1)北半球副热带高值带在中南半岛地区(100°～110°E,10°～20°N)发生断裂(以 280 线作为特征线),从苏门答腊的低中心有显著的低槽伸向这一地区,与北半球中纬度低值带相通;(2)南半球澳大利亚高值区显著北移(轴线移至 20°S),新几内亚伸向澳大利亚北部的低槽因此而消失。这些变化特征表明,随着太阳辐射的季节变化以及这些地区海陆对其响应的差异,整个系统北移,澳大利亚夏季风消失,标志着亚澳地区中低纬度大尺度环流的季节转换已经开始。

5 月,北半球副热带 T_{BB} 东部高值中心迅速东撤,与之相联系,菲律宾到我国南海地区的对流开始活跃。苏门答腊的低中心向西北移动,强度增强,范围扩大,显著的低槽占据中南半岛,

图 1　多年平均(1980—1988 年)的 3 月 (a)、4 月(b)、5 月(c)、6 月(d)T_{BB}水平分布图
(虚线为高值带脊线;粗实线为低值带轴线或 280 等值线)

原孟加拉湾东北部的高值中心迅速向西北移动至(80°E,25°N)附近。这表明中南半岛孟加拉湾地区夏季风对流已充分建立。此外,赤道东太平洋的高值带显著西伸到 150°E 附近。

6 月,西太平洋的 T_{BB} 高值中心继续东撤,且伴随着高值轴线的显著北跳(22°N 附近),该高值带北侧的低值带上有 2 个低值中心,分别位于我国长江下游和日本南部,它对应我国的梅雨和日本的 Baiu。我国南海—菲律宾地区已成为很强的低值中心,并且中东太平洋赤道北侧的低值带也随赤道高值带的西伸显著向西扩展,与南海—菲律宾地区的低值区相接,形成一条较为完整的低值带,它表明南海—西太平洋地区的夏季风已完全建立。冬季位于苏门答腊的低值中心已基本消失(为弱槽区),在孟加拉湾东北部出现了一个范围很大的强低中心,与冬季的高中心正好相对应,表明孟加拉湾到印度地区的夏季风形势已完全建立。

应该指出,冬季位于苏门答腊的低值中心在由冬向夏的季节转换中是系统地向龙移动的,5—6 月显著地北跳,7—8 月位于最北位置(20°N 以北),9 月开始回撤。后面还要讨论这一点。

综上所述,亚澳季风区的季节转换在 4 月就已开始,5—6 月是季节突变的时期,6 月夏季风形势完全建立。可以认为,我们的分析结果与 Yasunari 将 4—5 月作为"季风年"的开始时段是一致的。

3.2　时间—纬度剖面图上夏季风建立的特征

为了讨论各个地区夏季风建立的特征及其差异,我们计算了多年平均的逐候 T_{BB} 资料并分别制作了沿 80°E、100°E、120°E 和 140°E 的时间纬度剖面(图 2a,b,c,d)。

图 2a 基本上展示了印度东部地区夏季风建立的特征。由图可见,1—5 月印度(东部)地区 20°N 纬带附近受 T_{BB} 高值带控制,2—3 月最强,对应着印度旱季晴好天气;到 6 月上旬急剧减弱转受低值带控制,标志着印度夏季风爆发。7 月下旬低值带最强,影响范围最北,夏季风在印度全区达到盛期。以后低值带逐渐南移,于 9 月下旬开始逐渐转为高值带控制,标志着印度夏季风的撤退和冬季风的建立。值得注意的是,在 5 月上旬有低值带从赤道向北持续传播,从而于 6 月上、中旬在印度地区形成 T_{BB} 的低值中心,标志着对流加强,印度夏季风建立。上述建立过程表明,印度地区 T_{BB} 高值带向低值带的转换即夏季风的建立虽然带有爆发性,但明显地可以看出,它是与赤道地区低值带的向北传播密切相关的。

图 2b 反映的是高原东部和东南亚地区的夏季风建立特征。与印度夏季风的建立相比,东南亚地区季风的季节循环与印度地区看来较为相似,但存在 3 个显著的不同点:(1)东南亚地区 5 月上中旬就开始由高值带转为低值带,这意味着东南亚夏季风建立比印度夏季风建立约早一个月;(2)东南亚经度上近赤道南侧并不像印度地区那样始终存在一个低值带,而是伴随着冬季风向夏季风的转换,赤道低值带呈现出系统稳定的向北移动,同样伴随着夏季风的撤退,低值带变为向南移动回到赤道南侧;(3)高原东部(32°N 附近)在 6—9 月也存在一条低值带,且与东南亚地区的低值带(15°N)相分离,其间为相对高值带。这表明高原东部的夏季风与东南亚的夏季风是相对独立的系统,不像高原西部的夏季风是印度夏季风的向北延伸。

图 2c 展示了印尼—北澳季风区、中国南海季风区和中国大陆东部副热带季风区的演变特征。可以看出,南海地区(12°N 附近)5 月中旬高值带急速转变为低值带,标志着南海夏季风的建立。但必须指出的是,南海夏季风的建立与印度和东南亚地区相比,没有近赤道低值带向北传播的特征,似乎高值带向低值带的突然转变是在原地发生的,这是为什么?我们在后面将

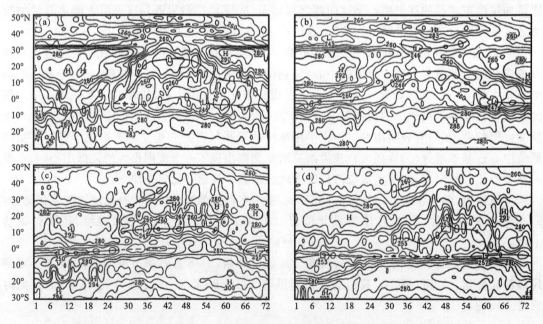

图 2 沿 80°E(a)、100°E(b)、120°E(c)、140°E(d)多年平均 T_{BB} 时间—纬度剖面图
(图中横坐标为时间,以候计,粗实线表 280 等值线或低值带轴线)

要重点讨论这一问题。另外,在我国大陆东部存在一条低值带,它对应着中国大陆的副热带夏季风云雨带,其倾斜的形式表明大陆夏季风随着季节向北推移。在这一低值带与南海低值带之间为一高值带(相应于西太平洋副热带高压)。12月—次年2月,在赤道附近有强的低值中心并向南伸展至北澳西部,这对应着南半球澳大利亚夏季风盛行期。

图2d显示了日本、西太平洋和北澳东部季风的季节循环。在这一经度上,北半球副热带高值带呈现出显著的南北进退的季节循环。1—5月,T_{BB}高值带脊线位于18°N左右,6月迅速北移,8月底到达最北位置30°N附近,以后逐渐南移,12月回到冬季位置。有趣的是,这一经度上10°N附近低值带的建立在经向剖面图上同样具有突然出现的特征,并伴随副热带高值带北移,于8月到达最北位置18°N附近,而后南移。可以看出,与上述低值带相关的北半球西太平洋热带夏季风的活动仅出现在6—10月。在日本南部的纬带(35°~40°N),6—7月有一个显著的低值带,它与日本的梅雨相对应,与我国东部低值带相连接,构成我国大陆—日本的副热带夏季风低值带。南半球澳大利亚北部,12月—次年3月为低值带控制,这是北澳夏季风盛行的反映。

综上所述,我们可以发现两个重要的特征:(1)印度和东南亚地区夏季风的爆发均伴随着赤道低值带的迅速向北移动,而南海地区夏季风低值带的建立则表现为局地突然生成的特征,西太平洋地区夏季风低值带的建立特征与南海地区类似,但较其缓。(2)东南亚和南海地区夏季风的建立较印度和西太平洋地区要早。是什么样的过程和机制形成了上述特征?下面我们将作进一步的讨论。

3.3 夏季风建立的过程和机制

根据多年平均的T_{BB}逐月演变(图略)和已有研究[5,7],我们已经可以看到,冬季位于苏门答腊岛上空(100°E,5°S)的对流中心由冬入夏是向北移动的,且不是严格沿经圈方向而是沿海洋大陆的"大陆桥"和中印半岛地区移动的,盛夏期间稳定在孟加拉湾北部到印度地区。尔后随着夏季风的撤退,这一对流中心又会南移,回到苏门答腊岛上空。为了追踪和证实苏门答腊对流中心这一连续演变特征,我们用多年平均的逐候T_{BB}资料特意制作了沿"大陆桥"中印半岛(从100°E,4°S到90°E,24°N)的时间剖面图,如图3所示。

图3中的低值轴线及其上的低中心反映了上述对流中心位置和强度的连续演变及年循环。可以看出,4月,对流中心已移至3°~5°N附近,5—6月有一次迅速北移,7—8月达到最北位置,9月撤至近赤道纬度。这表明,冬季位于苏门答腊的对流中心沿"大陆桥"和中印半岛西海岸的移动具有很规则的年循环特征。正是这一对流中心由冬入夏的向北偏西的移动导致了东南亚和印度地区夏季风先后建立的特征。

为了搞清南海夏季风T_{BB}低值带在经向剖面图上为什么具有局地突然生成的特征,我们绘制了北半球副热带高值带特征线(取280线)3—6月的连续演变,如图4所示。由图可以看出,3月北半球副热带高值带横贯整个研究区域(如实线所示),在4月首先在中印半岛地区(100°~110°E,10°~20°N)发生断裂,而后,5—6月,西部高值带西移消失,而东部西太平洋高值带连续东撤,并于6月有显著北进,它对应了西太平洋副高的活动。正是这一高值带的连续东撤并伴有北进导致了南海—菲律宾—西太平洋地区对流活动的依次活跃和夏季风的先后建立,这就解释了为什么在120°E经向剖面上南海地区夏季风低值带的突然建立没有伴随赤道低值带向北传播的原因。

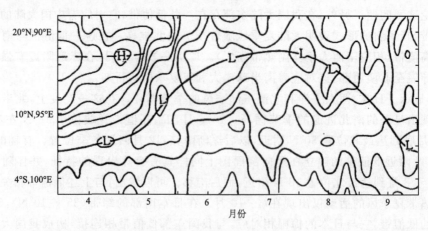

图3 沿海洋大陆"大陆桥"和中印半岛西海岸(100°E,4°S 到 90°E,24°N)的时间剖面图
(粗实线表示 T_{BB} 低值轴线或 260 等值线;虚线表示高值轴线;H(L)分别表示高(低)值中心;等值线间隔为 5)

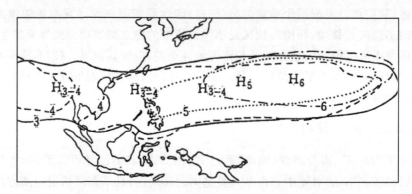

图4 北半球副热带 T_{BB} 高值带外围特征线(280 线)及有关中心的逐月动态图
(H、L 分别表示高、低值中心,下标表示相应的月份;等值线中所标注的数字表示其所在月份)

基于前面的分析,我们提出,苏门答腊对流中心由冬入夏向北偏西的系统性移动以及由此而造成的大尺度环流的变化是导致东南亚和印度地区夏季风建立的直接过程。而中印半岛地区对流的活跃,以及随后南海—西太平洋地区 T_{BB} 高值带的连续东撤则是导致该地区夏季风建立的主要过程特征,而它也与苏门答腊对流中心沿"大陆桥"和中印半岛的移动紧密相关。至于是什么机制导致了上述对流中心的系统性移动,这是一个非常有意义的科学问题。已有的研究表明[1],从苏门答腊到中南半岛的狭长陆地在由冬入夏时的迅速增温及相应的对大气的感热加热在亚澳地区大尺度环流季节转换中具有先兆和触发作用。

4 结 论

(1)亚澳季风区的季节转换在 4 月就已开始。主要特征表现为澳大利亚 T_{BB} 高值带的北

---

[1] Li C, Yanai M. The onset and Interannual variability of the Asian summer monsoon in relation to differential heating. Submitted to J. Climate."关于亚澳季风气象学和海洋学国际研讨会"报告论文,澳大利亚墨尔本,1995 年 4 月 10—12 日。

移(北移 5 个纬距)及北澳夏季风的消失,中印半岛地区对流的活跃以及副热带 T_{BB} 高值带在该区的断裂。5—6 月是季节突变的时期,主要表现为苏门答腊对流中心迅速北移至孟加拉湾和南海—西太平洋地区 T_{BB} 高值带的连续东撤。6 月,夏季风形势完全建立。

(2)东南亚和印度地区夏季风的建立与冬季位于苏门答腊的对流中心由冬入夏沿"大陆桥"和中印半岛的系统性移动直接相关,而南海—西太平洋地区夏季风的建立则与 T_{BB} 高值带(相应于西太平洋副高)的连续东撤关系密切。

(3)从苏门答腊到中印半岛的狭长陆地在由冬入夏时的迅速增温及相应的感热加热可能在亚澳地区大尺度环流季节转换中具有先兆和触发作用。苏门答腊对流中心移动的年循环具有重要的科学意义。

致谢:本文在定稿过程中得到陶诗言先生亲切指点,谨表衷心谢意。

参考文献

[1] Tao S, Chen L. A review of recent research on the east Asian summer monsoon in China[C]// Chang C-P, Kishnamurti T N. Monsoon Meteorology. Oxford: Oxford University press, 1987, 60-92.

[2] Zhu Qiangen, He Jinhai, Wang Panxing. A study of circulation differences between East-Asian and Indian summer monsoon with their interaction[J]. Adv Atmos Sci, 1986, 3: 466-477.

[3] 叶笃正, 陶诗言, 李麦村. 在 6 月和 10 月大气环流的突变现象[J]. 气象学报, 1958, 29: 249-263.

[4] Webster P J, Yang S. Monsoon and ENSO: Selectively interactive systems[J]. Quart J R Meteor Soc, 1992, 118: 877-926.

[5] Yasunari T. "The monsoon year"——a new concept of the climatic year in the tropics[J]. Bull Amer Meteor Soc, 1991, 72: 1331-1338.

[6] Lau K-M, Chan P H. Aspects of the 40-50-day oscillation during the Northern summer as inferred from outgoing longwave radiation[J]. Mon Wea Rev, 1986, 114: 1354-1367.

[7] Meehl G A. The annual cycle and its relationship to interannual variability in the tropical Pacific and Indian Ocean regions[J]. Mon Wea Rev, 1987, 115: 27-50.

北太平洋海温异常与东亚夏季风相互作用的年代际变化

李峰　何金海

(南京气象学院,南京 210044)

摘要: 主要利用英国气象局提供的海温资料和 NCAR/NCEP 提供的 40 年高度场及风场资料分析了东亚夏季风与北太平洋 SSTA 的关系,指出两者之间相互作用存在着年代际变化特征,20 世纪 70 年代中期以前,北太平洋海温异常通过一大圆波列作用于东亚夏季风,造成我国华北地区夏季降水偏多,1976 年以后,北太平洋海温异常使大圆波列减弱,与东亚夏季风关系淡漠,不再影响华北降水。由此指出:影响东亚天气气候的海温关键区,并不总固定在某一海区,随着海气作用的年代际变化关键区是会发生转移的。

关键词: 北太平洋海温;东亚夏季风;海陆温差指数;大圆波列

1 引　言

东亚背靠欧亚大陆,东、南毗邻广阔的太平洋,地跨热、温、寒三种气候带,海陆热力差异巨大,冬夏气候截然相反,冬季干燥而寒冷,夏季暖而湿润,属于典型的季风气候。由于特殊的地理环境,我国气象学者很早就注意到周围海洋异常状况对我国天气气候的影响,往往会造成我国局部地区灾害性天气和气候异常,如长江洪水、华北干旱。通过前人广泛而深入的研究,已经发现一些影响东亚气候的海温关键区,如葛孝贞等[1]指出冬春中纬度黑潮区海温偏高持续 2 个月以上,则夏半年副高脊线较平均位置南压 1~2 个纬度,江淮流域汛期将偏涝。许金镜[2]指出西北太平洋海温与副高脊线第二次北跳有关联。吴仁广等[3]和毛天松等[4]又强调了北太平洋海温的重要作用,指出 PNA 型环流的变化与北太平洋海温的联系较之赤道太平洋海温更密切,北太平洋海温距平造成的异常热成风影响,使西太平洋东西风带强度和副高位置发生异常,从而导致长江中下游夏季风的异常。何金海等[5]在研究东亚季风区旱涝年季节转换的不同特征时,指出涝年日本南部海区海温偏高,根据热成风原理,在此区域上空易形成一反气旋,叠加在西太平洋副高上,使副高强度偏大,脊线稳定在 25°N 附近,使江淮多雨。近几年,对赤道中东太平洋海温和 ENSO 的研究逐渐重视和深入,许多学者建立了 ENSO 与东亚夏季风气候关系模型。杨修群等[6]指出 4—6 月赤道中东太平洋海温偏高(偏低)则 6 月长江流域偏旱(偏涝)。杨修群等[7]利用海气耦合模式得到:ENSO 增暖在热带地区主要伴随着赤道中西太平洋 Walker 环流的减弱,中东太平洋气压降低以及表层辐合上升运动的增强;夏季表现为印度夏季风环流的显著减弱和东亚夏季风的增强。邹力等[8]诊断表明,El Niño 和 La

---

* 本文发表于《热带气象学报》,2000 年第 16 卷第 3 期,260-271。

Niña 事件中亚洲夏季风成员均发生不同程度的变化,我国东部地区夏季降水与 ENSO 循环的发展阶段密切相关,且在长江中下游地区和华南地区最为显著。最近,吴国雄等[9]又提出赤道印度洋—太平洋地区海气系统的齿轮式耦合和 ENSO 关系密切的概念。也有人[10]提出热带印度洋和太平洋协同作用对大气环流影响更符合观测事实。

从以上研究我们可以看到,不同的学者利用不同资料和方法可以提出不同观点,影响东亚天气气候的海温关键区就太平洋地区至少就有 2 个,哪一个才是影响东亚气候最重要、最关键、最显著的区域,研究其异常是否就能解决我国灾害性天气的预测问题?所得到海温关键区对东亚夏季风的影响,为什么在某些年份甚至更多的年份与观测的事实不符,这是什么原因呢?通过分析发现,前人所做的研究,都是建立在季节、年际变化的基础之上,而海温场和气候系统存在着显著的年代际变化,如果忽略这种气候背景,仅仅进行普遍计算,所得到的结果必然在某段时期是准确的,而在另一段时期,另一种气候态下,并不显著。许多研究表明[11],20 世纪 70 年代中期全球气候以及海温场都发生了一次突变,因此分别研究 70 年代中期之前和之后,不同海温背景下,海温与东亚夏季风的相互关系是必要的,本文的主要目的是要分析在不同的年代际背景下,海温与东亚夏季风之间的联系及其机制的变化,以及海温关键区是否会发生转移及如何转移,从而为我国旱涝预报提供新的启示。

2 资料与方法

主要利用英国气象局提供的 $1°×1°$ 的海温场资料和 NCAR/NCEP 提供的 1958—1997 年(共 40 年)$2.5°×2.5°$ 高度和风场资料,另外降水资料是中国气象局整编的全国 160 标准站 45 年资料;使用的主要方法有 EOF、一点相关、合成分析等。

3 20 世纪 70 年代中期海温关键区的转移

3.1 一种新的东亚夏季风的表示方法

过去的研究已经给出了许多表示东亚夏季风强度的方法,归纳起来主要有两种观点:一种是从环流的角度定义季风指数,可以较好地反映季风环流的状况,但却未能反映出季风另一重要特征即季风性降水状况,例如 Webster 等[12]的指标。另一种是从海陆热力差异的角度出发定义东亚季风强度指数,以郭其蕴[13]的工作为代表,取 $10°\sim 50°N$ 范围内,每隔 10 纬度(110°E 减 160°E 上)的月平均气压差 <-5 hPa 的所有值之和,然后各年值与多年值求比值,称为夏季风强度指数。该指数后来经施能、朱乾根(1996)改进,效果更好。它通过海陆气压系统的差异,一定程度上反映了决定东亚季风的海陆热力差异,并在反映降水方面有所改进。

本文引用了孙秀荣等①提出的定义东亚夏季风强度年际异常的一种新方法,直接用东亚地区海陆热力差异表示东亚夏季风强度,既包含纬向海陆热力差异,也包含经向海陆热力差异。用东亚季风区($27°\sim 35°N$,$105°E$ 以东的大陆)范围内的地表气温和副热带西北太平洋($15°\sim 30°N$,$120°\sim 150°E$)的海表温度之差表示东西向热力差异,用华南地区($27°N$ 以南、

---

① 孙秀英.东亚海陆热力差指数的年际变化及其与环流和降水的关系[D].南京:南京气象学院.

105°E以东的大陆)的气温和南海(105°~120°E,5°~18°N)海表温度之差表示南北向海陆热力差,然后加权求和,称为东亚海陆温差指数,简称为 LSTD(Land-Sea Thermal Difference)或(Land-Sea Temperature Difference),表示为:$LSTD = 4/5 \times (T_{EC} - SST_{STNWP}) + 1/5 \times (T_{SC} - SST_{SCS})$。该指数能很好地反映东亚夏季风的环流形势以及与夏季季风降水异常的关系,具体方法和意义请参考文献[12]。

图1给出了1958—1993年的东亚夏季海陆温差指数 LSTD 的年际变化。由图可见,LSTD 不但有明显的年际变化,而且具有显著的年代际变化趋势如虚线所示。20 世纪 70 年代中期前,LSTD 正值偏多,数值偏大,反映了东亚夏季风偏强,70 年代中期以后,多为负值,绝对值偏大,反映了东亚夏季风偏弱。根据孙秀荣的定义选取 LSTD>1.0 为强 LSTD 年,表示东亚强夏季风,LSTD<−1.0 为弱 LSTD 年,表示东亚弱夏季风,则 1959、1961、1967、1971、1978、1990 年为强 LSTD 年共 6 年,其中 4/6 的年份发生在 70 年代中期之前;1980、1982、1983、1987、1989、1992、1993 年为弱 LSTD 年共 7 年且全部发生在 70 年代中期以后,9 点平滑曲线更好地反映了年代际特征,如图从 1958 年到 70 年代初,9 点平滑曲线全为正值,70 年代末到 90 年代初 9 点平滑曲线全部转为负值,70 年代中期为过渡期;为了更好地确定东亚夏季风在 70 年代中期的气候转折点,利用 M-K 方法进行突变点检验,突变点约发生在 1975/1976 年间,1958—1975 年东亚夏季风总体表现为偏强,1976—1993 年东亚夏季风偏弱。

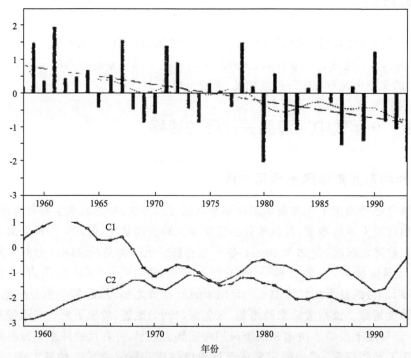

图 1　东亚夏季海陆热力差指数(LSTD)年际演变直方图

(长虚线为趋势线,点虚线为 9 点平滑曲线;C1:M-K 检验 C1 曲线;C2:M-K 检验 C2 曲线)

3.2　与东亚夏季风关系密切的海温关键区及其年代际转移

首先,我们合成强弱海陆温差指数年的太平洋海温异常分布如图 2 所示:海陆温差指数强年时,海温场表现为北太平洋区域为明显的大片正距平控制,中心位于41°N、170°E,西北太平洋也有两个SSTA正中心,而热带中东太平洋为负SSTA区。弱海陆温差指数年北太平洋地区变成了负SSTA区,而赤道中东太平洋为强大正SSTA控制,中心值达到1℃,整个太平洋海温场结构呈现出El Niño型。由此可见,东亚夏季风的异常确实对应着海温场的明显不同,夏季风强时对应着北太平洋区域海温异常偏高,赤道中东太平洋海温偏冷;夏季风弱时,海温场形势恰相反,因此北太平洋及赤道中东太平洋都是与东亚夏季风密切相关的海区。

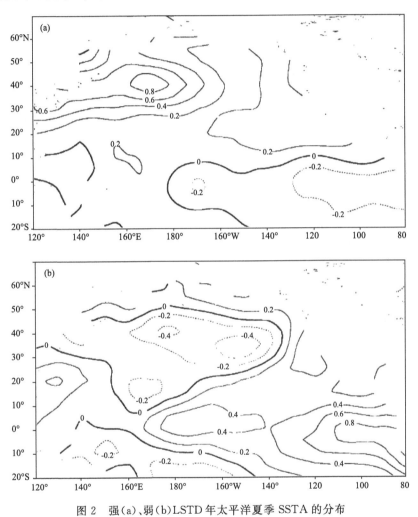

图2 强(a)、弱(b)LSTD年太平洋夏季SSTA的分布

东亚夏季风和太平洋海温在70年代中期都发生明显的变化,那么它们之间有什么联系?表1给出了LSTD指数与太平洋各海区的相关表。由表可见,1958—1993年东亚夏季风(LSTD)与北太平洋(30°~50°N,180°~150°W)区域,赤道中东太平洋(10°S~10°N,180°~90°W)海区关系都较好,相关分别达到0.423和-0.371,都通过0.05的信度检验。但是我们发现,东亚夏季风与这两个海区的关系随着年代际变化,1958—1975年,LSTD与北太平洋海区关系较好,相关达到0.401,而与赤道中东太平洋基本没有关系,相关仅为-0.0514,1976—1993年,东亚夏季风与北太平洋关系变得不明显,而与赤道中东太平洋关系变得非常密切,信度可达0.01。由此可见,与东亚夏季风关系密切的海温关键区在1976年前后是可能发生了

转移。

表1 东亚夏季海陆温差指数LSTD与太平洋各海区SSTA相关表

	东亚夏季海陆热力差异指数(LSTD)		
	1958—1993	1958—1975	1976—1993
北太平洋区域SSTA	0.423	0.401	−0.047
赤道中东太平洋SSTA	−0.371	−0.0514	−0.558
信度检验	$\|r\|_{0.08}=0.3246$	$\|r\|_{0.02}=0.3810$	$n=36$
	$\|r\|_{0.08}=0.4438$	$\|r\|_{0.02}=0.5155$	$n=18$

图3给出了东亚夏季海陆温差指数与太平洋SSTA的相关图。由图可见，在1958—1975年东亚夏季海陆温差指数与海温较大的相关区位于北太平洋，与赤道中东太平洋的关系很差。而1976—1993年大的相关区转移到赤道中东太平洋，与北太平洋的关系不明显。这直观地说明1976年前后海温关键区的转移问题。

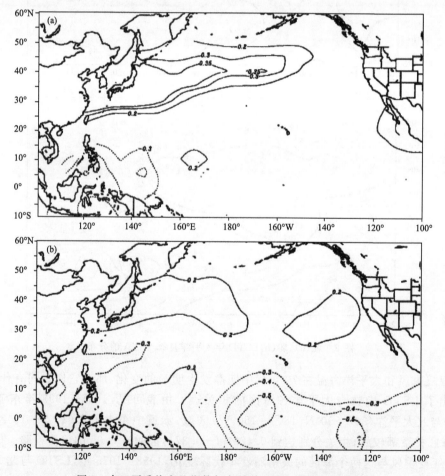

图3 东亚夏季热力差指数与太平洋SSTA同期相关分布
a. 1958—1975；b. 1976—1993（阴影区为通过0.05信度显著区）

4 北太平洋海区与东亚夏季风关系年代际转移的原因

4.1 东亚夏季风年际、年代际变化对太平洋 SSTA 的影响

根据前面强弱海陆温差指数年的定义,发现强温差指数年中有 4 年是出现在 1976 年以前,而 7 个弱海陆温差指数年全部发生在 1976 年以后,因此分析强弱海陆温差指数年的海气状况既包含了年际变化的信息,更主要反映了年代际的变化,具体的反映为 1976 年前后的海气作用情况。

图 4 给出了强弱海陆温差指数年对流层低、高层的流场差值分布,该图为弱 LSTD 年流场形势减强 LSTD 年流场形势。850 hPa(图 4a)上,经由阿拉伯海、印度半岛、孟加拉湾、中南半岛后转向北进入东亚大陆的西南夏季风减弱,欧亚大陆中高纬度纬向西风加强;北太平洋有一异常反气旋生产,中心位于 50°N、175°W 附近,气旋西面寒冷的北风在我们选取的北太平洋

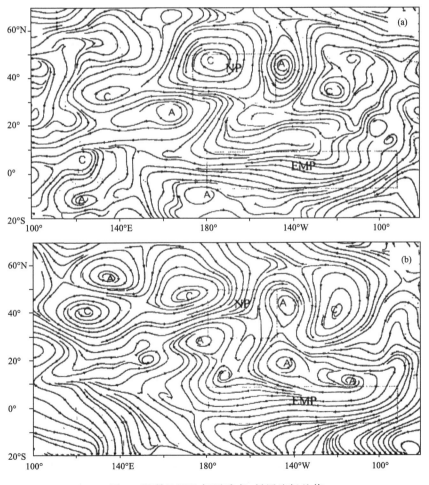

图 4　强弱 LSTD 年夏季高、低层流场差值
a. 850 hPa;b. 200 hPa

区域转为西风并与东亚大陆吹来的西风汇合。从而使该地区的西风异常加强,一方面使北太平洋区域海水潜热释放加快,另一方面使冷水上翻,海温持续降低;同时热带的赤道太平洋附近西风异常加强,东风减弱,从而驱动赤道中东太平洋海温持续升高。200 hPa 上(图 4b)北太平洋上空的异常反气旋中心较之低层偏移了约 10°,朝鲜半岛上空有一反气旋,鄂霍次克海和中国南部及中南半岛上空为反气旋控制,因此朝鲜半岛上空气旋西南转向的西风与北太平洋上空由北而来的转向冷气流在北太平洋地区汇合,使该地区西风异常增强。而热带地区,在赤道东太平洋上空出现了一强大的反气旋,反气旋南面的东风沿赤道异常强大。结合高低空流场异常形势,发现在北太平洋上空从低层到高层西风都增强,系统稳定,这样可以持续驱动北太平洋区域海温下降;而在赤道中东太平洋低层西风增强,高层东风增强,有利于 Walker(沃克)环流减弱,驱动该地区海温持续升高。

4.2 北太平洋海温年际、年代际变化尺度上与东亚夏季风的相互作用

在前面我们已经从流场角度讨论了东亚夏季风异常对太平洋海温的驱动作用,使不同海区的海温下降或升温。下面将从高度场形势来研究东亚夏季风与海温的相互作用,以找出海温关键区转移的原因。

根据强弱海陆温差指数年的分布特点,强 LSTD 年多发生在 1958—1975 年期间,弱 LSTD 年全部发生在 1976—1993 年,而前一阶段为东亚夏季风总体偏强时期,后一阶段为弱夏季风时期,因此我们用强、弱 LSTD 极端异常年份的合成场,代表强弱夏季风不同年代际气候背景场,这样可以突出东亚热力差因子的作用。然后,考虑在这种背景场下,夏季北太平洋海温的年际异常影响,以此来分析东亚夏季风与太平洋海温相互作用的年代际、年际变化。

图 5 是强弱 LSTD 年夏季 500 hPa 高度场异常形势。由图可见,强 LSTD 年(图 5a)从菲律宾到中南半岛、我国南部,经由我国江淮地区、朝鲜半岛、日本南部、鄂霍次克海、北太平洋、一直到北美西岸存在一个明显的波列,菲律宾到中南半岛、我国南部,鄂霍次克海分别为负异常,位势偏低,朝鲜半岛与北太平洋上空分别为正异常,位势偏高,东北太平洋为负异常,北美西海岸为正异常,这种分布形势减弱了鄂霍次克海高压,减弱中高纬经向环流,使西太平洋副高位置偏北,强度减弱,西风带北推,有利于夏季风北进,与偏北的冷空气相互作用,使梅雨锋位置偏北,从而使华北地区降水偏多,另外,菲律宾上空的高度场偏低,有利于华北地区对流增强,从而使华北降水偏多,而江淮流域处于单一强劲的夏季风气流控制下容易发生干旱。而弱 LSTD 年(图 5b)该波列形势依然明显,但从菲律宾、中南半岛到北美西海岸成"＋－＋－－"中心相间分布。这说明,夏季东亚东西、南北向海陆热力差异与北太平洋海温异常是通过这一波列相互作用,当东亚夏季海陆热力差异大(即东亚夏季风强),从菲律宾、中南半岛到北美西海岸成"－＋－＋－＋"型分布,我们把这种分布形势称为波列正位相,而当东亚夏季海陆热力差异小(东亚夏季风弱),该波列位相变为反位相,从菲律宾,中南半岛到北美西海岸成"＋－＋－＋－"型分布,这种分布形势加强了鄂霍次克海高压,加强中高纬度的经向环流,副高位置偏南,西风扰动偏南,有利于北方冷空气南下与夏季风交汇于江淮流域,使江淮降水偏多,华北旱,另外菲律宾周围上空位势偏高也有利江淮流域降水。而该波列与我国学者所说的 EAP 型(东亚太平洋型)很相似,其路径成一大圆,因此我们称之为"大圆波列"。

既然,在东亚热力差异(夏季风)与北太平洋区域之间存在着这样一个波列联系,那么,为什么北太平洋区域海温在 20 世纪 70 年代中期以前与东亚夏季风,华北地区降水关系好,而

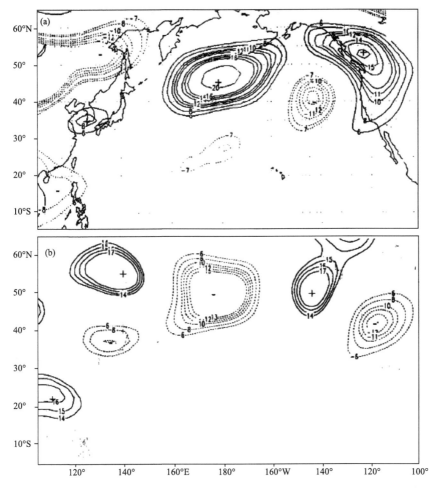

图 5　强(a)、弱(b)东亚夏季海陆热力差指数年亚太地区夏季 500 hPa 高度场异常分布型

70 年代中期以后关系又不明显了呢?为了解决这一问题,还需要分析海温异常对大气环流及东亚夏季风的影响。

上面我们已经给出了强弱夏季风的年代际背景场,下面我们将分析北太平洋夏季海温年际异常对这种背景场的作用。根据北太平洋区域(30°~50°N,180°~150°W)夏季平均海温年际演变曲线,合成峰值年减谷值年 500 hPa 高度场形势,作为北太平洋海温升高对 500 hPa 高度场的强迫。这样选取的年份可以通过 0.05 的 t 检验。如图 6 所示,在菲律宾、中南半岛为负异常中心,中国东部、朝鲜半岛、日本南部为正异常,鄂霍次克海上空为负异常,北太平洋上空为正异常中心控制,东北太平洋为负,北美西海岸为正。这种高度场异常分布与强东亚夏季风的气候背景场形势(图 5a)是一致的,而与弱东亚夏季风的气候背景场形势(图 5b)相反。这说明北太平洋夏季海温增温激发的大气异常,和 1958—1975 年强夏季风背景场存在的大圆波列的位相一致,异常形势叠加,该时期大圆波列加强,加强了东亚夏季风与北太平洋海温的联系;而 1976—1993 年阶段,北太平洋夏季海温对大气的强迫,与弱夏季风气候背景场形势相反,减弱了大圆波列,因此也减弱了北太平洋与东亚夏季风的联系,使两者关系变得淡漠。

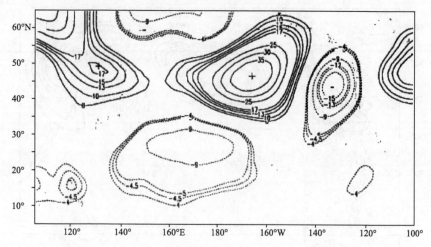

图6 北太平洋夏季海温年际异常激发的500 hPa高度场异常形势

4.3 赤道中东太平洋海区与东亚夏季风关系的年代际变化

近年来,国内外对ENSO事件的研究日渐深入,许多研究结果把赤道中东太平洋海温异常与东亚气候通过某种纽带联系起来,其中Walker环流上升支处于西太平洋至菲律宾附近,下沉支位于赤道中东太平洋,将赤道东太平洋海温异常与东亚大气环流异常密切联系起来,因此研究赤道中东太平洋海温与东亚季风的关系,分析Walker环流的异常状况是必要的。

由于Walker环流是一个横贯热带太平洋上空的纬向闭合环流,高层盛行西风,低层为东风,所以我们选取热带太平洋区域10°S~10°N,140°E~120°W的高低纬向风之差表示Walker环流的异常情况,称为沃克环流指数,简称WCI(Walker Circulation Index)。表2给出了WCI与东亚夏季风及赤道中东太平洋SSTA的相关表。由表可见,沃克环流异常与赤道中东太平洋海温异常几乎成反位相,关系异常密切,而沃克环流与东亚夏季风的关系却存在着年代际的变化,1958—1975年,相关系数为-0.0184,几乎没有关系,而1976—1993年相关达到0.4786,并通过了0.05的信度检验。因此1976年以后东亚夏季风与赤道中东太平洋关系异常密切,我们认为可能是通过沃克环流来实现的。

表2 沃克环流指数与赤道中东太平洋SSTA及东亚夏季海陆温差指数相关表

	沃克环流指数 WCI		
	1958—1993	1958—1975	1976—1993
赤道中东太平洋 STA	−0.8245	−0.8958	−0.7138
东亚夏季海陆温差指数 LSTD	0.2959	−0.0184	0.4786

5 结 语

(1)引用了一种新的表示东亚夏季风强度的方法和指数,很好地反映了东亚夏季风的年际、年代际异常形势。东亚夏季风与太平洋海温在20世纪70年代中期都发生了显著变化,并使与东亚夏季风密切相关的海温关键区发生了转移,70年代中期以前,东亚夏季风与北太平洋关系密切,70年代中期以后,东亚夏季风转而与赤道中东太平洋的关系强烈非常密切。

(2)东亚夏季风与北太平洋关系由强转弱的原因为:70年代中期以前,由强的东亚夏季海陆热力差激发的,存在于从菲律宾、中南半岛经北太平洋到北美西海岸的大圆波列,与由夏季北太平洋海温异常激发的波列同位相,加强两者之间的联系;1976年以后,由弱的东亚夏季海陆热力差激发的,存在于从菲律宾,中南半岛经北太平洋到北美西海岸的大圆波列,与由夏季北太平洋海温异常激发的波列反位相,使波列不明显,减弱了两者之间的联系。

(3)70年代中期以后海温关键区转移到赤道中东太平洋,可能是沃克环流发生了显著变化,加强了中东太平洋与东亚地区的联系,这有待进一步研究。

参考文献

[1] 葛孝贞,余志豪.海温变化与副热带高压活动的数值模拟[J].热带气象,1986,2:109-117.
[2] 许金镜.西北太平洋温度与西太平洋副热带高压脊线活动关系的初步分析[J].热带气象,1986,2:282-288.
[3] 吴仁广,陈烈庭.PNA流型的年际变化及温、热带太平洋海温的作用[J].大气科学,1992,16:583-590.
[4] 毛天松,许乃道.北太平洋海温距平对6月长江中下游夏季风影响及其机制分析[J].热带气象,1988,4:327-335.
[5] 何金海,温敏,罗京佳.东亚季风区旱涝年季节转换的不同特征[C]//何金海.亚洲季风研究的新进展——中日亚洲季风机制合作研究论文集.北京:气象出版社,1996:82-87.
[6] 杨修群,谢倩,黄士松.赤道中东太平洋海温和北极海冰与夏季长江流域旱涝的关系[J].热带气象,1992,8:261-266.
[7] 杨修群,谢倩,黄士松.全球大气/热带太平洋耦合模式中由ENSO增暖引起的全球大气环流异常[J].大气科学,1996,20:129-137.
[8] 邹力,倪允琪.ENSO对亚洲夏季风异常和我国夏季降水的影响[J].热带气象学报,1997,13:306-314.
[9] 吴国雄,孟文.赤道印度洋—太平洋地区海气系统的齿轮式耦合和ENSO事件I资料分析[J].大气科学,1998,22:470-481.
[10] 李永平,秦曾灏,孙照渤.冬季热带太平洋和印度洋SSTA对大气协同作用的数值实验[J].气象学报,1996,54:612-619.
[11] Nitta T,Yanmada S. Recent warming of the tropical sea surface temperature and its relationship to northern hemisphere circulation[J]. J Moteor Soc Jap,1989,67:375-383.
[12] Webster P J,Yong S. Monsoon and ENSO: selectively interactive systems[J]. Quart R Moteor Soc,1992,118:877-926.
[13] 郭其蕴.东亚季风强度指数及其变化分析[J].地理学报,1983,38:207-216.

关于南海夏季风建立的大尺度特征及其机制的讨论[*]

何金海 徐海明 周兵 王黎娟

(南京气象学院,南京 210044)

摘要:使用1998年南海季风试验期间高质量资料和NCEP/NCAR40年再分析资料分析了南海季风建立前后的大尺度环流特征和要素的突变及爆发过程。发现南亚高压迅速地从菲律宾以东移到中南半岛北部,印缅槽加强,赤道印度洋西风加强并向东向北迅速扩展和传播,以及相伴随的中低纬相互作用和西太平洋副高连续东撤是南海夏季风建立的大尺度特征,与此同时,亚洲低纬地区的南北温差和纬向风切变也发生相应的突变。数值实验结果指出,印度半岛地形的陆面加热作用在其东侧激发的气旋性环流对于印缅槽的加强有重要作用,并进而有利于南海夏季风先于印度夏季风爆发。

关键词:南海夏季风;大尺度;印缅槽;数值实验

1 引 言

亚洲季风系统由东亚季风和南亚季风两个既相互独立又相互联系的子系统组成[1],而东亚季风又可划分为南海—西太平洋热带季风和中国大陆东部—日本的副热带季风[2]。这是近20年来季风研究的主要成果之一。

南海夏季风由于其地理位置的特殊性,不仅是印度季风和东亚季风连接的媒介,也是亚澳季风相互联系、相互作用的表现。Murakami等[3]提出,由亚洲和澳大利亚热力差异所造成的气压梯度的大尺度季节转换比海陆热力差异所造成的大尺度季节转换敏感和强烈得多,南海地区正是这种变化敏感的区域。因此,关于南海夏季风建立特征的研究对于亚洲冬季风形势向夏季风形势的转换具有一种信号作用。

陶诗言和陈隆勋[1]首先指出,亚洲夏季风的爆发最早开始于南海北部,然后向西向北分阶段推进。He等[4]则发现亚洲夏季风的建立存在两个明显的阶段:第一阶段出现于5月,南海低层西风建立;第二阶段发生在6月,印度西海岸出现降水和阿拉伯西南风。何金海等[5]通过对多年TBB资料的分析指出,亚澳季风区的季节转换在4月就已开始,主要特征表现为澳大利亚TBB高值带(对应澳大利亚高压)迅速北移和中南半岛地区对流的活跃以及副热带TBB高值带(对应副热带高压)在该地区的断裂,伴随着东部TBB高值带的连续东撤,南海夏季风于5月第4候建立;而西部TBB高值带的向西退缩则伴随着亚洲季风建立自东向西的推进过程;东亚地区,伴随着TBB高值带和西太平洋副高的季节性北移,南海—西太平洋热带夏季风和中国大陆—日本的副热带夏季风也自南向北推进,但前者推进的经向距离远比后者小。

---

[*] 本文发表于《气候与环境研究》,2000年第5卷第4期,333-344。

上述研究表明,亚洲夏季风最早于南海地区建立,且其爆发性特点特别明显。然而,在南海季风试验之前,由于南海地区处于诸项国际科学试验范围的空白区,直接观测资料缺乏,致使有关南海夏季风问题的研究还不够多或者其研究结果还缺少可靠资料的支撑,同时南海夏季风是否最早建立尚有争议,其爆发性特征需要解释,爆发性过程需要进一步揭示。南海季风试验的高质量资料及近年来可能提供的其他资料为我们进一步揭示南海夏季风建立的大尺度突变特征及其机制提供了可能。关于南海夏季风特别是1998年南海夏季风建立的大尺度特征已有许多研究[6~9],然而本文提出的相关问题尚需进一步深入讨论。

2 资料说明

本文所用资料主要包括:南海季风试验期提供的1998年5—8月基本要素场逐日全球格点资料;CDC提供的1998年NCEP/NCAR每6小时一次的风场资料;ECMWF提供的1979—1993年850 hPa风场逐日全球格点资料;1959—1998年NCEP/NCAR 40年月平均再分析资料集;NOAA提供的1979—1993年向外长波辐射(OLR)资料。上述格点资料水平分辨率为$2.5°\times2.5°$经纬度。

3 南海夏季风建立的大尺度特征及过程

3.1 南海夏季风建立前后的突变特征

资料分析已经表明,南海夏季风最早建立于4月底,最迟建立于6月上旬,平均来说在5月中下旬(公认5月第4候)建立。这就是说,就月际尺度而言,4月属于南海季风建立前,6月整个亚洲夏季风形势建立,属于季风建立后,而5月是南海夏季风建立的时期。为了揭示南海夏季风建立前后的大尺度特征,我们给出了40年平均4月和5月的高低空流场资料(图1)。如图1a所示,4月南亚高压(或称季风高压)位于菲律宾以东洋面上;500 hPa副热带高压沿15°N基本呈带状分布,其上有两个高中心,一个位于菲律宾上空,一个位于阿拉伯以西地区,孟加拉湾为相对低值区;850 hPa上副热带地区仍为高压控制,且孟加拉湾低值区更加明显,西太平洋副高脊线仍控制着中南半岛—南海地区。印度以西的阿拉伯地区为一反气旋中心控制。东非越赤道气流已经可见,但强度不大,虽然澳大利亚高压已显著存在,但105°~120°E地区的越赤道气流却未出现。这种大尺度流场形势表明南海夏季风尚未建立。在图1b中,5月200 hPa南亚高压从菲律宾以东移到中南半岛北部,其东部的偏北偏东气流控制着南海地区;500 hPa和850 hPa上最明显的特征是,西太平洋副高脊线位置已明显东撤,印度以东的孟加拉湾地区出现了显著的低槽(印缅槽),与印缅槽相对应,赤道南半球一侧有气旋,低槽和气旋之间有较强的赤道西风。这支赤道西风看来是由来自阿拉伯海地区的中纬度西北气流和源自南半球的偏西气流组成。赤道西风流向中南半岛—南海地区,与西太平洋副高南侧的转向气流和源自澳大利亚高压北侧的越赤道气流汇合。显然,南海地区低层处于三支偏南气流的汇合处,高空位于南亚高压东部偏北气流控制下。这种形势表明就月际尺度而言,5月应是南海夏季风建立期。但值得指出的是,印度地区仍为阿拉伯地区反气旋前部的西北气流控制,这表明印度夏季风尚未爆发。

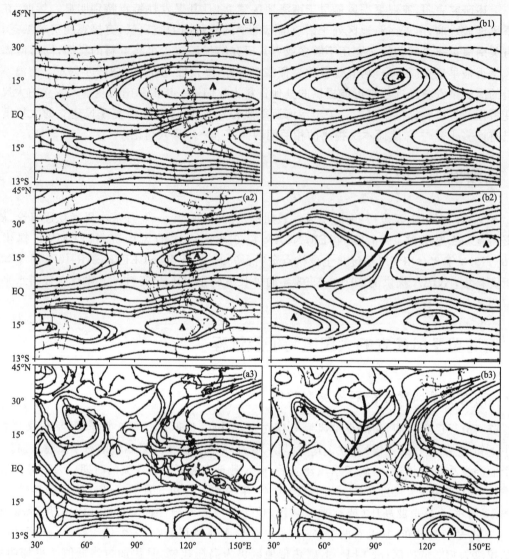

图 1 40 年(1959—1998)平均的 4 月和 5 月平均流场
(a1)、(a2)、(a3) 和 (b1)、(b2)、(b3) 分别为 4 月和 5 月 200、500、850 hPa
A:反气旋;C:气旋;粗实线为槽线

与上述大尺度流场 4 月到 5 月的突变相对应,亚洲中低纬地区南北温差和纬向风切变(低层减高层)也发生了显著变化。由图 2 可见,平均来说(图 2 中实线),90°E 以东的中低纬地区,南北温差在 5 月中下旬(27—29 候)都发生了稳定的符号转变(由正到负),这种转换最早发生在中南半岛地区(100°～110°E)依次是南海(110°～120°E)和孟加拉湾东部(90°～100°E)地区。孟加拉湾西部到印度地区(80°～90°E)南北温差由正到负的转换则发生在 6 月上中旬(32—33 候),这与印度夏季风最迟建立相对应。图 2 中虚线指出了 1998 年南北温差的演变,可以看出中南半岛和南海地区南北温差符号的转变均发生在 28 候附近,显然与南海夏季风建立日期基本一致。图 3 显示了纬向风切变的演变特征,与图 2 不同的是,纬向风切变由负到正转变的时间比南北温差转变要略早,且西早东迟。这种特征可能与加强的印缅槽槽前西南风和赤道西风自西向东扩展相联系。另外,1998年各个区域纬向风切变由负到正的转变均较多年平均转换时间要迟,这与1998年南海夏季风建立日期偏迟相吻合。

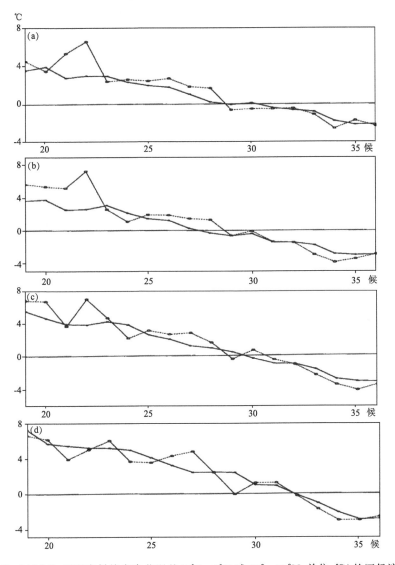

图 2　500 hPa 亚洲中低纬度南北温差(5°S～5°N 减 20°～30°N,单位:℃)的逐候演变
(a)110°～120°E;(b)100°～110°E;(c)90°～100°E;(d)80°～90°E
实线:1990—1998 年平均;虚线:1998 年

综上所述,与大尺度流场突变相联系,亚洲中低纬若干地区的南北温差(纬向风切变)在5月中下旬(上中旬)均发生转变,它与这些地区夏季风的建立有内在联系。然而,分析表明,单用某一个参数来确定特定年份南海夏季风的建立日期是困难的。

3.2　南海夏季风建立过程的分析

为了揭示南海夏季风建立过程的特征,我们采用合成分析方法制作了季风建立前后沿赤道地区的 850 hPa 纬向风的时间—经度剖面(图4)。图 4 中纵坐标 0 表示南海夏季风建立日

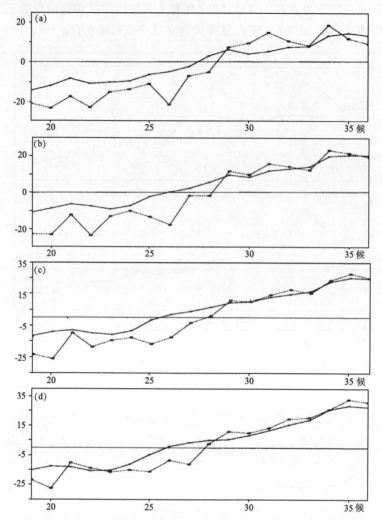

图 3 亚洲低纬(10°~20°N)各区域纬向风切变(850 hPa 减 200 hPa,单位:m·s^{-1})的逐候演变
(a)110°~120°E;(b)100°~110°E;(c)90°~100°E;(d)80°~90°E
实线:1990—1998年平均;虚线:1998年

期,其确定方法与文献[10]相同,只是采用日平均资料,所得到的南海夏季风建立日期亦与其基本相同。由图 4 可以看出,南海夏季风建立前一直为东风控制,约前 15 天(80°~90°E)地区的赤道西风有一个风速稳定增长的过程,且在前 12 天达到最大值,随后迅速东扩。显然,这一过程对南海夏季风的建立或爆发有重要作用。

为了进一步揭示南海夏季风建立的过程特征,我们还制作了从赤道印度洋到中国华南沿海的西南—东北向的剖面图(图 5)。由图 5a 可以看出,南海夏季风建立前约 15 天,赤道印度洋(80°E,EQ)西风不仅稳定增长,还明显地向东北扩展和传播,当它扩展到南海地区(110°E,15°N)时,南海夏季风建立。伴随赤道西风向东北的传播,亦有一个 OLR 的低值区向东北传播(图 5b)。1998 年南海季风试验期间类似的降水剖面图(图 6)也表明,南海夏季风建立前期,从赤道印度洋地区也有明显的降水向东北传播,当它到达中南半岛地区时,南海夏季风建立。另外一个值得指出的现象是,与赤道印度洋西风、对流活动和降水向东北传播的同时,均

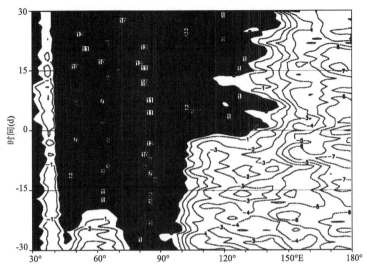

图 4　南海夏季风爆发前后 2.5°～5.0°N 平均的 850 hPa 纬向风的时间—经度剖面时间纵坐标 0 表示南海夏季风建立日期，正(负)值表示季风建立后(前)；阴影区为纬向风大于 0 的区域；等值线间隔 1.0 m·s^{-1}

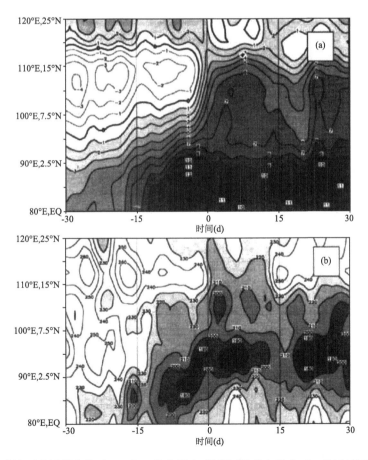

图 5　南海夏季风爆发前后 850 hPa 纬向风(a,阴影区为纬向风大于 0 的区域)和 OLR
(b,阴影区为 OLR 值小于 230 W·m^{-2} 的区域)沿(80°E,EQ)点至(110°E,15°N)点的时期演变图

有来自中国华南地区的西风、对流和降水向南扩展。这一现象可能与华南静止锋的南扩相联系。Chang 等[11]曾强调中纬度锋系对南海夏季风建立的触发作用,与我们的结果是一致的。

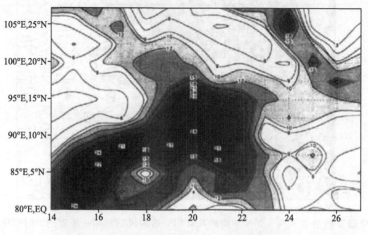

图 6 1998 年 5 月 14—27 日日降水剖面图

前面的分析表明,赤道印度洋西风一方面迅速东扩至南海的赤道地区,尔后向北扩展;另一方面又向东北传播和扩展,并伴随着对流和降水,与来自中纬度的西风和降水汇合。上述形式的中低纬相互作用导致了南海夏季风在很短时间内在南海地区南北大范围内迅速建立。这或许是导致南海夏季风建立的"爆发性"特征的原因之一。然而,我们不禁要问,赤道印度洋西风为什么在南海夏季风建立前有稳定增长的过程?它与印缅槽的加强有何联系?在下文中我们将要讨论这一问题。

为了揭示临近南海夏季风建立的流场变化特征,我们制作了 1998 年 5 月中旬与 5 月上旬流场较差图(图 7)。由图 7 可以看出,在 200 hPa,孟加拉湾北部地区有一显著的较差反气旋环流,其东部和南部的较差东风气流控制着中国南海地区和亚洲低纬度地区。这一较差反气旋与南亚高压自东南向西北的迅速移动相联系;在 850 hPa,从赤道印度洋经中南半岛直至中国南海和菲律宾地区为一强劲的偏西南较差气流,在赤道印度洋的南北两侧有一显著的涡旋对。根据前面的分析,夹在涡旋对之间的强西南风较差气流应是触发南海夏季风建立的直接原因之一。然而,这一涡旋对又是什么原因造成的?这仍是一个需要回答的问题。

4 关于南海夏季风为什么早于印度夏季风建立的讨论

亚洲夏季风究竟最早在哪里建立,这关系到冬季风形势向夏季风形势转换的最早信号问题。已有的研究[1]认为,夏季风最早在南海爆发,但有的作者认为夏季风最早在孟加拉湾或中南半岛地区爆发。我们通过分析要素场的突变和 θ_{se} 垂直结构的变化指出,1998 年亚洲夏季风首先于南海地区爆发[12]。这说明关于南海夏季风是否最早建立尚有争议,然而南海夏季风早于印度夏季风建立却是无可争议的事实。那么,这又是什么机制造成的呢?下面就来讨论这一问题。

从图 1 我们可以知道,4 月到 5 月对流层中低层大气环流的最大变化就是副热带高压的断裂,印缅槽的加强。有关合成分析的结果也指出,南海夏季风建立前期伴有印缅槽加强的过程。这就是说,孟加拉湾经中南半岛至南海地区处于印缅槽槽前西南气流的控制之下,而印度地区则处于印缅槽槽后西北气流的控制下。这一基本流场的特征有利于南海夏季风偏南气流

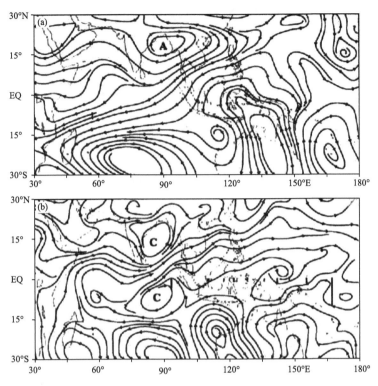

图 7　1998 年 5 月中旬与上旬流场较差图
A:较差反气旋;C:较差气旋。(a)200 hPa;(b)850 hPa

的建立,而不利于印度地区夏季风偏南气流的建立。只有当印度地区海陆温差转换引起的偏南风足够大到能抵消上述基本流场的偏北气流后,印度夏季风偏南气流才得以建立。因此,南海夏季风早于印度夏季风建立应该是印缅槽引起的大尺度流场造成的结果。这与前面指出的纬向风切变转换日期较南北温差转换要早的特征也是一致的。然而,印缅槽为什么在 5 月得以建立和加强呢? 为了回答这一问题,我们用 CCM3 模式[13]做了相关的控制试验(CTL)和变印度半岛地形为海洋的敏感性试验(NIT)。

CTL 的初始场采用经过多年积分以后得到的 9 月 1 日的模式场,由此开始积分 330 天,每 5 天取一次平均作为一次结果输出。在 NIT 试验中,将 20°N 以南的印度半岛地形变为海洋,同时将其变为海洋后的海温值用同纬度邻近海域的海温来代替,其余处理均与 CTL 相同。

CTL 的结果指出,南海夏季风平均于 5 月中旬建立。图 8a 和 b 表明,模拟的 5 月流场与图 1 中 5 月的气候场基本相似,即印缅槽已明显存在,槽前西南气流与赤道印度洋西风气流汇合于中南半岛和南海地区。6 月,从印度到中国南海直至菲律宾以东,已经出现了一条近似西北—东南向的亚洲季风槽,夏季形势已完全建立。这表明模式的控制试验能较好地再现亚洲夏季风的建立。图 8c 是 CTL 与 NIT 在 5 月的差值流场,它反映了印度半岛地形对大气环流的影响。由图 8c 可清楚地看出,在印缅槽地区有一个明显的气旋差值环流中心(85°E,18°N),其低槽可一直伸至赤道附近。这一差值气旋环流和低槽可使印缅槽和赤道印度洋西风得到加强,从而有利于南海夏季风的建立而不利于印度夏季风的建立。

进一步的分析表明,这一气旋性差值环流中心与印度半岛地形在由冬入夏时的陆面感热加热相联系。根据吴国雄等[14]的理论研究结果,当陆面加热和温度升高成同位相配置时,在大陆东部将有气旋式环流发展。上述数值实验证实了这一观点。

一般认为,印缅槽的存在与青藏高原大地形的作用有关[15],然而我们的 NIT 试验结果指出印度半岛的存在及其相应的陆面加热作用对印缅槽的加强有重要作用,正是印缅槽的加强有利于南海夏季风爆发,而不利于印度夏季风爆发。

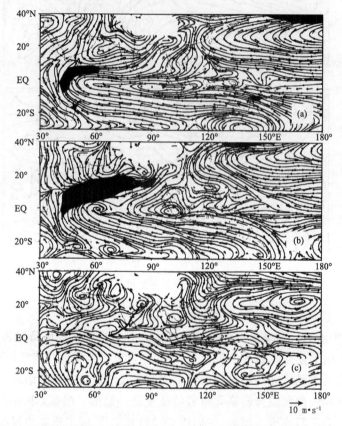

图 8　CCM3 控制试验 5 月(a)、6 月(b)850 hPa 流场和 5 月控制试验与无印度半岛试验 850 hPa 差值流场(c)阴影区为风速大于 8 m·s^{-1}

5　结果讨论

根据前面的分析,我们可以得到以下结论:

(1)有利于南海夏季风建立的大尺度形势是南亚高压迅速从菲律宾以东(4月)移到中南半岛北部(5月),印缅槽加深加强,其槽底偏西气流与源自南半球的赤道偏西气流汇合形成加强的赤道印度洋西风气流,然后流向中南半岛和南海地区。

(2)夏季风建立前后,南海地区处于来自赤道印度洋的西南气流、西太平洋副高南侧和源自澳大利亚高压北侧的越赤道气流汇合处,这给讨论南海夏季风的建立特征和过程带来了复杂性和更多的困难。赤道印度洋西风迅速东扩和向东北扩展以及相伴随的中低纬相互作用和西太平洋副高东撤是南海夏季风爆发的特征过程,至少 1998 年表现得特别清楚。与此同时,亚洲低纬地区的南北温差和纬向风垂直切变符号的转变也相应地发生，并且西风急流和湿度场也有明显的突变，但是用单一的要素场确定南海夏季风建立日期是有困难的。

（3）数值实验结果表明，印度半岛的存在对于亚洲夏季风的建立有重要影响。如果没有印度半岛，南海夏季风与印度夏季风的建立日期的差别远没有现在这样显著。印度半岛在由冬入夏时的陆面加热作用在其东侧激发出的气旋性环流加强了印缅槽，或许有利于夏季风建立前期赤道印度洋涡旋对的形成，正是印缅槽槽后的西北气流延缓了印度夏季风的爆发，槽前的西南气流加快了南海夏季风的爆发。

参考文献

[1] Tao Shiyan, Chen Longxun. A review of recent research on East Asian summer monsoon in China[M]. Monsoon Meteorology. Oxford: Oxford University Press, 1987, 60-92.

[2] Zhu Qiangen, He Jinhai, Wang Panxing. A study of circulation difference between East-Asian and Indian summer monsoons with their interactions[J]. Adv Atmos Sci, 1986, 3(4): 466-477.

[3] Murakami T, Chen Longxun, Xie An. Relationship among seasonal cycles, low-frequency oscillations and transent disturbances as revealed from outgoing long wave radiation data[J]. Mon Wea Rev, 1986, 114: 1456-1465.

[4] He Haiyan, Mcginnis J W, Song Zhengshen, et al. Onset of the Asian summer monsoon in 1979 and the effect of the Tibetan Plateau[J]. Mon Wea Rev, 1987, 115: 1966-1995.

[5] 何金海，朱乾根，Murakami M. TBB资料所揭示的亚澳季风区季节转换及亚洲夏季风建立的特征[J]. 热带气象学报, 1996, 12: 34-42.

[6] 丁一汇，薛纪善，王守荣，等. 1998年亚洲季风活动与中国的暴雨/洪涝[C]//丁一汇，李崇银. 南海季风爆发和演变及其与海洋的相互作用. 北京：气象出版社, 1999: 1-4.

[7] 李崇银，吴静波. 1998南海夏季风的分析研究[C]//丁一汇，李崇银. 南海季风爆发和演变及其与海洋的相互作用. 北京：气象出版社, 1999: 18-24.

[8] 罗会邦. 南海夏季风爆发及相关雨带演变特征[C]//丁一汇，李崇银. 南海季风爆发和演变及其与海洋的相互作用. 北京：气象出版社, 1999: 25-29.

[9] 邵慧，钱永甫. 1998年南海夏季风爆发前后区域环流变化的主要特征[C]//丁一汇，李崇银. 南海季风爆发和演变及其与海洋的相互作用. 北京：气象出版社, 1999: 34-42.

[10] 谢安，刘霞，叶谦. 南海夏季风爆发的气候特征[C]//何金海，丁一汇，陈隆勋. 亚洲季风研究的新进展. 北京：气象出版社, 1996: 132-142.

[11] Chang C P, Chen T J. Tropical circulation associated with southwest monsoon onset and westerly surge over South China Sea[J]. Mon Wea Rev, 1995, 123: 3254-3267.

[12] 王黎娟，何金海，徐海明，等. 1998年南海夏季风建立前后的突变特征及爆发过程[J]. 南京气象学院学报, 1999, 22(2): 135-140.

[13] Kiehl J T, Hack J J, Bonan G B, et al. Description of the NCAR Community Climate Model(CCM3)[R]. NCAR Tech Note, 1996, NCAR/TN-420+STR: 152.

[14] 吴国雄，刘屹岷，刘平. 空间非均匀加热对副热带高压带形成和变异的影响Ⅰ：尺度分析[J]. 气象学报, 1999, 57: 257-263.

[15] 吴国雄，张永生. 青藏高原的热力和机械强迫作用以及亚洲季风爆发Ⅰ：爆发地点[J]. 大气科学, 1998, 22: 825-838.

东亚季风指数及其与大尺度热力环流年际变化关系*

祝从文[1] 何金海[2] 吴国雄[3]

(1. 中国气象科学研究院,北京 100081;2. 南京气象学院气象系,南京 210044;3. 中国科学院大气物理研究所大气科学和地球流体力学数值模拟国家重点实验室(LASG),北京 100080)

摘要：将东西向海平面气压差与低纬度高、低层纬向风切变相结合,定义了东亚季风指数,该季风指数较好地反映了东亚冬、夏季风变化。其中,夏季风指数年际异常对西太平洋副热带高压南北位置变化和长江中下游旱涝具有较强的反映能力。分析表明：东亚夏季风年际变化与印度洋-西太平洋上空反 Walker 环流及夏季越赤道南北半球间的季风环流呈显著正相关关系。在强、弱异常东亚夏季风年份,异常的 Walker 环流在西太平洋上的辐合(辐散)中心在垂直方向不重合,高层(200 hPa)速度势与东亚夏季风显著相关区域位于西北太平洋上,该异常环流的高层的辐合(辐散)通过改变低层空气质量而影响夏季 500 hPa 西北太平洋副热带高压。采用 SVD 分析进一步发现：与海温耦合的异常 Walker 环流在西太平洋上空的上升支表现出南北半球关于赤道非对称结构,亚澳季风区受该异常 Walker 环流控制。因而,东亚季风与热带海气相互作用可直接通过这种纬向非对称的 Walker 环流发生联系。

关键词：季风指数；Walker 环流；ENSO；年际变化

1 引 言

亚洲是世界的主要季风区之一,由于海陆热力分布和青藏高原大地形的影响不同,东亚与印度季风存在本质差异[1]。夏季降水是夏季风活动导致的最直接后果,由于夏季西太平洋副热带高压南北位置年际差异,东亚降水异常主要表现为南北向正、负、正(或反向)雨型分布,特别是 20 世纪 80 年代以来更是这样,长江中下游基本位于这种雨型分布的中心[2-3]。因而,就东亚季风系统而言,西太平洋副热带高压是最关键的环流系统之一,其中长江中下游夏季降水在反映东亚夏季风年际变异方面具有较强的指示意义。

季风主要是海—陆热力差异导致的大气环流现象,海陆分布的非对称性造成的太阳辐射对大气加热非对称结果是季风地理分布不同和差异的主要原因[4]。东亚大陆东邻西太平洋,南部是中国南海和海洋大陆,不仅存在东西向的海陆热力对比,同时也存在南北向的海陆热力差异。因此,对东亚季风的整体变化研究应该考虑上述特点。

本文的目的是以东亚特殊的海陆分布为背景,寻求一种对东亚季风的客观描述物理量——东亚季风指数,讨论其反映东亚季风环流及旱涝年际变化的能力,并在此基础上分析它与大尺度热力环流及热带海气相互作用之间的年际变化关系和联系途径。

---

* 本文发表于《气象学报》,2000 年第 58 卷第 4 期,391-402。

文中的主要资料取自美国的 NCEP/NCAR 月平均再分析数据集,时间为 1980—1995 年,网格分辨率为 $2.5°×2.5°$。SST 来自英国气象局(U.K),网格分辨率为 $5°×5°$,时间为 1980—1994 年。西太平洋副热带高压脊线指数取自中国气象局国家气候中心。

2 东亚季风指数和东亚季风年际异常变化

季风指数是描述季风活动的简单物理量。海-陆热力差异的季节变化是季风的主要成因之一,考虑东西向海陆热力差异,郭其蕴[5]采用 $10°~50°N$ 内各纬度上月平均 $110°E$ 海平面气压减 $160°E$ 海平面气压,对差值 $≤-5$ hPa 的各纬度值累加,取绝对值然后标准化,即为夏季风强度指数(简称 SMI)。Webster 等[6]根据南亚纬向季风环流,采用 $0°~20°N;40°~110°E$ 区域平均纬向 850 hPa 与 200 hPa 切变风定义了一个大尺度南亚季风指数(简称 SAMI),夏季风指数年际距平大于(小于)零表示这一年夏季风强(弱)。此外,何敏曾采用热带高、低层纬向风切变将南海季风和 Walker 环流结合定义了一个热带环流指数(TCIX)。比较可以发现[7]:SMI 虽然物理意义清楚,但夏季指数年际异常反映东亚旱、涝年际变化能力较弱;SAMI 反映长江中下游旱涝的年际变化特征相对好些,但对 1991、1992 和 1993 年典型年份长江流域降水反映失真;TCIX 反映长江流域降水年际变化效果最好,TCIX 高(低)指数年份往往对应长江流域相对旱(涝)年。由于该指数未考虑中高纬度影响,因此,从完整意义上讲,TCIX 并未真正地反映东亚地区特殊的海陆热力差异年际变化。

分析可以发现,SMI 能较好地描述对流层低层(850 hPa)经向风和地转经向风季节演变,相对地表征了大陆上气压及其气压反映的热力状况的季节转换特征和东西向海陆热力差异,而 SAMI 主要反映的是南北向海陆热力状况季节转换及年际变化(图略)。如前所述,东亚不仅存在东西向海陆热力差异,而且还存在南北向海陆热力对比。考虑到海平面气压在反映南北向海陆热力差异的局限性,根据热带大气行星尺度运动具有准地转这一性质[8],分别计算了东亚季风在热带的主要活动区域的 $100°~130°E$ 范围内的 $0°~10°N$ 和 $2.5°~10°N$ 区域平均的 850 hPa 和 200 hPa 纬向风和纬向地转风切变,后者表示由于南北热力梯度导致的热成风变化。对比可以发现(图 1),该区域平均的实际风与地转风切变的季节变化趋势完全一致,所不同的是前者的量级大于后者,地转风切变表现出更清楚的季节转换特征,该指数反映的冬、夏季风转换的大尺度海陆热力背景在 4 月就已显示出来。通过比较看出,实际风切变基本反映了这种南北热力状况的季节变化特征。

考虑到将反映东西和南北向热力差异的季风指数相结合,以及所要定义的季风指数简单化,借鉴以往作者经验,首先将 $0°~10°N,100°~130°E$ 区域的 850 hPa 与 200 hPa 的纬向风切变($U_{850}-U_{200}$)和 $10°~50°N$ 内各纬度上月平均 $160°E$ 海平面气压减去 $110°E$ 海平面气压差分别作归一化处理,然后将经过标准化的两个指数相加,表示东亚特殊的海陆热力对比对东亚季风的影响,称为东亚季风指数(简称 EAMI),其值大于零表示夏季风(西南风)占优,小于零表示冬季风(东北风)占优。

东亚季风指数存在明显的季节变化(图 2),夏季风盛行期为 5—9 月,与以往指数不同,该夏季风指数 8 月达到峰值,冬季风 1 月最强。与西太平洋副热带高压脊线位置的季节变化对比,可以发现:东亚季风指数的季节变化能较好地反映副高的南北移动季节演变和春夏季节的突变特征。西太平洋副热带高压的活动是影响东亚天气和气候的最关键系统之一,尤其是夏

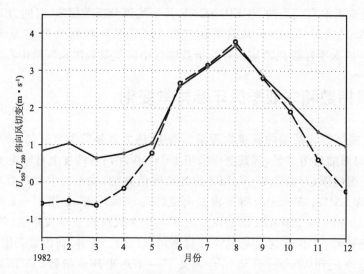

图 1 0°~10°N,100°~130°E 平均 $U_{850}-U_{200}$ 纬向风切变季节变化(实线)与 2.5°~10°N,100°~130°E 平均地转纬向风切变季节变化(虚线)(纬向地转风切变被缩小 6 倍,单位:m·s^{-1},资料取自 NCEP/NCAR)

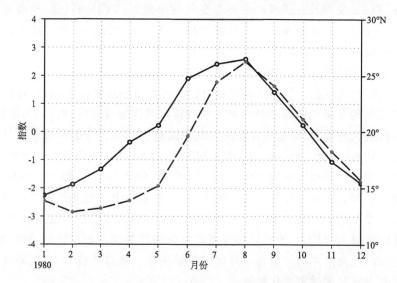

图 2 东亚季风指数(实线)和西太平洋副热带高压脊线指数(虚线)的季节变化(横坐标表示月份,纵坐标分别表示季风指数强度和副高脊线纬度)

季的南北位置异常是导致东亚旱、涝区域分布不均的主要原因之一。分析东亚夏季风指数(6—8月指数平均)年际异常可以发现(图3),除1988—1989年外,该指数与西太平洋副高脊线位置年际异常变化规律基本一致,对应强(弱)东亚夏季风指数年份,副热带高压容易偏北(偏南),长江中下游夏季容易旱(涝)。1—2月是东亚寒潮活动频繁的月份,该季节的季风指数在反映冬季风方面具有代表性。考虑到最初定义的季风指数是以夏季风方向出发点,因而冬、夏季风指数在符号上相反。取1—2月指数平均之后乘以-1来反映冬季风强度,通过比较冬、夏季风强度年际异常变化可以发现,除1990年之外,东亚冬季风与随后的夏季风强度表现出明显的反向变化关系,强(弱)冬季风之后往往伴随的是弱(强)夏季风,这与通常的理解比较一致。将冬季风指数与同期的 500 hPa 高度和温度场做相关,分析可以发现(图略),冬季的东亚大槽以及中国大陆东部地区的高度和温度场与指数呈显著负相关,即强(弱)冬季风指数异常与强(弱)冬季异常环流具有一致关系。对夏季风指数与同期 500 hPa 位势高度相关分析发现(图 4),相关场基本以 30°N 为纬向轴,从孟加拉湾到日界线呈东北—西南倾斜状的南负北正分布,对应夏季西太平洋副热带高压活动区,最大相关系数达到 -0.7 以上,超过 0.001 信度检验,即:强(弱)东亚夏季风年副热带高压易偏北(偏南),这与上面分析结果是一致的,上述结果进一步证明东亚季风指数定义的合理性和反映东亚大气环流变化的客观性。

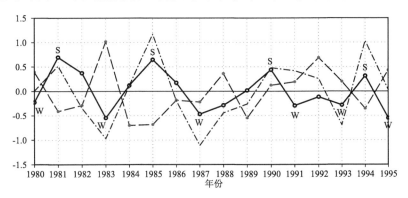

图 3　1980—1995 年东亚夏季(6—8 月平均)风指数(实线)、冬季(1—2 月平均)风指数(虚线)和西太平洋副热带高压脊线位置(点虚线)年际标准化异常变化(横坐标表示年,纵坐标表示指数距平)

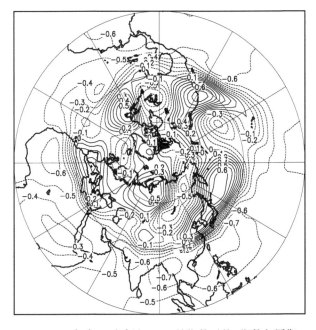

图 4　1980—1995 年东亚夏季风(6—8 月指数平均)指数与同期 500 hPa 位势高度相关系数分布(等值线间隔:0.1)

3 东亚夏季风与大尺度热力环流年际变化关系

东亚夏季风既独立于印度季风又与之发生联系,此外,南北半球环流也对其发生作用,上述环流主要表现为纬向型和南北半球热力差异导致的经向跨赤道气流。为了分析东亚夏季风年际变化与大尺度热力驱动的纬向环流和南北半球越赤道季风环流关系,对 850 hPa 和 200 hPa 速度势采用对称和反对称分解,对称和反对称模分别代表纬向 Walker 环流和南北半球越赤道季风环流[6]。

取夏季风指数极端峰值的 1981、1985、1990 和 1994 年以及极端谷值的 1980、1983、1987、1991、1993 和 1995 年为强、弱夏季风年份,分别对强、弱东亚夏季风指数年份夏季的对称和反对称模年际距平做合成和相关分析。合成分析结果表明(图 5):在强东亚夏季风年份(图 5a),

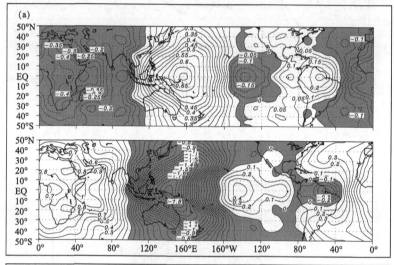

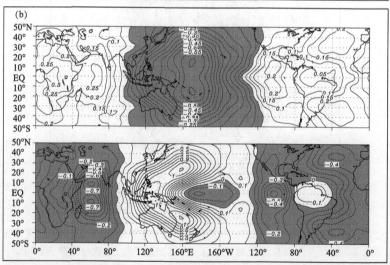

图 5 强(a)和弱(b)东亚夏季风年份 850 hPa(上图)和 200 hPa(下图)速度势对称模夏季(6—8 月)平均年际异常合成(阴影区表示负异常区等值线间隔 0.1,单位:10^6 m²·s⁻¹)

夏季 Walker 环流表现较强特征。速度势对称模异常表现为关于赤道对称和高、低层符号相反的 Walker 环流 2 波型分布,其中低层(850 hPa)的辐散中心主要位于赤道西印度洋和赤道东太平洋上空,辐合中心主要位于赤道西太平洋和赤道美洲上空。高层(200 hPa)的辐散和辐合中心与低层分布刚好相反。在东亚弱夏季风年份(图 5b),同期的 Walker 环流表现较弱特点。异常速度势对称模反映的 Walker 环流主要表现为 1 波型,表现为以赤道西太平洋暖池为中心的低层辐散、高层辐合特征,而在印度洋中部和东太平洋上空分别存在低层辐合和高层辐散中心。与较强季风年情形相比,印度洋、西太平洋及东太平洋上空的高低层的辐散(或辐合)中心整体位置明显东移,除此以外,西太平洋上空以赤道对称分裂成南北对称的低层辐散和高层辐合两个中心,其中北半球高层的辐合中心轴呈东北—西南向分布。

反对称模合成分析表明(图 6):东亚 100°E~160°W 范围内,对应强东亚夏季风年份(图 6a),

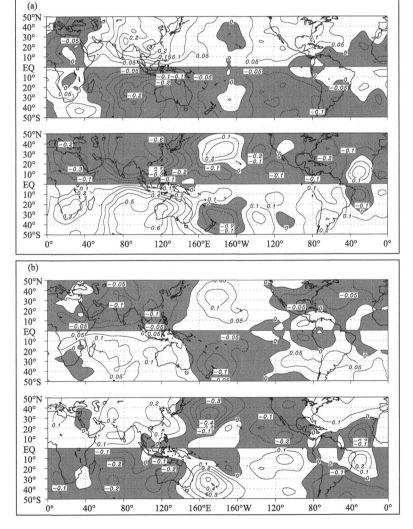

图 6 强(a)和弱(b)东亚夏季风年份 850 hPa(上图)和 200 hPa(下图)速度势关于赤道反对称模夏季(6—8 月)平均年际异常合成(阴影区表示负异常区,等值线间隔 0.1,单位:10^6 $m^2 \cdot s^{-1}$)

速度势反对称模表现为南(北)半球低层的辐散(辐合)和高层的辐合(辐散)特点,此外,南北半球 160°E~160°W 中、低纬度表现为高、低层局地反向变化,高层变化明显。弱夏季风年份(图 6b),除 120°W 以东外,速度势关于赤道反对称模异常分布与强季风年基本相反。反对称模的异常分布表明:强东亚夏季风年,东亚经度范围,低层南半球具有大范围辐散气流,北半球辐合,高层相反,南北半球季风经向越赤道气流较强;弱夏季风与之相反,越赤道气流较弱。160°E~160°W 范围内,强、弱东亚夏季风年份速度势反对称模差异除表现出高、低层南北半球分布反向外,还表现为中低纬度分布反向特点,由此可以猜测,该区域的大尺度速度势反对称模的分布特征除与南北半球热力差异有关外,与中、低纬度热力对比也存在一定联系。

对东亚夏季风指数与同期的速度势对称模相关分析表明(图 7),相关系数分布大致以 120°E 为界限,低层表现为东西反向变化,最大正、负相关系数达到 0.8、-0.6,中心分别位于 160°E 左右的赤道西太平洋和 60°E 左右印度洋上空;高层除与低层符号相反之外,在 120°E~180°的北半球范围内相关-0.8 以上的中心轴还表现出明显的东北—西南走向特征。同样,与前面弱季风年合成结果一致,最大相关系数中心轴不在赤道上空。夏季风指数与速度势反对称模的显著相关主要表现为 80°~160°E 南亚到亚澳季风区的南北半球和高低层反向分布,即东亚夏季风强度与上述区域南北半球高低层辐散(辐合)气流一致,强夏季风年对应南半球低层辐散、北半球辐合,高层变化相反。弱季风年份与强季风年变化相反。合成分析也反映类似结果。

众所周知,由于太阳辐射季节变化,夏季大气热源由海洋大陆移向北半球东亚和南亚季风区,东亚盛行夏季风,东西向辐合(辐散)环流结构反映了大尺度东亚与印度西南季风的联系,同时,东亚经度范围的速度势反对称模的分布大体上反映亚澳越赤道季风环流。因此,由上述相关分析结果可得出:东亚夏季风与南亚季风热力环流、南北半球越赤道气流呈显著的正相关关系,从相关系数大小可以得出,印度—西太平洋纬向热力环流与东亚夏季风关系更加密切。

速度势表现的异常环流与西太平洋副热带高压具有密切关系。从合成和相关分析可以看出:强夏季风年,西太平洋速度势对称模表现为低层(850 hPa)辐合和高层(200 hPa)辐散特征,印度洋上空则与之相反,上述两个区域环流构成反 Walker 环流。又由于弱季风年份西北太平洋 200 hPa 速度势异常中心远离赤道位于副热带高压主体上空,因此可以认为这个反 Walker 环流中心在高层的变化通过改变西太平洋上空的大气质量影响西太平洋副热带高压,即:强(弱)东亚夏季风年,西北太平洋高空的异常辐散气流(辐合气流),可能会导致西太平洋副热带高压的异常减弱(加强),其中弱季风年份上述特点表现更加明显。

4 东亚季风与热带海气相互作用联系的可能途径

由上面分析可以发现,东亚强冬季风(弱夏季风)年份多数发生在 El Nino 年中(1980、1995 年除外),很自然地将这种变化与 ENSO 联系在一起。最近,吴国雄等[9]的研究表明,印度洋纬向季风环流与太平洋 Walker 环流通过齿轮式耦合在一起,并指出印度洋上空纬向环流异常可以通过上述齿轮式耦合而影响中东太平洋的海气相互作用并触发 ENSO 事件发生。热带环流(尤其是纬向环流)的年际变化在很大程度上反映了热带海气耦合信号。从夏季风指数与热力环流关系中可以看出:东亚夏季风以及西太平洋副热带高压的年际变化与印度洋和西太平洋暖池的局地海气相互作用关系密切,而与热带海气相互作用最强烈的赤道东太平洋

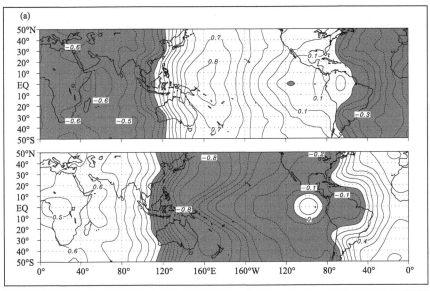

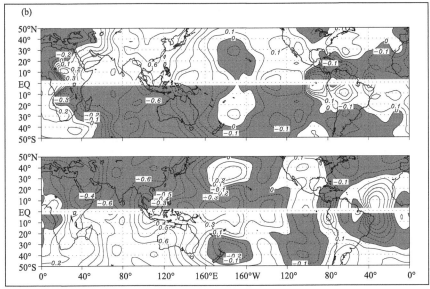

图 7 东亚夏季风指数与同期(6—8 月)850 hPa(上图)和 200 hPa(下图)速度势关于赤道对称模(a)和反对称模(b)相关分布(阴影区表示负相关区,等值线间隔 0.1)

不存在显著的相关关系。尽管如此,东亚季风年际变化中反映了 ENSO 信号,因此不能轻易否定东亚夏季风与 ENSO 关系的存在。多数研究者认为:El Nino 年,正常的 Walker 环流减弱东移,太平洋海温的东暖西冷异常分布容易导致暖池上空的对流减弱,从而造成局地 Hadley 环流减弱和西太平洋副热带高压的偏南。反之,则相反。然而,臧恒范等[10]研究表明,在 El Nino 年,西太平洋副热带高压较常年偏强、面积大和西伸明显;在 La Nina 年则相反。由于 El Nino(La Nina)年,西太平洋暖池海温异常偏低(高),假如局地 Hadley 环流存在,则必将导致 Hadley 环流减弱(加强),从而采用异常的 Hadley 无法解释西太平洋副高强度加强(减弱)。那么,东亚季风的年际异常是否与热带海气相互作用发生联系,它们之间的联系途径是什么?

为讨论上述问题,接下来将整个印度洋和太平洋作为研究对象,采用 SVD(Singular Value Decomposition)方法,着重分析热带海气同步线性耦合在东亚区域的大尺度热力环流表现形式。首先,将原始的 850PhPa 速度势减去 200PhPa 速度势,差值表示大尺度辐合(辐散)热力环流场,正值区表示低层辐合、高层辐散。反之,则相反。为方便计算,将速度势格点资料分别处理成为 10°×5°较粗分辨率数据,这样做的结果对研究大尺度环流不会造成大影响。之后将经过处理后的速度势与 SSTA 进行 SVD 分析。其中,分析的起始时间为 1980—1994 年,分析的范围为印度洋—太平洋地区。

图 8a、b 分别表示速度势反映的热力环流与 SSTA 第一对线性耦合模及其随时间变化。分析结果表明,第一对 SVD 模相关系数达到 0.73,它们联合解释原始方差为 85%,左(热力环流场)、右特征场(海表温度)占原始场方差百分比分别为 44.8% 和 18.0%。左、右异性(heterogeneous)相关分布型分别反映了反 Walker 环流和 El Nino 型海温异常分布特征。对 SVD 特征模分析可以发现(图 8a),与海温原始场显著相关的左特征场相关中心有两个,它们分别位于近赤道东太平洋上空和 100°~160°E 的澳大利亚到东亚 30°N 的亚澳季风区,后者表现出明显的纬向非对称特征,最大相关系数分别为 0.7 和 −0.6。印度洋上的相关系数虽表现与亚澳季风区同符号特点,但相关系数并不高(不到 −0.4)。与热力环流场相匹配,海温异性相关分布在太平洋表现为东西向反相关特征,最大相关系数分别位于中东太平洋近赤道地区,相关系数分别为 0.6 和 −0.6。其中,超过 0.5 以上的大值区呈赤道对称分布特征,西部最大异性相关系数正好位于西太平洋暖池地区。对比不难发现:与热力环流相类似,西太平洋海温异性相关分布同样表现出南北半球关于赤道的非对称性。印度洋海温异性相关系数虽与中东太平洋表现为同符号,但从数值上可以看出,该海域海温与热力环流相关不显著。SVD 左、右奇异向量对应的时间系数年际变化趋势表现出较好的一致性(图 8b)。结合特征向量场可以看出:对应 El Nino(La Nina)年,Walker 环流表现为弱(强)特征。

综上分析,与热带海表温度相耦合的异常 Walker 环流在亚澳季风区域表现出关于赤道的非对称特点,亚澳季风区受该异常环流的西支影响。因而,东亚季风可直接通过这种异常 Walker 环流与太平洋海气相互作用发生联系,其中,西太平洋暖池是这种联系的关键环节之一。热带海气相互作用对东亚季风的影响可以通过下面途径完成:1)异常 Walker 环流通过改变西太平洋上空的上升(下沉)气流,使低层的空气质量产生流出(流入)直接影响西太平洋副热带高压的强度、位置;2)在东亚中低纬度,热带海气相互作用导致的辐合(辐散)气流通过异常 Walker 环流与东亚正常的季风环流相叠加,从而加强(减弱)东亚季风夏季风(或冬季风)。例如,处于 El Nino(La Nina)位相,受这种异常 Walker 环流影响,在东亚中低纬度上空存在下沉(上升)气流,若此时正处于夏季,该下沉(上升)流向低纬度气流与西南季风相叠加,其结果可导致夏季风减弱(加强)。若此时为冬季风季节,如上道理,El Nino(La Nina)位相时的热带海气相互作用则将加强(减弱)东亚冬季风,热带海气相互作用对东亚季风影响和作用的时间取决于它的时间演变位相与东亚季风的季节变化之间的锁相关系。

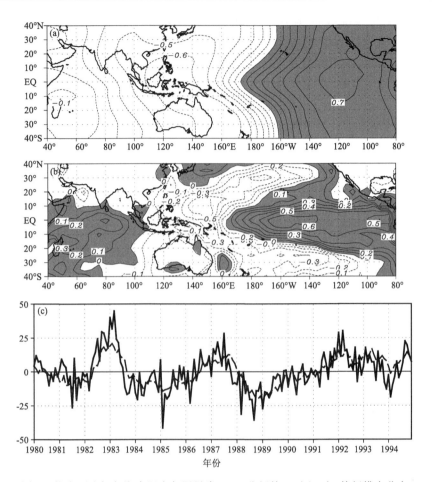

图 8 热力环流场与海表温度年际异常 SVD 分析第一对左(右)特征模态分布
(a. 左特征场:热力环流场特征模;b. 右特征场:海表温度特征模);c. SVD 分析的
第一对左、右特征场对应的时间系数(实线表示左场(热力环流)对应的时间
演变,虚线表示右场(海表温度)对应的时间演变)

5 结论与讨论

(1)反映东亚东西和南北海陆热力差异的季风指数无论其季节和年际变化均能较客观地描述东亚季风活动。夏季风指数异常对长江中下游旱涝和西太平洋高压具有较好的反映,对应东亚强(弱)夏季风年,长江中下游易旱(涝),西太平洋副高容易偏北(偏南)。此外,东亚冬季风指数对东亚冬季风也具有较强的反映,并表现与夏季风强度呈反向变化特征。

(2)东亚夏季风年际变化与同期反 Walker 环流呈显著正相关,显著相关中心位于西太平洋和赤道印度洋上空。对应强东亚夏季风年份,同期 Walker 环流表现 2 波型,西太平洋为辐合上升运动,Walker 环流强;弱季风年份反 Walker 环流以 1 波型分布为主,西太平洋上空为辐散下沉运动,Walker 环流弱。此外,东亚夏季风与 Walker 环流在 200 hPa 西北太平洋上空速度势呈东北—西南向显著正相关,该区域空气的辐合上升(辐散下沉)通过对 500 hPa 层空气质量的输出(输入)从而减弱(加强)夏季西太平洋副热带高压,进而影响夏季风强度。

(3)热带海表温度异常激发的 Walker 环流中心主要位于太平洋上空。在东亚中低纬度及邻近海域的 Walker 环流辐合上升(辐散下沉)支与西太平洋暖池海温相互耦合,表现为关于赤道非对称结构。因此,东亚季风可直接通过上面关于赤道非对称的 Walker 环流与热带海气相互作用发生联系。El Nino 对东亚夏(冬)季风的影响取决于 El Nino (La Nina)位相与季风的季节变化锁相关系:El Nino(La Nina)位相,受异常 Walker 环流影响,在东亚中低纬度上空存在下沉(上升)气流,若此时正处于夏季,该下沉(上升)流向低纬度气流与西南季风相叠加,其结果可导致夏季风减弱(加强)。对应冬季风季节,如上道理,El Nino(La Nina)位相时的海气相互作用则将加强(减弱)东亚冬季风。

本文主要讨论了东亚季风指数及其与热带海气相互作用年际变化关系,指出热带海气相互作用可直接通过南北向非对称的纬向 Walker 环流与季风发生作用。由于季风与 ENSO 之间作用是相互的,因而东亚季风对 ENSO 同样存在影响。限于篇幅,关于东亚季风(主要是冬季风)对 ENSO 的影响及其可能途径作者将在另文中加以阐述。

参考文献

[1] Tao S Y,Chen L X. A review of recent research on the East Asian Summer monsoon in China [C]//Chang C-P,Kishnamurti T N. Monsoon Meteorology. Oxford:Oxford University press,1987:60-92.

[2] 黄荣辉,孙凤英.热带西太平洋暖池的热状况及其上空的对流活动对东亚夏季气候异常的影响[J].大气科学,1994,18(2):141-151.

[3] 孙柏民,孙淑清.海温在东亚冬季风影响江淮流域旱涝中的作用[C]//灾害性气候过程及诊断.北京:气象出版社,1995:46-53.

[4] Krishnamurti T N,Ramanathan Y. Sensitivity of monsoon onset of differential heating[J]. J Atmos Sci,1982,39(6):1290-1306.

[5] 郭其蕴.东亚夏季风强度指数及其变化的分析[J].地理学报,1983,38(3):207-216.

[6] Webster P J,Yang S. Monsoon and ENSO:Selectively interactive systems[J]. Q J Roy Meteor Soc,1992,118:877-926.

[7] 祝从文,何敏,何金海.热带环流指数与夏季长江中下游旱涝的年际变化[J].南京气象学院学报,1998,21(1):15-22.

[8] 李崇银.热带大气运动的特征[J].大气科学,1985,9:366-376.

[9] 吴国雄,孟文.赤道印度洋—太平洋地区海气系统的齿轮式耦合和 ENSO 事件:Ⅰ资料分析[J].大气科学,1998,22(4):470-480.

[10] 臧恒范,王绍武.赤道东太平洋海温对低纬大气环流的影响[J].海洋学报,1984,6:16-24.

40 a 南海夏季风建立日期的确定*

高辉　何金海　谭言科　柳俊杰
(南京气象学院大气科学系,南京 210044)

摘要：对 1958—1997 年 NCEP/NCAR 4—6 月逐候资料的分析表明,将($10°\sim 20°N,110°\sim 120°E$)区域内面积平均的 850 hPa 层上稳定地有 $\theta_{se} \geqslant 335$ K 且纬向风由东风稳定地转变为西风的时刻为南海夏季风爆发时间具有较好的指示意义。所谓稳定是指从该时刻起,这一状况必须持续 3 候且其后间断不超过 2 候,或持续 2 候后间断 1 候但立刻又回到间断前状态。

关键词：南海夏季风；建立日期；突变

已有的研究表明,南海夏季风爆发是亚洲大气环流由冬季型向夏季型转换的最早征兆,也是亚澳季风区季节转换的一个阶段。因此,南海夏季风建立日期的确定是南海季风研究中一个至关重要的基本问题。

在具体确定南海夏季风建立日期时,通常是基于热力学和动力学的考虑,采用南海地区一定区域能反映南海夏季风建立特征的某些特征参数(例如,风、温度、湿度、散度、涡度、降水、OLR、TBB、海温等)作为指标,以一定的临界值确定夏季风建立日期。由于选择的区域和参数不同,人为规定的临界值有异。因此,尽管对于多数年份,多数作者对爆发日期早晚的定性判别基本一致,但对于爆发日期的定量确定上是有分歧的。

文献①较为详细地介绍了各种确定南海夏季风建立日期的标准,并认为将热、动力学结合起来考虑是一种较为合理的方法。但是由于早期 OLR 及 TBB 资料的缺乏,许多标准只适用于近 20 年左右,这也是各种标准确定的时间主要集中于 20 世纪 80—90 年代的根本原因。为了系统全面研究南海季风特性,必须有长时间的季风建立时间序列。这就要求我们寻求一种能够有效地反映南海夏季风爆发前后要素场的热、动力变化特征且所用资料序列时间较长的标准,以确定逐年南海夏季风建立日期。

何金海等②、冯瑞权等③提出将低空假相当位温 θ_{se}(此后简称"位温")和风场结合起来确定季风建立日期的方法,认为这种方法能较好地反映南海夏季风爆发前后要素场的热、动力变化特征。本文将在文献①的基础上,利用 NCEP/NCAR 40 a 逐日再分析资料,建立一种将低层位温和纬向风结合起来的标准,以确定 1958—1997 年逐年南海夏季风建立日期。

---

* 本文发表于《南京气象学院学报》,2001 年第 24 卷第 3 期,379-383。
① 何金海,丁一汇,高辉,等. 南海夏季风建立日期的确定及季风指数. 北京:气象出版社,2001.
② 何金海,谭言科. 1990—1998 年南海夏季风的建立日期. SCSMEX 研讨会. 北京:2000.
③ 冯瑞权,王安宇,吴池胜,等. 南海夏季风建立日期. SCSMEX 研讨会. 北京:2000.

1 资料和方法

采用南京大气资料中心提供的美国气候诊断中心(CDC)的 NCEP/NCAR 40 a(1958—1997年)逐日再分析全球资料。本文使用的是 850 hPa 纬向风场、比湿场及温度场,200 hPa 纬向风场,网格距 2.5°×2.5°。考虑到历年南海夏季风均在 4—6 月爆发,文中只取每年 4—6 月时段,且将其处理成逐候资料,其中 5 月第 6 候规定为 5 月 26—31 日。将温度场和比湿场处理成位温场可参见气象学基础教材。

文中使用 Yamamoto 信噪比突变检验法[1]。对于时间序列 x,人为设置某一时刻为基准点,基准点前后样本量分别为 n_1 和 n_2 的两段子序列 x_1 和 x_2 的均值为 $\overline{x}_1$ 和 $\overline{x}_2$,标准差为 s_1 和 s_2,定义信噪比为

$$SNR=|\overline{x}_1-\overline{x}_2|/(s_1+s_2)$$

假定两段子序列样本相同,即 $n_1=n_2=IH$,当 $t>SNR$,$\overline{IH}>t^T$ 且 $SNR>1.0$ 时,说明两段子序列的均值存在显著性差异,则认为在基准点发生了突变。其中 t^T 为给定显著水平 T,查 t 分布表得到的临界值。

2 南海夏季风建立日期确定标准的制定

2.1 区域的选择

如何选择合理的南海季风区域?气象学家们为此提出了各自的看法,也提供了各自选取的区域。虽然这些区域均在(0°~20°N,100°~120°E)范围内,但仍有较大差异。而区域的不同会直接影响到要素临界值的不同,从而使各自确定的日期差异较大,即使对于同一种标准而言亦是如此。例如,谢安等[2]最初将确定标准定义为南海区域(0°~20°N,105°~120°E)平均的 OLR 下降到 235 W·m^{-2},同时平均纬向风由东风转变为西风。后谢安又将南海区域改为(5°~20°N,105°~120°E),其他条件不变,但两种标准确定的日期在若干年份有较大分歧。由此可见,选择合理的区域并进而选择合适的临界值以便能较精确地确定南海夏季风的爆发时间是非常重要的。

具体到以位温和高低空风场为指标的确定方法上,何金海等①将 850 hPa 的区域选定为(7.5°~20°N,110°~120°E),200 hPa 纬向风为(10°~22.5°N,100°~120°E)。冯瑞权等②将 850 hPa 上 θ_{se} 及纬向风区域选定为(5°~20°N,105°~120°E)。因此,有必要考虑要素临界值对区域的敏感性。

在参考众多气象学家给出的各种定义基础上,将重点对以下 4 个界定区域进行讨论,即 1 区:(5°~20°N,105°~120°E);2 区:(10°~20°N,105°~120°E);3 区:(5°~20°N,110°~120°E);4 区:(10°~20°N,110°~120°E)。

在 4 个区面积平均的 40 a 平均 850 hPa θ_{se} 及纬向风逐候演变图上(图 1)可以看出,根据

---

① 何金海,谭言科.1990—1998 年南海夏季风的建立日期.SCSMEX 研讨会.北京;2000.
② 冯瑞权,王安宇,吴池胜,等.南海夏季风建立日期.SCSMEX 研讨会.北京;2000.

上述要素确定的南海季风建立日期对区域有着很强的敏感性,即不同区域同一要素相同临界值确定的爆发日期都会有较大的差异,对于每年情况(图略),这一点尤为突出。

进一步对整个南海区域(0°~20°N,100°~120°E)逐一格点的 850 hPa θ_{se} 及 u 作了突变检验(图2)。对于气候平均风场,若格点 i 上的 u 在 5 月第 3 候为负值而在 5 月第 4 候已变为正值,则认为该格点上的纬向风在 5 月第 4 候有突变。由图 2 发现,南海地区 40 a 平均要素场在 5 月第 4 候(目前公认的南海夏季风爆发候)发生突变的区域主要位于 4 区,即(10°~20°N,110°~120°E)。因此,我们认为在此区域的基础上,选择参数以确定南海夏季风的爆发日期具有较好的指示意义。

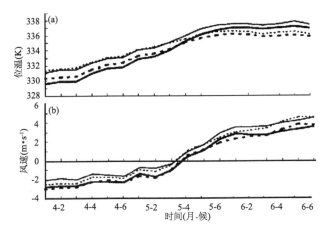

图 1　4 个区域面积平均的 40 a 平均 850 hPa θ_{se}(a)及纬向风(b)逐候演变
(其中粗实线为 4 区,细实线为 2 区,粗虚线为 3 区,细虚线为 1 区)

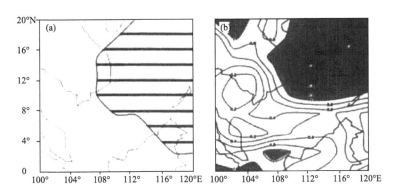

图 2　40 a 平均的 850 hPa 纬向风 u(a)及 θ_{se}(b)在 5 月第 4 候存在突变
的区域(阴影区)(灰色区域表示已通过 0.01 的信度检验)

2.2　临界值的选取

由上面的分析可知,区域(10°~20°N,110°~120°E)对确定南海夏季风爆发时间有着较好的指示意义。因此,作了该区域内面积平均的 40 a 平均 850 hPa 纬向风、位温及 200 hPa 纬向风均值随时间演变曲线(图 3)。考虑到南海夏季风爆发时间均集中于 4—6 月,故分析时段取为 4 月 1 候—6 月 6 候。

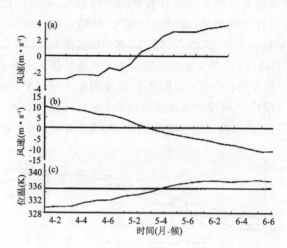

图3 (10°~20°N,110°~120°E)区域内面积平均的850 hPa 纬向风(a)、
200 hPa 纬向风(b)及850 hPa θ_{se}(c)逐候演变曲线

从图3可见,对流层低层(850 hPa)位温值在5月4候以前均小于335 K,在5月4候以后均大于335 K;低空纬向风在5月4候以前均为负值(东风),5月4候以后均为正值(西风);而对于高层,200 hPa 纬向风在5月3候以前均为正值(西风),5月3候以后均为负值(东风)。这表明,南海夏季风爆发前后,对流层低层温湿特性和高低层环流状况都经历了一个相当明显的突变,可以以位温是否大于335 K及纬向风风向的转变作为确定标准,但高空风的突变时间比低层要提前1候左右。

2.3 季风建立日期标准的确定

以上讨论的是40 a平均情况,但逐年情况如何呢?在具体分析了40 a逐年温湿及环流形势(图略)后,规定:若(10°~20°N,110°~120°E)区域内面积平均的850 hPa上稳定地有$\theta_{se}\geqslant$335 K且纬向风由东风稳定地转变为西风的时刻为南海夏季风爆发时间。所谓稳定是指从该

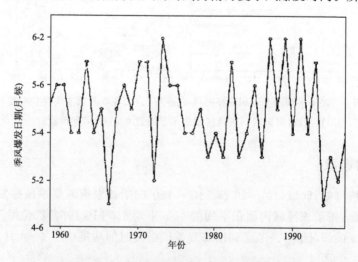

图4 40 a逐年南海夏季风建立日期变化曲线

时刻起,这一状况必须持续3候且其后间断不超过2候,或持续2候后间断1候但立刻又回到间断前状态。此外发现,200 hPa纬向风由西风转为东风的时间比南海夏季风爆发时间提早1~2候。因此,可以将其作为一个辅助参考或预测信号。

基于以上标准确定的40 a逐年南海夏季风爆发时间见图4和表1。由图4可见,南海夏季风的建立时间除有年际变化外,还具有明显的年代际变化特征,70年代中期以前爆发偏晚,70年代中期至80年代中期爆发较早,其后爆发又偏晚直至90年代中期。

表1 1958—1997年逐年南海夏季风爆发时间表(月-候)

年份	1958	1959	1960	1961	1962	1963	1964	1965	1966	1967
日期	5-5	5-6	5-6	5-4	5-4	6-1	5-4	5-6	5-1	5-5
年份	1968	1969	1970	1971	1972	1973	1974	1975	1976	1977
日期	5-4	5-5	6-1	6-1	5-3	6-2	5-6	5-5	5-4	5-4
年份	1978	1979	1980	1981	1982	1983	1984	1985	1986	1987
日期	5-5	5-3	5-3	5-3	5-6	5-3	5-4	5-6	5-2	6-2
年份	1988	1989	1990	1991	1992	1993	1994	1995	1996	1997
日期	5-5	6-2	5-4	6-2	5-4	5-6	5-1	5-3	5-2	5-4

3 结 论

对整个南海区域($0°\sim20°$N,$100°\sim120°$E)逐一格点(共9×9个格点)的4—6月850 hPa逐候θ_{se}及u作了突变检验分析,发现40 a平均要素场在5月第4候(目前公认的南海夏季风爆发候)发生突变的区域主要位于($10°\sim20°$N,$110°\sim120°$E),进而规定此区域内面积平均的850 hPa上θ_{se}稳定地大于等于335 K且纬向风由东风稳定地转变为西风的时刻为南海夏季风爆发时间。所谓稳定是指从该时刻起,这一状况必须持续3候且其后间断不超过2候,或持续2候后间断1候但立刻又回到间断前状态。此外发现,200 hPa纬向风由西风转为东风的时间比南海夏季风爆发时间提早1~2候,可将此作为预报信号。

参考文献

[1] 魏凤英.现代气候统计诊断与预测技术[M].北京:气象出版社,1999:66-68.
[2] 谢安,刘霞,叶谦.赤道涡旋与南海夏季风爆发[J].气象学报,1997,55(5):611-619.

黑潮地区海温影响南海夏季风爆发日期的数值试验

王黎娟 何金海 徐海明

(南京气象学院大气科学系,南京 210044)

摘要:采用合成分析和相关分析等方法讨论了季节转换时期(4—6月)黑潮地区海温异常对南海夏季风爆发时间和西太平洋副高位置与强度的影响。数值模拟结果进一步表明,该地区海温正异常将导致西太平洋副高位置偏南,强度偏强,南海夏季风建立较晚,强度偏弱,江淮流域偏涝;反之亦然。

关键词:黑潮海温异常;南海夏季风爆发;年际变化;数值试验

东亚季风区可分为南海—西太平洋热带季风区和中国大陆东部—日本副热带季风区[1-4]。南海夏季风爆发于北半球春末夏初,春季过渡季节对于确定其后的季风演变和海陆气耦合系统的年际变化是非常重要的。在这一时期,由于内部和外部强迫的作用,耦合系统很容易失去"记忆"功能,同其他季节相比,春季的可预报性较小,成为所谓的年际预报的可预报性障碍[5]。南海夏季风爆发的迟或早,受整个热带、副热带地区海陆气状况的制约,与海温的关系尤为密切。本文主要讨论黑潮地区海温对南海夏季风爆发时间和西太平洋副高位置与强度的影响。

1 资料及方法

(1)欧洲中期数值预报中心(ECMWF)提供的逐日全球网格点资料,格距为 2.5°×2.5°。我们将其处理为候平均的网格资料。(2)日本气象研究所提供的 GMS 观测的逐候黑体辐射温度(T_{BB})资料(范围为 80°E~160°W,60°S~60°N,分辨率为 2°×2°)。(3)国家气象中心整编的北半球月平均 500 hPa 高度场资料,格距为 10°×5°(经度×纬度),范围为 10°~85°N。(4)英国气象局提供的全球月平均海表温度资料,分辨率为 1°×1°。资料的时段均为 1980—1993 年。

研究工作主要采用合成分析、相关分析及数值试验方法。

2 南海夏季风爆发指数的定义

关于季风爆发时间的确定,不同区域有不同的标准。在陆上多是根据雨量记录的演变[6],而在海洋上多是根据低层流场的演变[7]。对南海地区,一系列试验发现[8,9],南海夏季风爆发时伴随着南海区域的纬向风从东风向西风的转变以及整个南海地区深对流的爆发性增长。本文从这两个方面综合定义南海夏季风的爆发。设

---

* 本文发表于《南京气象学院学报》,2000年第23卷第2期,211-217。

$$I_{ms} = \left| \frac{u}{\bar{u}} \right| - \left| \frac{T_{BB} - 275}{\overline{T_{BB}} - 275} \right| \tag{1}$$

式中,u 为南海区域(5°~20°N,110°~120°E)候平均纬向风,$\bar{u}$ 为南海地区纬向风的多年平均值,T_{BB} 为南海区域(5°~20°N,110°~120°E)候平均的黑体辐射温度,$\overline{T_{BB}}$ 为南海地区 T_{BB} 的多年平均值。定义 T_{BB} 下降至 275 K 时,对流爆发。因此,当 I_{ms} 由负转为正,并持续 2 候,则定义该候为南海夏季风爆发候。而且,I_{ms} 越大,表明南海夏季风越强。

表 1 是根据以上标准得到的 1980—1993 年各年南海夏季风爆发的时间。需要指出,这个指数是对整个南海区域平均而言,但南海地区各部分并非均匀发展。

表 1 1980—1993 年各年及多年平均南海夏季风爆发时间

年份	爆发时间(候)	年份	爆发时间(候)	年份	爆发时间(候)
1980	28	1985	22	1990	27
1981	27	1986	27	1991	32
1982	28	1987	32	1992	30
1983	32	1988	29	1993	29
1984	24	1989	28	平均	28

注:平均季风爆发时间是对 14 年平均的资料进行分析得到的季风爆发时间。

由表可见,南海夏季风的爆发具有明显的年际差异。最早的是 1985 年 4 月第 4 候,最晚的是 1983、1987、1991 年,均为 6 月第 2 候,早、晚相差近两个月。平均爆发时间是 5 月第 4 候(第 28 候),这与通常认为的南海夏季风爆发于 5 月中旬相一致,说明用(1)式来定义南海夏季风的爆发时间是合理的。定义 4 月季风爆发的年份为偏早年,即 1984、1985 年;6 月季风爆发的年份为偏晚年,包括 1983、1987、1991 年。根据文献[10,11]对旱涝年的定义可以发现,夏季风爆发早(晚)对应江淮流域偏旱(涝)。

3 夏季风爆发早、晚年副高指数的差异

图 1 为南海夏季风爆发早、晚年副高脊线位置和面积指数的逐月变化曲线。其中副高面积是指 588 线在 110°~180°E,10°N 以北范围内所围的格点数(经纬格距为 10°×5°);副高脊线位置是指脊线的平均位置,即 110°~150°E 内脊线与每隔 10°的经线交点的纬度平均值。由图可见,夏季风爆发早年副高脊线北进速度快,北抬明显,可达 30°N 以北;而晚年相反,7 月最北位置在 24°N 以南。南海夏季风爆发早年各月的副高面积指数均远小于晚年,表明副高明显弱于晚年。

4 南海夏季风爆发早晚与季节转换期 SST 的关系

在南海夏季风爆发晚年与早年 4—6 月海温较差分布图(图略)上,从菲律宾以东的近赤道地区向东北伸展至中太平洋地区为一显著的负较差区;从南海经我国东海及日本南部洋面伸向中太平洋地区为一显著的正较差区;黑潮地区始终维持一正的较差中心。何金海等[10]根据热成风原理认为,这种海温较差分布通过对大气的加热有利于大气温度较差形成相应的分布

型,正较差带(中心)南侧高空应有较差东风,北侧应有较差西风。因此,在正较差区域,南海夏季风爆发晚年应有一个异常的反气旋环流叠加在西太平洋副高之上,西太平洋副高较早年强;并且上述海温正中心的稳定也导致晚年异常反气旋的稳定,中心在(140°E,25°N),从而有利于西太平洋副高脊线在25°N附近持续停留,形成江淮梅雨以致洪涝。

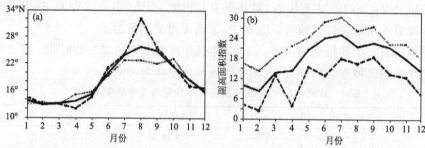

图1 南海夏季风爆发早、晚年及15年平均西太平洋副高脊线位置(a)和面积指数(b)的逐月变化曲线图
― ― ―:早年;……:晚年;———:平均

由南海夏季风爆发早迟与季节转换时期(4—6月)海温相关分布(图2)可以发现,赤道中东太平洋、南海、赤道印度洋、黑潮地区为显著的正相关区,达到0.01信度,说明季节转换时期(4—6月)黑潮地区海温对南海夏季风爆发早迟有一定影响。当该地区海温距平为正(负)时,南海夏季风爆发偏晚(早),长江流域偏涝(旱)。研究还发现,江淮梅雨期的长短与该地区同期(6—7月)海温的相关也十分密切,为显著的正相关区(图3)。也就是说,当该地区海温为正(负)距平时,梅雨期长(短),江淮流域偏涝(旱)。

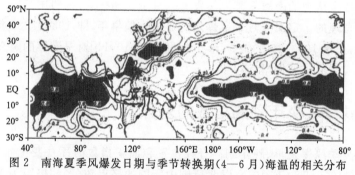

图2 南海夏季风爆发日期与季节转换期(4—6月)海温的相关分布
(等值线间隔为0.2,阴影区达到0.01信度)

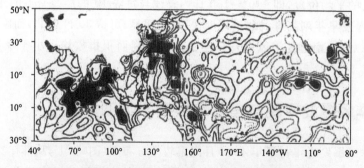

图3 江淮梅雨期长短与同期(6—7月)海温的相关分布
(等值线间隔为0.1,阴影区达到0.05信度)

5 数值试验

上述分析说明,季节转换时期黑潮地区海温异常对南海夏季风爆发时间及西太平洋副高的位置和强度有一定影响。当该地区海温为正(负)距平时,南海夏季风爆发偏晚(早),副高位置偏南(北),强度偏强(弱),江淮流域偏涝(旱)。下面将通过数值试验来进一步证实这种关系。

5.1 数值试验设计方案

采用菱形截断 15 个波的 9 层谱模式(L9R15)。此模式最初由 W Bourke 等设计,后经 Simmonds[12]、林元弼①以及吴国雄等[13-14]的进一步修改,成为现在所用的模式。

为了研究需要,设计了 3 个数值试验。

(1)CEX(控制试验):使用吴国雄等[13]给出的 1979—1988 年 10 年平均的逐月气候海温资料,其他强迫因子如雪盖、海冰、臭氧等用随季节变化的同期气候平均值,从 1 月 31 日开始积分 243 d,大约至 9 月底结束。

(2)EX+(异常试验一):在黑潮地区(即以 140°E,25°N 为中心的地区)加一正海温距平场(图 4),叠加于 4—6 月的气候海温场上,其他月份的海温和其他强迫因子都与控制试验完全一样。

(3)EX−(异常试验二):叠加负海温距平场,其他同 EX+。

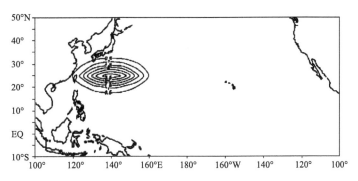

图 4　敏感性试验所加的海温距平分布

在 EX+和 EX−中,假设海温距平的分布为

$$\Delta \widetilde{T}(\lambda,h) = T_0 \left[\sin\frac{c(h-h_1)}{(h_2-h_1)} \sin\frac{c(\lambda-\lambda_1)}{(\lambda_2-\lambda_1)} \right]^2$$

$$\lambda_1 < \lambda < \lambda_2; h_1 < h < h_2$$

海温异常中心位于(140°E,25°N),即取 $\lambda_1=115°E, \lambda_2=165°E, h_1=17.5°N, h_2=32.5°N, T_0=4.0$。

---

① 林元弼. 南京大学大气科学系大气环流实验讲义. 1987.

5.2 控制试验结果

控制试验(CEX)所模拟的 6—8 月平均 500 hPa 和 850 hPa 流场见图 5。对流层中层(500 hPa)孟加拉湾地区为一气旋性环流,西太平洋副热带高压的脊线位置在 30°N 左右,与实际情况相一致,说明该模式对西太平洋副热带高压这一影响我国夏季降水的重要环流系统的位置具有较强的模拟能力,能基本满足研究的需要。

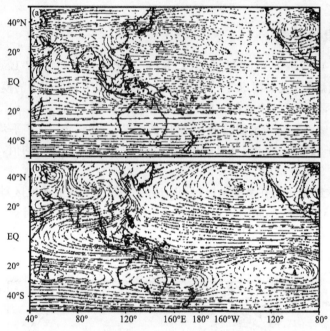

图 5　CEX 控制试验高、低空 6—8 月平均流场分布
(a. 500 hPa;b. 850 hPa)

东亚季风低空(850hPa)环流系统的主要成员,如澳大利亚冷性反气旋、东亚地区向北越赤道气流、南海夏季西南季风、南海季风槽、西太平洋副热带高压、副热带西南季风等系统与气候场相当接近;马斯克林高压、索马里越赤道气流、印度西南季风等系统也模拟得相当好,说明 L9R15 模式对夏季亚洲低层环流和热带印度洋—太平洋地区的环流也具有较强的模拟能力。

总的来说,L9R15 模式对夏季大气环流具有较强的模拟能力。

5.3 异常试验结果

从图 6 可见,5 月,台湾岛北部为一反气旋性差值环流,中心比叠加的海温距平中心偏西,但仍处于所叠加的海温距平区域范围之内。在这样的环流形势下,西太平洋副高位置偏南,强度偏强,南海大部分地区处于副高南侧的偏东气流之下,阻碍南半球越赤道气流的东伸北抬,导致南海夏季风爆发偏晚偏弱。

图 7 上,一反气旋性差值环流位于台湾岛附近。在这样的环流形势下,西太平洋副高位置偏南,江淮流域为西南气流控制,表明黑潮地区正海温异常将导致西太平洋副高位置偏南,强度偏强,江淮流域偏涝,反之亦然。这与前面的诊断分析结果非常一致。

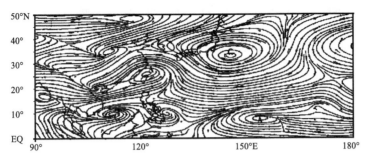

图 6　EX+与EX-试验的 5 月 850 hPa 差值环流分布

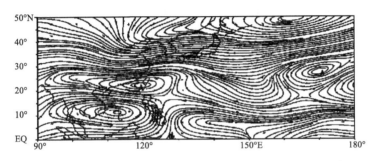

图 7　EX+与EX-试验的 6—8 月平均 500 hPa 差值环流分布

6　结　论

(1)南海夏季风爆发早(晚)年,西太平洋副高东移迅速(缓慢),位置偏北(南),强度偏弱(强),江淮流域偏旱(涝)。

(2)季节转换时期(4—6月)黑潮地区海温对南海夏季风爆发早迟有一定影响。数值模拟结果进一步表明:该地区海温正异常将导致西太平洋副高位置偏南,强度偏强,南海夏季风建立较晚,强度偏弱,江淮流域偏涝,反之亦然。

参考文献

[1]　Tao Shiyan,Chen Longxun. A review of recent research on the east Asian summer monsoon in China [C]//Chang C-P,Krishnaurti T N. Monsoon Meteorology. Oxford:OxfordUniversityPress,1987,60-92.

[2]　陈隆勋,朱乾根,何金海,等. 东亚季风[M]. 北京:气象出版社,1991.

[3]　何金海,朱乾根,Murakami M. TBB 资料揭示的亚澳季风区季节转换及亚洲夏季风建立的特征[J]. 热带气象学报,1996,12(1):34-42.

[4]　朱乾根,何金海,Murakami M. TBB 资料所揭示的亚澳季风之季节循环特征和年际异常[J]. 应用气象学报,1996,7(2):129-137.

[5]　Webster P J,Yang S. Monsoon and ENSO:selectively interactive system[J]. Quart J R Meteor Soc,1992,118(507):877-926.

[6]　Ananthakrishnan R,Soman M K. The onset of the southwest monsoon over Kerala:1901—1980[J]. J Climate,1988,8(3):283-296.

［7］ Holland G J. Interannual variability of the Australian summer monsoon at Darwin—1952/1982［J］. Mon Wea Rev,1986,114(3):594-604.

［8］ 郭品文.亚洲热带夏季风爆发的特征、机制及南海海气耦合低频振荡［D］.南京:南京气象学院,1997:22-27.

［9］ 谢安,刘霞,叶谦.南海夏季风爆发的气候特征［C］//何金海,等.亚洲季风研究的新进展.北京:气象出版社,1996:132-142.

［10］ 何金海,温敏,罗京佳.东亚季风区旱涝年季节转换的不同特征［C］//何金海,等.亚洲季风研究的新进展.北京:气象出版社,1996:82-86.

［11］ 金祖辉,村上胜人.东亚季风区对流活动的年际变异及与江淮地区旱涝关系的研究［C］//何金海,等.亚洲季风研究的新进展.北京:气象出版社,1996:88-97.

［12］ Simmonds I. Analysis of the "spinup" of a general circulation model［J］. J Geoph Res,1985,90(D3):5637-5660.

［13］ Wu Guoxiong, Liu Hui, Zhao Yucheng, et al. A nine-layer atmospheric general circulation model and its performance［J］. A A S,1996,13(1):1-18.

［14］ Liu Hui, Wu Guoxiong. Impacts of land surface on climate of July and onset of summer monsoon: A study with an AGCM plus SSIB［J］. A A S,1997,14(3):289-308.

中南半岛影响南海夏季风建立和维持的数值研究*

徐海明[1]　何金海[1]　温敏[1]　董敏[2]

(1. 南京气象学院,南京 210044；2. 国家气候中心,北京 100081)

摘要：利用美国大气研究中心研制的第三代公共气候模式（CCM3）模拟了中南半岛对南海夏季风的建立和维持的影响,数值试验结果表明,中南半岛对南海夏季风的建立和维持起了非常重要的作用。同时还就中南半岛影响南海夏季风建立和维持的机制进行了讨论。

关键词：中南半岛；南海；夏季风；数值模拟

1　引　言

研究表明[1],亚洲夏季风的爆发最早出现在南海,然后向北伸展到中国大陆及日本以南西太平洋地区,向西北伸展到孟加拉湾和印度,因而南海季风区被认为是东亚夏季风爆发的源地。因此,南海夏季风爆发特征和机制的研究对于正确认识整个亚洲夏季风的爆发及其随后的发展是至关重要的,同时也是东亚地区夏季降水准确预报的关键因素之一。近年来,国内外有关南海夏季风爆发的机制和原因已有了一些研究,但还没有明确的结论。大量的研究表明,南海表层海温在 5—6 月的增暖[2]、基本气流的对称不稳定[3]和低频振荡及中纬度系统的触发作用[4-5]等均可能导致南海夏季风的爆发。何金海等[6]采用黑体辐射温度（TBB）资料对南海夏季风爆发的过程进行了研究,结果表明南海夏季风爆发于 5 月中旬,它较印度季风要早一个月,这与文献[1]的结果是一致的。同时他们指出,4 月 TBB 高值区首先在中南半岛地区开始断裂为东、西两部分（意味着中南半岛对流开始活跃）。随后,分别向东、向西撤退。正是 TBB 高值带东段（相应于西太平洋副热带高压）的迅速东撤并伴随着北移,导致了南海—西太平洋地区对流的活跃和夏季风的建立。究竟是什么原因引起对流首先在中南半岛地区活跃以及随后 TBB 高值带的迅速东撤？引起夏季风建立和维持的根本原因是海陆的热力对比即海陆温差,随着北半球太阳辐射从冬到夏的逐渐增强,由于海洋和陆地热力特性的差异,陆地增暖快于海洋,导致海陆温差梯度先反转然后再加大,当该温差梯度达到某一临界值时即可引起夏季风的建立或爆发。因此,我们认为中南半岛在由冬到夏的迅速增温引起的对大气的感热加热,可能对亚洲夏季风最早在南海爆发起着非常重要的作用。实际情况是否如此,还有待用数值模拟来验证,这对于揭示南海夏季风爆发的原因和维持的机制是十分有意义的。

本文利用美国大气研究中心（NCAR）研制的第三代公共气候模式（CCM3）来研究中南半岛在南海夏季风建立和维持中所起的作用,并探讨其影响南海夏季风建立和维持的机制。

---

* 本文发表于《大气科学》,2002 年第 26 卷第 3 期,330-342.

2 数值模式和试验方案

本文所用模式为 CCM3,它是在第二代公共气候模式(CCM2)的基础上改进而成的,采用 σ-P 混合坐标的全球谱模式,垂直分为 18 层,大气顶取在 2.9 hPa,水平分辨率为 T42(相当于经纬度间距 2.8°左右,全球共 128×64 个格点)。模式包括了辐射对流、垂直扩散边界层及陆面过程等各种物理过程,同时还包括了日变化。辐射通量每隔 1 h 计算一次,在这之间辐射通量保持不变。时间积分步长为 20 min。它与 CCM2 相比在云的参数化、晴空长波辐射、深对流、边界层过程和陆面过程都作了改进。经过这些改进,模式的辐射偏差已大为减少,减弱了过强的水循环,使潜热释放及降水率均接近观测,辐射收支已接近平衡。关于模式的详细说明和改进可参阅 Kiehl 和 Hack 等人的工作[7-8]。

CCM3 模式中所用的海温为多年平均海温资料模式在运行过程中保持海温的季节变化而没有年际变化。为了研究中南半岛对南海夏季风的影响,我们设计如下两个数值试验:

试验 1 采用经过多年积分后得到的 9 月 1 日模式场作为本试验的初始场,然后再积分 330 天,每 5 天取一次平均作为一次结果输出,该试验我们称之为控制试验,用 CTL 表示。

试验 2 为了研究中南半岛对南海夏季风爆发及其维持的影响,我们将 20°N 以南的中南半岛改为海洋,同时将中南半岛变为海洋后的海温值同纬度邻近海域的海温来代替,然后,再采用与试验 1 同的初始场和积分时间以及相同的结果输出频率。该试验我们称之无中南半岛试验,用 NZT 表示。

3 试验结果分析

3.1 CCM3 揭示的南海夏季风建立特征

能否成功模拟中南半岛对南海夏季风建立及其维持的影响,关键在于 CCM3 控制试验(CTL)中能否很好地模拟出南海夏季风的主要特征。图 1 给出了控制试验中南海所在区域

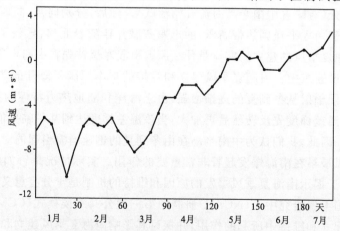

图 1 控制试验中(CTL)南海区域(5°~20°N,105°~120°E)
850 hPa 平均纬向风随时间的演变图

(5°~20°N，105°~120°E)平均的 850 hPa 纬向风随时间的演变。从图中可以看到,区域平均纬向风从冬到夏的总体特征是逐渐增加的,即东风减小、西风增强,但在东风减小、西风增强的过程中明显存在 30~60 天的低频振荡。从图中还可以看到,在 5 月的第 2 候平均纬向风由东风开始转为西风,且该西风一直维持到 5 月底。从 850 hPa 纬向风的时间—经度和时间—纬度的剖面图上(图 2、图 3),同样也清楚地反映出南海 5 月中上旬明显有一次西风的增强过程。在区域平均的降水量图上(图略),5 月第 3 候降水量也有一次明显的增强过程,结合降水量和纬向风两个要素的演变特征,可确定南海夏季风在 5 月 3 候爆发,这与用实际观测资料所确定的南海夏季风平均在 5 月中旬建立是比较一致的。

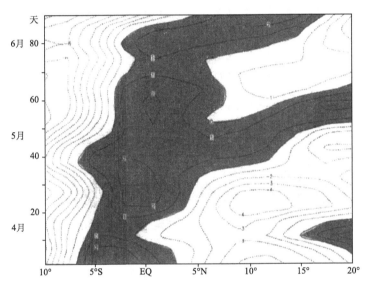

图 2　控制试验中(CTL)850 hPa 纬向风沿 105°~120°E 的时间—纬度剖面图
阴影区为风速大于 0 的区域

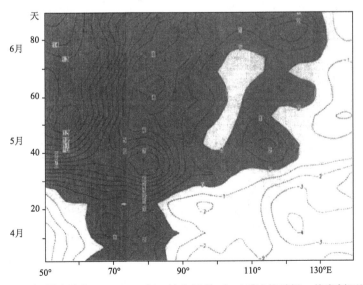

图 3　控制试验中(CTL)850 hPa 纬向风沿 5°~15°N 的时间—纬度剖面图
阴影区为风速大于 0 的区域

下面分析 CCM3 控制试验中南海夏季风建立前后对流层低层大气环流的演变特征。图 4 给出了控制试验中 CCM3 模拟的 4—6 月 850 hPa 风场的演变情况。由图可见：4 月，西太平洋上为强大副热带高压所控制，西太平洋副热带高压的西部脊向西伸至我国南海和中南半岛上空，南海上空盛行副热带高压南侧的偏东气流，即南海夏季风还没有建立；印度半岛东部上空为一近似南北走向的槽，即天气学上所谓的印缅槽，印度半岛则主要受到槽后西北气流所控制，此时，索马里越赤道气流和赤道印度洋上西风还未出现。5 月，索马里附近已出现了较强的向北越赤道气流，该气流越过赤道转向后，在赤道印度洋的北侧形成了较强的赤道西风，该西风一直向东扩展至中国南海南部；同时，西太平洋副热带高压西部脊已明显减弱东撤，原先位于印度半岛东部上空的印缅槽已东移至印度东部沿海上空，槽前的西南风气流相应增强并向东经孟加拉湾东部、中南半岛扩展至我国南海上空，表明夏季风已开始在南海建立；此时，印度半岛中北部仍受到印缅槽后的西北气流所控制，印度夏季风还没有建立。6 月，索马里越赤道气流进一步加强，该气流越过赤道转向后，在阿拉伯海上空形成了强的西南气流并向东扩展

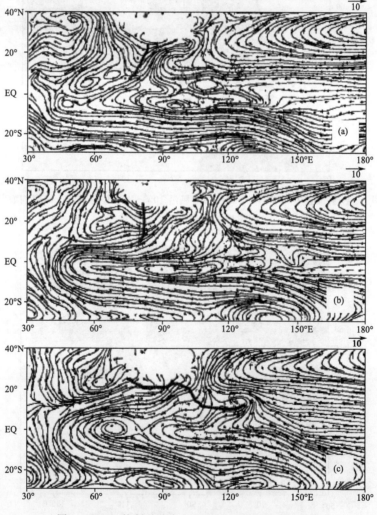

图 4 CCM3 控制试验中(CTL)4—6 月 850 hPa 风场
(a)4 月；(b)5 月；(c)6 月

至整个印度半岛上空,从印度半岛北部经中南半岛至我国南海上空为一近似东南—西北走向的季风槽所控制,表明 6 月印度夏季风已经建立,而南海夏季风则进入盛期。图 5 给出 40 年平均的 4—6 月 850 hPa 风场分布,所用资料取 NCEP/NCAR40 年再分析数据集。比较图 4 和图 5,我们不难发现 CCM3 所模拟的 4—6 月的大尺度环流场,除了太平洋副热带高压的西部脊在 6 月略微偏北和 7 月偏西偏北外,与实际的观测结果是一致的。这表明 CCM3 对亚洲夏季风,特别是南海夏季风的一些主要特征还是具有比较强的模拟能力。

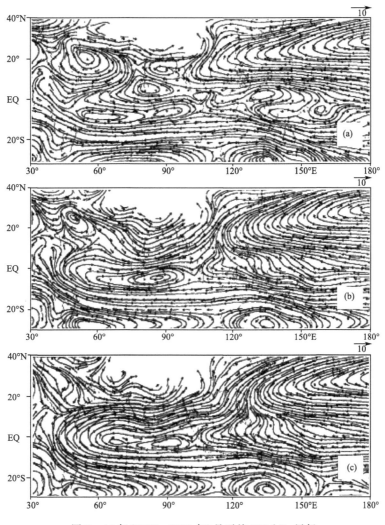

图 5　40 年(1958—1997 年)月平均 850 hPa 风场
(a)4 月;(b)5 月;(c)6 月

3.2　中南半岛对南海夏季风的影响

图 6 为无中南半岛试验(NZT)中,南海区域(5°~20°N,105°~120°E)平均的 850 hPa 纬向风随时间的演变。从图中可以看到,从冬到夏南海区域的东风分量也是逐渐减小的,但在整个过程中都没有出现西风。这从 850 hPa 纬向风的时间—经度和时间—纬度的剖面图上

(图7、图8),也同样反映出5—6月南海区域没有出现西风分量。这表明当CCM3模式中仅仅去除了中南半岛以后便不再出现南海夏季风,而由以上的分析结果知道CCM3的控制试验能很好地模拟出5月中下旬南海夏季风建立过程,由此可见,中南半岛的存在对南海夏季风建立及其维持起了非常重要的作用。

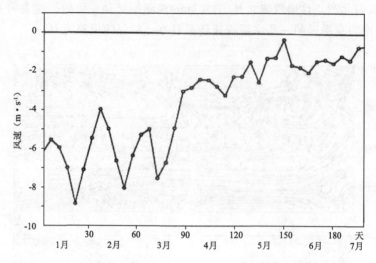

图6　无中南半岛试验中南海区域(5°~20°N,105°~120°E)850 hPa平均纬向风随时间的演变图

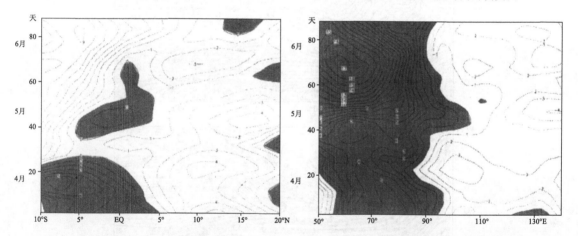

图7　无中南半岛试验中(NZT)850 hPa纬向风沿　　图8　无中南半岛试验中(NZT)850 hPa纬向风沿
　　　105°~120°E的时间—纬度剖面图　　　　　　　　　　5°~15°N的时间—经度剖面图
　　　阴影区为风速大于0的区域　　　　　　　　　　　　阴影区为风速大于0的区域

图9给出了无中南半岛试验中,4—6月对流层低层850 hPa的水平流场。由图可见,在无中南半岛试验(NZT)中,4—6月西太平洋副热带高压的西部脊较控制试验中西太平洋副热带高压脊要强得多。西太平洋副热带高压的西部脊,自台湾以东的西北太平洋地区一直向西伸展至南海北部和中南半岛上空,整个南海基本上受到副热带高压脊南侧的偏东风气流控制。由此可见,当CCM3中不考虑中南半岛时,西太平洋副热带高压脊较控制试验中的明显增强西伸并稳定地控制整个南海和中南半岛,从而阻碍了南海夏季风的建立和维持。

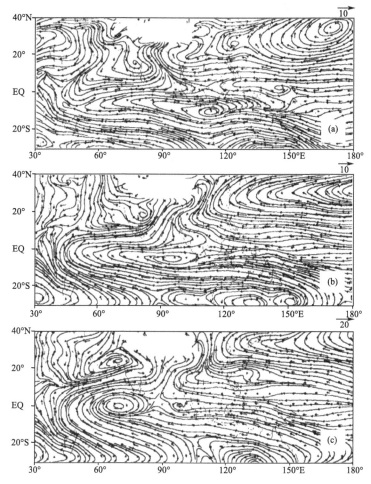

图 9 无中南半岛地形试验中(NZT)4—6 月 850 hPa 风场
(a)4 月;(b)5 月;(c)6 月

综上所述,当模式中包含了中南半岛时,CCM3 能很好地模拟出 5 月中下旬南海夏季风的爆发以及 6—7 月南海夏季风的维持;而当模式中仅仅去除中南半岛以后,南海夏季风无论在 5 月还是 6—7 月都不再出现。可见,中南半岛对南海夏季风的形成和维持起了非常重要的作用。

3.3 中南半岛对南海夏季风建立和维持影响的机理分析

由上面的分析,我们知道中南半岛地形对南海夏季风的形成和维持起了非常重要的作用,那么中南半岛又是通过何种机理影响南海夏季风的建立和维持的呢?本节就这一问题进行讨论。

地形对大气环流的影响主要起到两方面的作用,一是地形的动力强迫抬升作用,另一个是地形的热力作用。由于中南半岛都为平原和丘陵地区,它的动力强迫抬升作用很小,对大气环流的作用主要表现在它的热力作用。在采用数值模拟结果讨论中南半岛的热力作用通过何种途径影响南海夏季风的建立和维持之前,首先来看实际观测的结果。图 10、图 11 分别给出了

10 年平均的地表感热通量和地球同步卫星观测的 T_{BB} 沿 $10°\sim 20°N$ 的时间—经度剖面图,其中地表感热通量取自 NCEP/NCAR 逐日再分析数据集,T_{BB} 由日本气象研究所提供。由图可见,在 2 月中南半岛上空便出现了 $80\ W\cdot m^{-2}$ 以上的感热通量,并随时间增加逐渐增强,至 3 月中旬感热通量值达到了 $100\ W\cdot m^{-2}$ 以上,并一直维持到 4 月中旬,之后便逐渐减弱消失。与之相对应,随中南半岛上空感热加热的增强,T_{BB} 高值带 3 月首先在中南半岛上空出现断裂(对应于副热带高压带断裂),4 月出现了对流活动,5 月出现了较强的对流($T_{BB}<260\ K$)。由此可见,2—3 月中南半岛上空出现的强感热加热,对副热带高压带首先在中南半岛上空断裂起了非常重要的作用,而副热带高压带的断裂又为中南半岛上空对流的产生创造了条件。4—5 月中南半岛对流的出现,一方面阻碍了太阳辐射到达地面,使地表感热通量减小;另一方面又可能通过释放凝结潜热对南海夏季风的建立和维持产生影响。那么,中南半岛是否通过以上过程对南海夏季风的建立和维持产生影响?我们将用两个对比试验结果来加以分析讨论。

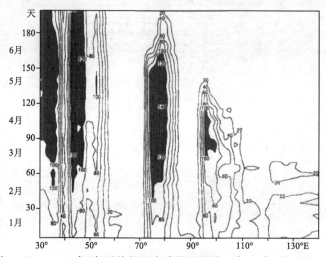

图 10　10 年(1982—1991 年)候平均的地表感热通量沿 $10°\sim 20°N$ 的时间—经度剖面图
阴影区为数值大于 $100\ W\cdot m^{-2}$ 的区域

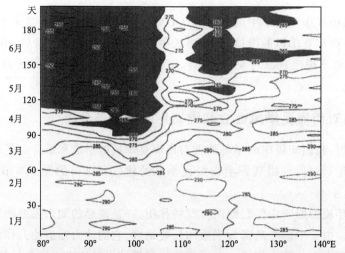

图 11　10 年(1982—1991 年)候平均的 T_{BB} 沿 $10°\sim 20°N$ 的时间—经度剖面图等直线间隔:5 K,阴影区为数值小于 265 K 的区域

图 12 为控制试验和无中南半岛试验中地面感热通量沿 10°～20°N 的时间—经度剖面图。由图可见，在控制试验中由于中南半岛的存在，早在 2 月中南半岛的上空便出现了强感热通量（100 W·m^{-2} 以上），感热通量在 3 月达到最强（120 W·m^{-2}），并一直维持到 4 月底，其变化特征与实际观测结果（图 10）是非常一致的，只是 CCM3 模拟的地表感热通量略大于观测值。在无中南半岛试验中，由于在 CCM3 中去除了中南半岛，所以，中南半岛上空的强感热通量便不再出现。由此可见，由于中南半岛的存在，在其上空 2 月便出现了强的感热加热，并一直维持到 4 月底，这样强而持久的感热加热势必会对中南半岛上空及其邻近地区的温压场和大气环流产生影响。

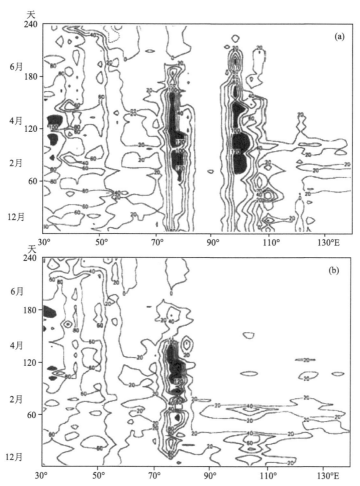

图 12　10°～20°N 平均的感热通量的时间—经度演变图
(a)控制试验(CTL)；(b)无中南半岛地形试验(NZT)；等值线间隔：200 W·m^{-2}
阴影区为数值大于 100 W·m^{-2} 的区域

图 13 分别给出了控制试验和无中南半岛地形试验中，10°～20°N 纬度带内平均的 850 hPa 温度在 50°～140°E 范围内的纬向偏差随时间—经度的演变情况。由图可见，在控制试验中，4 月之前印度半岛上一直维持正的温度偏差，这说明印度半岛上的气温比邻近的阿拉伯海和孟加拉湾的气温都要高；而在中南半岛所在经度上（105°E 附近），2 月之前一直维持强

的负偏差,这说明在北半球冬季,由于中南半岛的存在使中南半岛上空的气温比邻近的孟加拉湾和南海及西太平洋上的气温要低得多。到了3月,由于受到中南半岛地表强的感热加热的影响,在中南半岛的西部首先出现正的温度偏差,随后偏差迅速向东扩展至整个中南半岛,在3月中旬至4月中旬之间气温正偏差达到最强,其中心值达到2℃以上。进入5月中旬即南海夏季风爆发以后,该正偏差也随之消失。在无中南半岛地形试验中,由于在模式中去除了中南半岛地形,中南半岛所在区域便不再出现强感热加热,与之相对应,在中南半岛上空的气温与其邻近的南海和孟加拉湾上空的气温不再出现明显的温差(图13b)。由此可见,在南海夏季风爆发前2个月左右,中南半岛已通过强的感热加热使其上空的温度迅速增暖,从而使其与南海之间的东西向温度梯度在3月便已产生反转,这可能是造成亚洲夏季风在南海地区最早爆发的根本原因。

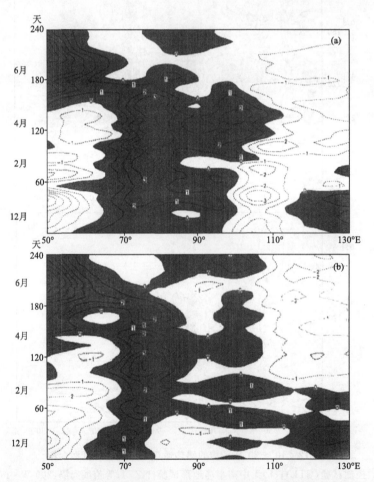

图13 10°～20°N间850 hPa平均温度在50°～140°E范围内纬向偏差的时间—经度演变图
(a)控制试验(CTL);(b)无中南半岛地形试验(NZT);阴影区为偏值大于0的区域

图14分别给出了4—6月控制试验(CTL)与无中南半岛地形试验(NZT)之间的850 hPa风场差。由图可见,4月孟加拉湾上空为一反气旋性的差值环流中心,中南半岛北部上空则为一闭合性的气旋性差值环流,该气旋性差值环流沿中南半岛向西南方向一直延伸到南海南部,而在巴士海峡以东的西太平洋上又为一个闭合的反气旋性差值环流,这种东西向的反气旋

旋和反气旋性差值环流的配置,明显有利于西太平洋副热带高压脊在中南半岛上空断裂并向东移动;5月,在中南半岛北部形成一个气旋性的差值环流并向东扩展至南海北部和我国华南沿海(图14b),在其南侧的整个南海和中南半岛产生一致的西南偏西差值气流,明显有利于南海夏季风的建立;中南半岛同样对6月的环流也产生一定的影响,在中南半岛经我国南海中部于菲律宾上空激发出一个气旋性的差值环流,并在菲律宾上空形成一个气旋性差值环流中心,明显有利于南海季风槽的维持。

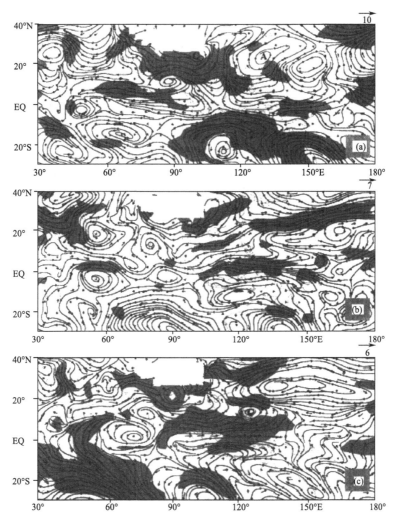

图14 控制试验与无中南半岛地形试验之间850 hPa风矢量差
(a)4月;(b)5月;(c)6月。阴影区通过0.05信度 t 检验区域

最近,吴国雄等[9]从动力角度研究了空间非均匀加热对副热带高压带的形成和变异的影响。下面,就利用吴国雄等的研究结果,从理论上来探讨一下3—4月中南半岛上空的感热加热对副热带高压带断裂所产生的影响。由文献[9]中(11)式知,水平非均匀加热与垂直涡度分量变化之间满足:

$$\left(\frac{\partial \zeta_z}{\partial t}\right)_x^Q \varpropto -\frac{g}{f\theta_z}\frac{\partial T}{\partial x}\frac{\partial Q}{\partial x} \tag{1}$$

$$\left(\frac{\partial \zeta_z}{\partial t}\right)_y^Q \varpropto -\frac{g}{f\theta_z}\frac{\partial T}{\partial y}\frac{\partial Q}{\partial y} \tag{2}$$

式中,Q 为非绝热加热率,T 为温度,θ 为位温。$(\partial \zeta_z/\partial t)_x^Q$ 表示由于温度(T)和加热场(Q)的纬向分布不均匀而引起的垂直涡度变化,而$(\partial \zeta_z/\partial t)_y^Q$ 则表示由于温度(T)和加热场(Q)的经向分布不均匀而引起的垂直涡度变化。由图 12a 和图 13a 知,3—4 月,中南半岛东西向感热加热和温度分布,在中南半岛以东的南海地区满足$(\partial T/\partial x)<0$、$(\partial Q/\partial x)<0$;而在中南半岛以西的孟加拉湾则满足$(\partial T/\partial x)>0$,$(\partial Q/\partial x)>0$。由(1)和(2)式可知,在南海和孟加拉湾地区都有利于反气旋涡度的发展,即$(\partial \zeta_z/\partial t)_x^Q<0$。由此可见,中南半岛 3—4 月出现的强感热加热以及由感热加热引起的对流层低层的增暖,可使南海和孟加拉湾的反气旋性环流增强,正如图 14a 所示,在南海和孟加拉湾分别出现反气旋性差值环流中心,而热力适应[10]的结果使得感热加热局地上空出现气旋式环流,从而可引起副热带高压带在中南半岛上空首先出现断裂。

综上所述,中南半岛影响南海夏季风建立和维持的机理可概述为:中南半岛首先通过感热加热对其上空温压场产生影响,使副热带高压带首先在中南半岛上空出现断裂,而高压带的断裂又为中南半岛上空对流的产生提供了一个切变辐合的有利条件。一旦中南半岛上空出现较强的对流,其一方面阻碍了太阳辐射到达地面,使地表感热通量迅速减小;另一方面又可能通过释放凝结潜热对西太平洋副热带高压产生影响,从而对南海夏季风的建立和维持产生影响。

4 结 论

本文利用 CCM3 模拟了中南半岛对南海夏季风的建立和维持所产生的影响,数值试验结果表明,中南半岛地形对南海夏季风的建立和维持起了非常重要的作用,当 CCM3 中考虑了中南半岛时,模式能很好地模拟出 5 月南海夏季风爆发和维持过程;而当模式中仅去除了中南半岛后,南海夏季风便不再出现。数值试验结果进一步表明,在 2 月中南半岛上便产生了强的感热加热中心,至南海夏季风爆发前的 2 个月左右,中南半岛便通过感热加热使其上空的气温迅速升高增暖,并使其与邻近的南海和孟加拉湾的温度梯度产生反转。中南半岛上空出现的强感热加热及对流层低层的迅速增暖,则导致原先控制南海和中南半岛的西太平洋副热带高压的西部脊在中南半岛上空出现断裂并东撤,而高压带的断裂又为中南半岛上空对流的产生提供了一个切变辐合的有利条件。一旦中南半岛上空出现较强的对流,其一方面阻碍了太阳辐射到达地面,使中南半岛地表感热通量迅速减小;另一方面又可能通过释放凝结潜热对西太平洋副热带高压脊的减弱东撤产生影响。随西太平洋副热带高压脊的减弱东撤、南海夏季风也随之建立和维持。

参考文献

[1] Tao Shiyan, Chen Longxun. A review of recent research on east Asian summer monsoon in China[C]// Monsoon Meteorology. Oxford:Oxford University Press,1987,60-92.

[2] 何有海,关翠华,甘子钧.南海南部海洋上层的热振荡[J].海洋学报,1992,14(3):19-28.

[3] Lau K M, Yang Song. Seasonal variation, abrupt transition and intraseasonal variability association with the summer monsoon in the GLA GCM[J]. J Climate, 1996, 9:965-985.

[4] Murakami T, Chen Longxun, Xie An. Relationship among seasonal cycles low-frequency oscillations and transent disturbances as revealed from outgoing long wave radiation data[J]. Mon Wea Rev, 1986, 114: 1456-1465.

[5] Chang C P, Chen T J. Tropical circulation association with southwest monsoon onset and westerly surge over South China Sea[J]. Mon Wea Rev, 1995, 123:3254-3267.

[6] 何金海, 朱乾根, Murakami M. TBB 资料揭示的亚澳季风区季节转换及亚洲夏季风建立的特征[J]. 热带气象学报, 1996, 12:34-42.

[7] Kiehl J T, Hack J J, Bonan G B, et al. Description of the NCAR Community Climate Model (CCM3)[R]. NCAR Tech. Note, 1996, NCAR/TN-420+STR, 152.

[8] Kiehl J T, Hack J J, Bonan G B, et al. The National Center for Atmospheric Research Community Climate Model: CCM3[J]. J Climate, 1998, 11:1131-1150.

[9] 吴国雄, 刘屹岷, 刘平. 空间非均匀加热对副热带高压带形成和变异的影响 I: 尺度分析[J]. 气象学报, 1999, 57:257-263.

[10] 吴国雄, 刘屹岷. 热力适应、过流、频散和副高 I. 热力适应和过流[J]. 大气科学, 2000, 24:433-446.

南海夏季风建立期间副高带断裂和东撤及其可能机制[*]

何金海[1]　温敏[1]　施晓晖[2]　赵巧华[1]

(1. 南京气象学院,KLME,南京 210044；2. 云南省气象台,昆明 650034)

摘要：使用1998年南海季风试验(SCSMEX)资料和日本气象研究所(MRI)提供的 T_{BB} 资料,分析了南海夏季风建立期间副高带断裂和东撤过程的主要特征及其可能机制。发现北半球副热带高压带的断裂(低层早于高层)和印缅槽(或孟加拉湾槽)的形成是南海夏季风建立的重要前期征兆之一,也为副高东撤和南海夏季风的建立提供了重要条件。斯里兰卡附近低涡的持续北移是副高断裂和印缅槽建立过程的显著特征。分析表明,南海夏季风建立之前,印度半岛的感热加热和中南半岛的潜热加热所激发的气旋性流场在孟加拉湾地区是相互叠加的,这有利于孟加拉湾低涡活动和低槽的形成,这可能是副高断裂和印缅槽活跃的机制。伴随着印缅槽前西南气流和赤道印度洋西风扰动的东传以及南海地区中低纬的相互作用,南海夏季风建立。南海夏季风爆发期间,副高东撤、季风加强和对流加热之间存在一种正反馈作用,这导致了副高的连续东撤和南海季风的"爆发性"以及各种要素的突变特征。当南海季风对流减弱后,西太平洋副高则会西伸。

关键词：南海夏季风建立；印缅槽和涡旋对；副高断裂和东撤；机制分析

亚洲季风区是全球最显著的季风区,可分为印度季风区和东亚季风区[1],而东亚季风又可划分为南海—西太平洋热带季风和中国大陆东部—日本的副热带季风[2]。南海夏季风由于其地理位置的特殊性,它不仅是印度季风和东亚季风连接的媒介,也是亚澳季风相互联系、相互作用的表现。Murakami等[3]提出,由亚洲和澳大利亚大陆热力差异所造成的气压梯度的大尺度季节转换比海陆热力差异所造成的季节转换要敏感和强烈得多,南海地区正是变化敏感的区域。因此南海夏季风的建立应是亚洲冬季风形势向夏季风形势转换的一种信号。关于南海夏季风特别是1998年南海夏季风建立的大尺度特征已有许多研究[4-10],并取得了令人鼓舞的进展。副热带高压(简称副高)是季风系统的重要成员,它的变动直接联系着夏季风的建立、活跃和进退[11-12]。何金海等[9]通过对多年 T_{BB} 资料的分析指出,亚澳季风区的季节转换在4月就已开始,主要特征表现为澳大利亚 T_{BB} 高值带(对应澳大利亚副热带高压)迅速北移和中南半岛地区对流的活跃以及北半球副热带 T_{BB} 高值带(对应副热带高压)在该地区的断裂,伴随着东部 T_{BB} 高值带(对应西太平洋副高)的连续东撤,南海夏季风于5月4候建立；而西部 T_{BB} 高值带的向西退缩则伴随着亚洲夏季风建立自东向西的推进过程。何金海等[4]最近又强调副热带高压带在孟加拉湾的首先断裂和印缅槽的建立对南海夏季风爆发的重要作用。因此我们自然要问,副高带为什么会在孟加拉湾地区首先断裂？它断裂的过程和机制如何？断裂

---

[*] 本文发表于《南京大学学报(自然科学)》,2002年第38卷第3期,318-330.

后副高连续东撤的机制又如何？本文试图对上述问题作一初步的分析。

1 资料及方法说明

本文使用的资料是1998年南海季风试验（SCSMEX）提供的逐日再分析资料和日本气象研究所（MRI）提供的T_{BB}资料。主要是：

(1)5月1日—8月31日,基本要素资料（u,v,t,z,q）和地表感热、潜热通量；格距为2.5°×2.5°经纬度；(2)T_{BB}资料：5月1日—8月31日,每天8次,处理成逐日资料（5月19日缺,经内插补齐）。

为了诊断加热场和流场（涡度场）之间的相互关系（或相互作用）,我们使用了吴国雄[10]得到的仅考虑外热源强迫的全型涡度方程：

$$\frac{\partial \zeta_z}{\partial t} = (1-\kappa)(f+\zeta_z)\frac{\omega}{p} - (f+\zeta_z)\frac{Q}{\theta} + \frac{f+\zeta_z}{\theta_z}\frac{\partial Q}{\partial z} - \frac{1}{\theta_z}\frac{\partial v}{\partial z}\frac{\partial Q}{\partial x} + \frac{1}{\theta_z}\frac{\partial u}{\partial z}\frac{\partial Q}{\partial y} - \mathbf{V}\cdot\nabla\zeta_z - \beta v \quad (1)$$

并用倒算法计算了大气视热源,其公式如(2)：

$$Q = c_p\left[\frac{\partial T}{\partial t} + \mathbf{V}\cdot\nabla T + \left(\frac{P}{P_0}\right)^\kappa \omega \frac{\partial \theta}{\partial p}\right] \quad (2)$$

2 季风建立前副高断裂过程及其机制分析

资料分析已经表明,南海夏季风最早建立于4月底,最迟建立于6月上旬,平均来说建立于5月中下旬（公认为5月第4候）。这就是说,就月际尺度而言,4月属于南海夏季风建立前的冬季风形势,6月整个亚洲夏季风形势建立,而5月属于南海夏季风建立时期。在多年月平均图上,4月500 hPa副热带高压沿15°N基本呈带状分布,5月副高带在印度东部的孟加拉湾地区开始断裂,850 hPa面上情况类似。

为了分析副高带的断裂过程,我们以1998年为例,采用候平均流场及其趋势流场分析南海夏季风建立前副高带的断裂过程。由图1可知,1998年直到5月3候,500 hPa副高带仍基本维持,850 hPa上5月2候副高带亦基本维持,但5月3候副高带状已严重变形,这表明副高变化低层要早于高层。5月4候,850 hPa上孟加拉湾地区已为一显著的低槽（印缅槽）,槽内有一较强的低涡,它与南半球赤道附近的低涡构成一明显的"涡旋对"。副高带已清楚地断裂为二部分,其西部主体停留在阿拉伯海上,东部的西太平洋副高脊仍控制着南海大部分地区,表明南海地区夏季风尚未建立。但是,孟加拉湾地区槽前西南气流已能越过中南半岛伸展至南海北部并伴有显著降水,这一特点导致有些作者认定南海夏季风已经爆发。5月5候,西太平洋副高脊已撤出南海并北抬,南海北部地区已建立一显著季风槽,105°E的越赤道流和源自南半球的赤道印度洋西南气流以及副高西南侧的转向偏南气流在南海—菲律宾地区汇合成偏西南气流,这表明整个南海地区夏季风已经建立。

由上分析可知,副高带的断裂及其伴随的印缅槽的建立对于南海夏季风的建立是至关重要的。然而,究竟是什么机制导致副高断裂,进而导致印缅槽的活跃呢？从形态上分析我们发现,5月2候在斯里兰卡附近有一低涡在游移,自12日起连续地向北移动,直至20日抵达孟加拉湾北部及高原南侧(如图1d粗实线所示),20日以后,这一低涡移向高原减弱。逐候趋势流场也清楚地显示出这一涡旋的向北移动(图2)。图2c清楚表明,5月5候上述涡旋移至高原南侧后减弱,印缅槽也减弱。但从4候到5候,沿东亚季风环流系统路径(见文献[1]),有偏南风和西南风的显著增强,如图中自南向北的阴影区所示,这反映了源自澳大利亚的扰动对南海夏季风的建立也有重要作用。综上所述,可以说斯里兰卡附近低涡的持续北移对于副高的断裂和印缅槽的活跃有重要作用,可是这一低涡又为什么会北移呢?何金海、徐海明等[4]所做的数值试验指出,由冬至夏,印度半岛的存在及其相应的陆面感热作用能够在印度半岛以东的孟加拉湾地区激发一个异常的槽,并在(18°N,85°E)附近出现一个低值环流中心,这或许能使斯里兰卡附近的低涡沿异常槽槽前西南气流北上并获得加强。

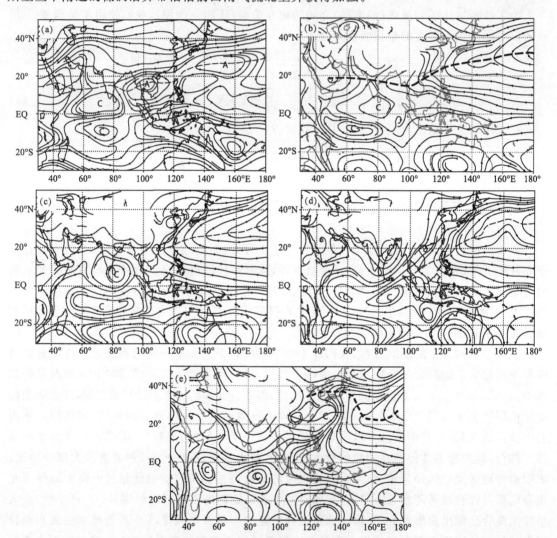

图1　1998年5月候平均流场形势(a.5月3候500 hPa;b.5月2候850 hPa;c.5月3候850 hPa;d.5月4候850 hPa;e.5月5候850 hPa,其中粗虚线为副高脊线,粗实线为南海季风槽槽线,黑色圆点及其连线为赤道印度洋北侧低涡中心及其移动路径,数字为移动的起止日期)

为了进一步讨论上述问题,我们绘制了印度半岛和中南半岛地区感热和潜热通量的时间

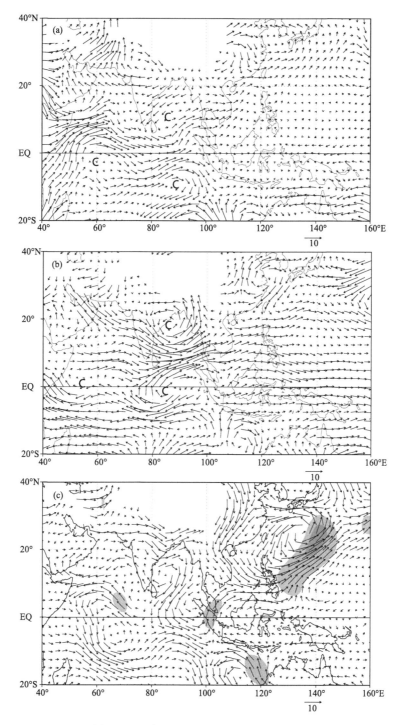

图 2　1998 年 5 月 850 hPa 候平均趋势风场
(a. 3 候减 2 候；b. 4 候减 3 候；c. 5 候减 4 候。阴影区为 $\Delta v > 4 \text{ m} \cdot \text{s}^{-1}$)

演变图(图 3)。由图 3 可以看出,印度半岛的地面感热通量 5 月均维持在 120 W·m^{-2}以上,而潜热通量始终很小,这是印度旱季的特征。中南半岛地区感热通量 5 月 11 日前虽有波动,但基本维持在 100 W·m^{-2} 附近或以上。可是 5 月 11 日以后，地面感热通量迅速下降至很小的数值，同时潜热通量迅速升至 100 W·m^{-2} 以上，与此相联系，5 月 11 日后该区有明显的 T_{BB} 低值区（参阅图 4）发展。这表明 5 月 11 日后中南半岛地区有显著的对流降水，其释放的潜热加热随高度增加（最大加热在 400 hPa 附近）。根据吴国雄[10]关于全型涡度方程的讨论可知，当地面附近有 100 W·m^{-2} 的感热加热时，在一天之内在最大加热上方的东侧（西侧）可强迫出气旋（反气旋）环流；而潜热加热作用则与感热加热作用相反，在潜热最大加热（400 hPa 附近）下方的东侧（西侧）可强迫出反气旋（气旋）环流。这就是说，在 5 月 11 日前，印度半岛和中南半岛地区的感热加热作用所激发的流场对孟加拉湾地区气旋性流场的贡献是相互抵消的，前者稍大；整个时段内印度半岛潜热的作用可以忽略，而中南半岛地区的潜热（特别是 5 月 11 日以后）作用对孟加拉湾地区气旋性流场有重要贡献，但 5 月 11 日后感热作用很小。综合感热和潜热的作用，可以看出 5 月 11 日以后，印度半岛感热和中南半岛的潜热激发的气旋性流场在孟加拉湾地区达到有效叠加。我们有理由认为，这一有效叠加的气旋性流场对于副高带的断裂，以及印缅槽的活跃和涡旋的活动有重要作用。

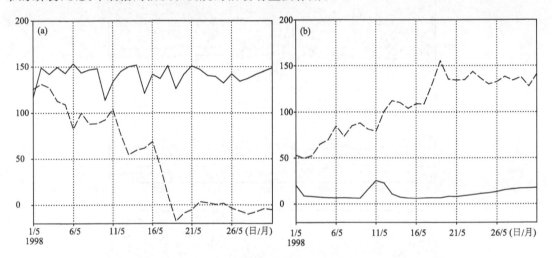

图 3　印度半岛（15°~22.5°N，75°~85°E，实线）和中南半岛（15°~22.5°N，90°~100°E，虚线）陆地地面感热通量（a）和地面潜热通量（b）（单位：W·m^{-2}）

3　副高东撤与南海夏季风建立

一般认为，南海夏季风建立的大尺度特征之一是西太平洋副高连续向东撤出南海地区，然而究竟是什么原因导致副高东撤？是副高东撤导致夏季风建立，还是夏季风建立使得副高东撤呢？它们之间的关系仍是一个需要探究的问题，下面就来讨论相关的问题。

图 4a 展示了 10°~20°N 纬带内平均的 T_{BB} 和 850 hPa 风场的时间—经度剖面。图中阴影区为（$T_{BB}=280$ K）等值线所围的低值区域，它反映了对流的演变。由图我们首先可以发现，如"箭矢"样的阴影区中，基本为偏西南气流控制，特别是在 5 月中旬以后，这表明了对流活跃和偏西南气流的有机联系。阴影区的东部边界基本上为偏西南风和偏东南风的分界线，因此东部边界的走向揭示了西太平副高脊西边界的移动，也就是说在 5 月 23 日前后，西太副高西脊点东撤至

150°E 附近(阴影区的最东边缘),23 日前,西太平洋副高连续东撤,23 日之后西太平洋副高表现为西伸。注意到 1998 年南海夏季风建立日期为 5 月 23 日(或 5 月 21 日和第 5 候)[4,6],可见南海夏季风建立前(后),西太副高有一次连续东撤(西伸)的过程。图 4b 显示了南海地区 T_{BB} 和低层纬向风的时间演变,可以看出,南海地区纬向东风在 5 月 11 日(A 点)以后迅速减小,至 5 月 16 日转为西风,5 月 20 日左右(图中 B 点)纬向西风迅速增加到 6 m·s^{-1} 左右并一直维持;而 T_{BB} 值(相应于对流降水)在 5 月 14 日前后(A' 点)开始迅速下降至 270 K,稳定一段时间后于 23 日前后又迅速下降到 250 K 以下。这表明 23 日以后,南海地区季风对流降水迅速增加。由以上讨论可以看出,南海夏季风的来临是先有风后有雨。这或许可以说副高的东撤对季风雨的来临是重要的。

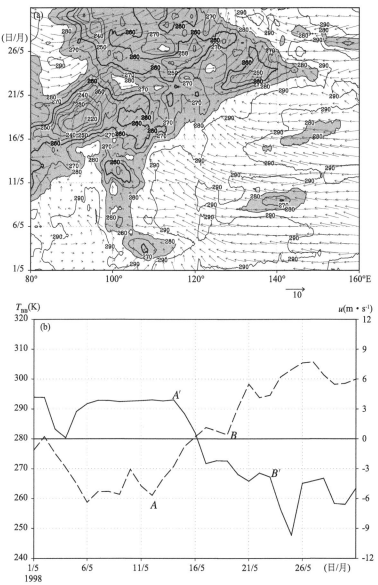

图 4 (a)1998 年 5 月 10°~20°N 平均的 T_{BB} 和 850 hPa 风场时间—经度剖面图,其中阴影区为 T_{BB}<280 K;
(b)南海地区(10°~20°N,110°~120°E)T_{BB}(实线)与 850 hPa 纬向风(虚线)的时间变化曲线,
其中横线为 T_{BB}=280 K 和 u=0 m·s^{-1}

为了进一步讨论副高东撤的原因,我们制作了经过南海地区110°~120°E的T_{BB}时间—纬度剖面和澳大利亚北部扰动北传剖面,如图5所示。由图5a可见,5月23日之前,10°~15°N的南海地区基本为大于280 K的T_{BB}高值区控制,这表明南海夏季风尚未建立,其两侧为T_{BB}

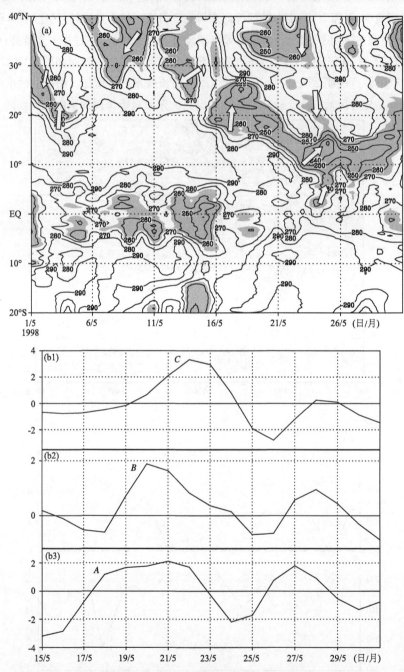

图5 (a)沿110°~120°E逐日T_{BB}时间—纬度剖面图,其中阴影区标示T_{BB}<265 K,加粗线为$T_{BB}=280$ K等值线,箭头为地面风向示意;(b)各区域南风扰动($v(t)-v(t-1)$)时间变化曲线,其中(b1):117.5°~122.5°E, 20°~12.5°S;(b2):100°~105°E,2.5°S~2.5°N;(b3):137.5°~145°E,15°~25°N

的低值带,阴影区显示了对流的活跃,其中箭头表示该区域相应的地面风向。我们还注意到,280 K 线所围的 T_{BB} 高值区的南北宽度随时间是减小的,这与副高连续向东撤出南海地区是一致的。也就是说南、北 T_{BB} 低值带随时间向高值带的挤压展示了中纬度或低纬度及其相互作用对副高东撤的影响。首先,让我们来讨论中纬度的冷空气强迫对 T_{BB} 高值带(相应于副高)的影响。我们不难发现,中纬度先后有 4 次冷空气强迫(伴有偏北风的低值区)冲击 T_{BB} 高值区,分别出现在 5 月初,5 月 2 候,5 月 3 候和 5 月 5 候。显然前 3 次冲击均未能从根本上动摇 T_{BB} 高值区在南海地区的存在,只能造成其北界的摆动。但第 4 次冷空气冲击则获得了动摇副高的"机遇"。一方面印缅槽前西南气流(伴有对流降水)在 5 月 4 候进入南海北部造成 T_{BB} 低值扰动(5 月 16 日以后的 T_{BB} 低值区向南扩展);另一方面 5 月 18 日前后来自澳大利亚北部南风扰动的北传(如图 5b 中 A、B、C 所示),在 5 月 23 日前后造成中低纬和南北半球相互作用的有利态势,将 T_{BB} 高值区挤出南海地区,而代之以显著的低值区,南海夏季风得以在整个南海地区建立。由此可以看出,中低纬和南北半球相互作用在副高东撤进而导致南海夏季风建立过程中担当了重要角色。

4 关于副高东撤过程中加热场作用的讨论

根据吴国雄等[10]推导的全型涡度方程(见(1)),可知非绝热加热的垂直变化对于涡度场有重要影响。当时间尺度很短时,凝结潜热加热将加强气旋的发展,计算结果表明,它的作用在一天之内即可使涡度场发生明显的变化。前文讨论的冷空气冲击和印缅槽前西南气流汇合于副高西部(西北部)边缘,形成对流降水释放潜热,引起气旋性涡度发展,导致副高东撤;另一方面,副高东撤又为对流发展和气旋涡度性发展提供了大尺度条件,这就是短时间尺度内副高东撤和对流降水(或潜热加热场)之间的相互作用。当时间尺度较长时,全型涡度方程(1)主要表现为 $\frac{f+\zeta_z}{\theta_z}\frac{\partial Q}{\partial z}$(记为 Q_z)、β_v 和 $\mathbf{V}\cdot\nabla\zeta_z$ 三大项的平衡。计算结果表明,候平均尺度基本满足"长时间尺度"条件,因此我们就以西太平洋地区(15°~25°N,135°~145°E)对流加热中心为例来讨论它们之间的平衡及其对副高的影响。图 6 展示了 5 月第 5 候平均的 T_{BB} 和风场分布。实际分析已经表明[9],T_{BB} 低值中心与对流加热中心一致,因此由图 6a 可知,在西太平洋地区有一个很强的对流加热中心(阴影区),与其相配合的是强西南风异常区(见图 6b)。在强西南风的西侧为强的气旋性流场(东亚季风槽或南海季风槽),并有一低涡中心;在其东侧则为反气旋高压环流。显然,这一季风对流加热中心对于季风槽的形成和维持以及副高的东撤是有重要贡献的。

为了揭示西太平洋地区全型涡度方程中平衡项的时间演变,我们还绘制了图 7,由图 7a 可见,5 月 21 日前后,纬向风和经向风急剧上升,T_{BB} 急剧下降,23 日前后达到极值,这就是说 23 日前后,西南风和对流加热达到最大,这是南海—西太平洋地区夏季风建立的显著特征,由图 7b 可以看出,5 月 21 日之前,全型涡度方程中的三大项数值均很小,但 21 日之后,加热作用项(Q_z)、β 效应项(β_v)和涡度平流项($\mathbf{V}\cdot\nabla\zeta_z$)均急剧增加,于 23 日前后达到最大,尔后减小。可以看出,加热项的峰值略有超前,这是否表示加热场对季风加强的主导作用值得进一步研究。

综上所述,我们可以看到,在副高东撤、季风加强和对流加热之间在季风爆发期间存在一种正反馈的相互作用。当季风对流减弱后,副高则又会西伸。

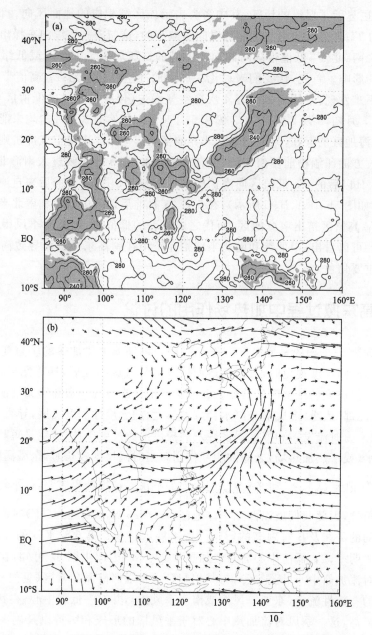

图 6 1998 年 5 月第 5 候 T_{BB}(a)和 850 hPa 风场(b)分布图
阴影区 $T_{BB} \leqslant 265$ K

5 结 论

(1)副热带高压的断裂(低层早于高层)和印缅槽的建立是南海夏季风建立的重要前期征兆之一,也为副高东撤和南海夏季风的建立提供了重要条件。斯里兰卡附近低涡的持续北移是副高断裂和印缅槽建立过程中的显著特征。分析表明,南海夏季风建立之前,印度半岛的感

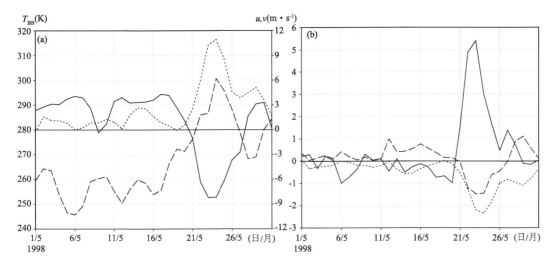

图 7 (15°～25°N,135°～145°E)区域平均各物理量的时间变化(a. T_{BB}(实线)和 850 hPa 纬向风(虚线)、经向风(点线);b. 850 hPa $\frac{f+\zeta_z}{Q_z}\frac{\partial Q}{\partial z}$(实线)、$\beta_v$(虚线)和 $\mathbf{V}\cdot\nabla\zeta_z$(点线),单位:$10^{-10}\,\text{s}^{-2}$)

热加热和中南半岛的潜热加热所激发的气旋性流场在印度东部的孟加拉湾地区是相互叠加的,这可能是副高断裂和印缅槽活跃的重要机制。

(2)副高断裂和东撤以及印缅槽的建立为副高西部(或西北部)边缘的暖湿气流和冷空气交汇提供了必要条件,中低纬相互作用导致的对流降水加热又促使副高东撤,在北传的澳大利亚南风扰动进入南海后,导致整个南海地区夏季风爆发。

(3)南海季风爆发期间,副高东撤、季风加强和对流加热之间存在一种正反馈作用,这导致了南海季风的"爆发性"和各种要素的突变特征。

参考文献

[1] Tao Shiyan,Chen Longxun. A review of recent research on East summer monsoon in China[C]//Monsoon Meteorology. Oxford:Oxford University Press,1987,60-92.

[2] Zhu Qiangen, He Jinhai, Wan Panxing. A study of circulation difference between East-Asia and India summer monsoons with their interactions[J]. Adv Atmos Sci,1986,3(4):466-477.

[3] Murakami T,Chen Longxun, Xie An. Relationship among seasonal cycles, low−frequency oscillations and transient disturbances as revealed from outgoing long wave radiation data[J]. Mon Wea Rev,1986, 114:1456-1465.

[4] 何金海,徐海明,周兵,等.关于南海夏季风建立的大尺度特征及其机制的讨论[J].气候与环境研究, 2000,5(4):333-343.

[5] 丁一汇,薛纪善,王守荣,等.1998年亚洲季风活动与中国的暴雨/洪涝[C]//南海季风爆发和演变及其与海洋的相互作用.北京:气象出版社,1999,1-4.

[6] 李崇银,吴静波.1998南海夏季风的分析研究[C]//南海季风爆发和演变及其与海洋的相互作用.北京:气象出版社,1999,18-24.

[7] 罗会邦.南海夏季风爆发及相关雨带演变特征[C]//南海季风爆发和演变及其与海洋的相互作用.北京:气象出版社,1999,25-29.

[8] 邵慧,钱永甫.1998年南海夏季风爆发前后区域环流变化的主要特征[C]//南海季风爆发和演变及其与海洋的相互作用.北京:气象出版社,1999,34-42.

[9] 何金海,朱乾根,Murakami M. TBB资料所揭示的亚澳季风区季节转换及亚洲夏季风建立的特征[J].热带气象学报,1996,12:34-42.

[10] 吴国雄,刘屹岷,刘平.空间非均匀加热对副热带高压形成和变异的影响 I:尺度分析[J].气象学报,1999,57(3):257-263.

[11] Qian Yongfu, Zhang Qiong, Zhong Xuehong. The South Asia High and its effect on China's Mid-summer climate abnormality[J]. J Nanjing Univ (Nat Sci), 2002, 38(3):295-307.

[12] Zhang Yaocun, Zhou Tianjun. Diagnostic and numerical study of evolutions of regional circulation system associated with the South China Sea monsoon onset[J]. J Nanjing Univ(Nat Sci), 2002, 38(3):331-340.

东亚副热带季风特征及其指数的建立*

周兵[1]　何金海[2]　吴国雄[1]　韩桂荣[2]

(1. 中国科学院大气物理研究所大气科学和地球流体力学数值模拟国家重点实验室,北京 100029;
2. 南京气象学院大气科学系,南京 210044)

摘要:基于大气对流活动和非绝热加热的分析,给出东亚夏季风区域变化特征,客观地确定东亚副热带季风及热带季风对流强度,并由此建立其与大尺度季风环流的内在联系,提出分别用所对应区域经向风垂直切变来构造季风指数,对比分析显示,上述季风指数在反映夏季环流和降水等方面效果显著,能很好地刻画东亚夏季风强度。同时指出,东亚副热带季风指数与西太平洋副热带高压及长江中下游降水密切相关,在空间场上表现出东亚/太平洋型(EAP 型)遥相关特征,高指数年副高偏南,长江中下游为涝;低指数年则相反。

关键词:对流指数;经向风垂直切变;季风指数;相关分析

1 引 言

由于东亚季风环流系统的异常活动会导致季风区降水发生异常,季风强弱与旱涝问题密切相关,而长江流域降水的多寡一直是我国汛期普遍关注的焦点问题,它直接与东亚副热带季风强弱密切相关。为此,人们一直在努力寻找能有效地度量季风变化的指标,即定义季风指数或环流(对流)指数。全印度夏季降水量指数(I_{AIR})就是这样一个广泛应用的指标[1]。由于东亚季风系统的复杂性,我国气象学家对季风指数的构造方法集中在热力学要素或动力学要素体制上[2-5]。

亚洲夏季风(ASM)变化可以是内部动力强迫或诸如海表温度(SST)和陆面边界强迫引起,具有区域变化特征,季风气候包括快过程、中间过程和慢过程,大气内动力快过程和陆面及近海中间过程是通过包含 Madden 和 Julian 30～60 天振荡(简称 MJO)和其他低频振荡的季节内变化相连接的,南亚季风与东(南)亚季风存在着不同的动力学过程,为此,Lau 等[6]给出了区域季风指数(南亚和东亚分别为 RM1 和 RM2)概念;同时指出 Webster 等[7]定义的南亚夏季风指数(简称为 WYI)存在一定的适用性;Ailikun 等[8]也认为 WYI 与西太平洋对流活动相联系而不是南亚季风区的对流活动;Goswami[9]发现广义的印度降水指数与 850 hPa 和 200 hPa 之间平均经向风切变的季风 Hadley 环流指数(MHI)关系很好,但缺乏动力基础;Wang 等[10]定义了南亚对流指数(CI1)/环流指数(WSI1),及东亚对流指数(CI2)/环流指数(WSI2),并进一步分析了季风指数反映的环流特征。李崇银等[11]在分析南海夏季风特征后提出用对流层散度差能有效地描述南海夏季风。

---

* 本文发表于《大气科学》,2003 年第 27 卷第 1 期,123-135.

如何定量描述东亚季风或东亚副热带季风的强度和年际变化是一个重要的科学问题。显然我们期望用一种简单而有意义的指数去表征东亚副热带季风的强度和变率。应该说，季风指数的定义是有随机性的，但选择一种合适的季风指数要有利于寻求季风变率与其他环流系统和气候变率之间的相关；有利于建立季风变率和内、外强迫因子之间的联系；有利于客观估价数值模式再现季风变率的能力。应该指出，东亚季风既包含热带季风，又包含副热带季风，它所占据的纬度范围非常广，可达 40～50 个纬距，受中高纬的影响又很显著，是否对热带季风和副热带季风分别选择指数也是一个必须考虑的问题。

本文在分析大尺度季风环流与对流活动存在内在的联系的基础上，针对东亚副热带季风关键区——长江中下游地区降水主要集中在夏季 6—7 月，考虑高低空越赤道气流在东亚季风环流系统中的重要性，以及季风区水汽经向输送等事实，提出能否采用经向风垂直切变来构造东亚区域季风指数，用于反映该地区季风环流特征。

2 资料来源

本文所用资料包括 NCEP/NCAR 1958—1998 年月平均常规资料集和大气顶向上长波辐射通量(OLR)资料，以及国家气候中心提供的 1951—1998 年 160 个站月平均降水资料。同时使用文献[12]方法直接计算 1958—1997 年大气非绝热加热。

3 东亚区域夏季风的特征

夏季 6—7 月 850 hPa 和 200 hPa 与 ASM 相关联的大尺度环流特征(图略)显示，低层大尺度流场主要特征表现为南北半球涡旋环流(IGC)，在南印度洋盛行偏东风，通过索马里急流转为偏西风，穿越印度次大陆；在 105°E 以东南半球印度尼西亚东南气流在赤道转向，使得南海附近和东亚东部地区盛行西南风，到达大陆后以南风分量为主，印度西南季风也能影响东亚地区。另一重要环流型是与西太平洋副热带高压相关联的反气旋环流型，低层热带西太平洋盛行偏东风。在 200 hPa，季风环流由南亚高压所控制，6—7 月间高压脊线在孟加拉湾北部，高压脊线呈纬向分布，西脊点向西伸展到阿拉伯海北部，东脊点向东扩展到长江中下游一带。在南海区域附近，高空东风急流和低空偏西气流间存在很强的东风切变；而东亚副热带区域出现强的西风切变。高空散度分布的气候特征反映出在赤道印度洋和西太平洋菲律宾对流区以外，辐散中心还出现在南海、东阿拉伯海和孟加拉湾。流函数和势函数场分别清楚地反映了南亚高压位置和菲律宾以北西太平洋洋面大气辐散特征。

夏季 6—7 月季风降水气候特征(图略)显示，从菲律宾暖池经南海向西扩展到孟加拉湾海域的大范围内降水率大于 6 mm·d^{-1}，印度西部邻近的阿拉伯海域日平均降水量达到 18 mm，我国东部长江中下游地区也为一强度在 10 mm·d^{-1} 的降水中心，这是通常所指的梅雨带，进一步分析表明该雨带以对流性降水为主。同时注意到印度尼西亚陆地为相对干季，这与文献[6]中用卫星降水产品微波探测装置(MSU)6—9 月资料得到的结果一致。我国 1951—1998 年 160 站月平均降水的分析也表明，6 月 150 mm 的等雨量线北界越过长江到达 32°N 附近，最大雨量中心在长江中下游南部及华南南部一带；7 月东部雨带前沿一直向北推到华北地区，江淮之间出现 200 mm 的雨量中心。

由于东亚夏季风和南亚夏季风分别具有不同的热源和对流活动中心,且两者年际变化相关性较弱,因此,寻找有代表性、能客观定量地描述东亚夏季风的年际变化的季风指数时需要考虑季风区域气候特征。一般,季风指数应当反映对流或者环流最强的季风区(即活动中心)的变化。山崎信雄等[13]采用黑体温度 TBB 定义对流指数,本文与文献[10]一样,采用 OLR 来估计对流强度,同时考虑大气整层非绝热加热过程。

图 1 给出了夏季 6—7 月 850 hPa 和 200 hPa 之间纬向和经向风的垂直切变以 OLR 的地理分布。从环流上看,季风主要由对流层中层释放的潜热所激发的最低阶斜压模态控制,用 850 hPa 减去 200 hPa 的风场表示的垂直切变是低阶斜压模的一级近似。从图 1a 上看出,最强的纬向风负垂直切变中心位于南亚高压北侧与西风急流轴之间,正切变中心位于南亚高压南侧北印度洋一带。经向风切变的情形则与纬向切变很不一样。从 850 hPa 和 200 hPa 经向风的分析发现,低层强南风的形成受到地形和海陆分布的影响,而高空强北风的形成与强大的南亚高压有关。图 1b 中 3 个最大北风切变中心分别位于我国东部、孟加拉湾和赤道非洲东部,而南海附近南风切变相对较弱。

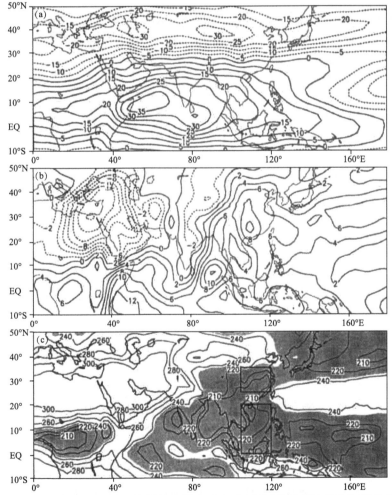

图 1　6—7 月 850 hPa 和 200 hPa 之间纬向风(a)和经向风(b)的垂直切变以及 OLR 的地理分布(c,阴影区为 OLR≤235 W·m^{-2},粗实线框为本研究对流区)

6—7月OLR分布(图1c)显示季风区强对流活动(≤220 W·m^{-2})主要集中在孟加拉湾向东经南海推进到我国东部的大范围区域、阿拉伯海东部、加里曼丹岛、菲律宾暖池。同时看到,WYI指数区除西侧没有对流发展外,包含了3个对流活动中心;RM1指数区也包含了2个对流活动中心,但其西北侧为弱对流区;显然用对流中心来评判RM2是不合适的,因为该指数主要考虑200 hPa副热带急流两侧平均纬向风切变;CI1指数作为南亚对流强度指标似乎比RM1更合理;CI2在描述6—7月菲律宾对流时可能会引起不合理。我国东部东亚副热带区域(25°～35°N,105°～120°E,粗实线框)主要反映了长江中下游及江淮流域对流活动和副热带急流南侧负纬向风切变,与此相对照的是南海季风区域(0°～20°N,105°～120°E)。

由图2a可知,在东亚季风区存在两个独立的视热源中心,分别位于南海和长江中游地区,加热强度均达到200 W·m^{-2},与印度季风区的相当;亚洲季风区内从青藏高原到孟加拉湾北部为最强的热源区,最大值超过600 W·m^{-2}。由视水汽汇的分布图2b可见,南海和长江中下游地区分别为两个视水汽汇加热中心,强度在150 W·m^{-2},它们与上述的视热源中心相对应,而印度地区的视水汽汇加热仅有100 W·m^{-2},较东亚季风区的要明显偏小。同样,青藏高原东南到孟加拉湾北部为最强的水汽汇中心,量值达到500 W·m^{-2}。上述分析与叶笃正等[14]用ECMWF资料得到的结果在形态上相近,但在副热带季风区可得到视热源和视水汽汇中心,且非绝热加热强度显著,与实际降水分布相一致,因此更趋合理。

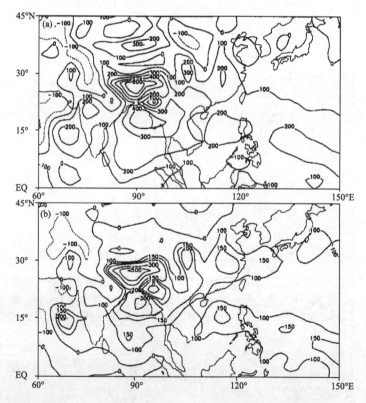

图2 夏季6—7月40年平均〈$Q1$〉(a)和〈$Q2$〉(b)分布(单位:W·m^{-2})

对比视热源和视水汽汇的分布可见:在东亚副热带季风区和南海地区,各主要热源与水汽潜热汇中心位置和强度相近,说明这些热源主要是凝结潜热释放形成的,与该地区存在强大的对流活动相关联。分析中同时兼顾了高低空越赤道气流正西风切变和对流加热。

4 大尺度环流与大气对流活动的内在联系

根据大气顶向上长波辐射通量再分析资料,对图1c中实线框区域分别计算副热带季风对流距平指数(I_e)和南海季风对流距平指数(I_s),以建立对流距平和环流距平之间的关系。I_e为(25°~35°N,105°~120°E)内OLR距平,I_s为(0°~20°N,105°~120°E)内OLR距平。

由于OLR是逆向变化量,因此$I_e>0$表示对流减弱,$I_e<0$表示对流增强;I_s也是如此。850 hPa和200 hPa的纬向风和经向风距平场与I_e的线性相关分布显示(图略),对于副热带季风区对流增强,850 hPa上东亚大槽槽后我国东北地区西风减弱,热带115°E以东越赤道西风也减弱,而长江中下游西风增强;200 hPa的情况在东亚南海北部以北地区也有相同的变化特征,且效果更显著。但在南海南部及其以东菲律宾暖池一带,西风增强,即越赤道东风减弱。东亚副热带季风区对流的增强也与南海北部850 hPa的南风增强和200 hPa的北风增强耦合在一起。

相应于东亚副热带季风的对流增强,30°N附近50°~130°E的大片区域内高低空纬向风垂直切变($u_{850}-u_{200}$)减弱(图3a),且在长江流域减弱最明显。40°以北中高纬度纬向切变得到加强,赤道印度洋纬向切变增强;与此同时,南海到华南东部高低空经向风垂直切变($v_{850}-v_{200}$)也得到增强(图3b)。根据上述两图可以定义两个环流指数:东亚西风环流指数I_{ew}定义为(25°~35°N,100°~125°E)平均的纬向风垂直切变;东亚南风环流指数I_{es}定义为(10°~25°N,100°~125°E)区域的平均经向风垂直切变。

$$I_{ew} = u_{850}-u_{200}, 在(25°\sim 35°N,100°\sim 125°E)$$
$$I_{es} = v_{850}-v_{200}, 在(10°\sim 25°N,100°\sim 125°E)$$

图3c反映了纬向风切变场和经向风切变场在对流距平场I_e上的线性回归分布,矢量是环流切变的回归,从图中可以看到,南海季风区主要为显著的经向风切变,副热带季风区南部也为经向风切变,但在其北部以纬向风切变为主。

对于南海地区对流距平指数I_s的相关分布是很不一样。对于$I_s<0$,即南海地区对流活动增强,20°N以南到赤道地区850 hPa西风增强(图略),而西风减弱的区域主要集中在80°E以东20°~30°N一带;200 hPa东风增强的区域分布在10°N以南的近赤道附近以及80°E以东20°~30°N附近。相应加强的850 hPa的南风与200 hPa的北风主要分布在以加里曼丹岛为中心的东南亚地区;加强的850 hPa北风出现在中南半岛。因此,对于南海地区对流的加强,纬向风切变在赤道附近印度洋增强(图4a),而经向风切变在与200hPa上北风加强的东南亚区域得到加强(图4b)。根据上述分析,相应于对流指数I_s,环流指数可以如下定义,南海西风环流指数I_{sw}是区域(0°~10°N,90°~120°E)内平均的纬向风垂直切变;南海南风环流指数I_{ss}为区域(10°S~10°N,100°~140°E)内平均的经向风垂直切变。

$$I_{sw} = u_{850}-u_{200}, 在(0°\sim 10°N,90°\sim 120°E)$$
$$I_{ss} = v_{850}-v_{200}, 在(10°S\sim 10°N,100°\sim 140°E)$$

环流切变在I_s上的回归矢量分布图4c与图3c的最大不同是南海地区为一气旋性切变,这样的季风环流距平的空间结构可以用Gill[15]的简单大气模式来理解。垂直切变场上所显示的气旋性距平环流主要是对距平热源的Rossby波响应。

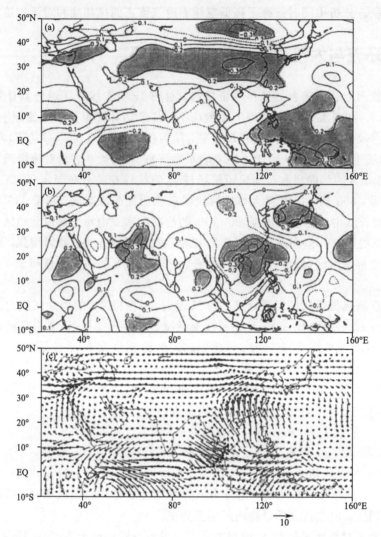

图3 纬向风(a)和经向风(b)的垂直切变相对于I_e线性相关场以及风垂直切变在I_e的回归分布(c)

    风场垂直切变和对流之间的内在动力学联系表明，可以用对流指数I_e和I_s来分别作为东亚副热带季风和南海季风指数，这两个指数不仅能代表相应对流中心的强度变化，而且能反映以纬向和经向风垂直切变为特征的大尺度环流的变化。表1给出了东亚地区对流指数和环流指数之间的线性关系。不难发现，I_e与副热带季风区纬向风垂直切变I_{ew}和南海及中南半岛经向风垂直切变I_{es}有显著的相关性，相关系数分别为－0.29和－0.49；I_s与北半球赤道附近东南亚地区纬向风垂直切变I_{sw}和东南亚越赤道气流带经向风垂直切变I_{ss}也有很好的相关，相关系数分别为－0.36和－0.47。

    I_e和I_{es}以及I_s和I_{ss}的高相关性表明，I_{es}和I_{ss}分别是描述副热带夏季风和南海夏季风的有意义的环流指数。对流增强，目标区域的经向风垂直切变增强；反之亦然。但正如I_e和I_s的相关性$r=-0.07$一样，I_{es}和I_{ss}的线性相关系数仅为－0.12。这说明在这两个季风区内，对流和环流的年际变化并不同步。进一步分析还发现，I_{ew}和I_{sw}的相关系数高达0.37，表明副热带地区与近赤道区域纬向垂直切变又是相互联系的；同一区域的平均纬向风和经向风垂

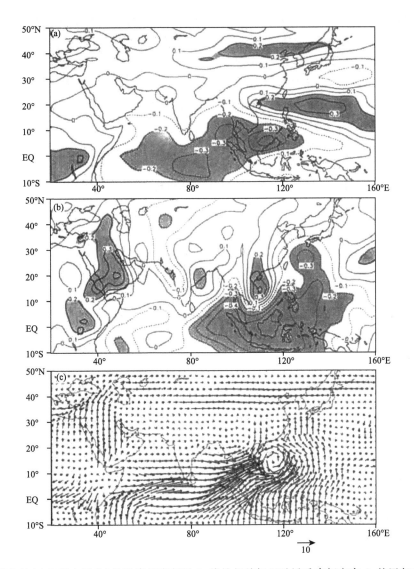

图 4 纬向风(a)和经向风(b)的垂直切变相对 I_s 线性相关场以及风垂直切变在 I_s 的回归分布(c)

直切变具有很好的关系,均能通过 99% 的信度检验,所不同的是副热带地区 $r=-0.34$,而南海地区 $r=0.31$,因此,变化趋势相反。

表 1 东亚夏季风对流和环流指数之间的线性相关系数

	I_e	I_s	I_{ew}	I_{es}	I_{sw}	I_{ss}
I_e	1.00	−0.07	−0.29	−0.49	0.08	−0.04
I_s		1.00	0.01	0.01	−0.36	−0.47
I_{ew}			1.00	−0.34	0.37	0.16
I_{es}				1.00	−0.27	−0.12
I_{sw}					1.00	0.31

注:$f=102$, $r_{0.01}=0.25$, $r_{0.001}=0.32$

对流指数（I_e，I_s）和环流指数（I_{es}，I_{ss}）分别客观、定量地反映出东亚副热带夏季风和南海夏季风的强度变化。因此，选择经向风垂直切变指数 I_{es} 和 I_{ss} 刻画东亚副热带夏季风和南海夏季风的年际变化。表2列出上述两个夏季风指数与 NCEP/NCAR 夏季6、7月平均日降水率 P_e（27.5°~32.5°N，110°~120°N）和 P_s（0°~20°N，105°~120°N）的线性相关系数。为了对照，同时分析了与不同种类季风环流指数 WYI、RM1（MHI）、RM2、WSI1 和 WSI2 等的关系。I_{es} 和 P_e 以及 I_{ss} 和 P_s 的相关系数分别为 0.35 和 0.31，均能通过 99% 的信度检验，而 I_{es} 和 P_s 以及 I_{ss} 和 P_e 的交叉相关系数接近于0，从而进一步说明 I_{es}/P_e 和 I_{ss}/P_s 确实反映了东亚夏季风明显的区域和动力学现象。P_e 和 P_s 的相关系数仅为 0.15，表明两者相关程度很弱。

表2 各种季风指数及其区域降水相互间的线性相关系数

	I_{es}	I_{ss}	WYI	RM1	RM2	WSI1	WSI2	P_e	P_s
I_{es}	1.00	−0.12	−0.2555	−0.13	−0.31	−0.30	−0.36	0.35	−0.05
I_{ss}		1.00	0.24	0.05	0.17	0.24	0.36	−0.01	0.31
WYI			1.00	0.65	0.63	0.95	0.68	−0.06	0.22
RM1				1.00	0.62	0.77	0.17	−0.05	0.07
RM2					1.00	0.73	0.31	−0.15	0.26
WSI1						1.00	0.53	−0.06	0.20
WSI2							1.00	−0.33	0.12
P_e								1.00	0.15

I_{es} 和东（南）亚夏季风环流指数 WSI2、东亚区域季风指数 RM2 有显著的负相关，与南亚季风指数 WYI、南亚区域季风指数 RM1 的负相关均能通过 $\alpha=0.01$ 信度检验；I_{ss} 和 WSI2 有显著的正相关，并能通过 99.9% 的信度检验。从这点可以认为 WSI2 是一个很好的指数，它与 P_e 的负相关系数为 −0.33，但与 P_s 的相关系数仅为 0.12。在 1949—1999 年夏季 6—7 月 WYI 与 WSI1 的相关系数达到 0.95，略高于 Wang 等[10]用 NCEP/NCAR 40年6—9月季平均资料得到的结果（$r=0.91$）。表2也能看出 WYI 与 RM1、RM2、WSI2 的相关系数高于 0.6，能有效地反映南亚夏季风及其扩展区域的环流特征。综上分析，可以认为：东亚副热带区域和南海区域分别选择经向风垂直切变 I_{es} 和 I_{ss} 来描述相应区域夏季风环流和降水是合适的。

5 I_{es} 与西太平洋副高及长江中下游降水的关系

分析夏季 6—7 月平均环流强度（I_{es}）距平发现，东亚副热带夏季风存在显著的年际变化（见图5），诸如 1962、1991、1998 年南风垂直切变极强，而 1966、1969、1975、1980、1983、1995 年偏强；1967、1988 年极弱，而 1964、1972、1974、1978、1985、1990 年偏弱。20世纪80年代经向风垂直切变强度偏弱，90年代则明显偏强。I_{es} 与长江中下游平均降水的相关系数为 0.44，显然可以通过 $\alpha=0.01$ 的信度检验。

图6进一步显示了 I_{es} 与同期我国降水相关的空间分布，不难发现，能够通过 $\alpha=0.05$ 信度检验的区域主要集中在长江中下游流域（阴影区），华南西江流域也为正相关显著区。由此可见：东亚副热带地区对流活动增强（$I_e<0$），其所选择的区域经向风垂直切变强度 I_{es} 增强，长江中下游地区降水明显增多；反之亦然。上述特征表明：副热带强（弱）环流指数年

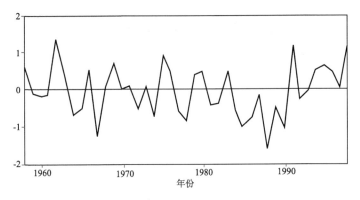

图 5 夏季 6—7 月平均东亚副热带季风指数(I_{es})距平年际变化（单位：m·s^{-1}）

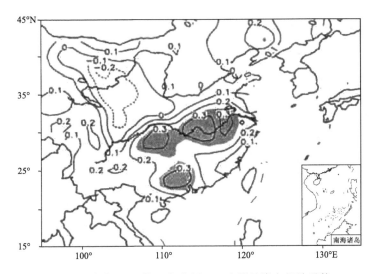

图 6 夏季 6—7 月 I_{es} 与我国 160 个测站降水相关系数
$f=49, r\ 0.05=0.27$

长江中下游地区容易导致涝（旱），从而证明该指数对研究区域降水有明确的表征能力。

为了分析强指数年与弱指数年高空环流的差异，选择 1980、1983、1991、1998 年为强指数代表年份，选择 1967、1985、1988、1990 年为弱指数代表年份，它们分别与长江中下游涝年和旱年相对应。制作涝年和旱年环流差值图，不难发现，我国东部 40°N 附近为 500 hPa 高度场差值负中心(7)，而 20°N 附近为正中心，由此可见，涝年较旱年副热带高压位置偏南；在差值流场（图略）上同样可看到相应的气旋性和反气旋性环流中心。

在 200 hPa 上，发现全风速正差值中心在 35°～40°N（图略），这表明西风急流位置在涝年要更偏南；同样，差值环流特征表现为我国长江以南大部分地区为高空差值反气旋区，有显著的辐散，在长江中下游地区存在很强的反气旋式切变，高空多个气旋式环流中心为高空辐散的发展提供有利条件。因此，可以认为西风急流位置和副高位置的异常是强弱指数年环流的主要差异。

为进一步识别西太平洋副高在强弱副热带环流指数年的显著差异，建立 I_{es} 与表征西太平洋副高的一些特征量的线性关系（见表 3）。

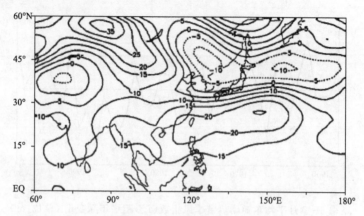

图 7　500 hPa 强指数（I_{es}）与弱指数年合成高度场差值分布

表 3　夏季 I_{es} 与西太平洋副高诸参数间的线性相关系数

特征量	副高脊线	副高面积	副高强度	副高北界	西伸脊点
相关系数	0.13	0.38	0.41	0.38	−0.33

注：副高参数说明同国家气候中心气候公报定义；$|r|_{0.05}=0.28$，$|r|_{0.01}=0.36$。

由表 3 可以看出，I_{es} 与夏季 6—7 月西太平洋副高脊线位置之间的关系并不好，这可能与夏季副高脊线存在短期变动直接有关；I_{es} 与副高 110°～180°E 平均面积和强度指数及 110°～150°E 平均北界位置的相关系数都能够通过 $\alpha=0.01$ 的信度检验；与 588 线平均西脊点经度间的负相关也能达到 $\alpha=0.05$ 的置信度。这进一步证明副高在东亚季风环流系统中的重要性，同时说明副热带环流指数不仅可以描述长江中下游的降水，也能充分反映与此直接相关联的西太平洋副高特征。

图 8a 给出了 5 个强指数年 500 hPa 高度距平场合成分布，可以看到长江流域以南的东亚地区有显著的正距平，表明副热带高压位置偏南，强度增强。东亚到北美高度距平呈现"＋－＋－"的分布特征，与 Huang 等[16] 的 EAP 型相一致。图 8b 给出了 5 个弱指数年 500 hPa 高度距平场合成分布，可以看到我国东部西太平洋海域为一显著的负距平区，长江流域北部为一正距平相对高值带，副热带高压位置偏北。菲律宾以东到北美高度距平呈现"－＋－＋"的分布特征，与反 EAP 型相一致。这些结果表明：北半球夏季环流异常存在东亚—太平洋遥相关型，行星尺度的扰动波列能够从东亚向北美传播，它严重地影响东亚夏季风的年际变化与旱涝发生。

图 8c 为 I_{es} 与 500 hPa 高度场的线性相关系数分布，对于副热带地区 I_{es} 减弱（即对流活动增强），亚洲东北部的阻塞高压系统加强，西太平洋副热带高压强度减弱，位置偏东，与此相关联的是印度西北部反气旋脊区向北扩展到非洲北部，I_{es} 遥相关波列呈 EAP 型。

6　结论讨论

根据前面的分析，我们可以得到以下主要结论：

（1）多年平均夏季 6—7 月东亚副热带季风区存在 200 W·m^{-2} 的视热源加热中心，其强度与南海及印度季风区相当；同时存在 150 W·m^{-2} 的水汽汇中心，其量值较印度地区的要

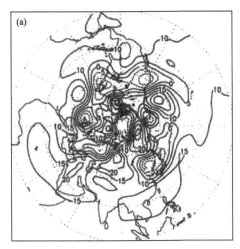
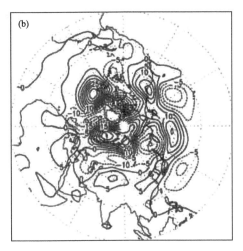
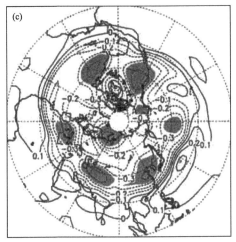

图 8　I_{es}高指数(a)和低指数(b)年 500 hPa 位势高度合成距平场空间分布及遥相关型(c)

大。利用 OLR 与大气整层非绝热加热相结合,可以较好地确定和估计对流活动区域及强度。

(2)对流强度与东亚季风环流特别是高低空经向风垂直切变间存在密切联系,由此可分别建立东亚副热带季风指数 I_{es} 和南海季风指数 I_{ss},并与其他的季风指数 WYI、WSI1 和 WSI2、RM1 和 RM2（MHI）进行比较,发现 I_{es} 和 I_{ss} 对副热带地区降水和南海地区降水分别具有最好的相关性。

(3)I_{es}具有显著的年际变化,高指数年长江中下游涝,低指数年则为旱,在空间场上表现出 EAP 遥相关型,旱涝年高度场正负距平中心相反。涝年高空西风急流位置偏南,副热带高压位置也偏南;反之,则偏北。I_{es} 与长江中下游降水及副高诸参数都有很好的线性相关,副高北侧的副热带夏季风雨带与副高之间存在一种相互联系或相互作用,它反映了由夏季风降水产生的凝结潜热与副高年际变动之间的内在联系。

参考文献

[1] Shukla J, Paolina D A. The southern oscillation and long range forecasting of summer monsoon rainfall over India[J]. Mon Wea Rev,1983,111:1830-1837.

[2] Tao Shiyan, Chen Longxun. A review of recent research on the summer monsoon China[C]// Chang C P, Krishnamurti T N. Monsoon Meteorology. Oxford: Oxford University Press,1987,60-92.

[3] 郭其蕴. 东亚夏季风的变化与中国降水[J]. 热带气象学报,1985,1(1):44-52.

[4] Zeng Qingcun, Zhang Banglin, Liang Youlin, et al. East Asian summer monsoon-a case study[J]. Proc Indian Natri Acad,1994,60,Part A,1:81-96.

[5] 谢安,刘霞,叶谦. 南海夏季风气候特征[C]//陈隆勋,等. 亚洲季风机制研究新进展. 北京:气象出版社,1999,132-142.

[6] Lau K M, Kim K M, Yang S. Dynamical and boundary forcing characteristics of regional components of the Asian summer monsoon[J]. J Climate,2000,13:2461-2482.

[7] Webster P J, Yang S. Monsoon and ENSO: selectively interactive systems[J]. Quart J Roy Meteor Soc,1992,118:877-926.

[8] Ailikun B, Yasunari T. On the two indices of Asian summer monsoon variability and their implications extended abstracts[R]. Int Conf on Monsoon and Hydrologic Cycle, Kyong, Korea, Korea Meteorological Society,1998,222-224.

[9] Goswami B N, Krishnamurthy V, Annaumalai H. A broad scale circulation index for the interannual variability of the Indian summer monsoon[J]. Quart J Roy Meteor Soc,1999,125:611-633.

[10] Wang B, Fan Z. Choice of South Asian summer monsoon idices[J]. Bull Amer Meteor Soc,1999,80:629-638.

[11] 李崇银,张利平. 南海夏季风特征及其指数[J]. 自然科学进展,1999,9(6):536-541.

[12] 周兵,徐海明,谭言科,等. 1998年武汉大暴雨切变涡度及非绝热加热垂直结构分析[J]. 气象学报,2001,59(6):707-718.

[13] 山崎信雄,井上丰志郎,田中实,等. 大尺度对流活动和降水系统的分析[C]//陈隆勋,等. 亚洲季风机制研究新进展. 北京:气象出版社,1999,160-171.

[14] 叶笃正,黄荣辉,等. 长江黄河流域旱涝规律和成因研究[M]. 济南:山东科学技术出版社,1996,387.

[15] Gill A E. Some simple solutions for heat induced tropical circulation[J]. Quart J Roy Meteor Soc,1980,106:447-462.

[16] Huang Ronghui, Sun Fengying. Impact of the tropical western Pacific on the East Asian summer monsoon [J]. J Meteor Soc Japan,1992,70:243-256.

有关东亚季风的形成及其变率的研究[*]

何金海[1,2]　宇婧婧[1]　沈新勇[1]　高辉[1]

(1. 南京信息工程大学大气科学系,南京 210044;
2. 中国气象局广州热带海洋气象研究所,广州 510080)

摘要:季风是全球气候系统中一种重要的大气环流系统,随着气候学研究及季风动力学研究的深入,国内外学者对季风的形成及其变率问题的研究取得了很大进展。回顾了近年来国内外关于亚洲季风的认识与形成机制的研究,特别是对东亚季风爆发及其过程的影响因子的认识。此外,对于东亚季风的季节性变化、低频振荡、年际和年代际变化及其可能机制也进行了分析、讨论。最后提出了在季风动力学研究方面需要进一步探讨的问题。

关键词:东亚季风;成因;变率

1 引 言

季风是一个古老的气候学概念,是全球气候系统中极其重要的一员,其变异严重影响着季风区的天气与气候。传统的观点认为,季风是近地面层冬、夏盛行风向接近相反且气候特征迥异的现象,最典型的季风区为亚澳季风区、非洲季风区和北美季风区。近年来有一些研究者提出,季风是一个三维环流系统,在对流层底层除经典的热带、副热带季风区外,还存在温寒带季风区和行星季风区等。我国处于东亚季风区,天气和气候深受其严重影响。早在 20 世纪 30 年代,竺可桢[1]就对夏季风进退及其与我国汛期降水关系进行过研究。之后,涂长望等[2]研究了东亚季风的进退对中国气候季节内变化的影响。继他们之后,陶诗言等[3-4]、陈隆勋[5]和丁一汇[6]对东亚夏季风的结构、性质作了系统的研究。最近几年,中国学者对亚洲季风的形成机理及其季节、季节内、年际和年代际等多时间尺度变率的研究取得了重要进展。

本文将首先对中国学者对东亚季风的研究作一简单介绍;第二部分分析亚洲季风着重分析东亚季风)的形成机理;第三部分着重分析东亚季风的季节、季节内、年际和年代际等多时间尺度变率的研究,并初步探讨季风的变化机制;最后提出在季风动力学研究方面有待进一步探讨的问题。

2 关于东亚季风的表述

东亚季风环流系统是一个与印度季风既独立又相互联系的季风环流系统。东亚季风区又

---

[*] 本文发表于《热带气象学报》,2004 年第 20 卷第 5 期,449-459。

可分为南海－西太平洋热带季风区和大陆－日本副热带季风区[7-8]，因而比印度季风要复杂。根据陶诗言和陈隆勋[7]的研究，东亚夏季风系统的成员（图1）包括：南海和赤道西太平洋的季风槽（或ITCZ）、印度的西南季风气流、沿100°E以东的越赤道气流、西太平洋副高和赤道东风气流、中纬度的扰动、梅雨锋以及澳大利亚的冷性反气旋。这个系统不仅受到印度西南季风环流系统的影响，而且还受到副热带高压和中纬度扰动系统的影响。每年亚洲季风首先在南海地区爆发，然后分别向西和向北扩展，使其印度夏季风和东亚夏季风逐渐建立。

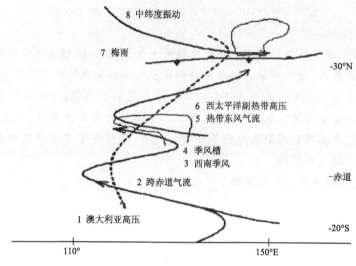

图1　东亚夏季风系统成员的配置（引自文献[7]）

3　亚洲季风的形成机理

早在1686年Halley[9]就指出，季风是由太阳对海洋和陆地加热差异形成的。他认为，海陆加热差异导致大气中气压存在差异，使得风由高压吹向低压。Hadley[10]引入地转偏向力的概念，对Halley的季风模型进行了补充和修正，很好地解释了南亚季风的风向及其变化特征。由海陆分布的经、纬向非对称导致的太阳辐射加热的非对称结果是形成季风地理分布差异的重要原因。陶诗言等[11]从大气环流的动力、热力变化来研究季风的季节变化，指出季风的建立不仅依赖于海陆季节温差，还受辐射变化、大气环流调整和地形等各种因素的综合作用。陶诗言、何诗秀等[3]曾提出季风爆发起因于东亚南半球的越赤道气流。从气候学上看，在低空，南亚夏季风以及部分东亚夏季风的源头，正是南半球热带印度洋的越赤道气流；而东亚夏季风的另一源头则是在印度尼西亚从苏门答腊到苏拉威西的越赤道气流；在高空，则与此方向近于相反的越赤道急流，这已成为气象界的常识[12]。季风是南北半球太阳辐射的季节性差异（即两半球大气的相互作用）所致，这一成因也成为季风本质研究中的重要内容。Chao[13-14]利用数值实验结果，提出季风起源于ITCZ的季节性移动，而海陆热力差异与海表温度差异一样仅决定ITCZ带的经向位置，其中地形的改变比简单的海陆热力差异对季风的作用更大。最近，曾庆存等[15]明确提出，行星热对流环流（由南到北的太阳辐射的季节性差异所致）是热带季风的"第一推动力"，地表面特性差异（包括海陆热力特性差异和地形高度等）所导致的准定常行星波是"第二推动力",二者共同作用使热带季风的表现具有局域性,第二推动力在亚澳季风区与第一推动力合拍,使热带季风在亚澳区内最明显,而各经圈环流圈的上下及其南北关联及与中高纬准定常行星波的配置则使全球范围内从低纬到高纬、从低空到高空有地域性的明显季节变化区,从而构成三度空间的全球季风系统,而由于第一推动力与第二推动力的效果相消,在热带大西洋和中东太平洋盛行"信风",环流季节变化很小。此外,大气中的湿过程也是驱动季风的机制之一[16]。空气中水汽的相变及其输送过程能够储存和重新分配热带和副热带大部分地区接收到的太阳能,并且有选择地释放这些能量,从而决定季风环流以及季风降雨的强度和地域,不妨可以说,湿过程是形成季风的第三推动力。

实际上,南亚季风就是主要由行星风系季节性移动而引起的,即南亚季风爆发是由ITCZ带的季节性移动或突跃引起,体现了南北半球相互作用的重要性;而对于东亚季风而言,其形成和性质与南亚季风不同,包含热带与副热带两种不同的季风系统,不仅与南半球越赤道气流有关,而且受副热带高压和中纬度扰动系统的影响,其爆发过程中南北半球和中低纬相互作用特征明显,因而比印度季风复杂。东亚季风则主要由纬向和经向海陆热力差异形成[17]。

亚洲季风区由冬到夏的季节变化以环流和天气型的突变为主要特征,它通常发生在5—6月,与亚洲夏季风爆发相联系。亚洲夏季风5月中旬首先在南海爆发,然后向西、向北推进,印度夏季风迟至6月中旬才爆发。因而南海夏季风爆发的成因研究对于其后季风推进过程研究具有重要意义。关于南海夏季风爆发过程和机制的研究很多,但有许多不同的看法。Murakaimi等[18]提出,由亚洲和澳大利亚的热力差异所造成的气压梯度的大尺度季节性转换比海陆热力差异所造成的大尺度季节性转换要敏感和强烈得多,且南海是这种变化的敏感区,当适当的低频振荡湿位相位移到这一地区时,将导致夏季风的爆发。Chang等[19]则利用1980—1986年的ECMWF资料研究了南海北部上空850 hPa西南气流涌的发展,发现5月份西南气流的加速依赖于中纬度斜压系统,该系统向南移到华南沿海,并保留一个准静止的位置,在其南侧出现西南气流,这种中纬度系统可能是南海夏季风爆发的触发机制。何金海等[20]则利用TBB资料的分析强调苏门答腊地区的对流中心沿着中南半岛移动的年循环对亚洲季风特别是南海夏季风建立有重要作用。关于南海夏季风爆发的成因,总结起来可能有以下几种:(1)海温的影响;(2)热带的强迫(例如热带对流影响);(3)青藏高原以及印度半岛、中南半岛的热力作用;(4)两半球的相互作用;(5)亚洲温带系统的激发。这些爆发机制的假设还有待于进一步验证。

何金海等[21]分析了亚洲季风地区特别是东亚地区的流场、降水场以及非绝热加热场对澳大利亚地区有、无冷空气活动的不同响应,着重讨论了澳大利亚地区冷空气活动影响东亚夏季风加强北进的传播过程。在该过程中,经向风扰动呈现出由南向北的传播,流场和降水场也有相应的向北移动,其时间滞后约12天。周学鸣等[22]发现乌拉尔阻高作为一个激发源能激发出一支东南—西北向的定常波列,其高空反气旋辐散环流(低空为槽区)刚好位于长江中下游地区,同时它增强东亚西风急流,有利于扰动的发展,且在急流入口区诱导出附加的次生环流,增强东亚季风上升气流。因此,乌拉尔阻高的存在及其激发的定常波列(即中高纬对东亚季风的影响)是导致江淮洪涝的大尺度关键因子和影响机制。

由于东亚地区所处的特殊地理位置,地形的热力、动力作用对东亚季风产生重要作用。正如Chao[14]的数值试验结果所表明,地形对季风产生的作用比简单的海陆热力对比还重要。亚洲夏季风的爆发是大气对海陆热力差异的季节改变的响应,其中青藏高原的加热对亚洲夏季风环流的形成和维持起着重要的作用。高原的加热可使其上大气柱每天增温 3℃左右,引起空气上升以及周围大气向高原的汇合,这有利于热带暖湿气流北上,最终引起夏季风爆发[23]。而吴国雄[24]对青藏高原的热力和机械强迫作用与亚洲季风爆发关系进行了研究,表明正是由于青藏高原的热力和机械强迫作用才使亚洲季风首先在孟加拉湾地区出现,并为南海季风爆发提供有利背景条件,最后随着亚洲热带流型的西移,印度季风爆发了。

骆美霞等[25]利用一个 σ 坐标的三层初始方程热带球带数值模式,以全球夏季平均纬向风场为初始场,研究了地形的纯动力作用和不同地区的大气热源、冷汇对亚洲夏季两支相对独立的季风环流系统形成的影响,发现对于东亚季风环流系统的形成,大气非绝热热力作用比地形的纯动力作用更重要;而对于印度季风环流系统的形成,地形的纯动力作用和大气的非绝热热力作用同等重要。徐海明等[26]利用 CCM3 模式模拟了印度半岛对亚洲热带夏季风进程的影响,研究表明印度半岛对亚洲热带夏季风的建立进程起到了至关重要的作用,印度半岛的地形主要通过地表的感热加热使其上空增温并与其相邻的阿拉伯海和孟加拉湾之间产生显著的海陆温差,而导致亚洲低纬副热带高压首先在孟加拉湾地区断裂。而 Zhang 等[27]分析了 1998 年和多年平均情况下南海夏季风爆发期间中南半岛地区的热力特征,讨论中南半岛热状况的异常与南海夏季风爆发之间的联系,结果发现南海季风爆发前中南半岛地区存在着较强的持续地面感热加热并具有显著的低频振荡特征,且出现有利于降水加强的低层强辐合、高层强辐散的垂直配置,进而触发南海季风的爆发。

4 亚洲季风的多时间尺度变率

4.1 亚洲季风的季节性变化

Tao 等[28]指出亚洲季风的季节变化首先表现在东亚夏季风的爆发上。平均而言,东亚夏季风在 5 月中旬建立,南亚夏季风在 6 月上旬建立,主要表现为低层西南风和高层东风分别在东亚和南亚的建立。伴随着季风,季风区大气环流包括降水在 6 月发生季节性突变[29]。亚洲季风的另一次明显的季节变化发生在 9 月中旬至 10 月。在低层,变化主要表现在夏季风的撤退和冬季风的建立。研究表明[30]这种由夏转秋的变化亦具有阶段性,即酝酿期与突发期,并从东亚地区先开始。东亚夏季风的撤退是非常迅速的,与印度季风的缓慢撤退不同,这是东亚季风的另一重要特征[5]。一般从 8 月中旬以后,夏季风从华北迅速南撤,不到 2 周就很快撤退到南海北部,并一直停留在那里到 10 月中旬。在 10 月中旬以后,东亚冬季风沿东亚沿岸向南吹到了东海和南海,尔后再折向西南吹向中印半岛和东南亚地区,这种冬季风可以维持到第二年的 4—5 月。东亚夏季风与冬季风交替盛行,年年如此,周而复始一直循环下去。亚洲冬季风对于夏季风也有重要影响,孙淑清等[31]指出,强的东亚冬季风(EAWM)能影响之后的东亚夏季风(EASM)。

4.2 亚洲季风的低频振荡

亚洲季风现象不仅有明显的季节变化,而且有季节内时间尺度的变化,即呈明显的 40~60 天及 10~20 天低频振荡。Murakami 等[18]利用 OLR 资料系统地研究了季节变化、低频变化与季风建立和撤退之间的关系,发现无论是南半球的澳大利亚还是北半球的孟加拉湾,夏季风的建立和撤退是在季节变化具备之后(即干位相转为湿位相或相反),由低频振荡触发(即低频振荡的第1个湿位相或相反)而产生的。亦即夏季风建立或撤退的具体日期决定于低频振荡的位相。Yasunari[32],Murakami等[33]、He等[34]、Wang等[35]、陈隆勋等[5]、李崇银[36]等详细讨论了亚洲季风区低频振荡的经向和纬向传播特性,低频振荡的起源与维持及其与我国天气的关系,这些工作大大拓宽了亚洲季风的研究领域。

东亚夏季雨带的进退与东亚季风前沿的进退有着密切关系[34]。因此,东亚夏季风的季节内变化(低频振荡)也可以表现为东亚夏季风雨带的进退和移动,从气候平均状况来说,东亚夏季风雨带在5月至6月上旬一般位于长江流域以南;在6月中旬很快北跳,并位于江淮流域、朝鲜半岛和日本。这就是我国梅雨季节的开始,即朝鲜半岛的Changma和日本的Baiu季节的开始;在7月中、下旬雨带又一次北跳到华北,即江淮流域梅雨季节结束,华北与东北的雨季开始。由此看来,我国大陆夏季风雨带(副热带季风)在向北推进过程中具有显著的阶段性,它和低频振荡有密切的关系。何金海等[38]指出南海地区向北水汽输送的每一次增强均对应着夏季风雨带的变动。朱乾根等[39]利用我国东部夏季风北界纬度为参数,进行功率谱分析,发现我国大陆副热带季风在北进过程中存在明显的准40天、双周和22天左右的低频振荡周期,其中以准40天振荡最为显著。他们指出,准40天振荡与副热带季风及其相伴雨区的2次北跳和3次停滞有关。当副热带季风准40天振荡的北进期与季节变化的北进期相叠加时,就可使副热带季风及相伴雨区产生北跃,当处于周期的其他阶段时,则北移缓慢或停滞。从以上结论可看出,我国大陆副热带季风的低频振荡主要表现在夏季风向北推进过程中的阶段性。当南海水汽输送的振荡处于加强过程时,我国大陆副热带季风处于迅速向北推进的阶段,反之处于停滞阶段。

林爱兰[40]分析了南海夏季风的低频特征,指出南海低频振荡在夏季风期间比冬季风期间明显增强,南海夏季风一般在初夏第一个较强低频振荡的负值位相(即湿位相)开始建立;南海夏季风期间低频振荡的实体是ITCZ的南北振荡与西太平洋高压西脊点的东西摆动,低频振荡与南海季风的活跃和中断(或减弱)密切相关。

东亚冬季风是北半球极为活跃的环流系统,其低频振荡的特点也受到了注意。金祖辉等[41]对东亚大陆冬季风低频振荡特征作了分析,发现低频振荡有明显的地区性,30～60天波一般在华北和长江流域(约28°N以北)比南部地区强;冬季风期间,东亚大陆的30～60天波的经向传播主要是由北向南传播的。大多数年份可以传播至东亚大陆的南端,历时3～6天,但是这种传播在长江流域以北地区表现得更加明显,在长江以南地区,一方面受到了冬季风本身的影响,另一方面还受到来自热带地区以致南半球夏季风低频振荡等因素的影响,使得经向传播往往变得复杂。

Lau等[42]指出,亚洲夏季风变化可以是内部动力强迫或者诸如海表温度(SST)和陆面强迫等外部强迫引起,具有区域变化特征,季风气候包括快过程、中间过程和慢过程,大气内动力快过程和陆面及近海中间过程是通过包含Madden和Julian30～60天振荡(MJO)和其他低频振荡的季节内变化相连接的,而南亚季风和东亚季风存在着不同的动力学过程。He[43]提出,南半球中纬度的低频振荡可以通过经圈方向环流的侧向耦合向北传播,即马斯克林高压和澳大利亚高压南侧低频冷空气活动伴随着西风的增强,可导致高压北部东南信风和越赤道气流的增强,进一步引起北半球夏季风的增强。许多学者研究指出,热带西太平洋海温及其上空对流活动(CISK机制)可能通过30～60天低频振荡影响东亚夏季风环流的季内变化。也许南半球中纬度的低频冷空气涌的外界强迫和波动 CISK 机制的结合可以作为亚洲季风区低频振荡的一种经向传播机制。最近的研究表明,热带大气低频振荡的活动对亚洲季风的建立和异常都有重要的作用[44-45]。然而由于亚洲夏季风的复杂性,其低频振荡机制还没有完全搞清楚。

4.3 亚洲季风的年际变率

季风的一个主要特点是年际变化大。为反映大范围季风的年际变化,定义一个或几个有代表性的表征季风强弱变化的指数是十分必要的。但是如何客观和定量地描述大范围季风的变化是一个难题,特别是由于东亚季风的复杂性,至今仍无比较一致的见解和定义。但无论是冬季还是夏季,季风活动的强弱和降水与温度状况有明显的联系,东亚夏季的旱涝气候灾害发生频率较高,尤其是从长江流域、淮河流域、韩国南部到日本南部的地区。因此,研究降水和温度的年际变化对研究季风的变化十分重要。最近的许多研究表明了东亚夏季风降水存在准2年周期振荡,特别在江淮流域、黄河流域和华北地区尤为明显[46-47]。准2年振荡(QBO)是亚洲季风环流系统的主要年际变化,这一周期不仅出现在季风本身[48],也出现在季风环流系统包括我国的降水和气温场中[28,49]。除此以外,亚洲季风还有着明显的准3.5年(QTO)、准5~6年等年际变化周期。

关于东亚夏季风年际变化的原因,以前许多研究强调青藏高原的热力作用,并指出了青藏高原的热力异常对亚洲夏季风变异有很大的影响[50-53]。亚洲季风的年际变化还受到欧亚积雪以及海陆热力差异的影响,Hahn 等[54],Dickinson[55]研究了印度季风降水和欧亚积雪的关系,表明这两者之间具有相反的关系,Khandekar[56]应用更新的资料也证实了这种关系。陈烈庭等[57-58],韦志刚和罗四维[59]的研究还表明青藏高原冬春季积雪能影响长江流域中、下游初夏的降水。

许多学者的研究表明,热带西太平洋暖池上空的对流活动对东亚夏季风的年际变化也起着重要作用[60-61]。黄荣辉等[62-63]借助观测资料和动力理论分析系统研究了暖池热状态及菲律宾附近的对流活动对东亚夏季风环流年际异常的影响,研究表明北半球夏季环流异常存在着一个遥相关型,即东亚-太平洋遥相关型(EAP 型),Lau 等[47]把这个遥相关型称为东亚-北美型。这个遥相关型表明了行星尺度扰动波列在北半球夏季能够从东亚向北美西部沿岸传播,它严重地影响着东亚夏季风的年际变化与旱涝发生。Nikaido[64]利用数值模拟证实了这种北半球东亚-太平洋遥相关型与热带西太平洋对流活动有关,而且表明在菲律宾周围对流活动强时,EAP 遥相关型较显著 Lu[65]指出太平洋上空的弱(强)对流对应着从热带西太平洋向西伸展到孟加拉湾的东风(西风)异常,而在对流弱(强)的夏季,热带西太平洋上空的对流和南海低层纬向风均表现出弱(强)的季节演变特征。

ENSO 循环是热带最引人注目的现象,许多研究表明了 ENSO 循环严重影响亚洲季风活动[66-67],并指出在 El Niño 年印度夏季风常常偏弱。符淙斌和滕星林[68]也指出东亚夏季风降水异常可能与 ENSO 循环的阶段有关。在 ENSO 发展阶段,我国江淮流域、日本南部和韩国常常出现洪涝灾害,而华北容易出现干旱;相反,在 ENSO 衰减阶段,我国的江淮流域、日本南部和韩国易出现干旱,而我国华北的降水可能正常或者偏多。Huang 等[69]也指出,ENSO 循环的阶段对东亚季风环流有很大的影响。穆明权、李崇银[44]利用1950—1989年40年的全球资料,讨论了东亚冬季风的年际变化特征,分析结果表明东亚冬季风年际变化中包含明显的ENSO 信号,表明两者之间有明显的相互作用。

综上,EASM 年际变化的原因是复杂的,其示意框图如图 2 所示。

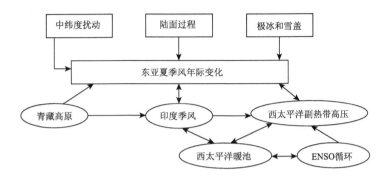

图 2　影响东亚夏季风年际变化的物理因子示意图(引自文献[70])

对于东亚冬季风,Chen 等[71]也研究了其年际变化,他利用东亚沿岸的风场定义了一个 EAWM 数(图 3)。可以清楚地看到 EAWM 的年际变化非常明显,图中指数若是负的,则代表强 EAWM,它说明该冬季沿东亚大陆是冷的,且东亚沿岸盛行强北风,在这种情况下,位于欧亚大陆对流层的高压是强的,位于中国东北的 500 hPa 的东亚大槽也是强的;相反,若指数是正的,则东亚大陆是暖的,东亚沿岸盛行弱北风,东亚对流层低层的高压和 500 hPa 的东亚大槽均较弱。结果表明亚洲冬季风的准 2 年振荡在年际变化中也较为突出。

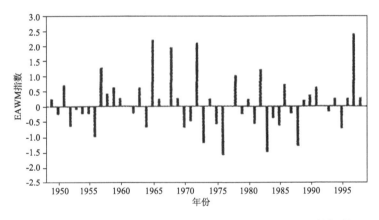

图 3　1951—2000 年东亚冬季风指数的年际变化(引自文献[71])

4.4　亚洲季风的年代际变率

亚洲季风大气环流和海温场还表现出一定的年代际变化[72],由于各作者选用的指标和资料长度有异,导致谱分析得出的这一尺度周期值差别较大,一般认为 11 年是其一个主要周期。分析资料表明[73-76],1976 年前后大气环流和北太平洋海温均发生了突变。与 1976 年突变相对应,东亚夏季风也发生了显著突变。1976 年以后,亚洲中高纬地区经向环流异常显著,从低层到高层呈相当正压结构。来自高纬地区的东亚异常偏北气流可以抵达孟加拉湾、澳洲西北部和中西太平洋,从而使 1976 年以后东亚夏季风明显减弱,热带太平洋出现西风异常,夏季西太平洋副热带高压偏强,位置偏南。李峰和何金海[77]分析了东亚夏季风和北太平洋 SSTA 的关系,指出两者之间相互作用存在着年代际变化特征,20 世纪 70 年代中期以前,北太平洋海温异常通过一大圆波列作用于东亚夏季风,造成我国华北地区夏季降水偏多,1976年以后,北太平洋海温异常使大圆波列减弱,而与东亚夏季风联系不紧密,不再影响华北降水。而黄刚[78]和Wang[79-80]的研究指出,由于亚洲季风从20世纪70年代后期减弱从而导致东亚地区降水发生年代际变化。如图4所示,1977年后长江流域降水明显增加,经常发生洪涝;而华北和黄河流域则有相反的情况出现,出现持续性的严重干旱。

另一方面,在图3中还可以看出,东亚冬季风也存在明显的年代际变化,20世纪70年代后期至80年代后期东亚冬季风趋于增强,而到90年代后期东亚冬季风又似乎趋于减弱。

而Nitta曾揭示了热带太平洋SST的年代际变化,而Huang[81]最近又对此做了分析,表明:赤道中、东太平洋SST异常有明显的年代际变化,即在70年代热带中、东太平洋的海水是冷的,而80—90年代海水变暖了;这些研究还指出,从1977年到现在,热带中、东太平洋发生了年代际的El Niño现象,这个年代际的El Niño事件减弱了亚洲季风。关于这些机制的研究,还有待以后进一步的探讨。

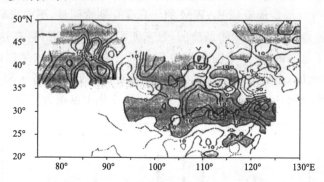

图4 中国1977—2000年与1967—1976年夏季(6—8月)平均降水距平百分率之差 阴影区为正值,虚线区为负值(引自文献[70])

5 结束语

通过以上回顾与总结,我们对东亚季风的结构、形成以及不同时间尺度的特点有了进一步的认识。但是,在东亚季风研究中还有许多问题仍不清楚,需要进一步进行探讨与研究。

(1)亚洲季风的建立与维持是多种热力因子相互作用及大气内部非线性变化的结果。而青藏高原及邻近地区的热力作用可能是决定亚洲季风爆发呈阶段性和区域性变化的一个重要因子,而这些因子影响亚洲季风的动力机制是什么,以及这些热力因子如何综合地影响亚洲季风?各个因子的相对重要性如何?决定亚洲季风爆发在不同地区和不同阶段的主要因子如何判别?这种影响的复杂性并未得深入研究。

(2)为了进一步理解季风动力学,必须把季风这一特殊现象置于海-陆-气的耦合系统中,通过研究各个系统的相互作用去认识其新的特征及有关的机制。例如,可以通过考察季风区海-气、陆-气相互作用的基本事实与动力机制,探讨决定亚洲季风爆发性和建立的阶段性的物理过程。而对于ENSO—季风的关系,青藏高原对季风爆发的动力、热力作用的更细致的动力机制问题还需进一步的研究。

(3)前面讨论了东亚季风的变率问题,即不同时间尺度的特征,而这些不同时间尺度系统之间的相互作用对季风有何重要的反馈作用,以及影响不同时间尺度的动力机制问题,有待进一步的研究。

参考文献

[1] 竺可桢.东南季风与中国之雨量[J].地理学报,1934,1:1-27.
[2] 涂长望,黄仕松.夏季风之进退[J].气象杂志,1944,18:82-92.
[3] 陶诗言,何诗秀,杨显芳.1979年季风实验期间东亚地区亚季风爆发时期的观测研究[J].大气科学,1983,7:347-355.
[4] 陶诗言,张庆云.亚洲冬夏季风对ENSO现象的响应[J].大气科学,1998,22:399.
[5] 陈隆勋,朱乾根,罗会邦,等.东亚季风[M].北京:气象出版社,1991:362.
[6] Ding Yihui. Monsoon over China[M]. Dordrecht-Boston-London:Kluwer Academic Publishers,1994,550.
[7] Tao Shiyan,Chen Longxun. The East Asian summer monsoon[R]. Tokyo Proceedings of International Conference on Monsoon in the Far East,1985,5-8:1-11.
[8] Zhu Qiangen,He Jinhai,Wang Panxin. A study of circulation differences between East-Asian and Indian monsoons with their interactions[J]. Adv Atmos Sci,1986,3:466-477.
[9] Halley E. An historical account of the trade winds and the monsoons,observable in the seas between and near the tropics,with an attempt to assign the physical cause of the said winds[J]. Phil Trans Roy Soc London,1686,16:153-168.
[10] Hadley G. Concerning the cause of the general trade-winds[J]. Phil Trans,1735,29:58-62.
[11] 陶诗言,陈隆勋.亚洲夏季大陆上空大气环流的结构[J].气象学报,1957,28:234-247.
[12] Lau K M,Ding Y,Wang J T,et al. A report of the field operations and early results of the South China Sea Monsoon Experiment (SCSMEX)[J]. Bull Amer Meteor Soc,2000,81:1261-1270.
[13] Chao W C. Multiple quasi-equilibria of the ITCZ and the origin of monsoon onset[J]. J Atmos Sci,2000,57:641-652.
[14] Chao W C. The origin of Monsoon[J]. J Atmos Sci,2001,58:3497-3507.
[15] 曾庆存,李建平.南北两半球大气的相互作用和季风的本质[J].大气科学,2002,26:433-448.
[16] He Jinhai,li Jun. Sensitivity experiments on summer monsoon circulation cell in East Asia[J]. Adv Atmos Sci,1989,6:120-132.
[17] 何金海.亚洲季风体系和研究进展以及爆发特征和可能机制[C]//现代大气科学前沿与展望.北京:气象出版社,1995:51-55.
[18] Murakami T,Chen Longxun,Xie An. Relationship among seasonal cycle,low-frequency oscillation and transientdisturbances as revealed from outgoing long wave radiation data[J]. Mon Wea Rev,1986,114:1456-1465.
[19] Chang C P,Chen T J. Tropical circulation associated with southwest monsoon onset and westerly surge over South China Sea[J]. Mon Wea Rev,1995,123:3254-3267.
[20] 何金海,朱乾根,Murakami T. T_{BB}资料揭示的亚澳季风区季节转换及亚洲夏季风建立的特征[J].热带气象学报,1996,12:36-42.
[21] 何金海,李俊,李永平.澳大利亚冷空气活动影响东亚夏季风的过程——数值试验[J].气象学报,1991,49:162-169.
[22] 周学鸣,何金海,叶榕生.乌拉尔阻塞高压影响亚洲夏季风环流和我国东部旱涝的数值试验[J].南京气

象学院学报,1995,18:25-32.
[23] Li Chengfeng,Yanai M. The onset and interannual variability of the Asian summer monsoon in to land-sea thermal contrast[J]. J Climate,1996,9:358-375.
[24] 吴国雄. 青藏高原的热力和机械强迫作用以及亚洲季风的爆发:1. 爆发地点[J]. 大气科学,1998,22:825-838.
[25] 骆美霞,张可苏. 大气热源和大地形对夏季印度季风和东亚季风环流形成的数值模拟[J]. 大气科学,1991,15:41-51.
[26] 徐海明,何金海,董敏. 印度半岛对亚洲夏季风进程影响的数值研究[J]. 热带气象学报,2001,17:117-124.
[27] Zhang Yaocun,Qian Yongfu. Mechanism of thermal features over the Indo-China Peninsula and possible effects on the onset of the South China Sea Monsoon[J]. Adv Atmos Sci,2002,19:885-900.
[28] Tao Shiyan,Chen Longxun. A review of recent research on the East Asian Summer monsoon in China [M]. Monsoon Meteorology,oxford University press,1987:60-92.
[29] 叶笃正,陶诗言,李麦村. 在六月和十月大气环流的突变现象[J]. 气象学报,1958,29:249-263.
[30] 陆菊中,林春育. 中国东部夏季风南撤与东亚环流季节变化的联系[C]//1981 全国热带季风议论文集. 昆明:云南出版社,1981:45-55.
[31] 孙淑清,孙伯民. 东亚冬季风环流异常与中国江淮流域夏季旱涝天气的关系[J]. 气象学报,1995,57:513-522.
[32] Yasunari T. A quasi-stationary appearance of 30-40 day period in the cloudiness fluctuations during the summer monsoon over India[J]. J Meteo Soc Japan,1985,65:67-102.
[33] Murakami T,Nakazawa T,He Jinhai. On the 40-50 day oscillation during the 1979 Northern Hemisphere summer,Part I:phase propagation[J]. J Meteo Soc Japan,1984,62:440-468.
[34] He Jinhai,Yang Song. Meridional propagation of East-Asian low frequency oscillation and mid-latitude low-frequency waves[J]. A M S(special Issue),1990,4:51-59.
[35] Wang Bin,Ding Yihui. An overview of the Madden-Julian oscillation and its relation to monsoon and mid-latitude circulation[J]. Adv Atmos Sci,1992,9:93-111.
[36] 李崇银. 大气低频振荡[M]. 北京:气象出版社,1993:40-46.
[37] Wang A Y,Fang S K,Ding Y H,et al. Onset,maintenance and retreat of Asian summer monsoon[C]// Fong S K,Wang A Y. Climatological ATLAS for Asian Summer monsoon. Macau:Macau Foundation,1989:251-318.
[38] 何金海,于新文. 1979 年夏季我国东部各纬度水汽输送周期振荡的初步分析[J]. 热带气象,1986,2:9-15.
[39] 朱乾根,杨松. 东亚副热带季风的北进及其低频振荡[J]. 南京气象学院学报,1989,12:249-258.
[40] 林爱兰. 南海夏季风的低频振荡[J]. 热带气象学报,1998,14:113-118.
[41] 金祖辉,孙淑清. 东亚大陆冬季风的低频振荡特征[J]. 大气科学,1996,20:101-111.
[42] Lau K M,Peng L. Origin of low frequency (intraseasonal) Oscillations in the tropical atmosphere,Part I:Basic theory[J]. J Atmos Sci,1987,44:950-972.
[43] He Jinhai. Discussion of meridional propagation mechanism of quai-40-day oscillation[J]. Adv Atmos Sci,1990,7:78-86.
[44] 穆明权,李崇银. 1998 年南海夏季风的爆发与大气季节内振荡的活动[J]. 气候与环境研究,2000,5:375-387.
[45] 李崇银,龙振夏,热带大气季节内振荡的异常与 1997 年 El Niño 事件的发生[J]. 大气科学,2001,25:589-595.

[46] 缪锦海,刘家铭. 东亚季风降水的年际变化[J]. 应用气象,1990,1:377-382.

[47] Lau K M, Shen S H. Biennial oscillation associated with the East Asian summer monsoon and tropical sea surface temperature[C]//Ye Duzheng, et al. Climate Variability. Beijing: China meteorological press, 1992:53-58.

[48] Guo Qiyun. The East Asian monsoon and the southern oscillation (1871-1980)[M]. Beijing: China Ocean Press, 1987:249-255.

[49] Chen Longxun, Chen Duo, Shen Rugui. Interannual oscillation of rainfall in China and its relation to air-sea system[J]. Acta Meteorological Sinica, 1990, 4:598-612.

[50] 叶笃正,高由禧. 青藏高原气象学[M]. 北京:科学出版社,1979:279.

[51] Nitta T S. Observational study of heat sources over the eastern Tibetan Plateau during the summer monsoon[J]. J Meteor Soc Japan, 1983, 61:590-605.

[52] Luo H B, Yanai M. The large-scale circulation and heat sources over the Tibetan Plateau and surrounding areas during the early summer of 1979[J]. Mon Wea Rev, 1984, 108:1849-1853.

[53] Huang Ronghui. Numerical simulation of the three-dimensional teleconnections in the summer circulation over the Northern Hemisphere[J]. Adv Atmos Sci, 1985, 2:81-92.

[54] Hahn D J, Shukla J. An apparent relationship between snow cover and the Indian monsoon rainfall[J]. J Atmos Sci, 1976, 33:2461-2462.

[55] Dickinson R R. Eurasian snow cover versus Indian rainfall-An extension of the Hahn-Shukla results[J]. J Clim Appl Meteor, 1984, 23:171-173.

[56] Khandekar M L. Eurasian snow cover, Indian monsoon and El Niño/Southern Oscillation-a synthesis[J]. Atmos Ocean, 1991, 29:636-647.

[57] 陈烈庭,阎志新. 青藏高原冬春积雪对大气环流和我国南方汛期降水的影响[C]//中长期水文气象预报文集Ⅱ. 北京:水利电力出版社,1979:185-195.

[58] 陈烈庭,阎志新. 青藏高原冬春异常雪盖影响初夏季风的统计分析[C]//中长期水文气象预报文集. 北京:水利电力出版社,1981:133-141.

[59] 韦志刚,罗四维. 中国西部积雪对我国汛期降水的影响[C]//"灾害性气候预测及其对农业年景和水资源调配的影响"项目论文集Ⅱ——灾害性气候的过程及诊断. 北京:气象出版社,1996:137-140.

[60] Nitta T S. Convective activities in the tropical western Pacific and their impact on the Northern Hemisphere summer circulation[J]. J Meteor Soc Japan, 1987, 64:373-390.

[61] Kurihara K. A climatological study on the relationship between the Japanese summer weather and the subtropical high in the western northern Pacific[J]. Geophy Mag, 1989, 43:45-104.

[62] Huang Ronghui, SUN Fengying. Impact of the tropical western Pacific on the East Asian summer monsoon[J]. J Meteor Soc Japan, 1992, 70(B):243-256.

[63] 黄荣辉,孙凤英. 热带西太平洋暖池的热状态及其上空的对流活动对东亚夏季气候异常的影响[J]. 大气科学,1994,18:141-151.

[64] Nikaido Y. The P-J like north-south oscillation-found in the 4-month integration of the global spectral model T42[J]. J Meteor Soc Japan, 1989, 67:587-604.

[65] Lu Riyu, Chan-Su Ryu, BUWEN Dong. Associations between the Western North Pacific Monsoon and the South China Sea monsoon[J]. Adv Atmos Sci, 2002, 19:12-24.

[66] Rassm13sson E M, Carpenter T. The relationship between eastern equatorial sea surface temperature and rainfall over India and Sri Lanka[J]. Mon Wea Rev, 1983, 111:517-527.

[67] Khandekar M L, Neralla V R. On the relationship between the sea surface temperature in the equatorial Pacific and the Indian monsoon rainfall[J]. Geophys Res Let, 1984, 11:1137-1140.

[68] 符淙斌,滕星林. ENSO 与中国夏季气候的关系[J]. 大气科学,1988(特刊):133-141.

[69] Huang Ronghui, Wu Yifang. The influence of ENSO on the summer climate change in China and its mechanisms[J]. Adv Atmos Sci,1989,6:21-32.

[70] 黄荣辉,陈文,丁一汇,等. 关于季风动力学以及季风与 ENSO 循环相互作用的研究[J]. 大气科学,2003,27:484-502.

[71] Chen Wen, Graf H F. The interannual variability of East Asian winter monsoon and its relationship to global circulation[R]. Max-Planck-Institute for Meteorologic report,1998:250.

[72] Xue Feng. Interannual to interdecadal variations of East Asian summer monsoon and its association with the global atmospheric circulation and sea surface temperature[J]. A A S,2001,18:567-575.

[73] 王绍武,朱锦红. 几个方兴未艾的气候学问题[J]. 应用气象学报,1999,10(增刊):104-113.

[74] Nitta T S, Yamada S. Recent warming of the tropical sea surface temperature and its relationship to the Northern Hemisphere temperature and its relationship to the Northern Hemisphere circulation[J]. J Meter Soc Japan,1989,67:373-375.

[75] 李崇银,李桂龙,龙振夏. 中国气候年代际变化的大气环流形势对比分析[J]. 应用气象学报,1999,10(增刊):1-8.

[76] Graham N E. Decadal-scale climate variability in the tropical and North Pacific during the 1970s and 1980s Observations and model results[J]. Climate Dyn,1994,9:135-162.

[77] 李峰,何金海. 北太平洋海温异常与东亚夏季风相互作用的年代际变化[J]. 热带气象学报,2000,16:260-271.

[78] 黄刚. 东亚夏季风环流指数与夏季气候变化关系的研究[J]. 应用气象学报,1999,10:61-69.

[79] Wang Huijun. The weakening of the Asian monsoon circulation after the end of 1970s[J]. Adv Atmos Sci,2001,18:376-386.

[80] Wang Huijun. Instability of the East Asian summer monsoon-ENSO relation[J]. Adv Atmos Sci,2002,19:1-11.

[81] Huang Ronghui. Decadal variability of the summer monsoon rainfall in East Asian and its association with the SST anomalies in the tropical Pacific[J]. CLIVAR Exchange,2001,2:7-8.

Possible Mechanism of the Effect of Convection over Asian-Australian "Land Bridge" on the East Asian Summer Monsoon Onset[*]

He Jinhai[1] Wen Min[2] Ding Yihui[3] Zhang Renhe[2]

(1. Jiangsu Key Laboratory of Meteorological Disaster (KLME), Nanjing Institute of Meteorology, Nanjing University of Information Science and Technology, Nanjing 210044, China;
2. State Key Laboratory of Severe Weather (LaSW), Chinese Academy of Meteorological Sciences, Beijing 100081, China;
3. National Climate Center, Chinese Meteorological Administration, Beijing 100081, China)

Abstract: The Asian-Australian "land bridge" is an area with the most vigorous convection in Asian monsoon region in boreal spring, where the onset and march of convection are well associated with the onset of East Asian summer monsoon. The convection occurs over Indo-China Peninsula as early as mid-April, which exerts critical impact on the evolution of monsoon circulation. Before mid-April, there are primarily sensible heatings to the atmosphere over Indo-China Peninsula and Indian Peninsula, so the apparent heating ratios over them decrease with height. However, after mid-April, it changes into latent heating over Indo-China Peninsula due to the onset of convection, and the apparent heating ratio increases with height in mid- and lower-troposphere. The vertical distribution of heating ratio and its differences between Indo-China Peninsula and Indian Peninsula are the key factors leading to the splitting of boreal subtropical high belt over the Bay of Bengal. Such mechanism is strongly supported by the fact that the evolution of the vertical heating ratio gradient above Indo-China Peninsula leads that of 850 hPa vorticity over the Bay of Bengal. Convections over Indo-China Peninsula and its surrounding areas further increase after the splitting. Since then, there is a positive feedback lying among the convective heating, the eastward retreat of the subtropical high and the march of monsoon, which is a possible mechanism of the advance of summer monsoon and convection from Indo-China Peninsula to South China Sea.

Key words: Asian-Australian "land bridge"; tropical convection; Indo-China Peninsula; the splitting of boreal subtropical high belt; East Asian summer monsoon onset.

---

[*] 本文发表于《Science in China Series D: Earth Sciences》,2006 年第 49 卷第 11 期,1223-1232. 其中文版发表于《中国科学 D 辑 地球科学》,2006 年第 36 卷第 10 期,959-967.

The formation and onset mechanisms have always been the highlight in the study of monsoon. Scholars suggest many kinds of possible mechanisms, but controversies still exist. Based on data and theoretic analysis, Zeng and Li[1] summarized that there are two major driving forces of monsoon: one is planetary thermal convection, which is caused by the seasonal variations of atmospheric general circulation and the cross-equatorial air flows (i. e. interactions between the Northern and Southern Hemispheres) due to the seasonal variation of incoming solar radiation; the other is the semi-permanent planetary waves due to the differences of the earth surface (such as the thermal differences between land and sea, topography, etc.). They pointed out that the two forces are in phase in the Asian-Australian monsoon region, resulting in pronounced monsoon prevails. The moist process in the atmosphere, the physical process associated with water-vapor phase change, plays important roles not only in the genesis and development of micro-and meso-scale, but also in those of synoptic and planetary scale systems[2-4]. It is indicated that the moist process is also a critical factor in the formation of monsoon, especially the summer monsoon, characterized by large range convective activities accompanying with strong latent heat release[5]. On one hand, the moist process may increase (decrease) the intensity of monsoon circulation by enhancing (weakening) land-sea thermal differences; on the other hand, it can store and redistribute solar energy gained in most areas of tropics and subtropics, and then conditionally release the energy in a narrow area to modify the location and intensity of the monsoon rainbelt. Therefore, the moist process in the atmosphere, i. e. convective activities, is the third forcing of monsoon[6].

The maritime continent (MC), including Sumatra, Kalimantan, etc. , is unique in Asian monsoon region. There is strong convection above MC in a whole year, in which the seasonal march of the convection over Sumatra is closely associated with the onset of summer monsoon over the Indo-China Peninsula (ICP)[7]. Located to the north of MC, the ICP connects the MC with the Asian continent and has important impacts on the onset of South China Sea (SCS) summer monsoon, even the weather and climate in East Asia. When considered together, the ICP and MC can be called as the Asian-Australian "land bridge"[6, 8, 9]. Studies on the "land bridge" in early years are scarce because of the limitation of observational methods and data[10]. But since the mid-1990s, with the comprehensive application of satellite remote sensing and the abundance of observational data, due attention has been paid to this region. It is indicated that the sensible heating over ICP is the key factor triggering the SCS summer monsoon[11-12]. He et al. [7] suggested that the movement of convections in Sumatra along the "land bridge" results in the onset of summer monsoon over Southeast Asia and Indian, which is supported by many other researches[10, 13]. Recent studies showed that convections over ICP also play a vital role during the onset of East Asian summer monsoon[14-15]. However, our knowledge for the convective characteristics of Asian-Australian "land bridge" is still not enough, and conclusions mentioned above are only abecedarian, the mechanisms of which needing further discussion.

In present study, the data of meteorological fields from the National Centers for Environmental Prediction-Nation Center of Atmospheric Research (NCEP-NCAR) reanalysis, the outgoing long-wave radiation (OLR) from the National Oceanic and Atmospheric Administration (NOAA) and the temperature of black body on the top of cloud (T_{BB}) from the Meteorological Research Institute (MRI) of Japan Meteorological Agency (JMA) during 1980-2001, are used to figure the convective characteristics of Asian-Australian "land bridge" with the focus on discussing the relationship between convective activities and the onset of summer monsoon and the possible mechanisms. Unless special explained, all results are derived from the climatological fields from 1980 to 2001.

1 Characteristics of convection over Asian-Australian "land bridge"

The flare-up of convection over Asian-Australian "land bridge" in boreal spring is the most significant characteristics of seasonal transitions in East Asian monsoon region, well associated with the onset of East Asian summer monsoon[7]. Fig. 1 shows the distribution of OLR in each stage of convective activities. In the first pentad of May (Fig. 1a), the northern and southern ITCZs converge at MC, with convective centers over each island and the strongest one over northern Sumatra. Many studies define this pentad as the onset pentad of summer monsoon in northern Sumatra and its surrounding areas[10, 16]. Notably, high OLR band has broken over ICP, and there is a low center below 230 W · m^{-2} over its southern areas. Usually, 240 W · m^{-2} is regarded as the threshold of active convection. Therefore, convection starts early over ICP, which is consistent with the analysis results of T_{BB} given by literature[7]. However, the summer monsoon over ICP hasn't fully established until the third pentad of May[10, 16] (Fig. 1b), when OLR decreases to lower than 220 W · m^{-2} over the southeast of the Bay of Bengal (BOB) and major ICP with the lowest center below 210 W · m^{-2} over the south. At the same time, convections over SCS is still weak with the OLR above 240 W · m^{-2}. In the fourth pentad of May, the OLR over SCS decreases rapidly to below 230 W · m^{-2}, indicating the onset of SCS summer monsoon (Figure Omitted).

During the march of monsoonal convection, the strong convection originated form MC shifts northward remarkably. In order to illustrate this process and to understand the primary moving tracks of convective centers, Hovmöller diagrams of climatological T_{BB}, from Sumatra to the northeast and northwest respectively, are plotted in Fig. 2. The lower panels of Fig. 2a and Fig. 2b are the same: from the southeast of northwest along Sumatra (Route AO in Fig. 1b), but the upper panels are northeastward to ICP (Route OB in Fig. 1b) and northwestward to northern BOB (Route OC in Fig. 1b) respectively. It is shown in Fig. 2 that the main convective region locates to the south of the equator in boreal winter, jumps to the north of the equator in early April, and strengthens in late April and early May indicating the onset of summer monsoon over Sumatra. In the middle of May, the convection extends northward to subtropics, resulting in the onset of summer monsoon over the east of BOB and ICP. The strong convective center occurs over northern BOB in June, which is the east march of the Indian summer monsoon onset[17]. It is notable during the whole process that T_{BB} over ICP drops to lower than 275 K in mid-April, indicating that the convection begins to flare up over the part of ICP. The similar characteristics have been pointed out by Matsumoto[18]. However, the convection begin to set up over the east of BOB in mid-May, with the deep convection ($T_{BB}<260$ K) occurring simultaneously with that over ICP. Both of them are the results of the northward movement of the tropical convection. That is to say, the tropical summer monsoon to the east of 90°E over East Asia is truly established until then.

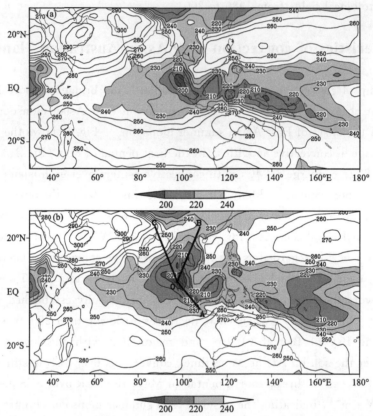

Fig. 1 Distributions of climatological OLR on (a) the first pentad of May, and (b) the third pentad of May. Shadings indicate OLR<240 W·m^{-2}. The thick lines in (b) are the cross routes in Fig. 2

2 Heating differences to the atmosphere between ICP and Indian Peninsula

As mentioned above, the convection over ICP starts to erect from mid-April. What role does the released heat exert on the onset of East Asian summer monsoon? It has been shown that it will directly induce the break of the subtropical high belt[14], but that research is a case study and only analyzes the change of surface heat flux. Although reflecting the change of rainfall in some ways, the surface heat flux is inherently different from the condensation latent heat in the atmosphere. In fact, the real factor resulting in the variation of the subtropical high is the inhomoge-

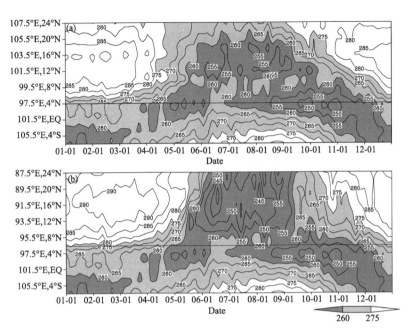

Fig. 2 Cross section of T_{BB} from Sumatra to (a) the northeast (AOB) and (b) the northwest (AOC). Shadings are $T_{BB} < 275$ K

neous vertical distribution of heating. Therefore, to further discuss the mechanism, the climatological atmospheric apparent heat source and apparent moist sink are calculated and their vertical structure is analyzed to realize the change of heating over ICP and Indian Peninsula (IP).

The atmospheric apparent heat source and apparent moist sink are defined as[19]:

$$Q_1 = c_p \left[\frac{\partial T}{\partial t} + \mathbf{V} \cdot \nabla T + \left(\frac{p}{p_0} \right)^{\kappa} \omega \frac{\partial \theta}{\partial p} \right] \quad (1)$$

$$Q_2 = -L \left(\frac{\partial q}{\partial t} + \mathbf{V} \cdot \nabla q + \omega \frac{\partial q}{\partial p} \right) \quad (2)$$

where Q_1 and Q_2 are called apparent heat source and apparent moist sink respectively, and $P_0 = 1000$ hPa. Points (100°E, 15°N) and (77.5°E, 15°N) are chosen to represent ICP and IP. Fig. 3 shows the evolution of vertical distribution of the heating rate Q_1/c_p and the drying rate Q_2/c_p (hereafter Q_1 and Q_2 briefly) of the two points.

Before mid-February, it is cloudless over ICP, so Q_1/c_p is negative in the mid-and upper troposphere, i.e. radiative cooling, and is positive in the lower troposphere caused by the turbulent transfer. The heating center in low level starts to strengthen from February, which results from the increase of solar radiation, the rapid rise of land surface temperature and the intensification of turbulence. A strong positive center of Q_2/c_p also exists in low level over ICP due to the turbulent transfer. In the meantime, Q_1/c_p in the mid-troposphere starts to increase too, but its intensity is lower than that in lower level. The maximal heating center lies still near the ground, i.e. sensible heating is dominant over ICP.

After mid-April, Q_1/c_p has a center of 3 K·d^{-1} in the mid-troposphere near 400 ~

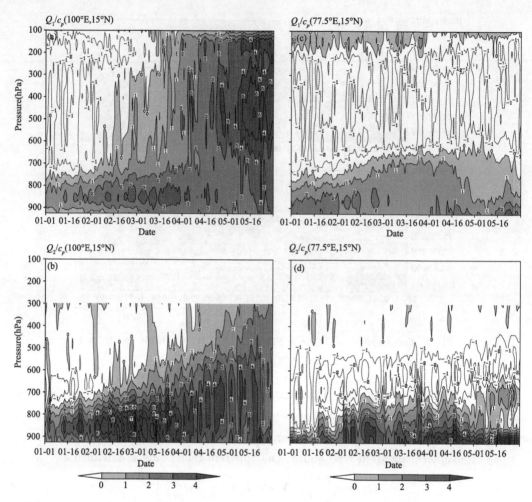

Fig. 3 Time-height sections of Q_1/c_p (above) and Q_2/c_p (below) over ICP (100°E, 15°N, left) and IP(77.5°E, 15°N, right). Shadings denote positive values. Unit: K·d^{-1}

500 hPa, which is the characteristic height of deep-convective condensation latent heating. At the same time, because of the flourish of convection, the cloud cover increases, thus reduces the solar radiation reaching the ground surface and decrease the land surface temperature. The heating center in the lower troposphere weakens with the turbulence. Fig. 3b shows the height of maximal Q_2/c_p is lifted to near 700 hPa. Here, the peaks of Q_1/c_p and Q_2/c_p are significantly in the different layer with the former is higher than the latter, indicating that convection is the primary forcing to produce the heating center[20]. The maxima of Q_1/c_p and Q_2/c_p are nearly equivalent, showing that condensation latent heating is the main contributor of apparent heating[20]. All the above characteristics present that the heating to atmosphere over ICP changes to the condensation latent heating associated with convection. After the onset of summer monsoon over ICP in mid-May, the convection strengthens and the apparent heating center reaches as much as 5 K·d^{-1}.

Comparing with ICP, the evolutions of Q_1/c_p and Q_2/c_p over IP are much simpler (Fig. 3c and Fig. 3d). Indian summer monsoon hasn't been established until mid-June, so the near-surface temperature over IP from January to May has a continuously increasing trend (Figure omitted). The centers of Q_1/c_p and Q_2/c_p are located in the lower troposphere, i.e. sensible heating is dominant. Calculation result shows that advection over IP is stronger than that over ICP, and its absolute value is equivalent to that of the vertical transfer, but their signs are opposite in the mid-and lower troposphere, thus offset each other. That the upward extension of the heating area of Q_1/c_p over IP disappears suddenly in early May and that the maximum of Q_2/c_p maintains near the ground seem to be related to the horizontal advection. As to the east of the subtropical high over Arabian Sea before the onset of summer monsoon, the IP is overlaid by northwesterly. The temperature gradient in tropics in boreal spring points to the north, so there is cold advection in northwesterly. Similarly, the specific humidity gradient points to the north due to the large humidity in the ocean, so there is dry moist advection in northwesterly. It is generally considered that the thermal low produced by land surface sensible heating is one of the preconditions for the convection developing. However, the above characteristics show that the advection over IP will suppress the development of vertical movement. Therefore, although surface temperatures start to increase in spring over both IP and ICP, the convection over IP is hard to set up. This might be an important cause for the later onset of summer monsoon over India than over East Asia.

In a word, after mid-April, the heating to the atmosphere over ICP changes progressively to latent heating, while the strong sensible heating maintains over IP. The heating characteristics between the two Peninsulas are significantly different. Therefore, the heating differences over ICP and IP reflected by the climatological atmospheric apparent heat source are in agreement with the conclusions of He et al.[14], indicating that the surface thermal flux may reflect the upper heating in a certain extent and that differences between heating over ICP and over IP in spring is ubiquitous.

3 Impact of the convection over ICP on the break of the subtropical high belt and the onset of East Asian tropical summer monsoon

According to Wu et al.[21], different heating will have different effects on the subtropics, especially near the ridge of the subtropical high. The heating ratio caused by sensible heating decreases with height, while that caused by latent heating increases with height in the mid-and lower troposphere. Therefore, in the mid-and lower troposphere, a cyclonic circulation is triggered to the east (west) of the sensible (latent) heating, while an anticyclonic circulation is triggered to the west (east) of the sensible (latent) heating. As discussed above, before mid-April sensible heating is dominant over ICP and IP, so their effects on BOB offset. After mid-April the convection over ICP starts to erect, so the heating to the atmosphere changes gradually to latent heating with a strong center occurring in late April, while the sensible heating is still dominant over IP. Thus, the effects of these two forcing superpose to trigger an abnormal cyclonic circulation over BOB. Attracting by this circulation, the vortex near Sri Lanka migrates to the north along the east coast of IP, leading to the break of the subtropical high belt, the formation of the BOB trough and the prevailing of southwesterly over ICP[14]. The strengthening of southwesterly will help the tropical convection flow into ICP, i.e. the north jump of convection in mid-May in Fig. 2. Therefore, the onset of convection over ICP might be the primary inducement of the break of the subtropical high belt and the onset of summer monsoon.

In order to illustrate the relationship of the convective heating over ICP with the break of the subtropical high belt, the vertical gradient of heating ratio is defined as the differences of apparent heating ratio Q_1/c_p between 400 hPa and 850 hPa to represent the evolution of character of heating to the atmosphere. It is compared with 850 hPa relative vorticity over BOB (Fig. 4) which is a good token of the status of the subtropical high belt[15].

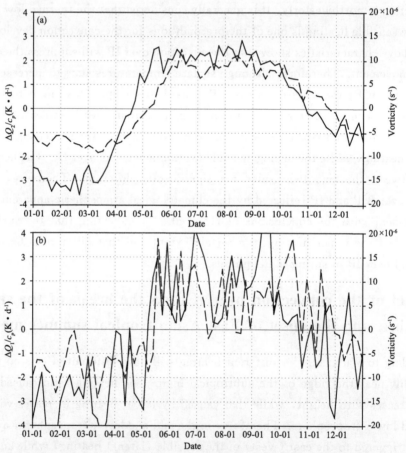

Fig. 4 Evolutions of the vertical gradients of Q_1/c_p over ICP (100°E, 15°N) (400~850 hPa, solid line, unit: K·d^{-1}) and 850 hPa relative vorticity over BOB (80°~90°E, 10°~20°N) (dotted line, unit: 10^{-6} s^{-1}) for (a) climatological mean and (b) 1998

Fig. 4a is for climatological situation. Consistent with the annual cycle of the convection over ICP, the vertical gradient of heating ratio is negative in winter half year and positive in summer. It begins to transit rapidly from negative to positive in mid-March, and becomes positive in mid-April which is simultaneous with the appearance of the Q_1 heating center near 400 hPa. The evolution of vorticity over BOB lags obviously behind that of the vertical gradient of heating ratio. The vorticity begins to develop toward positive in early April, and becomes positive in early May, which is concurrent with the break of the subtropical high belt. Fig. 4b shows the situation in the year of 1998. The time when the vertical gradient of heating ratio changes to positive is consistent with the onset of convection (refer Fig. 6) and earlier than the time when the subtropical high belt breaks. It has been shown that the above scenario can been found in most years [22]. The above analysis testifies further the critical role of heating to the atmosphere over ICP in the variation of the subtropical high belt. Interestingly, the transition of the heating ratio toward negative in boreal autumn also leads before that of relative vorticity over BOB, so the evolution of the heating profile over ICP may influence the subsequent change of the circulation field over BOB.

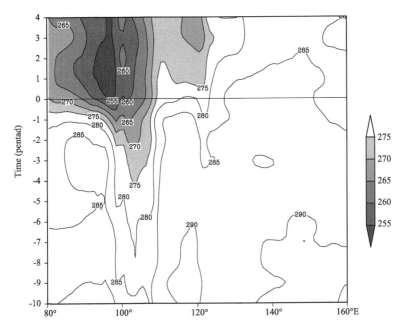

Fig. 5   Composed T_{BB} time-longitude section along 10°~20°N with the break time of the subtropical high belt as reference point. T_{BB}<275 K in shadow areas

After the break of the subtropical high belt, the tropical convection enters ICP strengthening the intensity of convection there. Fig. 5 is the composite time section of T_{BB} along 10°~20°N during 1980—2001 with the splitting date of the subtropical high belt [15] as the reference point, therein Pentad 0 is the pentad when the subtropical high belt breaks, Pentad -1 and $+1$ are the pentads before and after the break of the subtropical high belt respectively, and so on. It can be seen that the convection appears first over ICP, and is already active (T_{BB}<270 K) before the subtropical high belt breaks. After the break, the convection expands rapidly and covers the entire BOB and ICP, with the intensity of the center rises sharply. One or two pentads after the break of the subtropical high belt, T_{BB} over SCS drops to 275 K which is a symbol of the onset of summer monsoon in this region although it is much weaker than over BOB and ICP.

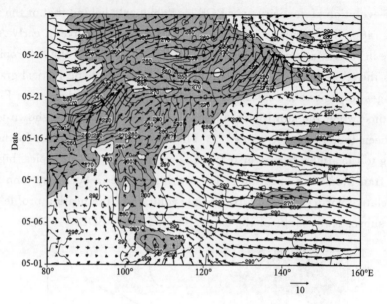

Fig. 6 Time-longitude section of T_{BB} and 850 hPa wind along $10°\sim20°$N in May 1998. $T_{BB}<280$ K in shadow areas

Many studies show that the SCS summer monsoon is "explosive" in the Asian summer monsoon region, which, as is shown in Fig. 5, is established in a wide latitude within one pentad. But we also find that convection sometimes advances gradually eastward, which may result from the interaction of convection and circulation. Fig. 6 shows the evolution of T_{BB} and 850 hPa wind in the latitude of $10°\sim20°$N in 1998. It is approximately southerly in the arrow-like shadow zone, indicating the connection of active convection with southwesterly. The process of eastward extension of the convection is also the process of the eastward withdrawal of West Pacific subtropical high and the advance of the tropical monsoon into SCS, which can be explained by the balance among three terms in the complete vorticity equation (CVE). A scaling analysis of CVE gives the following result [21]:

$$\beta v + \mathbf{V} \cdot \nabla \zeta_z \propto \frac{f+\zeta}{\theta_z}\frac{\partial Q_{LH}}{\partial z} \tag{3}$$

Below the maximal heating center in the convective region, $\frac{f+\zeta}{\theta_z}\frac{\partial Q_{LH}}{\partial z}>0$. As a response to the heating, the southwesterly strengthens under the convective center. To the west of southwesterly, there is a cyclonic circulation, and to the east there is an anticyclonic circulation. The exit region of the strong southwesterly is the region of new convection intensifying. Therefore, there is a positive feedback among the convective heating, the eastward withdrawal of the subtropical high and the strengthening of monsoon during the onset of summer monsoon, which is also a possible mechanism of the eastward advance of summer monsoon and convection from ICP to SCS.

4 Conclusions

The Asian-Australian "land bridge" is an area with the most vigorous convection in the Asian monsoon region in boreal spring. The onset process of convection is also the march of East Asian summer monsoon. With the seasonal transition, the strong convective center first jumps from the south to the north of equator, then reaches strongest in early May which corresponds to the onset of summer monsoon over northern Sumatra and its surrounding areas. The summer monsoon is entirely established over the east of BOB and ICP in the third pentad of May, represented by the rapid northward shift of the convective center originated from northern Sumatra. Then the convection advances to both east and west, enters SCS in the fourth pentad of May and northern IP in early June, respectively.

During the evolution of monsoonal circulation, differences between the heating to the atmosphere over ICP and IP are the key to the break of the subtropical high belt. The heating ratio in the mid-and low-troposphere over IP decreases with height from January to May, i. e. sensible heating is dominant. But over ICP, with the establishment of convection in mid-April, the trend of the heating ratio against height in the mid-and low-troposphere change from decrease to increase, i. e. the heating to atmosphere alter from sensible to latent heating. This change produces positive vortex in BOB leading to the break of the subtropical high belt there. Therefore, the convection over ICP plays a leading role in the break of subtropical high belt. Such mechanism is supported by the fact that the evolution of the vertical gradient of the heating ratio over ICP leads before that of the vorticity over BOB. In addition, the convection interacts with the circulation during its eastward migration: the convection strengths the southwesterly in the mid- and lower troposphere whose exit region is a new vigorously convective region and forces the subtropical high to retreat to the east. Therefore, the positive feedback among the convective heating, the eastward withdrawal of the subtropical high and the advance of monsoon is a possible mechanism of the march of summer monsoon and convection from ICP to SCS.

Acknowledgement. The authors would like to thank CDC, NCEP/NCAR, NOAA and MRI of Japan for providing the data. We are also sincerely grateful to anonymous reviewers for their comments and to Chen Hua for her kind assistance. This work was supported by the National Natural Science Foundation of China under Grant 40305005 and 40225012.

References

[1] Zeng Q C, Li J P. Interaction between the northern and southern hemispheric atmospheres and the essence of monsoon[J]. Chinese J Atmos Sci,2002,26(4):433-448.

[2] Gao S T,Lei T,Zhou Y S. Moist potential vorticity anomaly with heat and mass forcing in torrential rain systems[J]. Chinese Phys Lett,2002,19(6):878-880.

[3] Gao S T,Zhou Y S,Lei T. Structural features of the Meiyu front system[J]. Acta Meteorol Sin,2002,16(2):195-204.

[4] Zhou Y S,Gao S T,Shen S. A diagnostic study of formation and structures of the Meiyu Front system over East Asia[J]. J Meteorol Soc Jpn,2004,82(6):1565-1576.

[5] He J H,Li J. Sensitivity experiments on summer monsoon circulation cell in East Asia[J]. Adv Atmos Sci,1989,6(1):120-132.

[6] He J H,Yu J J,Shen X Y,et al. Research on mechanism and variability of East Asian monsoon[J]. J Trop Meteorol,2004,20(5):44-459.

[7] He J H,Zhu Q G,Murakami M. T_{BB} data-revealed features of Asian-Australian monsoon seasonal transition and Asian summer monsoon establishment[J]. J Trop Meteorol,1996,12(1):34-42.

[8] Chang C P,Harr P A,McBride J,et al. Maritime continent monsoon:annual cycle and boreal winter variability[C]//Chang C P. East Asian monsoon. Singapore:World Scientific Publishing Bo Pte Ltd,2004:107-152.

[9] Wang L J,He J H,Guan Z Y. Characteristic of convective activities over Asian-Australian "land bridge" areas and its possible factors[J]. Acta Meteorol Sin,2004,18(4):441-454.

[10] Lau K M,Yang S. Climatology and interannual variability of the Southeast Asian summer monsoon[J]. Adv Atmos Sci,1997,14(2):141-162.

[11] Zhang Y C,Qian Y F. Mechanism of thermal features over the Indo-China peninsula and possible effects on the onset of the South China Sea monsoon[J]. Adv Atmos Sci,2002,19(5):885-900.

[12] Wang S Y,Qian Y F. Basic characteristic of surface heat field in 1998 and the possible connections with the SCS summer monsoon onset[J]. Acta Meteorol Sin,2001,59(1):31-40.

[13] LinHo,Wang B. The time-space structure of the Asian-Pacific summer monsoon:A fast annual cycle view[J]. J Climate,2002,15(15):2001-2019.

[14] He J H,Wen M,Shi X H,et al. Splitting and eastward withdrawal of the subtropical high belt during the onset of the South China Sea summer monsoon and their possible mechanism[J]. J Nanjing University (Natural Sci),2002,38(3):318-330.

[15] Wen M,He J H,Xiao Z N. Impact of the convection over the Indochina peninsula on the onset of SCS summer monsoon[J]. Chinese J Atmos Sci,2004,28(6):864-875.

[16] Zhang Z Q,Chan J,Ding Y H. Characteristics,Evolution and Mechanisms of the summer monsoon onset over Southeast Asia[J]. Inter J Climat,2004,24:1461-1482.

[17] Wang B,LinHo. Rainy season of the Asian-Pacific summer monsoon[J]. J Climate,2002,15(4):386-398.

[18] Matsumoto J. Seasonal transition of summer rainy season over Indochina and adjacent monsoon region[J]. Adv Atmos Sci,1997,14(2):231-245.

[19] Yanai M,KEsbensen S,Chu J H. Determination of bulk properties of tropical cloud clusters from large-scale heat and moisture budgets[J]. J Atmos Sci,1973,30(4):611-627.

[20] Luo H,Yanai M. The large-scale circulation and heat sources over the Tibetan plateau and surrounding areas during the early summer of 1979. Part II:Heat and moisture Budgets[J]. Mon Wea Rev,1984,112(5):966-989.

[21] Wu G X,Liu Y M,Liu P. The effect of spatially nonuniform heating on the formation and variation of subtropical high I. Scale analysis[J]. Acta Meteorol Sin,1999,57(3):257-263.

[22] Wen M. The features of the convective activities over Asian-Australian "land bridge" and their effects on the East Asian summer monsoon onset[R]. CAMS postdoctoral report,2005.

Characteristics of the Onset of the Asian Summer Monsoon and the Importance of Asian-Australian "Land Bridge" *

He Jinhai   Wen Min   Wang Lijuan   Xu Haiming

(1. Jiangsu Key Laboratory of Meteorological Disaster (KLME) and School of Atmospheric Sciences, Nanjing University of Information Science & Technology, Nanjing 210044;

2. State Key Laboratory of Severe Weather (LaSW), Chinese Academy of Meteorological Sciences, Beijing 100081)

Abstract: Based on summarizing previous achievements and using data as long and new as possible, the onset characteristics of Asian summer monsoon and the role of Asian-Australian "land bridge" in the onset of summer monsoon are further discussed. In particular, the earliest onset area of Asian summer monsoon is comparatively analyzed, and the sudden and progressive characteristics of the onset of summer monsoon in different regions are discussed. Furthermore, the relationships among such critical events during the onset of Asian summer monsoon as the splitting of subtropical high belt over the Bay of Bengal (BOB), the initiation of convection over Indo-China Peninsula, the westward advance, reestablishment of South Asian High, and the rapid northward progression of convection originated from Sumatra in early summer are studied. The important impact of the proper collocation of the latent heating over Indo-China Peninsula and the sensible heating over Indian Peninsula on the splitting of the subtropical high belt, the deepening of BOB trough, the activating of Sri Lanka vortex (twin vortexes in Northern and Southern Hemispheres), and the subsequent onset of South China Sea summer monsoon are emphasized.

Key words: Asian summer monsoon onset, Asian-Australian "land bridge", splitting of subtropical high belt

1 Introduction

The onset characteristics and their possible mechanisms have always been the highlight in the study of Asian summer monsoon (ASM). In particular, the region where the earliest ASM onset occurs has been the focus of studies[1], which is also one of the scientific goals of the South China Sea Monsoon Experiment (SCSMEX)[2]. However, there is still no solid

---

* 本文发表于《Advances in Atmospheric Sciences》,2006 年第 23 卷第 6 期,951-963.

conclusion as to the initial onset of ASM. Currently, there exist four main viewpoints. First, the ASM initially establishes in the South China Sea (SCS), and then proceeds northward and westward[3]. Second, in eastern Bay of Bengal (BOB)[4]. Third, in Indo-China Peninsula (ICP) or its southern surrounding areas[5-11]. And fourth, simultaneously over the whole area in BOB, ICP and SCS. He et al.[12] suggested that the northward progression of the low TBB center over Sumatra and the occupation of a TBB trough in ICP lead to the onset of monsoonal convection in ICP and BOB, which actually belongs to the third viewpoint. In a word, where, on earth, is the earliest onset place of ASM needs further study.

The Asian monsoon is the most significant monsoon in the world. The Asian summer (winter) monsoon and Australian winter (summer) monsoon are so closely associated with each other that they can even be jointly called the Asian-Australian monsoon system. Therefore, the seasonal transition of the Asian monsoon, the interaction between the Northern and Southern Hemispheric atmospheres, and the seasonal migration of the tropical convection are indivisible. Zeng and Li[13] suggested that the seasonal migration of planetary thermal convection (the primary driving forcing) in phase with the quasi-stationary planetary waves (such as the land-ocean thermal contrast, topographic height, etc.) (the secondary forcing) is the main reason for the tropical monsoon being most obvious in the Asian-Australian monsoon region. The Maritime Continent (MC), including Sumatra and Kalimantan etc., has the strongest tropical convection in the world, and the seasonal migration of Sumatra convection is well associated with the onset of summer monsoon in ICP[12]. If called the ICP and MC as Asian-Australian "land bridge"[12,14-15], the seasonal migration of the Sumatra convection along "land bridge" is exactly the manifestation of the primary driving forcing of summer monsoon. Therefore, it is of importance to fully comprehend the role of Asian-Australian "land bridge" in the onset of ASM.

In the Asian monsoon region, there are coexistent latitudinal and longitudinal land-sea distributions (the Asian continent with the Indian Ocean, and the Asian continent with the Pacific Ocean, respectively). In addition, the largest plateau in the world, the Tibetan Plateau (TP), is located in the middle of Asian continent, together with several subcontinents such as the Indian Peninsula and ICP in the south, The surface thermal differences among them and their seasonal transition not only bring about complex, strong and sensitive monsoon in Asia[16-17], but also lead to the regional characteristics of the onset of ASM. Besides, due to the moist process known as the tertiary forcing[18], the onset of the summer monsoon is always accompanied with the convective rainfall. Therefore, studies on the onset characteristics of summer monsoon and their associated mechanisms attract much attention in the community of monsoon researchers.

In a word, where and how is the ASM first established? What are the onset characteristics and their associated mechanisms? And as a key factor to summer monsoon onset, what kind of roles does the Asian-Australian "land bridge" play? This paper aims at summarizing previous results and giving an integral image of the ASM onset process.

2 Data

The primary data sets used in this study are as follows:

1) The daily temperature of black body on the top of cloud (TBB), provided by the Japan Meteorological Agency, derived from 3-hourly Geostationary Meteorological Satellite (GMS) data with the horizontal resolution of 1°×1° spanning 80°E~160°W, 60°S~60°N;

2) The daily National Centers for Environmental Prediction-Nation Center of Atmospheric Research (NCEP-NCAR) reanalysis Products, with a global coverage and a 2.5° latitude-longitude resolution;

3) The pentadly Climate Prediction Center (CPC) Merged Analysis of Precipitation (CMAP), with a global coverage and a 2.5°×2.5° resolution;

4) The daily outgoing long-wave radiation (OLR), provided by National Oceanic and Atmospheric Administration (NOAA), also with a global coverage and a 2.5°×2.5° resolution.

The GMS came into use in 1980 and disused in 2002. In order to coincide with the time span of satellite data, 22 years from 1980 to 2001 are chosen as research period. All climatological fields are the mean of this period if no special announce is declared.

3 The earliest onset of ASM and the regional characteristics

3.1 The earliest onset area of ASM

As the onset of summer monsoon is always accompanied with convective rainfall, and the lower TBB can approximately represents convective clouds and heavy rainfall in the tropical and subtropical regions, thus TBB data were here used to discuss the characteristics of seasonal transition in Asian-Australian monsoon region and the initial onset of ASM[12]. Considering that there are many disputes on this issue, the latest TBB data are employed to make further analysis. Figure 1 shows the horizontal distribution of the climatological monthly TBB from March to June.

The distribution of TBB in March (Fig. 1a) is still the same as in January (figure omitted). There is a high TBB belt in subtropics in Southern Hemisphere (SH) with its ridgeline at 25°S and core in central Australia, corresponding to the austral subtropical high belt and the Australian high. There is also a high TBB belt in boreal subtropics, with its ridgeline around 15°~20°N and three high centers in West Pacific, SCS and northern BOB, corresponding to the boreal subtropical high belt and three anticyclonic centers, respectively. There is a low TBB between the two high TBB belts, with two low centers in MC and Sumatra. Moreover, a TBB trough extends from the low center over the New Guinea to northern Australia, in correspondence with summer monsoon there.

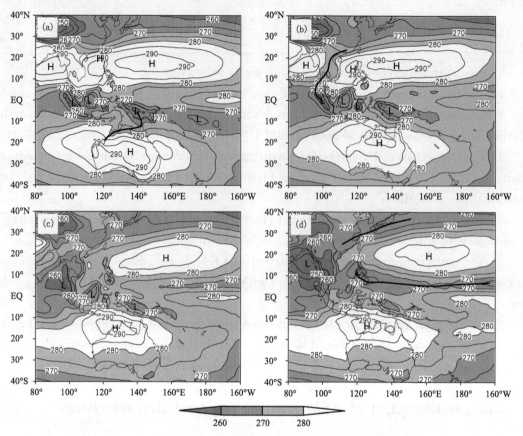

Fig. 1 Distributions of climatological T_{BB} in (a) March, (b) April, (c) May, and (d) June. The thick solid line is the axis of low belt and areas of $T_{BB} < 280$ K are shaded (Units: K)

However, the TBB characteristics change significantly in April (Fig. 1b): (1) the high TBB belt in subtropics in Northern Hemisphere (NH) breaks in ICP (100°~110°E, 10°~20°N) (280 K is the threshold), and an obvious trough extends from the low center in Sumatra to this region, connecting to the low TBB belt in mid-latitude; (2) the high TBB in Australia moves northward notably (the axis moves to 20°S), so the trough from New Guinea to northern Australia disappears. Those changes imply the northward movement of the entire circulation system, i. e., the first northward crush of the tropical convection into the subtropical high belt along ICP, and the disappearance of summer monsoon in Australia along with the seasonal change and the regional response to solar radiation, signifying the starting of the seasonal transition of large-scale circulation in mid and low latitude in Asian-Australian monsoon region.

In May (Fig. 1c), it is noticeable that: (1) the east part of TBB high belt in boreal subtropics retreat eastward rapidly, and convection from Philippine to SCS begins to flare up; (2) the low TBB center in Sumatra moves northwestward, strengthens and expands to occupy ICP, with the high center in northern BOB shifting northwestward rapidly to (80°E, 25°N). These changes indicate that the monsoonal convection has been fully established over

ICP and BOB in May.

In June (Fig. 1d), the high TBB belt in West Pacific continues to retreat eastward along with the obvious jump of the ridge line to around 22°N. There are two low centers in the low belt to the north of the high belt, lying in the lower reaches of Yangtze River and southern Japan respectively. They are in correspondence with Meiyu in China and Baiu in Japan, whose frontal structures and characteristics have been extensively studied[19]. The SCS-Philippine is dominated by a strong low center, showing the summer monsoon has fully established in SCS-West Pacific. The low center over Sumatra in winter has disappeared. A large strong low center appears in northeastern BOB in place of a high center in winter, indicating that summer monsoon has fully established from BOB to India.

To sum up, it is the northward progression of the tropical convection in Sumatra that lead to the break of the high TBB belt in ICP and a series of succeeding events resulting in the onset of ASM.

Figure 2a shows the time-longitude section of TBB along (10°~20°N). It can be seen that the convection illustrated by lower TBB flares up initially over ICP, and then extends

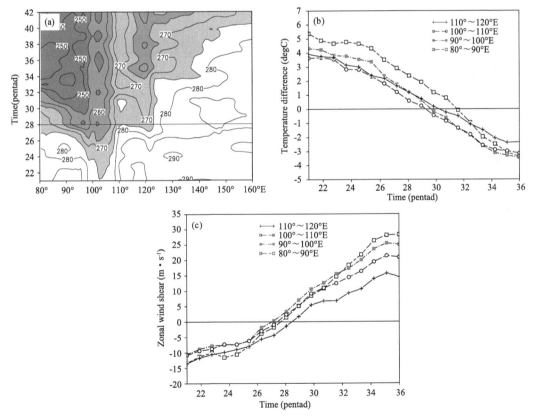

Fig. 2 (a) Time-longitude section of climatological T_{BB} along (10°~20°N). Areas of $T_{BB} \leqslant 275$ K are shaded. (Units: K) (b) Climatological evolution of the meridional temperature difference (5°S~5°N minus 20°~30°N) at 500 hPa. (Units: ℃) (c) Climatological evolution of the zonal wind shear (850 hPa minus 200 hPa) in low-latitude (10°~20°N). (Units: m·s^{-1})

eastward and westward. If $T_{BB}<275$ K (shaded areas) is taken as the signal of active convection, the onset of convection over eastern BOB (east of 90°E) is one pentad earlier than in SCS. In Fig. 2b, the meridional temperature gradient reverses its sign first over ICP (100°~110°E), then over eastern BOB and SCS, and finally over western BOB. Seen from Fig. 2c, the vertical shear of zonal wind reverses its sign almost simultaneously over ICP and BOB, earlier than that over SCS. This course is in agreement with the development of BOB trough and the eastward extension of the southwesterly during the onset of the Southeast Asian summer monsoon. Above results support the viewpoint of the simultaneous onset of summer monsoon over BOB, ICP and SCS in the 27th—28th pentad[20].

According to the above results, it is true that convection first flares up over ICP, which is closely associated with the seasonal migration of tropical convection along the Asian-Australian "land bridge". It will be further discussed in the following section.

3.2 Regional characteristics of the onset of summer monsoon

In order to discuss the characteristics of the summer monsoon onset and their differences in various regions, the time-latitude sections of climatological pentadly T_{BB} along 80°E, 100°E, 120°E and 140°E are plotted (Fig. 3), respectively.

The characteristic of the summer monsoon onset in East India (west of BOB) is roughly shown in Fig. 3a. It can be seen that the low TBB belt propagates northward gradually from the equator in early May, and a low center is formed in the east of India in early and mid June signaling the onset of Indian summer monsoon. Figure 3b illustrates the onset of summer monsoon in the east of TP and Southeast Asia. It can be seen that this region turns to be a low TBB belt from a high TBB belt in early and mid May, indicating that the onset of summer monsoon in Southeast Asia is one month earlier than in India. In addition, there is also a low belt in June-September in the east of TP (around 32°N), separated from the low belt in Southeast Asia (15°N). This means the summer monsoon over the east of TP is relatively independent of that over Southeast Asia, unlike the summer monsoon over the west of TP which is the northward extension of Indian summer monsoon. Figure 3c shows the evolutions of summer monsoons over Indonesia-Northern Australian, SCS and the subtropical region in eastern China. The rapid transition from a high TBB belt to a low TBB belt in mid May over SCS (12°N) denotes the onset of summer monsoon over SCS. Furthermore, there is a low belt over eastern China in June, corresponding to the subtropical summer monsoon rainfall belt in China (Meiyu). The high belt between them corresponds to the West Pacific subtropical high. A strong low center appears around the equator in December-February, and extends southward to the west of northern Australia, denoting to the prevalence of Australian summer monsoon in SH. Figure 3d shows the seasonal cycle of monsoon over Japan, West Pacific and the east of northern Australia. Along this longitude, the subtropical high shows an obvious seasonal cycle of meridional fluctuation. There are low belts on both sides of the subtropical high, with the northern one corresponding to Baiu in Japan and the south-

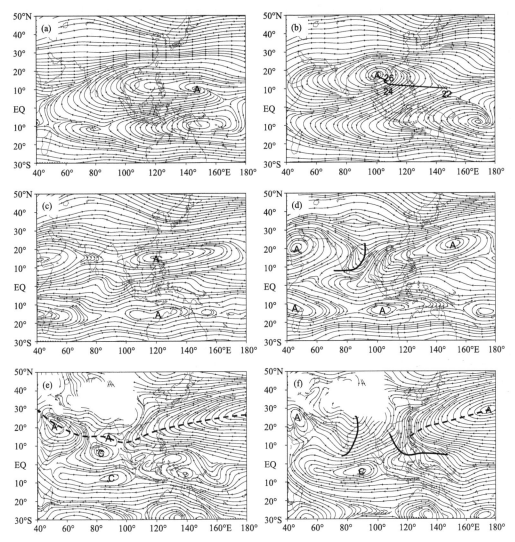

Fig. 3 Time-latitude sections of climatological TBB along
(a) 80°E, (b)100°E, (c)120°E, and (d)140°E. (Units: K)

ern one to the West Pacific summer monsoon. A low belt controls northern Australia in December-March, which embodies the prevailing of northern Australian summer monsoon.

In general, three crucial features can be drawn as follows: (1) The summer monsoons over East India and Southeast Asia are established progressively along with the rapid northward migration of the low belt at equator, reflecting the seasonal cycle of tropical convection. However, the onset of summer monsoon over SCS and West Pacific are quite different from above. The establishment of SCS summer monsoon is simultaneous in a wide range of 20 latitudes, that is, its abrupt behavior is much more obvious than Indian and Southeast Asian summer monsoon. This is directly associated with the rapid eastward retreat of the western Pacific subtropical high belt after its break[21]. (2) Besides the low TBB belt over SCS-West Pacific, there is another low belt to the east of 100°E, i.e. the subtropical monsoon rainfall belt in China-Japan (Meiyu in China and Baiu in Japan). The East Asian monsoon system includes not only tropical summer monsoon, but also subtropical monsoon[17], which is more complicated than Indian monsoon system. (3) The tropical summer monsoon is initially established in ICP, and then advances eastward and westward, respectively. What kind of processes and mechanisms result in such characteristics? Further discussion will be presented in the following sections.

4 Large-scale characteristics of Asian summer monsoon onset

According to Qian et al.[20], east of 90°E, the tropical Asian summer monsoon bursts simultaneously and abruptly first over the whole area in BOB, ICP and SCS in the $27^{th}-28^{th}$ pentad. West of 90°E, the onset of summer monsoon over Indian peninsula and Arabian Sea is relatively late, it first occurs to the south of 10°N and then proceeds northward gradually. Thereby, from a large-scale viewpoint, the onset of the tropical summer monsoon occurs earliest over SCS and its surrounding areas. After that, the large-scale circulation, water vapor transportation and convection over Asian-Australian monsoon regions have changed significantly[22]. We have ever discussed the climatological characteristics of the onset of SCS summer monsoon[23], and further details will be provided below.

4.1 Large-scale characteristics before and after onset

Figure 4 gives the upper and lower circulation fields before the onset of SCS summer monsoon (4th pentad of April) and during the onset (4th pentad of May). The South Asian high (SAH) center at 200 hPa lies on the ocean to the east of Philippine on the 4th pentad of April, then moves rapidly to southern ICP on the 6th pentad of April. It jumps from south of 15°N to the north, and extends westward on the $2^{nd}-4^{th}$ pentad of May with its main body over SCS, ICP, BOB, Indian peninsula and Arabian Sea at 10°~25°N and the northeasterly winds to the east of the high overlaying SCS. After the 5th pentad of May, the SAH moves northwestward to the north of 20°N.

The subtropical high at 500 hPa stretches along 15°N with two high centers (Fig. 4c), with one near Philippine and the other in the west of Arabian Sea. The relatively low value region is in BOB. The pattern begins to change on the 2nd pentad of May (Figure omitted): the subtropical high belt tends to break, and the trough in northern BOB strengthens. On the 4th pentad of May, i.e. the onset pentad of SCS summer monsoon (Fig. 4d), The subtropical high belt breaks completely with its east part retreating eastward and its west part controlling areas west of Arabian Sea, so India is controlled by northwesterly.

At 850 hPa (Fig. 4e, f), the splitting of the subtropical high belt, the rapidly eastward (westward) withdrawal of its east (west) part, and the formation and deepening of the BOB trough are similar to those at 500 hPa, but more complicated and a little earlier. In particular, in the mid April, there is a cyclone in the SH, forming twin cyclones straddling the

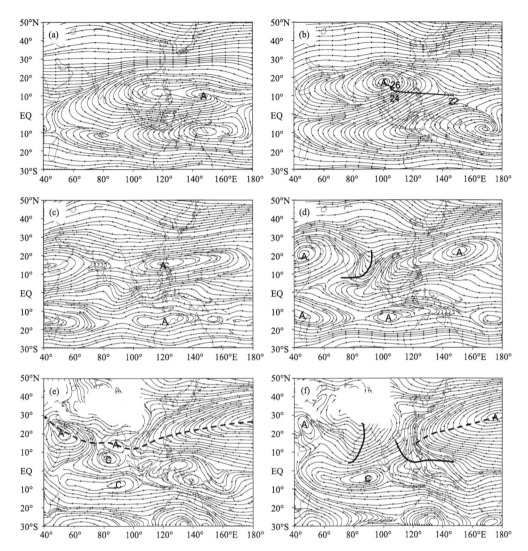

Fig. 4 Climatological circulation fields before and after the onset of South China Sea summer monsoon at (a),(b) 200hPa, (c),(d) 500hPa, and (e),(f) 850hPa. (a), (c) and (e) for the 4th pentad of April; (b), (d) and (f) for the 4th pentad of May. The thick solid line in (b) is the movement of South Asia High center, the numbers on it is the pentad when the center is there. The thick solid lines in other figures are trough lines, and dash line is the ridge line

equator with the Sri Lanka vortex. In between, the equatorial westerly accelerates. On the 4th pentad of May, the Sri Lanka vortex moves northward into the BOB trough with its center of cell disappearing. The equatorial westerly between BOB trough and the cyclone in SH is much stronger, originated from mid-latitude northwesterly over Arabian Sea and cross-equator flow from Somali. The westerly reaches the ICP-SCS region and converges at SCS with the cross-equatorial flow from northern Australia and the turning flow from the southern West Pacific subtropical high. Probably it is the frame of three flows that leads to the complexity of the onset course of SCS summer monsoon, furthermore arises controversies.

The meridional temperature and vertical shear of zonal wind in mid- and low- latitudes in Asia have also converted corresponding to changes of large-scale circulation (see Fig. 2). Therefore, it may be considered that the onset of SCS summer monsoon is not a local phenomenon, but the prominent large-scale event that happened earliest during the seasonal transition of the Asian-Australian monsoon region and the onset course of Asian summer monsoon. In this sense, it is reasonable that the Asian summer monsoon is established earliest over SCS, as claimed by Tao and Chen[3].

The relationship between the movement of SAH and the onset of ASM is widely accepted. Qian et al.[20] specially emphasized the correlation between the position of SAH center and the onset time of ASM. We can see from Fig. 4 that the SAH advances westward rapidly to southern ICP during the 4th — 6th pentad of April, then moves northward, and summer monsoon is established over SCS afterwards. Why does the SAH advance westward rapidly? Why does it move northward along ICP?

Figure 5 shows the variation of SAH at 150 hPa on the 5th and 6th pentad of April and the vertical section of apparent heating ratio on the 5th pentad of April. The SAH disintegrates into two centers on the 5th pentad of April, lying on the east and west of Philippine respectively. On the 6th pentad of April, the center on the east weakens and disappears, while the other one on the west strengthens and moves westward to southern ICP. Hence, the rapid

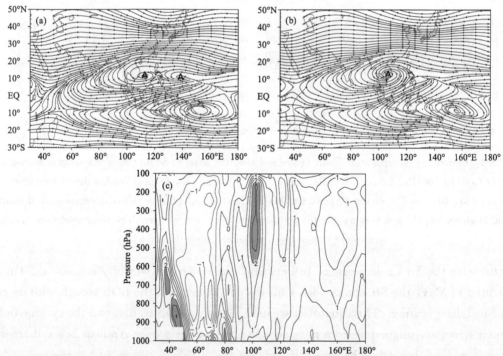

Fig. 5 Climatological circulation fields at 150 hPa on the (a) 5th pentad and (b) 6th pentad of April. (c) Vertical section of heating ratio on the 5th pentad of April averaged in (7.5°~15°N), and areas of heating ratio greater than 3degK · d^{-1} are shaded. (Units: degK · d^{-1})

westward progression of SAH on the $4^{th}-6^{th}$ pentad of April pentad is actually the process of disintegration and reestablishment. It can be seen from Fig. 5c that there is a heating center (the apparent heating ratio is greater than $2\ K\cdot d^{-1}$) at middle and upper troposphere above southern ICP (7.5°~15°N, 105°E) on the 5th pentad of April. The SAH center at 150 hPa is located exactly above the heating center on the next pentad, implying an important role of the upper latent heating in the reestablishment or the rapid westward movement of SAH. It is noticed that the convection over Sumatra strengthens and proceeds northward rapidly in late April and early May. Therefore, we hypothesize with venture that there are some interconnections among the rapid northward progression of convection over Sumatra, the flourish of convection over ICP and the rapid westward movement of SAH.

4.2 Characteristics of the onset course of SCS summer monsoon

Figure 6 shows the composite circulation at 850 hPa with the splitting date of the subtropical high belt[24] as reference point. Two pentads before the belt breaks, there is still a zonal subtropical high belt over southern Asia, and twin cyclones on both sides of the equator near 80°E. On the pentad of the subtropical high belt break (i.e. pentad 0), Sri Lanka vortex migrates into the trough region, leading to the deepening of the BOB trough. The southwesterly in front of the trough arrives at ICP, but the summer monsoon has not yet

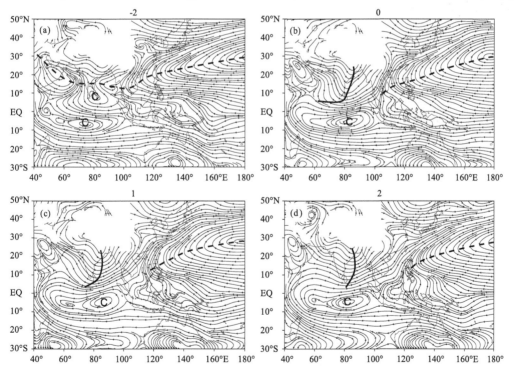

Fig. 6 Composite circulation fields at 850hPa with the splitting dates of the subtropical high belt as reference point. (a) Pentad 2, (b) Pentad 0, (c) Pentad +1, and (d) Pentad +2. The dashed line is the ridge line of the subtropical high, solid line is the trough line

established over SCS where is still controlled by the western Pacific subtropical high (WPSH). As the WPSH withdraws eastward rapidly to the east of Philippine two pentads later, the summer monsoon is fully established over SCS. It is seen that a series of events, such as the appearance of twin cyclones, the northward movement of Sri Lanka vortex, the formation and development of BOB trough, the splitting of zonal subtropical high belt and the rapid eastward withdrawal of WPSH, lead to the onset of summer monsoon over SCS. It is also clear that the onset of SCS summer monsoon is the process southwesterly proceeds from west to east rather than from south to north(at least this is the case in climatology), which helps to explain why SCS summer monsoon is established simultaneously in a wide range of 20 latitudes.

The composite time sections of zonal wind at 850 hPa and OLR with the onset date of SCS summer monsoon as reference point are shown in Fig. 7[23]. Before (after) the onset of monsoon, the SCS is controlled by easterly (westerly) and high (low) OLR, exactly the onset characteristics of summer monsoon. In addition, the westerly and low OLR propagate from the equatorial Indian Ocean (80°E) to ICP and SCS, which are associated with the activating of the BOB trough and the strengthening of the westerly in the equatorial Indian Ocean. It is remarkable that the westerly and low OLR from South China propagate southward during the onset of monsoon, which might be the manifestation of the southward movement of South China stationary front triggering the onset of SCS summer monsoon[24]. Liu et al.[25] stressed that the convective latent heat release may trigger two-dimensional asymmetric Rossby wave train after the onset of BOB summer monsoon. This Rossby wave train is also favorable to the southward movement of South China stationary front. In a word, there are interactions between mid-latitude and low-latitude systems during the onset of SCS summer monsoon, resulting in the abrupt features of SCS summer monsoon onset.

5 Seasonal cycle of the tropical convection along Asian-Australian "land bridge" and its relationship with other events

We have discussed a series of processes such as the northward progression of the tropical convection over Sumatra, the rapid westward movement of SAH to southern ICP, the activating of convection over ICP and the subsequent splitting of subtropical high belt, the establishment and deepening of BOB trough and the onset of SCS summer monsoon. How are they connected with each other? What is the possible mechanism? We'll discuss these issues in the following.

Figure 8 shows the time section of climatological (1980—1997) low TBB and CMAP rainfall along Asian-Australian "land bridge". It is seen that the low TBB over Sumatra (103°E, 3°S) in winter moves to 3°N in April, proceeds northward rapidly in May-June, reaches its northernmost position in July-August, and retreats to near the equator in September. This seasonal cycle of the movement of the tropical convection is the manifestation of

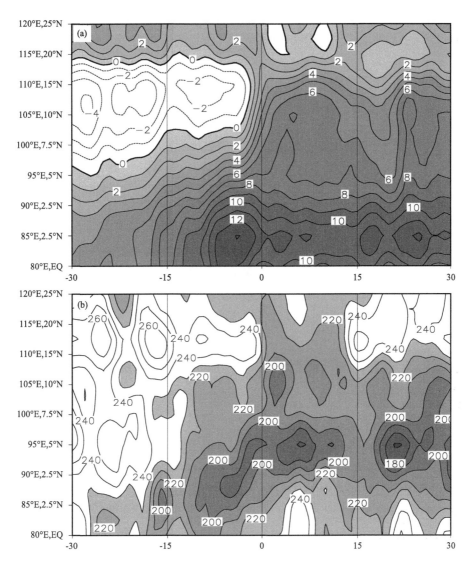

Fig. 7 Time sections of (a) 850 hPa zonal wind (units: m·s^{-1}) and (b) OLR (units: W·m^{-2}) before and after the onset of South China Sea summer monsoon from (80°E, EQ) to (120°E, 25°N). Areas of zonal wind at 850 hPa greater than zero are shaded in (a); areas of OLR≤230 W·m^{-2} are shaded in (b). (adopted from He et al.[23])

the primary driving forcing of monsoon. It is the movement of the convective center along "land bridge" from winter to summer that denotes the onset of summer monsoon over ICP, east of BOB, SCS and India afterwards.

The tropical convection over Sumatra begins to proceed northward rapidly and strengthens in late April and early May when the SAH advances westward rapidly to southern ICP. Thereby, it is reasonable to consider the rapid northward progression of convection over Sumatra as a critical event of the onset of ASM. Time series of the rapid northward shifting of tropical convection were determined to discuss its relationship with the activating of convec-

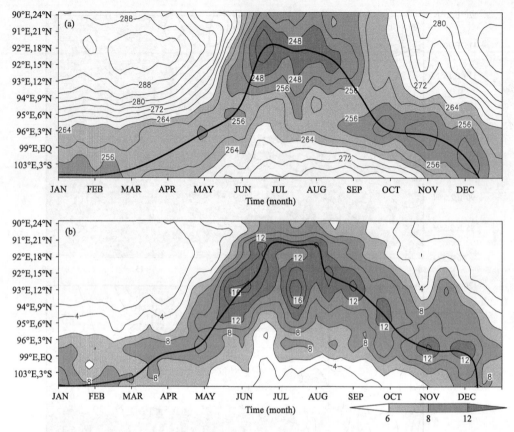

Fig. 8 Climatological (1980—1997) time sections of (a) TBB (units: K) and (b) rainfall (units: mm · d^{-1}) along "Asian-Australian land bridge" from (105°E, 5°S) to (90°E, 25°N). In (a), areas of $T_{BB} \leqslant 265$ K are shaded, the thick solid line is the axis of low TBB. In (b), areas of rainfall more than 6 mm · d^{-1} are shaded, and the thick solid line is the axis of high rainfall(adopted from Wang et al. [24])

tion in ICP, the splitting of subtropical high belt and the onset of SCS summer monsoon[15]. The results indicate the interannual variation of the tropical convection proceeding northward rapidly is in agreement with that of the flourish of convection over ICP with their correlation coefficient is 0.5243. Additionally, in most years, the beginning date of the later tends to lag behind that of the former. Consequently, the rapid northward shifting of convection may have an influence upon the activating of convection in ICP. Furthermore, the activating of convection in ICP, the splitting of subtropical high belt and the onset of SCS summer monsoon are so well correlated that the coefficient of the former two is 0.7263 and the coefficient of the latter two is 0.7206. We can conclude that these events have some inner connections and temporal sequences, and the seasonal movement and rapid northward progression of tropical convection may be an important triggering mechanism.

6 Conclusions and discussions

Wu et al. [4] emphasized the mechanical and thermodynamic effects of TP and pointed out that the summer monsoon is initially established over the east of BOB. Qian et al.[20] stressed that the surface sensible heat flux in mid- and high- latitudes of East Asia in spring and early summer encourages the tropical low belt to break through the high belt first at 110°~120°E, resulting the earliest onset of southwesterly to the east of 90°E. Summarizing their results, it can be found that they all focus on the thermodynamic effect of the earth surface (including TP), which is the most vigorous factor in land-ocean contrast as the secondary forcing. The fact emphasized in this study, however, is the seasonal movement of tropical convection along "land bridge", which leads to the activating of convection in ICP, the splitting of subtropical high belt first in BOB, and the onset of summer monsoon in related regions. Therefore, the disputes on the earliest onset area of ASM are coordinate instead of antagonistic. It can be summarized as the following:

1) Within the range of southern Asia (10°~20°N), the monsoonal convection flares up initially over ICP, then over BOB and SCS, and last over India. The flare-up of convection over ICP is related to the underlying surface sensible heating, but more importantly, it is the result of the northward progression of tropical convection and the arrival of SAH at this place.

2) The latent heating over ICP is favorable for the formation of a cyclonic circulation on the west side of the heating[26], while the sensible heating over Indian Peninsula favors the existence of a cyclonic circulation or trough on the east side[27]. The effective combination of both[18], along with the mechanical and thermodynamic effects of TP and the thermodynamic effects in mid- and high-latitudes in East Asia, leads the subtropical high belt to break first over BOB. Accompanied by the formation and deepening of BOB trough, eastern BOB is controlled by tropical southwesterly in front of the trough, so that the convection flares up, and the summer monsoon is established.

3) In the meantime, the convective latent heat over BOB may trigger an asymmetric Rossby wave train which encourages the overturning of meridional temperature gradient in SCS, thus favors the onset of SCS summer monsoon[27]. The rapid eastward withdrawal of the eastern subtropical high after the belt breaks results directly in the onset of SCS summer monsoon and its sudden characteristics.

4) As India is overlaid by northwesterly in front of the ridge of high (i.e. behind BOB trough) after the high belt breaks, it is unfavorable for the onset of summer monsoon over India. Therefore, the summer monsoon is established last over India[28].

As a matter of fact, the splitting of subtropical high belt, the onset of BOB trough, the eastward withdrawal of subtropical high and the onset of SCS summer monsoon are completed rapidly and accompanied by the seasonal abrupt transition of the large-scale circulation and the water vapor transportation in Asia. Therefore, Qian et al.[20] pointed out that the summer monsoons are established suddenly and simultaneously over BOB, ICP and SCS. However, seen from the source of the onset of summer monsoon, the rapid northward progression of tropical convection over Sumatra in late April and early May is the earliest sign of the onset of ASM. Accordingly, it is reasonable to determine the earliest onset area of ASM in the surrounding areas from southern ICP to northern Sumatra, and Asian-Australian "land bridge" plays an important role during the onset of ASM. Above academic viewpoints can be represented as the following diagrams (Fig. 9).

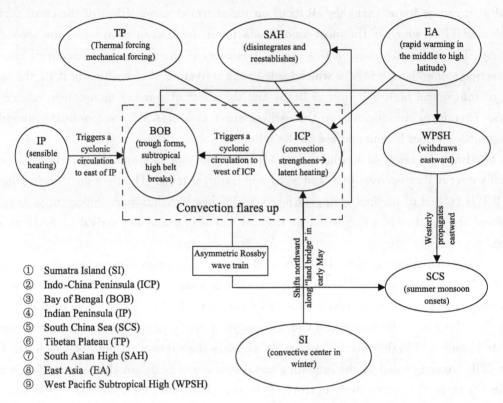

Fig. 9　Schematic illustration of the scenario of the East Asian summer monsoon onset

Acknowledgments. The authors would like to thank CDC, NCEP/NCAR, NOAA and MRI of Japan for providing the data. We also wish to thank CHEN Hua, ZHAN Ruifen, QI Li and LI Fei for their kind assistance. This work was jointly supported by the National Key Program for Developing Basic Sciences under Grant 2006CB403607, and the National Natural Science Foundation of China under Grant 40305005 and 40135020.

References

[1]　Wang B, LinHo. Rainy season of the Asian-Pacific summer monsoon[J]. J Climate, 2002, 15: 386-398.
[2]　Ding Yihui, et al. South China Sea Monsoon Experiment (SCSMEX) and the East-Asian monsoon[J]. Acta Meteor Sinica, 2004, 62(5): 561-586. (in Chinese)
[3]　Tao Shiyan, Chen Longxun. A review of recent research on East summer monsoon in China[C]//Chang C P, Krishramurti T N. Monsoon Meteorology. Oxford: Oxford University Press, 1987: 60-92.
[4]　Wu Guoxiong, Zhang Yongsheng. Tibetan plateau forcing and the timing of the monsoon Onset over South Asia and the South China Sea[J]. Mon Wea Rev, 1998, 126: 913-927.

[5] Li Chongyin, Qu Xin. Characteristics of Atmospheric Circulation Associated with Summer monsoon onset in the South China Sea[C]//Ding Yihui, Li Chongyin, et al. Onset and Evolution of the South China Sea Monsoon and Its Interaction with the Ocean. Beijing: Chinese Meteorological Press, 1999: 200-209.

[6] Zhang Zuqiang, Chan J, Ding Yihui. Characteristics, evolution and mechanisms of the summer monsoon onset over Southeast Asia[J]. Inter J of Climat, 2004, 24: 1461-1482.

[7] Lau K M, Yang S. Climatology and interannual variability of the Southeast Asian summer monsoon[J]. Adv Atmos Sci, 1997, 14: 141-162.

[8] Matsumoto J. Seasonal transition of summer rainy season over Indochina and adjacent monsoon region [J]. Adv Atmos Sci, 1997, 14: 231-245.

[9] Webster P J, Magaña V O, Palmer T N, et al. Monsoons: processes, predictability, and the prospects for prediction[J]. J Geo Res, 1998, 103: 14451-14510.

[10] Wang B, Fan Z. Choice of South Asian summer monsoon indices[J]. Bull Amer Meteor Sci, 1999, 80: 629-638.

[11] Lu Junmei, Zhang Qingyun, Tao Shiyan, et al. The onset and advance of the Asian summer monsoon[J]. Chinese Sci Bull, 2006, 51(1): 80-88.

[12] He Jinhai, Zhu Qiangen, Murakami M. T_{BB} data-revealed features of Asian-Australian monsoon seasonal transition and Asian summer monsoon establishment[J]. J Trop Meteor, 1996, 12(1): 34-42. (in Chinese)

[13] Zeng Qingcun, Li Jianping. Interaction between the northern and southern hemispheric atmospheres and the essence of monsoon[J]. Chinese J Atmos Sci, 2002, 26(4): 433-448. (in Chinese)

[14] Chang C P, Harr P A, McBride J, et al. Maritime continent monsoon: annual cycle and boreal winter variability[C]//Chang C P. East Asian Monsoon. World Scientific Publishing Co Pte Ltd, 2004: 107-152.

[15] Wang Lijuan, He Jinhai, Guan Zhaoyong. Characteristic of convective activities over Asian-Australian "land bridge" areas and its possible factors[J]. Acta Meteor Sinica, 2004, 18: 441-454.

[16] Murakami T, Chen Longxun, Xie An. Relationship among seasonal cycles, low- frequency oscillations and transient disturbances as revealed from outgoing long wave radiation data[J]. Mon Wea Rev, 1986, 114: 1456-1465.

[17] Zhu Qiangen, He Jinhai, Wang Panxing. A study of circulation differences between East Asian and Indian summer monsoon with their interaction[J]. Adv Atmos Sci, 1986, 3: 466-477.

[18] He Jinhai, Yu Jingjing, Shen Xinyong, et al. Research on mechanism and variability of East Asian monsoon[J]. J Trop Meteor, 2004, 20(5): 449-459. (in Chinese)

[19] Gao Shouting, Zhou Yushu, Lei Ting. Structural features of the Meiyu front system[J]. Acta Meteor Sinica, 2002, 16(2): 195-204.

[20] Qian Y, Jiang J, Zhang Y, et al. The earliest onset area of the tropical Asian summer monsoon and its mechanisms[J]. Acta Meteor Sinica, 2004, 62: 129-139. (in Chinese)

[21] He Jinhai, Wen Min, Shi Xiaohui, et al. Splitting and eastward withdrawal of the subtropical high belt during the onset of the South China Sea summer monsoon and their possible mechanism[J]. J Nanjing University (Natural Sciences), 2002, 38(3): 318-330. (in Chinese)

[22] Ding Yihui. Seasonal march of the East-Asian summer monsoon[C]// Chang C P. East Asian Monsoon. World Scientific Publishing Co Pte Ltd, 2004: 3-53.

[23] He Jinhai, Xu Haiming, Wang Lijuan. Climatic features of SCS summer monsoon onset and its possible mechanism[J]. Acta Meteor Sinica, 2003, 17(suppl): 19-34.

[24] Chang C P, Chen G T J. Tropical circulations associated with southwest monsoon onset and westerly surges over the South China Sea[J]. Mon Wea Rev, 1995, 123: 3254-3267.

[25] Liu Yimin, Chan J, Mao Jiangyu, et al. The Role of Bay of Bengal convection in the onset of the 1998 South China Sea summer monsoon[J]. Mon Wea Rev, 2002, 130: 2731-2744.

[26] Wen Min, He Jinhai, Xiao Ziniu. Impact of the convection over the Indochina peninsula on the onset of SCS summer monsoon[J]. Chinese J Atmos Sci, 2004, 28(6): 864-875. (in Chinese)

[27] Xu Haiming, He Jinhai, Wen Min, et al. A numerical study of effects of the Indo-China peninsula on the establishment and maintenance of the South China Sea summer monsoon[J]. Chinese J Atmos Sci, 2002, 26(3): 330-342. (in Chinese)

[28] He Jinhai, Xu Haiming, Zhou Bing, et al. Large scale features of SCS summer monsoon onset and its possible mechanism[J]. Climatic Environ Res, 2000, 5(4): 333-344. (in Chinese)

关于东亚副热带季风和热带季风的再认识[*]

何金海[1]　祁莉[1]　韦晋[1]　池艳珍[1,2]

(1. 江苏省气象灾害重点实验室,南京信息工程大学,南京 210044；
2. 福建省气候中心,福州 350001)

摘要：利用 NCEP/NCAR 再分析数据集和 CMAP(Climate Prediction Center Merged Analysis of Precipitation)降水资料,分析了东亚副热带夏季风与热带夏季风的区别和联系,以及两者相互作用问题,深入讨论了东亚副热带季风的本质。分析发现东亚副热带夏季风建立早于热带夏季风,于 3 月中旬已经开始建立。两者是相互独立的两个过程,前者并非后者向北推进的结果；相反,前者建立后的突然南压有利于后者的爆发。副热带夏季风为渐进式建立,但撤退迅速；热带夏季风爆发突然,但撤退缓慢。副热带夏季风的建立以偏南风的建立为特征,而热带夏季风的建立以偏东风向偏西风转变为特征。热带夏季风的建立时间取决于经向海陆热力差异转向,而东亚副热带夏季风则更依赖于纬向海陆热力差异的逆转。亚洲大陆(含青藏高原)与西太平洋之间的纬向海陆热力差异的季节逆转无论对东亚副热带夏季风还是热带夏季风均有重要作用。

关键词：东亚副热带夏季风；热带夏季风；青藏高原；纬向海陆热力差异

1　引　言

东亚季风既包含热带季风又包含副热带季风,它们之间存在相互作用,并直接影响我国大范围的旱涝。Zhu 等[1]明确提出东亚季风系统分为南海热带季风与副热带季风。陶诗言先生在给《东亚季风》一书[2]作序时写到"东亚季风与印度季风主要不同之一是东亚季风包括有热带季风和副热带季风的两个子系统,而印度季风是热带季风性质"。李建平和曾庆存[3-4]构造动态标准化季节变率指数,划分了全球季风系统的地理分布,指出东亚季风从南至北由"热带季风""副热带季风"和"温寒带季风"组成。不少学者[5-7]通过建立季风指数来讨论夏季东亚副热带季风与东亚降水的关系。但东亚副热带季风本质及其与热带季风的相互关系仍是一个尚未引起重视的科学问题。究竟什么是副热带季风？或者说副热带季风的本质是什么？人们通常有两方面的含混：一是基于地域的考虑,认为东亚副热带地区盛行的季风即称"东亚副热带季风",而不问究竟是什么机制形成了副热带季风。二是将东亚地区热带季风的向北延伸看成副热带季风,同样是忽视了副热带季风自身存在的原因。

东亚地域纵跨高、中、低纬,且广袤的大陆濒临最宽广的太平洋,海陆热力差异季节变化的

---

[*] 本文发表于《大气科学》,2007 年第 31 卷第 6 期,1257-1265.

影响十分突出。欧亚大陆(包括青藏高原)对亚洲夏季风的爆发和强度有重要影响。吴国雄[8-9]和梁潇云[10]等提出高原的"热力滑轮"对亚洲夏季风的爆发地点有"锚定"作用。由于海陆热力差异是形成季风的关键因子,大多季风的研究[2-4,11-17]强调东亚地区经向海陆(包括青藏高原)热力差异是形成热带季风的重要推动力。然而,在副热带地区,高原作为抬高了的热源(或冷源),增强了纬向海陆(亚洲大陆－太平洋)热力差异,使其季节转换更具敏感性和特殊性。张艳等[18]和钱永甫等[19]首先讨论了纬向海陆热力差异对夏季风的影响,但没有与副热带季风相联系。祁莉等[20]首次讨论了东亚副热带季风环流与纬向海陆热力差异的关系,得到了一些新的认识。然而很多问题仍然未知,例如,纬向海陆热力差异的转向对东亚副热带夏季风的建立十分关键,其在热带夏季风的建立过程中起何作用?东亚副热带夏季风与热带夏季风的建立特征有什么不同?两者如何相互作用?本文拟就上述相关问题展开讨论,进一步探讨东亚副热带夏季风的本质。

2 资料与方法

本文使用 1950—2000 年 NCEP/NCAR(National Centers for Environmental Prediction/National Center for Atmospheric Research)再分析数据集[21]及 1979—2000 年 CMAP(Climate Prediction Center Merged Analysis of Precipitation)[22]降水资料来分析讨论东亚副热带夏季风和热带夏季风的区别和联系。

3 纬向海陆热力差异季节逆转的特征

图 1 给出了气候平均场上北半球中低纬度各纬带上 500 hPa 温度纬向偏差(定义为各经度上的温度与 80°~150°E 平均温度之差,引自文献[20])逐候的演变。可见,温度纬向偏差的量值随纬度从南至北增大;除 35°N 外各纬度上其东西向差异的季节转换特征明显:西太平洋地区冬春为暖区,春末夏初转为冷区;而西侧东亚大陆上冬春为冷区,春末夏初转为暖区;即冬春季节为西冷东暖,夏秋季节则转变为西暖东冷。但各纬带发生季节逆转的时间及特征有所不同。

15°N 纬带上,温度纬向偏差以南海经度(115°E,图 1 中竖线所示)为界,东西向截然相反;24 候左右南海以东(以西)几乎同时由暖转冷(由冷转暖),各经度上几乎同步发生季节逆转,具有爆发性和同时性。20°N 上也基本以南海经度为分界线;南海以东大洋上 24 候后转为冷区,以西地区最早增温的为中南半岛,约 18 候,后逐渐向西推进,印度半岛东部 30 候以后才转为暖区。25°N 纬带的特征与 20°N 相似,所不同的是其分界线东移至 120°E(我国东部海岸线附近);120°E 以西增暖最早的同样是大陆东侧(华南地区),14 候后已转为暖区。相比较而言,30°N 上青藏高原热源的加热作用造成该纬带上温度纬向偏差量值最大,此处季节逆转也最早;东海以东的大洋上 24 候前就已经转冷,而高原东侧(90°E 以东)东亚大陆上空各处 12 候均转为暖区。35°N 纬带上西太平洋上 50 候前均为冷区,50 候后转暖,而 110°E 以西大陆上基本没有季节变化,常年均为暖区。

由图 1 还可看出,东亚各纬度基本以 120°E 附近为冷暖区分界线,因此选取 120°~150°E 代表西太平洋地区,80°~120°E 代表东亚大陆,求得各纬带上的纬向海陆热力差异如图 2 所示。35°N海陆热力差异基本维持负值,仅10—11月转为正值,季节变化不明显,Zhao等[23]提出该地区没有纬向风的季节逆转;35°N以南各个纬带海陆热力差异均有显著的季节变化,春末夏初由正转负,表明东亚大陆和西太平洋地区由东暖西冷转为东冷西暖,形成向西的温度梯度;秋季则由负转正,温度梯度转向东。图2最显著的特征为各纬带相比转向最早的在副热带地区(30°N),3月末4月初已经转负;热带地区各纬度24候同时转向,具有同时性和爆发性。热带和副热带地区相差一个月,这一特点造成了副热带季风和热带季风建立时间的差异。

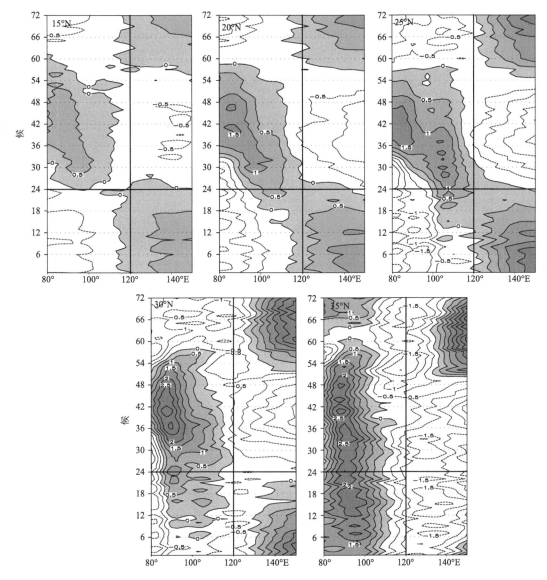

图1　500 hPa各纬带温度纬向偏差(单位:K)随时间逐候的演变,阴影为暖区
(引自文献[20])

根据图1,由春入夏西太平洋地区各纬度均于24候前后转冷,但东亚大陆及孟加拉湾地区转暖时间差异较大。因此,纬向海陆热力差异季节转向的早晚可能更多地取决于东亚大陆及其南侧地区增暖的早晚。钱永甫等[19]指出由春入夏高原东侧(90°E以东)我国东部大陆地

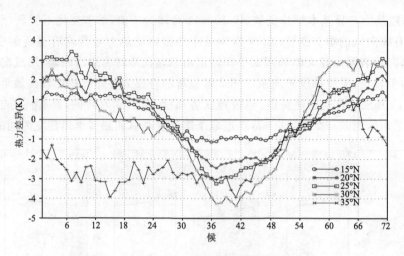

图2　500 hPa纬向海陆热力差异逐候的演变

区迅速增温,其中30°N纬带为增温和降压最快的地区。这使得纬向海陆热力差异最早于高原所在纬度的副热带地区(30°N)发生逆转。另一方面,3月中旬高原热源开始加强[2],15候左右由冷源转为热源(图3),作为抬高的热源直接加热对流层中层大气,加剧了该处的纬向海陆热力差异,使其强度最强。因此,副热带地区纬向海陆热力差异的季节转换不但具有其敏感性、独特性,而且具有超前性。

不但如此,东亚副热带纬向海陆热力差异的季节逆转还超前于东亚经向海陆热力差异的季节变化(图略)[20]。经向海陆热力差异季节逆转最早发生在中南半岛地区(105°E),约为27候由南暖北冷转为南冷北暖。此后季节逆转同时向东西两侧推进,南海(115°E)与孟加拉湾(90°E)同时于30候发生逆转(对流层低层更早些,约28候),印度次大陆(70°~80°E)及130°E附近地区随后于32候逆转,最后为西北太平洋地区(150°E)。整个经向海陆热力差异的季节逆转过程与东亚热带夏季风的建立与推进的进程基本一致。

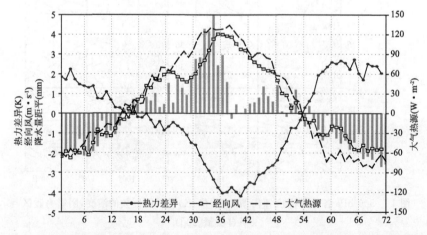

图3　气候平均场上副热带地区(27.5°~32.5°N平均)500 hPa纬向海陆热力差异,副热带地区
(27.5°~32.5°N,110°~140°E平均)850 hPa经向风、降水量距平(去除年平均,柱形图)
及青藏高原(27.5°~37.5°N,80°~100°E)整层积分大气热源的逐候演变

4 对流层低层环流

东亚经度平均(110°～130°E)的 850 hPa 流场、涡度场及降水距平(去除年平均值)如图 4 所示。降水正距平最早出现在 25°～32.5°N 的副热带地区,约 15 候(3 月中旬),与该区域纬向海陆热力差异季节逆转的时间一致(图 2),该地区相应的 850 hPa 风场由西北风转为西南风,涡度也转为气旋性涡度。祁莉等[20]提出,基于海陆热力差异是形成季风的基本推动力的观点,上述副热带地区从偏北风转为偏南风的过程表明东亚副热带夏季风已经建立(图 3)。因此该雨带实为副热带季风雨带,相应正涡度带对应于副热带季风槽。与此同时,25°N 以南的热带地区均为较弱偏东风,涡度为负,降水为负距平,即热带夏季风仍未爆发,明显晚于东亚副热带夏季风。随后副热带季风雨带继续维持,偏南风也有所加强;热带地区风场基本没有变化。

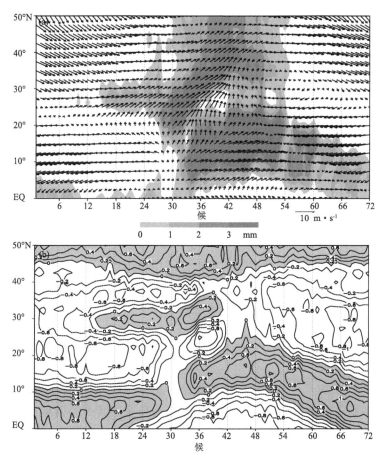

图 4 气候平均东亚地区(110°～130°E 平均)850 hPa 风场和降水距平(去除年平均值,阴影)(a)及 850 hPa 涡度(b,单位:10^{-5} s^{-2},阴影为气旋性涡度)的逐候演变

随着热带地区纬向海陆热力差异于 24 候转向(图 2),副高南侧转向气流加强,热带地区南风分量也开始加强;副热带季风雨带加强并迅速地向南扩展,相应的异常南风分量也显著南压(去除年平均值,图略)。陈隆勋[24]和江志红[25]等也发现南海夏季风爆发前副热带雨带有

向南传播的特征,但其突然南传之原因还未有人给出合理解释。27 候副热带季风雨带到达 20°N 以南地区,该地区已由较弱南风控制;28 候 10°~20°N 的南海地区降水同时转为正距平,对流层低层由偏东风转为偏西风,表明南海夏季风爆发,具有爆发性。

随后,副热带夏季风雨带及相应正涡度带迅速加强北抬,6 月到达长江中下游,正涡度达到最强,表明此时进入梅雨期。7 月底 8 月初,副热带季风雨带抵达华北,华北雨季开始,正涡度带减弱转为负值,25°~40°N 地区均由较强的负涡度带控制,相应降水带和偏南风也开始减弱,表明副热带夏季风开始撤退。随着副热带纬向海陆热力差异于 54 候由负转正(图 2),25°N 以北的广大副热带地区的降水同时由正距平转为负距平,相应的偏南风也减弱转为偏北风。这意味着东亚副热带夏季风已经完全撤退,秋高气爽天气来临。与上述特征相比较,在 20°N 以南的热带地区,随着南海夏季风于 28 候建立,热带地区出现降水正距平,低层 850 hPa 风场由偏东风转为偏西风。30 候后热带季风雨带发展加强,偏西风大大加强,涡度场上也出现了相应的正涡度带。值得注意的是,此时东亚地区从南至北显著地存在两条正涡度带,两者由一较强的负涡度带隔离;相应的降水场上也出现两支降水大值带,分别对应于副热带夏季风雨带和热带夏季风雨带,这表明副热带夏季风与热带夏季风是相互独立的,前者并非是后者向北延伸的结果。相反,热带夏季风(南海夏季风)的建立与副热带夏季风雨带的突然南压密切相关。1995 年 Zhang 等[26]就曾提出华南静止锋的南压对南海夏季风的建立有促发作用。54 候后,热带地区各纬度上纬向海陆热力差异相继由负转正(图 2),热带夏季风雨带也明显减弱,并缓慢南退,相应的对流层低层风场也由偏西风转为偏东风,热带夏季风开始撤退。但其撤退是缓慢的,10 月后雨带已撤至较南纬度。

因此,(1)东亚副热带夏季风建立早于热带夏季风。(2)东亚副热带夏季风与热带夏季风是两个独立的过程,这在涡度场上表现得更为明显些(图 4b)。前者不是后者向北推进的结果。相反,热带夏季风的爆发可能与副热带夏季风建立后突然南压密切相关。(3)副热带夏季风为渐进式建立,但撤退迅速;热带夏季风爆发突然,但撤退缓慢。这与王黎娟等[27]的分析结果一致。(4)副热带夏季风的建立以偏南风的建立为特征,而热带夏季风主要表现为偏东风向偏西风的转变。(5)南海夏季风爆发于 28 候,恰好对应于南海地区经向海陆热力差异转向[20]。可见,热带夏季风的建立时间取决于经向海陆热力差异转向的早晚,而东亚副热带夏季风则更依赖于纬向海陆热力差异的逆转。

5 对流层高层环流

南海夏季风建立的大尺度特征在对流层高层表现为,南亚高压自菲律宾以东迅速向西移动,在到达中南半岛南部后沿中南半岛西北行最终上高原。其中的西移过程发生在 24 候前后。事实上,南亚高压中心的位置变化并非简单的西进,而是一个重建和替代的过程[28]。

多年平均的 150 hPa 流场上(图 5),18 候南亚高压中心位于 160°E 以东的洋面上,21 候高压缓慢西扩,与之相对应的南半球的高压也相应地西扩。22 候南亚高压中心仍在 150°E 附近。有意思的是在 23 候南亚高压在 120°E 附近发生分裂,生成两个反气旋环流中心,分别位于 105°E 以东(中南半岛东岸)和 135°E;到 24 候南亚高压偏西中心继续西移,移上中南半岛南部,23 候时位于 135°E 的南亚高压中心强度大大减弱,已基本消失不见。随后南亚高压沿中南半岛逐步西北行,26 候移至 15°N 以北,最终登上高原。

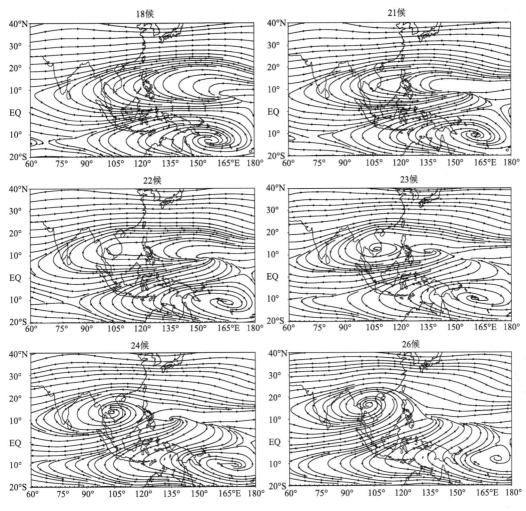

图 5 气候平均场 150 hPa 流场的演变

可见南海夏季风建立前(22—24 候)存在南亚高压分裂为两个中心,随后西侧中心增强,东侧中心消亡的这样一个过程,可以称之为南亚高压的分裂重建过程。这个过程是流场的重新调整,整体表现为南亚高压中心迅速由菲律宾以东移动到中南半岛南部,其移动具有突发性。

胡鹏等[28]提出南亚高压中心分裂重建与苏门答腊地区对流北推有密切联系,深对流引发的凝结潜热释放可能是造成南亚高压重建的重要原因。值得注意的是 24 候热带地区纬向海陆热力差异发生了季节性逆转(图 2),15°N 南海以东/以西几乎同时由暖转冷/由冷转暖(图 1),其爆发性和同时性显然与南亚高压的分裂重建和迅速西移密切相关。但两者孰因孰果? 是纬向海陆热力差异的逆转触发了南亚高压的重建西移,还是南亚高压的西移加速了纬向海陆热力差异的季节逆转? 抑或是其中存在某种正反馈机制? 这需要我们进一步的深入分析。

6 结果和讨论

东亚季风既包含热带季风又包含副热带季风,它们之间存在相互作用,并直接影响我国大范围的旱涝。然而副热带季风由于仍存在很多未知而一直颇受争议,尤其是其本质问题及其与热带季风的相互作用是一个尚未引起重视的科学问题。本文立足于气候平均场,以文献[20]先前对纬向海陆热力差异季节逆转与东亚副热带夏季风环流关系的分析为基础,通过进一步分析东亚副热带夏季风与热带夏季风的区别和联系以及两者的相互作用,深入讨论东亚副热带季风的本质。

分析发现,东亚副热带夏季风建立早于热带夏季风,3月中旬已经建立。两者是相互独立的两个过程,前者并非后者向北推进的结果;相反,前者建立后的南压促发了后者的爆发。副热带夏季风为渐进式建立,但撤退迅速;热带夏季风爆发突然,但撤退缓慢。副热带夏季风的建立以偏南风的建立为特征,而热带夏季风主要表现为偏东风向偏西风的突然转变。

南海夏季风爆发于28候,恰好对应于南海地区经向海陆热力差异转向[20]。可见热带夏季风的建立时间取决于经向海陆热力差异转向的早晚,而东亚副热带夏季风则更依赖于纬向海陆热力差异的逆转。但无论是副热带夏季风还是热带夏季风,纬向海陆热力的重要影响是不言而喻的。东亚纬向海陆热力差异的影响并非是局地性的。在中纬度地区,其影响主要表现在对流层低层,偏南风的增强和对流降水的出现预示着东亚副热带夏季风的建立(图6)。而在热带地区,其影响主要体现在对流层高层,南亚高压分裂重建,西跃上大陆(图6),这一特征已经成为南海夏季风建立前的重要征兆[29]。此时还伴有副热带夏季风的突然南压。因此,

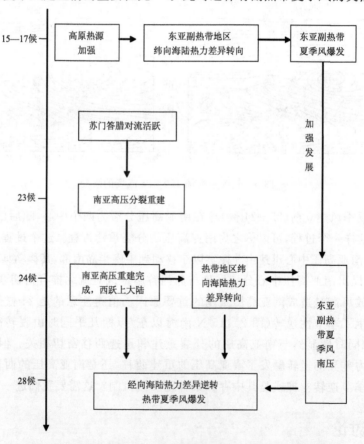

图6 东亚副热带夏季风与热带夏季风爆发过程中的关键事件

亚洲大陆(含青藏高原)与西太平洋之间的纬向海陆热力差异对东亚夏季风的影响贯穿于整个对流层,更涉及两支夏季风——副热带夏季风和热带夏季风的爆发和传播进程。吴国雄等[30,31]通过数值模拟发现高原的存在加强了欧亚大陆东南部的西南气流,使东亚夏季风向北发展至50°N。结合本文的研究结果,这可能表明东亚季风的向北推进正是东亚热带季风和东亚副热带季风相互配合,相互作用的结果。

同时,根据图6我们发现,24候前后东亚地区中低纬度纬向海陆热力差异均已发生季节逆转,东亚副热带夏季风建立并加强南压,南亚高压分裂重建并抵达中南半岛南端。大气环流从对流层低层至高层均发生了调整。这可能意味着东亚地区夏季环流开始建立。

参考文献

[1] Zhu Qiangen, He Jinhai, Wang Panxing. A study of circulation differences between East-Asian and Indian summer monsoons with their interaction[J]. Adv Atmos Sci,1986,3(4):466-477.

[2] 陈隆勋,朱乾根,罗会邦,等. 东亚季风[M]. 北京:气象出版社,1991.

[3] 曾庆存,李建平. 南北半球相互作用和季风的本质[J]. 大气科学,2002,26(4):433-448.

[4] Li Jianping, Zeng Qingcun. A new monsoon index and the geographical distribution of the global monsoons[J]. Adv Atmos Sci,2003,20(2):299-302.

[5] 张庆云,陶诗言. 夏季东亚热带和副热带季风与中国东部汛期降水[J]. 应用气象学报,1998,9:17-23.

[6] 周兵,何金海,吴国雄,等. 东亚副热带夏季风特征及其指数的建立[J]. 大气科学,2003,27(1):123-135.

[7] 赵平,周自江. 东亚副热带夏季风指数及其与降水的关系[J]. 气象学报,2005,63(6):933-941.

[8] 吴国雄,张永生. 青藏高原的热力和机械强迫作用以及亚洲季风的爆发[J]. 大气科学,1998,22(6):825-838.

[9] 梁潇云,刘屹岷,吴国雄. 青藏高原对亚洲夏季风爆发位置及强度的影响[J]. 气象学报,2005,63(5):799-805.

[10] 吴国雄,王军,刘新,等. 欧亚地形对不同季节大气环流影响的数值模拟研究[J]. 气象学报,2005,63(5):603-612.

[11] Tao Shiyan, Chen Longxun. A review of recent research on the East Asian summer monsoon in China [C]// Chang C-P, Krishnamurti T N. Monsoon Meteorology. Oxford:Oxford University Press,1987:60-92.

[12] 何金海,丁一汇,高辉,等. 南海夏季风建立日期的确定与季风指数[M]. 北京:气象出版社,2001.

[13] Qian Yongfu, Wang Shiyu, Shao Hui. A possible mechanism effecting the earlier onset of South westerly monsoon in the South China Sea compared to the Indian monsoon[J]. Metero & Atmos Physics,2001,76(3-4):237-250.

[14] Huang Ronghui, Zhou Liantong, Chen Wen. The progress of recent studies on the variability of the East Asian monsoon and their causes[J]. Adv Atmos Sci,2003,20(1):55-69.

[15] He Haiyan, Sui Chung-Hsiung, Jian Maoqiu, et al. The evolution of tropospheric temperature field and its relationship with the onset of Asian summer monsoon[J]. J Meteor Soc Japan,2003,81(5):1201-1223.

[16] Zhang Zuqiang, Chan J C L, Ding Yihui. Characteristics, evolution and mechanisms of the summer monsoon onset over Southeast Asia[J]. Int J Climatol,2004,24:1461-1482.

[17] 丁一汇,李崇银,何金海,等. 南海季风试验与东亚夏季风[J]. 气象学报,2004,62(5):561-586.

[18] 张艳,钱永甫. 青藏高原地面热源对亚洲季风爆发的热力影响[J]. 南京气象学院学报,2002,25(3):298-306.

[19] 钱永甫,江静,张艳,等.亚洲热带夏季风的首发地区和机理研究[J].气象学报,2004,62(2):129-139.
[20] 祁莉,何金海,张祖强.纬向海陆热力差异的季节转换与东亚副热带季风环流[J].科学通报,2007,52(24):2895-2899.
[21] Kalnay E, et al. The NCEP/NCAR 40-Year Reanalysis Project[J]. Bull Amer Meteor Soc,1996,77:437-471.
[22] Xie P, Arkin P A. Global Precipitation: A 17-Year Monthly Analysis Based on Gauge Observations, Satellite Estimates and Numerical Model Outputs[J]. Bull Amer Meteor Soc,1997,78:2539-2558.
[23] Zhao Ping, Zhang Renhe, Liu Jiping, et al. Onset of southwesterly wind over eastern China and associated atmospheric circulation and rainfall[J]. Clim Dyn,2006,DOI:10.1007/s00382-006-0212-y.
[24] 陈隆勋,李薇,赵平,等.东亚地区夏季风爆发过程[J].气候与环境研究,2000,5(4):345-355.
[25] 江志红,何金海,李建平,等.东亚夏季风推进进程的气候特征及其年代际变化[J].地理学报,2006,61(7):675-686.
[26] Chang C P, Chen G T J. Tropical circulation associated with southwest monsoon onset and westerly surges over the South China Sea[J]. Mon Wea Rev,1995,123:3254-3267.
[27] 王黎娟,何金海,管兆勇.东亚副热带夏季风槽的气候特征及其与南海夏季风槽的比较[J].气象学报,2006,64(5):583-593.
[28] 胡鹏.南海夏季风建立和南亚高压重建及其数值模拟[D].南京:南京信息工程大学,2005.
[29] He Jinhai, Wen Min, Wang Lijuan, et al. Characteristics of the onset of the Asian summer monsoon and the importance of Asian-Australian "Land Bridge"[J]. Adv Atmos Sci,2006,23(6):951-963.
[30] Wu Guoxiong, Mao Jiangyu, Duan Anmin, et al. Recent progress in the study on the impacts of Tibetan Plateau on Asian summer climate[J]. Acta Meteorologica Sinica,2004,62(5):528-540.
[31] 吴国雄,刘屹岷,刘新,等.青藏高原加热如何影响亚洲夏季的气候格局[J].大气科学,2005,29(1):47-56.

Seasonal Transition Features of Large-Scale Moisture Transport in the Asian-Australian Monsoon Region[*]

He Jinhai[1] Sun Chenghu[2] Liu Yunyun[1]
Jun Matsumoto[3] Li Weijing[2]

(1. Department of Atmospheric Sciences, NUIST, Nanjing, China;
2. LCS, National Climate Center, CMA, Beijing, China; 3. Department of Earth and Planetary Science, Graduate School of Science, The University of Tokyo, Tokyo, Japan)

Abstract: Using NCEP/NCAR reanalysis data for the period of 1957-2001, the climatologically seasonal transition features of large-scale vertically integrated moisture transport (VIMT) in the Asian-Australian monsoon region have been investigated in this paper. The basic features for the seasonal transition of VIMT from winter to summer are establishment of the summertime "great moisture river" pattern (named the GMR pattern) and its eastward expansion, associated with a series of climatological events occurred in some "key periods", which include the occurrence of notably southerly VIMT over the Indochina Peninsular in mid-March, the activity of low VIMT vortex around the Sri Lanka in late April, and the onset of the South China Sea summer monsoon in mid-May, etc. However, during the transition from summer to winter the characteristics are mainly exhibited by establishment of the easterly VIMT belt located to the tropical area, accompanying with some events occurred in "key periods". Our further analyses disclose a great difference between the Indian and East Asian monsoon region when viewed from the meridional migration of the westerly VIMT during the seasonal change process, according to which the Asian monsoon region can be easily divided into two parts along the west of the Indochina Peninsula and it may also denote different forming mechanisms between those two regions.

Key words: seasonal transition; moisture transport; Asian-Australian monsoon

1 Introduction

The Asian-Australian Monsoon (A-AM) is an integral component of the earth's climate system, involving complex interactions of the atmosphere, the hydrosphere and the bio-

---

[*] 本文发表于《Advances in Atmospheric Sciences》,2007 年第 24 卷第 1 期,1-14.

sphere. It covers a vast domain: 40°~160°E and 30°S~45°N[1-6], which is home for more than half of the world's population. The monsoon transports abundant water vapor from the Pacific and Indian Ocean to the A-AM region, which greatly affects rainfall and water budget in the region[7-11], which is crucial to agricultural production and economy of the region. So research of moisture transport has both scientific and social value.

The seasonal transition of the Earth climate is the result of solar forcing, i. e. driven by the "march" of the sun back and forth across the equator. In the monsoon domain, the seasonal march always encompasses the alternation between wet and dry phases, associated closely with transition of two monsoon's sectors: summer and winter monsoon (in this paper the seasonal transition is always depicted from the Northern Hemisphere perspective, unless a specific term is added). Thus, the interests in rather complex migration process of monsoon motivate the research on seasonal transition of the A-AM region. So far, much work has been contributed to it by a great variety of ways, such as: precipitation or satellite-observed convection, circulation criteria and dynamical or thermodynamic methods[12-22]. The investigators have shown that onset and retreat of monsoon exhibit several stages and significant transitions in the large-scale atmospheric and oceanic circulations[23-28], owning to large-scale interactions between surface heating and atmospheric dynamic, thermal, and hydrologic processes[26-27,29-32]. However, these studies have been largely concerned with summer phase, rather than the whole year. Therefore, it is worth comprehensively investigating the seasonal transition over the great A-AM domain by using a more reliable dataset containing enough samples.

To diagnose the seasonal transition of monsoon, usually there are several variables to be chosen such as: the rainfall or satellite-observed convection and circulation criteria, etc. In this paper, we choose the moisture transport to diagnose it, due mainly to the close relationship between monsoon precipitation and moisture transport as emphasized by some researches[17,33-46]. Besides, there are other motivations to base our diagnostic investigations on the moisture transport. Firstly, hydrologic fields, such as the vertically integrated moisture transport (VIMT), are linked directly to the lateral and transverse heating gradients, which are the basic monsoon forcing[5,47-48]. Secondly, analysis of the VIMT field can avoid influence of mountainous terrain, thus gets a rather continuous analysis field than the circulation field on a certain level (e. g. , 850-hPa). Thirdly, to diagnose the onset and withdrawal over a broad scale, the VIMT is generally well modeled and observed in contrast to the rainfall, which is poorly measured and modeled and has caused great trouble to monsoon researches[49]. For example, the serious scarcity and inaccuracy in rainfall historical datasets, particularly around ocean, often cause large disagreements in the climatological onset date of Asian summer monsoon [49].

Furthermore, with regard to moisture transport, some kinds of improper treatment may confuse the results. For example, some studies employ the moisture transport only on one level (e. g. , 850-hPa level) or just integrate from 1000-hPa level to get the VIMT dataset.

Actually, these methods essentially contain serious errors over mountainous terrain, thus can not provide true information (as mentioned by Zhou and He et al.[50]). And, others utilize monthly averaged horizontal wind vector and specific humidity data directly to calculate the monthly mean moisture flux vector dataset, instead of utilizing the daily mean data. Thus, they may neglect the contribution of the transient components, which are also of importance sometimes[51]. To address the above problems, this paper calculates the day by day VIMT dataset integrated from the surface to 300-hPa for the 1957-2001 period, on which the pentad and monthly averaged dataset are based. And on the basis of these datasets, the seasonal transition of the VIMT over the A-AM region is discussed. The data used in present study are described in Section 2. The climatological features of seasonal distribution and seasonal transition process are investigated in Section 3 and Section 4, respectively. In Section 5, the sub-regional characteristics of the VIMT and corresponding rainfall in the A-AM region are discussed, and the conclusions follow in Section 6.

2 Data and analysis methods

The data employed in the present study are NCEP/NCAR daily data[52] from 1957 to 2001 with a spatial resolution of 2.5° and 8 levels (from surface to 300-hPa), including horizontal wind and specific humidity. Now, the total vertically integrated water vapor flux vector Q can be expressed as:

$$Q = -\frac{1}{g}\int_{p_s}^{p} q U \mathrm{d}p \qquad (1)$$

Where g is gravity acceleration, p_s represents surface pressure, p stands for top pressure, which is chosen as 300-hPa in this study. Above 300-hPa, specific humidity amounts are poorly known and are therefore not part of the reanalysis[52]. Parameter q is special humidity, and U is horizontal wind vector. the unit of Q is $kg \cdot m^{-1} \cdot s^{-1}$. Above 300-hPa specific humidity in the Tropics is at least two orders of magnitude smaller than near the surface, and moisture transports are therefore of negligible influence to the calculation of total VIMT[49]. Further, the pentad and monthly mean are calculated by the daily mean data. We should remind the reader that the VIMT pentad mean are constructed by the principle that the last pentad in any month always covers the period from the 26th to the end of the month. Thus, there are always contains 6 pentads in a calendar month (i.e. 72 pentads for a year), and is quite convenient to match the month corresponding to the pentad mean.

The main precipitation data analyzed in this study are the Climate Prediction Center Merged Analysis of Precipitation (CMAP: Xie and Arkin[53]) data derived by merging rain gauge observation, microwave and infrared satellite images, and numerical model outputs. The pentad mean CMAP rainfall data cover the period from January 1979 to December 2001 with global coverage having a spatial resolution of 2.5°. However due to the different pentad mean definition by Xie and Arkin[53], it contains 73 pentads for a calendar year.

3 The seasonal distribution of VIMT

The majority of previous work focused mainly on the summertime features of moisture transport, while less attention has been paid in other seasons that are also of importance. Hence, in this section we will firstly exhibit the VIMT condition in the four seasons. The most salient features in winter are four zonally oriented VIMT belts existing over the great A-AM region (Fig. 1a): (i) The easterly VIMT belt located to the tropical area (5°~10°N) is the strongest of the four. (ii) Another easterly one confined to the 15°~25°S latitudinal band exhibits discontinuous feature. (iii) The westerly VIMT belt in the 5°~10°S latitudinal band is associated with three prominent inter-hemisphere VIMTs (around the 80°E, 105°E and 130°E, respectively) emanating from the aforementioned easterly VIMT. (iv) The last one is transported by the mid-latitude westerly, progressively eastward from the northern Arabian Sea to Western Pacific, and is the weakest of the four.

During summer time, significant differences are presented in Fig. 1c: a very strong VIMT flow originated from the Southern Hemisphere gets across the equator along the Somali coast and down to the Arabian Sea, India and the Bay of Bengal (BOB), then extends northeastward from the SCS to the Eastern China, Korea, Japan and Northwestern Pacific (The distribution bears great similarity to that of Chen[54]; Park and Schubert[55]; Simmonds et al.[41]). Due to the great intensity and merging other two notably cross-equatorial moisture exchanges (from 80°E and 105°E) and southeasterly VIMT from the southern periphery of WPSH into it, here we call this VIMT pattern the "great moisture river" (hereafter referred to as the GMR).

In two transitional seasons, the main features have more or less similarities by very weak inter-hemisphere moisture exchanges, which indicate weakly coupling between the two hemispheres (Fig. 1b and Fig. 1d). Also noted are the asymmetric features between the two transitions. For example, the mid-latitude westerly VIMT in the Northern Hemisphere is much pronounced in spring, while the westerly one only exists in the central tropical Indian Ocean in autumn. And the southwesterly (northeasterly) VIMT prevails over the Indochina region in spring (autumn) transition.

4 The seasonal transition features of VIMT

The preceding section has described the distributions of seasonal mean VIMT while details of the sub-seasonal change are intact. To this end, we have to examine the evolving VIMT pattern associated with seasonal march of the monsoon in this section.

4.1 VIMT features in the migration from winter to summer

In this section, pentad-to-pentad VIMT from January to March are analyzed in Fig. 2 to

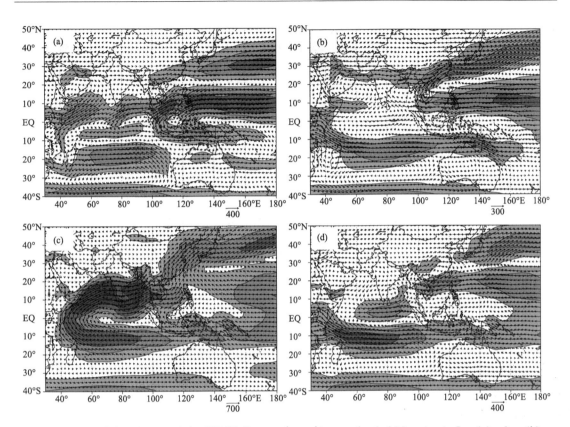

Fig. 1　Seasonal distribution of the VIMT (kg·m^{-1}·s^{-1}) over the A-AM region in Jan (a), Apr (b), Jul (c) and Oct (d). The climatology for VIMT is based on NCEP/NCAR data from 1957 to 2001 and the absolute value of VIMT>100 kg·m^{-1}·s^{-1} is shaded, and the shading interval is 100 kg·m^{-1}·s^{-1}

depict the transitional process from boreal winter to summer. From the Fig. 2, it can be noted that the whole process contains several stages as follows: Before the 3rd pentad of Feb, the VIMT pattern in A-AM region already has reached the peak stage of winter pattern mainly characterized by the four pronounced southward cross-equatorial VIMTs along the Somali coast, 80°E, 105°E and 130°E (Fig. 2a). In the 4th (figure omitted) and 5th pentad of Feb (Fig. 2b), the southward cross-equatorial VIMT along 80°E and the Somali coast nearly disappear, thus leading to serious weakness of the westerly VIMT over the tropical southern Indian Ocean (Fig. 2c). In the 3rd pentad of Mar, four dramatic changes occur simultaneously, heralding the collapse of winter pattern and the commencement of transition from winter to summer (see changes from Fig. 2c to Fig. 2d): the southerly VIMT exceeding 100 kg·m^{-1}·s^{-1} appears over the Indochina Peninsula; the southward cross-equatorial VIMT along approximately 105°E and 130°E weaken strikingly, and a continuous easterly VIMT belt forms in the 15°~25°S latitudinal band. Thus, in late Mar (Fig. 2e-f) the basic features of the spring transition are established (as shown in Fig. 1b), with the enhancement of the southerly VIMT established over the Indochina region, the disappearance of the southward cross-equatorial VIMT and the westerly VIMT over the north of the Australian

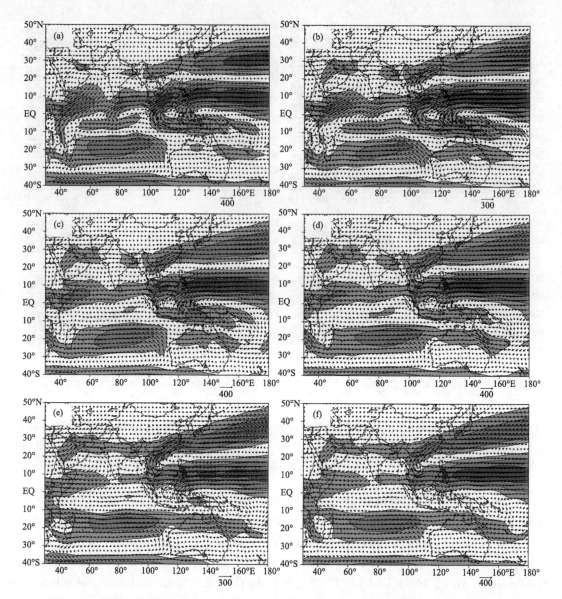

Fig. 2 The VIMT distribution in the seasonal migration from winter to summer. Figures from (a) to (e) exhibit the distribution in the 3rd pentad of Jan, the 5th pentad of Feb, the 2nd, 3rd, 5th and 6th pentad of Mar, respectively

continent, and also the split of the Asian tropical easterly VIMT belt over the Sri Lanka.

However, in this paper, the end date of Asian winter monsoon defined by the VIMT is slightly different from Matsumoto[14], which is one pentad earlier than our definition (7-11 March for Matsumoto and 11-15 March for our definition). Probably, it results from the different definition standard or dataset coverage period from which the climatological data were derived.

4.2 Features before the onset of the SCS summer monsoon

The large-scale Asian summer monsoon commences each year around early and middle May over the BOB, the SCS and the Indochina region[13-14,16-18,20,25-27,56-59], hence examining the evolution of moisture circulation before the onset of SCS summer monsoon has a very important meaning for understanding the migration of Asian monsoon.

Base on the pentad to pentad evolution of VIMT, the chain of remarkable events before the onset of SCS summer monsoon may be identified bellow: In the late-Apr (Fig. 3a), a moisture low vortex around the Sri Lanka (hereafter referred to as the "Sri Lanka low vortex") becomes active, which may greatly favor acceleration and eastward propagation of the southwesterly VIMT in the tropical Indian Ocean along the southern periphery of it. Around early and middle May (Fig. 3b, c), with initiation of the northward cross-equatorial VIMT off the Somali coast (i.e. the Somali VIMT jet), expansion and northward-migration of the "Sri Lanka low vortex", the southwesterly VIMT further eastward propagates to cross the Indochina region, the Southern China and the northern SCS, and merges with other two VIMTs, namely, the mid-latitude westerly one and the southerly one flowing along southern

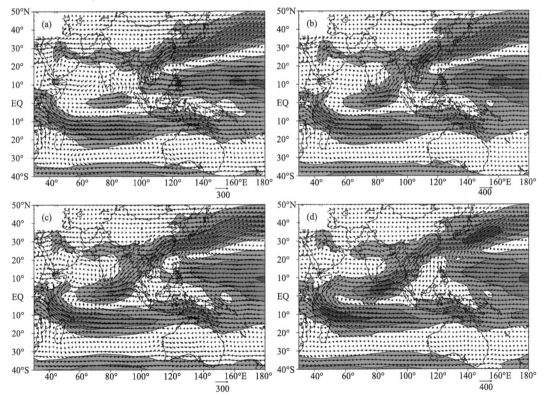

Fig. 3 Same as Fig. 2, but for the period before the onset of the SCS summer monsoon. Figures from (a) to (d) exhibit the distribution in the 6th pentad of Apr, the 2nd, 3rd and 4th pentad of May, respectively

periphery of the WPSH from the Indochina region to the Southern China. However, the typical summertime features as establishment of the "GMR" pattern are not integrated in this period, for that the westerly VIMT over the equatorial 50°~60°E longitudes still keeps disconnection, the SCS region is under the control of WPSH ridge yet and the SCS summer monsoon is still not onset. Just one pentad later, in the 4th pentad of May (Fig. 3d), with rapidly eastward retreat of the WPSH from the SCS, and with the dominion of the southerly VIMT over the whole SCS, the "GMR" transport pattern initially comes into being and manifests an impressive onset feature of the SCS summer monsoon, although the inter-hemisphere VIMT near 105°E is still very weak.

4.3 VIMT in the peak of the Asian summer and Australian winter monsoon

As shown in Fig. 4, after breaking out of the SCS summer monsoon, associated with a rapid intensification of the Somali VIMT jet, the southwesterly VIMT in the tropical Indian Ocean undergoes a considerable intensification and expansion along both northward and eastward routes to overcast the northern part of the Arabian Sea, the Indian Peninsular and the BOB around the 1st pentad of Jun (Fig. 4a), thus leading to forming a climatological onset moisture transport feature over these regions. Subsequently, with pronounced reinforcement of the cross-equatorial VIMT around 105°E in mid-Jun (see change between Fig. 4b and Fig. 4c), and rapid expansion of the strong "GMR" along northward and eastward routes simultaneous to reach the regions north of 50°N and east of 140°E and also enhancement of the cross-equatorial VIMT around 130°E in late Jul (Fig. 4d), the VIMT situation of the Asian summer and Australian winter monsoon gradually reach their peak phases. Five pentads later after reaching the peak phase, Asian-Australian monsoon enter the decaying period, as the strong "GMR" suddenly moves from the Southeastern China to the Western Pacific in the 4th pentad of Aug (Fig. 4e), and the northward inter-hemisphere VIMT at approximately 130°E weakens strikingly as manifested as the disappearance of VIMT belt exceeding 100 kg · $m^{-1} \cdot s^{-1}$ (Fig. 4f).

4.4 VIMT features in the summer to winter transition

In this period, there are two events mark the commencement of seasonal transition from summer to winter, which are the weakening of the inter-hemisphere VIMT around 130°E after mid-Aug (Fig. 4e), and breaking of the monsoon southwesterly VIMT belt over the western North Pacific in early September (Fig. 5a). After these key periods, a series of dramatically stepwise changes as described subsequently happen leading to establishing of the boreal winter and Australian summer VIMT stage: Firstly, in the 2nd pentad of Sep (figure omitted), accompanied by westward and southward retreat of the southwesterly VIMT, and westward expansion of the easterly VIMT along the southern periphery of WPSH, the westerly VIMT is replaced by the easterly one in the northern part of SCS (the similar phenomenon can be found in Matsumoto 1997 but for wind vectors), which manifest the SCS summer

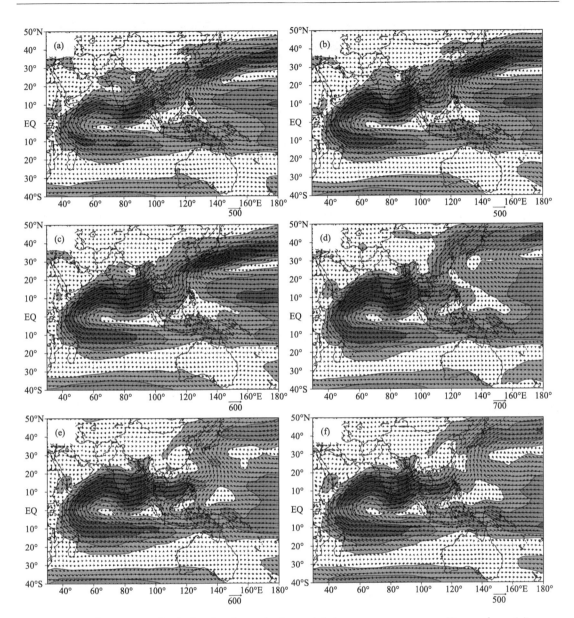

Fig. 4 The same as Fig. 2, but for the peak phase of Asian summer monsoon. Figures from (a) to (f) depict the distribution in the 1st, 3rd and 4th pentad of Jun, the 5th pentad of Jul, the 4th and 5th pentad of Aug, respectively

monsoon transport commences its withdrawing process. Then, after the 4th pentad of Sep (Fig. 5b and c), the northward cross-equatorial VIMT around 105°E weakens remarkably, while the easterly VIMT along the southern periphery of WPSH further advances westward to reach the region west of 100°E. In mid-October, the former monsoon southwesterly VIMT retreats more westward and becomes rather weak marked by the decaying of the northward Somali VIMT Jet in Fig. 5d, thus the typical autumn VIMT situation is shaped at this time (as exhibited in Fig. 1d).

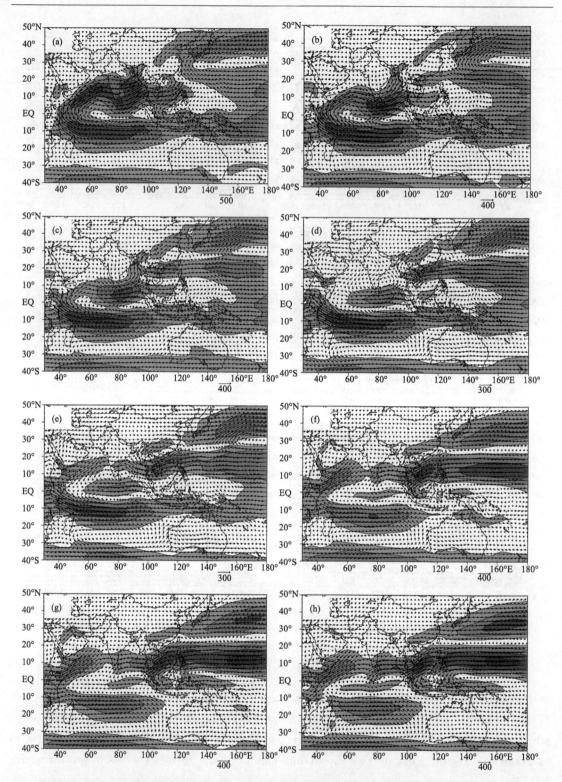

Fig. 5 The same as Fig. 2, but for the transition from summer to winter. Figures from (a) to (h) represent the 1st, 5th pentad of Sep, the 1st, 3rd, 6th pentad of Oct, the 6th pentad of Nov, the 2nd and 4th pentad of Dec, respectively

In subsequent pentads, the VIMT undergoes its transition to the winter phase as follows: In late Oct, with reactivity of the "Sri Lanka low vortex", the easterly VIMT along the southern periphery of WPSH extends more westward, then the wintertime strong low-latitude easterly VIMT is elementarily established, indicating the end of Indian summer monsoon transport (Fig. 5e). Afterward, accompanied by the intensification of the aforementioned easterly VIMT, decaying of the "Sri Lanka low vortex", and splitting of the Southern hemisphere mid-latitude easterly VIMT belt over the northern Australian continent, a pronounced southward-cross-equatorial VIMT at 105°E is established in the late November (Fig. 5f), the 130°E one in early December (Fig. 5g), heralding the commencement of the Southern Hemisphere summer monsoon. However, the VIMT signatures for entering mature stage of the Asian winter and the Southern Hemisphere summer monsoon transport, and also accomplishment of the seasonal transition for VIMT in the A-AM region, are indicated by the establishment of the cross-equatorial VIMTs along the Somali Coast and 80°E in the 4th pentad of December (Fig. 5h). Actually, the former cross-equatorial VIMT has crossed the equator in late November but still not stretched southward to approximately 10°S latitude that the VIMT in the domain: 30°~40°E, 0°~10°S is almost northerly.

5 Features in different A-AM sub-regions

In this section, in order to clarify the regional VIMT features, the zonal component of VIMT and precipitation are applied simultaneously in different parts of the A-AM region according to the great differences between the South Asian and East Asian monsoon[13,60].

5.1 The VIMT features in the BOB

In the western BOB region (80°~90°E), the westerly VIMT originated from 5°S starts its northward march after Pentad 15(mid-March) and crosses the equator at Pentad 22(mid-April). Therefore, before the 6th pentad of April (P24), the easterly VIMT covers the western BOB while the westerly one dominates north and south of it. After that, from south to north of the region, the VIMT reverses its direction from easterly to southwesterly and intensifies notably over the whole domain in the 1st pentad of June (P31), which is consistent well with the climatological onset date of the Indian summer monsoon. Around late July (P40) the southwesterly VIMT gets to its maximum intensity and expands to the northern most latitude of 30°N. Hereafter, following southward retreat of the southwesterly VIMT, the easterly one occupies northern part of the region once again in early October (around P55). However, the whole region is not covered by the easterly VIMT until the westerly VIMT retreats into Southern Hemisphere around late October (P60). From above transition process, we may find that the onset of summer monsoon over the western BOB is closely related to northward propagation of the equatorial westerly VIMT, although it seems to result from a sudden eruption of southwesterly VIMT as a local perspective. Actually, when we investigate the evolving figures in Section 4.2, we may find that the northward migration of westerly VIMT is closely associated with the activity and northward march of the "Sri Lanka low vortex" and the enhancement of the Somali northward cross-equatorial VIMT.

Corresponding to the VIMT, the similarities in rainfall can also be found from Fig. 6. The rainbelt originated from south of 5°N also moves northward after the 6th pentad of April (P24) and intensifies sharply on early June (P31), reflecting the onset characteristics of the Indian summer monsoon VIMT. It can also be noted that after rainbelt reaches its maximum intensity and extends to its northernmost location in late July (P41), it progressively withdraws southward and back to its cradle land around late October (P60: 23-27September).

Compared with the characteristics of seasonal change over the western BOB, the one over the eastern part (90°~100°E) exhibit similarities to some extent, but obviously different features can also be noted. First, we can find the intensity of westerly VIMT belt over the eastern BOB is weaker than the western one, while the mid-latitude southwesterly VIMT is stronger and can expand more northward. Second, after monsoon onset around early-May (P26), the southerly VIMT in the eastern BOB is more conspicuous than that in the western BOB, which exhibits more westerly component after onset around early Jun. Another noted feature in the eastern BOB is the occurrence of two maximum westerly VIMT centers in late-Jun and mid-Aug independently, and with the first occurrence of maximum westerly VIMT, the southwesterly VIMT propagates to the north of 25°N synchronously. As for the features of precipitation, over the eastern BOB, the main difference from the western one is monsoon rainband located at 5°~15°N latitudes, which is located nearly ten degrees latitude southward in contrast to the western one.

5.2 The features in the Indochina Peninsula

In the Indochina peninsula region (100°~110°E), the most interesting feature compared to the BOB region is the westerly VIMT from the Southern Hemisphere is unable to unceasingly migrate northward to cross the equator into the Northern Hemisphere during winter to summer transition. Before P15 (mid-March), the northeasterly VIMT domains the 3°~12°N latitudinal band but very strong westerly VIMT occupies the south and north of it. Afterward, the southerly VIMT occurs at 15°N then propagates northward, implying the dayspring of seasonal transition from winter to summer for both the Indochina Peninsula and the whole A-AM region. After mid-May (P27), caused by the quickly eastward-expansion of the tropical-strong-westerly VIMT from the eastern Indian Ocean (as depicted by Fong and Wang[61]; Ding and sun[62] but for wind vectors), the southern Indochina region (south of 15°N) undergoes a striking change that the VIMT shifts from southeasterly to southwesterly, while to the BOB that results from northward migration of the Southern Hemisphere westerly VIMT. In the same period, the notable mid-latitude westerly VIMT expands more northward than that over the BOB region to merge with the high-latitude westerly VIMT. However, the westerly VIMT does not expand to its northern edge of 35°N until summer monsoon

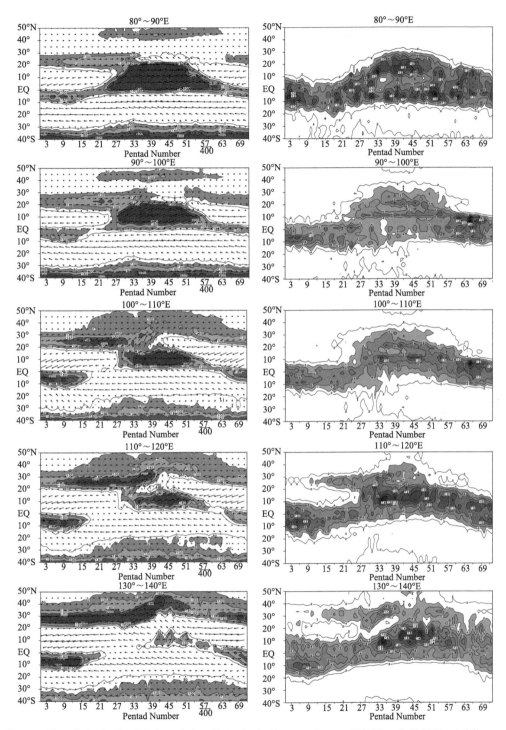

Fig. 6 Time-latitude distribution of the climatologically pentad mean VIMT and CMAP rainfall over different regions. The climatolology for rainfall was made by using CMAP data from 1979 to 2001. The shading for VIMT indicates the westerly component exceeding 50 kg·m^{-1}·s^{-1}; for rainfall it denotes the precipitation rate exceeding 4 mm·d^{-1}. And, the shading interval is 50 kg·m^{-1}·s^{-1} for VIMT, and 2 mm·d^{-1} for rainfall

reaches its prosperous epoch around P40 (late July). Hereafter, with withdrawal of the summer monsoon, the westerly VIMT also gradually retreats southward and eventually back into the Southern Hemisphere in early November.

Corresponding to the VIMT, both similar and different features are exhibited in the precipitation. In general, the characteristics of rainfall and VIMT are rather similar in summer to winter transition as depicted as a continuously southward withdrawing into the Southern Hemisphere. But different features can also be found as follows: Firstly, in winter to summer transition, the rainbelt marches northward unceasingly from the Southern to Northern Hemisphere, while the VIMT does not. This phenomenon is primarily due to the march of rainbelt is so closely associated with migration of the VIMT convergence center (the figures omitted) that although the westerly VIMT can't cross the equator and supply moisture for precipitation in the region, the easterly one can also do it, its convergence will cause enough precipitation over the region. Secondly, the notable mid-latitude westerly merges with the southwesterly VIMT from the eastern Indian Ocean, while the rainbelt does not. Thirdly, the rainy season reaches its maximum intensity in autumn[14-15,18], while the VIMT does in summer.

5.3 The features in the SCS

Over the SCS longitudes (110°~120°E), the main VIMT feature in contrast to the BOB is exhibited by sudden arrival of the westerly VIMT in the 4th pentad of May (P28), when the SCS summer monsoon breaks out. In contrast, over the BOB longitudes the feature along with the onset process is characterized by a gradually northward propagation of westerly VIMT. On the other hand, over the SCS longitudes, the abrupt feature is also displayed by the pronounced increase of rainfall after P28 (as mentioned by Wang and LinHo[58]). The above results may demonstrate that the arrival of monsoon rainfall in the SCS concurs with the sudden advent of the southwesterly VIMT. After that, around P33 with expansion of the southwesterly VIMT, the rainband suddenly migrates from 23°N to 30°N indicating the onset of Mei-yu/Baiu season in East Asian (this feature is more obvious over the 130°~140°E longitudes). After reaching summer peak phase (around P42), the westerly VIMT begin retreating southward, and around the 1st pentad of September (P49) it withdraws from the northern SCS, thus the easterly VIMT establishes once again in the region (Matsumoto[15]). In late October (around P59), as the westerly VIMT returns to Southern Hemisphere, the rainbelt also retreats to south of 5°N. The westerly VIMT and rainbelt become vigorous again when they retreat to the 5°~10°S latitudinal bands during the period from January to February of subsequent year, corresponding to peak phase of Australian summer monsoon. Moreover, a remarkable invasion process from middle latitude before the onset of SCS summer monsoon (as emphasized by Chan et al[63]) can also be found both in rainfall and VIMT figures, i.e., both the mid-latitude westerly VIMT and rainbelt march southward before the onset of SCS summer monsoon. Thus, compared with the characteristics over the BOB region, the intrusion of mid-latitude events may contribute as a major factor to the SCS summer monsoon onset [63-66].

5.4 The features of the western North Pacific and Australian region

In this section, the longitudinal band between 130°E and 140°E is chosen as a representative zone of the western North Pacific (WNP) and Australian monsoon region. For the evolution of VIMT, it shows a significantly asymmetric feature, which is similar to that in the SCS region. But what is worth noting is the more remarkably northeastward expansion of southwesterly VIMT after P32 near 30°N than that in the SCS, and it agrees quite well with the northeastward migration of rainbelt after the onset of Meiyu/Baiu season as emphasized in Section 5.3. From the figures, we can also find that after P33 from 5°N to north of it, the VIMT gradually shifts from easterly to westerly, which signifies that summer monsoon is under its northward propagation. Meanwhile, the precipitation increases sharply implying the onset of WNP summer monsoon. Hereafter, the WNP summer monsoon reaches peak phase around P45 then withdraws after P57 agreeing well with those defined by Wang and LinHo[18]. Corresponding to the equatorward withdraw of the WNP summer monsoon, around P68, the VIMT at 10°S latitude shifts from easterly to westerly. Concurrently, the precipitation strengthens markedly, denoting the commencement of Southern Hemisphere summer monsoon. However, in the Australian sector, the onset date of summer monsoon is just around P1, characterized by a suddenly increasing of the rainfall and enhancing of the westerly VIMT (as described by Matsumoto 1992 but for wind vectors). Around P6, it reaches its peak stage with occurring of the maximum precipitation. After P18, the Australian summer westerly VIMT begins retreating, while it does not disappear completely from Southern Hemisphere until P24, which is coincident with the previous findings by Murakami and Matsumoto from wind vectors[56].

Actually, there are several views on the demarcation of the Indian, East Asian and western North Pacific monsoon regions in the A-AM region. For example, Marakami and Matsumoto[56] regarded the boundary between the Indian and western North Pacific monsoon as being east of the Indochina Peninsula, while Wang and LinHo[18] treated the Indochina region as a kind of buffer or transitional region between the two. But in this paper, we regard the west of the Indochina region as the boundary between the Indian and East Asian-western North Pacific monsoon when viewed from seasonal transition of the moisture transport pattern, which can be easily summarized from Fig. 6 as follows: in the region west of the Indochina Peninsular that is in the Indian monsoon sector, there are distinct symmetries in the meridional migration of westerly VIMT characterized by poleward advancement and equatorward withdrawal unceasingly across the equator in the all of the seasonal transitions. However, in the region east of the Indochina Peninsular, it has a pronounced asymmetry by the impossibility of marching poleward to cross the equator during winter to summer transition, but the possibility of withdrawal equatorward to cross the equator during transition from summer to winter.

6 Conclusions

As shown above, over the whole A-AM region, the basic seasonal transition features of moisture transport from winter to summer are characterized by establishment of the "GMR" pattern and its eastward expansion associated with a series of climatological events such as: the occurrence of the notably southerly VIMT over the Indochina Peninsular, the activity of the "Sri Lanka low VIMT vortex", the occurrence of the strong northward cross-equatorial VIMT, the eastward retreat of the WPSH and the onset of the SCS summer monsoon, etc. It is also found that the precursory signals in seasonal transition from winter to summer pattern are mainly revealed by the establishment of strong southerly VIMT exceeding 100 kg · m^{-1} · s^{-1} over the Indochina Peninsula in mid-March (the 3rd pentad in March). However, the commencement of large-scale summertime moisture transport pattern over the A-AM region is demonstrated by the initial establishment of the "GMR" pattern in the 4th pentad of May.

On the other hand, the chief characteristics in summer to winter transition are the formation of the tropical Asian easterly VIMT belt and subsequent occurrence of the southward cross-equatorial VIMT, accompanied by several stepwise changes as: the breakup and westward retreat of the "GMR" pattern, the westward expansion of easterly VIMT emanating from the southern periphery of the WPSH, etc. Among these events, the early signal of summer to winter transition is indicated by the event that the "GMR" pattern splits over the Western Pacific in early September.

Our results further show that great differences exist in the sub-regions as viewed from the meridional migration of the VIMT and precipitation during seasonal transition process. In the region west of the Indochina peninsula, a particular phenomenon associated with the seasonal transition is the westerly VIMT can advance northward to cross the equator into the Northern Hemisphere during winter to summer transition, while to the east of the region in the same period, it can not advance northward to cross the equator at all. Therefore, these different migration features of the westerly VIMT can not only easily divide the sub-monsoon region for the great Asian monsoon region, but also reflect the different formation mechanisms between the Indian monsoon and East Asian monsoon.

Acknowledgements. We wish to thank Dr. Xiang Lin, Ms. Dong-Yan Wang and Ms. Li-Wei Jia for their help with the revision of the text, and two anonymous reviewers for their comments and suggestions to improve the quality of this paper. We also thank the National Centers for Environment Prediction and National Center for Atmospheric Research for providing the reanalysis data, and the Climate Prediction Center for providing the Merged Analysis of Precipitation data. This research was supported by the National Natural Science Foundation of China (No. 40475021 & No. 40375025), and the Natural Science Foundation of Guangdong province, China (No. 0400391).

References

[1] Ramage C. Monsoon meteorology[M]. International Geophysics Series Vol. 15. San Diego: Academic Press,1971.

[2] Krishnamurti T N,Surge N,Manobianco J. Annual cycle of the monsoon over the global tropics[R]. WMO world climate research programme publications Ser. 4,WMO TD65,Part IV-I-IV-21,1985.

[3] Wang B. Climate regimes of the tropical convection and rainfall[J]. J Climate,1994,7:1109-1118.

[4] Wang B,Clements S C,Liu P. Contrasting the Indian and East Asian monsoons: implications on geologic timescales[J]. Marine Geology,2003,201:5-21.

[5] Webster P J,Magana V O,Palmer T N,et al. Monsoons: Processes,Predictability,and the prospects for prediction[J]. J Geophys Res,1998,103:14451-14510.

[6] Trenberth K E,Stepaniak D P,Caron J M. The global monsoon as seen through the divergent atmospheric circulation[J]. J Climate,2000,13:3969-3993.

[7] Starr V P,White R M. Direct measurement of the hemispheric poleward flux of water vapor[J]. J Mar Res,1955,14:217-225.

[8] Xu Shuying. Water-vapor transfer and water balance over the eastern China[J]. Acta Meteorologica Sinica,1958,29: 33-43. (in Chinese)

[9] Shen J Z. Atmospheric water vapor budget in summer monsoon over China[C]//Proceedings of the symposium on the summer monsoon in Southeast Asia. People's Press of Yunnan Province,1983: 147-157. (in Chinese)

[10] Ding Yihui. Monsoon over China [M]. Kluwer Academic publishers, Dordrecht/Boston/London, 1994:419.

[11] Qiao Yunting,Luo Huibang,Jian Maoqiu. The temporal and spatial characteristics of moisture budgets over Asian and Australian monsoon regions[J]. J Trop Meteor,2002,18:203-210. (in Chinese)

[12] Murakami T,Chen L,Xie A. Relationships among seasonal cycles,low frequency oscillations and transient disturbances as revealed from outgoing longwave radiation data[J]. Mon Wea Rev, 1986, 114: 1456-1464.

[13] Tao S Y,Chen L X. A review of recent research on the East Asian summer monsoon in China[C]// Monsoon Meteorology. Chang C P,Krishnamruti T N. Oxford:Oxford University Press,1987: 60-92.

[14] Matsumoto J. The seasonal changes in Asian and Australian monsoon regions[J]. J Meteor Soc Japan, 1992,70: 257-273.

[15] Matsumoto J. Seasonal transition of summer rainy season over Indo-China and adjacent monsoon region [J]. Adv Atmos Sci,1997,14:231-245.

[16] He Jinhai,Zhu Qiangen,Murakami M. TBB data revealed features of Asian-Australian monsoon seasonal transition and Asian summer monsoon establishment[J]. J Trop Meteor,1997,3:18-26. (in Chinese)

[17] Lau K M,Yang S. Climatology and interannual variability of the Southeast Asian summer monsoon[J]. Adv Atmos Sci,1997,14:141-162.

[18] Wang B,Lin Ho. Rainy seasons of the Asian-Pacific monsoon[J]. J Climate,2002,15: 386-398.

[19] Qian W,Lee D K. Seasonal march of Asian summer monsoon[J]. Int J Climatol,2000,20: 1371-1386.

[20] Qian W,Yang S. Onset of the regional monsoon over southeast Asian[J]. Meteor Atmos Phys,2000,75: 29-38.

[21] Qian W,Kang H S,Lee D K. Distribution of seasonal rainfall in the East Asian monsoon region[J]. Thero Appl Climatol,2002,73:151-168.

[22] Zhang Z Q, Chan J C L, Ding Y H. Characteristics, evolution and mechanisms of the summer monsoon onset over southeast Asia[J]. Int J Climatol, 2004, 24: 1461-1482.

[23] Rao Y P. Southwest monsoon: Synoptic Meteorology[J]. Meteor Monogr, 1976, 1, India Meteorological Department, 367.

[24] Murakami T, Nakazawa T. Transition from the Southern to Northern Hemisphere summer monsoon[J]. Mon Wea Rev, 1985, 113: 1470-1486.

[25] Lau K M, Wu H T, Yang S. Hydrologic processes associated with the first transition of the Asian summer monsoon: a pilot satellite study[J]. Bull Amer Meteor Soc, 1998, 79: 1871-1882.

[26] Wu G, Zhang Y. Tibetan Plateau forcing and timing of the monsoon onset over South Asian and the South China Sea[J]. Mon Wea Rev, 1998, 126: 913-927.

[27] Hsu H H, Terrg C T, Chen C T. Evolution of large scale circulation and heating during the first transition of the Asian summer monsoon[J]. J Climate, 1999, 12: 793-810.

[28] DingYihui. Seasonal march of the East Asian summer monsoon[C]// Chang C-P. East Asian Monsoon. Singapore: World Scientific Publisher, 2004: 3-53.

[29] Li C, Yanai M. The onset and interannual variability of the Asian summer monsoon in relation to land-sea thermal contrast[J]. J Climate, 1996, 9: 358-375.

[30] Kumar K K, Kumar K R, Pant G B. Pre-monsoon maximum and minimum temperatures over India in relation to the summer monsoon rainfall[J]. Int J Climatol, 1997, 17: 1115-1127.

[31] Ueda H, Yasunari T. Role of warming over the Tibetan Plateau in early onset of the summer monsoon over the Bay of Bengal and the South China Sea[J]. J Meteor Soc Japan, 1998, 76: 1-12.

[32] Takagi T, Kimura F, Kono S. Diurnal variation of GPS precipitable water at Lhasa in premonsoon and monsoon periods[J]. J Meteor Soc Japan, 2000, 78: 175-180.

[33] Murakami T. The general circulation and water vapor balance over the Far East during the raining season [J]. Geophy Mag, 1959, 29: 131-171.

[34] Akiyama T. The large-scale aspects of the characteristics of Baiu front[J]. Pap Meteor Geophys, 1973, 24: 157-188.

[35] Asakura T. Transport and source of water vapor in the northern hemisphere and monsoon Asian[C]// Yoshino M M. Water balance of Monsoon Asian. Tokyo: Univ Tokyo press, 1973: 131-151.

[36] Murakami T, Nakazawa T, He J. On the 40-50 day oscillation during the 1979 Northern Hemisphere summer. Part II: Heat and moisture budget[J]. J Meteor Soc Japan, 1984, 62: 469-484.

[37] Chen T C, Yen M C, Murakami M. The water vapor transport associated with the 30-50 day oscillation over the Asian Monsoon regions during 1979 summer[J]. Mon Wea Rev, 1988, 116: 1983-2002.

[38] Huang Ronghui, Zhang Zhenzhou, Huang Gang, et al. Characteristics of the water vapor transport in East Asian monsoon region and its difference from that in South Asian monsoon region in summer[J]. Chinese J Atmos Sci, 1998, 22: 469-479. (in Chinese)

[39] Ninomiya K, Kobayashi C. Precipitation and Moisture Balance of the Asian Summer Monsoon in 1991 Part I: Precipitation and Major Circulation Systems[J]. J Meteor Soc Japan, 1998, 76: 855-877.

[40] Ninomiya K, Kobayashi C. Precipitation and Moisture Balance of the Asian Summer Monsoon in 1991 Part II: Moisture Transport and Moisture Balance[J]. J Meteor Soc Japan, 1999, 77: 77-99.

[41] Simmonds I, Bi D H, Hope P. Atmospheric water vapor flux and its association with rainfall over China in summer[J]. J Climate, 1999, 12: 1353-1367.

[42] Zhang Renhe. The role of Indian summer monsoon water vapor transportation on the summer rainfall anomalies in the northern part of China during the El Nino mature phase[J]. Plateau Meteorology, 1999,

18：567-574. (in Chinese)

[43] Zhang Renhe. Relations of water vapor transport from Indian monsoon with that over East Asia and the summer rainfall in China[J]. Adv Atmos Sci,2001,18：1005-1017.

[44] Xie An,Song Yanyun,Mao Jiangyu,et al. Climatological characteristics of moisture transport during summer monsoon over South China Sea[J]. Climatic and Environmental Research,2001,6：425-434. (in Chinese)

[45] Tian S F,Yasunari T. Climatological aspects and mechanism of spring persistent rains over central China [J]. J Meteor Soc Japan,1998,76：57-71.

[46] Zhang Renhe,Akimasa S. Moisture circulation over East Asia during El Nino Episode in Northern Winter,Spring and Autumn[J]. J Meteor Soc Japan,2002,80：213-227.

[47] Webster P J. Mechanisms of monsoon transition：Surface hydrology effects[J]. J Atmos Sci,1983,40：2110-2124.

[48] Webster P J. The role of hydrological processes in ocean-atmosphere interaction[J]. Rev Geophys,1994,32：427-476.

[49] Fasullo J,Webster P J. A hydrological definition of Indian Monsoon onset and withdraw[J]. J Climate,2003,16：3200-3211.

[50] Zhou ChangYan,He Jinhai,Li Wei,et al. Climatological characteristics of water vapor transfer over East Asian in summer[J]. J Nanjing Inst Metero,2005,28：18-27. (in Chinese)

[51] Shi Huimin,Weng Dumin,Lin Shoulian. Comparison of different climatological calculation of methods for water transportation in the atmosphere with discussion[J]. J Nanjing Inst Metero,1994,17：372-378.

[52] Kalnay E,et al. The NCEP/NCAR 40-Year Reanalysis Project[J]. Bull Amer Meteor Soc,1996,77：437-471.

[53] Xie P,Arkin P A. Global Precipitation：A 17-Year Monthly Analysis Based on Gauge Observations,Satellite Estimates and Numerical Model Outputs[J]. Bull Amer Meteor Soc,1997,78：2539-2558.

[54] Chen T C. Global water vapor flux and maintenance during FGGE[J]. Mon Wea Rev,1985,113：1801-1819.

[55] Park C-K,Schubert S D. On the nature of the 1994 East Asian summer drought[J]. J Climate,1997,10：1056-1070.

[56] Murakami T,Matsumoto J. Summer monsoon over the Asian continent and western north Pacific[J]. J Meteor Soc Japan,1994,72：719-745.

[57] Wang B. On the annual cycle in the tropical eastern-central pacific[J]. J Climate,1994,7：1926-1942.

[58] Wang B,LinHo,Zhang Y,et al. Definition of South China Sea monsoon onset and commencement of the East Asia Summer monsoon[J]. J Climate,2004,17：699-710.

[59] Wang Shiyu,Qian Yongfu. Diagnostic study of apparent heat sources and moisture sinks in the South China Sea and its adjacent areas during the onset of 1998 SCS monsoon[J]. Adv Atmos Sci,2000,17：285-298.

[60] Lau K M,Li M T. The monsoon of East Asia and its global association —A survey[J]. Bull Amer Meteor Soc,1984,65：114-125.

[61] Fong S K,Wang A Y. Climatological Altas for Asian summer monsoon,Macau Meteorological and Geophysical Bureau and Macau Foundation,2001：318.

[62] DingYihui,Sun Ying. Some aspects of climatology of Asian summer monsoon[J]. To be submitted to Acta Meteorological Sinica,2001.

[63] Chan J C L,Wang Y G,Xu J J. Dynamic and thermodynamic characteristics associated with the onset of

the 1998 South China Sea summer monsoon[J]. J Meteor Soc Japan,2000,78:367-380.

[64] Chang C P,Chen G T J. Tropical circulation associated with southwest monsoon onset and westerly surges over the South China Sea[J]. Mon Wea Rev,1995,123:3254-3267.

[65] DingYihui,Liu Yanju. Onset and the evolution of the summer monsoon over the South China Sea during SCSMEX field experiment in 1998[J]. J Meteor Soc Japan,2001,79:255-276.

[66] Hung C W,Yanai M. Factors contributing to the onset of the Australian summer monsoon[J]. Quart J Roy Meteor Soc,2004,130:739-758.

A Review of Recent Advances in Research on Asian Monsoon in China[*]

He Jinhai[1] Ju Jianhua[2] Wen Zhiping[3] Lü Junmei[4] Jin Qihua[1,4]

(1. College of Atmospheric Sciences, Nanjing University of Information Science & Technology, Nanjing 210044; 2. China Meteorological Administration Training Centre, Beijing 100081; 3. Department of Atmospheric Science, Sun Yat-Sen University, Guangzhou 510275; 4. Chinese Academy of Meteorological Sciences, Beijing 100081)

Abstract: This paper reviews briefly advances in recent research on monsoon by Chinese scholars, including primarily: (1) the establishment of various monsoon indices. In particular, the standardized dynamic seasonal variability index of the monsoon can delimit the geographical distribution of global monsoon systems and determine quantitatively the date of abrupt change in circulation. (2) The provision of three driving forces for the generation of monsoon. (3) The revelation of the heating-pump action of the Tibetan Plateau, which strengthens southerlies in the southern and southeastern periphery of the Plateau and results in a strong rainfall center from the northern Bay of Bengal (BOB) to the Plateau itself. (4) Clarification of the initial onset of the Asian Summer Monsoon (ASM) in the BOB east of 90°E, Indochina Peninsula (ICP) and the South China Sea, of which the rapid northward progression of tropical convection in the Sumatra and the rapid westward movement of the South Asia High to the Indochina Peninsula are the earliest signs. (5) The provision of an integrated mechanism for the onset of the East Asian Summer Monsoon (EASM), which emphasizes the integrated impact of sensible heat over Indian Peninsula, the warm advection of the Tibetan Plateau and the sensible heat and latent heat over the Indochina Peninsula on the one hand, and the seasonal phase-lock effect of the northward propagation of low frequency oscillation on the other. (6) The revelation of the "planetary-scale moisture transport large-value band" from the Southern Hemisphere through to the Asian monsoon region and into the North Pacific, which is converged by several large-scale moisture transport belts in the Asian-Australian monsoon regions and whose variation influences directly the temporal and spatial distribution of summer rainfall in China. (7) Presenting the features of the seasonal advance of the EASM, the propagation of intraseasonal oscillation, and their relationship with rainfall in China; indicating that the intraseasonal oscillation of the EASM propagates in the form of a wave-train along the coast and behaves as monsoon surge propagating northward. (8) Describing the interannual and interdecadal variation of Asian monsoon, revealing the factors affecting it, and possible mechanisms of the variation of Asian monsoon. An elementary outlook on the existing problems and future direction of monsoon research is also provided.

Key words: recent research; monsoon; China; review; advances; existing problems; prospects

---

[*] 本文发表于《Advances in Atmospheric Sciences》,2007 年第 24 卷第 6 期,972-992.

1 Introduction

Monsoon is an important component of the global climate system, whose variation plays a critical role in weather and climate in the monsoon regions. China is located in the East Asian monsoon region, and therefore research on monsoon has always been high on the agenda of Chinese meteorologists. Research on monsoon in China has a history of nearly 80 years, and many achievements have been made by Chinese meteorologists in research on the East Asian Monsoon, producing significant influences in China and abroad. As early as the 1930s, Zhu[1] studied the advancement and withdrawal of summer monsoon and its relationship with rainfall in China. Following this, Tu et al.[2] explored the advancement and withdrawal of East Asian monsoon and its impact upon the intraseasonal variation of climate in China. Ye et al.[3] investigated the seasonal transition of the atmospheric circulation over the Northern Hemisphere. Later, Chen et al.[4] and Ding[5] proposed the structure and seasonal variation of the East Asian Summer Monsoon (EASM). Recently, Chinese meteorologists have made important progress in research on the forming mechanisms, onset characteristics, and the multi-scale variability of Asian monsoon. This paper comments briefly on recent advances in related research.

2 Description of monsoon and monsoon indices

The monsoon index is a critical parameter for quantitatively describing and studying monsoon. In order to reflect the essence of monsoon, investigators put forward various monsoon indices.

In order to define the East Asian Winter Monsoon (EAWM) index, some factors—for example, differential heating between land and sea, meridional wind regionally averaged over East Asia, and zonal horizontal wind shear—have been considered in previous research[6-9]. It is well known that meridional wind is a crucial aspect of the EAWM. However, zonal wind also plays an important role, as strong zonal winds over mid and high latitudes can obstruct cold air outbreaks southward. Considering two-dimensional wind vectors, two distinct modes of the EAWM have been identified[10], resulting in a complete understanding of the EAWM.

Based on the contrary trend between the intensity of the tropical monsoon trough and that of the mei-yu Front and the variation of their wind anomalies, Zhang et al.[11] defined an EASM index as the difference of the zonal wind anomaly at 850-hPa between the tropical area (10°~20°N, 100°~150°E) and the subtropical area (25°~35°N, 100°~150°E) in East Asia. This had synoptic meaning, was easy to calculate, and reflected well the interannual variation of the wind and rainfall in East Asia. Utilizing the method of combining dynamic and thermodynamic factors, Ju et al.[12] dealt with the southwesterlies and outgoing longwave radiation (OLR) in the East Asian monsoon region to build synthetically an EASM index which could reflect appropriately not only the interannual variation of the East Asian Subtropical Monsoon, but also the essence of the monsoon system. He et al.[13] set up a new monsoon index by combining the zonal wind shear between the upper and lower levels with the geopotential height at 850-hPa. They investigated the relationship between temperature in the troposphere and the Asian Summer Monsoon (ASM) and found that the reverse of the meridional temperature gradient in the upper level is generally earlier than or synchronous with the onset of summer monsoon in both the East Asian and Indian monsoon regions.

Li and Zeng[14] presented a standardized dynamic seasonal variability index which can not only describe the seasonal variation and interannual variation in various monsoon regions, as well as their relationship with precipitation, but also outline the distribution of the global monsoon system (Fig. 1). Therefore, this index was named the "unified monsoon index" [15-16] and represents significant progress in the study of monsoon indices.

Zeng et al. [17] and Zhang et al.[18] adopted the "normalized finite temporal variation" method to determine quantitatively the critical day of abrupt transition in atmospheric circulation, which is just 2-4 days earlier than the so-called "onset date" of monsoon, i. e. the "presage date" of monsoon onset. Whether in the tropical or subtropical monsoon region, the seasonal adjustment and transition of atmospheric circulation have already finished in the stratosphere before summer monsoon breaks out in the lower level. It then extends suddenly downward to the troposphere to trigger the onset of monsoon in the lower level, which indicates that the seasonal transition of the atmospheric circulation from winter to summer commences initially in the Southern Hemisphere and in the stratosphere.

There are three aspects in the research of monsoon indices so far: (1) the intensity of monsoon as described by winds (zonal or meridional), divergence, vorticity or moisture transport averaged over an area, which derives directly from the monsoon circulation itself; (2) the strength of monsoon as depicted by the combination of circulation and rainfall or convective parameters (e. g. , OLR) averaged over an area; and (3) the EASM index defined using the meridional land-sea thermal contrast, the zonal land-sea thermal contrast, or both. Whichever index is employed, it is appropriate as long as it distinguishes floods from droughts and relates significantly to the circulation or parameters of the circulation systems (e. g. , the position and area of the subtropical high), and also if the relation can be maintained as the database of observations is prolonged. If such an index can reflect the date of abrupt seasonal transition of circulation as well, it can be used to determine the onset date of monsoon. Monsoon indices are generalized in a booklet by the research group of the South China Sea Monsoon Experiment (SCSMEX)[19].

3 Mechanisms in forming monsoon

The fundamental mechanism for the formation of monsoon is generally considered to be land-sea thermal contrast. In the meantime, the formation of monsoon is associated with

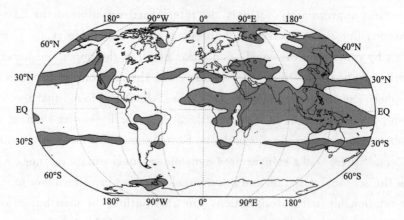

Fig. 1 Distribution of global surface monsoon. Red, green, and blue areas represent the tropical monsoon region, subtropical monsoon region, and extratropical monsoon region, respectively. The red and blue bold solid lines show the positions of the ITCZ in summer and winter respectively[14]

seasonal differences in solar radiation between the Northern and Southern Hemispheres. However, if the surface is covered entirely by oceans, monsoon would maybe still exist. In such an ocean model, the sea surface temperature (SST) may play an important role in the onset and existence of monsoon. When the forcing of seasonal variation in SST was introduced, numerical experiments simulated not only the vivid monsoon circulation, but also the abrupt out-break and active-break features[20]. Monsoon is probably the outcome of the seasonal movement of the ITCZ[21-22]. The land-sea thermal contrast, like the difference in SST, determines the meridional position of the ITCZ band, but orographic change has more influence on monsoon than land-sea thermal contrast. The formation of monsoon results from the cooperation between various driving forces. Zeng and Li[23] suggested two driving forces for monsoon generation: (1) planetary thermal convective circulation (induced by seasonal differences in solar radiation between the Northern and Southern Hemispheres); and (2) the quasistationary planetary wave induced by differences in surface characteristics (including land-sea thermal contrast, orographic height etc.). He et al. [24-25] considered moist processes as a third driving force of monsoon. The phase-change of moisture in the air and its transporting process can store and redistribute solar energy in the tropics and subtropics, release the energy selectively, and determine the intensity and location of monsoon circulation and rainfall. Research on the role of the Tibetan Plateau in the ASM has achieved new progress recently. Numerical experiments have shown that in "no-orography experiments" (Fig. 2), monsoon appears in Africa, South Asia and East Asia, and the rough distribution of monsoon regions is similar to the actual situation, indicating that the distribution of land and sea is primary in the formation of Asian monsoon. After introducing an ideal plateau in the experiment, northerlies occur round the west side of the plateau in the lower troposphere, and monsoon on the southwest side of the plateau withdraws southward; thus, the monsoon rainfall in Africa to the west of the plateau is separated from that in Asia. The heating pump action of the plateau strengthens the southerlies in the southern and southeastern periphery and results in a strong rainfall center from the northern Bay of Bengal (BOB) to the south of the plateau. Seen from the difference of two experiments, the low-level cyclonic circulation triggered by the plateau in summer strengthens significantly the southwesterlies in the southeast of the Eurasian continent and brings the EASM to Northeast China at 50°N, which greatly strengthens the East Asian monsoon[26-27].

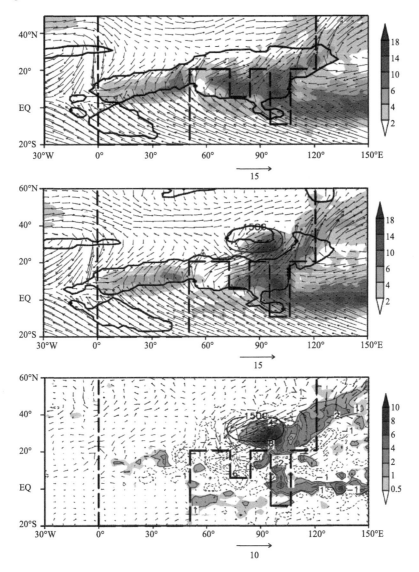

Fig. 2 Distribution of the surface wind vectors (units: m·s^{-1}) and rainfall (units: mm·d^{-1}) in July in ideal experiments: (a) no-orography experiment; (b) orography experiment; and (c) difference of (b)−(a). The bold dashed line is the land boundary, the curve covers the region where wind diverges more than 120° in January and July, and the ellipse is the contour line of the 1500-m plateau[26]

The thermal contrast between the African continent or Indian subcontinent and the surrounding areas (including the Indian Ocean, Arabian Sea and BOB) might be the main mechanism maintaining the Indian monsoon circulation. In particular, the former has more influence. However, the thermal contrasts between the Indochina Peninsula and the South China Sea (SCS), Australia and the West Pacific play a crucial role in forming the tropical monsoon of the EASM, which indicates it is reasonable to divide the Asian monsoon system into East Asian monsoon and South Asian monsoon[28].

4 Onset of the ASM and its mechanisms

4.1 Onset time and characteristics

Each year the ASM in the 10°~20°N latitudinal band breaks out initially in the BOB east of 90°E, ICP and SCS, then expands westward and northward, with the EASM and Indian Summer Monsoon establishing themselves one after the other. The Indian Summer Monsoon breaks out later than the EASM[29-30]. The rapid northward progression of tropical convection in the Sumatra in late April and early May is the first sign of such a process[31-35].

The Asian tropical summer monsoon is established initially in the equatorial East Indian Ocean and Sumatra in late April (Pentad 24) and then in the east of the BOB and Indochina Peninsula in the 2nd pentad of May (Pentad 26), signifying the outbreak of summer monsoon in these regions. The tropical summer monsoon proceeds northeastward to the central SCS in the 4th pentad of May (Pentad 28), signifying the outbreak of the South China Sea Monsoon (SCSM). It then advances northwestward in late May and early the June, with its front arriving at the southern Indian peninsula. Two tropical summer monsoons from the west coast of India and BOB converge in the central Indian Peninsula in the 1st and 2nd pentads of June (Pentads 31 and 32), which signifies the outbreak of the Indian Monsoon. They then arrive in the north-west of India in early July, signifying the establishment of tropical summer monsoon in the whole of India.

The process of summer monsoon onset in East Asia seems complex. It is well known that tropical summer monsoon has proceeded to the Yangtze River and its southern areas in the 1st and 2nd pentads of June after it has initially broken out in the SCS. However, Zhao et al.[36-37] pointed out that persistent low-level southwesterly winds occur at their earliest over southeastern China and gradually advance northward, resulting in the onset of large-scale southwesterly winds over eastern China. Moreover, southerly winds extend southward to the SCS. This suggests the EASM breaks out at its earliest over southeastern China.

Accompanied by the northward march of the EASM, summer monsoon rainfall occurs in South China in the first flood period in mid and late May, moves northward rapidly in early June, and brings the mei-yu front to the Yangtze River valley and the mei-yu front to Japan in mid June. The mei-yu front lasts in the Yangtze River valley and then proceeds rapidly northward to North China in early July when the North China rainy season begins. In mid July, front of summer monsoon arrives in Northeast China, which is the northernmost position of the ASM. The entire process is represented by three still phases and two abrupt jumps, which are closely associated with the activity of the Western Pacific Subtropical High (WPSH)[38]. Jiang et al.[39] distinguished between the mei-yu front in the Huaihe-River and south of the Yangtze River, and pointed out that the rain belt has three jumps northward.

It can be seen from the differences of 500-200-hPa thickness between 20°N and 5°N that the difference turns from negative to positive in early May, first in the longitude of 100°E, i. e., the ICP, and expands east-ward to the SCS in the 4th pentad of May (Pentad 28) and westward to the Indian subcontinent in the 1st and 2nd pentads of June (Pentads 31 and 32), which corresponds to the onset of summer monsoon in these regions. In addition, the axis of maximal thermal contrast tilts obviously to the west with time, indicating the maximal thermal gradient moves regularly west-ward. Therefore, the reverse of the meridional thermal gradient and the westward propagation of the maximal gradient are the manifestation of Asian monsoon advancing from east to west after breaking out in the tropical east Indian Ocean and Indochina Peninsula.

In summarizing the SCSMEX, Ding[38] revealed a series of important phenomena accompanying the outbreak of monsoon as follows: cross-equatorial flow develops in the equatorial East Indian Ocean and Somali; heat sources in the Indochina Peninsula, South China, Tibetan Plateau and surrounding areas strengthen rapidly in seasonal transition; lower-level westerlies accelerate in the equatorial East Indian Ocean; the Subtropical High belt splits in the BOB and a monsoon low or cyclonic circulation forms; tropical southwesterlies expand eastward from the tropical East Indian Ocean; rainy season arrives in the BOB and Indochina Peninsula; under the influence of mid-latitudinal systems, southwesterlies expand further to the SCS; the main body of the Subtropical High weakens significantly and withdraws eastward; convective cloud, rainfall, low-level southwesterlies and upper-level northeasterlies develop abruptly in the SCS, and so on. These achievements provide favorable references for researching Asian and East Asian monsoon.

4.2 Onset mechanisms

4.2.1 Impact of the Tibetan Plateau on the onset position and intensity of the ASM

The "heating pulley" action of the Plateau strengthens the southerlies, increases rainfall, and strengthens the latent heat over the land to the southeast of the Plateau; meanwhile, it produces northerlies, decreases rainfall, and strengthens the sensible heat over the land surface to the southwest of the Plateau. The Plateau anchors the location of the ASM onset. In the background of the tropical land-sea distribution, the ASM breaks out initially on the east side of the ocean and west side of the continent to the southeast of the Plateau, and thus the monsoon rainfall in Asia is redistributed. The initial outbreak of the ASM in the east of the BOB is associated with obvious heating in the south side of the Plateau in spring. It has been proven, after analyzing the evolution of the common boundary of easterlies and westerlies, that in the lower troposphere in the Asian monsoon region the transition from the prevalence of easterlies in winter to southwesterlies in summer occurs firstly in the east of the BOB due to heating of the Plateau in spring, accompanied by drastic convective rainfall in its east. Therefore, the region of "east of the BOB-west of the Indochina Peninsula" may be where Asian monsoon initially breaks out[26-27,40-41].

4.2.2 Impact of the onset of the BOB Monsoon on the onset of the SCSM

The introduction of convective latent heating over the BOB results in vigorous ascending motion and onset of the BOB Monsoon, as well as the development of westerlies and vertical ascent over the northern SCS due to an asymmetric Rossby-wave response. Together with low-level moisture advection, convection is induced over the northern SCS. It is the condensation heating over the northern SCS that causes the overturning of the meridional gradient of temperature. Consequently, the vertical slope of the ridge of the subtropical high over the SCS turns from a winter pattern to a summer pattern according to the relation of the thermal wind, i. e. , the subtropical high in the low level weakens and moves southward. Eventually, as convection develops over the entire SCS domain, the subtropical high moves out of the region and the SCSM breaks out [42-43].

4.2.3 The antecedent effect of the "Asian-Australian land bridge" on the onset of the ASM

Within the South Asian region (10°~20°N), con-vection flares up initially in the ICP, then in the BOB and SCS, and lastly in India. The convection activity in the ICP is related to sensible heating of the Earth's surface there, but more importantly, it is the result of the northward progression of the tropical convection in Sumatra and the arrival of the South Asia High in this region. The latent heating in the ICP is favorable for triggering a cyclonic circulation on its west side, while the sensible heating in the Indian Peninsula favors triggering of a cyclonic circulation or trough on its east side. The effective match of these two, along with the around-flow and thermodynamic effects of the Plateau and the thermodynamic effect in the middle and high latitudes in East Asia, helps the subtropical high belt in South Asia to split first in the BOB. Accompanied by the formation and strengthening of the BOB trough, the eastern BOB is controlled by tropical southwesterlies in front of the trough, then large quantities of moisture are transported to the SCS where upon they converge with the flow from the west side of the subtropical high. Therefore, the atmosphere is destabilized, convection perks up, and summer monsoon is established. In the meantime, convective latent heating in the BOB triggers two-dimensional asymmetric Rossby Waves, which generate diversion of the meridional gradient of temperature in the SCS and favor the onset of the SCSM. The rapid eastward retreat of the eastern subtropical high after the belt splits results directly in the onset of the SCSM and its explosive characteristics. As India is controlled by northwesterlies in front of the ridge (i. e. , behind the BOB trough) after the high belt breaks, it is unfavorable for the onset of summer monsoon in India. Therefore, summer monsoon is established lastly in India. As a matter of fact, the break of the subtropical high belt, the occurrence of the BOB trough, the eastward retreat of the subtropical high, and the onset of the SCSM are completed rapidly and accompanied by the seasonal abrupt change of large-scale circulation and moisture transport in Asia. However, in terms of the source of the onset of summer monsoon, the rapid northward progression of the tropical convection in Sumatra in late April and early May is the earliest sign of the onset of the ASM[24,33,41-42,44-54]. This process is represented in Fig. 3.

4.2.4 Impact of the thermal anomalies of the oceans on the onset of the SCSM

A series of studies have been carried out on the impact of anomalous SSTs in the tropics on the onset of the SCSM. Many researchers have suggested that the preceding SST anomalies (SSTAs) in the tropical Indian Ocean and Pacific Ocean have a good relation with the onset of the SCSM. Wen et al.[55] indicated that the sign of the SSTAs in the tropical Indian Ocean and Pacific Ocean are opposite in some years, so their impacts on the early or late onset of the SCSM are different. The negative (positive) SSTA in the tropical Indian Ocean is favorable (unfavorable) to the early establishment of the anti-Walker Circulation over South Asia; while the negative (positive) SSTAs in the tropical Pacific Ocean favor (disfavor) the early strengthening of the Walker Circulation over the Pacific. The SSTAs in the tropical Indian

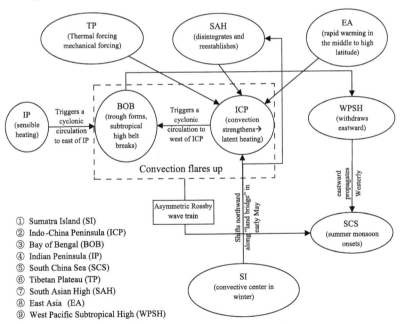

Fig. 3 Relationships among the northwestward movement of convection in the Sumatra along the "land bridge", the onset of convection in the Indochina Peninsula, the break of the subtropical high belt, the formation of the BOB trough, and the onset of the SCS Summer Monsoon[47]

Ocean and Pacific Ocean have an impact on the early or late onset of the SCSM through influencing the Walker Circulation.

Warm SST in the SCS in winter and spring is favor-able for the formation of monsoon circulation throughout all levels of the atmosphere over the sea, which hastens the onset of the SCSM. The effect of cold SST is generally the opposite. The local land-sea thermal contrasts in the SCS are one of the possible reasons for the SCSM onset [56].

4.2.5 Impacts of anomalous atmospheric circulation in different latitudes on the onset of the SCSM

The SCSM onset is also influenced by low-latitude circulation and anomalous atmospheric circulation in the mid and high latitudes. Through studying the impacts of mid-and high-latitude atmospheric circulation anomalies, and the activities of 30-60-day low-frequency convection over low latitude areas on the onset of the SCSM, Wen et al. [57] discovered that when there exists an anomalous wave train with negative anomalies of geopotential height field (low-frequency cyclone) over the Ural Mountains and its western region and along the seacoast of eastern China, along with positive anomalies (low-frequency anticyclone) over the mid-latitude continent and Sea of Okhotsk during 1-15 May, the ridge of the subtropical high will withdraw earlier from the SCS. At the same time, the low-frequency convection over the eastern part of the BOB is active and moves eastward, develops and moves westward around the Philippines, is active and moves southward over South China, and is also active and moves northward over Kalimantan. In this case, the establishment of the SCSM is earlier.

4.3 Integrated mechanism of the onset of the EASM

As shown in Fig. 4, due to the existence of the Indian Peninsula and its corresponding surface sensible heating, the BOB trough strengthens, the convection and precipitation strengthen initially in this area, and the subtropical high weakens and breaks first here in the seasonal progression of the atmospheric circulation. The ascending motion is evidently set up in the tropical eastern Indian Ocean-Indochina Peninsula and SCS region, subsequently under the effect of the positive vortex advection (PVA) in front of the BOB trough, while the northwesterlies behind the trough are unfavorable for the establishment of southerlies in India and the development of ascending motion (there is descent behind the trough). On the other hand, the cooperation of the warm westerly advection over the eastern Plateau and the sensible and latent heat over the Indochina Peninsula reverses the temperature gradient in the area, builds the upper-level easterlies, and forces the subtropical high to withdraw eastward. The weakening and withdrawal of the subtropical high conduce to the further development and strengthening of convection and precipitation, and then the tropical eastern Indian Ocean-Indochina Peninsula summer monsoon breaks out. Correspondingly, the upper-level westerlies weaken and are entirely substituted by tropical easterlies; the low-level westerlies strengthen and advance eastward to the SCS area. In the meantime, the 30-60-day and 10-20-

day low-frequency oscillations from the east and west in the tropics propagating to the north are phase-locked in the SCS and its surrounding areas around mid May, which triggers the rapid onset of the SCSM[58]. Cold air from the north can also trigger the onset of summer monsoon in some years[18,59].

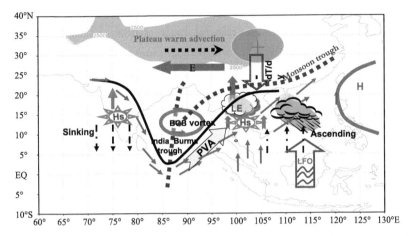

Fig. 4 Sketch map of the onset mechanism of the ASM based on the outcome of the South China Sea Monsoon Experiment (SCSMEX) [58]

5 Moisture transport of Asian monsoon

The activities of the ASM are closely associated with moisture transport. Based on vertically integrated moisture transport, Fasullo and Webster[60] employed the Hydrologic Onset and Withdrawal Index (HOWI) to investigate the interannual variation of the onset date of Indian summer monsoon. The results showed that there is a close relationship between moisture transport and the onset of summer monsoon. A planetary-scale water vapor transport band of high values is formed through the converging of large-scale water vapor transport bands in the Asian-Australian regions in summer. This starts from the Southern Hemisphere, crosses the Asian monsoon region, and then flows to the North Pacific. The north border of moisture transfer by southerlies is near 50°N in Northeast China, and the west border of moisture transfer by southeasterlies from the south side of the WPSH is near 100°E in the southeast of Gansu Province[61]. The relations between moisture transport from the Indian monsoon region and that over East Asia and their influences on summer rainfall in China have been investigated. It is found that the distribution and transport of moisture in the East Asian monsoon region differed greatly from those in the Indian monsoon region[62]. That is to say, the former has larger meridional transport, with moisture convergence and divergence resulting primarily from moisture advection, while the latter has more zonal transport and moisture convergence and divergence resulting primarily from convergence and divergence of the wind field. Moreover, the moisture transport from the Indian monsoon region is inverse to that over East Asia, i. e. more Indian monsoon moisture transport corresponds to less moisture transport over East Asia and less rainfall in the middle and lower reaches of the Yangtze River valley [63].

Further studies have shown that there are four moisture corridors influencing summer rainfall in China: the southwest corridor, SCS corridor, southeast corridor from the low latitudes, and the weak northwest corridor from the high latitudes. These represent the impacts of the South Asian Monsoon, the SCSM, subtropical monsoon and mid-latitude westerlies on summer rainfall in China, respectively. The region affected by the EASM is located to the east of 100°E. The southwest corridor is the moisture source for central South China, Southwest China and Northwest China; the SCS corridor influences rainfall directly in South China; the southeast corridor transfers moisture to the Yangtze River valley; and the northwest corridor transports moisture to the mid and upper reaches of the Yellow River and eastern North China [64-65].

There are bifurcations on the primary corridor of moisture transport to the Yangtze River valley. For instance, Xu et al. [66] considered that the moisture transport of the mei-yu belt has a structure of multi-sources, with the BOB, SCS and the tropical western Pacific the primary sources. There is also a moisture corridor in the mid and high latitudes in the western Pacific, which converges with the former in the Yangtze River valley. The moisture flows from the SCS and the Indian Ocean, converges in the BOB, then transfers northward and turns eastward by the dynamic forcing of the Qinghai-Tibetan Plateau to the Yangtze River valley, which forms the principal moisture corridor of the mei-yu rain belt in the Yangtze River valley.

The water vapor transport in the SCS is closely related to strong rainfall in China. The moisture over the SCS is mainly from the West Pacific before the monsoon onset, while it is from the tropical eastern Indian Ocean and BOB during the onset pentad. After the SCSM onset, moisture from the west side through the Indochina Peninsula to the SCS increases significantly, which becomes the primary water vapor source, and forms an obvious moisture source region in the SCS where large quantities of moisture are accumulated and transferred northward to South China and the Yangtze River valley, providing necessary moisture conditions for strong precipitation [34,59,67].

6 Variability of monsoon and its relationship with other circulation systems

6.1 Intraseasonal oscillation of East Asian monsoon

There is a close relationship between the onset of summer monsoon and intraseasonal oscillation (ISO). Low-frequency zonal westerlies appear two days earlier than the SCSM onset; strong development of at-mospheric ISO to the east of the Philippines and its westward extension play an important role in the ISO of the atmosphere in the SCS and the onset of summer monsoon[68]. The intraseasonal oscillation of the EASM is arranged in a pattern of wave trains along the seacoast of East Asia, and is represented by the northward propagation of monsoon surge with time. The monsoon surge consists of several ISO wet phases, and summer monsoon breaks out when the ISO wet phase is introduced or developed[69-70]. Quasi-biweekly oscillation of the tropical atmosphere has a critical role in establishing the processes of summer monsoon. Investigations have found that coherent adjustments of the convective disturbance and wind fields associated with atmospheric Rossby-wave response may be an important maintaining mechanism of the tropical quasi-biweekly oscillation[53].

Intraseasonal oscillation is one of the factors affect-ing the early or late onset of summer monsoon. Using 30-60-day filtered data of OLR averaged over the SCS for 1981-1996, Wen et al.[71] defined an index to describe the "low-frequency convection outbreak" of the SCS and analyzed ten "low-frequency convection outbreak" years synthetically. They pointed out that the dry phase of ISO in the low latitudes inhibits the development of southwesterlies and cyclonic circulation in the SCS before the outbreak of low-frequency convection; together with the northeastward propagation of ISO in the tropical Indian Ocean areas, there appears anomalous low-frequency cyclonic circulation in the SCS, and the low-frequency southwesterlies and convection strengthen rapidly before the SCSM breaks out.

Two preferential modes (30-60 days and 10-20 days) may play a significant part in the adjustment of the EASM. Analysis of data from 1998 indicates that the activity of the SCS is mainly controlled by 30-60-day oscillation, but adjusted by the 10-20-day mode. The low-frequency oscillation waves propagating northward in the form of a wave train and monsoon surge along the East Asia coast connect activities in the tropics and subtropics, which results in contrary phases in these two regions. Quasi-30-60-day oscillation is obvious in strong monsoon surge years, producing more rainfall in the mid and lower reaches of the Yangtze River; while it weakens in weak monsoon surge years when 10-20-day oscillation is the primary period, producing drought in the mid and lower reaches of the Yangtze River[12,38,69]. Analysis has shown that no matter whether the EASM is strong or weak, the westward propagation of the atmospheric ISO in the Pacific is stronger in every flood summer for several regions of East Asia, while it is weaker in every drought summer, indicating that strong or weak westward propagation of the ISO in the Pacific is the necessary condition for the precipitation amount in the EASM region[72].

ISO also has two significant modes in the western North Pacific (WNP) monsoon region, in which the 30-60-day mode is predominant[73]. Synthetic analysis of different phases has pointed out that the low-frequency convection and westerlies in the WNP propagate to the west and to the north. The monsoon rainfall, convection and active-break cycle in the WNP are adjusted by 30-60 and 10-20-day low-frequency oscillation to a great extent. However, 30-60 day oscillation in the SCS and tropical WNP in summer is also influenced by ENSO in the preceding winter[74].

The strongest center of kinetic energy in the trop-ical ASM region is located over 75°∼

95°E, with a sec-ondary center over the Somali jet channel, around 50°E. The disturbances of both kinetic energy and meridional wind are observed east of 90°E, mainly coming from the western Pacific and propagating westward to the BOB through the SCS. However, the propagation directions of both kinetic energy and meridional wind are rather disorderly between the BOB and the Somali jet channel. Therefore, the EASM and Indian Summer Monsoon are different in the propagation features of the disturbances of kinetic energy and meridional wind. The above facts indicate that the East Asian monsoon system undoubtedly exists even in the equatorial region, and quite distinct from the Indian monsoon system, it is mainly affected by the disturbances coming from the tropical western Pacific rather than from the Indian monsoon region. The boundary of the two monsoon systems is around 95°~100°E, which is more westward (5°~10°E) than its counterpart, as proposed in earlier studies[75-76]. In addition, Wu et al. [77-78] proposed a new concept of the droughts-floods coexistence during the normal summer monsoons which has close association with the EASM ISO anomalies.

6.2 Factors affecting the interannual variation of East Asian monsoon and its possible mechanisms

Huang et al. [79-80] summarized that the reasons for the interannual variation of East Asian monsoon are complicated and affected by many factors. The interannual variation of East Asian monsoon is jointly influenced by the Indian Monsoon, the WPSH, mid-latitude disturbances, the ENSO cycle, the warm pool in the tropical western Pacific, the thermal conditions of the tropical Indian Ocean, land surface processes, snow cover in Eurasia, the dynamic and thermal effects of the Tibetan Plateau, sea ice in the Arctic Ocean, and so on.

6.2.1 Warm pool in the western Pacific, its convective activity and interannual variation of the EASM

Huang et al. [81] perfected and developed view-points on the characteristics of the interannual variation of the onset and advance of the EASM and their associations with thermal states of the tropical western Pacific. Figure 5 shows that when the tropical western Pacific is in a warming (cooling) state in spring and summer, convective activities are intensified (weakened) around the Philippines. In this case, there is an anomalous cyclonic (anticyclonic) circulation in the lower troposphere over the SCS, and the WPSH shifts eastward (westward); thus, an earlier (later) onset of the SCSM can be caused. Moreover, since the WPSH shifts abruptly northward in mid June and early July, respectively, when the tropical western Pacific is warm, the abrupt northward shift of the EASM rain band from South China to the Yangtze River and the Huaihe River valleys is obvious in mid June and this monsoon rain band again jumps northward to the Yellow River valley, North China and Northeast China in early July. As a result, summer monsoon rainfall is below normal and drought may occur in the Yangtze River and the Huaihe River valleys, but summer rainfall is normal or above normal in the Yellow River valley, North China and Northeast China. On the other hand, when the tropical western Pacific is in a cooling state in spring and summer, the EASM rain band can be maintained in the Yangtze River and Huaihe River valleys. That is to say, the summer monsoon rainfall is above normal and flooding may occur in these two valleys, but summer rainfall is below normal and drought may occur in North China.

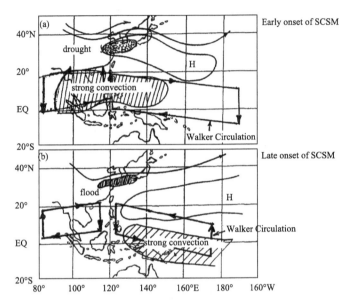

Fig. 5 Sketch map of the relationships among the thermal states of the tropical western Pacific (warm pool), the convective activities around the Philippines, the early or late onset of the SCSM, the Western Pacific Subtropical High, and the distribution of drought and flood in the Yangtze River and Huaihe River valleys: (a) warm pool in a warm state; and (b) warm pool in a cold state[81]

6.2.2 El Niño

Observational analyses and numerical experiments found that El Niño has a certain impact on the early or late onset of southwesterly monsoon in the SCS, which is less significant than the anomalous SST in the western Pacific has. In the preceding periods of strong (weak) SCSM years, the tropical SST is distributed as a La Niña (El Niño) pattern, in which the distribution of SST in December relates mostly to the intensity of the coming SCSM[82-84].

Zhang and Sumi [85] investigated features of the moisture circulation over East Asia during different El Niño episodes. It was found that in the El Niño mature phase, the anomalies of precipitation in China, water vapor transport and moisture divergence over East Asia differ from those in the rest of the phases, and the impact of El Niño on the East Asian climate is significant. The physical process through which El Niño affects the climate in East Asia was also discussed.

6.2.3 SST in the Indian Ocean

The Indian Ocean Dipole (IOD) has significant im-pacts on weather and climate in the East Asian monsoon region, especially in summer and in El Niño periods. In the positive phase of the IOD, southwesterly monsoon in East Asia breaks out later, with its intensity strengthening and rainfall increasing in China; while in the negative phase of the IOD, southwesterly monsoon in East Asia breaks out earlier, with its intensity weakening and plentiful rainfall in Southeast China[86]. Furthermore, the SSTA in the equatorial east Pacific and the IOD exert cooperative influences on climate variation over the East Asian monsoon region. The possible processes, through which the Indian Ocean SSTA impacts on the onset of the SCSM, are also different when the ENSO signal is included or removed[55,87]. Lu et al.[88] pointed out that SST in the Atlantic also has important impacts on the EASM and rainfall.

6.2.4 Snow state over the Tibetan Plateau

Numerical simulation shows that the increase of both snow cover and snow depth over the Tibetan Plateau can delay the onset and weaken the intensity of summer monsoon obviously, resulting in a decrease of precipitation in South China and North China and an increase in the Yangtze River and Huaihe River basins. The influence of winter snow depth is more substantial than that of both winter snow cover and spring snow depth. The snow anomalies over the Tibetan Plateau firstly change the soil moisture and surface temperature through snow melting processes; in the meantime, heat, moisture and radiation fluxes from the surface to the atmosphere are altered. Abnormal circulation conditions induced by changes of surface fluxes may in turn affect the underlying surface properties. Such a long period of interaction between the wetland and atmosphere is the key process resulting in later climate changes[89].

6.2.5 Somali jet

The Somali jet, the predominant cross-equatorial flow, plays a key role in water vapor transport between the two hemispheres. It transports water vapor through the equator from the Southern Hemisphere to the Northern Hemisphere during boreal summer time, and from the Northern Hemisphere to the Southern Hemisphere during boreal winter time. The interannual variation of the Somali jet is found to be linked with many changes around the globe, including the wave pattern along the East Asia coast, the South Asian High, the dipole pattern to the southeast of Australia, and the SSTA in the northern Indian Ocean in spring. Results have also revealed that interannual variation of the Somali jet in boreal spring has significant influences on East Asian summer rainfall and atmospheric circulation[90].

6.2.6 Circulations in the Southern Hemisphere

Studies on the interannual variation of the Mas-carene High and Australian High indicated that the former is controlled by Antarctic Oscillation (AAO), while the latter is related to both ENSO and AAO. In spring and summer, especially in spring, the intensity of the Mascarene High and Australian High is closely associated with summer rainfall in East Asia. When the Mascarene High strengthens in spring and summer, there is more rainfall from the Yangtze River valley to Japan and less rainfall from South China to the western Pacific east of Taiwan and the mid latitudes in East Asia. The influence of the Australian High on summer rainfall in East Asia is restricted in specific area. When the Australian High strengthens, rainfall is less in South China. The influence of the Australian High on East Asia monsoon is weaker than that of the Mascarene High, which plays a crucial role in the interactions between the atmospheric circulations of the two hemispheres. Studies have proven that the AAO is another strong interannual signal influencing summer rainfall in East Asia[91-92].

Nan and Li[93] found that there is a significantly positive correlation between the boreal spring southern hemisphere annular mode (SAM) and the following summer precipitation in the mid and lower reaches of the Yangtze River valley. While there is a strong SAM in spring, a pair of anomalous anticyclones exist in the Mongolian Plateau and Tianshan Mountains, respectively. Meanwhile the anomalous northerlies prevail from Northeast China to the mid latitudes of South China. These anomalous circulations may persist until the following summer and weaken the EASM; the west ridge of the WPSH strengthens and extends westward in summer following the spring of a strong SAM. These circulation anomalies are related to more precipitation in the mid and lower reaches of Yangtze River valley.

6.2.7 The activities of stationary planetary waves and East Asian winter monsoon

Chen et al.[94] investigated the relationship between the activities of stationary planetary waves and East Asian winter monsoon and suggested that in winter when the stationary planetary waves act frequently, the upward propagations of the waves from troposphere to stratosphere weaken, along with small polar vortex disturbance, which results in chilling and strengthening of the polar vortex. In the meantime, the westerly jet in East Asia, the East Asian trough, the Siberian High and the Aleutian Low become significantly weakened, which also weakens the northeasterlies over East Asia and warms this area. Further studies have pointed out that the planetary waves of zonal two-wave patterns play a leading role in the variability of East Asia winter monsoon.

6.3 Interdecadal variation of the Asian monsoon

There exists interdecadal variation in Asian monsoon circulation[14,26,36,95-97]. The EASM has been weakening since the 1960s on a decadal scale, with two abrupt interdecadal changes having appeared in the mid 1960s and late 1970s, but there is still dispute over whether or not such decadal weakening is related to global warming induced by human activity. Obvious interdecadal variation in the position and intensity of the WPSH also occurred in the mid 1970s. As a result, summer southwesterlies influence directly the areas from the lower reaches of the Yangtze River to Korea after the mid 1970s when rainfall was lower in North China and higher in the Yangtze River to Korea; while before the mid 1970s, summer southwest-erlies could reach North China, and then rainfall was higher in North China and lower in the Yangtze River[98-100]. The East Asian Winter Monsoon also has obvious interdecadal. The temperature in the north of East Asia has been rising significantly since the mid 1970s, which is directly influenced by the interdecadal variation of the East Asian Winter Monsoon[101].

The interdecadal variation of Asian monsoon may be associated with temperature decrease in the troposphere and the interdecadal variation of the sea thermal states. Wu et al.[26] found that in the two weakening processes of Indian summer monsoon in the mid 1960s and late 1970s, the thermal contrast in the troposphere decreases between East Asia and the tropical regions from the east Indian Ocean to the tropical western Pacific, which weakens the Indian Summer Monsoon circulation. In the last 20 years, the impact of El Niño on East Asian summer circulation has been increasing, and the warm phase of the Pacific Decadal Oscillation (PDO) has basically synchronized with the interdecadal variation of the atmospheric circulation in East Asia[102]. Yang et al.[103] pointed out that the states in the upper layer in the Pacific, and further revealed the related mechanism.

Another possible affecting factor is the Arctic Oscillation (AO)[104]. In the last 20 years, the Asian continent in winter and spring warms in mid and high latitudes and cools in lower latitudes, along with the trend in the AO toward its high-index polarity after the late 1970s. In the meantime, rainfall increases in the Tibetan Plateau and South China, which increases the soil moisture. Due to the memory of soil moisture, the cooling of the southern continent is maintained in summer and the warming of the Asian continent in summer slows down. Moreover, the Pacific to the east and the Indian Ocean to the south of Asia are warming from winter to summer; therefore, the land-sea thermal contrast decreases in summer, which results in a weakening of the EASM circulation.

6.4 Relationship between the ASM and other circulation systems

6.4.1 Relationship between summer rainfall in East Asia and the WPSH

The northward progression of summer monsoon in East Asia is closely associated with the variation of the subtropical high. The seasonal northward advancement of the subtropical high in March-July is represented mainly by three abrupt changes, corresponding temporally to the onset of the SCSM, the occurrence of the mei-yu and the end of the mei-yu, respectively[105].

When the ridge line of the WPSH shifts more southward than normal, or the ridge point shifts more west than normal, the EASM circulation is weaker. Correspondingly, there is an anti-cyclonic circulation in the anomalous wind field at 850-hPa over the tropical area in East Asia, and there is a cyclonic circulation over the subtropical area. The anomalous ascending motion at 500-hPa weakens over the tropical area in East Asia, while the ascending motion over the mei-yu frontal area strengthens. Meanwhile, at 500-hPa there is a blocking situation over the Sea of Okhotsk, which is located in the high latitudes of East Asia. The cold air from the high latitudes reaches the mid latitudes and strengthens the disturbance of the mei-yu Front. As a result, the rainfall in the Yangtze River valley is above normal. On the contrary, when the ridge line of the subtropical anticyclone over the western Pacific shifts more northward than normal, or the ridge point shifts more east than normal, the EASM circulation is stronger. The activities of the circulation systems appear opposite in anomaly patterns, and the rainfall in the Yangtze River valley is below normal[11]. Studies have also shown that the WPSH sometimes has two ridge lines, influencing the distribution of rainfall in eastern China and the genesis and lysis of mei-yu [106-107].

At the seasonal scale, the intensity of the WPSH, the position of its western boundary, and the intensity of the South Asia High are closely associated with the strength of the EASM, while the meridional position of the northern boundary of the WPSH and the blocking situation over the Urals area are related to the rainfall amount in East Asia [108].

Lu and Dong[109] pointed out that the monsoon rainfall in East Asia is influenced by not only the meridional, but also the east-west movement of the subtropical high. Lu also showed that the anomalous atmospheric convection over the Philippines impacts the east-west movement of the subtropical high through Rossby Waves, which modifies Gill's theory[110] and makes new progress.

6.4.2 Multi-scale conditions of strong summer monsoon rainfall

The east-west movement and meridional shift of the subtropical high determines the position of the EASM rain belt; the SCSM surge transports large amounts of warm and wet air from sea to land, ensuring abundant provision of moisture for continuous rainfall; cold air in the mid and high latitudes increases the moisture contrast of the air in the north and south, maintains and strengthens the mei-yu Front; and the meso-σ scale convective systems in the Plateau propagate eastward to the Yangtze River valley, promoting the formation and development of the meso-σ scale systems in the mei-yu Front. Such interactions among different scale systems provide the circulation conditions for the occurrence of continuous strong rainfall in the mid and lower reaches of the Yangtze River valley. When the above systems are best combined (locked), i.e., they are all in an active phase, it is common to see large-scale and long-lasting rainstorms, which may cause severe floods [111]. Gao et al. [112] and Zhou et al. [113] found "double fronts" (dew-point front and mei-yu front) structure of the summer monsoon rainfall in East Asia, and further suggested the conception of the mei-yu front system, which sheds light on the structure of summer monsoon rainfall.

6.4.3 Relationship between monsoon circulation and meso-scale systems

There is a kind of positive feedback mechanism between large-scale circulations and meso-scale convective systems. At the early stage of monsoon onset, the wide range background provides favorable synoptic and dynamic conditions for the summer monsoon onset and the formation of meso-scale convective activities; whereas after the summer monsoon onset, occurrence of persistent and large-scale meso-scale convective activities produces an obvious feedback effect on large-scale circulation. Because of the release of latent heat driven by enhanced convective activity, intense atmospheric heating appears over the northern SCS, which results in the meridional temperature gradient over the SCS reversing from the upper level to the lower level, and then the large-scale circulations become changed seasonally. Correspondingly, the surface pressure over the northern SCS deepens continually and forms broad monsoon troughs and obvious pressure-reducing areas, thus causing the subtropical high to eventually move out of the SCS. With the development of low-pressure circulations in the mid and lower troposphere, the meso-scale convective systems further enhance and extend southward, which is favorable for the onset of the SCSM and its maintenance over the central and southern SCS. The deepening of the monsoon trough promotes strengthening of monsoon flow and moisture transport on the southern side of it, and consequently the onset of monsoon prevails[34,114].

6.4.4 Relationship between monsoon and the westerly jet

Liao et al.[115] analyzed the activities of the subtropical westerly jet in boreal summer and the association with the distribution of anomalous SST in the equatorial-central-Pacific-subtropical-North-Pacific-extra-tropical-central-North-Pacific, which exert influences on the intensity of the South Asia High at 200-hPa, the anomaly of the EASM and the anomaly of rainfall in East China.

The upper westerly jet jumps northward twice during the transition process from winter to summer, which are periods closely related to EASM activities. Li et al.[116] emphasized that the northward jump of the upper westerly jet in East Asia occurs for the first time on average around 8 May, and it is about seven days earlier than the onset date of the SCSM (mean date is 15 May). The northward jump of the upper westerly jet over East Asia for the second time occurs on average around 7 June, and is about 10 days earlier than the beginning date of the mei-yu rainfall in the Yangtze River and Huaihe River basins (mean date is 18 June) and can be the forewarning of the commencement of mei-yu rainfall. These two-time northward jumps of the upper westerly jet are related to two-time reverses of meridional temperature gradient in the upper middle troposphere (500~200-hPa), respectively. During the seasonal transition, the continent is heated quickly, so that the meridional temperature gradient in the upper-middle troposphere will be reversed at $5°\sim25°N$ in South Asia. Then, through the geostrophic adjustment, the flow field adjusts to the pressure field (temperature field) and it will lead to the northward jumps of the upper westerly jet location. Analyses have also shown that sometimes the enhancement and northward movement of the upper westerly jet in the southern hemispheric subtropics can also influence the first northward jump of the upper westerly jet in East Asia.

6.4.5 Correlation with the anomalous circulation in the western Pacific

There is key relation between the tropical western Pacific circulation anomaly in winter and the following ASM. The winter anti-cyclonic (cyclonic) circulation anomaly in the tropical western Pacific moves gradually northeastward and expands westward, and the anomalous easterlies (westerlies) to the south of the anti-cyclonic (c-clonic) circulation extend westward to the Indian Peninsula, leading to the weakening (strengthening) of South Asian summer monsoon[79-80,117].

Li and his research group [118-119] paid special attention to the impact of the East Asian Winter Monsoon on the anomalous circulation in the western Pacific, and pointed out that an anomalous pattern of pressure exists for a long period in the western Pacific and Southeast Asia (Fig. 6). The different pressure gradients in the equatorial western Pacific region induced by strong (weak) East Asian winter monsoon play an important role in the formation of anomalous westerlies (easterlies) in the equatorial western Pacific; meanwhile, the anomalous westerlies (easterlies) formed in the equatorial western Pacific and the anomalous easterlies (westerlies) triggered in and around 20°N cooperate with the anomalous northerlies (southerlies) along the East Asian coast, favoring the anomalous cyclonic (anti-cyclonic) circulation to the east of the Philippines, which is a key factor influencing monsoon weather and climate and provides a critical signal for the prediction of weather and climate in China.

7 Problems and outlook

(1) How to describe quantitatively the intensity of the East Asian Monsoon and its interannual variability is still a basic scientific problem in the study of monsoon. Obviously, we expect using a simple and effective index to represent the intensity and variability of monsoon. The Southern Oscillation Index is the best example of employing one parameter to describe a complex phenomenon. An appropriate monsoon index is favorable to the quest of correlation among monsoon variability, other circulation systems and climate variability, to the establishment of association among monsoon variability and the inner and outer forcing factors, and to the objective evaluation of numerical models' potential to reproduce monsoon variability. Various monsoon indices so far have certain effects, but they are not conformable to all wishes. The key problem lies in that the East Asian Monsoon contains both the SCS western Pacific tropical monsoon and the China mainland/Japan subtropical monsoon. It covers a distance of 40-50 latitudes and is significantly influenced by mid and high latitudes. Therefore, the choice of East Asian monsoon index is much more difficult than the choice of Indian monsoon index. The circulation intensity and rainfall amount in the East Asian monsoon region have the characteristic of "distribution" rather than "unanimity", and so indices defined by different sub-areas and different parameters have no "comparability", and even

attain contrary conclusions. We believe a "common area" and a "common parameter" should be chosen to define an East Asian monsoon index that can reflect both the wind and rain, thus allowing scholars to discuss it in a "common language". Hence, further exploration is needed to obtain an accepted monsoon index.

(2) The essence of East Asian subtropical monsoon and its interaction with the tropical monsoon is a scientific problem that is not paid much attention. As mentioned above, East Asian monsoon contains both the tropical monsoon and subtropical monsoon, which interact and influence directly large-scale flooding and drought in China. But what actually is subtropical monsoon? Or what is the essence of subtropical monsoon? There are generally two misunderstandings: one is based on regions, i.e. monsoon prevalent in the subtropical areas in East Asia is called "East Asian subtropical monsoon"; the other considers the northward extension of tropical monsoon in East Asia as subtropical monsoon. Both neglect the reason why subtropical monsoon exists. We believe that the meridional land-sea thermal contrast in East Asia is the key driving force of tropical monsoon, while the seasonal cycle formed by the thermal contrast between Asian continent (including the Plateau) and the western Pacific may be an independent driving force of subtropical monsoon. Obviously, if there was no land-sea contrast between the Asian continent and western Pacific, the subtropical high belt would not break, and the East Asian tropical monsoon could not extend to Northeast China at around 50°N. We may imagine that the northward progression of East Asian monsoon is the result of coordination and interaction between the East Asian tropical monsoon and the East Asian subtropical monsoon. Therefore, we have to study thoroughly the essence of East Asian subtropical monsoon and its interaction with tropical monsoon, which will improve the prediction of flooding and drought in China.

(3) If the East Asian Subtropical Monsoon is an in-dependent system from tropical monsoon, what members are involved in the East Asian subtropical monsoon system? What about the seasonal cycle? When is the East Asian subtropical monsoon established and when does it end? It is related directly to the seasonal cycle in East China, so it is a scientific problem worthy of study.

(4) The thermal effect of the Tibetan Plateau strengthens the meridional land-sea (Asian continent and Indian Ocean) thermal contrast and the zonal land-sea thermal contrast (Asian continent and the Pacific). Especially, as an uplifted heat source (or cold source), the Plateau makes the seasonal transition of thermal contrast more sensitive and more advanced.

That is to say, the Plateau is a key and sensitive area for monitoring and predicting monsoon. Therefore, studies on the Plateau are very important.

(5) The East Asian subtropical summer monsoon rain belt locates on the forward side of summer monsoon, so the rain belt moves northward as the summer monsoon marches northward. Rainfall is relatively less in the controlling region of summer monsoon, i.e. the activity on the forward side of summer monsoon is directly related to the progression of the rain belt. Hereby, the research on monsoon edge is also a meaningful problem.

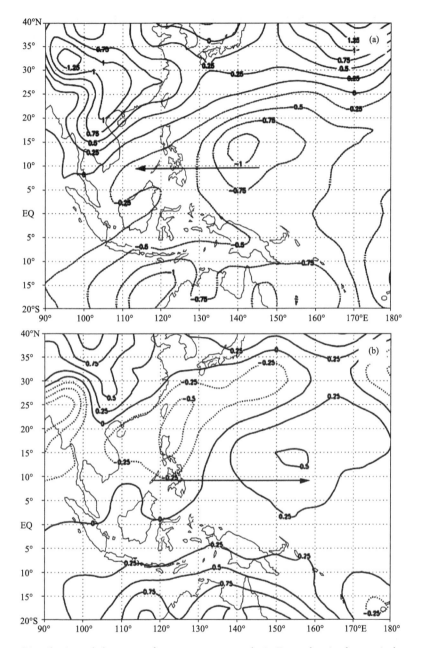

Fig. 6 Distribution of the sea surface pressure anomaly in December in the tropical western Pacific region in (a) strong and (b) weak East Asian Winter Monsoon[118]

(6) Asian monsoon is the outcome of interactions among the Earth, ocean, atmosphere, hydrosphere, biosphere, and cryosphere, and the evolving rule and variability of monsoon influences greatly the plantation, bio-Earth, chemistry, economy and society in the entire Asian monsoon region. Therefore, it is necessary to place monsoon in the coupled system of land-sea-atmosphere and to know its characteristics and mechanisms through studying the interactions among different spheres, layers, systems and scales.

(7) The seasonal prediction of monsoon activity is critical to national economy and social development, but the predicting ability of the present coupled model of land-sea-atmosphere has many difficulties. A proper description of the Tibetan Plateau in the model, a correct introduction of land-atmosphere processes, and a reasonable parameterization of physical processes are critical to upgrade the ability to predict the East Asian Monsoon. Thus, progress in the study of models is highly necessary.

(8) The change of the coverage on land and the atmospheric components at both regional and global scales induced by human activity may influence the future of Asian monsoon to a great extent. Research on this aspect helps to regulate human activity, realize the harmony between human and nature, and protect the living environment of human beings.

To sum up, the mechanisms of variation of East Asian monsoon, the mechanisms of East Asian monsoon influencing weather and climate in China, especially flooding and drought, and the theories and methods of predicting East Asian monsoon are still the primary aims in the study of monsoon in future.

Acknowledgements. The authors would like to thank all the authors of the references for their research achievements. We also wish to thank Chen Hua, Wei Jin and Zhu Xiaying for their kind assistance. This work was jointly supported by the National Natural Science Foundation of China (Grant Nos. 40633018; 40675056) and the key project A of the State Ministry of Science and Tech-nology of China "South China Sea Monsoon Experiment (SCSMEX)".

References

[1] Zhu K. Southeast monsoons and rainfall in China[J]. Acta Geographica Sinica,1934,1:1-27. (in Chinese)

[2] Tu C,Huang S,Gao Y. The advance and retreat of the summer monsoon[J]. Meteor Mag,1944,18:1-20.

[3] Ye D,Tao S,Li M. The abrupt change of circulation over the Northern Hemisphere during June and October[M]. The Atmosphere and the Sea in Motion,Rockefeller Institute,New York,1959:249-267.

[4] Chen L,Zhu Q,Luo H,et al. East Asian Monsoon[M]. Beijing:China Meteorological Press,1991:362. (in Chinese)

[5] Ding Y. Monsoons over China. Kluwer Academic Publishers,Dordrecht,Boston,London,1994:420.

[6] Chen W,Graf H-F,Huang R. The interannual variability of East Asian winter monsoon and its relation to the summer monsoon[J]. Adv Atmos Sci,2000,17:48-60.

[7] Wu B,Wang J. Winter Arctic Oscillation,Siberian high and East Asian winter monsoon[J]. Geophys Res Lett,2002,29:1897,doi:10.1029/2002GL015373.

[8] Yang S,Lau K-M,Kim K M. Variations of the East Asian jet stream and Asian-Pacific-American winter climate anomalies[J]. J Climate,2002,15:306-325.

[9] Jhun J-G,Lee E-J,2004. A new East Asian winter monsoon index and associated characteristics of the winter monsoon[J]. J Climate,17:711-726.

[10] Wu B,Zhang R,D'Arrigo R. Distinct modes of the East Asian winter monsoon[J]. Mon Wea Rev,2006,134:2165-2179.

[11] Zhang Q,Tao S,Chen L. The interannual variability of East Asian summer monsoon indices and its association with the pattern of general circulation over East Asia[J]. Acta Meteorologica Sinica,2003,61(4): 559-568. (in Chinese)

[12] Ju J,Zhao E. Impacts of the low frequency oscillation in East Asian summer monsoon on the drought and flooding in the middle and lower valley of the Yangtze River[J]. Journal of Tropical Meteorology,2005, 21(2):163-173. (in Chinese)

[13] He H,Sui C-H,Jian M,et al. The evolution of tropospheric temperature field and its relationship with the onset of Asian summer monsoon[J]. J Meteor Soc Japan,2003,81(5):1201-1223.

[14] Li J,Zeng Q. A new monsoon index, its interannual variability and relation with monsoon precipitation [J]. Climatic and Environmental Research,2005,10(3):351-365. (in Chinese)

[15] Li J,Zeng Q. A new monsoon index and the geographical distribution of the global monsoons[J]. Adv Atmos Sci,2003,20(2):299-302.

[16] Li J,Zeng Q. A unified monsoon index[J]. Geophys Res Lett,2003,29(8):1151-1154.

[17] Zeng Q,et al. The abrupt seasonal transitions in the atmospheric general circulation and the onset of monsoons. Part I :Basic theoretical method and its application to the analysis of climatological mean observations[J]. Climatic and Environmental Research,2005,10(3):285-302. (in Chinese)

[18] Zhang M,Zhang D,Zuo R,et al. The abrupt seasonal transitions in the atmospheric general circulation and the onset of monsoons. Part II :The onset of summer monsoon in South China Sea region[J]. Climatic and Environmental Research,2005,10(3):303-314. (in Chinese)

[19] He J,Ding Y,Gao H,et al. The Definition of Onset Date of South China Sea Summer Monsoon and the Monsoon Indices[M]. Beijing:China Meteorological Press,2001:123. (in Chinese)

[20] Yano J-I,McBride J L. An Aquaplanet Monsoon[J]. J Atmos Sci,1998,55:1373-1399.

[21] Chao W C. Multiple quasi equilibria of the ITCZ and the origin of monsoon onset[J]. J Atmos Sci,2000, 57:641-651.

[22] Chao W C,Chen B. Multiple quasi equilibria of the ITCZ and the origin of monsoon onset. Part II :Rotational ITCZ attractors[J]. J Atmos Sci,2001,58:2820-2831.

[23] Zeng Q,Li J. Interactions between the Northern and Southern Hemispheric atmospheres and the essence of monsoon[J]. Chinese J Atmos Sci,2002,26(4):433-448. (in Chinese)

[24] He J,Wang L,Xu H,et al. Climatic features of SCS summer monsoon onset and its possible mechanism [J]. Bimonthly of Xinjiang Meteorology,2004,27(1):1-7. (in Chinese)

[25] He J,Yu J,Shen X,et al. Research on mechanism and variability of East Asian monsoon[J]. Journal of Tropical Meteorology,2004,20(5):449-459. (in Chinese)

[26] Wu G,Liu Y,Liu X,et al. How the heating over the Tibetan Plateau affects the Asian climate in summer [J]. Chinese J Atmos Sci,2005,29(1):47-56. (in Chinese)

[27] Wu G,Mao J,Duan A,et al. Recent progress in the study on the impacts of Tibetan Plateau on Asian summer climate[J]. Acta Meteorologica Sinica,2004,62(5):528-540. (in Chinese)

[28] Jin Q,He J,Chen L,et al. Impact of ocean-continent distribution over southern Asian on the formation of summer monsoon[J]. Acta Meteorologica Sinica,2006,20(1):95-108.

[29] Mooley D A,Shukla J. Variability and forecasting of the summer monsoon rainfall over India[C]// Monsoon Meterorlogy. Chang C-P,Krishnamurti T N. Oxford:Oxford University Press,1987:26-59.

[30] Tao S,Chen L. A review of recent research on the East Asian summer monsoon in China[C]// Monsoon Meterorlogy. Chang C-P,Krishnamurti T N. Oxford:Oxford University Press,1987:60-92.

[31] He J,Zhu Q,Murakami M. TBB data revealed features of Asian-Australian monsoon seasonal transition

and Asian summer monsoon establishment[J]. Journal of Tropical Meteorology,1996,12(1):34-42. (in Chinese)

[32] Ding Y,Li C,He J. South China Sea monsoon experiment (SCSMEX) and the East-Asian monsoon[J]. Acta Meteorologica Sinica,2004,62(5):561-586. (in Chinese)

[33] Qian Y,Jiang J,Zhang Y,et al. The earliest onset area of the tropical Asian Summer monsoon and its mechanism[J]. Acta Meteorologica Sinica,2004,42(2):129-139. (in Chinese)

[34] Liu Y,Ding Y,Zhao N. A study on the meso-scale convective systems during summer monsoon onset over the South China Sea in 1998, I: analysis of large-scale fields for occurrence and development of meso-scale convective systems[J]. Acta Meteorologica Sinica,2005,63(4):431-442. (in Chinese)

[35] LÜ J,Zhang Q,Tao S,et al. The onset and advance of the Asian summer monsoon[J]. Chinese Sci Bull, 2006,51(1):80-88.

[36] Zhao P,Zhang R. Relationship of interannual variation between an Eastern Asia-Pacific dipole pressure pattern and East Asian monsoon[J]. Chinese J Atmos Sci,2006,30(2):307-316. (in Chinese)

[37] Zhao P,Zhang R,Liu J,et al. Onset of southwesterly wind over eastern China and associated atmospheric circulation and rainfall[J]. Clim Dyn,2007,28:797-811.

[38] Ding Y. Seasonal march of the East-Asia summer monsoon[C]//Chang C P. East Asian Monsoon. World Scientific Publishing,Singapore,2004,562-563.

[39] Jiang Z,He J,Li J,et al. Northerly advancement characteristics of the East Asian summer monsoon with its interdecadal variations[J]. Acta Geographica Sinica,2006,61(7):675-686. (in Chinese)

[40] Wu B. Weakening of Indian summer monsoon in recent decades[J]. Adv Atmos Sci,2005,22(1):21-29.

[41] Liang X,Liu Y,Wu G. Effect of Tibetan Plateau on the site of onset and intensity of the Asian summer monsoon[J]. Acta Meteorologica Sinica,2005,63(5):799-805. (in Chinese)

[42] Liu Y,Chen Z,Mao J,et al. Impacts of the onset of the Bay of Bengal monsoon on the onset of the South China Sea monsoon. Part I :A case study[J]. Acta Meteorologica Sinica,2003,61(1):1-10. (in Chinese)

[43] Liu Y,Chen Z,Wu G. Impacts of the onset of the Bay of Bengal monsoon on the onset of the South China Sea monsoon. Part II :Numerical experiments[J]. Acta Meteorologica Sinica,2003,61(1):11-19. (in Chinese)

[44] Xu H,He J,Dong M. Numerical study of the effect of Indian peninsula on South Asian summer monsoon process[J]. Journal of Tropical Meteorology,2001,17(2):117-124. (in Chinese)

[45] Xu H,He J,Wen M,et al. A numerical study of effects of the Indo-China Peninsula on the establishment and maintenance of the South China Sea summer monsoon[J]. Chinese J Atmos Sci,2002,26(3):330-342. (in Chinese)

[46] He J,Wen M,Shi X,et al. Splitting and eastward withdrawal of the subtropical high belt during the onset of the South China Sea summer monsoon and their possible mechanism[J]. Journal of Nanjing University (Natural Sciences),2002,38(3):318-330. (in Chinese)

[47] He J,Wen M,Wang L,et al. Characteristics of the onset of the Asian Summer Monsoon and the importance of Asian-Australian "land bridge" [J]. Adv Atmos Sci,2006,23(6):951-963.

[48] Zhang Y,Qian Y.:Mechanism of thermal features over the Indo-China Peninsula and possible effects on the onset of the South China Sea monsoon[J]. Adv Atmos Sci,2002,19(5):885-900.

[49] Luo M,Li C. A study of the onset of the summer monsoon over the South China Sea[J]. Climatic and Environmental Research,2004,9(3):495-509. (in Chinese)

[50] Wang L,He J,Guan Z. Characteristics of convection activities over Sumatra area and its relationship with SCS summer monsoon onset[J]. Journal of Nanjing Institute of Meteorology,2004,27(4):451-460. (in

Chinese)

[51] Wen M, He J, Xiao Z. Impact of the convection over the Indo-China Peninsula on the onset of SCS summer monsoon[J]. Chinese J Atmos Sci, 2004, 28(6):864-875. (in Chinese)

[52] Zhang Z, Chan J C L, Ding Y. Characteristics, evolution and mechanisms of the summer monsoon onset over Southeast Asia[J]. Int J Climatol, 2004, 24:1461-1482.

[53] Wen M, Zhang R. Possible maintaining mechanism of climatological atmospheric quasi-biweekly oscillation around Sumatra[J]. Chinese Sci Bull, 2005, 50(10):1054-1056.

[54] Zhou W, Chen Z, Li C. South China Sea summer monsoon onset in relation to the off equatorial ITCZ[J]. Adv Atmos Sci, 2005, 22(5):665-676.

[55] Wen Z, Liang Z, Wu L. The relationship between the Indian Ocean sea surface temperature anomaly and the onset of South China Sea summer monsoon, II: mechanism analysis[J]. Chinese J Atmos Sci, 2006, 30(6):1138-1146. (in Chinese)

[56] Ren X, Qian Y. Numerical simulation experiments of the impacts of local land-sea thermodynamic contrasts on the SCS summer monsoon onset[J]. Journal of Tropical Meteorology, 2003, 9(1):1-8.

[57] Wen Z, Huang R, He H, et al. The influences of anomalous atmospheric circulation over mid-high latitudes and the activities of 30-60d low frequency convection over low latitudes on the onset of the South China Sea summer monsoon[J]. Chinese J Atmos Sci, 2006, 30(5):952-964. (in Chinese)

[58] Liu Y, Ding Y. Analysis of the Basic Features of the onset of the Asian Summer Monsoon[J]. Acta Meteorological Sinica, 2007, 21(3):257-276.

[59] Ding Y, Chan J C L. The East Asian summer monsoon: An overview[J]. Meteor Atmos Phys, 2005, 89:117-142.

[60] Fasullo J, Webster P J. A hydrological definition of Indian monsoon onset and withdrawal[J]. J Climate, 2003, 16:3200-3211.

[61] Zhou C, He J, W Li, et al. Climatological characteristics of water vapor transfer over East Asia in summer [J]. Journal of Nanjing Institute of Meteorology, 2005, 28(1):18-27. (in Chinese)

[62] Huang R, Zhang Z, Huang G, et al. Characteristics of the water vapor transport in East Asian monsoon region and its difference from that in South Asian monsoon region in summer[J]. Scientia Atmospherica Sinica, 1998, 22(4):460-469. (in Chinese)

[63] Zhang R. Relations of water vapor transports from Indian monsoon with those over East Asia and the summer rainfall in China[J]. Adv Atmos Sci, 2001, 18:1005-1017.

[64] Tian H, Guo P, Lu W. Characteristics of vapor inflow corridors related to summer rainfall in China and impact factors[J]. Journal of Tropical Meteorology, 2004, 20(4):401-408. (in Chinese)

[65] Wang B, Huang Y, He J, et al. Relation between vapor transportation in the period of East Asian summer monsoon and drought in Northwest China[J]. Plateau Meteorology, 2004, 23(6):912-918. (in Chinese)

[66] Xu X, Chen L, Wang X, et al. Moisture sink and source for the mei-yu rain band in the Yangtze River valley[J]. Chinese Sci Bull, 2003, 48(21):2288-2294. (in Chinese)

[67] Ding Y, Hu G. A study on water vapor budget over China during the 1998 severe flood periods[J]. Acta Meteorologica Sinica, 2003, 61(2):129-145. (in Chinese)

[68] Li C. Recent progress in atmospheric intraseasonal oscillation research[J]. Progress in Natural Science, 2004, 14(7):734-741. (in Chinese)

[69] Ju J, Qian C, Cao J. The intraseasonal oscillation of East Asian summer monsoon[J]. Chinese J Atmos Sci, 2005, 29(2):187-194. (in Chinese)

[70] Qian W H, Kang H S, Lee D K. Seasonal march of Asian summer monsoon. Int J Climatol, 2000, 20(11):

1371-1378.

[71] Wen Z, He H, Huang R. The influences of 30-60d oscillation on the development of the South China Sea summer monsoon[J]. Acta Oceanologica Sinica, 2004, 23(4):568-579.

[72] Han R, Li W, Dong M. The impact of 30-60 day oscillations over the subtropical Pacific on the East Asian summer rainfall[J]. Acta Meteorologica Sinica, 2006, 64(2):149-163. (in Chinese)

[73] Wang H, Ding Y, He J. The climate research of summer monsoon over the western North Pacific[J]. Acta Meteorologica Sinica, 2005, 63(4):418-430. (in Chi-nese)

[74] Lu R, Ren B. The influence of ENSO on the seasonal convection evolution and the phase of 30-60-day oscillations during boreal summer[J]. J Meteor Soc Japan, 2005, 83:1025-1040.

[75] Chen L, Gao H, He J, et al. Zonal propagation of kinetic energy and convection in the South China Sea and Indian monsoon regions in boreal summer[J]. Science in China (Series D), 2004, 34(2):171-179. (in Chinese)

[76] Gao H, Chen L, He J, et al. Characteristics of zonal propagation of atmospheric kinetic energy at equatorial region in Asia[J]. Acta Meteorologica Sinica, 2005, 63(1):21-29. (in Chinese)

[77] Wu Z W, Li J P, He J H. The occurrence of droughts and floods during the normal summer monsoons in the mid- and lower reaches of the Yangtze River[J]. Geophys Res Lett, 2006, 33, L05813, doi:10.1029/2005GL024487.

[78] Wu Z W, Li J P, He J H. The large-scale atmospheric singularities and the summer long-cycle droughts-floods abrupt alternation in the middle and lower reaches of the Yangtze River[J]. Chinese Sci Bull, 2006, 51(16):2027-2034.

[79] Huang R, Chen W, Ding Y, et al. Studies on the monsoon dynamics and the interaction between monsoon and ENSO cycle[J]. Chinese J Atmos Sci, 2003, 27(4):484-502. (in Chinese)

[80] Huang R, Zhou L, Chen W. The progress of recent studies on the variability of the East Asian monsoon and their causes[J]. Adv Atmos Sci, 2003, 20(1):55-69.

[81] Huang R, Gu L, Xu Y, et al. Characteristics of the interannual variations of onset and advance of the East Asian summer monsoon and their associations with thermal states of the tropical western Pacific[J]. Chinese J Atmos Sci, 2005, 29(1):20-36. (in Chinese)

[82] Liang J, Wu S. A study of southwest monsoon onset date over the South China Sea and its impact factors [J]. Chinese J Atmos Sci, 2002, 26(6):829-841. (in Chinese)

[83] Liang J, Wu S. The study on the mechanism of SSTA in the Pacific Ocean affecting the onset of summer monsoon in the South China Sea-Numerical experiments[J]. Acta Oceanologica Sinica, 2003, 25(1):28-41. (in Chinese)

[84] Zhang, Y, Xie A, Dai N. Preceding features associated with the anomalous summer monsoon over the South China Sea[J]. Journal of Tropical Meteorology, 2004, 20(5):460-471. (in Chinese)

[85] Zhang R, Sumi A. Moisture circulation over East Asia during El Niño episode in northern winter, spring and autumn[J]. J Meteor Soc Japan, 2002, 80:213-227.

[86] Yan X, Zhang M. A study of the Indian Ocean dipole influence on climate variations over East Asian monsoon region[J]. Climatic and Environmental Research, 2004, 9(3):427-436. (in Chinese)

[87] Yan X, Zhang M. Numerical simulation of the Indian Ocean dipole influence on climate variations over East Asian monsoon region during equator east Pacific SSTA[J]. Journal of Tropical Meteorology, 2004, 20(4):376-382. (in Chinese)

[88] Lu R, Dong B, Ding H. Impact of the Atlantic multidecadal oscillation on the Asian summer monsoon[J]. Geophys Res Lett, 2006, 33, L24701, doi:10.1029/2006GL027655.

[89] Qian Y, Zhang Y, Zheng Y. Impacts of the Tibetan Plateau snow anomaly in winter and spring on precipitation in China in spring and summer[J]. Arid Meteorology, 2003, 21(3): 1-7. (in Chinese)

[90] Wang H, Xue F. Interannual variability of Somali jet and its influences on the inter-hemispheric water vapor transport and on the East Asian summer rainfall. Chinese Journal of Geophysics, 2003, 46(1): 18-25. (in Chinese)

[91] Xue F, Jiang D, Lan X, et al. Influence of the Mascarene high and Australian high on the summer monsoon in east Asia: Ensemble simulation[J]. Adv Atmos Sci, 2003, 20(5): 799-809.

[92] Xue F, Wang H, He J. Interannual variability of Mascarene high and Australian high and their influences on summer rainfall over East Asia[J]. Chinese Science Bulletin, 2003, 48(3): 287-291. (in Chinese)

[93] Nan S, Li J. The relationship between the summer precipitation in the Yangtze River Valley and the boreal spring Southern Hemisphere annular mode. I: Basic facts[J]. Acta Meteorologica Sinica, 2005, 63(6): 837-846. (in Chinese)

[94] Chen W, Yang S, Huang R. Relationship between stationary planetary wave activity and the East Asian winter monsoon[J]. J Geophys Res, 2005, 110, D14110, doi: 10.1029/2004JD005669.

[95] Fong S, Wang A, Tong T, et al. Climatological characteristics of the maintaining period of South China Sea summer monsoon, II: Interdecadal Variation[J]. Journal of Tropical Meteorology, 2005, 21(2): 123-130. (in Chinese)

[96] Jiang D, Wang H. Natural interdecadal weakening of East Asian summer monsoon in the late 20th century[J]. Chinese Sci Bull, 2005, 50(20): 2256-2262. (in Chinese)

[97] Zhao P, Zhou Z. East Asian subtropical summer monsoon index and its relationships to rainfall[J]. Acta Meteorologica Sinica, 2005, 63(6): 933-941. (in Chinese)

[98] Dai X, Wang P, Chou J. Multiscale characteristics of the rainy season rainfall and interdecadal decaying of summer monsoon in North China[J]. Chinese Sci Bull, 2003, 38(23): 2483-2487. (in Chinese)

[99] LÜ J, Ren J, Ju J. The interdecadal variability of East Asian Monsoon and its effect on the rainfall over China[J]. Journal of Tropical Meteorology, 2004, 20(1): 74-80. (in Chinese)

[100] Qian W. Review of variations of the summer monsoon from seasonal to interannual and interdecadal scales[J]. Journal of Tropical Meteorology, 2005, 21(2): 199-206. (in Chinese)

[101] Ju J, Ren J, LÜ J. Effect of interdecadal variation of Arctic Oscillation on temperature increasing in North of East Asian Winter[J]. Plateau Meteorology, 2004, 23(4): 429-434. (in Chinese)

[102] Wang Y, Kiyotoshi T. Decadal climate variability of rainfall around the middle and lower reaches of Yangtzs River and atmospheric circulation[J]. Journal of Tropical Meteorology, 2005, 21(4): 351-358. (in Chinese)

[103] Yang X, Xie Q, Zhu Y, et al. Decadal-to-interdecadal variability of precipitation in North China and associated atmospheric and oceanic anomaly patterns[J]. Chinese Journal of Geophysics, 2005, 48(4): 789-797. (in Chinese)

[104] Ju J, J LÜ, J Cao, et al. Possible impacts of the Arctic Oscillation on the interdecadal variation of summer monsoon rainfall in East Asia[J]. Adv Atmos Sci, 2005, 22(1): 39-48.

[105] Shu T, Luo H. Abrupt change of the west Pacific subtropical high and its interannual variation during the later spring and early summer[J]. Journal of Tropical Meteorology, 2003, 19(1): 17-26. (in Chinese)

[106] Zhan R, Li J, He J. Statistical characteristics of the double ridges of subtropical high in the Northern Hemisphere[J]. Chinese Sci Bull, 2005, 50(20): 2336-2341.

[107] Qi L, He J, Zhan R, et al. Characteristics of the western Pacific subtropical high double ridges process in 1962[J]. Chinese J Atmos Sci, 2006, 30(4): 682-692. (in Chinese)

[108] Liu C, Wang H, Jiang D. The configurable relationships between summer monsoon and precipitation over East Asia[J]. Chinese J Atmos Sci, 2004, 28(5): 700-712. (in Chinese)

[109] Lu R, Dong B. Westward extension of North Pacific subtropical high in summer[J]. J Meteor Soc Japan, 2001, 79: 1229-1241.

[110] Gill A E. Some simple solutions for heat-induced tropical circulation[J]. Quart J Roy Meteor Soc, 1980, 106: 447-462.

[111] Zhang S, Tao S, Zhang Q, et al. Multi-scale conditions of torrential rain causing floods in the mid-and lower-reaches of the Yangtze River[J]. Chinese Sci Bull, 2002, 47(6): 467-473. (in Chinese)

[112] Gao S, Zhou Y, Lei T. Structural features of the mei-yu front system[J]. Acta Meteorologica Sinica, 2002, 16(2): 195-204. (in Chinese)

[113] Zhou Y S, Gao S T, Shen S. A diagnostic study of formation and structures of the mei-yu front system over East Asia[J]. J Meteor Soc Japan, 2004, 82(6): 1565-1576.

[114] Liu Y, Ding Y. A study on the meso-scale convective systems during summer monsoon onset over the South China Sea in 1998, II: Effect of the meso-scale convective systems on large-scale fields[J]. Acta Meteorologica Sinica, 2005, 63(4): 443-454. (in Chinese)

[115] Liao Q, Gao S, Wang H, et al. Anomalies of the extratropical westerly jet in the North Hemisphere and their impacts on East Asian summer monsoon climate anomalies[J]. Chinese Journal of Geophysics, 2004, 47(1): 10-18. (in Chinese)

[116] Li C, Wang Z, Lin S, et al. The relationship between East Asian summer monsoon activity and northward jump of the upper westerly jet location[J]. Chinese J Atmos Sci, 2004, 28(5): 641-658. (in Chinese)

[117] Wu B, Wang D, Huang R. Relationship between sea level pressures of the winter tropical western Pacific and the subsequent Asian summer monsoon[J]. Adv Atmos Sci, 2003, 20(4): 496-510.

[118] Li C, Pei S, Pu Y. Dynamical influence of anomalous East Asian winter monsoon on zonal wind over the equatorial western Pacific[J]. Chinese Science Bulletin, 2005, 50(11): 1136-1141. (in Chinese)

[119] Pu Y, Pei S, Li C, et al. Influence of anomalous East Asian winter monsoon on zonal wind anomalies over the equatorial western Pacific[J]. Chinese J Atmos Sci, 2006, 30(1): 69-79. (in Chinese)

Discussion of Some Problems as to the East Asian Subtropical Monsoon[*]

He Jinhai[1]   Zhao Ping[2]   Zhu Congwen[2]   Zhang Renhe[2]
Tang Xu[3]   Chen Longxun[2]   Zhou Xiuji[2]

(1. Jiangsu Laboratory of Meteorological Disasters, Nanjing University of Information Science & Technology, Nanjing 210044; 2. CMA Academy of Meteorological Science, Beijing 100081; 3. Shanghai Meteorological Bureau, Shanghai 200030)

Abstract: Based on NCEP/NCAR gridded reanalysis, TRMM precipitation data, CMAP, and rainfall observations in East China, a study is conducted with focus on the timing and distinctive establishment of the rainy season of the East Asian subtropical monsoon (EASM) in relation to the South China Sea (SCS) tropical summer monsoon (SCSM). A possible mechanism for the EASM is investigated. The results suggest that 1) the EASM rainy season begins at first over the south of the Jiangnan region to the north of South China in late March to early April (i.e., pentads 16-18), and then the early flooding period in South China starts when southerly winds enhance and connective rainfall increases pronouncedly; 2) the establishment of the EASM rainy season is earlier than that of its counterpart, the SCSM. The EASM and the SCSM each is featured with its own independent rain belt, strong southwesterly wind, intense vertical motion, and robust low-level water vapor convergence. The SCSM interacts with the EASM, causing the EASM rainy belt to move northward. The two systems are responsible for the flood/droughts over the eastern China; and 3) in mid-late March, the eastern Asian landmass (especially the Tibetan Plateau) had its thermal condition changing from a cold to a heat source for the atmosphere. A reversal of the zonal thermal contrast and related temperature and pressure contrasts between the landmass and the western Pacific happens. The argument about whether or not the dynamic and thermal effects of the landmass really act ae a mechanism for the earlier establishment of the EASM rain belt is discussed and to be further clarified. Finally, the article presents some common understandings and disagreements regarding the EASM.

Key words: East Asian subtropical monsoon; South China Sea tropical monsoon

---

[*] 本文发表于《Acta Meteorologica Sinica》,2008 年第 22 卷第 4 期,419-434. 其中文版发表于《气象学报》,2008 年第 66 卷第 5 期,683-696.

1 Introduction

The land-sea distribution of East Asia with the Tibetan Plateau as its prominent terrain causes huge meridional and zonal thermal contrasts, resulting in a complex and unique East Asian monsoon system. One of the features of this monsoon system is that there coexist tropical and subtropical elements.

As indicated in Chen and Jin [1], as a tropical monsoon system, the South China Sea (SCS) monsoon circulation is independent of the Indian counterpart. Subsequently, Zhu et al. [2] showed that there is an SCS-western Pacific tropical monsoon system to the south of the western Pacific subtropical high (denoted as sub-high hereinafter) and also a subtropical monsoon circulation north of the sub-high that stretches from the China mainland to Japan. Tao and Chen[3] presented a complete and explicit picture of the East Asian subtropical monsoon system (see Fig. 1), of which the sub-high and Meiyu front are the principal East Asian subtropical monsoon members. This widely cited result contributes greatly to the development of the East Asian monsoon research. In the following decade and more, through the Sins-US and Sins-Japan cooperative monsoon projects such as the SCS Monsoon Experiment (SCSMEX), Chinese scientists achieved encouraging progress in this research field [4-13]. He et al. [14] made a review on the advances of recent monsoon research by Chinese investigators, finding that not many studies have been carried out on the subtropical monsoon. It is worth noting that the concept of the East Asian subtropical monsoon zone has been proposed on a geographical and biological basis. Gao [15], for example, delimited clearly the zone over $20°\sim50°N$ of eastern China. Later, Zhang et al. [16] demarcated the zone in latitude and longitude extent, and Li et al. [13] broadly determined the zone by means of a normalized monsoon index. Qian and Lee [17] further divided the subtropical monsoon zone from China to Japan into East and Northeast Asian subtropical monsoon regions. Through analyzing monsoon properties at a range of regions, Wang et al. [18] presented a division of the Asian summer monsoon zone into the SCS-western Pacific zone called the "northwestern Pacific monsoon area", and China to Japan subtropical zone referred to as the "Eastern Asian summer monsoon area". This delineation of the monsoon regions has drawn much attention of the meteorological researchers internationally.

To sum up, the East Asian subtropical monsoon (EASM) is a monsoon system that differs from the tropical monsoon. It has its own approximate domain (Fig. 1), which has been accepted by the meteorological community (Fig. 1) at home and abroad. But there are two misleading ideas about the EASM. One is associated with geography, without understanding the intrinsic features of the system, holding that the monsoon prevailing over the East Asian subtropics is by name none other than the East Asian subtropical monsoon. The other maintains that the northward extension of the SCSM is just the subtropical monsoon. The present work is intended to address such issues as how the EASM rainy season is established, how it marches with season towards the north, how the EASM is related to the SCSM, and what drives the EASM, with an ultimate aim to promote in-depth knowledge of the EASM.

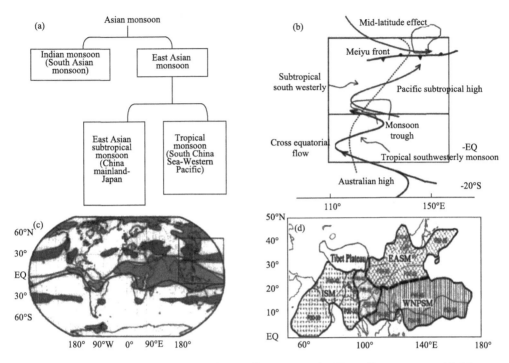

Fig. 1 The Asian monsoon regions from different sources. These figures are provided by (a) Zhu et al. [2], (b) Tao and Chen [3], (c) Li and Zeng [13], and (d) Wang et al. [18]

2 Data sources

The present study used the following datasets:

1) 1961-2000 NCEP/NCAR gridded reanalysis data with 2.5°×2.5° horizontal resolution;

2) 1998-2006 TRMM precipitation data with 2.5°×2.5° resolution;

3) 1981-2000 rainfall data at 693 stations in East China;

4) 1979-2000 precipitation data from Climate Prediction Center Merged Analysis of Precipitation (CMAP) with 2.5°×2.5° resolution.

Climatological means were performed with all the variables. The annual average was removed for some variables and "anomaly or departure" was used herein to reveal the seasonal cycle of the East Asian subtropical monsoon (EASM).

3 Establishment of the EASM rainy season

When and where is the EASM rainy season or rainbelt firstly established and what are its properties? These are the first and foremost problems. Many investigators, Chinese and foreign, have performed considerable studies. For example, the rainfall in late March to early April occurring in South China is denoted an initial stage of the early flooding period (see Ban[19]) and also called a springtime persistent precipitation stage[20]. Subsequently, Wan and Wu [21] made a pioneer study on the mechanism of the springtime rainfall. The April-June rainfall in South China is referred to as an early summer rainy season [5-6] and also as the South China-Taiwan Meiyu rainfall [22]. As indicated in Chen et al. [7], the EASM rainfall begins early in April in the north of South China and the Jiangnan region, spreading out south-and south-westward into the southern China seaboard and the Indo-China Peninsula by the end of April. Xu et al. [23] maintained that the EASM breaks out early in April over the central Jiangnan region. The southwest wind prevailing in East Asia to the Western Pacific happens first at subtropical latitudes, accompanied by the beginning of the subtropical rainy season[24]. Some other researchers pointed out that the seasonal reversal of land-sea thermal contrast between East Asia and the western Pacific takes place at first at subtropics in late March and early April, with the prevailing winter northerlies changed to summer southerlies in the lower troposphere, together with concurrently occurring connective precipitation, which possibly marks the EASM establishment (see Qi et al. [25]). As shown in Wang et al. [26-27], there are three rainbelts associated with the East Asian summer monsoon over East China happening, in order, in the Jiangnan zone, mid-lower Yangtze River Valley, and North-Northeast China, with their duration in the pentads 20-34, 35-39, and 40-44, respectively. This suggests that the earliest monsoon rainband starts early in April (pentad 20) over the Jiangnan region. For convenience in later use all the aforementioned viewpoints are collectively referred to as class A. In spite of the fact that the class A viewpoints hold that the EASM rainy season, i. e. , its rainbelt, or its sustained rainfall stage, is established first in late March to early April over the southern China-the Jiangnan zone, but the terms and connotations very greatly, as exemplified by such names as "early flooding season""Jiangnan spring rainy interval""early summer rainy season" "South China-Taiwan Meiyu precipitation period" "subtropical monsoon rainy stage", and "subtropical summer monsoon rainfall period".

However, Lau and Yang[28], Webster et al. [29], and Wang and Lin [18] indicated that the EASM rainband is established first around mid May over South China, followed by its northward advance. The dates of its appearance (figure omitted) show that only after the tropical summer monsoon starts over the SCS and its thereabouts is the EASM rainy season established. As seen from the northward moving contours of the establishment dates, the EASM establishment is thought to be the product of the SCSM traveling northward. In addition, Lian et al. [30] showed the EASM is established at pentad 26. Obviously, these viewpoints differ greatly from class A and are thus grouped as class B. Essentially, the key issue lies in how to define the rainy period lasting from late March to the time of the SCSM establishment. Could it belong to a part of the EASM rainy season? If this rainy period is cascaded with the South China rainy season, Jiang-Huai Meiyu precipitation, and North-Northeast

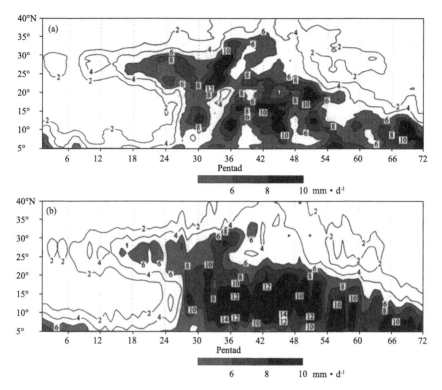

Fig. 2 Time-latitude sections of climatological precipitation averaged over 110°~120°E from TRMM (a) and CMAP (b)

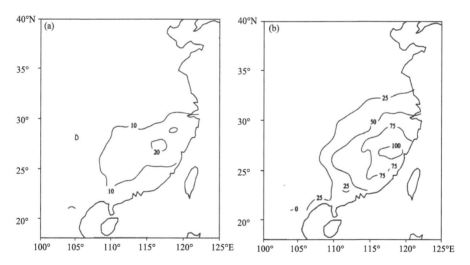

Fig. 3 Climatological station rainfall (0.1 mm · d^{-1}) for pentads 6 (a) and 16 (b) (from Zhao et al.[24])

China rainfall stages, those after the SCSM establishment around mid May, they constitute a complete seasonal cycle of the EASM. It is then natural to regard this late March to mid May rainy period as an early stage of the EASM rainy season. Thereby, the beginning of the

EASM rainy season should be shifted ahead from post mid-May to the end of March. This serves as a new line of thinking in predicting summer droughts and floods in the eastern China, so the study of the problem is of much significance in either the theoretical or the application perspective.

To further understand the above problems, we have to look at relevant facts. As shown in Fig. 2, rainfall over the southern China (25°~30°N) has been steadily over 6 mm · d^{-1} after pentad 16 (late in March), suggestive of the initiation of the rainband, with its maximum reaching 10 mm · d^{-1} (Fig. 3b), in comparison with a maximum of 2 mm · d^{-1} at pentad 6 (late January, Fig. 3a). It is of particular note that the rainband, after staying over for a certain time in South China, begins to move southward prior to the SCSM establishment, then advances northward in early June, reaches the Jiang-Huai Basin in mid-late June, and finally arrives in North-Northeast China in July-August. Therefore, the rainbelt produces, all the way to the north, an early flooding rainy reason in South China, the Meiyu rainfall over the Jiang-Huai Basin and a rainy period over North-Northeast China. Then the rainband starts to retreat southward rapidly in September. The aforementioned facts show that the EASM rainy season is established late in March (after pentad 16) in the Jiangnan region-southern China (25°~30°N), advancing northward after the SCSM establishment, and then retrogressing rapidly back to the south in September. The rainband experiences a complete seasonal cycle accompanied by a seasonal cycle of the high meridional gradient zone of θ_{se} (i.e., the frontal zone) and the sub-high, and the associated southerly winds undergo a similar seasonal cycle too, a statement that will be detailed later.

In summary, we come to a conclusion that from late March to the time before the SCSM establishment the persistent rainfall period (rainy episode) ought to be considered as a stage of the EASM rainfall. What are the properties of the precipitation? This is a problem to be answered next.

Fig. 4 presents the evolution of water vapor transfer around pentad 16 at lower levels, integrated from surface to 700-hPa. It is seen therefrom that the low-latitude water vapor transport by easterly winds is more intense at pentad 13, accompanied by southward cross-equatorial water vapor transfer, a situation that is typical for winter[31]. When it comes at pentad 16, a distinct water vapor passage is formed that stretches from the western Pacific to South China. And it is the convergence ahead of the strong water vapor passage that leads to the substantial increase of rainfall over the Jiangnan region-South China in late March. At that time, the SCSM and its related water vapor passage have not yet been established. At pentad 28 the water vapor transfer pattern undergoes a dramatic change: the low-latitude easterly water vapor transfer passage and the southward cross-equatorial transport disappear, instead another water vapor corridor is established that originates in the southern low-latitude ocean, crossing the equator in the neighborhood of Somali into the Arabian Sea, following the path through the southern Indian Ocean and Bay of Bengal into South China, a situation typical of summer water vapor transfer. Meanwhile, the SCSM breaks out.

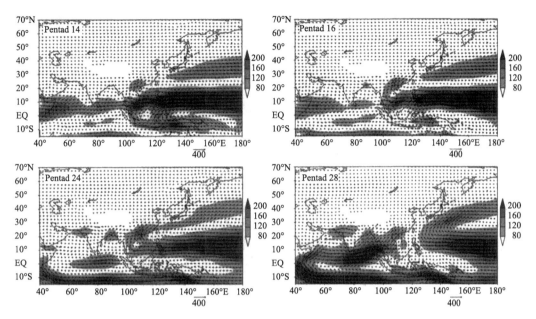

Fig. 4 Climatological low-level water vapor transport ($kg \cdot m^{-1} \cdot s^{-1}$) integrated from surface to 700-hPa, with shading for the transport $>80\ kg \cdot m^{-1} \cdot s^{-1}$

Fig. 5 gives latitude-height cross sections of time-dependent vertical velocity averaged between 110° and 120°E, indicating that the meridional distribution of the vertical motion at pentad 1 bears a case for winter, with subsidence everywhere except feeble updrafts observed at mid-lower levels over China in 25°~30°N; at pentad 16 the updrafts have covered the whole troposphere over these latitudes. The related precipitation intensity has arrived at 6 mm $\cdot$ d^{-1} (Fig. 2c), reaching the level of deep connective rainfall[28,32], during which sinking downdrafts are dominant over the SCS, suggesting that the SCSM has not yet been established there; at pentad 36 the deep convective band moves northward over the Jiang-Huai Basin, producing the so-called "Jiang-Huai Meiyu". At that time, the SCS zone is under the control of deep convection, indicative of the SCSM establishment.

In association with the deep convection establishment and development in the southern part of China, the latent heating height and strength are increased rapidly posterior to pentad 16 (Fig. 6). The foregoing analysis shows that the lasting precipitation of China in 25°~30°N after pentad 16 is of deeply connective character and is linked closely to the subsequent South China rainy season and Jiang-Huai Meiyu rainfall.

To summarize, we have sufficient reasons to give the statement that the sustained connective precipitation over the southern part of China in late March (after pentad 16) represents the beginning of the EASM rainy season, and the rainfall stage covering late March to the time before the SCSM establishment designates the incipient phase of the seasonal cycle of the EASM rainy season.

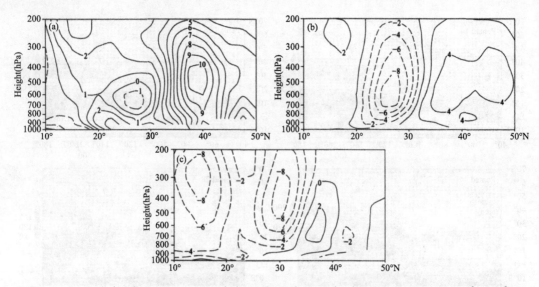

Fig. 5 Latitude-height cross sections of climatological vertical velocity averaged over 110°~120°E for pentad 6 (a), pentad 16 (b), and pentad 36 (c) (Zhao et al. [24])

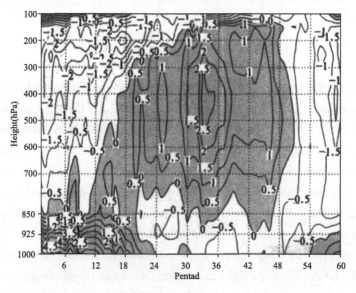

Fig. 6 Pentad-varying heating rate (°C · d^{-1}) in northern subtropical latitudes over 27.5°~32.5°N, 110°~140°E

4 Relationship of the EASM and SCSM

As mentioned earlier, the class B viewpoints hold that the EASM rainbelt is established after the tropical summer monsoon has been established over the SCS and its vicinity, and the EASM establishment is the product of the northward movement of the SCSM[3,18,28-29]. But the foregoing analyses in this study indicate that the persistent rainfall over South China in late March is of deep convective nature and can be classified into the cascade of rainfall events associated with the EASM. We suggest designating the late March persistent rainfall period as the beginning of the EASM rainy period and the initial stage of the EASM seasonal cycle. The establishment of the EASM rainy season is earlier than that of the SCSM rainfall. It is necessary to re-examine the relationship between the EASM and the SCSM.

The time-latitude sections of 850-hPa wind, precipitation anomalies, vorticity and moisture flux divergence are given in Fig. 7, showing that positive rainfall anomalies emerge first in $25°\sim32.5°$N around pentad 16 (late March), where the related northwesterly winds change to southwesterly flows, with the low-level water vapor flux convergence and cyclonic

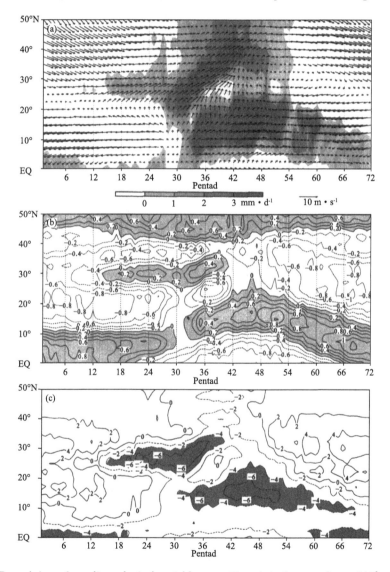

Fig. 7 Pentad-dependent climatological variables over East Asia (averaged over $110°\sim130°$E). (a) 850-hPa wind and rainfall anomalies (Shaded); (b) 850-hPa vorticity (units of 10^{-5} s^{-1}, cyclonic vorticity shaded); and (c) moisture flux divergence integrated from surface to 700-hPa

vorticity intensified greatly. Through the earlier analysis, the positive rainfall anomalies are associated with the EASM system, and the positive vorticity zone corresponds to the EASM trough. Wang et al. [26] indicated that the EASM trough differs in character from the SCSM trough; the former is formed earlier than the latter. When the EASM rainy season starts, to its south in the tropical latitudes easterly flows are maintained, water vapor flux divergence and negative vorticity dwell at lower levels, suggesting that the SCSM trough has not yet been established, and no precipitation appears. All these demonstrate that the SCSM has not broken out yet when the EASM already comes into play.

Afterwards, the EASM rainbelt continues to maintain, with southerly winds intensified somewhat, and the tropical wind filed remains nearly unaltered. The EASM rainband begins to spread out southward at pentad 24, reaching the zone south of 20°N at pentad 27; the rainfall anomaly in the SCS region (10°~20°N) changes to be positive at pentad 28, during which easterlies change to westerlies at lower levels, and the lower troposphere moisture fluxes change from divergence to convergence, suggesting the commencement of the SCSM.

Posterior to the SCSM commencement, the EASM rainbelt makes rapid northward movements, arriving at the mid-lower Yangtze River Basin in mid-late June to form the Jiang-Huai Meiyu period and at North-northeast China in July-August to form a North-Northeast China rainy season. When September comes, the rainbelt retreats southward rapidly. And at pentad 54 the rainfall anomalies change fast from positive to negative for the vast expanse north of 25°N and the related southerlies change into northerlies (Fig. 7a). This suggests that the EASM as a whole has retreated out of China, implying the advent of pleasant autumnal days. From the above analysis we see that the EASM rainbelt exhibits a progressive establishment and a swift retreat in sharp contrast to the SCSM, which is marked by an explosive establishment and a slow withdrawal.

During the span from the SCSM establishment to the EASM retreat, there are always two rainbands in the north and south with concentrated high rainfall values, corresponding to the rainbelt of the EASM and that of the SCSM. Two positive vorticity zones also exist, namely, the EASM trough and the SCSM trough; they are separated by a band of strong negative vorticity in between. A similar pattern shows up in the moisture flux convergence field as well. Moreover, two zones of upward motion of about the same intensity are separated by downward motion in the middle (Fig. 5c), with convective maximums located above 400-hPa. As shown in Yu and Mao[33] and Zhou et al. [34], the EASM has its own meridional circulation and heat source.

To sum up, the EASM rainband is established earlier than the SCSM rainbelt. After the SCSM establishment, there are two independent rainbands, each with its own related southwesterlies, strong vertical ascending areas and moisture flux convergence zones. The SCSM, after establishment, interacts with the EASM, causing the latter to make a seasonal northward advance. The two systems are responsible for the floods and droughts over the eastern China.

5 A possible mechanism for the EASM

The land-sea thermal contrast is the principal stimulus. Most researchers emphasize the longitudinal thermal contrast in East Asia (with the Tibetan Plateau) as an essential factor. However, in the subtropics, the plateau acting as an elevated heat (or cold) source increases also the zonal thermal contrast between the Asian landmass and the western Pacific, making the seasonal transition more sensitive and unique. Qian et al. [11] addressed first an important effect of the zonal thermal difference on summer monsoon, especially the tropical monsoon.

Fig. 8 gives the climatological pentad-varying 500-hPa zonal temperature difference (defined as the discrepancy between the temperature at a given longitude and the mean temperature averaged over 80°~150°E[25]) at some selected latitudes. We see that the difference increases poleward; south of 35°N the zonal thermal difference is characterized by distinct seasonal reversals: in late March the western Pacific changes from a warm into a cold region and the reverse is true for the East Asian landmass in the west; by the end of September an opposite reversal takes place in both regions. The timing and characteristics of the seasonal reversals at various latitudes differ from one latitude to another. Over 15°~35°N the warming of the East Asian landmass in the north becomes little by little earlier than that in the south, with a warm region appearing at pentad 12 (March) at 30°N, which may be related to the Tibetan Plateau heating at those latitudes. The seasonal evolution of the 20°~35°N mean zonal temperature difference is similar to that at 30°N, showing that the East Asian continent becomes already warm in March. Consequently, the zonal temperature difference at 30°N is selected to represent the East Asian subtropical thermal condition, and utilized to investigate a possible mechanism for the EASM.

Fig. 8 shows that the immediate neighborhood along 120°E serves roughly as a partition line between cold and warm areas, so we choose two longitudinal ranges: 120°~150°E (for the western Pacific) and 80°~120°E (for the East Asian continent) to calculate the zonal thermal contrast between land and sea at 30°N (Fig. 9a). The subtropical zonal thermal contrast changes from positive to negative in late March, forming a westward temperature gradient, while the tropical meridional thermal contrast is reversed at pentad 24 [35-36], lagged behind by one month relative to the above subtropical shift. This explains the discrepancy in the commencing time between the subtropical and tropical summer monsoons over East Asia. Besides, the difference of the vertically integrated heat source between land and ocean reveals that the thermal regime changes from heating (cooling) in the east (west) to the opposite in late March (Fig. 10). This demonstrates once again that the seasonal reversal happens first in late March.

As there occurs a seasonal reversal of zonal land-sea thermal contrast in East Asia in late March, the low-level sea level pressure (SLP) changes from a pattern with high (low) values in the west (east) into the opposite (Fig. 9b), accompanied by positive anomalies of precipi-

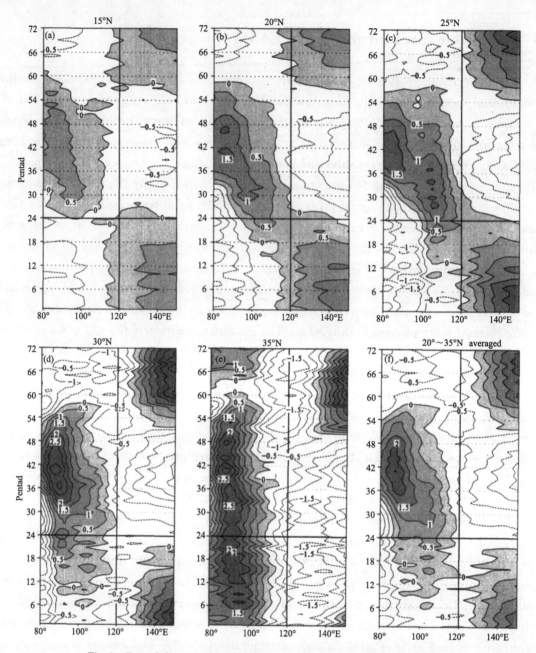

Fig. 8 Pentad-dependent zonal difference of 500-hPa temperature for the latitudes of East Asia (from Qi et al.[25])

tation in the subtropics (Fig. 9a). In the meantime, the lower-level winds veer from northerlies to southerlies. The southerlies originate from two branches of flows: 1) the southwesterly air flows deviated from the easterlies on the southern side of the sub-high over the western Pacific; and 2) the northwesterly air flows detouring along the southern side of the Tibetan Plateau into the subtropics [25].

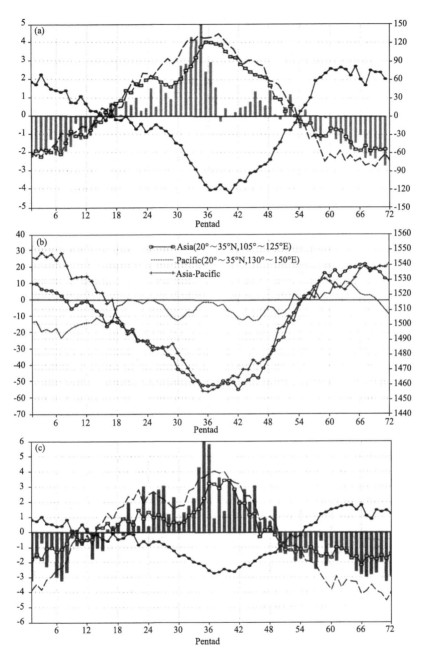

Fig. 9 (a) 500-hPa climatological zonal land-sea thermal contrast, averaged across 27.5°~32.5°N (see Sec. 5 for longitude ranges), denoted by solid circle chain; 850-hPa meridional wind in 27.5°~32.5°N, 110°~140°E (open square chain); precipitation anomalies (bars); and the vertically integrated heat source (dashed line) over the Tibetan Plateau in 27.5°~37.5°N, 80°~100°E. (b) Climatological 850-hPa geopotential height and its zonal difference. (c) Climatological 500-hPa zonal land-sea thermal contrast over 25°~35°N (see Sec. 5 for longitude ranges), denoted by solid circle chain; zonal SLP difference (long dash) (see Sec. 5 for longitude ranges); 850-hPa longitudinal wind anomalies averaged over 25°~35°N, 110°~120°E (annual mean removed, open circle chain); TRMM rainfall departures (bars)

In summary, as the seasonal reversal happens to the zonal land-sea thermal contrast, the EASM rainband is established and the lower-level winds veer from northerlies to southerlies. Such a scenario also holds true for other key regions of the Asian monsoon on the whole (Fig. 9c). Can the temporal consistency in the reversal of the zonal land-sea thermal contrast and low-level wind direction as well as the increase in convective precipitation attest that the East Asian zonal land-sea thermal contrast is the driving force of the EASM? This is a problem that deserves our further efforts.

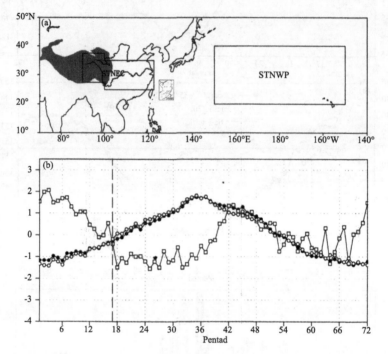

Fig. 10 (a) The land and sea domains selected to perform the domain average tropical (STNEC= subtropical NE continent and STNWP=subtropical NW Pacific). (b) Climatological evolution of the vertically integrated atmospheric heat source (normalized) over land (solid circles) and sea (hollow box), and their difference (normalized, hollow circles). This figure is finished by Chen Longxun

6 Conclusions and discussion

Based on the above analyses and discussion, the main conclusions in this paper are summarized as follows:

1) The EASM rainy season begins in late March over the Jiangnan region, with low-level southerlies intensified and connective rainfall enhanced greatly from then on.

2) The EASM rainband is persistent during March to May over the Jiangnan region, meanwhile expanding into South China prior to the SCSM establishment (mid-late May), thereby forming the initial phase of an early flooding season over South China. After the SCSM is established, the EASM marches northward, producing rainfall all the way, in

sequence, forming the posterior phase of the South China early rainy season, the Jiang-Huai Meiyu season and the rainy season of North-Northeast China.

3) The rainy season is established earlier for the EASM than for the SCSM. After the SCSM establishment, there are clear differences between the features of the two systems. Each system has its own rainbelt, strong southwestly winds, vigorous low-level moisture convergence and intense vertical motion. They also interact with each other. The SCSM initiation promotes the seasonal northward movement of the EASM rainband.

4) In March, the East Asian continent including the Tibetan Plateau turns from a cold to a warm heat source to the overlying atmosphere. The question about whether the temporal consistency in the reversal of the zonal land-sea thermal contrast and low-level wind direction as well as the increase in convective precipitation can attest that the East Asian zonal land-sea thermal contrast is the driving force of the EASM remains to be further elucidated.

The EASM is a key factor directly influencing the floods and droughts of China. To promote relevant studies and the application of research results, a panel meeting was held in Beijing on May 9, 2008, during which monsoon researchers discussed, debated, and achieved agreements on some issues, meanwhile disagreements and disputes also came forth[37].

The participants diverged regarding the rain-rich phase from late March to the time prior to the SCSM establishment, i.e., about whether the phased rainfall belongs to "spring rainfall" or "summer precipitation". Those who hold the view that it is the spring rains have their principal arguments as follows. 1) It is the frontal precipitation, 2) the summer atmospheric circulation pattern has not been established, and 3) the strengthened southerlies are caused mainly by the Tibetan dynamic effect.

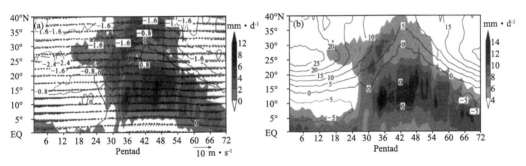

Fig. 11 Climatological evolutions of several variables averaged over the subtropics (100°~130°E). (a) Climatological $\partial \theta_{se}/\partial y$ (contour, unit: 10^{-6} K·m^{-1}; thick dotted line denotes $\partial^2 \theta_{se}/\partial y^2 = 0$); CMAP precipitation anomalies with annual mean removed (shaded, unit: mm·d^{-1}); and wind anomalies at 850-hPa with annual mean removed. (b) Climatological zonal wind (contour, unit: m·s^{-1} thick dotted line: $u=0$) and CMAP precipitation (shaded, unit: mm·d^{-1}). This figure is provided by Zhu et al.[2]

In our view, the above disagreement arises from a different understanding of the EASM. To solve the controversy, let us look at Fig. 11, where there is a stretch of high meridional gradient of θ_{se} over East Asia (the bold dashed line thereof), we may refer to as an East Asian frontal surface or frontal zone. It is worth noting that the frontal zone is on the northern side of the sub-high, with its annual cycle extremely consistent. This is in fact a manifestation of the meridional yearly cycle of the Meiyu front and sub-high as members of the East-Asian monsoon system, a result given by Tao and Chen[3]. Associated with the θ_{se} gradient cycle is the yearly cycle of the low-latitude rainband, which is actually a rejection of the meridional annual cycle of the ITCZ rainband. By contrast, the East Asian subtropical rainy zone is established late in March (pentads 16-18), undergoing a similar yearly cycle to that of the frontal zone and the robust southwesterlies. The EASM rainbelt is thus inferred to correlate with the frontal zone, and we can even further claim that the frontal rain is none other than an essential characteristic of the EASM rainband. If we assume that the tropical monsoon is caused by the ITCZ meridional shift driven by the seasonal change of the solar radiation and the meridional land-sea thermal contrast, then we deem it is justifiable as well to believe that the EASM and its meridional jumping rainband are driven by the seasonal variability of the solar radiation and the zonal land-sea thermal contrast. The interactions between the SCSM and the EASM determine the floods/droughts over China.

From the foregoing analysis we already know that the rainy season beginning in late March and the strengthening of the associated southerlies experience a progressive buildup and advance to the north, followed by a swift retreat equatorward, as opposed to an explosive onset and a slow retreat of the SCSM. This result agrees well with the findings of Wang[27]. In other words, the EASM establishment does not necessarily require an abrupt change of the large scale circulation pattern, nor a complete setup of the summer monsoon situation. The establishment means just the incipient phase of the seasonal cycle of the EASM rainy season. Only the Jiang-Huai Meiyu season is the prime stage of the summer monsoon rains, when the EASM is in full swing. For this reason, we maintain that the rain-rich phase over the Jiangnan region-South China between late March and the time prior to the SCSM onset is just a preceding stage of the early flooding rainfall period of South China (April-June). It is part of the seasonal cycle of the EASM rainy season and is thus the incipient phase or "pregnant" period of the monsoon rainy season.

Does this initial phase precipitation belong to "spring rains" or "summer precipitation"? If we refer to it as "summer rainfall", it seems unjustifiable because late March to early April (pentads 16-18) is the time shortly after the beginning of spring. The "spring rains" seems a more appropriate title. However, with the concept of "monsoon year", a year can be divided into two parts: winter (wetness) and prevalence of northerly (southerly) winds. During the summer half year, the EASM rainband is established in late March, developing and traveling northward till September, and retreating fast to the south in mid-late September, a period that covers just the summer half year. In the classic works, e. g., Great Britain Encyclopedia and Collier Encyclopedia, and from the meteorological glossary, we can find the definition of the reversal of large-scale meteorological conditions associated with the Asian monsoon on a half-year time frame. Therefore, the rainy period from late March to the time before the establishment of the SCSM is referred to as the start of the EASM rainy season or simply "early summer rainfall stage", which has been termed this way, for example, in Ding[5] and Wang[27].

Also, it needs to be noted that throughout the year there is a stretch of southerlies along the eastern flank of the Tibetan Plateau[24] where air flows detour commonly, and the southerlies have a significant seasonal variation. The rain-rich period in late March-early April in the Jiangnan region and southern China bears a close relation to the enhancement and eastward expansion of the southerly winds in the detouring zone, which may be the evidence of the Tibetan Plateau dynamic effect. In winter, the East Asian northerlies produced by the zonal thermal contrast, i. e., warm air in the east and cold in the west, and the zonal pressure difference, i. e., low air pressure in the east and high in the west, and the Tibetan Plateau cold source effect in combination, inhibit and suppress the detouring southerlies in the eastern Plateau. After late March (pentads 16-18), the reversal in the extensive thermal and pressure contrasts strengthens just the southerlies, thereby causing the so-called "spring rains". Evidently, the constantly detouring zone over the eastern Tibetan Plateau with southerlies prevailing all year long is a region that is less influenced by the large-scale winter circulations. Once the large-scale environmental conditions favor the formation of southerly winds, it can be expected that this region and the vicinity, the Jiangnan-south China is the first place that the summer monsoon and its precipitation should arrive. Hence, it is justifiable to claim that the dynamic effect of the Tibetan Plateau and the reversal of the zonal thermal contrast between the East Asian landmass (including the Tibetan Plateau in particular) and the western Pacific cause the earlier establishment of the EASM but not the SCSM. Nonetheless, which is predominant, the dynamic or the thermal effect remains a problem to be answered.

References

[1] Chen Longxun, Jin Zuhui. The medium-range variations of the summer monsoon circulation system over East Asia[J]. Adv Atmos Sci, 1984, 1: 224-233.

[2] Zhu Qiangeng, He Jinhai, Wang Panxing. A study of circulation differences between East-Asian and Indian summer monsoons with their interactions[J]. Adv Atmos Sci, 1986, 3(4): 466-477.

[3] Tao Shiyan, Chen Longxun. A review of recent research on the East Asian summer monsoon in China [C]//Monsoon Meteorology. Oxford: Oxford Univ Press, 1987, 60-92.

[4] Chen Longxun, Zhu Qiangen, Luo Huibang, et al. East Asian Monsoons[M]. Beijing: China Meteorological Press, 1991. (in Chinese)

[5] Ding Yihui, Mura Kamimikito. Monsoon in East Asia[M]. Beijing: China Meteorological Press, 1994, 74-92. (in Chinese)

[6] Ding Yihui, Chen Longxun, Murakami M. Asian Monsoon[M]. Beijing: China Meteorological Press, 1994. (in Chinese)

[7] Chen Longxun, Li Wei, Zhao Ping, et al. Onset process of East-Asian summer monsoon[J]. Clim & Environ Res, 2000, 5(4): 345-355. (in Chinese)

[8] Wu Guoxiong, Zhang Yongsheng. The Tibetan thermal effect and its mechanic forcing in relation to Asian summer monsoon onset. Part II: Onset timing[J]. Chinese J Atmos Sci,1999,23(1):51-56. (in Chinese)

[9] Huang Ronghui, Zhou Liantong, Chen Wen. The progresses of recent studies on the variabilities of the East Asian monsoon and their cause[J]. Adv Atmos Sci,2003,20:55-69.

[10] Wu Guoxiong, Liu Yimin, Liu Xin, et al. How does the Tibetan heating impact on the pattern of summer climate in Asia[J]. Chinese J Atmos Sci,2005,20(1):47-56. (in Chinese)

[11] Qian Yongfu, Jiang Jing, Zhang Yan, et al. On the region of the earliest onset of tropical summer monsoon in Asia with the mechanism studied[J]. Acta Meteorologica Sinica,2004,82(2):129-139. (in Chinese)

[12] Ding Yihui, Li Chongyin, He Jinhai, et al. SCSMEX and East-Asian Summer Monsoon[J]. Acta Meteorologica Sinica,2004,82(5):561-586. (in Chinese)

[13] Li Jianping, Zeng Qingcun. A new monsoon index and the geographical distribution of the global monsoons[J]. Adv Atmos Sci,2003,20(2):299-302.

[14] He Jinhai, Ju Jianhua, Wen Zhiping, et al. A review of recent advances in research on Asian monsoon in China[J]. Adv Atmos Sci,2007,24(6):972-992.

[15] Gao Youxi, Xu Shuying, Guo Qiyun, et al. Monsoon regions and climate over China, in some problems of East-Asian monsoon[M]. Beijing:Chinese Scientific Press,1962,49-63. (in Chinese)

[16] Zhang Renhe, Akimasa Sum, Masahido. Impacts of El Nino on the East Asian monsoon: A diagnostic study of the 86/87 and 91/92 events[J]. J Meteor Soc Japan,1999,74:49-62.

[17] Qian Weihong, Lee D-K. Seasonal march of Asian summer monsoon[J]. International Journal of Climatology,2000,20:1371-1386.

[18] Wang B, Lin H. Rainy season of the Asian-Pacific summer monsoon[J]. J Climate,2002,15:386-398.

[19] Ban Chenglan. Tropical Meteorological[M]. Beijing:Chinese Science Press,1980:27. (in Chinese)

[20] Tian S F, Yasunari T. Climatological aspects and mechanism of spring persistent rains over central China[J]. J Meteor Soc Japan,1998,76:57-71.

[21] Wan Rijin, Wu Guoxiong. Study of climate cause and mechanism for the Jiangnan springtime precipitation[J]. Chinese Sciences (D),2006,36(1):936-950. (in Chinese)

[22] Chen G T J. Reserch on the phenomenon of Meiyu during the past quarter century: An overview in the East Asian monsoon[C] // Chang C P. World Scientific Publishing Co, Singapore,2004,2:357-403.

[23] Xu Guoqiang, Zhu Qiangen, Ran Yufang. The characteristic onset of 1998 summer monsoon over the South-China Sea and its thereabouts with the mechanism explored[J]. J Appl Meteor,2002,13(5):535-549. (in Chinese)

[24] Zhao Ping, Zhang Renhe, Liu Jiping, et al. Onset of southwesterly wind over eastern China and associated atmospheric circulation and rainfall[J]. Clim Dyn,2007,28:797-811.

[25] Qi Li, He Jinhai, Zhang Zuqiang, et al. The seasonal reversal of zonal land-sea thermal contrast in relation to eastern Asian tropical monsoon circulation[J]. Chinese Sci Bull,2007,52(24):2895-2899. (in Chinese)

[26] Wang Lijuan, He Jinhai, Guan Zhaoyong. Comparison of the double summer monsoon troughs over East Asia[J]. Acta Meteorologica Sinica,2007,21(1):75-86.

[27] Wang Zunya. Climatological variability of Chinese summer rainfall with the possible mechanism investigated [D]. Journal of Chinese Academy Meteorological Sciences (in Chinese),2007.

[28] Lau K M, Yang Song. Climatology and interannual variability of the southeast Asian summer[J]. Adv Atmos Sci,1997,14:141-162.

[29] Webster P J, et al. Predictability and the prospects for prediction[J]. J Geophys Res, 1998, 103: 14451-14510.

[30] Lian Yi, Shen Baizhu, Gao Zongding, et al. A scheme for determining the establishment and region of eastern Asian-western Pacific subtropical summer monsoon[J]. Acta Meteorologica Sinina, 2007, 85(4): 503-510. (in Chinese)

[31] He Jinhai, Qi Li, Wei Jin, et al. Re-understanding of eastern Asian subtropical and tropical monsoons[J]. Chinese J Atmos Sci, 2007, 31(6): 1257-1265. (in Chinese)

[32] Qian Weihong, Kang H S, Lee D-K. Distribution of seasonal rainfall in the East Asian monsoon region [J]. Theor Appl Climatol, 2002, 73: 151-168.

[33] Yu Shihua, Mao Weiping. Characteristics of 1979 eastern Asian summer monsoon circulation in relation to precipitation[J]. J Trop Meteor, 1986, 2(4): 349-354 (in Chinese).

[34] Zhou Bing, He Jinhai, Wu Guoxiong, et al. Properties of East Asian subtropical monsoon and the establishment of its index[J]. Chinese J Atmos Sci, 2003, 27(1): 123-135. (in Chinese)

[35] He Jinhai, Wen Min, Wang Lijuan, et al. Characteristics of the onset of the Asian summer monsoon and the importance of Asian-Australian "Land Bridge" [J]. Adv Atmos Sci, 2006, 23(6): 951-963.

[36] He Jinhai, Sun Chenghu, Li Yunyun, et al. Seasonal transition features of large-scale moisture transport in the Asian-Australian monsoon region[J]. Adv Atmos Sci, 2007, 24(1): 1-14.

[37] Liu Zongxiu. Academic Meeting on Eastern Asian Subtropical Monsoon Held in Beijing[J]. Acta Meteorologica Sinica, 2008, 66(3): 478. (in Chinese)

Seasonal Cycle of the Zonal Land-Sea Thermal Contrast and East Asian Subtropical Monsoon Circulation[*]

Qi Li[1,2]   He Jinhai[1]   Zhang Zuqiang[2]   Song Jinnuan[3]

(1. Key Laboratory of Meteorological Disaster of Jiangsu Province, Nanjing University of Information Science and Technology, Nanjing 210044, China; 2. National Climate Center, Beijing 100081, China; 3. College of Remote Sensing, Nanjing University of Information Science and Technology, Nanjing 210044, China)

Abstract: Based on analysis of the climatic temperature latitudinal deviation on middle troposphere, its seasonal cycle suggests that due to the rapid warming from eastern China continent to the east of Tibetan Plateau and the heating of Tibetan Plateau in spring, seasonal transition of the thermal difference between East Asia continent and West Pacific first takes place in the subtropical region with greatest intensity. On the accompanying low troposphere, the prevailing wind turns from northerly in winter to southerly in summer with the convection precipitation occurring at the same time. This maybe indicates the onset of the East Asian subtropical summer monsoon. Consequently, we advice that the seasonal cycle formed by the zonal thermal contrast between Asian continent and West Pacific may be an independent driving force of East Asian subtropical monsoon.

Key words: zonal land-sea thermal contrast; East Asian subtropical monsoon; tropical monsoon; Tibetan Plateau

1 Introduction

Zhu et al.[1] definitely brought forward that the East Asian monsoon system can be divided into South China Sea (SCS) tropical monsoon and subtropical monsoon in 1986. Some scholars[2-4] have discussed subtropical monsoon impact on precipitation by using the monsoon indexes. But the essence of East Asian subtropical monsoon and its interaction with the tropical monsoon have not been paid much attention to. They interact and influence directly large-scale flood and drought in China. But what is subtropical monsoon on the earth? Or what is the essence of subtropical monsoon? There are generally two ambiguities: one is based on regions, i. e. monsoon prevalent in the subtropical areas in East Asia is called "East

Asian subtropical monsoon", and the other considers the northward extension of tropical monsoon in East Asia as subtropical monsoon. But both of them neglect the reason why the subtropical monsoon exists.

Eurasia plays an important role in Asian summer monsoon onset and intensity. Wu et al.[5-7] advanced that the heat pulley function made by Tibetan Plateau anchors the onset site of Asian summer monsoon. On the other hand, because the land-sea thermal contrast is generally considered as the fundamental mechanism of the formation of monsoon, at present, most research on monsoon[8-17] emphasized that the meridional land-sea thermal contrast in East Asia is the key driving force of tropical monsoon. However, over the subtropical region, the Tibetan Plateau, as an uplifted heat source (or cold source), strengthens the zonal land-sea thermal contrast (Asian continent and the West Pacific), making the seasonal transition of thermal contrast more sensitive and specific. Zhang et al.[18] and Qian et al.[19] firstly documented the effect of zonal land-sea thermal contrast on summer monsoon, but the subtropical monsoon was not involved. So the present paper focuses on the seasonal cycle of zonal land-sea thermal difference over East Asian subtropical region, and its relationship with East Asian subtropical monsoon.

2 Data

We utilize two datasets: the daily meteorological variables form NCEP/NCAR reanalysis from 1950 to 2000[20] and pentad precipitation from CMAP (Climate Prediction Center Merged Analysis of Precipitation)[21] from 1979 to 2000. The pentad-averaged values were calculated from the former daily data.

3 Discussion on subtropical monsoon essence

Fig. 1 depicts evolution of 500-hPa climatic air temperature latitudinal deviation (TLD, defined as the temperature difference between a certain longitude and the average of 80°~150°E) on Northern Hemisphere. We can find that the magnitude of TLD increases with the latitude from south to north. The seasonal transition of its zonal contrast is significant at all latitudes except 35°N (Fig. 1e): the West Pacific is warm during winter and spring, and turns cool in later spring and early summer with anomalous centers located east of 130°E. The Asian continent is cold during winter and spring, and becomes warm in later spring and early summer with centers situated from Bay of Bengal to Tibetan Plateau. That is, it is cold/warm in the west/east during winter and spring and vice versa during summer and autumn. However, the time and the features of transition at different latitudes are different.

At 15°N (Fig. 1a), the TLD in the west of SCS (115°E) is almost opposite completely to that in east (the boundary is showed with the vertical line in Fig. 1a); by pentad 24, the east/west to SCS turns from warm/cold to cold/warm; at all longitudes, transit occurs synchronously almost at the same time with explosiveness. 20°N (Fig. 1b) also is bounded by SCS; the ocean to the east of SCS turns cold in pentad 26, while on the other side, the Indochina Peninsula transits first in about pentad 18; then the seasonal transition advances westward, causing the east India to change warm later than pentad 30. The seasonal cycle at 25°N (Fig. 1c) is similar to that at 20°N (Fig. 1b) with the boundary little eastward to 120°E (the coastline of South China); on its east flank, the eastern continent (South China) warms first much earlier than pentad 14. Comparatively, the TLD at 30°N (Fig. 1d) is much greater due to the Tibetan Plateau heating; moreover, its seasonal cycle transits earliest, that is, the ocean east of East China Sea has cooled before pentad 24, while East Asian continent east of the plateau (east of 90°E) turns warm all around pentad 12. At 35°N (Fig. 1e), West Pacific is cool before pentad 50, and the continent west of 110°E is warm all the time with no seasonal cycle.

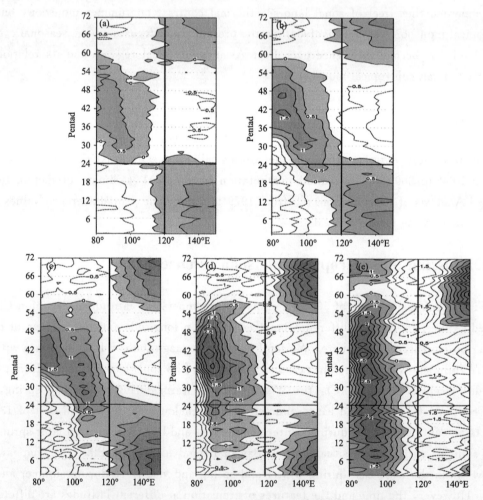

Fig. 1 Evolution of air temperature zonal deviation on 500-hPa. (a) 15°N; (b) 20°N; (c) 25°N; (d) 30°N; (e) 35°N. Warm zone is shaded, units: K

In order to calculate the zonal land-sea thermal con-trast at every latitude, the TLD anomalous centers on both sides are chosen as the key regions, namely 130°~150°E for West Pacific, 80°~100°E for East Asian con-tinent. As shown in Fig. 2a, the contrast in 35°N has no seasonal cycle leading to no seasonal transition of wind direction over this region[22]. However, it has significant seasonal variation at other latitudes. It turns from positive to negative in later spring and early summer, indicating that cold East Asian continent/warm West Pacific turns to warm continent/cool ocean forming westward temperature gradient. While it turns from negative to positive in autumn with eastward temperature gradient. Comparison between all latitudes suggests that the seasonal transition takes place earliest in subtropical region in later March. Later the tropical region transits at 15°N, 20°N, 25°N in turn in the first and middle ten days of May. Their interval isn't longer than 1~2 pentad. It is probably connected with the onset of the East Asian tropical summer monsoon and advance northward.

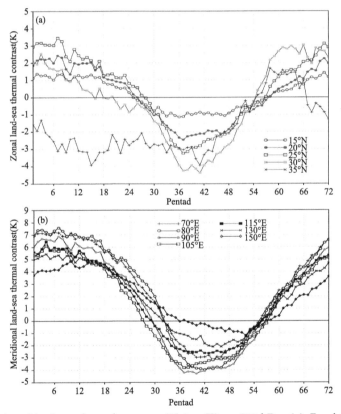

Fig. 2 Evolution of land-sea thermal contrast (Units: K) on 500-hPa. (a) Zonal (West Pacific: 130°~150°E, East Asia: 80°~100°E); (b) meridional (Tropics: 5°S~5°N, East Asia: 20°~30°N)

On an average, the zonal land-sea thermal contrast transits in pentad 27 over tropical region (15°~25°N (figure omitted)) and in pentad 18 over subtropics (27.5°~32.5°N (Fig. 3)). The time interval is greater than a month, proving that they are independent of each other. As shown in Fig. 1, the West Pacific turns cool in about pentad 24 at all latitudes, just a little ahead at 30°N. However, the time is evidently different for Bay of Bengal and East Asia to turn warm. On this account, when zonal land-sea thermal difference transits occur depends on when the East Asia and its south turn warm. From spring to summer, east China continent and the east of Plateau (east of 90°E) warm in short order, with the quickest temperature rising and air pressure lowering at latitude of 30°N, leading to seasonal cycle of the zonal land-sea thermal contrast fist transiting here, that is, the latitude of Tibetan Plateau. On the other hand, heating source over the plateau begins to enhance in later March, as an uplifted one heating the middle and upper troposphere directly, and intensify the zonal land-sea thermal contrast there, leading to maximum. Nevertheless, the contrast over tropics transits along with the solar radiation seasonal cycle extending from south to north. On this account, the zonal land-sea thermal difference over subtropical regions is not only sensitive and specific, but also more ahead.

Furthermore, the zonal land-sea thermal contrast over subtropics is also ahead relative to the East Asia meridional land-sea thermal contrast. The meridional land-sea thermal contrast seasonal transition first takes place in Indochina Peninsula (105°E) in about pentad 27, presenting warm in south and cold in north changing to warm in north and cool in south. Then the seasonal transition extends to both sides at the same time. SCS and Bay of Bengal transit simultaneously during pentad 30 (more ahead in low troposphere), so do Indian Peninsula and the neighborhood of 130°E in pentad 32. The contrast transits last in West Pacific (150°E). It is obvious that the seasonal cycle of the meridional land-sea thermal difference is basically consistent with the process of the onset of East Asian tropical summer monsoon and advances northward.

Generally speaking, the land-sea thermal contrast is considered as the fundamental mechanism of monsoon formation. The meridional land-sea thermal contrast in East Asia (including the Tibetan Plateau) is the key driving force of tropical monsoon. Could the seasonal transition of zonal land-sea thermal difference over subtropics be closely related with subtropical summer monsoon onset? From winter to summer, the zonal land-sea thermal contrast over subtropics turns from warm east and cool west to cold east and warm west in pentad 18 and persists (Fig. 3). Before the transition (pentad 12-17(Fig. 4a)) at the 850-hPa level, the India-Burma trough is much evident; a vortex is centered around equator 85°E; the subtropical high cells have formed, but the high belt dose not split; the north edge of the southerly to the east of 110°E is located at about 25°N, therefore the subtropics (27.5°~32.5°N) are controlled by northerly. After transition (pentad 19-24 (Fig. 4b)), the vortex moves to Sri Lanka; sub-tropical high belt is weakened but still orbicular; there is no cross-equatorial flow formed, suggesting that the interactions between the Northern and Southern Hemispheric atmospheres are not remarkable. On this account, the SCS summer monsoon does not set up here; the West Pacific subtropical high is strengthened with wide meridional span; the southerly advances northward to 35°N, that is, the wind over subtropics turns from northerly to southerly. Notable southerly sets up over subtropical region of East Asia, which mainly comes from the southwesterly from easterly of south West Pacific sub-tropical high over ocean and northwesterly from south Tibetan Plateau at mid and high latitudes, causing mixed feature of subtropics.

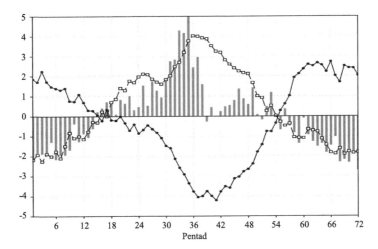

Fig. 3　Climatological evolution of subtropics (27.5°～32.5°N) zonal land-sea thermal contrast on 500-hPa (solid dotted line, units: K), subtropics (27.5°～32.5°N, 110°～140°E on average) meridional wind on 850-hPa (line with open square, units: m·s^{-1}) and anomaly precipitation (annual mean is taken off, histogram, units: mm)

Moreover, precipitation anomaly over subtropical regions (averaged over 27.5°～32.5°N, 110°～140°E, annual mean is taken out (Fig. 3)) turns from negative to positive as the zonal land-sea thermal contrast season-ally transits.

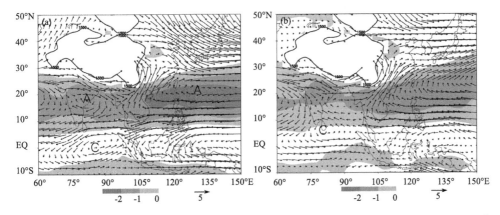

Fig. 4　The stream filed on 850-hPa (m·s^{-1}) and anticyclonic vorticity on 500-hPa (shaded, in 10^{-5} s^{-2}).
(a) Before zonal land-sea thermal contrast transits over subtropics from winter to summer (12-17P);
(b) after zonal land-sea thermal contrast transits over subtropics from winter to summer (19-24P)

Both wind and rain, two characteristics of the monsoon occur together, may indicate the onset of subtropical summer monsoon. It illuminates that the subtropical summer monsoon sets up earlier relative to the SCS summer monsoon.

4 Summary and discussion

Conclusively, based on analysis of the climatic temperature latitudinal deviation in middle troposphere, its seasonal cycle suggests that due to the rapid warming of east China continent and the east of Tibetan Plateau and heating of Tibetan Plateau in spring, seasonal transition of the thermal difference between East Asia continent and West Pacific first takes place over the subtropical region with greatest intensity. Accordingly, the prevailing wind turns from northerly in winter to southerly in summer in low tropospheres, which partly come from the ocean and have the mix feature of subtropics. Wan and Wu[23] pointed out that the precipitation to the east of plateau during spring belongs to spring persistent rain over southeastern China, essentially caused by mechanical forcing and thermal forcing of Tibetan Plateau on a climatic basis. Herein, the convention precipitation is thought to take place at the same time. That is, both the wind and rain appear in the meantime. Based on the general viewpoint that the land-sea thermal contrast is the fundamental mechanism of the formation of monsoon, we might as well consider that the process of wind over subtropics turning from northerly to southerly indicates East Asia subtropical summer monsoon onset.

In consequence, (1) Subtropical monsoon is independent of tropical monsoon; the former is not the northward extension of the latter. (2) Subtropical monsoon sets up earlier relative to tropical monsoon. (3) The seasonal cycle formed by the zonal thermal contrast between Asian continent and western Pacific on the background of solar radiation seasonal variation may be an independent driving force of East Asian subtropical monsoon. Subtropical monsoon sets up earlier than tropical monsoon in that the temperature gradient between East Asia continent and the ocean transits earlier than the meridional one over tropics.

The authors would like to thank the two anonymous reviewers for their valuable suggestions and comments.

References

[1] Zhu Q G, He J H, Wang P X. A study of circulation differences between East-Asian and Indian summer monsoons with their interaction[J]. Adv Atmos Sci, 1986, 3(4): 466-477.

[2] Zhang Q Y, Tao S Y. Tropical and subtropical monsoon over East Asia and its influence on the rainfall over Eastern China in summer[J]. Quarterly J Appl Meteorol (in Chinese), 1998, 9(Suppl): 17-23.

[3] Zhou B, He J H, Wu G X, et al. Characteristics of East Asian Subtropical Monsoon Index and its definition[J]. Chin J Atmosph Sci (in Chinese), 2003, 27(1): 123-135.

[4] Zhao P, Zhou Z J. East Asian Subtropical Summer Monsoon Index and its relationships to rainfall[J]. Acta Meteorol Sin (in Chinese), 2005, 63(6): 933-941.

[5] Wu G X, Zhang Y S. Thermal and mechanical forcing of the Tibetan Plateau and the Asian Monsoon Onset. Part I: Situating of the onset[J]. Chin J Atmosph Sci (in Chinese), 1998, 22(6): 825-838.

[6] Liang X Y, Liu Y M, Wu G X. Effect of Tibetan Plateau on the site of onset and intensity of the Asian

Summer Monsoon[J]. Acta Meteorol Sin (in Chinese),2005,63(5):799-805.

[7] Wu G X,Wang J. Numerical modeling of the influence of Eurasian Orography on the atmospheric circulation in different seasons[J]. Acta Meteorol Sin (in Chinese),2005,63(5):603-612.

[8] Tao S Y,Chen L X. A Review of Recent Research on the East Asian Summer Monsoon in China[C]// Chang C P,Krishnamurti T N. Monsoon Meterorlogy. Oxford:Oxford University Press,1987:60-92.

[9] Chen L X,Zhu Q G,Luo H B,et al. East Asian Monsoon (in Chinese) [M]. Beijing:China Meteorological Press,1991.

[10] He J H,Ding Y H,Gao H,et al. The South China Sea summer monsoon onset data and monsoon index (in Chinese) [J]. Beijing:China Meteorological Press,2001.

[11] Qian Y F,Wang S Y,Shao H. A possible mechanism effecting the earlier onset of South Weasterly Monsoon in the South China Sea compared to the Indian Monsoon[J]. Metero Atmos Physics,2001,76(3-4): 237-250.

[12] Zeng Q G,Li J P. Interactions between the Northern and Southern Hemispheric atmospheres and the essence of monsoon[J]. Chin J Atmosph Sci (in Chinese),2002,26(4):433-448.

[13] Huang R H,Zhou L T,Chen W. The progress of cecent studies on the variability of the East Asian Monsoon and their causes[J]. Adv Atmos Sci,2003,20(1):55-69.

[14] He H Y,Sui C-H,Jian M Q,et al. The evolution of Tropospheric Temperature Field and its relationship with the onset of Asian Summer Monsoon[J]. J Meterol Soc Japan,2003,81(5):1201-1223.

[15] Li J P,Zeng Q C. A new monsoon index and the geographical distribution of the global monsoons[J]. Adv Atmos Sci,2003,20(2):299-302.

[16] Zhang Z Q,Chan Johnny C L,Ding Y H. Characteristics,evolution and mechanisms of the Summer Monsoon onset over Southeast Asia[J]. Int J Climatol,2004,24:1461-1482.

[17] Ding Y H,Li C Y,He J H et al. South China Sea Monsoon experiment (SCSMEX) and the Ease-Asian Monsoon[J]. Acta Meteorol Sin (in Chinese),2004,62(5):561-586.

[18] Zhang Y,Qian Y F. Thermal effect of surface heat source over the Tibet Plateau on the onset of Asian Summer Monsoon[J]. J Nanjing Institute Meteorol (in Chinese),2002,25(3):298-306.

[19] Qian Y F,Jiang J,Zhang Y,et al. The earliest onset area of the Tropical Asian Summer Monsoon and its mechanism[J]. Acta Meteorol Sin (in Chinese),2004,62(2):129-139.

[20] Kalnay E,Kanamitsu M,Kistler R,et al. The NCEP/NCAR 40-Year reanalysis project[J]. Bull Amer Meteor Soc,1996,77:437-471.

[21] Xie P,Arkin P A. Global Precipitation:A 17-year monthly analysis based on Gauge Observations,satellite estimates and numerical model outputs[J]. Bull Amer Meteor Soc,1997,78:2539-2558.

[22] Zhao P,Zhang R H,Liu J P,et al. Onset of southwesterly wind over Eastern China and associated atmospheric circulation and rainfall[J]. Clim Dyn,2006,doi:10.1007/s00382-006-0212-y.

[23] Wan R J,Wu G X. Mechanism of the spring persistent rains over Southeastern China[J]. Sci China Ser D-Earth Sci,2007,50(1):130-144.

东亚副热带季风雨带建立特征及其降水性质分析*

任珂 何金海 祁莉

(南京信息工程大学江苏省气象灾害重点实验室,南京 210044)

摘要:利用 1961—2006 年 NCEP/NCAR 再分析数据集和 TRMM、CMAP 多年平均逐候降水资料,分析了我国东部东亚副热带季风雨季的起始时间、建立特征及其降水性质。结果表明,第 16—18 候,在中国江南南部和华南北部地区(25°～30°N)日降水率达到 6 mm·d^{-1},且范围较大,在低层该雨带的水汽主要来源于西太平洋副热带高压南侧转向的西南水汽输送,其源地即为西太平洋副热带雨季开始。雨带建立同时,东亚副热带地区—中东太平洋的纬向海平面气压梯度首先在中纬度发生反转,即西低东高(相应于西暖东冷)。中国东部副热带地区出现加热中心并伴有上升运动,强度逐渐增强,并伸展至对流层顶,其强度及对流高度与热带地区相当,对流层中低层大气呈对流不稳定,降水已具有对流性降水性质。与此同时,南海—西太平洋地区仍在副热带高压控制之下,盛行下沉运动,无降水产生,南海夏季风及其相应的水汽输送尚未建立。东亚副热带季风雨带的建立(3 月底 4 月初)早于热带夏季风雨带,两雨带分别具有独立的热源中心和上升运动。南海夏季风即将爆发之际,赤道地区加热中心快速北移至南海地区,与副热带地区热源相互作用。

关键词:东亚副热带季风雨季;南海夏季风;低层水汽输送;大气加热率

1 引言

中国地处东亚季风区,天气气候主要受季风影响。陈隆勋等[1-3]提出了东亚季风系统的概念。Zhu 等[4]明确提出东亚季风区可分为南海—西太平洋热带季风区(简称热带季风区)和大陆—日本副热带季风区(简称东亚副热带季风区)。喻世华等[5-6]也指出东亚季风环流由热带季风环流和副热带季风环流组成,正是环流特征的多样性,引起了东亚季风降水的复杂性。因此,深入认识不同时期降水的性质及其影响系统将有助于我国旱涝预测水平的提高。

以往的研究大多关注 6—7 月东亚副热带夏季风降水盛期发生的江淮梅雨[7-9]。然而,4—6 月的雨季是除了江淮梅雨外又一个多雨时段,同时也是大范围洪涝灾害的频发时段,近年来对其研究也受到了广泛的重视。陈隆勋等[10]指出副热带季风雨季于 4 月初开始于华南北部和江南地区。陈邵东等[11-12]指出:江南汛期降水始于 3 月,雨量集中在 4—6 月。池艳珍等[13]分析了华南前汛期不同降水时段的特征,把华南前汛期降水分为锋面降水(副热带季风降水)和热带夏季风降水两个时段。郑彬等[14]对华南前汛期的副热带季风降水和热带夏季风

---

* 本文发表于《气象学报》,2010 年第 68 卷第 4 期,550-558.

降水日期的划分进行了研究。强学民等[15]对华南前汛期降水开始和结束日期的确定方法进行了综述。以上作者均指出副热带季风雨季开始于3月底4月初,但尚未涉及此时降水的性质。何金海等[16]发现16—18候东亚大陆(含高原)与西太平洋地区之间纬向温度梯度开始转向并伴随高原由冷源转为热源,与此同时,东亚副热带南风和降水异常(减去年平均值)首先在我国东部的副热带地区(江南)建立,从季风的基本推动力的观点出发,这一现象可以被认为是东亚副热带季风及其雨带的开始。Zhao等[17]指出高原东南部快速增温使得东亚与西太平洋的副热带东西向热力差异较早反转,纬向海陆热力差异的转变意味着大气环流形势由冬向夏的转变,此时的东亚副热带季风雨带是否已具有夏季风雨带的性质尚不甚清楚,甚至存有争议,因此,本文对副热带季风雨季开始前后的大气环流特征和降水性质进行初步分析。

2 资料与方法

本文使用的资料是(1)1961—2006年NCEP/NCAR逐日再分析资料,水平分辨率2.5°×2.5°;(2)1998—2006年热带测雨卫星(TRMM)降水资料,水平分辨率为0.25°×0.25°。TRMM是1997年底发射的卫星,目前资料时段相对较短,但资料的准确性较好、时空分辨率较高,能反映更多的降水细节,有利于进一步认识副热带季风雨带的建立及其移动特征;(3)1979—2000年CMAP(Climate Prediction Center Merged Analysis of Precipitation)降水资料,水平分辨率为2.5°×2.5°。

本文计算了气候平均的垂直大气加热率 Q_1 及假相当位温垂直梯度,以分析讨论东亚副热带季风雨季大气热状况特征。

利用"倒算法"计算了整层大气的视热源 Q_1

$$Q_1 = c_p \left[\frac{\partial T}{\partial t} + \mathbf{V} \cdot \nabla T + \left(\frac{p}{p_0} \right)^k \omega \frac{\partial \theta}{\partial p} \right]$$

式中,T 为温度;ω 为 p 坐标的垂直速度;$p_0 = 1000$ hPa;$k = R/c_p$;$\mathbf{V}$ 为水平风矢量。

假相当位温的计算公式[18]为

$$\theta_{se} = T_k \left(\frac{1000}{p} \right)^{0.2854(1-0.28r)} \times \exp\left[\left(\frac{3376}{T_L} - 2.54 \right) r(1+0.81r) \right]$$

式中,T_k、p、r 分别为起始面上绝对温度(K)、气压(hPa)、比湿(g·g^{-1}),T_L 为凝结高度的绝对温度。

$$T_L = \frac{2840}{3.5\ln T_k - \ln e - 4.805} + 55$$

3 副热带季风雨季初期降水特征

祁莉等[19-20]提出在东亚副热带地区,纬向海陆热力差异在第16—18候由东暖西冷转为东冷西暖,东亚副热带地区同时出现降水正异常。第15候,中国东部副热带地区的降水分布较为零散(图1a),且日降水率为4 mm·d^{-1},东部沿海存在分布零散的降水率为6 mm·d^{-1}降水中心,整个副热带地区没有形成较大范围的雨带。第16候,我国东部副热带地区出现大范围降水(图1b),日降水率达到6 mm·d^{-1},与东部沿海降水形成一条十分明显的西南—东北走向的雨带。随后,该大值雨带一直维持且降水量逐渐增大(图略)。Lau 等[21]、Qian 等[22]均以降水率大于 6 mm·d^{-1} 定义了南海季风爆发日期,在此,我们同样以日降水率大于 6 mm·d^{-1} 为副热带季风雨季开始的标准,可见,从气候态来说,东亚副热带季风雨季于 3 月底开始。从 TRMM 降水和 CMAP 降水时间—纬度剖面(图 2)同样可以看到:16 候左右在副热带地区(25°~30°N)日降水率首先达到 6 mm·d^{-1}。比较图 2a、图 2b,TRMM 降水资料反映了更多的降水细节:副热带雨带开始并持续稳定一段时间后,同时向南、向北扩展。第 24 候开始,副热带雨带明显地向南扩展,形成华南前汛期前期降水,第 28 候,南海地区出现降水正异常,热带夏季风雨季开始,形成热带夏季风雨带,第 30 候左右,华南前汛期强降水开始季节性地向北推进。可见,副热带季风雨季的开始要早于热带季风雨季的开始。

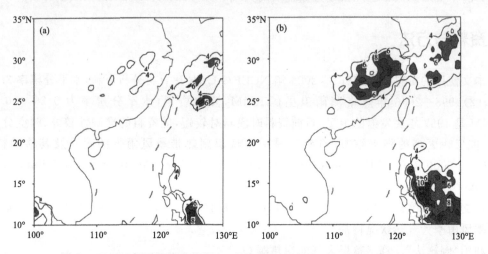

图 1　TRMM 候平均降水分布(a. 第 15 候,b. 第 16 候;单位:mm·d^{-1})

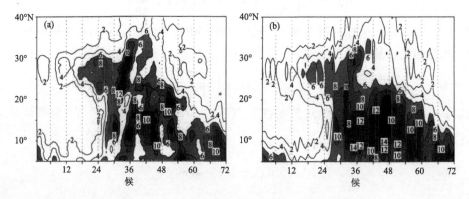

图 2　气候平均 TRMM 降水(a)和 CMAP 降水(b)的时间(候)—纬度剖面(110°~120°E 平均)
(单位:mm·d^{-1})

4　水汽通量输送及其散度

以往对水汽输送的研究主要着眼于整层积分的水汽输送,而低层水汽通量的敏感度要比整层积分的水汽通量敏感度高,图 3 给出了多年平均的低层水汽输送通量(从地面积分到

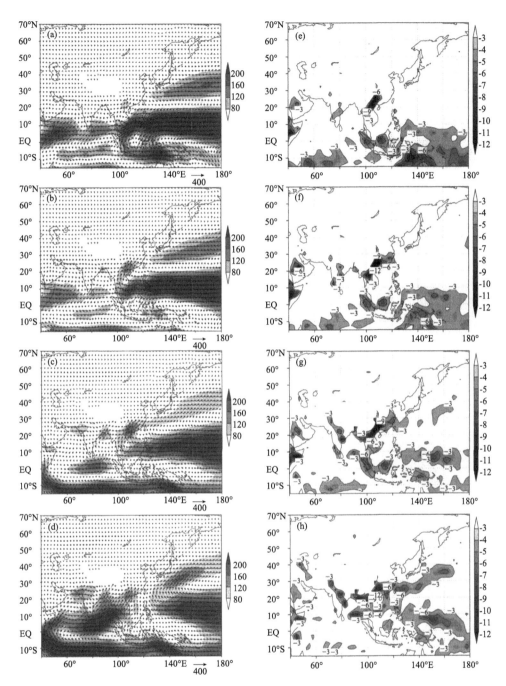

图3 气候平均低层水汽输送(a、b、c、d;阴影区表示水汽输送大于 80 kg·(m·s)$^{-1}$)及其散度(e、f、g、h;阴影区表示水汽通量散度小于 -3×10^{-5} g·(kg·s)$^{-1}$)(a、e. 第13候, b、f. 第16候, c、g. 第24候, d、h. 第28候)

700 hPa)及其散度。第13候高原东南侧有一小范围的水汽输送大值区,有零星的辐合,无论是水汽输送还是水汽通量辐合均没有伸展到110°E以东的副热带地区。第16候的水汽输送较第13候最大的不同在于其水汽得到了来自西太平洋的补充,即西太平洋副热带高压南侧转向的西南水汽输送强度增强,范围扩大,逐步向110°E以东扩展,同时水汽辐合加强,东亚副热带地区形成大范围的西南—东北走向雨带。可见,此时维持东亚副热带季风降水的水汽主要来自西太平洋。24候西太平洋的西南水汽输送通道更加宽广,强水汽输送抵达我国东亚副热带地区,与副高北侧日本以南的强水汽输送带打通,该水汽输送带上水汽辐合增强,华南地区降水量增多。28候,副高东撤,强盛的越赤道西南气流,经孟加拉湾、中南半岛把充沛的水汽输送到南海并辐合,形成强降水,南海夏季风建立。热带地区低层水汽输送的季节转换特征与已有的大量研究结果一致[23-24],在此不再重复叙述。

可见,在南海夏季风爆发前,中国东亚副热带地区已出现较强雨带,其水汽主要来自西太平洋副热带高压南侧转向的西南水汽输送,即来自于西太平洋。南海夏季风爆发后,强盛的越赤道西南气流,经孟加拉湾、中南半岛把充沛的水汽输送到南海及华南地区,中国东亚副热带地区水汽更加充沛,季风雨带雨量增多,而此时南海夏季风雨季刚刚开始。东亚副热带季风雨带和南海夏季风雨带水汽来源的不同再次证明前者的建立早于后者。

5 海平面气压转换特征

王绍武等[25]指出3—4月大气环流与气候异常持续性最小,即在3—4月大气环流变化激烈,并指出这可能在于大气冷热源的强烈改变,下垫面热力状况很难具体表征。钱永甫等[26]指出海平面气压与地面气温的时间演变特征是一致的,或者说海平面气压与下垫面热力状况是一致的,即高压冷低压暖。因此用海平面气压表征地表海陆热力差异,分析在由冬到夏的季节转换期间,海平面气压(海陆热力差异)的季节转换特征。

从海平面气压的纬向偏差分布(图4)可见。第6候,在中高纬度海平面气压纬向偏差是典型的冬季环流形势,亚洲大陆到北太平洋海平面气压纬向偏差呈正、负分布型,即西高东低(相应于西冷东暖),且正负中心在较高维度(50°N左右);中国东亚副热带地区、南海及我国东部沿海海域均为正的海平面气压纬向偏差,为冷源。第16、17候,高原南部的海平面气压纬向

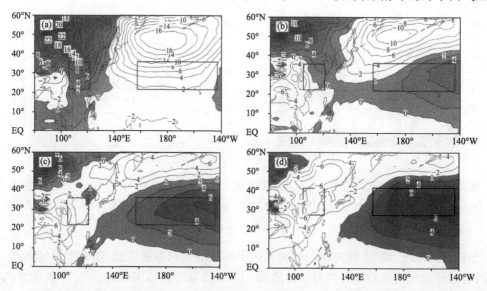

图4 海平面气压纬向偏差(a.第6候;b.第16候;c.第20候;d.第24候;单位:hPa)

偏差负中心强度不断增强且向东扩展,中国东部副热带地区逐渐转为负中心,同时,中东太平洋出现纬向偏差正中心,且向西扩展,东亚副热带地区—中东太平洋之间纬向海平面气压梯度首先在中纬度发生反转。反转之后的西低东高(相应于西暖东冷)海平面气压纬向偏差中心不断增强,且向北扩展,使得高纬度东高西低的海平面气压纬向偏差中心强度减弱、范围减小,至24候,20°～50°N的中高纬度海平面气压纬向偏差已完全呈西低东高型,即夏季的环流形势。选取海平面气压纬向偏差首先发生反转的区域:西区(20°～35°N,105°～120°E),东区(20°～35°N,160°E～140°W),做区域平均的海平面气压纬向之差(图5),同样看到:在13候,东亚副热带地区与中东太平洋之间的海平面气压相差2 hPa,在15候,两者之间的纬向海平面气压之差已转为−1.5 hPa,即中纬度副热带地区海平面气压纬向梯度发生反转。而南海海平面气压纬向偏差的季节转换并不明显,正是由于这种热力性质的转变导致副热带季风雨季的开始早于南海夏季风雨季的开始。

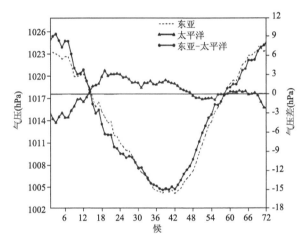

图5 区域平均海平面气压的逐候演变
(左纵坐标:海平面气压值;右纵坐标:区域平均的海平面气压纬向之差)

6 大气加热率及其垂直环流

从中国东亚副热带季风雨季开始到南海夏季风爆发时期各个关键候大气垂直运动及其加热率的纬度−高度剖面(图6)可见,第1候(图6a),赤道地区大气整层为热源区,中心最大增温率达到3 ℃·d^{-1},南海及副热带地区大气整层为冷源区。相应的在赤道地区有强盛的上升运动,南海及副热带地区为下沉运动,这表明大气垂直环流及大气热状况维持冬季形势。16候(图6b),赤道地区的加热区继续稳定,但强度有所减弱,中心最大增温率为2～2.5 ℃·d^{-1};南海地区仍为冷却区,强度有所减弱,然而,副热带对流层中低层出现大气加热中心并伴有上升运动,其位置正好与这个时期的副热带季风雨带相重叠(图1,图2)。随后,该大气加热率及上升运动不断增强,在22候(图6c)中心最大增温率已达到2 ℃·d^{-1},其强度大于赤道地区的1.5 ℃·d^{-1},此时,副热带地区的上升运动高度也已达到对流层顶。就其发生、发展来看,副热带地区的加热中心(上升运动)并不是赤道加热中心(赤道上升运动)向北移动而产生,两者中间隔着南海冷源区(下沉运动),我国东部的副热带季风雨带有其独立的热源中心。南海夏季风即将爆发前(27、28候),赤道地区加热中心快速北移至南海地区,南海地区上空转变为热源,中心强度达到 2 ℃·d^{-1},同时,赤道上升运动中心北移至 10°N。副热带地区加热中心一直稳定在 20°~30°N,强度逐渐加强,中心强度达到 2.5 ℃·d^{-1},可见,即使在南海夏季风爆发时期,南海地区热源中心及我国东部的副热带地区热源中心仍是相互分离的,副热带地区热源中心强度甚至比南海地区热源中心强度要强。

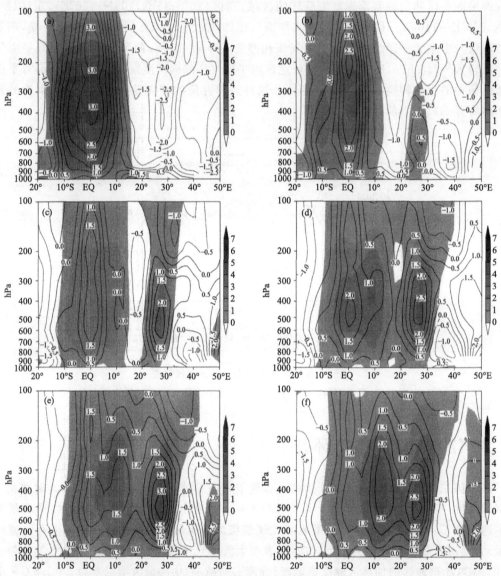

图 6 气候平均场垂直速度(阴影区表示上升运动,单位:10^{-2} m·s^{-1})及大气加热率
(等值线,单位:10^{-5} ℃·d^{-1})纬度—高度剖面(110°~140°E)
(a.第 1 候;b.第 16 候;c.第 22 候;d.第 26 候;e.第 27 候;f.第 28 候)

综上所述,可以认为,中国东部的副热带季风雨带与南海热带季风雨带分别有其独立的热源中心、上升运动与之相配合,副热带地区加热中心(上升运动)是在副热带地区局地发展起来的,其位置正好与这个时期的副热带季风雨带相吻合;而南海夏季风的爆发与赤道加热中心

(上升运动)迅速向北移至南海地区有关。我国东亚副热带地区大气加热中心,上升运动的出现均早于南海—西太平洋地区。

7 大气不稳定度

当大气低层的水汽和热量聚集达到一定的量值时,冷暖气流交汇影响大气稳定度,大气就会形成强烈的不稳定层结,为对流和降水提供温床[27]。

中层冷暖平流与降水的相关性较强,图 7 给出了 500 hPa 温度平流的时间演变。$-\mathbf{V}\cdot\nabla T<0$ 为冷平流;$-\mathbf{V}\cdot\nabla T>0$ 为暖平流。18 候之前,30°N 以北受十分强的冷平流控制,中心强度达到-3×10^{-5} K·s^{-1},20°~30°N 为较弱的暖平流,甚至为冷平流。3 月底,30°N 以北的冷空气变性增暖,冷平流减弱,同时,副热带地区的暖平流增强,强度能够达到 1×10^{-5} K·s^{-1}。可以看到,降水落在暖平流区(图 2),此时季风降水属暖平流降水。

假相当位温垂直梯度表示大气的对流不稳定度,当 $\partial\theta_{se}/\partial p>0$ 时,为对流不稳定,当 $\partial\theta_{se}/\partial p<0$ 时,为对流稳定。图 8 给出了气候平均场假相当位温垂直梯度($\partial\theta_{se}/\partial p$)时间—高度剖面,3 月底 4 月初,中国东部的副热带地区大气低层已有水汽的辐合及大气的加热(图 3、图 6),且对流层中层为暖平流(图 7),从图 8 又可以看到:在第 19 候,对流层中低层副热带地区大气呈对流不稳定,从副热带季风雨季初期到雨季盛期,大气对流不稳定强度逐渐增大,对流不稳定区已伸展至 700 hPa 以上。可见,3—4 月中国东部副热带地区大气已满足对流不稳定条件,此时季风降水具有对流性降水性质。

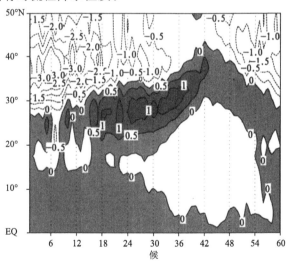

图 7 气候平均 500 hPa 温度平流(阴影区:$-\mathbf{V}\cdot\nabla T>0.0$,单位:$10^{-5}$ K·s^{-1})的逐候演变(110°~140°E)

8 总结与讨论

综上所述,可得如下结论:

(1)从气候态来说,3 月底 4 月初,中国东部副热带地区(25°~30°N)出现较大范围、强度较强的降水,日降水率达到 6 mm·d^{-1},我们通常以日降水率大于 6 mm·d^{-1} 为季风雨季开

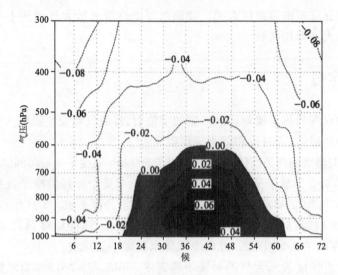

图8 气候平均θ_{se}垂直梯度时间—高度剖面(25°~30°N,110°~120°E平均)

始的标准,同时,其低层水汽主要来自西太平洋副热带高压南侧转向的西南水汽输送,即来自西太平洋,可以认为副热带季风雨季已经开始,相对于南海季风雨季的建立要早。

(2)此时,副热带地区海平面气压纬向梯度发生反转,而南海—西太平洋地区海平面气压纬向偏差变化不大。与东亚副热带季风雨带相配合,3月底4月初东亚副热带地区大气垂直加热中心(上升运动)也迅速发展起来,而南海夏季风的爆发与赤道加热中心(上升运动)迅速向北移至南海地区有关。中国东亚副热带地区大气加热中心,上升运动的出现均早于南海—西太平洋地区。

(3)在东亚副热带雨带建立时期,东亚副热带地区对流层中层温度平流为暖平流,大气呈对流不稳定,降水已具有对流性降水性质。

从对气候平均的海平面气压(相应于地表冷热源)、大气垂直加热率、温度平流及大气对流不稳定的分析结果,我们可以认为3月底4月初建立的东亚副热带季风雨带已具有初夏夏季风雨带的性质,当然,我们还需深入研究以进一步求证相关结论。

参考文献

[1] 陈隆勋,金祖辉,刘喜礼,等.夏季亚洲地区热带环流的中期振荡[J].海洋学报,1983,5(5):575-586.
[2] 陈隆勋.东亚季风环流系统的结构及其中期变动[J].海洋学报,1984,6(6):744-758.
[3] Chen Longxun, Jin Zuhui. The medium-range variations of the summer monsoon circulation system over East Asia[J]. Adv Atmos Sci,1984,1:224-233.
[4] Zhu Qiangen, He Jinhai, Wang Panxing. A study of circulation differences between East-Asian and Indian summer monsoons with their interaction[J]. Adv Atmos Sci,1986,3(4):465-477.
[5] 喻世华,茅卫平.1979年东亚地区的夏季风环流特征及其与降水的关系[J].热带气象,1986,2(4):349-354.
[6] 喻世华,杨维武.副热带季风环流圈的特征及其与东亚夏季环流的关系[J].应用气象学报,1991,2(3):242-247.
[7] 张庆云,陶诗言.夏季东亚热带和副热带季风与中国东部汛期降水[J].应用气象学报,1998,9(增刊):

17-22.

[8] 周兵,何金海,吴国雄,等.东亚副热带季风特征及其指数的建立[J].大气科学,2003,27(1):123-135.

[9] Zhu Xiaying, He Jinhai, Wu Zhiwei. Meridional seesaw-like distribution of the Meiyu rainfall over the Changjiang-Huaihe River Valley and characteristics in the anomalous climate years[J]. Chinese Sci Bull, 2007,52(17):2420-2428.

[10] 陈隆勋,李薇,赵平,等.东亚地区夏季风爆发过程[J].气候与环境研究,2000,5(4):345-355.

[11] 陈绍东,王谦谦,钱永甫.江南汛期降水基本气候特征及其与海温异常关系初探[J].热带气象学报,2003,19(3):260-268.

[12] 王谦谦,陈绍东.江南地区汛期降水与热带海温关系的SVD分析[J].干旱气象,2004,22(3):11-16.

[13] 池艳珍,何金海,吴志伟.华南前汛期不同降水时段的特征分析[J].南京气象学院学报,2005,28(2):163-171.

[14] 郑彬,梁建茵,林爱兰,等.华南前汛期的锋面降水和夏季风降水划分日期的确定[J].大气科学,2006,30(6):1207-1216.

[15] 强学民,杨修群,孙成艺.华南前汛期降水开始和结束日期确定方法综述[J].气象,2008,34(3):10-15.

[16] 何金海,祁莉,韦晋,等.关于东亚副热带季风和热带季风的再认识[J].大气科学,2007,31(6):1257-1265.

[17] Zhao Ping, Zhang Renhe, Liu Jiping,et al. Onset of southwesterly wind over eastern China and associated atmospheric circulation and rainfall[J]. Clim Dyn,2007,28:797-811.

[18] Bolton D. The computation of equivalent potential temperature[J]. Mon Wea Rev,1980,108:1046-1053.

[19] 祁莉,何金海,张祖强,等.纬向海陆热力差异的季节转换与东亚副热带季风环流[J].科学通报,2007,52(24):2895-2899.

[20] Qi Li, He Jinhai, Zhang Zuqiang,et al. Seasonal cycle of the zonal land-sea thermal contrast and East Asian subtropical monsoon circulation[J]. Chinese Science Bulletin,2008,53(1):131-136.

[21] Lau K M,Yang S. Cliamtology and interannual variability of the Southeast Asian monsoon[J]. Adv Atmos Sci,1997,14:141-162.

[22] Qian Weihong, Lee D. Seasonal march of Asian summer monsoon[J]. J Climatology,2000,20:1371-1378.

[23] 刘芸芸,何金海,梁建茵,等.亚澳季风区水汽输送季节转换特征[J].热带气象学报,2006,22(2):138-146.

[24] 陈际龙,黄荣辉.亚澳季风各子系统气候学特征的异同研究Ⅱ:夏季风水汽输送[J].大气科学,2007,31(5):766-778.

[25] 王绍武,赵宗慈,陈振华.月平均环流异常的持续性与韵律性和海气相互作用[J].气象学报,1983,41(1):33-42.

[26] 钱永甫,江静,张艳,等.亚洲热带夏季风的首发地区和机理研究[J].气象学报,2004,62(2):129-139.

[27] 李昀英,宇如聪,傅云飞,等.一次热对流降水成因的分析和模拟[J].气象学报,2008,66(2):190-202.

Seasonal Transition of East Asian Subtropical Monsoon and Its Possible Mechanism*

Zhu Zhiwei   He Jinhai   Qi Li

(Jiangsu Key Laboratory of Meteorological Disaster, Nanjing University of Information Science and Technology, Nanjing 210044 China)

Abstract: The NCEP/NCAR reanalysis datasets and Climate Prediction Center (CPC) Merged Analysis of Precipitation (CMAP) rain data are used to investigate the large scale seasonal transition of East Asian subtropical monsoon (EASM) and its possible mechanism. The key region of EASM is defined according to the seasonal transition feature of meridional wind. By combining the "thermal wind" formula and the "thermal adaptation" equation, a new "thermal-wind-precipitation" relation is deduced. The area mean wind directions and thermal advections in different seasons are analyzed and it is shown that in summer (winter) monsoon period, the averaged wind direction in the EASM region varies clockwise (anticlockwise) with altitude, and the EASM region is dominated by warm (cold) advection. The seasonal transition of the wind direction at different levels and the corresponding meridional circulation consistently indicates that the subtropical summer monsoon is established between the end of March and the beginning of April. Finally, a conceptual schematic explanation for the mechanism of seasonal transition of EASM is proposed.

Key words: East Asian Subtropical Monsoon; seasonal transition; mechanism

1 Introduction

As one of the Asian Monsoon members[1], the East Asian Monsoon has been further divided into SCS tropical monsoon (SCSTM) and the East Asian subtropical monsoon (EASM)[2-3]. Different East Asian monsoon members interact with each other and mutually result in the flood and drought in eastern China[4]. Through observational experiments and international cooperation studies, scientists have obtained basic knowledge about the characteristics of the Asian monsoon and its primary mechanisms[5-9]. However, the exact time of seasonal transition of EASM and its possible mechanism have not been clarified yet.

---

* 本文发表于《Journal of Tropical Meteorology》,2012 年第 18 卷第 3 期,305-313.

In recent years, Wu et al.[10] based on the "thermal adaptation" theory, have proposed an original view on the formation of subtropical monsoons. They pointed out that the long wave radiation cooling (LO) in eastern oceans, the sensible heating (SE) in western continents, the condensation heating (CO) in eastern continents and the dual type of heating (LO+CO) in western oceans comprised the so-called "Four Leaves LOSECOD" heating in boreal summer, which is related to the zonal land-sea thermal contrast in the extratropical region and is the main cause of the coexistence of monsoon and desert. However, when and how does the "LOSECOD" heating set up as well as the heating pattern in Asian-Pacific region change from the "west cold east warm" into the "east cold west warm"? Subsequently, a great number of relative issues still remain to be further investigated.

In fact, on the onset time of the East Asian subtropical summer monsoon, scientists have been in profound dispute for a long time, and two major considerable controversial views are as follows: Tian et al.[11] and Wan et al.[12] pointed out, by analyzing the features of spring rainfall in the south of China, that the spring rain does not belong to the East Asian summer monsoon rainfall. Wu et al.[10] also classified the spring rain into a winter type of the atmospheric circulation based on the F-GOALS model sensitivity experiments; On the contrary, Ding et al.[13] firstly called the persistent rainfall over the southern China during April to June an early summer rainy season. Chen et al.[6] suggested that the East Asian subtropical monsoon rainy season began in early April over the southern China first. Zhao et al.[14] pointed out that the southwest summer monsoon set up in the subtropics first, along with the beginning of a subtropical rainy season. Wang et al.[15] proposed that in late March and early October, along the east coast of Asia, not only the sensible heat flux but also the latent heat flux presents a distinct inverse of land-sea thermal contrast. He et al.[16] asserted that the seasonal transition of the zonal thermal contrast between the East Asian continent and the Western Pacific Ocean happened over the subtropical region first around the end of March and early April, with prevalence of northerly in winter turning into southerly in summer at lower troposphere, and the emergence of convective precipitation at the same time, which all marked the establishment of the East Asian subtropical summer monsoon. As mentioned above, one point of view held that the transition of East Asian atmospheric circulation from winter to summer occurred in May, and the EASM set up after the SCS summer monsoon; while the other point of view admitted that the EASM established at the end of March to the beginning of April. Therefore, when does the East Asian subtropical monsoon seasonal transition actually occur? And what is its possible mechanism? These are the urgent issues needed to be further researched and investigated.

In this paper, the key EASM region is defined according to the dominant seasonal transition feature of the meridional wind. Based on the "thermal wind" rule in the atmospheric dynamics and "thermal adaptation" theory, a new "thermal-wind-precipitation" relationship has been deduced. At last, the seasonal transition features of EASM and its possible mechanism have been illustrated.

2 Data and analysis methods

Daily reanalysis datasets from the NCEP/NCAR (National Centers for Environmental Prediction/National Center for Atmospheric Research, USA), which include the variables of air temperature, geopotential height, and 10-meter wind (with a horizontal resolution 2.5°× 2.5°), and the pentad data of the Climate Prediction Center (CPC) Merged Analysis of Precipitation (CMAP) (with a horizontal resolution 2.5°×2.5°) from 1979 to 2009 are applied. The pentad mean data were calculated from daily data.

3 Definition of EASM region

As we all know, there are two basic features over the monsoon regions: the first one is significant seasonal transition of the lower-layer wind, especially the wind direction reverses from winter to summer, while the second one is a dry period turning into a wet period corresponding with the wind changing from winter to summer. As indicated in Fig. 1, the easterly occurred mostly in the lower layers and the westerly dominated in the middle and higher layers. As to the meridional wind, there is south wind in lower layers from March to September, and northerly in the rest of the months; while in the upper layers there is north wind from April to October, and south wind in the remaining months, and all the levels are characterized by significant alternation of the meridional wind. Based on its significant seasonal reverse of meridional wind, the key EASM region is objectively defined by whether there is a robust change in the meridional wind at 925-hPa and a meridional vertical wind shear between 925-hPa and 200-hPa.

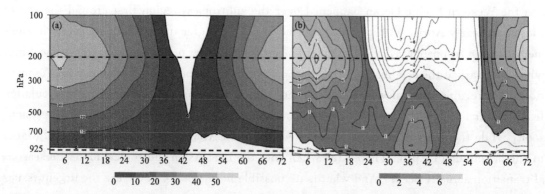

Fig. 1 Time-height cross-section of zonal (a) and meridional (b) wind over the EASM region (20°~35°N, 110°~120°N). y-coordinate: hPa, x-coordinate: pentad; The shades in (a) and (b) indicate the westerly and southerly, respectively. Units: m·s^{-1}

The pentad evolution of the meridional wind is illustrated in Fig. 2, which shows that no seasonal transition exists between 100°~110°E at the lower layer, and all the latitudes are characterized by south wind throughout the year. We conclude that the region over 100°~110°E is not the EASM region. In the region of 110°~120°E, 20°~35°N, there is significant change of the low-layer meridional wind and its vertical wind shear. To the north of 35°N, the change of the meridional wind is weak, which we identify as the edge of the EASM. Besides, the change of the meridional wind is also significant over the western Pacific Ocean (120°~150°E). Due to the distribution of observations and for the sake of convenience, the region (110°~120°E, 20°~35°N) is consequently defined as the key region of EASM.

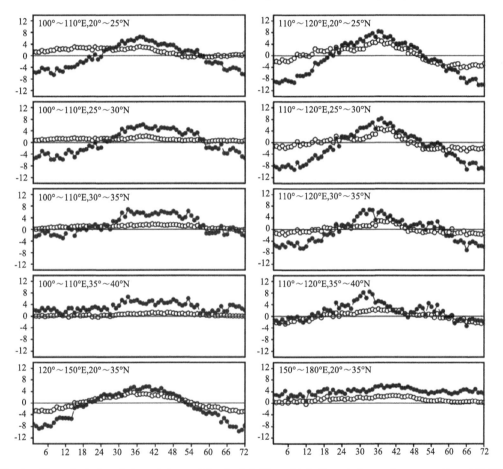

Fig. 2　Climatological evolution of the meridional wind (line with hollow circles, units: m·s^{-1}) at 925-hPa and its vertical shear between 925-hPa and 200-hPa (925~200-hPa, line with solid circles, units: m·s^{-1})

4　Thermal-wind-precipitation relation of EASM

By simplifying the quasi-geostrophic vorticity equation, the relationship between the vertical motion and vertical shear of the meridional wind can be deduced, i. e.

$$w \propto -\frac{\partial v}{\partial z} \qquad (1)$$

where w is the vertical motion, v is the meridional wind.

According to the thermal wind principle,

$$-\frac{\partial \mathbf{V}_g}{\partial p} = \frac{R}{fp} k \times \nabla_P \cdot T \tag{2}$$

where T is air temperature, f is geostrophic vorticity, and R is a constant parameter.

It is noticed that there is a strong transition of zonal land-sea thermal contrast from winter to summer, while there is no signal change of meridional land-sea thermal contrast. In addition, in the subtropical region, the observed wind can be approximated as the geostrophic wind. Therefore, Eq. (3) can be rewritten as:

$$\frac{\partial v}{\partial p} = -\frac{R}{f}\frac{1}{p}\left(\frac{\partial T}{\partial x}\right)_P \tag{3}$$

By combining the Eq. (3) and Eq. (1), a new relationship Eq. (4) is inferred as:

$$\left(\frac{\partial T}{\partial x}\right)_P \propto -\frac{\partial v}{\partial z} \propto w \tag{4}$$

Thus, in Eq. (4), a thermal-wind-precipitation relationship is established by combining the thermal and dynamical processes, which include zonal land-sea thermal contrast, vertical shear of meridional wind, and precipitation associated with the summer monsoon. To validate Eq. (4) and for the sake of convenience, we choose 500-hPa area mean air temperature contrast between the (80°~110°E, 20°~35°N) and (120°~150°E, 20°~35°N) region to stand for the whole-column thermal contrast between land and sea. For the whole troposphere, we take the meridional wind vertical shear as the parameter of monsoon meridional circulation of EASM, and the 300-hPa vertical motion as the inner response. Hence, Eq. (4) is deduced as

$$(T_L - T_S)_{500 \text{ hPa}} \propto (v_{200 \text{ hPa}} - v_{925 \text{ hPa}}) \propto -w_{300 \text{ hPa}} \tag{5}$$

where T_L is whole-layer mean air temperature over land, T_S is air temperature over the sea.

Based on Eq. (5), the thermal-wind-precipitation relationship is shown in Fig. 3.

As can be seen from Fig. 3, the zonal sea-land thermal contrast, the vertical shear of meridional wind, and the 300-hPa vertical movement along with the rainfall anomaly all presented a dominant seasonal reverse from winter (summer) to summer (winter). During late March and early April, the zonal land-sea thermal contrast and the vertical meridional wind shear both reversed from positive to negative, while the w and the anomaly rainfall changed from negative to positive.

5 Seasonal transition of EASM

5.1 The seasonal change of thermal flux

It has been known that the most fundamental definition of monsoon is the reverse of wind direction(e. g., from winter to summer). We first analyzed the wind direction on each of the layers at different periods of time.

Fig. 4 shows the averaged wind direction change with height in different pentads. The thermal wind principle shows that the wind direction rotates anticlockwise with height throughout the winter half-year, associated with cold advection. By contrast, in summer, the wind direction on the low level (e. g. , 925-hPa) is southeast or south, west wind dominates in the middle layer (e. g. , 500-hPa) of the troposphere, and there is either northerly or westerly in the upper layer (e. g. , 200-hPa). It is identified that the transition time of EASM happened right after the 12th pentad and before the 22nd pentad.

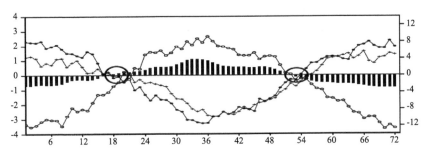

Fig. 3 Seasonal transition of the three factors in the "thermal-wind-precipitation" relationship and pentad mean CMAP anomaly over the key EASM region (20°~35°N, 110°~120°E) (cross: land-sea thermal difference (left vertical axis), units: deg K; solid circle: omega (left vertical axis), units: 0.01 m · s^{-1}; hollow circle: the high and low difference (right vertical axis), units: m · s^{-1}; columnar: pentad precipitation anomaly (right vertical axis), units: mm · p^{-1}; the transition time of the factors is indicated by two big circles)

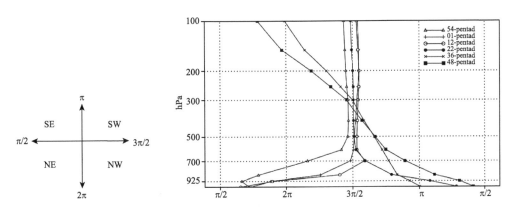

Fig. 4 Vertical profile of wind direction over the key region of EASM

During seasonal transition, the reverse of wind direction and thermal advection on the lower layer of the troposphere is most obvious (Table 1). As shown in the 925-hPa wind field (Fig. 5), in early January (1st pentad), the entire East Asia was covered by cold advection; in early March (12th pentad), the EASM region, except for a small area at the southwest side, was generally controlled by cold advection. However, in early April (the 22nd pentad), the EASM region was fully covered with warm advection, the EASM was completely established, and warm advection was centered in the Hunan, Guangdong, and Guangxi region. In late June and early July (36th pentad, i. e. during the SCS summer monsoon), warm advection was further developed, the maximum value center was located along the south of China to the north of China to the Huaihe River basin; while in late August (48th pentad), over the EASM region, warm advection decreased significantly, but the whole eastern China was still covered with warm advection; in early October (54th pentad), warm advection in the EASM region retreated, which was replaced by uniform cold advection. At this time, the mainland of the eastern China entered a rapidly cooling period in which the East Asian subtropical winter monsoon began to establish.

Table 1 Seasonal transition of wind direction and thermal advection

Season type	Time/pentad	Wind-direction on 925-hPa	Wind-direction on 500-hPa	Wind-direction on 200-hPa	R-type with height	Thermal advection
Winter	54	NE	NW	NW	AC	CA
	1	NE	NW	W	AC	CA
	12	NE	NW	SW	AC	CA
Summer	22	ES	SW	W	C	WA
	36	S	SW	WN	C	WA
	48	ES	SW	NE	C	WA

(NE for northeasterly, ES for southeasterly, S for southerly, NW for northwesterly, SW for southwesterly, W for westerly, AC for anticyclone, C for cyclone, CA for cold advection, WA for warm advection)

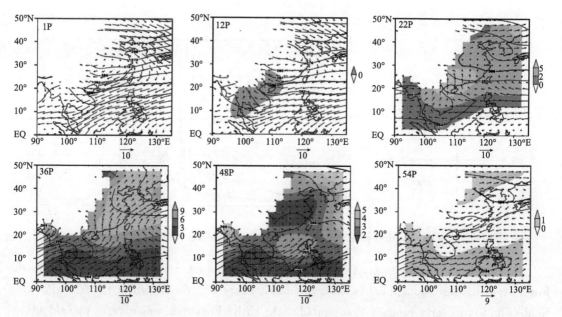

Fig. 5 Horizontal wind (vector), temperature (contour; units: K) and thermal flux (shaded, units: 10^{-5} K·s^{-1}) at 925-hPa (1P, 12P, 22P, 36P, 48P, and 54P represents the 1st, 12th, 22nd, 36th, 48th, and 54th pentad, respectively)

5.2 The rapid mutation of low-level wind direction

The above analysis confirmed that both the wind angle and the corresponding thermal flux characterized the seasonal transition of EASM from winter (summer) to summer (winter). In order to further clarify the exact time of seasonal transition of EASM, we designed a diagram to characterize the temporal evolution of area-averaged wind direction over the EASM region (Fig. 6).

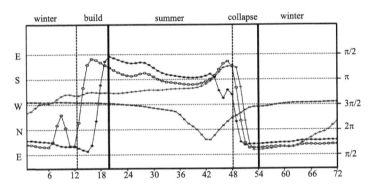

Fig. 6 The pentad mean evolution of wind direction angle over the key region of EASM (hollow circle: 10-m wind angle, solid circle: 925-hPa wind angle, cross: 850-hPa wind angle, fork: 200-hPa wind angle)

As shown in Fig. 6, at 925-hPa, the north wind suddenly changed into the south wind between the 14th pentad and the 16th pentad over the EASM region before changing into a southwesterly wind during summer (6th-7th pentad). In the 48th pentad, an abrupt change of wind occurred from southeasterly to northeasterly. The ground 10-m wind field also shows an obvious change from winter to summer: before the 16th pentad, and the wind direction is northeasterly, which changed into southerly around the 18th pentad and northeasterly before the 54th pentad. Although not as obvious as the 10-m wind field and 925-hPa wind field, a seasonal change at 850-hPa and 200-hPa appears: the wind angle rotated almost 180 degrees just in three pentads (half a month). The summer and winter types of EASM are stabilized in late March to early April and late September to early October, respectively, representing the transition of EASM from winter (summer) to summer (winter).

As the evolution of wind direction at each of the layers is characterized as significant meridional change, the transformation of wind angle in high and low layers roughly reflected a transition in EASM meridional circulation. From the meridional circulation field in different periods (Fig. 7) we can find that in early January (1st pentad), there are downward flows above the EASM region; the higher layers are with the southerly while the lower layers are with the northerly. In early March (12th pentad), a weak southerly appeared in the low layer, meanwhile week convection is generated at the middle-and lower-layer over the EASM region. In early April (22nd pentad), an obvious upward motion appeared over the EASM region. Southerly wind is prevalent at the low layer while there is weak northerly wind at the high layer. It is also noticed that at 200-hPa, a clear closed circulation circle appeared around 20°N; to its south (north), there are downward (upward) airflows. This closed circulation is an excellent indicator for the setup of EASM. At the end of May and beginning of June (36th pentad), a time when SCS summer monsoon has already fully established, although there is consistent upward motion from the equator to 35°N, subtropical summer monsoon circulation and tropical summer monsoon circulation are still separated. By this time, the high (low) level has been occupied by the northerly (southerly) wind. At the end of August to early September (48th pentad), the south wind at the low layer are decelerated and the subtropical summer monsoon convection came into a decay stage. From the end of September to early October (54th pentad), it turned to the northerly wind at the low level when the subtropical winter monsoon was established.

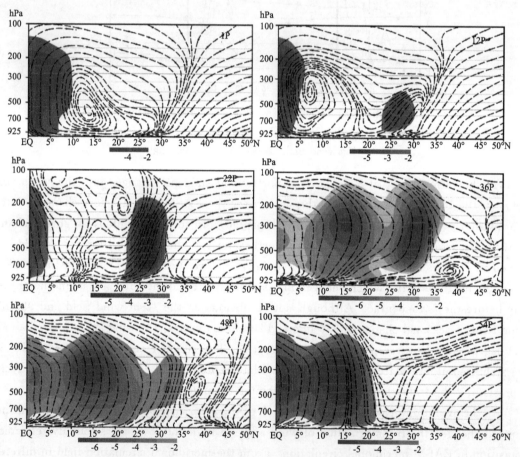

Fig. 7 The meridional circulation and vertical motion in different periods
(shaded areas stand for the rising motion, unit: 10^{-2} m·s^{-1})

5.3 The seasonal transition of low-level wind field

The seasonal transition of East Asian subtropical monsoon starts in the lower troposphere. In order to understand its building process and corresponding weather system, the evolution of the 10-m wind field is examined (Fig. 8). As shown in Fig. 8, in early January (1st pentad), the whole East Asia was controlled by a continental high. In early March (12th pentad), a mainland cold high moved eastward, when the northerly wind was replaced by northeasterly over eastern China; In early April (22nd pentad), the continental cold high moved towards the north of East China Sea, when eastern China was controlled by the southeasterly wind, which marked the establishment of the subtropical summer monsoon. At the end of June and early July (36th pentad), the cold high merged into the western Pacific subtropical high and the eastern China was entirely controlled by the southerly wind. In late September to early October (54th pentad), a cold high with a closed center was located over Shandong province, eastern China was controlled by northwesterly wind, and the subtropical monsoon circulation transformed from summer to winter. From the last panel of Fig. 7 that shows the moving tracks of the cold high from the 12th pentad to 22nd pentad, it is known that it was around the 18th pentad that the continental cold high moved eastward to the sea.

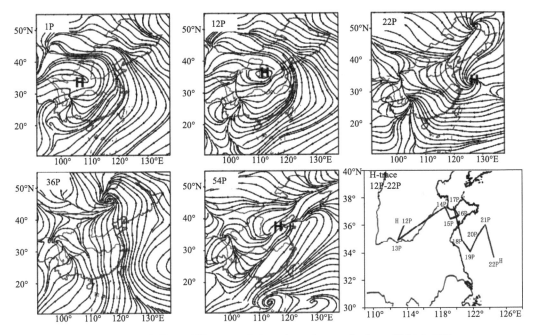

Fig. 8 10-m wind field in different periods and moving track of a cold high (H) center

6 Possible mechanisms of seasonal transition of EASM

Through the above research, we analyzed the seasonal transition features of EASM and clarified that the transition period of EASM (from the winter type to the summer type) is around 18th to 22nd pentad (from late march to early April). Besides, based on previous work[11], we preliminarily present here a mechanism (Fig. 9) for the establishment of the subtropical summer monsoon. As we can see from Fig. 9, when the Sun passes from the Southern Hemisphere to the Northern Hemisphere, the surface thermal contrast (including the effects of the Tibetan Plateau) causes sensible heat to distribute asymmetrically in the zonal direction, which changes the non-uniform distribution of heat from a pattern of "west cold - east warm" into one of "west warm-east cold". Due to the sensible heating forcing over the continent, a low-layer cold high moved eastward into the sea. As a result, the southerly wind at the low layer is stimulated over the subtropics. The southerly wind conveys heat from lower latitudes, inducing positive feedback in the "west warm - east cold" pattern. Meanwhile, the southerly wind at the low layer changed the sign of vertical wind shear, inducing rising motion and rainfall which ultimately released latent heat and warmed the high-level atmosphere. Through this "thermal adaptation" process, latent heating not only strengthened the southerly wind at the low layer, but also induced the northerly wind at the high layer.

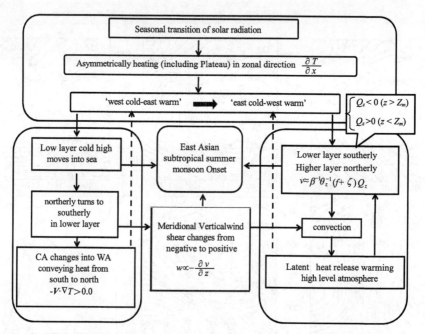

Fig. 9 The mechanism of East Asian subtropical summer monsoon onset

7 Summary and discussion

Based on the definition of the EASM region and the onset time of the subtropical summer monsoon, the formation and development mechanism of the subtropical summer monsoon is examined and discussed. Major conclusions are given as follows:

(1) According to the characteristics of changes in the meridional wind of EASM, the area (20°~35°N, 110°~150°E) is defined as the EASM area, and the 35°~40°N is a border area of EASM.

(2) A "thermal-wind-precipitation" relationship, i. e. the change phase of the land-sea temperature contrast, the high and low layer vertical wind shear, and the vertical movement associated with the monsoon rains, is identified for the EASM region. The wind direction experiences significant seasonal change: during the winter monsoon, wind direction rotates anticlockwise with height, representing cold advection; while in the summer monsoon, wind direction rotates clockwise with height, representing warm advection.

(3) In late march to early April, at the low level, the wind direction changes almost 180°. The seasonal transition of the EASM circulation occurs simultaneously.

(4) Due to the surface thermal contrast and the seasonal transition in solar radiation, the heating type over the EASM region changes from a pattern of being "west cold-east warm" to one of being "west warm-east cold". The latter induces the southerly wind at the low layer. As a result, the wind direction rotates clockwise with height, representing warm advection that conveys heat from the south to the north. Meanwhile, the "thermal adaptation" process produces the upward motion and rainfall, releasing latent heat and warming the upper layer. Both of the two processes have positive feedback effects on the thermal distribution pattern of being "warm in the west and cold in the east".

In this paper, we made a preliminary examination of the seasonal transition mechanism of EASM. However, the effects of various external forcing factors and internal positive feedback processes on the formation of EASM are still not quite clear. In our future work, we will utilize large-scale circulation models to gain better understanding of the formation mechanism of EASM.

References

[1] Tao Shiyan, Chen Longxun. A review of recent research on the East Asian summer monsoon in China [C]//Monsoon Meteorology. Oxford: Oxford University Press, 1987, 60-92.

[2] Zhu Qiangen, He Jinhai, WANG Panxing. A study of circulation differences between East Asian and Indian summer monsoon with their interaction [J]. Adv Atmos Sci, 1986, 3: 466-477.

[3] Wang Bin, Lin Ho. Rainy season of the Asian-Pacific summer monsoon [J]. J Climate, 15: 386-398.

[4] Zhang Qingyun, Tao Shiyan. Tropical and subtropical monsoon over East Asia and its influence on the rainfall over eastern China in summer [J]. Quart J Appl Meteor, 1998, 9: 17-23.

[5] Ding Yihui, Murakami M. The Asian Monsoon [M]. Beijing: China Meteorological Press, 1994: 1-38.

[6] Chen Longxun, Li Wei, Zhao Ping, et al. On the process of summer monsoon onset over East Asia [J]. Clim Environ Res, 2000, 5(4): 345-355.

[7] Ding Yihui, Li Chongyin, He Jinhai, et al. South China Sea Monsoon Experiment (SCSMEX) and the East-Asian Monsoon [J]. Acta Meteor Sinica, 2004, 62(5): 561-568.

[8] Qian Yongfu, Jiang Jing, Zhang Yan, et al. The earliest onset area of the tropical Asian Summer Monsoon and its mechanisms [J]. Acta Meteor Sinica, 2004, 62(2): 129-139.

[9] He Jinhai, Qi Li, Wei Jin, et al. Reinvestigations on the East Asian subtropical monsoon and tropical monsoon [J]. Chin J Atmos Sci, 2007, 31(6): 1257-1265.

[10] Wu Guoxiong, Liu Yimin, Yu Jinjin, et al. Modulation of land-sea distribution on air-sea interaction and formation of subtropical anticyclones[J]. Chin J Atmos Sci, 2008, 32(4): 720-740.

[11] Tian S F, Yasunari T. Climatological aspects and mechanism of spring persistent rains over central China [J]. J Meteor Soc Japan, 1998, 76(1): 57-71.

[12] Wan Rijin, Wu Guoxiong. On the mechanism of spring persistent rains[J]. Sci in China (Ser D), 2006, 36(1): 936-950.

[13] Ding Yihui, Chen Longxun, MURAKAMI M. The East Asian Monsoon[M]. Beijing: China Meteorological Press, 1994, 1-263.

[14] Zhao Ping, Zhou Xiuji, Chen Longxun, et al. Characteristics of subtropical monsoon and rainfall over eastern China and western North Pacific and associated reasons [J]. Acta Meteor. Sinica, 2008, 66(6): 940-954.

[15] Wang Tongmei, Wu Guoxiong, WAN Rijin. Influence of the mechanical and thermal forcing of Tibetan Plateau on the circulation of the Asian summer monsoon area[J]. Plateau Meteor, 2008, 27(1): 1-9.

[16] He Jinhai, Zhao Ping, ZHU Congwen, et al. Discussion of some problems as to the East Asian subtropical monsoon[J]. Acta Meteor Sinica, 2008, 66(5): 683-696.

[17] Zeng Qingcun, Li Jianping. Interactions between the Northern and Southern Hemispheric atmospheres and the essence of monsoon [J]. Chin J Atmos Sci, 2002, 26(4): 433-444.

东亚与北美东部降水和环流季节演变差异及其可能机理分析[*]

常炉子[1]　何金海[1]　祁莉[1]　温敏[2]

(1. 南京信息工程大学气象灾害省部共建教育部重点实验室,南京 210044；
2. 中国气象科学研究院灾害天气国家重点实验室,北京 100081)

摘要：利用 NCEP/NCAR 再分析资料和 CMAP、GPCP 卫星反演降水资料,对比分析了东亚与北美东部地区降水和大尺度大气环流季节演变特征的差异。结果显示,东亚和北美东部地区冬季环流形势较为相似,而夏季差异则较大,这正是东亚为季风区,北美为非季风区的表现。此外,基于季风的两大特征量"风"和"雨",分析了两地降水和低空风场季节变化的显著差异：东亚副热带地区降水季节变率大,呈"夏湿冬干"的季风降水特征,低层盛行风向随季节逆转,冬季盛行偏北风,夏季盛行偏南风,具有显著的副热带季风区特征。北美东部副热带地区全年雨量分配均匀,低层常年盛行偏西风,呈现非季风区特征。进一步的分析发现,作为季风基本推动力的海、陆热力差异在东亚和北美东部地区有着显著的区别：东亚地区的经向和纬向温度梯度随季节反转的特征显著；而北美东部地区虽有纬向温度梯度的季节反转但幅度很小,且经向海、陆热力差异随季节反转不明显。此外,与青藏高原和落基山脉相伴的纬向环流也存在显著差异。鉴于此,提出经向和纬向海、陆热力梯度反转特征的不同以及青藏高原和落基山脉地形的不同作用很可能是造成东亚副热带季风气候而北美东部非季风气候的主要原因,上述结论还有待于数值试验的进一步证实。

关键词：副热带夏季风；降水季节演变；副热带高压；海、陆热力差异

1　引　言

中国位于欧亚大陆的东部,每年的天气气候深受东亚季风活动的影响。作为亚洲季风的主要成员,东亚季风具有显著不同于印度季风的特征。东亚季风可划分为南海—西太平洋热带季风和中国大陆东部—日本的副热带季风[1],后者是直接影响中国旱涝的关键因子。它与前者相互作用,共同影响着中国的旱涝分布[2]。虽然迄今为止,对东亚副热带季风的研究已取得了很多成果,但对其形成原因的认识存在争议,一部分学者认为东亚副热带夏季风是南海热带夏季风的向北延伸[3],另一部分则认为东亚副热带夏季风早于热带夏季风建立,它更依赖于纬向海、陆热力差异的季节性反向[4-7]。与东亚不同的是,虽然美国东部大陆同样位于北美高原(落基山脉)东侧,且东面为海洋(大西洋),属于大陆东海岸亚热带湿润气候区,但却没有典型的季风气候特征。究竟"北美东部地区"为什么是非季风区至今并没有答案。早在20世纪

---
[*] 本文发表于《气象学报》,2013 年第 71 卷第 6 期,1074-1088.

50年代,叶笃正等[8]就指出"由于青藏高原与落基山脉的形状不同,对大气环流可能产生很多不同的效果。"近几十年来,也有不少科研成果展示了落基山和青藏高原对大气环流和天气系统影响的差别较大[9-11]。那么,究竟这种不同的海、陆分布和地形在东亚季风气候与美国东部非季风气候的产生中起何作用?鉴于此,深入对比分析东亚和北美东部的大尺度环流演变特征,并在此基础上探讨可能的物理成因,具有重要的科学意义和实际应用价值。

近年来,吴国雄等[12]从"热力适应"理论出发,提出与海、陆分布相关的长波辐射冷却(LO)、感热加热(SE)、凝结加热(CO)和双主加热(D)(LOS-ECOD)四叶型加热是形成副热带地区季风与沙漠共存的主要机制之一。冬季海洋上空的大气较暖,陆地上空的大气较冷,而夏季的情形与之相反。因此,夏季副热带地区的上升运动出现在大陆东部和海洋西部,而下沉运动发生在大陆西部和大洋东部。夏季副热带的海、陆(含高原)上空大气加热型决定了大尺度的经向风场和垂直运动及其相应的降水。这种理论为解释副热带环流的形成机制提供了理论基础,为研究副热带季风的形成机理提供了新思路。然而,据此理论北美东部地区应与欧亚大陆东部地区一样,同属副热带季风降水区("绿洲")。Li等[13]根据冬夏近地面盛行风向反转程度所定义的归一化季风指数,给出的全球季风区的分布图显示,欧亚大陆东部属于副热带季风区,但北美大陆的东部却是非季风区(图1)。中外诸多学者根据不同的标准所定义的全球季风区分布[14-15]在东亚地区均存在与图1类似的结论。可见,在北美东部地区,"沙漠-季风"共存机制无法解释该地区的非季风区特征。

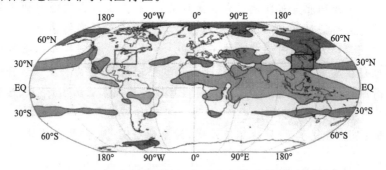

图1 全球季风区的划分[13](两个实线框分别表示北美东部和东亚地区)

众所周知,季风区通常具有两个基本特征[16]:(1)低层风的季节变化显著,特别是盛行风向的冬夏反向,(2)存在与风向变化相对应的明显干、湿季交替。Li等[13]把既有风向转变又有干湿转换的地区定义为标准季风区,把有风向变化而无干湿转换的地区称为普通季风区,而将有干湿转换却无风向变化的地区称为类季风区。那么东亚与北美东部地区的降水以及低层风向的季节演变特征有何差异?造成这种差异的可能机制是什么?回答这些问题有助于揭示季风的季节变化和旱涝转换[17-19]的机制,进而为深入理解东亚副热带季风形成的本质(推动力)提供线索,对改进中国东部降水的预测方法有着十分重要的科学价值和实际意义。本研究将通过对比分析,弄清东亚和北美东部地区降水和环流季节演变特征的差异,着重从海、陆热力差异的角度分析造成上述区别的可能原因。

2 资料与方法

使用的资料是1979—2008年NCEP/NCAR逐日,逐月再分析数据集[20],水平分辨率为

2.5°×2.5°,气象要素包括500和850 hPa风场,位势高度场以及大气温度资料等。同时还使用了CMAP(Climate Prediction Center Merged Analysis of Precipitation)[21]以及GPCP(Global Precipitation Climatology Project)[22]逐日,逐月降水资料。

本研究关注的是气候平均态,因此,首先求各变量的气候平均值(各变量1979—2008年的算术平均)。由于主要分析大尺度环流的特征,为了滤去高频扰动故将逐日资料处理为逐候(5d)平均。另外,由于对流层温度总体上自赤道向两极递减,存在明显的经向差异,为了更好地反映温度梯度在东西方向上的变化特征,本研究用温度的纬向和经向非对称分量(T')进行研究,其中,$T'=T-\bar{T}$(T是气温,$\bar{T}$是T的纬向平均),用海洋、陆地上空的温度梯度表征海陆热力梯度。

3 冬、夏季东亚、北美东部地区大气环流背景场和水汽输送特征分析

东亚和北美东部大陆同位于北半球高原(青藏高原和落基山脉)东侧,海洋(太平洋和大西洋)西侧,且所处纬度相近,海陆分布具有一定的相似性,但是气候特征却有着显著的差异(一个为季风气候,另一个为非季风气候)。为了比较,首先认识一下两个地区的大气环流背景有着怎样的异同,它们为各地气候的形成提供了怎样的背景条件?

3.1 冬、夏季东亚和北美上空平均槽、脊分布特征

图2给出了冬、夏季北半球大气环流分布特征。由图可见,冬季,东亚与北美对流层中层的大气环流形势较为相近:10°~20°N为副热带高压带,在中高纬度,东亚东海岸及北美东海岸各有一大槽(分别为东亚大槽和北美大槽),东亚大陆西部及北美西部各有一脊。由冬入夏,环流形势发生了显著变化:中低纬度,东亚南面的西太平洋副热带高压(副高)发展北移,副高不再维持带状,发生断裂,东亚大陆上空为气压相对低值区,其中,西太平洋副高单体中心位于海上,其东侧外围偏南气流影响东亚地区;而北美南面的亚速尔高压同样发展北移,但仍旧维持带状,其北侧的偏西环流影响北美东部地区。在中高纬度地区,东亚大陆西部的气流平直甚至由脊变为槽,东亚大槽减弱并东移,"东槽西脊"的环流形势发生改变,而北美环流形势与冬

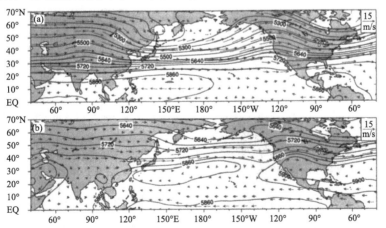

图2 (a)冬(1月)、(b)夏(7月)季北半球500 hPa位势高度场(等值线,gpm)以及风速场(箭头)分布特征

季相比变化不大,"东槽西脊"的环流形势稳定维持,只是强度略微减弱。可见,由冬入夏,西太平洋副高位置、强度和型态的改变以及中高纬度地区槽、脊的变化决定了东亚环流显著的季节变化特征,而北美地区,南面的亚速尔高压强度和位置虽与西太平洋副高一样加强和北移,但最显著的差别在于其型态并没有明显的季节变化,始终维持带状,且中高纬度"东槽西脊"的分布也并没有太大改变,故北美东部大陆环流形势并不存在明显的季节变化特征。

综上所述,虽然控制东亚和北美地区的冬季对流层中层的环流形势十分相似,但夏季差别较大,两个地区环流形势的季节演变情况存在显著差异。

3.2 冬、夏季东亚和北美高空急流的分布特征

图3给出了冬、夏季北半球200 hPa纬向风以及风矢量分布特征,清楚地反映了冬、夏季高空急流的特征。冬季,东亚和北美上空的急流分布特征较为相似:在青藏高原南北两侧各有一支西风急流,两支急流在东亚沿海上空交汇,并在日本岛东南侧上空达最强;北美落基山脉南北两侧同样各有着一支西风急流,并在北美东南大陆上空交汇达最强,但强度较东亚西风急流弱,这与谢义炳等[23]、叶笃正等[8]的研究结果一致。夏季,青藏高原南侧的西风急流消失,由冬入夏,整个急流轴北跳到40°N左右,同时,西风急流的中心西退至青藏高原、伊朗高原北部上空;同样,落基山脉南侧的西风急流也消失不见,北美西风急流轴也发生北跳,至45°N左右,位置较东亚西风急流更北,但是北美西风急流中心的位置在东西方向并无太大改变,这与东亚情况存在较大差异。除此差异以外,夏季,东亚热带地区出现东风急流。中心强度达30 m·s^{-1},然而北美热带地区虽有东风带,但风速较小(不到10 m·s^{-1})。

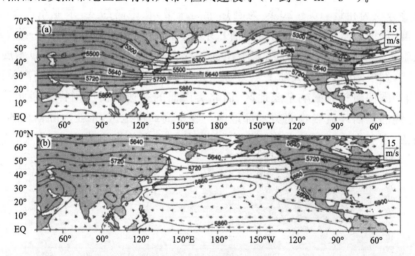

图3 (a)冬(1月)、(b)夏(7月)季200 hPa纬向风场(等值线,m·s^{-1})以及风矢量(箭头)分布特征

综上可见,冬季,东亚与北美高空的急流分布情况相似,东亚西风急流略强。夏季两地有较大差别,急流轴虽同是北跳但急流中心东西向的位置却存在很大差异。

3.3 冬、夏季东亚和北美副热带地区平均纬圈环流的特征

图4给出了冬、夏季沿32.5°N纬带的平均纬圈环流特征,由图可见,冬季,青藏高原和落基山脉及其周围地区的平均纬圈环流分布型态较为相似:在平直的西风带中,青藏高原和落基山脉上空及其东部存在下沉运动。夏季,两高原及其周边的纬圈环流发生了显著的变化,存在

较大差异:该纬度带上,75°E 以东的青藏高原及东亚大陆上空表现为一致的强上升运动,其对应的下沉支位于东太平洋至北美西海岸上空;相反,落基下沉运动隔断了落基山上空的上升气流以及大陆东山脉上空的上升运动较弱,且其对应的下沉支位于北美的中东部地区,这支下沉运动隔断了落基山上空的上升气流以及大陆东部上空的上升气流,不利于北美东部地区上升气流的发展。

综上所述,冬季,东亚与北美东部地区具有较为相似的大气环流形势,然而夏季差别很大,这为东亚季风气候、北美东部地区非季风气候的形成提供了大尺度的环流背景条件。

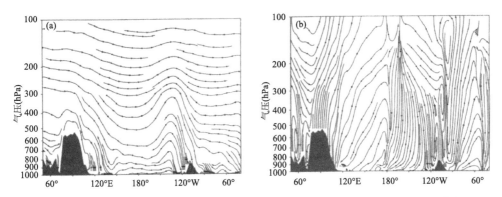

图 4　(a)冬(1 月)、(b)夏(7 月)季高原及周围地区(沿 32.5°N)的平均纬圈环流特征

3.4　冬、夏季东亚和北美东部地区水汽输送特征

图 5 给出了冬(1 月)、夏(7 月)季东亚和北美东部地区垂直积分(地面至 300 hPa)的水汽输送通量特征。可见无论冬、夏,东亚和北美东部地区的水汽输送特征均存在显著差异。冬季东亚大陆盛行冬季风,中国大部分地区盛行西北风,干冷气流自北向南输送的水汽量都很小。另有一股绕青藏高原南侧的偏西风水汽输送为中国南方冬季降水带来水汽。而北美东部地区上空以偏西风水汽输送为主,水汽主要来源于北美大陆西面的太平洋,另有一支在亚速尔高压西南侧转向的偏西南风水汽输送,将墨西哥湾的水汽输送到北美南岸的大陆上空。可见,即使在冬季,北美东部地区也有着较多的暖湿水汽,这与北美东部地区冬季也存在明显的降雨带相对应。夏季,东亚地区的环流形式发生了较大的改变,水汽输送量已然比冬季增大了几倍,这时来到中国大陆(100°E 以东)的水汽,主要来自西南,中国南海和东南沿海,以经向输送为主,这几股南来气流直往北进,可抵河套南部黄河下游、华北及松江平原,与之对应,东亚雨带随季节不断北进;然而,北美东部地区水汽输送量虽比冬季增大,但水汽输送形势并未有太大改变,且仍以西风输送为主,经向输送很弱,与之相对应,北美东部地区的降水带常年稳定少动,且降水量四季分配均匀。

综上可见,东亚地区水汽来源冬、夏季完全不同,冬季水汽输送量少,夏季水汽来源充沛,且以经向输送为主。相比而言,北美东部地区水汽输送形势较为稳定,且一直以纬向西风输送为主,夏季输送量增多,没有显著的季节变化,这与北美东部地区雨带位置稳定有关。

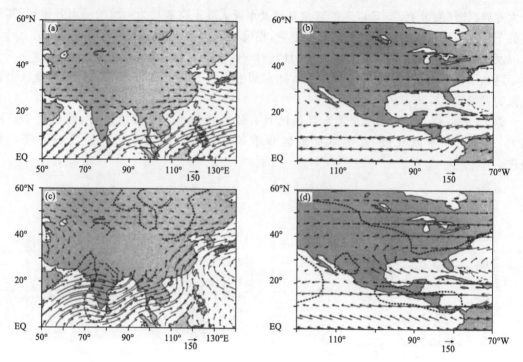

图5 (a、b)冬(1月)、(c、d)夏(7月)季东亚(a、c)和北美东部地区(b、d)垂直积分(地面至300 hPa)的水汽输送通量(单位:kg·(m·s)$^{-1}$)(虚线表示经向风速值为0的区域)

4 东亚与北美东部地区降水及环流季节演变特征的对比分析

4.1 降水季节演变特征差异

图6给出了两种降水资料东亚和北美东部地区降水的时间—纬度剖面。由图6a、c可见,东亚地区的降水最早于第15候在副热带地区(25°～32.5°N)增多,平均降水量达到6 mm·d^{-1}。随后,雨带于第24候迅速向南扩展并不断加强,这与陈隆勋等[24]、江志红等[25]、叶笃正等[8]的研究结果一致。第30—46候,雨带逐渐向北推进,先后抵达江淮流域、华北—东北地区。而在北美东部副热带地区,一条明显独立的降水带位于30°～40°N。与东亚副热带降水的明显年循环特征不同,北美东部地区降水四季分配比较均匀,稳定维持在3～4 mm·d^{-1}(图7)。另外,雨带位置稳定少动,无明显的南北摆动。同样的特征在GPCP给出的图6b、d中也得以体现,只是GPCP降水量普遍大于CMAP降水量且CMAP降水为双峰值(6与8月),而GPCP只有6月为峰值,这种差异主要是由资料处理的差别造成的[26-27],但不影响上述分析结果。

以上分析表明,东亚副热带地区降水的季节变化大,夏湿冬干,具有显著的季风降水的特征。相反,北美东部副热带地区全年雨量分配均匀,表现出非季风区降水的特点。此外,两地副热带雨带位置的季节演变特征也有很大差异:北美东部地区副热带高值(大于3 mm·d^{-1})雨带位置常年稳定,无明显的南北位移。

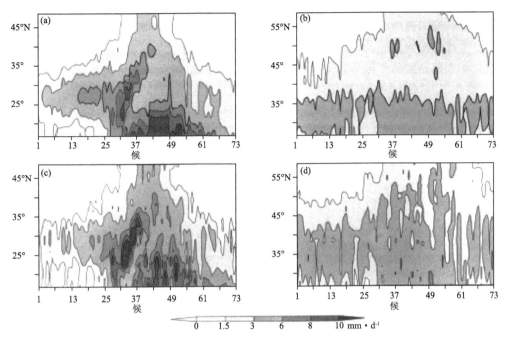

图6 (a、c)东亚(110°~130°E平均)、(b、d)北美东部地区(90°~70°W平均)
CMAP(a、b)与GPCP(c、d)降水的时间—纬度剖面

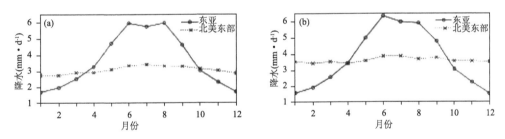

图7 东亚(20°~40°N,100°~140°E)和北美东部地区(30°~50°N,100°~60°W)
CMAP(a)、GPCP(b)降水量的季节演变

4.2 环流季节演变特征的差异

雨带的维持和移动与大气环流的演变密切相关。就东亚地区而言,对流层中、低层的西太平洋副高是夏季对流层下部最重要的天气、气候系统之一,它的活动与东亚天气、气候的变化密切相关[28]。而北美东部地区的天气、气候变化则与北大西洋副高的活动特征联系紧密[29-31]。因此,西太平洋副高和北大西洋副高季节演变特征的差异可能是造成东亚和北美东部地区降水季节变化不同的原因之一。故首先需要对比东亚和北美东部地区对流层中、低层位势高度场上主要环流系统的季节演变特征。

图8给出了东亚和北美东部地区对流层低层850 hPa风场的分布情况。冬季,东亚的副热带以南盛行东北风,以北盛行西北风,中国东部受大陆冷高压东侧干冷的偏北风控制,干旱少雨。由冬入夏,亚洲低纬度副热带高压带断裂(图略),东亚位于大陆热低压东侧与西太平洋

副高西侧的偏南气流控制下,同时越赤道气流不断增强,低空盛行风向发生明显季节性反转。中低纬度盛行越赤道输送至东亚地区的西南风、西太平洋副热带高压西侧外围的东南风以及来自南海南部的偏南气流,3股气流在东亚地区汇合,带来大量暖湿空气,有利于降水增多。而影响北美东部地区的环流形势冬、夏并未发生明显改变。冬季,北美东部地区位于强盛的西风带中,盛行偏西风。由冬入夏,虽然北美上空的副高略微加强北抬,同时副热带高压带在北美大陆西部断裂(图略),但是影响北美东部地区的环流仍以北大西洋副高北侧的中纬度偏西气流为主,说明北美东部地区的低空盛行风向并没有显著的季节变化。

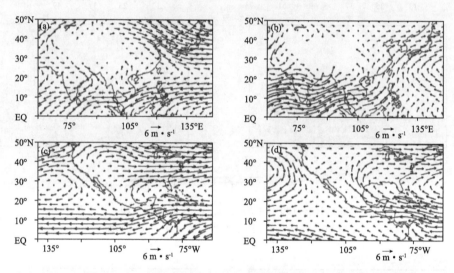

图 8 东亚(a、b)和北美东部地区(c、d)冬季(a、c)、夏季(b、d)850 hPa 风场
(白色区表征 1500 m 的地形)

以上分析表明,东亚地区低空环流形势有明显的季节变化特征,最突出的表现是由冬入夏,东亚低空盛行风向由偏北风转为偏南风;而影响北美东部地区的低空环流形势则较为稳定,常年盛行偏西风,经向风分量很小,没有明显的风向季节性反转特征。

图 9 给出了对流层中层(500 hPa)纬度平均(15°～25°N)的相对涡度的时间—经度剖面。如图 9 所示,由冬入夏,东亚上空的负涡度带转变为正涡度带,说明西太平洋副高在该地区上

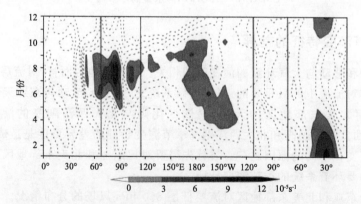

图 9 15°～25°N 纬带平均的 500 hPa 相对涡度的时间—经度剖面
(等值线虚线表示负涡度值,实线加阴影区表示正涡度值;两个实线框分别表示北美东部和东亚地区)

空发生断裂,与低空形态一致。而北大西洋副高始终盘踞在北美上空,表现为终年的负涡度区,说明北大西洋副高的型态无明显改变,北美大陆上空始终为带状分布,显示出比低空更强的稳定性。与涡度场对应,500 hPa 高度场上(图略):东亚上空,冬季,亚洲低纬度副高呈带状分布,其中心居于热带洋面上,副热带地区大陆受其北侧偏西环流控制。夏季,西太平洋副高强度变强,并向北移动,同时在东亚大陆上空断裂,中心居于海洋上,其西伸脊点可至东亚大陆,中国南方受其外围西侧偏南气流影响,有利于将对流层中层的大量海上暖湿空气向中国南方—日本一带输送。伴随着西太平洋副高的北进,中国主雨带明显北抬(图 2)。而北大西洋副高夏季位于北美 32°N 以南地区上空,呈带状分布,这种环流背景不利于南面墨西哥湾上空的低层暖湿空气向北输送。北美东部地区终年位于北大西洋副高北侧,盛行偏西风,其雨带位置稳定少变。

5 降水季节演变差异可能原因

上述结果表明,北美东部地区的低空盛行风场和降水都没有显著的季节变化特征,这与东亚地区完全相反。然而,东亚和北美东部地区同处于大陆东部,其东部和南部均为海洋,但它们的降水和环流的季节变化为什么有如此显著的差异呢? 对东亚地区而言,许多学者强调了经向海、陆热力差异对季风形成和推进的重要作用[32-38]。此外,近年来又有不少学者[39-41]认为纬向海、陆热力差异在副热带季风的形成中至关重要。因此,下文将从海、陆热力差异的角度,试图解释造成东亚和北美东部地区低空盛行风和降水季节演变差异的可能原因。

5.1 纬向海、陆热力差异的季节演变特征

赵平等[5]指出,用 500~200 hPa 平均值表示对流层平均热状况是合理的,因此,首先分析北半球中低纬度 500~200 hPa 平均温度纬向偏差(T')的逐候演变(图 10),以反映海、陆热力差异的季节变化特征。由图可见,东亚地区的纬向温度偏差具有显著的季节反转特征:西太平洋上空的气温在 1—3 月底为偏暖,4—9 月偏冷,而秋冬季又偏暖,其中,高值区基本位于 120°E 以东洋面上空;大陆上的情况完全相反,4—9 月偏暖,10 月—次年 3 月偏冷。总的看来,东亚及其邻近西太平洋地区,冬半年大陆冷、海洋暖,夏半年反转为大陆暖、海洋冷,令温度梯度发生明显的季节性反转,冬半年由海洋指向大陆,夏半年反转为大陆指向海洋。根据热成风关系,冬季对流层低层(高层)盛行偏北风(偏南风),夏季低层(高层)盛行偏南风(偏北风)(图 8,高空图略),表现为明显的季风区特征。而北美东部地区(图 10b)的对流层平均温度偏差也有从东暖西冷(冬)向东冷西暖(夏)的反转特征,但温度偏差的季节变化较小。同时计算了东亚和北美东部地区大陆和海洋的纬向温度差异(图 10c,80°~95°E 代表东亚大陆,125°~140°E 代表西太平洋;115°~105°W 代表北美大陆,65°~55°W 代表西大西洋),发现北美东部地区的纬向热力差异虽有反转,但其幅度较东亚地区要小得多。

然而为什么这种弱的纬向热力差异的转换没有形成冬夏盛行风向弱的季节性反转呢? 后面进一步讨论这个问题。

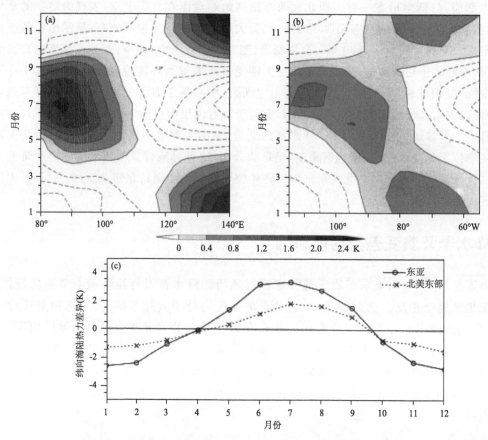

图 10　500～200 hPa 平均温度纬向偏差（沿 30°～32.5°N 平均纬带）的经度—时间剖面
（a. 东亚，b. 北美东部地区）；(c)两个地区纬向海陆热力差异季节演变

5.2　经向海、陆热力差异的季节演变特征

图 11 给出了冬、夏季东亚和北美东部大陆地区温度经向偏差的纬度—高度剖面。冬季（图 11a、c），东亚和北美东部大陆地区与各自相邻南部海域间的经向温度偏差特征大体相同：近于整个对流层内（200 hPa 以下），南面海洋上为相对暖区，其北部大陆上为冷区，北美东部地区经向温度偏差较大。夏季（图 11b、d）东亚和北美东部大陆地区与各自相邻南部海域间的经向温度偏差特征存在较大差异：东亚大陆及其南面海域上空的经向热力梯度方向发生了季节性的反转，呈现为"南冷北暖"，相应的低空气压场（高度场）也发生了反转（图略）。然而，北美东部大陆及其南面墨西哥湾上空的经向温度偏差虽然量值有所减小，但稳定地维持"南暖北冷"的分布形态（图 11d），相应的低空气压场（高度场）也是南高北低。这种稳定的南暖北冷海、陆热力分布为北美东部地区常年盛行的偏西风提供了稳定的大尺度热力背景。

综上可见，作为季风基本推动力的海、陆热力差异在东亚和北美东部地区有着显著的不同：东亚地区的经向和纬向海、陆热力差异随季节反转的特征显著；而北美东部地区的经向海、陆热力差异没有季节性反转，纬向热力差异随季节反转的幅度很小。两者之间如何配置？这可能是决定大陆东部是否为季风区的关键。

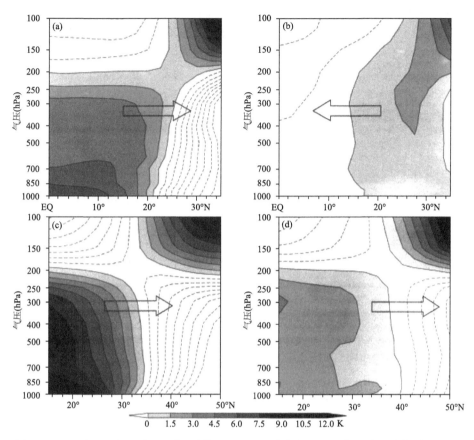

图 11 (a、b)东亚(沿 100°~120°E 平均)和(c、d)北美东部(沿 100°~80°W 平均)冬(a、c)、夏(b、d)季温度经向偏差的纬度—高度剖面(箭头表示经向热力梯度方向)

5.3 纬向海、陆热力梯度的反转与环流季节差异的联系

为进一步讨论为何弱的纬向海、陆热力差异的转换没有形成北美东部地区冬、夏盛行风向弱的季节性反转，根据图 10c，把 4—9 月定义为纬向海、陆热力差异的正位相期(此时大陆暖，海洋冷)，10 月—次年 3 月定义为负位相期，分析纬向海、陆热力差异负位相转为正位相(由冬半年到夏半年)前后的大气环流形势的变化。

图 12 说明在纬向海、陆热力性质由海洋暖大陆冷反转为海洋冷大陆暖后，沿 30°~32.5°N纬带的垂直运动发生了很大的变化：东亚地区表现为全区一致的上升运动区，下沉运动主要在东太平洋上。而北美大陆西侧落基山脉上空为上升运动，中部为下沉，东部为上升运动。可见，纬向海、陆热力梯度反向后(负位相转为正位相)，东亚地区上升运动显著增强，而北美东部地区则没有明显的垂直运动的变化，仅有弱的上升运动。根据热力适应理论[12]可知，当纬向海陆热力性质反转为大陆暖海洋冷后，会在大陆的东部激发上升运动。春季青藏高原感热快速增强，由于其高耸的地形且位于亚欧大陆的东部，能够直接加热对流层中部的大气[42]，加强了东亚大陆的夏季热源(纬向海陆热力对比较强)。然而，落基山脉位于北美大陆的中西部，且东西范围较为狭窄，其对大气的直接加热作用明显弱于青藏高原(纬向海、陆热力对比较弱)。因此，北美上空的大陆加热作用较小，相应的加热区东侧(北美东部地区)的上升

运动也较弱。吴国雄等[12]还指出,青藏高原和落基山脉本身强迫的垂直运动对上述由行星尺度海陆分布激发的大陆东部垂直运动有不同的贡献:青藏高原位于亚欧大陆东部,夏季,它强迫的环流在东亚地区为上升运动,与大尺度海陆分布在大陆东岸激发的上升运动叠加,增强了东亚地区的上升运动;然而,落基山脉位于北美大陆的中西部,且东西范围较为狭窄,它所强迫的下沉气流位于北美中东部地区,削减了大陆东岸垂直运动对行星尺度海陆分布所决定的加热响应,进而减弱了这种行星尺度海陆分布在大陆东岸激发的上升运动。

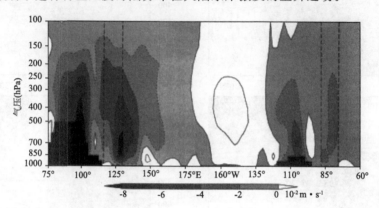

图12 纬向海、陆热力差异负位相转为正位相前后垂直速度变化的经度-高度剖面(沿 30°～32.5°N 平均纬带)(黑色阴影表示地形,粗虚矩形框表示东亚和北美东部大致经度范围)

图13a、b给出了东亚和北美东部地区对流层经向风在纬向海陆热力差异正(夏)、负(冬)位相期的差异。由图可见,东亚(110°～130°E)500 hPa 以下的对流层有较强的偏南风,高空有偏北风,结合气候平均图可知,东亚夏季低空有较强的偏南风,高空有较强的偏北风。这种配置有利于夏季在东亚地区产生较强的上升运动,导致降水,释放潜热,进而加强低空偏南风和高空偏北风,这种正反馈机制放大了纬向海、陆热力差异产生的经向风和降水的季节差异(下文简称为正反馈的放大效应)。而北美东部地区(90°～70°W)低空经向风季节变化很小,故上述正反馈放大作用亦很弱,且落基山脉强迫的下沉运动(图12)也抑制了这种正反馈作用,进而造成北美东部地区正(夏)、负(冬)位相期降水和经向风的差别不大。由纬向风的差异(图13c、d)可以看出,东亚 30°N 以南的低空有异常的偏西风,这是东亚热带副热带季风区的反映,而高空则呈大范围的偏东风,即夏季西风气流的减弱。值得注意的是,北美东部地区正(夏)、负(冬)位相期纬向风的差异较东亚要小得多,尤其是低空。这种纬向风在冬、夏的较小差异不足以改变北美东部地区南暖(高)北冷(低)的温(压)场决定的纬向西风气流(气压差值图略),进而北美东部地区常年盛行偏西风,没有风向的冬夏反转。

6 结论与讨论

东亚和北美东部地区冬季环流形势较为相似,而夏季差异则较大,这正是东亚为季风区,北美为非季风区的表现。此外,基于季风的两大特性——"风"和"雨",对比分析了东亚和北美东部地区气候平均降水和大尺度大气环流季节演变特征的差异,在此基础上,探讨造成这种差异的可能物理成因。主要结论如下:

(1)东亚副热带地区降水季节变率大,"夏湿冬干",且低层盛行风向随季节逆转,冬季盛行

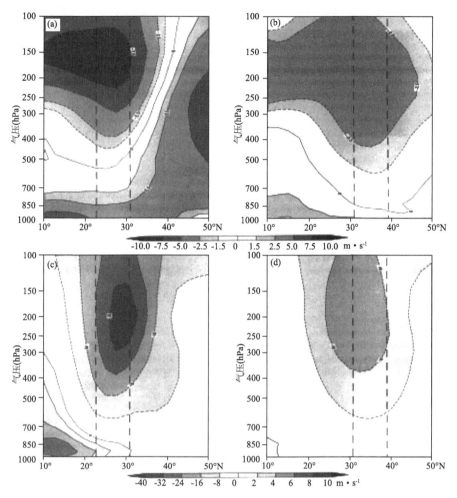

图 13 东亚(a、c)和北美东部地区(b、d)纬向海、陆热力差异负位相转为正位相前后水平风场变化的纬度—高度剖面(a、b.经向风;c、d.纬向风)

偏北风,夏季盛行偏南风,具有显著的副热带季风区特征。相比较而言,北美东部副热带地区全年雨量分配均匀,低层常年盛行偏西风,降水和风向均无明显的季节变化,呈非季风区的特征。

(2)由冬入夏,东亚地区的经向和纬向海、陆热力差异发生反转,亚欧大陆由冷高压转为热低压控制,且副高在大陆上空断裂,中心位于海上,东亚由冷高压东侧偏北风控制转而受副高西侧以及热低压东侧的偏南气流控制;然而,北美东部地区的经向海、陆热力差异没有季节性反转,纬向热力差异虽随季节有反转但幅度很小。北美大陆上空低层(850 hPa)副高虽断裂,但断裂处位于大陆西部,中层(500 hPa)仍呈带状分布,北美东部地区自低层至中层稳定受副高北侧偏西气流控制。

(3)由冬入夏,东亚与北美东部地区的纬向海陆热力差异都发生反转,但春季青藏高原感热快速增强,加强了东亚大陆的夏季热源(纬向海、陆热力对比较强),相应加热区东侧(即东亚地区)上升运动较强,低空有较强的偏南风发展,高空相反。而落基山脉对大气的直接加热作用明显弱于青藏高原(纬向海、陆热力对比较弱),相应的加热区东侧(即北美东部地区)上升运动较弱,低空经向风季节变化很小。另外,由于东亚经向海、陆热力差异也发生了季节性的反转,由冬向夏,偏西风环流明显减弱,有利于上述低空偏南风的发展,潜热释放正反馈的放大效应也得以进一步加强,它的作用与高原强迫的上升运动在东亚地区叠加,形成东亚显著的副热带季风气候,可见经向海、陆热力梯度的反转对于副热带季风的形成是必要的。然而,北美东部地区常年稳定的南冷北暖的经向热力分布决定了该地区四季都受副高北侧稳定的偏西风气流控制,限制了经向风的发展,正反馈的放大效应也很难进一步加强,此外,落基山脉强迫的下沉支又抵消了这种较弱的正反馈效应,进而北美东部大陆东部的降水量不会随季节有显著的改变,形成了北美东部非季风区气候。

综上所述,纬向和经向海、陆热力差异在东亚和北美东部地区的不同配置以及青藏高原和落基山脉地形的不同作用可能是造成这两个地区降水和环流季节演变特征差异的重要原因。这为进一步认识东亚副热带季风的成因和本质提供了参考依据。当然,上述结论还有待于数值模拟和理论分析的进一步验证。

致谢:感谢南京信息工程大学吴志伟教授、刘伯奇博士和朱志伟博士的有益建议和帮助。

参考文献

[1] 朱乾根,何金海.亚洲季风建立及其中期振荡的高空环流特征[J].热带气象学报,1985,(1):9-18.

[2] 张庆云,陶诗言,陈烈庭.东亚夏季风指数的年际变化与东亚大气环流[J].气象学报,2003,61(5):559-568.

[3] Wang B,Lin H. Rainy season of the Asian-Pacific summer monsoon [J]. J Climate,2002,15(4):386-398.

[4] 何金海,赵平,祝从文,等.关于东亚副热带季风若干问题的讨论[J].气象学报,2008,66(5):683-696.

[5] 赵平,陈军明,肖栋,等.夏季亚洲—太平洋涛动与大气环流和季风降水[J].气象学报,2008,66(5):716-729.

[6] 赵平,周秀骥,陈隆勋,等.中国东部—西太平洋副热带季风和降水的气候特征及成因分析[J].气象学报,2008,66(6):940-954.

[7] 赵平,蒋教平,周秀骥,等.春季东亚海 陆热力差异对我国东部西南风降水影响数值试验[J].科学通报,2009,54(16):2372-2378.

[8] 叶笃正,高由禧,刘匡南.1945-1946年亚洲南部与美洲西南部急流进退之探讨[J].气象学报,1952,28(1-2):1-32.

[9] 李维京.关于5、6月东亚大暴雨与北美强风暴天气气候背景的对比研究[J].高原气象,1985,4(S2):65-76.

[10] 盛华,陶诗言.青藏高原与落基山背风气旋的对比分析(一)天气结构[J].大气科学,1988,12(3):233-241.

[11] 李振军,赵思雄.东亚春季强冷锋结构及其动力学诊断研究 I:东亚锤击强冷锋结构[J].大气科学,1996,20(6):662-672.

[12] 吴国雄,刘屹岷,宇婧婧,等.海陆分布对海气相互作用的调控和副热带高压的形成[J].大气科学,2008,32(4):720-740.

[13] Li J P,Zeng Q C. A monsoon index and the geographical distribution of the global monsoon [J]. Adv Atmos Sci,2003,20(2):299-302.

[14] Qian W,Deng Y,Zhu Y,et al. Demarcating the worldwide monsoon [J]. Theoret Appl Climato,2002,71(1-2):1-16.

[15] Wang B, Ding Q H. Global monsoon: Dominant mode of annual variation in the tropics[J]. Dyn Atmos Ocean, 2008, 44(3-4):165-183.

[16] Murakami T, Matsumoto J. Summer monsoon over the Asian continent and western North Pacific [J]. J Meter Soc Japan, 1994, 72:719-745.

[17] Wu Z W, Li J P, He J H, et al. Occurrence of droughts and floods during the normal summer monsoons in the mid- and lower reaches of the Yangtze River [J]. Geophys Res Lett, 2006, 33 (5): L05813, doi: 10.1029/2005GL024487.

[18] Wu Z W, Li J P, He J H, et al. Large scale atmospheric singularities and summer long cycle droughts floods abrupt alternation in the middle and lower reaches of the Yangtze River [J]. Chin Sci Bull, 2006, 51(16):2027-2034.

[19] 吴志伟,李建平,何金海,等. 正常季风年华南夏季"旱涝并存、旱涝急转"之气候统计特征[J]. 自然科学进展,2007,17(12):1665-1671.

[20] Kalnay E, Kanamitsu M, Kistler R, et al. The NCEP/NCAR 40 year reanalysis project [J]. Bull Amer Meteor Soc, 1996, 77(3):437-471.

[21] Xie P P, Arkin P A. An intercomparison of gauge observations and satellite estimates of monthly precipitation [J]. J Appl Meteor, 1995, 34(5):1143-1160.

[22] Robert F A, George J H, Chang A, et al. The version-2 global precipitation climatology project (GPCP) monthly precipitation analysis (1979 present) [J]. J Hydro Meteor, 2003, 4(6):1147-1167.

[23] 谢义炳,陈玉樵. 冬季西太平洋及东亚大陆北部上空的温度场及流场[J]. 气象学报,1951,22(1):47-48.

[24] 陈隆勋,李薇,赵平,等. 东亚地区夏季风爆发过程[J]. 气候与环境研究,2000,5(4):345-355.

[25] 江志红,何金海,李建平,等. 东亚夏季风推进过程的气候特征及其年代际变化[J]. 地理学报,2006,61(7):675-686.

[26] 白建峰. 全球共享降水数据在我国的适用性研究[D]. 上海:东华大学,2002,112.

[27] Gruber A, Su X J, Kanamitsu M, et al. The comparison of two merged rain Gauge-satellite precipitation datasets [J]. Bull Amer Meteor Soc, 2000, 81(11):2631-2644.

[28] 孙淑清,马淑杰. 西太平洋副热带高压异常及其与1998年长江流域洪涝过程关系的研究[J]. 气象学报,2001,59(6):719-729.

[29] Henderson K G, Vega A J. Regional precipitation variability in the southeastern United States [J]. Phys Geogr, 1996, 17(2):93-112.

[30] Katz R W, Parlange M B, Tebaldi C. Stochastic modeling of the effects of large-scale circulation on daily weather in the southeastern US[J]. Climate Change, 2003, 60(1-2):189-216.

[31] Li W, Li L, Fu R, et al. Changes to the North Atlantic subtropical high and its role in the intensification of summer rainfall variability in the Southeastern United States[J]. J Climate, 2001, 24(5):1499-1506.

[32] 陈隆勋,朱乾根,罗会邦,等. 东亚季风[M]. 北京:气象出版社,1991,362.

[33] 何金海,丁一汇,高辉,等. 南海夏季风建立日期的确定与季风指数[M]. 北京:气象出版社,2001:123.

[34] Qian Y F, Wang S Y, Shao H. A possible mechanism effecting the earlier onset South Westerly Monsoon in the South China Sea compared to the Indian Monsoon [J]. Meteor Atmos Phys, 2001, 76(3-4):237-249.

[35] Huang R H, Zhou L T, Chen W. The progresses of recent studies on the variabilities of the East Asian Monsoon and their causes [J]. Adv Atmos Sci, 2003, 20(1):55-69.

[36] He H Y, Sui C H, Jian M Q, et al. The evolution of tropospheric temperature field and its relationship with the onset of Asian Summer Monsoon [J]. J Meter Soc Japan, 2003, 81(5):1201-1223.

[37] Zhang Z Q, Chan J C L, Ding Y H. Characteristics, evolution and mechanisms of the summer monsoon on-

set over Southeast Asia [J]. Int J Climat,2004,24(12):1461-1482.

[38] 丁一汇,李崇银,何金海,等.南海季风试验与东亚夏季风[J].气象学报,2004,62(5):561-586.

[39] 祁莉,何金海,张祖强,等.纬向海陆热力差异的季节转换与东亚副热带季风环流[J].科学通报,2007,52(24):2896-2899.

[40] 何金海,祁莉,韦晋,等.关于东亚副热带季风和热带季风的再认识[J].大气科学,2007,31(6):1257-1265.

[41] Zhu Z W,He J H,Qi L. Seasonal transition of east Asian Subtropical monsoon and its possible mechanism [J]. Tropic Meteor,2012,18(3):305-313.

[42] Wu Z W,Li J P,Jiang Z H,et al. Modulation of the Tibetan Plateau snow cover on the ENSO teleconnections:From the East Asian summer monsoon perspective [J]. J Climate,2012,25(7):2481-2489.

江南雨季地理区域及起止时间的客观确定*

詹丰兴[1,2]　何金海[1]　章毅之[3]　朱志伟[1]

(1. 南京信息工程大学气象灾害省部共建教育部重点实验室 大气科学学院,南京 210044;
2. 江西省气象局,南昌 330046;3. 江西省气候中心,南昌 330046)

摘要:本文基于通过国家气象信息中心整编的全国1675个台站观测资料以及NCEP/NCAR的再分析资料,定义了候降水指数,利用旋转正交经验函数分解(REOF)的方法对全国候降水的季节进程进行了诊断分析,得到了表征气候态降水逐候进程的南、北方模态及各自的时间系数,发现REOF第二模态对应降水季节进程中的江南雨季。综合考虑我国南方(31°N以南、110°E以东区域)气候态降水的候进程、降水季节进程(4—6月减去6—8月降水指数)年际变率以及雨季(4—6月降水指数)降水年际变率的一致性,客观定义了江南雨季的地理范围。利用客观划定区域内的降水指数、925 hPa经向风以及西北太平洋副热带高压500 hPa脊线位置这三个指标,制定了判定江南雨季起止时间的方法,进而对1961—2012年江南雨季起止时间进行了客观确定,给出了江南雨季起止时间序列。本文旨在为规范江南雨季的监测提供参考和借鉴,并为其预测提供科学基础。

关键词:江南雨季;地理区域客观界定;降水季节进程;起止时间

1　引　言

江南地区人口稠密、资源丰富、生态良好、社会发达,是我国经济、文化、生态和社会建设的重要区域。然而,受东亚热带季风及副热带季风系统的共同影响[1-2],江南地区旱涝等气象灾害频发、重发。东亚季风异常危及该区域人民生命财产及生态环境的安全,严重制约经济社会可持续发展,因此对江南地区的降水进行深入研究,进一步提高监测预测能力非常必要。由冬至夏的季节转换过程中(3—6月),我国南方存在一段持续显著的降雨期,这段降水期属于东亚副热带季风进程的初期,为东亚副热带季风孕育阶段。本文将此段降水期称之为"江南雨季"。但迄今为止,对江南雨季的研究中,地理区域和雨季起止时间没有客观、统一的确定方法,影响了对江南雨季的研究和监测、预测。作为东亚季风降水季节进程的一个重要环节[3],江南雨季是海陆气相互作用的产物,它既受热带、副热带地区下垫面影响,又受中高纬度天气系统作用。热带低纬度地区下垫面强迫通过遥相关作用影响江南雨季的降水,如西太平洋暖池热状况通过北传的EAP/PJ波列和改变局地经圈环流影响江南地区降水[4-7],中东太平洋和印度洋海表温度通过Gill激发产生西传的Rossby波和东传的Kelvin波影响江南地区降

---

* 本文发表于《海洋学报》,2015年第37卷第6期,1-11.

水[8-13];同时,中高纬度因子也影响江南地区降水:青藏高原感热加热作用通过在其东侧对流层低层激发气旋性涡度影响江南雨季[14-15],东亚副热带西风急流等对江南雨季也起到不同程度的作用[16]。不同时间尺度上诸多影响因子的共同作用,又使得江南雨季具有多时间尺度的变率特征。因此,对江南雨季时空分布及其变化规律的研究不仅具有巨大的现实意义,同时也具有十分重要的理论价值。江南雨季的研究受到专家学者的广泛关注和极大重视。

在已有研究中,江南雨季常被纳入华南前汛期[17]或长江以南梅雨期[18-19],但也有不少研究指出,江南雨季的季节内—季节以及年际变化有显著的特殊性,降水时段集中在4—6月,与华南南部和长江流域有着明显的不同。万日金等[20]依据高原感热和江南地区西南风速的气候态特征,将13—27候的降水定义为江南春雨期。陈菊英[21]将持续接近或超过"年平均日降水量的旬累计值"1.5倍的时段定义为雨季,并指出江南区域平均雨季从4月上旬开始,到6月下旬结束,是我国雨季开始最早也结束最早的区域,雨季集中期存在两个峰值。高波[22]等认为江南地区汛期集中于每年的3—6月,其中4—6月的降水量为717 mm,约占全年降水量的46%,且旱涝频发,因此应将4—6月作为江南地区的主要雨季进行研究。陈绍东[23]等发现江南地区雨量集中于春末夏初(4—6月),该区域内雨季开始于4月。如,江西省的雨季为4月到7月上旬[24]。

可见,在冬夏转换期内存在的江南地区持续降水期已被作为一段具有特殊性的独立雨季来分析。然而,由于各研究中对江南地区的区域划定差异较大,得到的江南雨季变率特征也就不尽相同,这给正确认识江南雨季变化规律,提高江南雨季的监测和预测水平带来较大困难。例如,陈菊英[21]利用浙江、江西、湖南、福建4省的16个观测站作为江南地区代表站;陈绍东等[22]利用聚类分析方法从长江下游和华南北部挑选14个站点构成江南地区代表站;Tian等[25]在研究春季持续降水(即江南春雨)时,将我国东部地区25°~30°N确定为江南地区;朱志伟和何金海[26]在研究东亚副热带季风降水(即江南春雨)时将东亚副热带季风关键区定义为20°~35°N,110°~120°E;尚可等[27]将气候平均的春季持续连阴雨期(SPR)降水的峰值区(大于等于5 mm·d^{-1})定义为江南春雨区(20°~30°N,110°~122°E);万日金等[20]甚至认为江南雨季范围的南界位置应该划至雷州半岛。除此之外,还有一些研究将江南雨季范围分成两部分,其南部划归为华南地区,而北部则纳入长江中下游地区。如池艳珍等[17]在研究华南前汛期降水时将华南区域定义为20°~27.5°N,106°~121°E;王遵娅和丁一汇[29]在研究我国雨季气候特征时,将我国南方的广大地区定义为华南区(20°~28°N,110°~122°E);蔡学湛[28]从福建、广东、广西及江西南部、湖南南部选取了25个站点作为华南代表站,上述定义的华南区域北部与江南南部区域大部分重合;王永光[30]将我国东部4-9月的降水做旋转经验正交函数(REOF)展开,定义南方的两个高荷载区分别为长江中下游地区(26°~31°N,106°~121°E)和华南区(20°~25°N,106°~120°E),这正好将江南雨季范围分隔开来,分别列入以上两个区域。

综上所述,华南北部与江南南部降水特征相似,最大降水期发生在4—6月,而华南南部连续三个月最大降水量出现在5—7月,与华南北部和江南地区降水特征有明显差异[21],而长江中下游地区的主要降水出现在6—8月。因此江南雨季是一个有别于华南南部和长江流域而独立存在的雨季,雨季起止时间等特征和其他地区有较大区别。江南地区气候平均的雨季为4—6月,与我国现行气象业务中的主汛期(6—8月)、江淮梅雨期(6—7月)以及华北—东北雨季(7—8月)在时间上有较大差异,是我国雨季开始最早、结束最早的区域。目前大多数研究将江南雨季泛称为华南前汛期。而即使将江南地区作为独立区域进行研究时,其具体区域的划分也往往根据简单的行政区划或工作经验来选取江南地区代表站,缺乏客观划分江南雨季具体区域范围的方法。对于江南雨季的开始期也仅仅从气候态出发,仅考虑降水变化,将江南雨季开始期定义在4月初,这显然带有相当的主观性和经验性。何金海[31-32]等在研究东亚副热带季风时指出,3月底4月初,以对流层低层偏南风的建立为标志,江南地区出现了较大范围、强度较强的降水,此降水属于东亚副热带季风降水性质。受此启发,我们认为,在考虑江南雨季开始期指标时,有必要结合降水形成的机制,如同时考虑对流层低层偏南风的建立来制定相应的雨季开始期指标,这样才能较为合理地确定江南雨季的起止时间。

如果一个区域内各个站点的降水期能够称为独立雨季,那么该区域各个站点的降水多时间尺度特征应该有许多相似性,如气候态降水逐候进程(下称候进程)存在一致性,降水季节进程的年际变率存在一致性,雨季降水的年际变率存在一致性。因此,本文将通过诊断我国气象台站日降水观测资料,从上述3个降水特征的相似性标准出发,客观划定江南雨季的具体地理范围。在此基础上分析区域内降水和风场变化特征,确定江南雨季起止时间,为规范江南雨季的监测提供指导和参考,为江南雨季的预测提供科学基础。

2 资料及方法

文中使用了我国气象台站逐日降水资料,资料时段为1961—2012年。降水资料来自国家气象信息中心。为保证资料的完整性,选取了我国1675个无降水缺测的气象台站降水记录。研究中还使用了NCEP/NCAR再分析资料的925 hPa风场、500 hPa高度场等环流资料。

文中研究的时间尺度为候。为消除各候天数不同的影响,定义候降水指数作为研究对象,其定义如下:

$$RI = \frac{R/N}{\bar{R}} \times 100\% \tag{1}$$

式中,RI为候降水指数(无量纲值),即表示候内日平均降水量占平均年降水量的百分比,R为候降水量,$\bar{R}$为台站年降水量的气候平均值(气候平均值的统计时段为1981—2010年),N为一候的天数。N值一般为5,在大月的最后一候为6,根据平年、闰年的不同,2月最后一候的天数分别为3、4。在分析月、季降水特征时,采用类似方法定义月、季降水指数。图1给出了研究雨季时分别采用降水指数、降水量进行分析的比较。假设候内日平均降水量大于等于5 mm·d^{-1},或者候降水指数$RI \geqslant 0.4\%$的候为雨季,可以看出用降水指数RI做指标(图1a)分析出的雨季由南向北推进表现较为连续,而用降水量作指标(图1b)时雨季由南向北推进时会出现突然跳跃,在41候时雨季从35°N跳跃到了40°N。这主要是由于我国东部各台站降水量差异较大,如果用降水量做指标定义雨季阈值,会导致不同台站之间的不可比较性和相应的误差,而降水指数RI是一个相对量,反映了候降水占年降水量的比例,用相对量定义雨季阈值,消除了区域内台站降水量的气候态差异,在区域内更具一致性。

本文采用REOF方法分析降水指数的时空变化。首先对降水指数场标准化,进行EOF分析,截取累计方差贡献达90%的前k个空间型进行几何旋转,REOF分析得到的前k个空间型累计方差解释原场总方差的百分率保持不变,而得到的单个空间型能够尽量反映各模态的局部相关结构。在绘制REOF空间特征场时,我们将得到的第k个特征场V_{jk}按下式进行了变换:

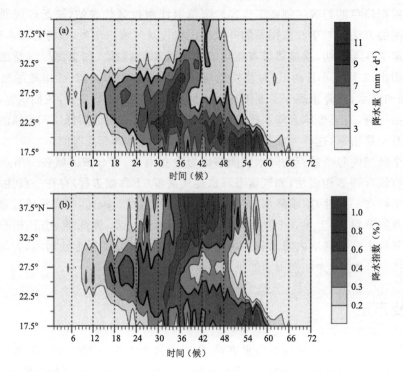

图1 我国东部（110°E以东）经向逐候日平均降水量（a）和逐候降水指数候进程演变图（b）

$$a_{jk}=V_{jk}\sqrt{\lambda_k} \tag{2}$$

式中，V_{jk}、λ_k 分别为 REOF 得到的第 k 个特征场和第 k 个特征值。通过变换得到的 a_{jk} 为各站点降水指数和第 k 个特征场对应的时间系数之间的相关系数，反映了区域内站点降水场和模态特征场的联系程度，并且特征场空间结构不会发生改变[34]。

此外，文中通过分析我国气象站气候态逐候降水指数，研究各台站气候态降水候进程的一致性；将各站 4—6 月降水指数与 6—8 月降水指数的差值作为降水的季节进程，进而研究各台站降水季节进程年际变率的一致性；对 4—6 月的降水指数进行 EOF 分析，研究雨季降水年际变率的一致性，进而根据上述 3 种一致性去客观确定江南雨季的地理范围。

3 江南雨季区域的客观确定

3.1 我国气候态降水的候进程分析

利用所选 1675 个气象台站逐候降水指数的气候平均值，构建我国逐候降水指数气候平均场（见下式）：

$$X_{m\times n}=\begin{bmatrix} x_{11} & x_{21} & \cdots & x_{1n} \\ x_{21} & x_{22} & \cdots & x_{2n} \\ \vdots & \vdots & & \vdots \\ x_{m1} & x_{m2} & \cdots & x_{mn} \end{bmatrix} \tag{3}$$

式中，$m=1675$，$n=72$。

对 X 标准化后进行 EOF 分解,前 7 个特征场共解释了 90% 以上的方差,将 EOF 得到的前 7 个特征场再进行旋转,得到了 REOF 分析的 7 个特征场。REOF 得到的前 2 个特征场的方差贡献分别为 51.89%、21.8%,均远大于其他特征场的方差贡献,因此我们重点对这两个特征场进行分析。

REOF 分析得到的时间系数反映了气候态降水候进程,特征场(已经过前述变换)代表的是各站点候降水时间序列与该模态时间系数的相关系数。图 2 为 REOF 分析得到的前 2 个特征场空间分布及其相应的时间系数。由于江南地区地处我国东部,文中对两个特征场中的我国东部地区进行分析。REOF 得到的第一特征场解释方差为 51.89%,对应的时间系数在 34—54 候为正值,7—8 月为最大区间。第一特征场中 0.8 以上的高值区(红色区域)分布在我国东北、华北一带,这说明我国东北、华北地区降水与这一降水候进程模态的相关性较好,34—54 候为雨季;而 0.3 以下的低值区(绿色区域)分布在江南地区,说明该区域降水与该模态相关性较差,其时间系数也不能反映江南地区主要雨季的时段。REOF 得到的第二特征场解释方差为 21.8%,对应的时间系数在 15—37 候为正值,在 24 候左右出现一个相对低谷,最大值出现在 6 月,4 月中有一个次大值。第二特征场中 0.7 以上载荷高值区主要分布在江南地区,而东北、华北地区为低值区,数值在 0.2 以下。这表明江南地区降水和降水候进程第二模态的相关系数在 0.7 以上,15—37 候对应着江南雨季的主要时段。

为了更进一步进行深入分析,我们关注图 2 中 REOF 分析得到的第一、二特征场中分别大于 0.6 的红色区域。在第一特征场中大于 0.6 的大值区主要分布在 31°N 以北,而在第二特征场大于 0.6 的大值区主要分布在 31°N 以南。因此根据气候态降水候进程可以将我国东部划分为两个区域:31°N 以北区域划分成一类,这个区域内气候态降水候进程呈现 REOF 得到

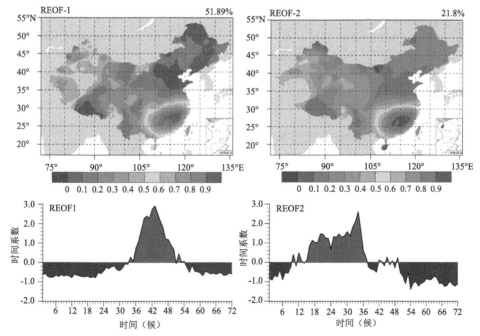

图 2 我国逐候气候态降水指数场 EOF 分解的前七个特征向量进行旋转
得到的第一、二特征场及其时间系数

的第一特征场时间系数的特点,雨季出现在34—54候;而31°N以南区域划分为另一类,这个区域气候态降水候进程表现为REOF得到的第二特征场的时间系数特征,15—37候为雨季的主要时段。可见,我国东部31°N以南和以北的雨季发生时段呈现出显著的不同特征。

3.2 区域降水季节进程的年际变率分析

根据我国气候态降水的候进程特点,以31°N为界,我国东部可以划分为两个区域,31°N以北雨季出现在6—8月,31°N以南雨季主要出现在4—6月。为分析降水季节进程的一致性,将各站4—6月降水与6—8月降水的差值(或者说是4—5月与7—8月的差值)作为降水季节进程,并且聚焦31°N以南区域。通过对区域内各站1961—2012年降水季节进程进行EOF分析,我们得到了EOF分解的第一特征场,该模态解释了25.8%的方差。空间载荷分布的高值区(0.6以上)位于江西、湖南、福建西部的大部分区域以及广东西北部,高值中心在江西中部。该模态有较明显的年代际特征,20世纪90年代以前,时间系数以正值为主,而90年代以后的时间系数负值出现更多,尤其在1990—2002年时间系数一直为负。

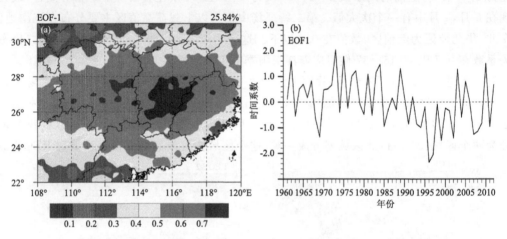

图3 1961—2012年区域降水季节进程EOF分析第一特征场空间分布(a)及时间系数图(b)

3.3 区域雨季降水的年际变率分析

由于31°N以南的雨季主要出现在4—6月,因此对31°N以南区域内1961—2012年4—6月总降水进行EOF分解,以分析区域内雨季各站点4—6月总降水年际变率的一致性。图4给出了EOF分析的第一特征场空间分布图。第一特征场解释了26.8%的方差,高载荷区(大于0.6)在江西、湖南、福建大部以及广东的西北部,高值中心在江西中部到福建北部28°N左右的狭窄区域,模态系数的年际变化波动性较大。

综合考虑气候态降水候进程、降水季节进程年际变化以及雨季降水年际变化的空间一致性,对以上3个特征场进行叠加(图5),选取0.6为阈值,将三者均大于0.6的重叠区域作为客观划定的江南雨季范围,可见该区域位于110°E以东、25°~28°N之间,包括江西、湖南大部、浙江西南部、福建西北部、广东北部、广西东北部,即南岭以北、28°N以南、110°E以东的我国东部地区。以下研究中,选取定义区域内的112个气象台站作为代表站,对江南雨季的起止时间进行分析和划定。

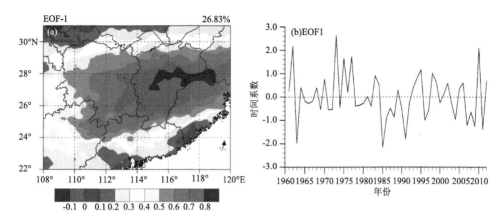

图 4　1961—2012 年区域雨季降水 EOF 分析第一特征场空间分布(a)和时间系数图(b)

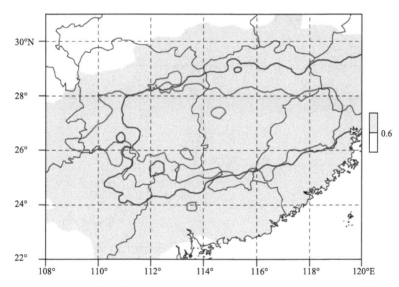

图 5　我国逐候气候态降水指数场 EOF 第二特征载荷向量大于 0.6 区域(阴影区),区域降水季节进程 EOF 第一特征载荷向量大于 0.6 区域(红色等值线)以及区域雨季降水 EOF 第一特征载荷向量大于 0.6 区域(蓝色等值线)

为了验证江南雨季地理区域划定的合理性,我们绘制了我国华南地区、江南地区、长江中下游地区区域平均气候态降水的候进程变化曲线(图 6)。其中,华南地区、长江中下游地区参考陈菊英[21]的划分,利用 20°～24.5°N,110°E 以东气象观测站计算了华南区域平均降水,利用 28°～31°N,115°E 以东气象台站的资料计算了长江中下游区域平均降水。江南区域平均降水的计算则利用了文中定义的 112 个台站资料。从图 6 可以看出这 3 个区域降水最大值分别出现在 32、34、36 候,呈现出由南向北推进的特点,这是因为南海季风爆发后,副热带高压由南向北推进,从而导致我国雨季由南向北推进,这正是众所周知的我国东部主雨带的推进特征。然而值得指出的是,江南地区降水逐候演变呈现出显著的不同特征。不难看出,江南地区的雨季开始最早,并呈双峰型特征,即在 16 候左右江南地区显著降水开始,并逐渐增强,到 20 候左右出现第一个峰值,然后降水略有下降,在 24 候左右出现低值,随后降水再逐渐增强,到 34 候左右出现第二个峰值,36候降水迅速减少,随后江南雨季结束。在江南地区出现第一个降水峰值时,华南、长江中下游地区的降水均较江南地区明显偏少,而华南地区在25候左右降水才明显增强,长江中下游地区大约在30候降水明显增强。因此,江南地区雨季出现的时间较其他两个地区都早。37候以后江南地区受副热带高压控制,降水明显减少,而华南、长江中下游地区的降水还在维持,可见江南地区的雨季结束也早于其他两个地区。因此,本文定义的江南雨季降水和毗邻的华南、长江中下游地区的气候态降水候进程有明显的不同,呈现出一定的特殊性,表明江南雨季地理范围的划分是合理而有意义的。

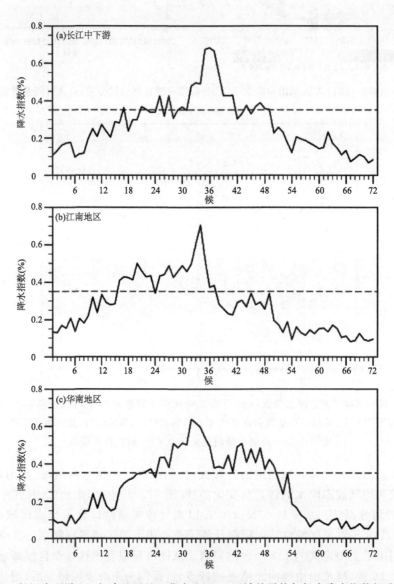

图6 长江中下游(a)、江南地区(b)、华南地区(c)区域的平均气候态降水指数候进程,图中虚线为降水指数参考阈值(0.35)

4 江南雨季起止时间的客观确定

上面我们客观划分了江南雨季的地理区域,那么江南雨季的起止时间又该如何确定?何金海等[31]及朱志伟和何金海[26]指出伴随着东亚纬向海陆热力差异的季节反转,对流层低层(925 hPa)风场也由偏北风转为偏南风,降水明显增强,东亚副热带季风雨季于3月底4月初首先在江南开始。为此,我们重点分析江南地区降水和经向风逐候演变特点。为了考虑江南降水和低层经向风的关系,文中将江南区域平均1961—2012年逐候降水指数 RI 和925 hPa 经向风 V 构建双时间序列矩阵 X(式4),并进行多变量 EOF 分析,这样可以得到降水和经向风候进程的主要模态。

$$X_{144\times 52} = \begin{pmatrix} RI_{1,1} & RI_{2,1} & \cdots & RI_{1,52} \\ RI_{2,1} & RI_{2,2} & \cdots & RI_{2,52} \\ \vdots & \vdots & & \vdots \\ RI_{72,1} & RI_{72,2} & \cdots & RI_{72,52} \\ V_{1,1} & V_{2,1} & \cdots & V_{1,52} \\ V_{2,1} & V_{2,2} & \cdots & V_{2,52} \\ \vdots & \vdots & & \vdots \\ V_{72,1} & V_{72,2} & \cdots & V_{72,52} \end{pmatrix} \quad (4)$$

多变量(降水指数和经向风)EOF 分析的第一模态能解释 47.08% 的方差,其降水指数和 925 hPa 经向风候进程曲线如图 7a 所示。由图可以看出,降水指数和经向风在 36 候之前均以波动形式不断增强,特别是在 3 月底 4 月初(18 候前后),降水和经向风有一个迅速增强的过程,这对应江南雨季的建立,体现了"季风"的建立既是"风"又是"雨",风雨交加的特点。然而在 36 候以后,降水迅速下降,但经向风却迅速增强并维持高值,这体现了"风"和"雨"的分离。由图 7b 可以看出,对应降水的迅速下降,副高脊线表现为迅速北抬,正是副高的迅速北抬导致了江南雨季的结束,这是符合广大天气气候工作者的预报实践的。因此我们选用降水和经向风的结合来确定江南雨季的起始日期,用降水和 500 hPa 副高脊线位置的结合来确定雨季结束日期。副高脊线位置的确定采用刘芸芸等[35]研究的客观方法。

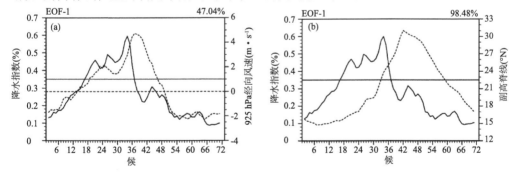

图 7 1961—2012 年江南区域逐候降水指数和 925 hPa 经向风多变量 EOF 分析第一模态(a),降水指数第一模态和气候态副高脊线位置候进程图(b),图中实线为降水指数,左图虚线为 925 hPa 经向风、右图虚线为副高脊线,水平虚实线代表各指标阈值

为了确定雨季起止日期，我们选择降水指数的阈值为 0.35%，其在多变量 EOF 第一模态中对应的经向风大于 0.5 m·s^{-1}。在区域内所选台站中，候降水指数大于 0.35% 相当于候内平均日降水量为 4.5~7.4 mm·d^{-1}，区域平均为 5.9 mm·d^{-1}。该阈值反映的日降水量与王遵娅和丁一汇[29]制定的主雨季标准—逐候降雨量标准化值超过 0.5[在东部季风区主雨季阈值 30~40 mm·(5d)$^{-1}$]，Lau 和 Yang[36]及 Qian 等[37]制定的深对流降水标准—6 mm·d^{-1} 较一致，而与 Wang[38]采用相对候平均降水率超过 5 mm·d^{-1} 的标准定义季风雨季的标准也十分接近。

因此，根据以上分析，给出江南雨季起止时间客观确定指标：从 3 月开始，当连续两候 925 hPa 经向风为南风，风速大于 0.5 m·s^{-1}，且降水指数之和大于等于 0.7%，则将该时段的第一候作为江南雨季的开始期；在江南雨季开始以后，当出现连续两候降水指数之和小于 0.7%，且副热带高压脊线位置在 25°N 以北时，则该时段的第一候为江南雨季的结束期。

表 1 为根据上述三指标客观确定的 1961—2012 年江南雨季的起止时间。从表中可以看出，江南雨季开始期多在 3 月下旬到 4 月上旬，最早的年份是 1980 年，13 候江南雨季开始，而 2007 年的江南雨季开始期为 30 候，是雨季开始最晚的年份。可见江南雨季开始日期的年际变率较大，最早最晚年相差 17 候；江南雨季大多在 6 月下旬到 7 月上旬结束，最早结束期为 33 候，分别出现在 1963、1971、1979、1997 年，而 1973 年雨季结束最晚，为 43 候。江南雨季结束日期存在明显的年际变率，但其年际变率不如开始日期显著。

表 1 1961—2012 年江南雨季开始和结束日期（单位：候）

年份	开始	结束	年份	开始	结束	年份	开始	结束	年份	开始	结束
1961	15	34	1974	21	41	1987	15	38	2000	22	36
1962	14	37	1975	19	37	1988	25	37	2001	24	35
1963	21	38	1976	20	39	1989	21	38	2002	21	38
1964	19	36	1977	23	37	1990	17	34	2003	20	37
1965	19	36	1978	21	35	1991	13	38	2004	19	35
1966	16	39	1979	22	33	1992	22	36	2005	18	39
1967	17	38	1980	13	40	1993	23	39	2006	19	40
1968	20	40	1981	16	38	1994	19	36	2007	30	38
1969	22	39	1982	22	39	1995	22	40	2008	23	40
1970	24	40	1983	16	38	1996	29	38	2009	21	38
1971	25	32	1984	17	34	1997	18	33	2010	18	39
1972	18	36	1985	19	34	1998	23	36	2011	24	35
1973	20	43	1986	14	38	1999	21	36	2012	23	37

5 总结与讨论

本文采用国家气象信息中心整编的 1675 台站资料以及 NCEP/NCAR 再分析资料，定义了能消除地区差异的候平均降水指数，综合考虑我国南方气候态降水的候进程、降水季节进程的年际变率以及雨季降水的年际变率的一致性，客观划定了江南雨季的地理区域，制定了判定

江南雨季起止时间的方法,确定了1961—2012年江南雨季起止时间。得到的主要结论如下:

(1)我国东部地区气候态降水逐候进程存在两个主模态,分别为北方模态和南方模态。北方模态高载荷区位于31°N以北,这个区域内显著降水主要出现在34—54候,对应东亚季风主汛期降水;而南方模态高载荷区位于31°N以南,区域内降水主要时段集中在15—37候,对应江南雨季。

(2)综合我国南方(31°N以南、110°E以东区域)的气候态降水逐候进程、降水季节进程的年际变率以及雨季降水年际变率的一致性特征,客观划定了江南雨季的地域范围,该区域位于110°E以东,25°~28°N之间,主要包括江西、湖南大部,浙江西南部,福建西北部,广东北部以及广西东北部,亦即南岭以北、28°N以南,110°E以东的江南地区。

(3)考虑江南地区对流层低层925 hPa经向风,西北太平洋副热带高压500 hPa脊线和江南地区降水指数之间的关系,用降水指数和经向风的结合来确定雨季起始期,用降水指数和副高脊线位置的结合来确定雨季结束期。对1961—2012年江南雨季的起止时间进行了确定,给出了江南雨季起止时间序列。江南雨季的起始期大多在3月底和4月初,而结束期主要集中在6月底和7月初。

值得指出的是,本文仅解决了江南雨季区域的客观划分以及江南雨季起止时间客观判定的相关问题。限于篇幅,没有直接给出与江南雨季起止时间相联系的海气背景。本文注意到江南雨季的气候态季节进程存在双峰型特征:第一个峰值出现在21候左右,第二个峰值出现在34候,第二峰值为江南雨季的主峰值。这种双峰型特征的成因是什么?此外,江南雨季的起止时间和降雨量之间是否存在相关关系?其物理过程如何?这些问题还有待进一步研究。江南雨季作为东亚副热带季风雨季的一个阶段,其形成是海陆气相互作用的产物,特别是西太平洋暖池热含量以及中东太平洋和印度洋海表温度异常等都是江南雨季建立及其降水量的重要影响因子,但如何影响,其物理过程是什么?都是我们将要深入探究的科学问题。

致谢:感谢国家气象信息中心提供的1675台站资料以及NCEP/NCAR提供的再分析资料。本文所有图为NCL软件所绘制。

参考文献

[1] Zhu Qiangen, He Jinhai, Wang Panxing. A study of circulation differences between East-Asian and Indian Summer Monsoons with their interaction [J]. Adv Atmos Sci,1986,3(4):466-477.

[2] 张庆云,陶诗言.夏季东亚热带和副热带季风与中国东部汛期降水[J].应用气象学报,1998,9:17-23.

[3] Wang B, Lin H. Rainy season of the Asian-Pacific summer monsoon[J]. J Climate,2002,15(4):386-398.

[4] Huang R H, Sun F Y. Impacts of the tropical western Pacific on the East Asian summer monson[J]. J Meteorol Soc Jpn,1992,70(1B):243-256.

[5] Nitta T. Convective activities in the tropical western Pacific and their impact on the Northern Hemisphere summer circulation[J]. J Meteorol Soc Jpn,1987,65(3):373-390.

[6] 尚可,詹丰兴,何金海,等.前期夏季西太平洋暖池热含量对江南春雨的影响及其可能机理[J].海洋学报,2014,36(1):86-97.

[7] 任雪娟,钱永甫.南海地区潜热输送与我国东南部夏季降水的遥相关分析[J].海洋学报,2000,22(2):25-34.

[8] 王磊,张文君,祁莉,等.两类La Nina季节演变过程的海气耦合特征对比[J].海洋学报,2014,36(1):

72-85.

[9] 陈迪,陈锦年,左涛. 西太平洋副热带高压变化与赤道太平洋海温场的联系[J]. 海洋学报,2013,35(6): 21-30.

[10] Wang B,Wu R,Fu X. Pacific-East Asia teleconnection:how does ENSO affect East Asian climate[J]. J Climate,2000,13(9):1517-1536.

[11] Xie S P,Hu K,Hafner J,et al. Indian Ocean Capacitor Effect on Indo-Western Pacific Climate during the Summer following El Nino[J]. J Climate,2009,22(3):730-747.

[12] Zhang R,Sumi A,Kimoto M. A diagnostic study of the impact of El Nino on the precipitation in China [J]. Adv Atmos Sci,1999,16(2):229-241.

[13] Zhang R,Sumi A,Kimoto M. Moisture circulation over East Asia during El Nino episode in northern winter,spring and autumn[J]. Journal Meteorological Society of Japan Series,2002,80(2):213-227.

[14] Wan R J,Wu G X. Mechanism of the spring persistent rains over southeastern China[J]. Sci China Ser D, 2007,50(1):130-144.

[15] Wu G,Liu Y,Zhang Q,et al. The influence of mechanical and thermal forcing by the Tibetan Plateau on Asian climate[J]. J Hydrometeorol,2007,8(4):770-789.

[16] 张耀存,郭兰丽. 东亚副热带西风急流偏差与中国东部雨带季节变化的模拟[J]. 科学通报,2005,50 (13):1394-1399.

[17] 池艳珍,何金海,吴志伟. 华南前汛期不同降水时段的特征分析[J]. 南京气象学院学报,2005,28(2): 163-171.

[18] 陈科艺,王谦谦,胡娟,等. 江南北部地区梅雨期降水与海温的 SVD 分析[J]. 南京气象学院学报,2006, 29(2):258-263.

[19] 黄琰,张人禾,龚志强,等. 中国雨季的一种客观定量划分[J]. 气象学报,2014,72(6):1186-1204.

[20] 万日金,吴国雄. 江南春雨的时空分布[J]. 气象学报,2008,66(3):310-319.

[21] 陈菊英. 中国旱涝的分析和长期预报研究[M]. 北京:气象出版社,1991:15-28.

[22] 高波,陈乾金,任殿东. 江南南部-华南北部前汛期严重旱涝诊断分析[J]. 应用气象学报,1999,10(2): 219-226.

[23] 陈绍东,王谦谦,钱永甫. 江南汛期降水基本气候特征及其与海温异常关系初探[J]. 热带气象学报, 2003,19(3):260-268.

[24] 陈双溪. 气象与领导[M]. 北京:气象出版社,1999.

[25] Tian S F,Yasunari T. Climatological aspects and mechanism of Spring Persistent Rains over central China [J]. J Meteorol Soc Jpn,1998,76(1):57-71.

[26] 朱志伟,何金海. 东亚副热带季风的季节转变特征及其可能机理[J]. 热带气象学报,2013,29(2): 245-254.

[27] 尚可,何金海,朱志伟,等. 西太平洋暖池区热含量和海表温度与江南春雨的相关性对比研究[J]. 地理科学,2013,33(8):987-992.

[28] 蔡学湛. 青藏高原雪盖与东亚季风异常对华南前汛期降水的影响[J]. 应用气象学报,2001,12:358-367.

[29] 王遵娅,丁一汇. 中国雨季的气候学特征[J]. 大气科学,2008,32(1):1-13.

[30] 王永,陆维松,龚赟. 近53a中国东部雨季降水的区域特征[J]. 南京气象学院学报,2005,28(5):609-616.

[31] 何金海,祁莉,韦晋,等. 关于东亚副热带季风和热带季风的再认识[J]. 大气科学,2007,31(6): 1257-1265.

[32] 何金海,赵平,祝从文,等. 关于东亚副热带季风若干问题的讨论[J]. 气象学报,2008,66(5):683-696.

[33] Zhao P,Zhang R,Liu J,et al. Onset of southwesterly wind over eastern China and associated atmospheric circulation and rainfall[J]. Clim Dynam,2007,28(7/8):797-811.

[34] 吴洪宝,吴蕾.气候变率诊断和预测方法[M].北京:气象出版社,2005:33-44.

[35] 刘芸芸,李维京,艾婉秀,等.月尺度西太平洋副热带高压指数的重建与应用[J].应用气象学报,2012,23(4):414-423.

[36] Lau K M,Yang S. Climatology and interannual variability of the Southeast Asian summer monsoon[J]. Adv Atmos Sci,1997,14(2):141-162.

[37] Qian W,Kang H S,Le D K. Distribution of seasonal rainfall in the East Asian monsoon region[J]. Theor Appl Climatol,2002,73(3/4):151-168.

[38] Wang B. Rainy season of the Asian-Pacific summer monsoon[J]. J Climate,2002,15(4):386-398.

The East Asian Subtropical Summer Monsoon: Recent Progress

He Jinhai[1,3]   Liu Boqi[2,4]

(1. Key Laboratory of Meteorological Disaster of Ministry of Education, Nanjing University of Information Science & Technology, Nanjing 210044;
2. State Key Laboratory of Severe Weather, Chinese Academy of Meteorological Sciences, Beijing 100081; 3. Laboratory of Research for Middle-High Latitude Circulation and East Asian Monsoon, Changchun 130062; 4. State Key Laboratory of Numerical Modeling for Atmospheric Sciences and Geophysical Fluid Dynamics, Institute of Atmospheric Physics, Chinese Academy of Sciences, Beijing 100029)

Abstract: The East Asian subtropical summer monsoon (EASSM) is one component of the East Asian summer monsoon system, and its evolution determines the weather and climate over East China. In the present paper, we firstly demonstrate the formation and advancement of the EASSM rainbelt and its associated circulation and precipitation patterns through reviewing recent studies and our own analysis based on JRA-55 (Japanese 55-yr Reanalysis) data and CMAP (CPC Merged Analysis of Precipitation), GPCP (Global Precipitation Climatology Project), and TRMM (Tropical Rainfall Measuring Mission) precipitation data. The results show that the rainy season of the EASSM starts over the region to the south of the Yangtze River in early April, with the establishment of strong southerly wind in situ. The EASSM rainfall, which is composed of dominant convective and minor stratiform precipitation, is always accompanied by a frontal system and separated from the tropical summer monsoon system. It moves northward following the onset of the South China Sea summer monsoon. Moreover, the role of the land–sea thermal contrast in the formation and maintenance of the EASSM is illustrated, including in particular the effect of the seasonal transition of the zonal land–sea thermal contrast and the influences from the Tibetan Plateau and midlatitudes. In addition, we reveal a possible reason for the subtropical climate difference between East Asia and East America. Finally, the multi-scale variability of the EASSM and its influential factors are summarized to uncover possible reasons for the intraseasonal, interannual, and interdecadal variability of the EASSM and their importance in climate prediction.

Key words: East Asian subtropical summer monsoon; rainbelt formation and advancement; precipitation property; zonal land-sea thermal contrast seasonal transition; midlatitude influence, multi-scale variability

---

* 本文发表于《Journal of Meteorological Research》,2016 年第 30 卷第 2 期,135-155。

1 Introduction

Monsoon, as an ancient concept in climatology, is not only associated with the annual cycle of solar radiation, but also one of the most typical seasonal variations of the atmospheric circulation. The basic characteristics of monsoon is the seasonal reversal of the wind direction between winter and summer, which results in wet flow from the cold ocean to the warm land in summer and dry flow from the cold land to the warm ocean in winter. Halley[1] was the first to treat monsoon as a scientific issue, pointing out that monsoon is the result of the land-sea thermal contrast in response to the seasonal changes in solar radiation. He was also the first to consider monsoon as a planetary-scale land-sea wind system. Later, Hadley[2] introduced the effect of earth's rotation into Halley's monsoon model, suggesting that the southwesterly monsoon is formed by earth's rotation, which deflects the wind to its right-hand side after its arrival over warm land from the Southern Hemisphere. In fact, the generation of monsoon is also evidently influenced by condensation heating, and the rainfall and circulation possess a mutual cause-effect relationship[3]. Thus, there are three factors driving the formation of monsoon: (1) seasonal changes in solar elevation angle; (2) non-uniform underlying conditions; and (3) moisture processes (i. e. , feedback between monsoon precipitation and circulation)[4-6].

The Asian monsoon is the strongest in the global monsoon system, and is composed of the South Asian monsoon (i. e. , Indian monsoon) and the East Asian monsoon[7-8]. The Asian monsoon can also be divided into tropical monsoon, subtropical monsoon, and temperate monsoon, based on climatic zones. Specifically, the tropical monsoon includes the Indian monsoon, the Bay of Bengal monsoon, and the South China Sea (SCS) —western North Pacific (WNP) monsoon[9]. The East Asian monsoon contains the tropical monsoon (including the SCS and WNP monsoon), the subtropical monsoon (including the rainbelt over South China, the Meiyu rainbelt over the Yangtze-Huaihe River, and the Baiu or Changma rainbelt from Korea to Japan), and the temperate monsoon, according to geographic latitude[10,11]. The East Asian subtropical summer monsoon (EASSM), as an important member of the East Asian summer monsoon, exhibits great effects on the weather and climate in China. Thus, studies on the dynamical mechanisms and multi-scale variability of the formation and development of the EASSM are essential for the improvement of weather forecasting and climate prediction in China.

With the aim to provide a scientific basis and clues for further development in our understanding of monsoon dynamics, in this paper we comprehensively review recent research on the EASSM, alongside an analysis of the latest datasets, with particular focus on its characteristic nature, formation mechanisms, and multi-scale variability (i. e. , intraseasonal, interannual, and interdecadal timescales), as well as their influences.

2 Data

Daily Japanese 55-yr Reanalysis (JRA-55) data, provided by the Japanese Meteorological Agency (JMA)[12], are employed to describe the characteristics of the EASSM and its formation mechanisms. We also use pentad-scale CPC Merged Analysis of Precipitation (CMAP)[13] and Global Precipitation Climatology Project (GPCP)[14] precipitation data, as well as daily Tropical Rainfall Measuring Mission (TRMM 3B42)[15] precipitation data, to validate the seasonal evolution of the EASSM rainbelt. The record length of the above datasets spans from 1981 to 2010, except for TRMM 3B42, which is from 1998 to 2010.

3 Intrinsic characteristics of the EASSM

3.1 Formation and advancement of the EASSM rainbelt

Chinese scientists proposed the concept of the EASSM in the mid 1980s and early 1990s[16-19]. They suggested that the East Asian monsoon could be divided into the SCS-WNP monsoon (i.e., tropical East Asian monsoon) and the continent-Japan subtropical monsoon (i.e., EASSM). Tao and Chen[7] described the climate features of the East Asian monsoon, pointing out that the EASSM rainbelt was accompanied by the Meiyu front, along with prevailing southerly winds to its south, but northerly winds to its north where mid-latitudinal influences were dominant, and also a typical feature of the EASSM is the northward movement of the Meiyu front to the Yangtze-Huaihe River. Recent studies have argued that the EASSM begins with the generation of a rainbelt to the south of the Yangtze River (generally referred to as South China) in early April[20-23].

Fig. 1 presents the seasonal evolution of the EASSM rainbelt and its related front. The daily mean precipitation is weak over the region to the south of the Yangtze River in mid-February, when the EASSM has yet to be established (Fig. 1a). The precipitation over this region enhances rapidly from March to April, along with a front over the southeastern coast of China settling to the south of the precipitation maximum (Figs. 1b and 1c). This indicates the beginnings of the formation of the EASSM rainbelt in situ. This rainbelt has been named the "early rainy season rainfall over South China"[24], the "spring permanent rainfall"[25], the "early summer rainy season"[26], and the "spring rainfall over South China" or "Meiyu over South China and Taiwan"[27]. The precipitation strengthens from April to May, with the front located to the south of the Yangtze River, where the EASSM rainbelt develops and lingers (Figs. 1c and 1d). Afterwards, when the SCS summer monsoon emerges in late May, the precipitation to the south of the Yangtze River further increases, accompanied by northward movement of the front to its south (Fig. 1e), indicating that the EASSM rainbelt starts to shift northward. During June-July, when the Meiyu rainfall begins, the EASSM

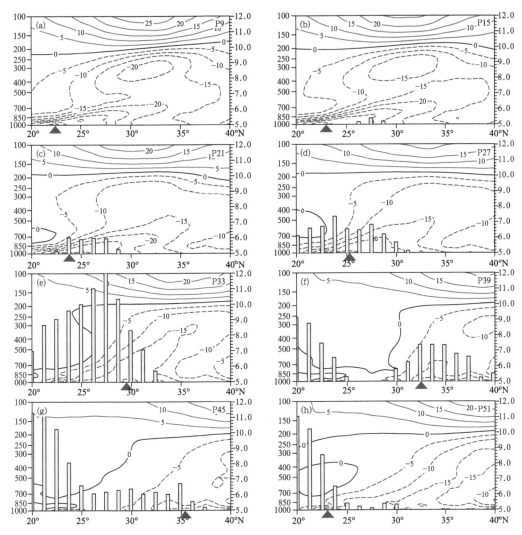

Fig. 1 Climatological (1981—2010) seasonal evolution of East Asian (averaged along 110°~130° E) precipitation (bars, mm·d^{-1}) and the pressure-latitudinal cross section of pseudo-equivalent potential temperature θ_{se} (contours, K). Blue triangles indicate the position of frontal zones defined by the maximum meridional θ_{se} gradient at 925-hPa. (a) Pentad 9; (b) Pentad 15; (c) Pentad 21; (d) Pentad 27; (e) Pentad 33; (f) Pentad 39; (g) Pentad 45; (h) Pentad 51

rainbelt and its associated front arrive at the Yangtze-Huaihe River, which marks the formation of the typical pattern of the EASSM rainbelt (Figs. 1e and 1f). The EASSM rainbelt is then displaced further northward North and Northeast China, inducing the local rainy season, while the Meiyu season ends and the second flood season involving tropical cyclones and typhoons begins over South China (Fig. 1g). Subsequently, the EASSM rainbelt withdraws southward rapidly and reaches the southeast coast of China in mid-September, marking the end of the EASSM (Fig. 1h). In addition, studies have reported noticeable interannual variation in the northward advancement of the EASSM rainbelt. For instance, Lian et al.[28,29]

defined an East Asian summer monsoon index based on NCEP reanalysis data and gauge-based observed rainfall data, and subsequently divided the northward advancement of the EASSM into a south path, middle path, and east path, according to their distinct features. Meanwhile, the low-level southerly flow has been found to be well correlated with the northward shift of the EASSM[30].

Importantly, two separate rainbelts are apparent over East Asia. One is the tropical monsoon rainbelt to the south of 20°N, and the other is the EASSM rainbelt to the north of the Nanling Mountains near 25°N. The seasonal evolutions of these two well-organized rainbelts are isolated from one another (Fig. 2). Climatologically, the East Asian tropical summer monsoon starts with the onset of the Bay of Bengal summer monsoon in early May, and then propagates eastward to the SCS in mid-May to result in the onset of the SCS summer monsoon in mid-late May[31-34]. However, the formation of the EASSM in early April is much earlier. Therefore, the formation of these two rainbelts over East Asia shows no direct association.

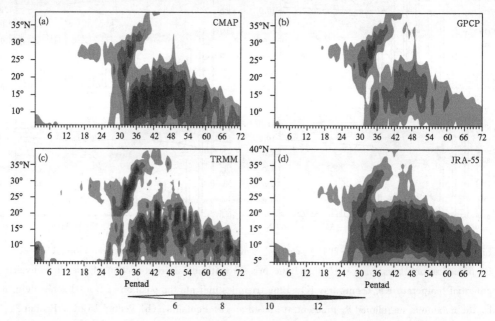

Fig. 2 Latitude-time cross section (averaged over 110°~130°E) of rainfall (mm · d^{-1}) in different datasets: (a) CMAP; (b) GPCP; (c) TRMM; (d) JRA-55

3.2 Properties of the EASSM rainbelt

The EASSM rainfall is always associated with a large-scale frontal system. This is in contrast to the tropical summer monsoon precipitation, which is largely composed of convective rainfall[7,20,35]. The EASSM rainfall constitutes both stratiform and convective precipitation, and their ratio presents a conspicuous seasonal evolution (Fig. 3a). During March-April, the EASSM rainbelt is preliminarily built up to the south of the Yangtze River, with the front located over the southeast coast of China (Fig. 1c). In the meantime, large-scale stratiform precipitation is dominant in the EASSM rainfall. Although the EASSM rain still lingers over South China in April, the lower troposphere of the East Asian subtropics becomes more convectively unstable in Pentad 19. Subsequently, this convective instability gradually develops during the early and mature phase of the EASSM rainy season, manifesting in the form of a gradual upward extension of a convectively unstable zone above 700-hPa (Fig. 3b). Thus, convective instability is a feature of the lower troposphere of the subtropics over East China during late March and early April, resulting in convective precipitation dominating the EASSM total rainfall (Fig. 3a). Therefore, the rainbelt located to the south of the Yangtze River in late March and early April is characterized by early summer monsoonal convection, and it can be treated as an indicator of the beginnings of the formation of the EASSM.

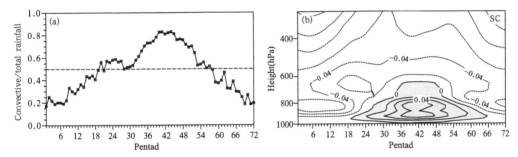

Fig. 3 (a) Time series of the ratio between convective and total precipitation over the region to the south of the Yangtze River (24°~29°N, 110°~130°E). (b) Vertical profile evolution of the convective instability ($\partial\theta_{se}/\partial p$) over the region to the south of the Yangtze River (shading indicates the convectively unstable region, 10^{-2} K · Pa^{-1})

3.3 Atmospheric circulation associated with the formation and development of the EASSM

The seasonal transition of the precipitation properties of the EASSM is closely associated with the general circulation. In late March, when the rainfall increases suddenly to the south of the Yangtze River, enhanced low-level in situ southerly merges with the southerly to the west of the western Pacific subtropical anticyclone, and northerly, due to the mechanical forcing of the Tibetan Plateau upon the westerly (Fig. 4a). The development of the southerly can be ascribed to the seasonal reversal of the zonal land-sea thermal contrast. In the meantime, the strengthening and advancement of the EASSM is separate from that of the Asian tropical summer monsoon. As shown in the evolution of the 850-hPa relative vorticity (Fig. 4b), a positive vorticity belt of the EASSM appears to the south of the Yangtze River in early March, whereas the SCS is controlled by a negative vorticity belt. In mid-May, the positive vorticity belt of the EASSM intrudes southward, followed by the onset of the SCS summer monsoon in Pentad 28, suggesting a possible triggering effect of the development of the EASSM on the onset of the SCS summer monsoon. The subtropical positive vorticity belt then becomes isolated from the tropical one over East Asia, presenting a separation of the circulation between the EASSM and the East Asian tropical summer monsoon (Fig. 4b). Similar characteristics can also be observed in the evolution of moisture convergence in the lower troposphere over East Asia (Fig. 4c). As a result, the evolution of the EASSM and the East Asian tropical summer monsoon should be considered as two relatively separate processes[21,36].

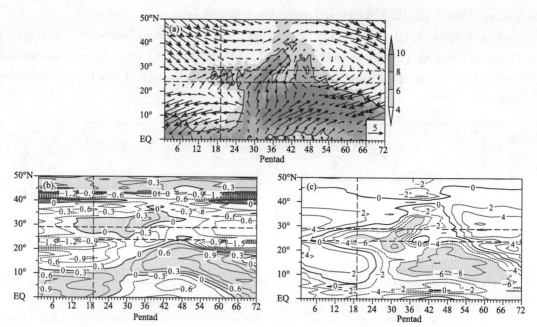

Fig. 4 Latitude-time cross section (averaged over 110°~130°E) of (a) precipitation (shading, mm d^{-1}) and the 850-hPa wind field (vectors, m·s^{-1}), (b) 850-hPa relative vorticity (10^{-6} s^{-1}), and (c) moisture flux divergence integrated from the surface to 700-hPa (10^{-5} kg·m^{-2}·s^{-1})

The intrinsic characteristics of the EASSM can be summarized as follows: (1) The formation of the EASSM is connected with the seasonal development of low-level southerly flow to the south of the Yangtze River, and its rainfall is always accompanied by a frontal system, which is absent in the tropical summer monsoon rainfall[10]. (2) The EASSM rainfall contains both stratiform and convective precipitation, and their ratio has apparent seasonal evolution. Convective precipitation becomes the dominant component after the formation of the EASSM. (3) The evolution of the EASSM is isolated from that of the Asian tropical summer monsoon. The EASSM establishes in mid-April, and its subsequent southward intrusion is beneficial for the onset of the SCS summer monsoon. As a result, the strengthening and northward advancement of the EASSM rainbelt and circulation does not result from the seasonal northward shifting of the Asian tropical summer monsoon.

4 Dynamical mechanisms involved in the formation and development of the EASSM

As mentioned above, the formation and development of the EASSM is associated with the seasonal enhancement of low-level southerly flow to the south over the Yangtze River. Also, it is one of the important differences between the EASSM and the Asian tropical summer monsoon. According to the thermal wind relationship, the formation and strengthening of the low-level southerly is relevant to the seasonal transition of the zonal land-sea thermal contrast over East Asia, which is a major forcing of the EASSM[23,36,37].

4.1 Seasonal reversal of the zonal land-sea thermal contrast over East Asia

On the planetary scale, LinHo et al.[38] linked the formation of the EASSM with the seasonal transition of global circulation from winter to summer, and argued that the meridional land-sea thermal contrast decreases in February to weaken the Asian winter monsoon, which favors the commencement of summer-type circulation over South China. Previous studies have pointed out that the zonal land-sea thermal contrast between the East Asian continent and the western Pacific reverses during late March and early April, when the low-level prevailing northerly turns to a southerly in situ (Fig. 5)[22,37].

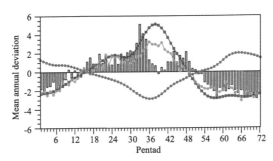

Fig. 5 Time series of the climate-mean (1981—2010) annual deviation of rainfall (bars, mm · d^{-1}), 850-hPa meridional wind (blue line, m · s^{-1}), zonal land-sea thermal contrast ($\partial \overline{T}/\partial x$) in the middle troposphere (red line, averaged between 600 and 400-hPa, 5×10^{-5} K · m^{-1}) and the zonal sea level pressure gradient ($\partial(SLP)/\partial x$, green line, 10^{-6} hPa · m^{-1}) over the region to the south of the Yangtze River (24°~29°N, 110°~125°E)

The seasonal reversal of the zonal land-sea thermal contrast originates from the seasonal evolution of the thermal conditions over the Tibetan Plateau, the Asian continent, and the western Pacific. Zhu et al.[39] used the thermal wind relationship and thermal adaptation theory to investigate the physical mechanism through which the formation of the EASSM is induced by the seasonal changes in zonal land-sea thermal contrast. They proposed that the land surface sensible heating enhances rapidly with the gradually northward shifting of the subsolar point, leading to a zonal asymmetric structure of the surface temperature field over

East Asia involving a "warm in the west but cold in the east" pattern. The land surface sensible heating would weaken the winter cold high over land that moves eastward to the western Pacific. Meanwhile, the southerly to the east of the high prevailing over South China brings the lower latitudinal warm and wet air northward to strengthen the "warm in the west but cold in the east" pattern. Afterwards, the low-level southerly is able to strength the vertical northerly shear, resulting in ascending motion and rainfall locally. Furthermore, the condensation heating released by the rainfall can further strengthen the low-level southerly. Such positive feedback can ultimately lead to the formation of the EASSM. Recently, Qi et al.[40] applied a high-resolution regional climate model in quantifying the influences of sensible heating over the Tibetan Plateau, the East Asian plain, and the northern West Pacific, on the formation of the EASSM. Their results suggested that the contribution of sensible heating over the WNP to the formation of the EASSM is small, whereas the sensible heating over the East Asian plain might enhance the low-level southerly over South China. However, their numerical experiment also validated the fact that the sensible heating over the Tibetan Plateau could prevent the establishment of the low-level southerly, thus inhibiting the formation of the EASSM. This finding is markedly distinct from those of other studies (e. g., Pan et al.[41]), which support the positive effect of sensible heating over the Tibetan Plateau on the pre-monsoonal rainfall over South China. Thus, further research is needed to further understand the dynamical mechanisms involved in the formation of the EASSM rainbelt.

4.2 Zonal land-sea thermal contrast in the subtropics of East Asia

On the global scale, the subtropical summer monsoon and its related precipitation firstly builds up to the south of the Yangtze River, but no monsoon is observed over East America at the same latitude. First of all, the formation of the EASSM is closely associated with the forcing of the Tibetan Plateau[42,43]. In boreal winter, when the subtropical westerly meets the Tibetan Plateau, the climbing flow and rounded wind generate a large-scale anticyclone to the north, and a cyclone to the south, of the Tibetan Plateau. This pair of large-scale circulation systems can strengthen the southward intrusion of higher-latitudinal cold air and the northward transport of tropical warm air over East Asia. The cold air and warm air then merge together over South China to benefit the formation of the EASSM rainbelt. Li and Zhang[44] also identified that the wind surrounding the Tibetan Plateau strengthens in March, giving rise to the spring permanent rainfall over Southeast China.

Moreover, a recent study has revealed a distinction between the land-sea thermal contrast of East Asia and East America[45]. A seasonal reversal of the zonal and meridional temperature gradient is evident over East Asia (Figs. 6a, 7a, and 7b); but in contrast, no seasonal transition of the meridional temperature gradient exists over East America (Fig. 6b), and the seasonal reversal of the zonal temperature gradient is very weak (Figs. 7a and 7b). Consequently, the seasonal changes in low-level southerly flow over East America are much

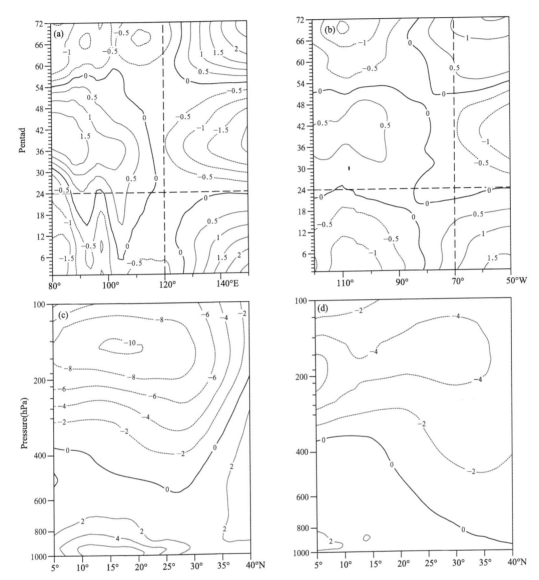

Fig. 6 Hovmüeller diagram of the 500-hPa zonal deviation of air temperature (K) over (a) East Asia (averaged along 24°~29°N) and (b) East America (averaged along 30°~35°N). Panels (c) and (d) show the seasonal change in the meridional wind (m·s^{-1}) over East Asia (110°~120°E) and East America (90°~70°W), respectively. The seasonal change is defined as the difference in mean values between April-September and October-March of the following year

weaker than those over East Asia (Figs. 6c and 6d). Furthermore, the different seasonal change in the zonal land-sea thermal contrast can be attributed to the difference between the thermal forcing of the Tibetan Plateau and that of the Rocky Mountains. In boreal spring, the sensible heating enhances rapidly over the Tibetan Plateau, strengthening the summer atmospheric heat source over the East Asian continent. The zonal land-sea thermal contrast is then large enough to develop strong ascending motion and low-level southerly flow over

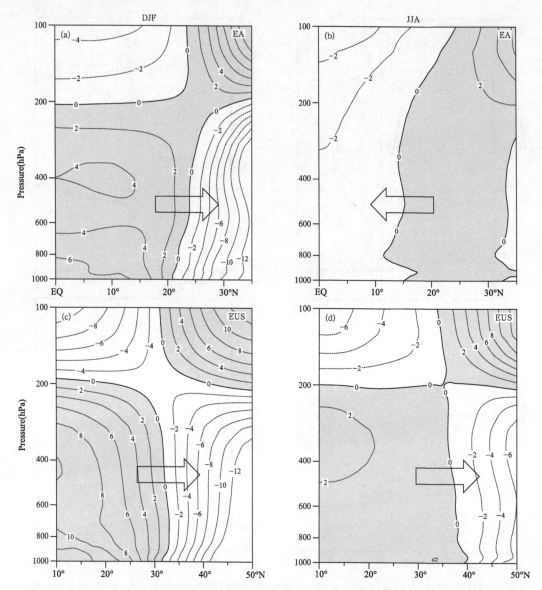

Fig. 7 Pressure-latitude cross section of air temperature without its meridional mean (K) over (a, b) East Asia (averaged along 100°~120°E) and (c, d) East America (averaged along 100°~80°W) in (a, c) boreal winter and (b, d) summer. The arrows indicate the thermal gradient direction

East Asia to the east of the heating region. In contrast, the direct heating effect of the Rocky Mountains on the local atmosphere is much weaker over East America, related to a small zonal land-sea thermal contrast and weak ascending motion and seasonal change in the low-level southerly flow. Therefore, the difference in the seasonal transition of the zonal land-sea thermal contrast is an important factor resulting in the distinct spring subtropical climates of East Asia and East America.

4.3 Influence from the mid-latitudes

Another prominent characteristic of the EASSM is the influence from the midlatitudes, which is the cold air source that maintains the EASSM front. This is also one of the key differences between the EASSM and the tropical summer monsoon. In the formation and development phase of the EASSM, the lower-tropospheric cyclone associated with the cold vortex over Northeast China is active over the mid-high latitudes of East Asia, suggesting a midlatitudinal influence on the seasonal advancement of the EASSM. The cold vortex over Northeast China occurs all year round, with high frequency from late spring to early summer in the Northern Hemisphere[46], and can be ascribed to the mid-high latitudinal Rossby wave activity[47]. Previous studies have indicated that the strength of the cold vortex over Northeast China is positively correlated with the precipitation over South China during its pre-flood rainy season from May to June. Also, significant correlation exists between the strength of the cold vortex over Northeast China and the Meiyu rainfall, indicating that the Meiyu rainfall increases/decreases with a stronger/weaker cold vortex over Northeast China from June to July[48-50].

Ding and Chan[51] reported a number of possible mechanisms involved in the influence of the mid-high latitudes on the EASSM. The mid-high latitudinal circulation might instigate the subtropical convection and rainfall through the effect of uplift. It could enhance the meridional pressure gradient to increase the low-level northeasterly, leading to the shear vorticity and cyclone near the shear lines. It could not only provide the potential energy available to the turbulence and mesoscale systems in the frontal zone, but also strengthen the low-level trough to withdraw the western Pacific subtropical anticyclone southeastward. Furthermore, the mid-latitudinal systems could affect the advancement of the EASSM. For instance, the formation of the Meiyu front is closely associated with the continuous southward intrusion of cold air, which is ascribed to the development and maintenance of a blocking high over the mid-high latitudes of Eurasia. The most favorable circulation pattern for persistent Meiyu rainfall is characterized by the development of two blocking highs over the Ural Mountains and the Okhotsk Sea, respectively[52,53,54]. Thus, the mid-latitudinal influence can be treated as a major difference between the East Asian monsoon and the South Asian monsoon, but it remains unclear as to how the synoptic systems over the mid-high latitudes modulate the advancement of the EASSM.

In conclusion, the formation and development of the EASSM is closely associated with the seasonal reversal of the zonal land-sea thermal contrast over East Asia. The particularity of the EASSM can also be attributed to the mechanical and thermal forcing of the Tibetan Plateau. In addition, synoptic systems (e.g., a cold vortex over Northeast China and a blocking high) can also influence the activity of the EASSM.

5 Multi-scale variability of the EASSM and relevant factors of influence

The EASSM also exhibits multi-scale variability, and the relevant factors of influence are different on specific timescales. Thus, a summary of the multi-scale variability of the EASSM and relevant factors of influence is of great scientific importance, both in a basic sense (for a deeper understanding of the nature of the EASSM) and in an applied sense (for developing seamless forecasting technologies).

5.1 Intraseasonal variability

The intraseasonal oscillation (ISO) of the EASSM varies over different regions in an individual season. In the pre-flood rainy season over South China (March to May), the EASSM rainbelt is situated over South China with a significant 10-20-day ISO, which can be attributed to the intraseasonal variability of surface sensible heating over the Tibetan Plateau with the same period in boreal spring[41]. When the spring surface sensible heating over the Tibetan Plateau is stronger (weaker), the surrounding low-level air can converge (diverge) under the influence of the sensible heating air pump. Since South China is affected by large-scale converging circulation to the southeast of the Tibetan Plateau, the lower-tropospheric convergence would enhance (weaken), corresponding to more (less) local rainfall.

After the EASSM rainbelt moves northward to the Yangtze-Huaihe River, both the bi-weekly and the 21-30-day ISO are significant for the monsoonal precipitation with similar strength[55]. Further studies have shown that the bi-weekly ISO is induced by the southward movement of vorticity anomalies in the mid-latitudinal westerly jet, whereas the 21-30-day ISO is associated with the westward extension of the western Pacific subtropical anticyclone in the lower troposphere[55]. Moreover, Mao et al.[56] pointed out that a 20-50-day ISO exists in the Meiyu rainbelt of the EASSM, which is produced by the ISO of the western Pacific subtropical anticyclone via the northward or northwestward propagation of a Rossby-wave-like coupling system between convection and circulation. Recently, He et al.[57] observed a 30-60-day ISO related to the MJO over the EASSM region, and subsequently constructed a relevant monitoring index that demonstrates a certain amount of predictability.

5.2 Interannual variability

As well as intraseasonal variability, the EASSM also demonstrates prominent interannual variability. The primary underlying conditions influencing the interannual variability of the EASSM include sea surface temperature anomalies (SSTAs; e.g., ENSO events and Indian Ocean SSTAs) and the land surface status (e.g., the snow anomalies over Eurasia and the Tibetan Plateau, and abnormal soil moisture over East Asia).

On the interannual timescale, the EASSM is weaker (stronger) following an El Niño

(La Niña) event, with a warmer (colder) SSTA in the equatorial tropical central-eastern Pacific and an enhanced (weakened) western Pacific subtropical anticyclone. The summer rainfall then tends to increase (decrease) over the Yangtze-Huaihe River. ENSO events in the previous winter may modulate the EASSM circulation and rainfall by changing the SST in the tropical Indian Ocean, the western Pacific, and the SCS in the following seasons[58,59]; also, the Walker circulation and the monsoonal cell will change subsequently[60-63]. Furthermore, ENSO events may modulate the monsoonal circulation and its onset process by simulating an anomalous anticyclone over the Philippine Sea[64-68] and changing the formation and evolution of the South Asian high[69-72]. The land surface conditions over Eurasia can also be affected by ENSO events to indirectly change the activity of the Asian summer monsoon[73-75]. Moreover, the role of ENSO in the EASSM depends on its specific phase. The decaying phase of ENSO is linked with the first mode of the EASSM, presenting out-of-phase rainfall anomalies between the northern SCS-Philippine Sea and southern Japan. However, the second mode of the EASSM, with negative correlation of abnormal rainfall between South China and North-Northeast China, is related to the developing phase of ENSO[76]. Specifically, colder SST in the tropical WNP occurs in the developing phase of ENSO, suppressing local convection and leading to the southward movement of the western Pacific subtropical anticyclone. Thus, abnormally high rainfall appears over the middle and lower Yangtze River, while abnormally low precipitation exists over South China due to the effect of the Pacific-Japan wave train[77].

Moreover, local SSTAs, snow anomalies over the Tibetan Plateau, and abnormal soil moisture over East Asia, can also affect the interannual variability of the EASSM. The effects of local SSTAs on the EASSM are distinct among regions. For anomalies of the EASSM rainfall, a warm SSTA in the WNP is able to induce a westward extension of the western Pacific subtropical anticyclone, which results in persistently strong rainfall events over the middle and lower Yangtze River[78]. Meanwhile, Indian Ocean SSTAs, including the Indian Ocean basin mode (IOBM) and the Indian Ocean dipole (IOD), can alter the rainfall and circulation of the Asian summer monsoon. The IOBM SSTAs act as a capacitor that prolongs the El Niño effect to enhance the subtropical anticyclone over the WNP via a Gill-type response and strengthened South Asian high, corresponding to increased rainfall over the mid and lower Yangtze River and decreased precipitation over South China[79]. Meanwhile, the East Asian jet is located to the north of its climate-mean position[80]. Furthermore, in the summer following a positive IOD event, an anomalous low-level easterly occurs over South China, accompanied by an anticyclonic anomaly settling over the Bay of Bengal, and there is abnormally high rainfall over South China but abnormally low rainfall over southern India[81-83]. Guan and Yamagata[84] proposed that the IOD SSTAs might influence the EASSM by inducing anomalous diabatic heating over the Bay of Bengal, stimulating a Rossby wave to affect the circulation over the Mediterranean to its west. Thereafter, the circulation anomalies would affect the EASSM via the "silk road" wave guide. Also, the IOD SSTAs could produce a northeastward-propagating Rossby wave train from the Indian Peninsula and Bay of Bengal. In addition, the interannual variability of sea ice could modulate the EASSM rainfall. Specifically, the sea ice over the Arctic and Greenland in spring could affect the EASSM rainfall via the Eurasian wave train at 500-hPa. More (less) spring sea ice over the Arctic and Greenland would correspond to increased (decreased) rainfall over northeastern and central China, and less (more) precipitation over South China[85-87]. Meanwhile, the sea ice over the Antarctic might affect the Mascarene and Australian high via anomalous Antarctic oscillation in the Southern Hemisphere, and then the cross-equatorial flow would change to influence the convection over the western Pacific and the westward extension of the western Pacific subtropical anticyclone, and ultimately the spatial distribution of the EASSM rainbelt would be modulated[88-89].

Many data analyses and numerical simulations have been conducted on the effects of Tibetan Plateau snow on the EASSM and climate in China[53,90-99]. Most results have suggested that the thermal effect of the Tibetan Plateau would be damped to weaken the EASSM in years with more and deeper snow over the Tibetan Plateau. Meanwhile, there would be more flood events over the Yangtze River. A recent study validated that anomalies of surface sensible heating over the Tibetan Plateau in spring could influence the local summer heating source via land-air interaction in situ, leading to interannual variability of the EASSM rainbelt[100]. However, further investigation on the physical process through which Tibetan Plateau snow influences the EASSM is still required, because of the non-uniform spatial distribution of Tibetan Plateau snow[101-102] and the uncertainty in observations and numerical models.

In addition, the East Asian summer monsoon rainfall can also be altered by soil moisture anomalies over East Asia in boreal spring. If the soil is wetter from the middle and lower Yangtze River to northern China, but drier over northeastern China, the land surface evaporation strengthens to cool the surface temperature over eastern China, which in turn decreases the land-sea thermal contrast to weaken the EASSM. Thus, the EASSM rainbelt would be located to the south of its climate-mean position, presenting abnormally high precipitation over the Yangtze River and abnormally low precipitation over northern and southern China[103-104].

5.3 Interdecadal variability

The EASSM has evidently weakened since the 1970s to aggravate the "flood in the south but drought in the north" rainfall pattern in summer over eastern China[105]. The detailed features of this change include: (1) more extreme rainfall over the Yangtze River but less precipitation over North China[106]; (2) weaker surface southerly flow in summer and a cooling trend of the air temperature in the mid and upper troposphere over East Asia[107-108]; (3) a southward extension of the WNP subtropical anticyclone and an increased zonal span of the South Asian high in the upper troposphere[106,109-110]; (4) southward movement of the

200-hPa East Asian subtropical westerly jet that induces the stronger westerly to its south but weaker one to its north[107,111]; and (5) weakened land-sea thermal contrast between the Asian continent and its surrounding oceans[112]. However, rainfall has been decreasing evidently to the south of the Yangtze River in late spring since the 1970s, along with a significant cooling of the upper troposphere over central China[113]. Studies show that the precipitation in the pre-flood season over South China and the rainy season over North China decreased from 1958 to 2000, when more Meiyu rainfall appeared over the Yangtze-Huaihe River[114]. During the late 1980s, summer rainfall increased over southern East China, which is closely associated with the decreased snow cover over the Eurasian continent in boreal spring and the warmer SSTAs in the WNP in boreal summer on the interdecadal timescale[115,116]. In the early 1990s, the East Asian summer monsoon started to recover, corresponding to a northward shifting tendency of the monsoonal rainbelt[117]. Meanwhile, the center of the WNP subtropical anticyclone started to withdraw eastward with a northward shifting of its ridgeline, which transported more moisture to the Huaihe River[118]. Also, the land-sea thermal contrast over East Asia enhanced in this period due to the increased sea level pressure over the western Pacific, ascribed to a phase transition of the Pacific Decadal Oscillation (PDO)[39]. The latest research suggests that the interdecadal variability of the EASSM over southern China in spring was opposite to that in summer in the 1990s, exhibiting more spring rainfall but less summer precipitation over southern China after 1994, which was attributed to the spring La Niña-like interdecadal cooling in the tropical eastern Pacific SST and the interdecadal summer warming in the IOBM SST[119].

Although we still have no consensus on the reason for the interdecadal weakening of the East Asian summer monsoon in the 1970s, mounting evidence suggests that the interdecadal warming in the tropical ocean may play a critical role. Such warming in the tropical ocean could decrease the meridional temperature gradient in the mid-upper troposphere of the East Asian monsoon region to weaken the monsoonal circulation. The weakening of the East Asian summer monsoon in the last 50 years can be considered as one aspect of the interdecadal variability of the global monsoon, which is altered by interdecadal changes in the tropical ocean, especially the tropical central-eastern Pacific and western Indian Ocean. Such interdecadal change in the tropical ocean is part of the PDO, belonging to natural variability[120-122]. Moreover, the more frequent summer floods over central-eastern China are associated with the warming SST trend in the western Pacific warm pool[123]. Nevertheless, one cannot neglect the role of air-sea feedback in the interdecadal variability of the East Asian summer monsoon[121,124]. Aside from the influence of SST, research also stresses the effect of land-air interaction on the interdecadal weakening of the East Asian summer monsoon. Specifically, the interdecadal decrease in spring sensible heating over the Tibetan Plateau and the increased snow cover or depth over Eurasia may attenuate the East Asian summer monsoon circulation, leading to a southward withdrawal of the monsoon rainbelt[101,98,125-127]. Since the warming in the higher latitudes of Eurasia is more rapid than that in the subtropics against the background of global warming, the East Asian westerly jet weakens, corresponding to a weaker EASSM[128].

6 Conclusions and discussion

The EASSM, as the main component of the Asian summer monsoon, determines the rainfall distribution over eastern China, and its multi-scale variability plays an important role in weather and climate anomalies over China as a whole. In the present paper, we revisit the nature of the EASSM by reviewing the literature and analyzing the latest datasets. In particular, the formation and advancement of the EASSM is demonstrated, as well as the precipitation properties of the EASSM and the related circulation. Additionally, we discuss the characteristics of, and the dynamical mechanisms involved in, the seasonal transition of the land-sea thermal contrast and mid-latitudinal influence, as well as their roles in the formation and maintenance of the EASSM. The characteristics of the EASSM on different timescales and the related factors of influence can be summarized as follows:

(1) The EASSM is characterized by the seasonal transition of zonal land-sea thermal contrast in the subtropics of East Asia, which generates the EASSM rainbelt. The EASSM rainfall features frontal precipitation, which is absent in the tropical monsoon rainfall. The tropical monsoon rainfall is composed of convective precipitation, while both large-scale stratiform and convective precipitation constitute the EASSM rainfall, with a distinct seasonal variation in the ratio between the two precipitation types. Specifically, the convective precipitation becomes dominant in the EASSM rainfall after the EASSM becomes established.

(2) The EASSM rainbelt firstly establishes over the region to the south of the Yangtze River from March to April, accompanied by seasonal enhancement of local low-level southerly flow associated with the seasonal transition of the zonal land-sea thermal contrast between the East Asian continent and the western Pacific. In addition, the mechanical forcing of the Tibetan Plateau provides a favorable background for the enhancement of low-level southerly flow over the region to the south of the Yangtze River, which directly results from the seasonal evolution of the thermal conditions over the Tibetan Plateau, East Asian continent, and western Pacific. Since the seasonal transition of land-sea thermal contrast gives rise to the monsoon, the establishment of the rainbelt over the region to the south of the Yangtze River in spring can be treated as the beginning of the EASSM, or at least its embryonic stage.

(3) The seasonal evolution of the EASSM is separate to the formation and propagation of the tropical summer monsoon, and it is also influenced by mid-latitudinal systems. The EASSM rainbelt builds up earlier than the onset of the tropical summer monsoon, while its southward intrusion in mid-May favors the onset of the tropical summer monsoon, followed by a strengthening of low-level southerly flow that further advances the EASSM rainbelt northward. The occurrence of a cold vortex over Northeast China is most frequent during

late spring and early summer, modulating the seasonal evolution of the EASSM rainbelt. Generally, the strength of such a cold vortex over Northeast China is significantly and positively correlated with the rainfall amount of the EASSM.

(4) The spatial difference in the land-sea thermal contrast leads to the distinct climates of East Asia and East America. The seasonal variation of rainfall is large in the subtropics of East Asia, exhibiting monsoonal precipitation characterized by a "wet in the summer but dry in the winter" pattern. Also, the low-level wind direction reverses from winter to summer, with northerly wind in winter but southerly wind in summer. Thus, a subtropical monsoon climate is observed over East Asia. In contrast, a non-monsoonal climate exists over East America, where the precipitation amount changes little all year round and there is a prevailing westerly in the lower troposphere. The differences originate from the distinct land-sea thermal contrast over the two regions. A seasonal reversal of both the meridional and zonal temperature gradient is evident over East Asia, but the seasonal change in the zonal temperature gradient is weak over East America and the seasonal reversal of the meridional temperature gradient is absent. Moreover, the zonal cell related to the Tibetan Plateau is substantially different to that induced by the Rocky Mountains. As a result, the primary reason for the distinct climates of East Asia and East America is the different seasonal transition of the land-sea thermal contrast and the differing roles of the Tibetan Plateau and Rocky Mountains.

(5) Multi-scale variability exists in the EASSM circulation and rainfall, and the factors of influence on specific timescales are different. On the intraseasonal timescale, the non-uniform spatiotemporal distribution of the EASSM ISO is ascribed to the intraseasonal variability of surface sensible heating over the Tibetan Plateau in spring, the movement of anomalous vorticity in the mid-latitudinal westerly jet, the short-term oscillation of the western Pacific subtropical anticyclone, and the meridional propagation of MJO. On the interannual timescale, the major forcings are ENSO events, Indian Ocean SSTAs (including the IOBM an IOD SSTAs), snow anomalies over Eurasia and the Tibetan Plateau in the previous winter and spring, and anomalous soil moisture over East Asia in spring. Meanwhile, on the interdecadal timescale, the EASSM weakened from the 1970s, but began to recover from the 1990s. The interdecadal variability of the EASSM can be attributed to the anomalous tropical convection over the Indian Ocean and Maritime Continent, and the long-term changes in the thermal conditions of the Tibetan Plateau, which are affected by both the natural variability of climate systems and anthropogenic greenhouse gas and aerosol emissions.

(6) The low-level southerly maximum is sometimes accompanied by the rainfall center in the EASSM, but it is separated from each other at other times. In the tropical summer monsoon, meanwhile, the wind maximum is always collocated with the rainfall center. In the formation and development stage of the EASSM, the southerly appears together with the rainfall, but it is isolated in the mature stage of the EASSM. This is because the EASSM rainfall is always closely associated with the large-scale frontal system, and the rainfall takes

place in the convergence near the front. Therefore, the wind and rainfall emerge together over the region to the south of the Yangtze River, where the front is located, in the formation and development stage of the EASSM. Subsequently, the southerly is enhanced to shift the front northward, consistent with the northward movement of the rainfall region. The southerly is then located to the south of the front with no precipitation, exhibiting a separation of wind from rainfall. Consequently, a combined occurrence of wind and rainfall exists over the forward position of the EASSM, but they are separated over the EASSM hinterland. Since the tropical summer monsoon rainfall is ascribed to local convergence without a large-scale frontal system, it is always accompanied by the maximum low-level southerly.

It is important to note that the EASSM rainfall is always accompanied by a frontal system. Since the position and strength of the front depend upon the multi-scale configuration between the warm-and-wet flow from lower latitudes and the cold-and-dry air from higher latitudes, it is possible to extract their activity and the related impact factors through scale separation technology. Ultimately, high-quality weather and climate prediction may then be achievable via hierarchical forecasting methods.

It is clear that many achievements have been made through studies of the EASSM, providing important clues for furthering our understanding of the Asian monsoon systems. However, some key scientific questions remain open. For instance, most previous studies have focused on the formation mechanisms of the EASSM rainfall and circulation in boreal summer, but little attention has been paid to the process of EASSM establishment from winter to summer, especially the seasonal transition from winter monsoon to summer monsoon and the relevant contributing factors over the East Asian subtropics. Also, it should be noted that the EASSM is closely associated with the thermal forcing of the Tibetan Plateau. Thus, improvements in the observation network over the Tibetan Plateau is required to obtain accurate surface heating flux information, which is of great importance to future studies on the interaction between the sub-systems of the Asian summer monsoon.

Acknowledgments. We thank the two anonymous reviewers for their insightful comments.

References

[1] Halley E. An historical account of the trade winds, and monsoons, observable in the seas between and near the tropics, with an attempt to assign the physical cause of the said winds[J]. Phil Trans, 1686, 16: 153-168.

[2] Hadley G. Concerning the cause of the general trade-winds[J]. Phil Trans, 1735, 39: 58-62.

[3] Eady E T. The cause of general circulation of atmosphere[J]. Centen Proc Roy Meteor Soc, 1950: 156-172.

[4] He Jinhai, Li Jun, Zhu Qiangen. Sensitivity experiments on summer monsoon circulation cell in East Asia [J]. Adv Atmos Sci, 1989, 6: 120-132.

[5] He Jinhai, Yu Jingjing, Shen Xinyong, et al. Research on mechanism and variability of East Asian mon-

soon[J]. J Trop Meteor,2004,20:449-459. (in Chinese)

[6] Li Jianping,Zeng Qingcun. A new monsoon index and the geographical distribution of the global monsoons[J]. Adv Atmos Sci,2003,20:299-302.

[7] Tao S Y,Chen L X. A review of recent research on the East Asian summer monsoon in China[C]// Chang C-P,Krishnamurti T N. Monsoon Meteorology. Oxford:Oxford University Press,1987:60-92.

[8] Huang Ronghui,Zhang Zhenzhou,Huang Gang,et al. Characteristics of the water vapor transport in East Asian monsoon region and its difference from that in South Asian monsoon region in summer[J]. Scientia Atmos Sinica,1998,22:460-469. (in Chinese)

[9] Lau N C,Ploshay J J. Simulation of synoptic-and subsynoptic-scale phenomena associated with the East Asian summer monsoon using a high-resolution GCM[J]. Mon Wea Rev,2009,137:137-160.

[10] Wang Lijuan,He Jinhai,Guan Zhaoyong. Climatological features of East Asian subtropical summer monsoon trough and its comparison with South China Sea summer monsoon trough[J]. Acta Meteor Sinica, 2006,64:583-593. (in Chinese)

[11] Tang Xu,Chen Baode,Liang Ping,et al. Definition and features of the north edge of the East Asian summer monsoon[J]. Acta Meteor Sinica,2010,24:43-49.

[12] Kobayashi S,Ota Y,Harada Y,et al. The JRA-55 Reanalysis:General specifications and basic characteristics[J]. J Meteor Soc Japan,2015,93:5-48.

[13] Xie P P,Arkin P A. Global precipitation:A 17-yr monthly analysis based on gauge observations,satellite estimates,and numerical model outputs[J]. Bull Amer Meteor Soc,1997,78:2539-2558.

[14] Adler R F,Huffman G J,Chang A,et al. The version-2 global precipitation climatology project (GPCP) monthly precipitation analysis (1979-present) [J]. J Hydrometeor,2003,4:1147-1167.

[15] Huffman G J,Bolvin D T,Nelkin E J,et al. The TRMM multisatellite precipitation analysis (TMPA): Quasi-global,multiyear,combined-sensor precipitation estimates on fine scales[J]. J Hydrometeor,2007, 8:38-55.

[16] Yu Shihua,Mao Weiping. The characteristics of the East Asian monsoon circulation in summer and its relation with the rainfall[J]. J Trop Meteor,1986,2:349-354. (in Chinese)

[17] Yu Shihua,Yan Jingrong. An analysis of the establishment processes of the East Asian summer monsoon circulation[J]. J Trop Meteor,1986,2:55-61. (in Chinese)

[18] Zhu Qiangen,He Jinhai,Wang Panxing. A study of circulation differences between East Asian and Indian summer monsoons with their interaction[J]. Adv Atmos Sci,1986,3:465-477.

[19] Yu Shihua, Yang Weiwu. The features of the subtropical monsoon circulation cell and its relationship with summer circulation over East Asia[J]. J Appl Meteor Sci,1991,2:242-247. (in Chinese)

[20] Chi Yanzhen, He Jinhai, Wu Zhiwei. Features analysis of the different precipitation periods in the pre-flood season in South China[J]. J Nanjing Inst Meteor,2005,28:163-171. (in Chinese)

[21] He Jinhai,Qi Li,Wei Jin,et al. Reinvestigations on the East Asian subtropical monsoon and tropical monsoon[J]. Chinese J Atmos Sci,2007,31:1257-1265. (in Chinese)

[22] Zhao Ping,Zhang Renhe,Liu Jiping,et al. Onset of southwesterly wind over eastern China and associated atmospheric circulation and rainfall[J]. Climate Dyn,2007,28:797-811.

[23] Zhu Congwen,Zhou Xiuji,Zhao Ping,et al. Onset of East Asian subtropical summer monsoon and rainy season in China[J]. Sci China (Ser. D),2011,54:1845-1853.

[24] Bao Chenlan. Tropical Meteorology [J]. Beijing:Science Press,1980:1-27. (in Chinese)

[25] Tian S F,Yasunari T. Climatological aspects and mechanism of spring persistent rains over central China [J]. J Meteor Soc Japan,1998,76:57-51.

[26] Ding Yihui. Chinese summer monsoon rainfall and its regional characteristics[C]// Ding Yihui, Kamimikito M. Monsoon in East Asia. China Meteorological Press, 1994:76-83. (in Chinese)

[27] Chen G T J. Research on the phenomena of Meiyu during the past quarter century: An overview[C]// Chang C P. East Asian Monsoon. World Scientific Publishing Co, Singapore, 2004:357-403.

[28] Lian Yi, Shen Baizhu, Gao Zongting, et al. A method describing the subtropical summer monsoon active domain in the East Asian-Northwest Pacific region[J]. Acta Meteor Sinica, 2007, 65: 503-510. (in Chinese)

[29] Lian Yi, Shen Bozhu, Liu Gang, et al. Preliminarily investigation on the advancing path of East Asian subtropical summer monsoon from May to July[R]. Assembly of the 5th workshop on the subtropical meteorological operations and studies, Shanghai, China, 2009, 16-17 Nov, Chinese Meteor Soc, 8-9. (in Chinese)

[30] Liu Gang, Shen Baizhu, Lian Yi, et al. Feature analysis of the main active path about East Asian summer monsoon[J]. J Jilin Univ (Earth Sci Ed.), 2008, 38:196-199. (in Chinese)

[31] Lau K M, Yang S. Climatology and interannual variability of the Southeast Asian summer monsoon[J]. Adv Atmos Sci, 1997, 14:141-162.

[32] Webster P J, Magana V O, Palmer T N, et al. Monsoons: Processes, predictability, and the prospects for prediction[J]. J Geophys Res, 1998, 103:14451-14510.

[33] Wang B, LinHo. Rainy season of the Asian-Pacific summer monsoon[J]. J Climate, 2002, 15:386-398.

[34] Liu Boqi, Liu Yimin, Wu Guoxiong, et al. Asian summer monsoon onset barrier and its formation mechanism[J]. Climate Dyn, 2015, 45:711-726.

[35] Ren Ke, He Jinhai, Qi Li. The establishment characteristics of the East Asian subtropical monsoon rain belt and an analysis of its nature of precipitation[J]. Acta Meteor Sinica, 2010, 68:550-558. (in Chinese)

[36] He Jinhai, Zhao Ping, Zhu Congwen, et al. Discussion of some problems as to the East Asian subtropical monsoon[J]. Acta Meteor Sinica, 2008, 22:419-434.

[37] Qi Li, He Jinhai, Zhang Zuqiang, et al. Seasonal cycle of the zonal land-sea thermal contrast and East Asian subtropical monsoon circulation[J]. Chinese Sci Bull, 2008, 53:131-136.

[38] LinHo, Huang X L, Lau N C. Winter-to-spring transition in East Asia: A planetary-scale perspective of the South China spring rain onset[J]. J Climate, 2008, 21:3081-3096.

[39] Zhu Zhiwei, He Jinhai, Qi Li. Seasonal transition of East Asian subtropical monsoon and its possible mechanism[J]. J Trop Meteor, 2012, 18:305-313.

[40] Qi Li, He Jinhai, Wang Yuqing. The terraced thermal contrast among the Tibetan Plateau, the East Asian plain, and the western North Pacific and its impacts on the seasonal transition of East Asian climate[J]. Chinese Sci Bull, 2014, 59:212-221.

[41] Pan Weijuan, Mao Jiangyu, Wu Guoxiong. Characteristics and mechanism of the 10-20-day oscillation of spring rainfall over Southern China[J]. J Climate, 2013, 26:5072-5087.

[42] Wan Rijin, Wu Guoxiong. Mechanism of the spring persistent rains over southeastern China[J]. Sci China (Ser. D), 2007, 50:130-144.

[43] Wu Guoxiong, Liu Yimin, Zhang Qiong, et al. The influence of mechanical and thermal forcing by the Tibetan Plateau on Asian climate[J]. J Hydrometeor, 2007, 8:770-789.

[44] Li Qiang, Zhang Renhe. Seasonal variation of climatological bypassing flows around the Tibetan Plateau[J]. Adv Atmos Sci, 2012, 29:1100-1110.

[45] Chang Luyu, He Jinhai, Qi Li, et al. A study of the different characteristics of seasonal variations of the precipitation and large-scale circulation between East Asia and eastern North America and its possible

mechanism[J]. Acta Meteor Sinica,2013,71:1074-1088. (in Chinese)

[46] Zhu Qiangen,Lin Jinrong,Shou Shaowen,et al. Principle and Methods of Synoptic Meteorology[J]. Beijing:China Meteorological Press,1992:186-187. (in Chinese)

[47] Lian Yi,Bueh Chaolaw,Xie Zuowei,et al. The anomalous cold vortex activity in Northeast China during the early summer and the low-frequency variability of the Northern Hemispheric atmosphere circulation [J]. Chinese J Atmos Sci,2010,34:429-439. (in Chinese)

[48] Miao Chunsheng,Wu Zhiwei,He Jinhai,et al. The anomalous features of the Northeast cold vortex during the first flood period in the last 50 years and its correlation with rainfall in South China[J]. Chinese J Atmos Sci,2006,30:1249-1256. (in Chinese)

[49] Miao Chunsheng,Wu Zhiwei,He Jinhai. Relationship among the Northern Hemisphere annual mode,the Northeast China cold vortex and precipitation during the first yearly rainy period in South China[J]. J Trop Meteor,2006,22:593-599. (in Chinese)

[50] He Jinhai,Wu Zhiwei,Jiang Zhihong,et al. "Climate effect" of the northeast cold vortex and its influences on Meiyu[J]. Chinese Sci Bull,2007,52:671-679.

[51] Ding Yihui,Chan J C L. The East Asian summer monsoon:an overview[J]. Meteor Atmos Phys,2005,89:117-142.

[52] Ding Yihui. Advanced Meteorology . China Meteorological Press,1991:792. (in Chinese)

[53] Zhang Shunli,Tao Shiyan. The influences of snow cover over the Tibetan Plateau on Asian summer monsoon[J]. Chinese J Atmos Sci,2001,25:372-390. (in Chinese)

[54] Wu Renguang. A midlatitude Asian circulation anomaly pattern in boreal summer and its connection with the Indian and East Asian summer monsoons[J]. Int J Climatol,2002,22:1879-1895.

[55] Yang Jing,Wang Bin,Bao Qing. Biweekly and 21-30-day variations of the subtropical summer monsoon rainfall over the lower reach of the Yangtze River basin[J]. J Climate,2010,23:1146-1160.

[56] Mao Jiangyu,Sun Zhang,Wu Guoxiong. 20-50-day oscillation of summer Yangtze rainfall in response to intraseasonal variations in the subtropical high over the western North Pacific and South China Sea[J]. Climate Dyn,2010,34:747-761.

[57] He Jinhai,Chang Luyu,Chen Hua. Meridional propagation of the 30-60-day variability of precipitation in the East Asian subtropical summer monsoon region:Monitoring and prediction[J]. Atmos Ocean,2015,53:251-263.

[58] Klein S A,Soden B J,Lau N C. Remote sea surface temperature variations during ENSO:Evidence for a tropical atmospheric bridge[J]. J Climate,1999,12:917-932.

[59] Alexander M A,Blade I,Newman M,et al. The atmospheric bridge:The influence of ENSO teleconnections on air-sea interaction over the global oceans[J]. J Climate,2002,15:2205-2231.

[60] Webster P J,Yang S. Monsoon and ENSO:Selectively interactive systems[J]. Quart J Roy Meteor Soc,1992,118:877-926.

[61] Ju Jianhua,Slingo J M. The Asian summer monsoon and ENSO[J]. Quart J Roy Meteor Soc, 1995,121:1133-1168.

[62] Wu Guoxiong,Meng Wen. Gearing between the Indo-monsoon circulation and the Pacific-Walker circulation and the ENSO. Part I:Data analysis[J]. Chinese J Atmos Sci,1998,22:470-480. (in Chinese)

[63] Kawamura R,Matsuura T,Iizuka S. Equatorially symmetric impact of the El Nino-Southern Oscillation on the South Asian summer monsoon system[J]. J Meteor Soc Japan,2003,81:1329-1352.

[64] Zhang Renhe,Sumi A,Kimoto M. Impact of El Nino on the East Asian monsoon:A diagnostic study of the '86/87 and '91/92 events[J]. J Meteor Soc Japan,1996,74:49-62.

[65] Zhang Renhe, Sumi A, Kimoto M. A diagnostic study of the impact of El Nino on the precipitation in China[J]. Adv Atmos Sci, 1999, 16: 229-241.

[66] Wang Bin, Wu Renguang Fu Xiuhua. Pacific-East Asian teleconnection: How does ENSO affect East Asian climate[J]. J Climate, 2000, 13: 1517-1536.

[67] Wang Bin, Zhang Qiong. Pacific-East Asian teleconnection. Part II: How the Philippine Sea anomalous anticyclone is established during El Nino development[J]. J Climate, 2002, 15: 3252-3265.

[68] Zhang Renhe, Sumi A. Moisture circulation over East Asia during El Nino episode in northern winter, spring and autumn[J]. J Meteor Soc Japan, 2002, 80: 213-227.

[69] Liu Boqi, Wu Guoxiong, Mao Jiangyu, et al. Genesis of the South Asian high and its impact on the Asian summer monsoon onset[J]. J Climate, 2013, 26: 2976-2991.

[70] Liu Boqi, Wu Guoxiong, Ren Rongcai. Influences of ENSO on the vertical coupling of atmospheric circulation during the onset of South Asian summer monsoon[J]. Climate Dyn, 2015, 45: 1859-1875.

[71] Guo Shuaihong, Wang Lijuan, Wang Miao. The timing of South-Asian high establishment and its relation to tropical Asian summer monsoon and precipitation over East-central China in summer[J]. J Trop Meteor, 2014, 30: 129-136. (in Chinese)

[72] He Jinhai, Liu Boqi, Wu Guoxiong. Formation of South Asian high from late spring to early summer and its association with ENSO events[J]. Chinese J Atmos Sci, 2014, 38: 670-684. (in Chinese)

[73] Meehl G A. Influence of the land surface in the Asian summer monsoon: External conditions versus internal feedbacks[J]. J Climate, 1994, 7: 1033-1049.

[74] Yang S. ENSO-snow-monsoon associations and seasonal-interannual predictions[J]. Int J Climatol, 1996, 16: 125-134.

[75] Yang S, Lau K M. Influences of sea surface temperature and ground wetness on Asian summer monsoon[J]. J Climate, 1998, 11: 3230-3246.

[76] Wang Bin, Wu Zhiwei, Li Jianping, et al. How to measure the strength of the East Asian summer monsoon[J]. J Climate, 2008, 21: 4449-4463.

[77] Huang Ronghui, Wu Yifang. The influence of ENSO on the summer climate change in China and its mechanism[J]. Adv Atmos Sci, 1989, 6: 21-32.

[78] Ren Xuejuan, Yang Xiuqun, Sun Xuguang. Zonal oscillation of western Pacific subtropical high and sub-seasonal SST variations during Yangtze persistent heavy rainfall events[J]. J Climate, 2013, 26: 8929-8946.

[79] Huang Gang, Qu Xia, Hu Kaiming. The impact of the tropical Indian Ocean on South Asian high in boreal summer[J]. Adv Atmos Sci, 2011, 28: 421-432.

[80] Qu Xia, Huang Gang. Impacts of tropical Indian Ocean SST on the meridional displacement of East Asian jet in boreal summer[J]. Int J Climatol, 2012, 32: 2073-2080.

[81] Li Chongyin, Mu Minquan. The dipole in the equatorial Indian Ocean and its impacts on climate[J]. Chinese J Atmos Sci, 2001, 25: 433-443. (in Chinese)

[82] Qian Wei, Guan Zhaoyong. Relationships between SSTA of tropical Indian Ocean and summer rainfall in southern China[J]. J Nanjing Inst Meteor, 2007, 30: 9-16. (in Chinese)

[83] Yan Hongming, Yang Hui, Li Chongyin. Numerical simulations on the climate impacts of temperature dipole in the equatorial Indian Ocean[J]. Acta Ocean Sinica, 2007, 29: 31-39. (in Chinese)

[84] Guan Zhaoyong, Yamagata T. The unusual summer of 1994 in East Asia: IOD teleconnections[J]. Geophys Res Lett, 2003, 30, doi: 10.1029/2002GL016831.

[85] Wu Bingyi, Zhang Renhe, Wang Bin, et al. On the association between spring Arctic sea ice concentration

and Chinese summer rainfall[J]. Geophys Res Lett,2009,36,L09501,doi:10.1029/2009GL037299.

[86] Wu Bingyi,Zhang Renhe,Wang Bin. On the association between spring Arctic sea ice concentration and Chinese summer rainfall:A further study[J]. Adv Atmos Sci,2009,26:666-678.

[87] Wu Bingyi,Zhang Renhe,D'Arrigo R,et al. On the relationship between winter sea ice and summer atmospheric circulation over Eurasia[J]. J Climate,2013,26:5523-5536.

[88] Xue Feng,Wang Huijun,He Jinhai. Interannual variability of Mascarene high and Australian high and their influences on East Asian summer monsoon[J]. J Meteor Soc Japan,2003,82:1173-1186.

[89] Xue Feng,He Juanxiong. Influence of the Southern Hemispheric circulation on east-west oscillation of the western Pacific subtropical high[J]. Chinese Sci Bull,2005,50:1532-1536.

[90] Chen Lieting,Yan Zhixin. Statistical analysis about the effects of winter-to-spring snow cover anomalies over Tibetan Plateau on the early summer monsoon[M]. Essays of Chinese Long-term Hydrological Prediction (Vol. II), Yangtze Valley Planning Office, Ed. ,China Water & Power Press,1981,133-141. (in Chinese)

[91] Guo Qiyun,Wang Jiqin. The snow cover on Tibetan Plateau and its effect on the monsoon over East Asia [J]. Plateau Meteor,1986,5:116-123. (in Chinese)

[92] Barnett T P,Dumenil L,Schlese U,et al. The effect of Eurasian snow cover on regional and global climate variations[J]. J Atmos Sci,1989,46:661-686.

[93] Luo Yong. Studies on the effect of snow cover over the Qinghai-Xizang Plateau in winter and spring on general circulation over East Asia in summer[J]. Plateau Meteor,1995,14:505-512. (in Chinese)

[94] Wu Guoxiong,Xue Jishan,Wang Zaizhi,et al. The modification of radiation effect of snowmelt sooner or later on the seasonal variation in the Tibetan Plateau[C]//Huang Ronghui. Numerical Simulation and Prediction of Disaster Climate. China Meteorological Press,1996:151-161. (in Chinese)

[95] Ding Yihui,Sun Ying. Interdecadal variation of the temperature and precipitation patterns in the East-Asian monsoon region[J]. Proc Int Symp on Climate Change (ISCC),Beijing,2003:66-71.

[96] Li Yueqing. Surface heating in the Tibetan Plateau and general circulation over it and their relations with the prediction of drought-flood at its eastern side[J]. Chinese J Atmos Sci,2003,27:107-114. (in Chinese)

[97] Qian Yongfu,Zhang Yan,Huang Yanyan,et al. The effects of the thermal anomalies over the Tibetan Plateau and its vicinities on climate variability in China[J]. Adv Atmos Sci,2004,21:369-381.

[98] Zhao Ping,Zhou Zhijiang,Liu Jiping. Variability of Tibetan spring snow and its associations with the hemispheric extratropical circulation and East Asian summer monsoon rainfall:An observational investigation[J]. J Climate,2007,20:3942-3955.

[99] Wu Zhiwei,Li Jianping,Jiang Zhihong,et al. Modulation of the Tibetan Plateau snow cover on the ENSO teleconnections:From the East Asian summer monsoon perspective[J]. J Climate,2012,25:2481-2489.

[100] Wang Ziqian,Duan Anmin,Wu Guoxiong. Time-lagged impact of spring sensible heat over the Tibetan Plateau on the summer rainfall anomaly in East China:Case studies using the WRF model[J]. Climate Dyn,2014,42:2885-2898.

[101] Wu Tongwen,Qian Zhi'an. The relation between the Tibetan winter snow and the Asian summer monsoon and rainfall:An observational investigation[J]. J Climate,2003,16:2038-2051.

[102] Qin D H,Liu S Y,Li P J. Snow cover distribution,variability,and response to climate change in western China[J]. J Climate,2006,19:1820-1833.

[103] Zuo Zhiyan,Zhang Renhe. The spring soil moisture and the summer rainfall in eastern China[J]. Chinese Sci Bull,2007,52:3310-3312.

[104] Zhang Renhe, Zuo Zhiyan. Impact of spring soil moisture on surface energy balance and summer monsoon circulation over East Asia and precipitation in East China[J]. J Climate, 2011, 24: 3309-3322.

[105] Zhou T J, Gong D Y, Li J, et al. Detecting and understanding the multi-decadal variability of the East Asian summer monsoon-Recent progress and state of affairs[J]. Meteorol Z, 2009, 18: 455-467.

[106] Hu Z Z. Interdecadal variability of summer climate over East Asia and its association with 500-hPa height and global sea surface temperature[J]. J Geophys Res, 1997, 102: 19403-19412.

[107] Yu Rucong, Wang Bin, Zhou Tianjun. Tropospheric cooling and summer monsoon weakening trend over East Asia[J]. Geophys Res Lett, 2004, 31, L22212, doi: 10.1029/2004GL021270.

[108] Yu Rucong, Zhou Tianjun. Seasonality and three-dimensional structure of the interdecadal change in East Asian monsoon[J]. J Climate, 2007, 20: 5344-5355.

[109] Zhang Qiong, Qian Yongfu, Zhang Xuehong. Interannual and interdecadal variations of the South Asian high[J]. Chinese J Atmos Sci, 2000, 24: 67-78. (in Chinese)

[110] Gong D Y, Ho C H. Shift in the summer rainfall over the Yangtze River valley in the late 1970s[J]. Geophys Res Lett, 2002, 29, doi: 10.1029/2001GL014523.

[111] Schiemann R, Luthi D, Schar C. Seasonality and interannual variability of the westerly jet in the Tibetan Plateau region[J]. J Climate, 2009, 22: 2940-2957.

[112] Ding Yihui, Wang Zunya, Sun Ying. Interdecadal variation of the summer precipitation in East China and its association with decreasing Asian summer monsoon. Part I: Observed evidences[J]. Int J Climatol, 2007, 28: 1139-1161.

[113] Xin X G, Yu R C, Zhou T J. Drought in late spring of South China in recent decades[J]. J Climate, 2006, 19: 3197-3206.

[114] Wu Zhiwei, Jiang Zhihong, He Jinhai. The comparison analysis of flood and drought features among the first flood period in South China, Meiyu period in the Yangtze River and the Huaihe River valleys and rainy season in North China in the last 50 years[J]. Chinese J Atmos Sci, 2006, 30: 391-401. (in Chinese)

[115] Zhang Renhe, Wu Bingyi, Zhao Ping. The decadal shift of the summer climate in the late 1980s over eastern China and its possible causes[J]. Acta Meteor Sinica, 2008, 22: 435-445.

[116] Wu Bingyi, Yang Kun, Zhang Renhe. Eurasian snow cover variability and its association with summer rainfall in China[J]. Adv Atmos Sci, 2009, 26: 31-44.

[117] Liu Haiwen, Zhou Tianjun, Zhu Yuxiang, et al. The strengthening East Asian summer monsoon since the early 1990s[J]. Chinese Sci Bull, 2012, 57: 1553-1558.

[118] Zhou Tianjun, Yu Rucong. Atmospheric water vapor transport associated with typical anomalous summer rainfall patterns in China[J]. J Geophys Res, 2005, 110, doi: 10.1029/2004JD005413.

[119] Zhu Zhiwei, Li T, He Jinhai. Out-of-phase relationship between boreal spring and summer decadal rainfall changes in southern China[J]. J Climate, 2014, 27: 1083-1099.

[120] Zhou Tianjun, Yu Rucong, Chen Haoming, et al. Summer precipitation frequency, intensity, and diurnal cycle over China: A comparison of satellite data with rain gauge observations[J]. J Climate, 2008, 21: 3997-4010.

[121] Zhou Tianjun, Wu Bo, Scaife A A, et al. The CLIVAR C20C Project: Which components of the Asian-Australian monsoon circulation variations are forced and reproducible[J]. Climate Dyn, 2009, 33: 1051-1068.

[122] Li H M, Dai A G, Zhou T J, et al. Responses of East Asian summer monsoon to historical SST and atmospheric forcing during 1950-2000[J]. Climate Dyn, 2010, 34: 501-514.

[123] Yang F L, Lau K M. Trend and variability of China precipitation in spring and summer: Linkage to sea surface temperatures[J]. Int J Climatol, 2004, 24: 1625-1644.

[124] Zhou Tianjun, Zhou Liwei. Understanding the predictability of East Asian summer monsoon from the reproduction of land-sea thermal contrast change in AMIP-type simulation[J]. J Climate, 2010, 23: 6009-6026.

[125] Zhao Ping, Chen Longxun. Climatic features of atmospheric heat source/sink over the Qinghai-Xizang Plateau in 35 years and its relation to rainfall in China[J]. Sci China (Ser. D), 2001, 44: 858-864.

[126] Gao Rong, Zhong Hailing, Dong Wenjie, et al. Impact of snow cover and frozen soil in the Tibetan Plateau on summer precipitation in China[J]. J Glaciol Geocryol, 2011, 33: 254-260. (in Chinese)

[127] Duan Anmin, Wang Meirong, Lei Yonghui, et al. Trends in summer rainfall over China associated with the Tibetan Plateau sensible heat source during 1980-2008[J]. J Climate, 2013, 26: 261-275.

[128] Molnar P, Boos W R, Battisti D S. Orographic controls on climate and paleoclimate of Asia: Thermal and mechanical roles for the Tibetan Plateau[J]. Annu Rev Earth Planet Sci, 2010, 38: 77-102.

Relationship Between the Seasonal Transition of East Asian Monsoon Circulation and Asian-Pacific Thermal Field and Possible Mechanisms[*]

Huang Jiaowen[1,2]　　He Jinhai[1,2]　　Xu Haiming[1,2]　　Jin Qihua[3]

(1. Key Laboratory of Meteorological Disaster, Ministry of Education/Collaborative Innovation Center on Forecast and Evaluation of Meteorological Disasters, Nanjing University of Information Science & Technology, Nanjing 210044, China;

2. College of Atmospheric Science, Nanjing University of Information Science & Technology, Nanjing 210044, China;

3. Chinese Academy of Meteorological Sciences, Beijing 100081, China)

Abstract: The NCEP/NCAR reanalysis, CMAP rainfall and Hadley Centre sea surface temperature (SST) datasets are used to investigate the relationship between the seasonal transition of East Asian monsoon and Asian-Pacific thermal contrast, together with the possible causes. Based on the 250-hPa air temperature over two selected key areas, the Asian-Pacific thermal difference (APTD) index is calculated. Results show that the APTD index is highly consistent with the Asian-Pacific Oscillation (APO) index defined by Zhao et al., in term of different key areas in different seasons. Moreover, the time point of the seasonal transition of the Asian-Pacific thermal contrast can be well determined by the APTD index, indicative of seasonal variation in East Asian atmospheric circulation from winter to summer. The transition characteristic of the circulation can be summarized as follows. The continental cold high at lower tropospheric level moves eastward to the East China Sea and decreases rapidly in intensity, while the low-level northerlies turn to southerlies. At middle tropospheric level, the East Asia major trough is reduced and moves eastward. Furthermore, the subtropical high strengthens and appears near Philippines. The South Asia high shifts from the east of Philippines to the west of Indochina Peninsula, and the prevailing southerlies change into northerlies in upper troposphere. Meanwhile, both the westerly and easterly jets both jump to the north. The seasonal transition of atmospheric circulation is closely related to the thermal contrast, and the possible mechanism can be concluded as follows. Under the background of the APTD seasonal transition, the southerly wind appears firstly at lower troposphere, which triggers the ascending motion via changing vertical shear of meridional winds. The resultant latent heating accelerates the transition of heating pattern from winter to summer. The summer heating pattern can further promote the adjustment of circulation, which favors the formation and strengthening of the low-level southerly and upper-level northerly winds. As a result, the meridional circulation of the East Asian subtropical monsoon is established through a positive feedback between the circulation and thermal fields. Moreover, the time point of this seasonal transition has a significant positive correlation with the SST anomalies over the tropical central-eastern Pacific Ocean, providing a basis for the short-term climate prediction.

Key words: East Asian monsoon; seasonal transition; Asian-Pacific Oscillation (APO); possible mechanism

1 Introduction

The East Asian subtropical monsoon (EASM) has a great impact on the weather and climate over China and surrounding areas, thus becoming one of the hot topics in the study of monsoon. However, the seasonal transition time of EASM circulation and the associated problems have not been clarified yet.

Rainy season onset over East Asian region has been extensively investigated in the literature[1-3]. For example, Ding et al.[1] firstly defined the precipitation in South China region in April as the early summer rainy season. From the perspective of the thermal condition over Tibetan Plateau (TP), some studies pointed out that this transition has already started in late March and early April[4-5]. Besides, the EASM is also considered to burst in late March and early April through investigating East Asian monsoon onset[6-7]. Based on the analysis of the seasonal transition of the zonal thermal contrast between East Asia and the western Pacific, it is suggested that seasonal transition happened over the subtropical region first in late March and early April[8-11]. The low-level northerly wind in winter turned into southerly in summer, accompanied by the emergence of convective precipitation, indicative of the establishment of the East Asian subtropical summer monsoon (EASSM). Obviously, many studies held that the transition of East Asian atmospheric circulation from winter to summer occurred in late March and early April with the onset of EASM, which is earlier than the South China Sea summer monsoon (SCSSM).

On the contrary, some studies argued that the spring rainfall in late March and early April is not related to the EASM[12-13]. Numerical simulation showed that the effect of Eurasian orography on the atmospheric circulation are seasonal dependent[14], which can be classified into winter (November to following April), summer (June to September) and transition (May and October) patterns. In other words, the spring rainfall is classified into the winter type. Clearly, these results admitted that the seasonal transition of East Asian atmospheric circulation occurred in May with the onset of SCSSM, which is earlier than the EASM.

Recently, Wu et al. [15-16] have proposed an original view on the main cause of the coexistence of subtropical monsoon and desert. The continent-scale heating over land and cooling over ocean can induce the ascent (descent) over the east (west) of continents and west (east) of oceans. Thus, local-scale sea-breeze forcing along coastal regions can lead to the formation of the well-defined summertime subtropical "LOSECOD" quadruplet-heating pattern across each continent and adjacent oceans. Moreover, Zhu et al. [17] proposed that the heating type over the EASM region changes from a pattern of being "east warm west cold" to one of being "east cold west warm" during the seasonal transition. Thus, the heating pattern and thermal field are different in winter and summer. Therefore, when and how will the "LOSECOD" heating pattern establish and the pattern of "east warm west cold" change to the "east cold west warm"?

Zhao et al. [18] revealed a zonal seesaw of tropospheric temperature over the Asian-Pacific region in the mid-latitudes during the boreal summer, named the Asian-Pacific Oscillation (APO). That is, the troposphere cooling occurred over the Asian continent in the mid-latitudes, while warming in the central and eastern North Pacific, and vice versa. Furthermore, the APO is also found in other seasons [19]. The APO plays an important role in the linkage between the Pacific Ocean and the Asian continent and is highly related to the Asian monsoon and rainfall [19-21], the tropical cyclone (TC) activities [22-25], and El Nino-Southern Oscillation (ENSO) [26-28]. Since the APO index indicates the zonal difference between the Asian and Pacific thermal fields, it is a good way to describe the thermal features over the Asian-Pacific region and study its interannual variation. However, problems referring to the seasonal transition cannot be focused on. As a result, whether or not the APO index in different seasons can be defined by selecting the same key regions to study the seasonal variation of East Asian monsoon circulation?

Toward this goal, the transition time point of the thermal fields in the Asian-Pacific region is calculated objectively in this study. The relationship between the seasonal transitions of East Asian monsoon circulation and Asian-Pacific thermal field is further discussed with focus on the transition from winter to summer, together with the possible mechanism involved.

2 Data and methods

The daily atmospheric reanalysis data, including air temperature, wind, geopotential height and sea level pressure (SLP) are taken from the National Centers for Environmental Prediction-National Center for Atmosphere Research (NCEP-NCAR) reanalysis with 2.5° horizontal resolution. The monthly SST from the Hadley Centre Sea Ice and Sea Surface Temperature dataset (HadISST) with 1.0° horizontal resolution and the pentad rainfall from the Climate Prediction Center (CPC) Merged Analysis of Precipitation (CMAP) with 2.5°

horizontal resolution are also used in this study. The datasets above are all available from January 1979 to December 2014, and the pentad atmospheric data is calculated from the daily NCEP-NCAR reanalysis.

An Empirical Orthogonal Function (EOF) analysis with area weighting is conducted to detect a teleconnection pattern. Correlation and regression analyses are applied to examine the relationships between different variables. The atmospheric heating is estimated using the NCEP/NCAR reanalysis data based on the thermodynamic equation as the following:

$$Q1 = c_p \left[\frac{\partial T}{\partial t} + \mathbf{V} \cdot \nabla T + \left(\frac{p}{p_0} \right)^k \omega \frac{\partial \theta}{\partial p} \right]$$

where T is air temperature, t is time, p is pressure, p_0 (=100 hPa) is the reference pressure, c_p is the specific heat at constant pressure, $k = R/c_p$ with R the gas constant of dry air, $\mathbf{V}$ is the horizontal wind vector, ω is vertical p-velocity, and θ is potential temperature. Column-integrated atmospheric heating $Q1$ is obtained by integrating the above equation from the surface (P_s) to 100-hPa.

3 Definition of Asian-Pacific thermal difference index

The eddy temperature, defined as the deviation of temperature from the zonal mean, is used to find the key regions of Asian-Pacific thermal difference (APTD) index. As shown in Fig. 1a and Fig. 1b, Asian continent is warmer (colder) than the Pacific with positive (negative) values over most Asian continent and negative (positive) values over the Pacific, forming an "east cold west warm" ("east warm west cold") pattern in summer(winter). Moreover, temperature anomalies over the continent and ocean are centered near 250-hPa in summer and winter. To select key regions for describing the seasonal transition, the differences of eddy temperature between summer and winter at 250-hPa are given in Fig. 1c. It can be clearly seen that temperature anomalies appear over the Asian continent while negative ones over the Pacific, indicating the largest differences between summer and winter thermal conditions. Thus, the anomalous positive center over Asia (60°~110°E, 15°~45°N) and negative center over Pacific (160°E~140°W, 15°~45°N) are chosen as two key regions (Fig. 1c). The difference of air temperature between these two key regions is defined as the APTD index:

$$I_{APTD} = T(60°\sim110°E, 15°\sim45°N) - T(160°E\sim140°W, 15°\sim45°N)$$

The APTD index is highly correlated with the APO index with significant correlation coefficients of 0.93, 0.88, 0.98 and 0.91 in spring, summer, autumn and winter respectively, indicating the APTD index agree well with the APO index (Fig. 2). Therefore, the modified APTD indexes in different seasons are simply calculated with the same key regions and can be used to monitor and predict the seasonal transition.

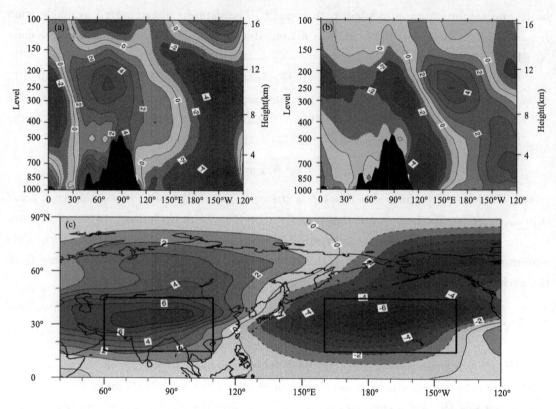

Fig. 1 Longitude-height cross section along 35°N of eddy temperature (°C) over the Asian-Pacific region in (a) summer and (b) winter (the black shaded areas denote mountains); (c) The differences of eddy temperature between summer and winter (°C) at 250-hPa (two boxes represent the key regions for Asia and Pacific, respectively)

4 Determination and features of the seasonal transition time

Since the APTD index indicates the zonal thermal difference between the Asian and Pacific thermal fields, the annually pentad APTD index can be also calculated. Then, the time point of seasonal transition can be ascertained by the corresponding cumulative curve. Take the year of 1979 for example to illustrate how to determine the seasonal transition time (Fig. 3). The APTD index in 1979 shows a dominant seasonal reverse, representing the shift of thermal pattern between "east warm west cold" and "east cold west warm". As a result, the time point of APTD index changing between positive and negative correspond to the seasonal transition time of thermal field and heating pattern. Because the pentad APTD index is unstable around zero during the transition, it is difficult to determine the seasonal transition time accurately. Therefore, the minimum (maximum) of the corresponding cumulative curve represent the APTD index shifting from negative (positive) to positive (negative) values, namely, the transition point from winter (summer) to summer (winter) labeled as TPWS (TPSW).

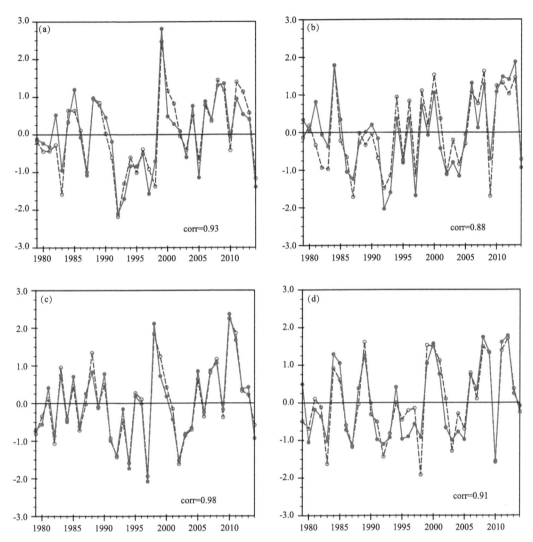

Fig. 2 Normalized time series of the APTD index (red solid line) and APO index (blue dashed line) in (a) spring, (b) summer, (c) autumn and (d) winter. The APO index was defined by Zhao et al. [19] as the time series of the first principal component of EOF which is performed on the anomalies of each season upper-tropospheric eddy temperature over the Asian-Pacific region

Accordingly, interannual variations of the TPWS and TPSW can be obtained (Fig. 4a). To further explore the features of the transition points, Fig. 4b portrays the climatological evolution of the APTD index and several meteorological indicators related to the EASM, including 925- and 100-hPa meridional winds and their vertical shear, column integrated atmospheric heating and pentad rainfall anomalies. The results show that the meridional winds exhibit an obvious seasonal cycle, that is, the 925-hPa northerly wind turns to southerly wind in pentad 14, while the opposite is found at 100-hPa in pentad 22. Correspondingly, the vertical shear of meridional winds shows an inverse change from winter to summer around pentad 18. Besides, the APTD index, atmospheric heating and pentad rainfall anomalies over

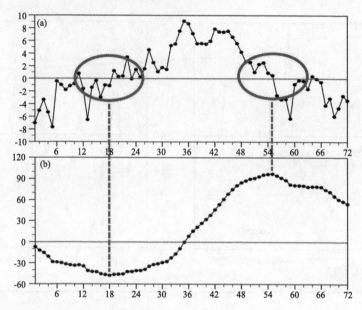

Fig. 3 The pentad evolution of (a) APTD index and (b) its related cumulative curve in 1979

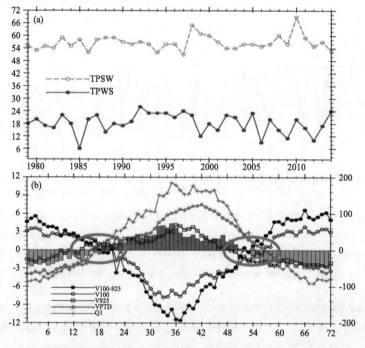

Fig. 4 (a) Time series of the TPWS (blue solid line) and TPSW (green dashed line) (the vertical axis corresponding to the transition time; pentad). (b) Climatological evolution of the meridional wind at 925-hPa (hollow square; m·s^{-1}), 100-hPa (hollow circle; m·s^{-1}) and their vertical shear (solid circle; m·s^{-1}), APTD index (red prismatic; ℃), column-integrated atmospheric heating (red cross; W·m^{-2}) and rainfall anomaly (column bar; mm) over the EASM region (20°~35°N, 110°~120°E)

the EASM region also show the seasonal variation, with negative values to positive around pentads 14-16. Comparing the seasonal transition time of atmospheric circulation with thermal field, it can be found that seasonal transition from winter to summer occurs in late March and early April. Note that, low-level southerly wind firstly appears around pentad 14, suggesting that the appearance of low-level southerly wind occurred earlier than that of thermal field and high-level northerly wind.

From the analysis above, the seasonal transition of thermal field is consistent with atmospheric circulation field. Therefore, how does the East Asian atmospheric circulation change? What is the connection between these changes and the thermal field during the transition process? Thus, we will focus on the transition period from middle March to early April, and discuss circulation features and processes of seasonal transition from winter to summer in the following section.

5 Circulation features and processes of seasonal transition

5.1 Circulation features

During the seasonal transition from winter to summer, the reverses of geopotential height and meridional winds at the 100-hPa and 500-hPa are obvious. As shown in Fig. 5a, southerly wind prevails over the East Asian subtropical region, which is located in the northwest of the anticyclone over Philippines. After the transition, this anticyclone strengthens and moves eastward to the west of Indochina Peninsula, leading to the southerly wind replaced by the northerly wind (Fig. 5b). Then, this anticyclone moves to the Tibet Plateau in late May and becomes the South Asia High[29]. From Fig. 5c and Fig. 5d, it can be observed that the East Asia major trough weakens and moves eastward in middle and high latitudes. In the low latitudes, the high pressure belt turns into anticyclonic circulation with a closed center and the Western Pacific subtropical high appears near the Philippines.

Similarly, SLP and low-level wind also represent a clear seasonal variation from winter to summer (Fig. 5e and Fig. 5f). The strong Aleutian Low and continental cold high in winter are weakened. This continental cold high moves eastward to the sea, and the subtropical high strengthens and extends westward. Therefore, the SLP over Asian-Pacific region shifts from "east low and west high" to "east high and west low", as a result, low-level northerly wind turns into southerly wind.

5.2 Processes

The process of seasonal transformation can be reflected in the seasonal variation of upper tropospheric easterly and westerly jet (Fig. 6). As shown in Fig. 6a, the westerly jet exhibits a feature of "strong in winter and weak in summer". It maintains near 30°N stably with its strength exceeding 30 m · s^{-1} before pentad 18. When the westerly jet pushes northward to

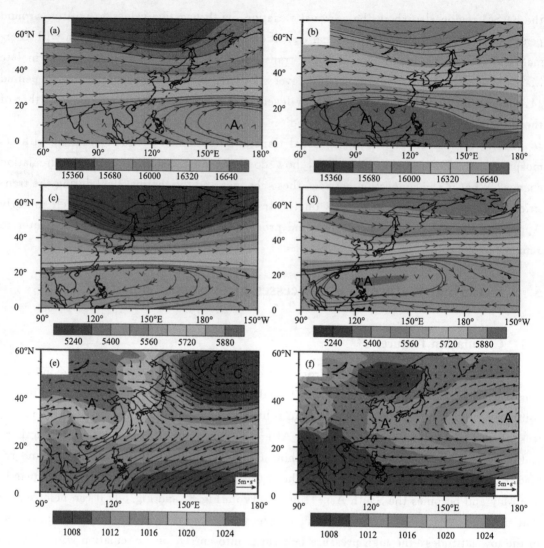

Fig. 5 The geopotential height (shaded; gpm) and horizontal winds (vector; m · s^{-1}) at (a, b) 100-hPa, (c, d) 500-hPa and (e, f) the SLP (shaded; Pa) and 10-m wind (vector; m · s^{-1})

40°N approximately in pentad 42, its strength decreases 15 m · s^{-1} and its southern boundary moves from 20°N to 30°N. Similarly, the easterly jet is located in the Southern Hemisphere with its intensity exceeding 15 m · s^{-1} before pentad 18 (Fig. 6b). Then it suddenly weakens and moves northward to 15°N after pentad 18. These results confirm that the northward shifting processes of westerly and easterly jets can be observed in late March and early April, which is consistent with previous studies [30].

The composite results (taking the TPWS as the reference point) can also reveal the similar processes of seasonal change of high-level jets (Fig. 6c and Fig. 6d). The westerly jet maintains near 30°N stably with its intensity greater than 30 m · s^{-1} before the TPWS. After then, it pushes northward to 40°N and weakens, together with its south boundary moving from 20°N to 30°N. Similarly, the easterly jet suddenly weakens and moves northward from the Southern Hemisphere to 15°N after the TPWS. As a result, the TPWS can well represent the northward shifting process of high-level jets in the transition period.

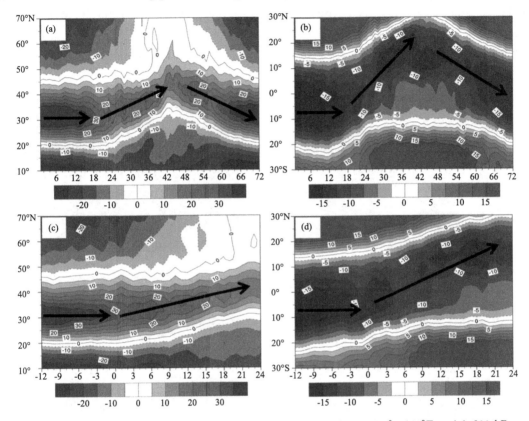

Fig. 6 Seasonal variation of meridional deviation of mean zonal wind over 110°~140°E at (a) 200-hPa, and (b) 100-hPa; The composite results of meridional deviation of mean zonal wind over 110-140°E at (c) 200-hPa, and (d) 100-hPa (Taking the TPWS as the reference point, zero of x-axis represents the reference point, the positive and negative valves represent the time early and later than that of reference point)

To further examine whether the TPWS can represent the seasonal transition of atmospheric circulation, SLP and 10-m wind are composited by using the TPWS as a reference point (Fig. 7). The atmospheric circulation pattern over Asian-Pacific region is a typical winter pattern in pentad-9, including the Aleutian low, the East Asian continental cold high and the East Pacific subtropical high. Consequently, SLP shows a spatial pattern of "east low and west high". In pentad-6, the continental cold high gradually weakens and moves eastward but its center still stays over the continent, thus the northerly wind over eastern China begin to weaken. When the mainland cold high moves eastward to the East China Sea in pentad-3, the northerly wind has already been replaced by the southerly wind. After the TPWS (pentad 0), the cold high merges into the western Pacific subtropical high and the southerly wind dominates the East Asian subtropical region. Meanwhile, the Aleutian low also weakens during the transition period and is divided into two parts moving to the west and east respectively.

In summary, the TPWS can well indicate the transformation processes of SLP and low-level circulation in the transition period. Besides, the composite results also shows that the northerly wind has already been replaced by the southerly wind before the TPWS, suggesting that the low-level southerly wind appears earlier than the seasonal transition of high-level thermal field. These features are generally consistent with the climatological transformation features shown in Fig. 4.

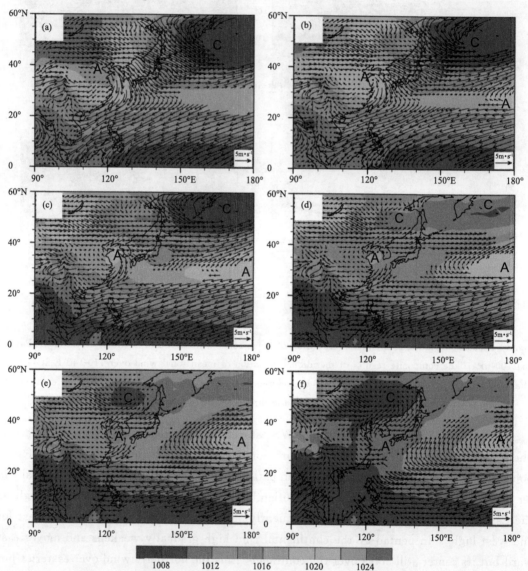

Fig. 7 The composite results of SLP field (shaded; hPa) and 10-m wind field (vectors significant at the 95% confidence level; m·s^{-1}) in different periods by taking the TPWS as the reference point. (a), (b), (c), (d), (e) and (f) represents the −9, −6, −3, 0, 3, and 6 pentad. This numbers indicate the average of three pentads nearby and the positive and negative values represent the time early and later than that of reference point

Fig. 8 exhibits the composites of vertical heating profiles over the East Asian subtropical region by taking the TPWS as the reference point. The profile in January (July) represents typical winter (summer) heating pattern. Both vertical profiles below 850-hPa are positive due to the sensible heat near the surface, and then rapidly weaken with height. The winter heating pattern generally exhibits atmospheric cooling above 850-hPa, with a value around $1 \sim 2$ ℃ · d^{-1}. On the contrary, the summer heating pattern shows atmospheric warming above 850-hPa, with a heating center of about 3 ℃ · d^{-1} at 400-hPa. According to the theory of Wu et al.[31-32], strong latent heating induces the northly (southerly) wind above (below) the heating center over the EASM region, indicating that the meridional wind is related to the vertical shear of atmospheric heating. Therefore, the eastablishment of summer heating pattern favors the formation and enhancement of meridional monsoon circulation.

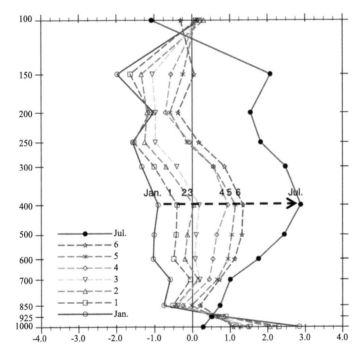

Fig. 8 As Fig. 7, but for the vertical heating profiles over the East Asian subtropical region (20°~35°N, 110°~120°E) (Blue solid line: Jan., red solid line: Jul., dashed line 1~6 represents the pentad $-9 \sim 6$; ℃ · d^{-1})

Furthermore, it can be seen that the atmosphere heating patterns in pentads -9 and -6 are similar to the winter pattern. Compared with the typical winter heating profile, the cooling rate is almost unchanged in the upper troposphere and it reduces to around zero in the middle and lower troposphere, indicating that the seasonal transformation does not start yet. In pentad -3, it can be found that the cooling rate weakens and shifts from negative to weak positive value at middle and lower layers. In pentad 0, the heating rate turns to positive below 300-hPa and increases rapidly with a center value of 1 ℃ · d^{-1} at 400-hPa, representing the seasonal transition of atmosphere heating pattern. Besides, in pentads 3 and 6, summer heating further strengthens and gradually develops to the typical summer pattern, and the upper-level heating rates turns into positive. These results suggest that the TPWS can well reflect the transformation process of atmosphere heating pattern in the transition period.

According to the theory of "thermal adaptation" proposed by Wu et al.[16], the quasi-geostrophic vorticity equation can be simplified due to the vorticity advection is small. Thus, the positive correlation between the vertical motion and vertical shear of meridional wind can be inferred as:

$$w \propto -\frac{\partial v}{\partial z}$$

where w is the vertical motion, v is the meridional wind.

In pentads -9 and -6 (Fig. 9a and Fig. 9b), the northerly wind prevails in the shallow lower layers while the southerly wind in the upper levels. Consequently, the descent appears over the EASM region, corresponding to the positive vertical shear of meridional wind. In pentad -3 (Fig. 9c), the surface northerly wind begin changing to the southerly wind and the vertical shear gradually weakens, favoring the enhancement of vertical velocity. Therefore, the descent weakens and the heat source in the lower troposphere shows a weak heating effect, indicative of the seasonal transition of heating pattern. In pentad 0 (Fig. 9d), the southerly wind dominates the entire troposphere and the descent further weakens. Meanwhile, the winter heating pattern obviously transforms into summer pattern with a heating center in the middle troposphere (pentad 0 in Fig. 8). In pentads 3 and 6, the upper (lower) troposphere has been controlled by the northerly (southerly) wind, and the descent appears over the EASM region, with the vertical shear from positive to negative. By this time, the atmospheric heat source in the troposphere shows a heating effect, suggesting the establishment of meridional monsoon circulation. Note that there are two separated heating center existing over the equatorial region and the EASM region, indicating that the EASM and tropical summer monsoon circulation are separated. The above analysis indicates that the meridional monsoon circulation has been gradually established followed by the seasonal transition of heating pattern from winter to summer.

5.3 Possible mechanism

As discussed above, a high relationship can be observed between the seasonal transition of the East Asian monsoon circulation and Asian-Pacific thermal field although they do not change simultaneously. When the sun passes from the Southern Hemisphere to the Northern Hemisphere, the surface thermal contrast induces rapidly increased sensible heat over the East Asian continent, which generates anomalous gradients of diabatic heating related to precipitation[31]. As a result, the low-level cold high is weakened and moves eastward into the sea, and the southerly wind firstly appears over the East Asian subtropical region. Meanwhile, the sign of vertical wind shear is changed due to the low-level southerly wind, inducing vertical velocity and rainfall. The resultant latent heat acts to further warms the upper

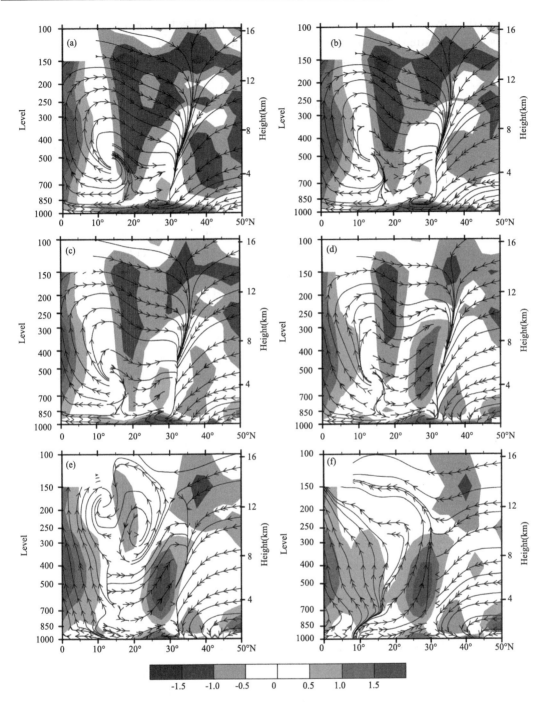

Fig. 9 As Fig. 7, but for the meridional circulation and atmospheric heat source (shaded; ℃·d^{-1}) along 110°~130°E

troposphere, leading to the heating pattern transforms from the winter type to the summer. Moreover, according to the theory of Wu et al.[32-33], the summer heating pattern with a center at middle troposphere not only induces the low-level southerly wind but also strengthens the upper-tropospheric northerly wind. These changes further shift the vertical wind shear from positive to negative to enhance the ascending motion, the resultant latent heat strengthens the summer heating pattern. Therefore, a positive feedback mechanism is formed between the thermal field and circulation. When the circulation is adjusted from the middle to upper layer, the upper-level southerly wind gradually is weakened and turns into the northerly wind, resulting in the upward motions appearing in the entire troposphere. The upper-tropospheric circulation also transforms from the winter type to the summer type. Finally, meridional monsoon circulation is built up, characterized by the southerly (northerly) wind at lower (upper) troposphere.

Besides, from the aspect of temperature and circulation, the transition of Asian-Pacific thermal field from "east warm west cold" to "east cold west warm" corresponds to that of pressure from "east low west high" to "east high west low". The shift of the air pressure gradient provides a large-scale background for the adjustment of circulation. Changes of the low-level wind from the southerly wind to the northerly wind are corresponding to the transition of advection from cold to warm. Thus, the southerly wind transports heat from the low latitudes, inducing positive feedback in the 'east cold west warm' pattern.

In summary, under the background of the APTD seasonal transition, the low-level southerly wind is firstly strengthened. Then, East Asian monsoon circulation and Asian-Pacific thermal field transform from the winter pattern to the summer pattern via the adjustment and adaption processes between thermal and circulation fields.

6 A predicted signal for the transition time from winter to summer

According to the analysis above, the TPWS can objectively represent the seasonal transition time and well represent the transformation of atmospheric circulation and thermal field. What's more, the seasonal transition time is related to rainy season in China, thus the TPWS is important to weather and climate prediction in China. Therefore, whether or not this time point can be predicted?

Fig. 10a exhibits the correlation coefficients between the TPWS and the previous winter SST. Significantly positive correlations can be found over the tropical eastern Pacific region, indicating that the early winter SST anomalies over this region may be considered as a preceding signal for the TPWS. The average of previous winter SST anomalies in tropical eastern Pacific region (180°~120°W, 10°S~10°N) is defined as a mean anomalous SST index. There is a positive relationship between that index and the TPWS (Fig. 10b) with a correlation coefficient of 0.56 (significant at the 99.9% confidence level). This high correlation indicates that when El Niño(La Niña) occurs in the previous winter with warm (cold) SST in tropical eastern Pacific, the seasonal transition of East Asian monsoon circulation would be later (earlier).

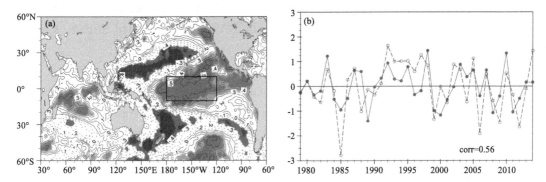

Fig. 10 (a) Correlation coefficients between the TPWS and the previous winter SST during 1979-2014 (Shaded exceeding the 95% and 99% confidence level; the black box representing the key region of the anomalous SST index). (b) Normalized time series of the TPWS (blue dashed line) and the mean anomalous SST index (red solid line) in previous winter over tropical eastern Pacific with their linear trends removed

7 Summary and discussion

(1) The areas of (60°~110°E, 15°~45°N) and (160°E~140°W, 15°~45°N) are selected to represent the key regions of the Asian continent and Pacific. The pentad Asian-Pacific thermal difference (APTD) indexes are calculated by the differences of 250-hPa air temperature between these two regions. The APTD index is highly correlated with the APO index defined by Zhao et. al using different key areas in different seasons. Therefore, this modified APTD index is simply calculated with the same key regions in different seasons and can be used to study the seasonal transition.

(2) The seasonal transition time point determined by the APTD index can well represent the transition of East Asian atmospheric circulation from winter to summer. The transition features of circulation can be summarized as follows. The continental cold high at lower troposphere moves eastward to the East China Sea and decreases rapidly in intensity, and the low-level northerly wind turns to the southerly wind. In the middle troposphere, the East Asia major trough is reduced and moves eastward. Furthermore, the subtropical high strengthens and appears near Philippines. The South Asia high shifts westward from the east of Philippines to the west of Indochina Peninsula, and the upper-level southerly wind changes into northerly wind. Meanwhile, both the westerly and easterly jets jump to the north.

(3) The seasonal transition of atmospheric circulation is closely related to the thermal contrast, and the possible mechanism can be concluded as follows. Under the background of the APTD seasonal transition, the southerly wind appears firstly at lower troposphere, which triggers the ascending motion via changing vertical shear of meridional winds. The resultant latent heating accelerates the transition of heating pattern from winter to summer. The summer heating pattern can further promote the adjustment of circulation, which favors

the formation and strengthening of the low-level southerly wind and the upper-level northerly wind. As a result, the meridional circulation of the EASM is established through a positive feedback between the circulation and thermal fields.

(4) There is a positive relationship between the TPWS and previous winter SST anomaly in tropical eastern Pacific, indicating that when El Niño (La Niña) occurs in the previous winter with warm (cold) SST in tropical eastern Pacific, the seasonal transition of East Asian monsoon circulation would be later (earlier). This high correlation provides a basis for the short-term climate prediction.

In this study, we analyzed the relationship between the seasonal transition of East Asian monsoon circulation and Asian-Pacific thermal field from winter to summer and possible mechanisms. However, the mechanism of positive relationship between the TPWS and previous winter SST anomaly in tropical eastern Pacific are not well addressed and the situation of seasonal transition from summer pattern to winter pattern is still not quite clear. In our future work, we will utilize atmosphere general circulation models to better understand the formation mechanism of the EASM and the mechanism of tropical SST affecting the seasonal transition.

References

[1] Ding Yihui, Chen Longxun, Murakami M. Asian monsoon [M]. Beijing: China Meteorological Press, 1994:1-263 (in Chinese).

[2] Chen Longxun, Li Wei, Zhao Ping, et al. On the process of summer monsoon onset over East Asia[J]. Clim Environ Res, 2000, 5(4):345-355 (in Chinese).

[3] Wang Zunya, Ding Yihui. Climatic characteristics of rainy seasons in China [J]. Chin J Atmosph Sci, 2008, 32(1):1-13 (in Chinese).

[4] Yao Yonghong, Wang Qianqian, Ge Zhaoxia. Climatological features of decade sensible heat flux over the Tibetan Plateau and its surrounding area [J]. J Nanjing Inst Meteorol, 2000, 23(3):404-411 (in Chinese).

[5] Wang Tongmei, Wu Guoxiong, Wang Rijin. Influence of the mechanical and thermal forcing of Tibetan Plateau on the circulation of the Asian summer monsoon area [J]. Plat Meteorol, 2008, 27(1):1-9 (in Chinese).

[6] Xu Guoqiang, Zhu Qiangen, Ran Yufang. Analyses of features and mechanisms of summer monsoon onsets over SCS and its vicinity in 1998 [J]. J Appl Meteorol Sci, 2002, 13(5):535-549 (in Chinese).

[7] Jian Maoqiu, Luo Huibang, Qiao Yunting. Seasonal variability of atmospheric heat sources over the Asian-Australian monsoon region [J]. Acta Sci Nat Univ Sunyatseni, 2004, 43(3):106-109 (in Chinese).

[8] Qi Li, He Jinhai, Zhang Zuqiang, et al. Seasonal cycle of the Zonal land-sea thermal contrast and East Asian subtropical monsoon circulation[J]. Chinese Science Bulletin, 2008, 53(1):131-136.

[9] He Jinhai, Zhao Ping, Zhu congwen, et al. Discussions of some problems as to the East Asian subtropical monsoon [J]. Acta Meteorol Sin, 2008, 22(4):619-634.

[10] He Jinhai, Qi Li, Liu Danni, et al. Transition of East Asian general circulation from winter to summer and its features [J]. Sci Meteorol Sin, 2010, 30(5):591-596 (in Chinese).

[11] Zhu Congwen, Zhou Xiuji, Zhao Ping, et al. Onset of East Asian subtropical summer monsoon and rainy season in China [J]. Sci in China (Ser D), 2011, 54(12):1845-1853.

[12] Tian S F, Yasunari T. Climatological aspects and mechanism of spring persistent rains over central China [J]. J Meteor Soc Japan, 1998, 76(1):57-71.

[13] Wan Rijin, Wu Guoxiong. Mechanism of spring persistent rains over southeastern China[J]. Sci in China (Ser D), 2007, 50(1):130-144.

[14] Wu Guoxiong, Wang Jun, Liu Xin, et al. Numerical modeling of the influence of Eurasian orography on the atmospheric circulation in different seasons [J]. Acta Meteorol Sin, 2005, 63(5):603-612 (in Chinese).

[15] Wu Guoxiong, Liu Yiming, Zhu Xiaying, et al. Muti-scale forcing and the formation of subtropical desert and monsoon[J]. Ann Geophys, 2009, 27:3631-3644.

[16] Wu Guoxiong, Liu Yiming, Yu Jingjing, et al. Modulation of land-sea distribution on air-sea interaction and formation of subtropical anticyclones [J]. Chin J Atmosph Sci, 2008, 32(4):720-740 (in Chinese).

[17] Zhu Zhiwei, He Jinhai, Qi Li. Seasonal transition of East Asian subtropical monsoon and its possible mechanism[J]. J Trop Meteor, 2012, 18(3):305-313.

[18] Zhao Ping, Zhu Yani, Zhang Renhe. An Asian-Pacific teleconnection in summer tropospheric temperature and associated Asian climate variability [J]. Clim Dyn, 2007, 29:293-303.

[19] Zhao Ping, Chen Junming, Xiao Dong, et al. Summer Asian-Pacific oscillation and its relationship with atmospheric circulation and monsoon rainfall[J]. Acta Meteorol Sin, 2008, 22(4):455-471.

[20] Zhou Botao, Zhao Ping. Modeling variations of summer upper tropospheric temperature and associated climate over the Asian-Pacific region during the mid-Holocene[J]. J Geophys Res, 2010, 115 (D20):898-907.

[21] Liu Ge, Zhao Ping, Chen Junming, et al. A precursory signal for June precipitation anomalies over the middle and lower reaches of Yangtze River-the Asian-Pacific Oscillation [J]. Acta Meteorol Sin, 2012, 70 (5):1064-1073 (in Chinese).

[22] Zhou Botao, Cui Xuan, Zhao Ping. Relationship between the Asian-Pacific oscillation and the tropical cyclone frequency in the Northwestern Pacific [J]. Sci in China (Ser D), 2008, 51(3):380-385.

[23] Zou Yan, Zhao Ping. Relation of summer Asian-Pacific oscillation to tropical cyclone activities over the coastal waters of China [J]. Acta Meteorol Sin, 2010, 24 (5):539-547.

[24] Zou Yan, Zhao Ping. A study of the relationship between the Asian-Pacific oscillation and tropical cyclone activities over the coastal waters of China during autumn [J]. Acta Meteorol Sin, 2011, 69 (4):601-609 (in Chinese).

[25] Zou Y, Zhao P, Lin Q. Asian-Pacific oscillation in autumn and its relationships with the subtropical monsoon in East Asia [J]. J Trop Meteorol, 2015, 21(2):143-152.

[26] Nan Sulan, Zhao Ping, YANG Song, et al. Springtime tropospheric temperature over the Tibetan Plateau and evolutions of the tropical Pacific SST[J]. J Geophys Res, 2009, 114(D10):895-896.

[27] Zhao Ping, Yang Song, JIAN Maoqiu, et al. Relative controls of Asian-Pacific summer climate by Asian land and Tropical-North Pacific sea surface temperature[J]. J Climate, 2011, 24(15):4165-4188.

[28] Liu Ge, Zhao Ping, CHEN Jun-ming, et al. Preceding Factors of Summer Asian-Pacific Oscillation and the Physical Mechanism for Their Potential Influences[J]. J Climate, 2015, 28(7):2531-2543.

[29] Liu BOqi, He Jinhai, WANG Lijuan. Characteristics of the South Asia high establishment processes above the Indo-China Peninsula from April to May and their possible mechanism [J]. Chin J Atmosph Sci, 2009, 33(6):1319-1332 (in Chinese).

[30] Chi Yanzhen, Liu Danni, He Jinhai, et al. Features of transformation on circulation and thermal fields over East Asian-Pacific regions from winter to summer and its possible mechanism [J]. Plat Meteorol, 2013, 32(4):983-992 (in Chinese).

[31] Ke Dan, Guan Zhaoyong. Variations in regional mean daily precipitation extremes and related circulation anomalies over central china during boreal summer [J]. J Meteor Res, 2014, 28(4):524-539.

[32] Wu Guoxiong, Liu Yiming, Liu Ping. The effect of spatially nonuniform heating on the formation and variation of subtropical high part I: scale analysis [J]. Acta Meteorol Sin, 1999, 57 (3):257-263 (in Chinese).

[33] Liu Yiming, Wu Guoxiong, Liu Hui, et al. The effect of spatially nonuniform heating on the formation and variation of subtropical high part Ⅲ: condensation heating and south Asia high and western Pacific subtropical high [J]. Acta Meteorol Sin, 1999, 57 (5):525-538 (in Chinese).

何金海论文选

第二部分 贰

季风气候预测

1979年6月东亚和南亚上空的水汽通量*

何金海   T. 村上多喜雄
（美国夏威夷大学气象系）

摘要：1979年随着印度季风的建立，亚洲季风区上空的水汽输送场经历了非常显著的变化。一条强水汽输送带从阿拉伯海经印度南部、孟加拉湾和南海，然后向北进入华南和日本。由于垂直积分的水汽通量（Q_2）的强烈辐合而形成明显的降水。6月16—20日，孟加拉湾和南海 Q_2 辐合突然加强。同时，在上述两地区和菲律宾，向外长波辐射（OLR）急剧下降，对流指数（I_c）迅速上升。这些变化表明，季风在印度、孟加拉湾、南海和菲律宾几乎是同时建立的。在东亚，6月5日左右，受西太平洋副高控制的强水汽输送带在日本南部建立。同时，Q_2 辐合加强，OLR下降，I_c 上升。这些变化表明，日本雨季早在印度季风建立以前约两个星期就开始了。季风前后两个时期 Q_2 的较差也指出有两条水汽通道。一条与南亚季风相联系，另一条则与东亚季风相联系，对 Q_2、OLR 和 I_c 的分析表明，1979年东亚与南亚季风是相对独立的系统。

1 引言

大气环流输送水汽，而水汽又通过释放潜热影响大气环流的型式。Sutcliffe[1]指出，正确估计水分平衡有利于进一步了解大气环流，分析水汽输送的演变可以提供有关大气环流的若干有用的信息。

不少作者对亚洲季风区域上空的水汽输送进行了研究。Flohn[2]研究指出，来自东亚大陆的水汽输送是很大的；在潮湿地区，陆地的蒸发比海洋上的蒸发略小些。Murakami[3]发现，东亚水汽输送在很大程度上受两股气流的控制，即700 hPa上的印度西南季风和环绕副高的近地面东南气流，前者在东亚季风的早期阶段起主要作用，后者则在晚期阶段起主要作用。徐淑英[4]研究了中国东部的水汽输送和水分平衡指出，冬、夏的水汽输送有显著的不同，但无论冬夏，南面流入的水汽均比从西面和北面来的重要得多。

Saitio[5]分析了亚洲季风区域的水汽输送发现，起源于印度的西南季风气流在850 hPa 或 700 hPa 最强，而副高周围的东南季风气流风速向上增加。Asakura[6]发现，潮湿气流通常起源于太平洋副高的内部，当它抵达副高外部边缘时变得更加强烈。

Saha[7]指出，在 42°~75°E 之间，有显著的穿越赤道进入阿拉伯海的向北水汽输送，然后随着盛行的低层西风带向东输送。Cadet 等[8]发现，在阿拉伯海，季风的建立伴随着地面水汽

---

* 本文发表于《南京气象学院学报》，1983年第2期，159-173。

通量的显著增加,穿越印度西海岸的水汽输送的大约70%起源于南半球,其余的由阿拉伯海上的蒸发供给。Rao等[9]对强季风和弱季风期间的水汽输送进行了对比分析,他们发现弱季风期间,纬向水汽输送有显著的减少,而强季风期间,沿着印度东北部的向北经向输送几乎增加一倍。

上面列举的关于大尺度水汽输送通量的研究,大多数都集中于水汽输送本身的长期或季节的平均状况,然而水汽输送通量的散度场对于较短周期的季风环流的降水和潜热的释放应有更大的重要性。我们应用FGGE资料研究了水汽输送通量本身的变化,还研究了相应的水汽输送散度场的变化特点,并着重研究了若干区域同季风建立紧密关联的水汽通量辐合的突然变化。这些区域包括印度东北部、菲律宾东海岸和日本等地区。

2 资料和计算程序

本文使用1979年6月1—30日一天一次的u, v, H, T和r在六个层次(即300、400、500、700、850和1000 hPa)上的资料。这些资料取自欧洲中期天气预报中心(ECMRWF)制作的FGGE Level Ⅲ b资料。计算范围是从56.25°E~176.25°W和18.75°S~52.5°N,格距为3.75°经纬度。

我们首先根据相对湿度r和温度T计算每一个标准等压面上的比湿q,然后计算各层各网格点上纬向和经向水汽通量qu和qv,最后计算各网格点上的垂直积分的水汽通量,即

$$\langle qu \rangle = \frac{1}{g}\int_{300}^{P_0} qu \, dP$$

$$\langle qv \rangle = \frac{1}{g}\int_{300}^{P_0} qv \, dP$$

式中,P_0是洋面或地面气压。

垂直积分的水汽通量散度由下式计算

$$Q_2 = \frac{\partial \langle qu \rangle}{a\cos\varphi \partial\lambda} + \frac{\partial \langle qv\cos\varphi \rangle}{a\cos\varphi \partial\varphi} \tag{1}$$

式中,φ和λ分别为纬度和经度。

本文还采用1979年6月每天9时(标准地方时)各网格点的向外长波辐射(OLR)资料,它由NOAA极轨卫星扫描辐射仪测得,一般在有云层的地区OLR较小,云层越厚。OLR越小,可惜某些地区,某几天的资料遗缺。

此外,也采用了6月逐日平均的日本卫星的"红外"资料,由这些资料组成对流活动强度指数(I_c), Murakami M (1983)将它定义成

$$I_c = \frac{T_{BB} - T_{400}}{T_{tr} - T_{400}} \times 100 \tag{2}$$

式中,T_{BB}表示云顶的黑体温度,T_{tr}和T_{400}分别表示对流层顶和400 hPa的温度。对流发展越旺盛,云顶越高,T_{BB}越低,从而I_c值越大。资料范围从80°~180°E和30°S~40°N,格距为2°经纬度。

3 季风区域水汽输送的变化

为了研究水汽输送场的特点,我们选择两个时段进行对比分析,即6月1—10日和6月

20—30 日。由于 1979 年印度中部的季风建立日期被确定为 6 月 19 日,为了方便起见,我们称前一个时段为"季风前位相",后一个时段为"季风后位相"。必需指出,季风的建立日期在亚洲的各个区域差别是很大的,这在后面还要进一步讨论。我们分别计算了"季风前位相"和"季风后位相"的垂直积分的平均水汽通量场(图 1 和图 2),同时还计算了两个时段平均垂直积分水汽通量的散度场(Q_2)、OLR 和 I_c 场(图略)。

在"季风前位相"图 1 上,季风区域有三股主要的水汽输送,第一股是来自蒙古的向南输送,第二股是从孟加拉湾南部向中国南部的输送,第三股则是沿着太平洋副高边缘的水汽输送。其中第一股输送最弱,而第三股输送最强。这三股输送在南海、华南和日本地区汇合,特别在华南和日本南部有显著的水汽输送的辐合中心。与辐合中心相配合有 OLR 的低值中心和 I_c 的高值中心,反映了这些区域的强烈对流活动。这表明华南和日本南部的雨季的建立要比印度中部季风的建立早得多。与此成为鲜明对照的是在印度却为水汽输送的辐散区,显著高值的 OLR 区和低值的 I_c 区显示了印度季风建立前的干旱天气。

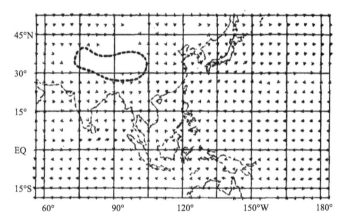

图 1 "季风前位相"(1979 年 6 月 1—10 日)平均的垂直积分的水汽通量向量分布图

在 110°E 以西的西太平洋,15°～20°N 之间有一 OLR 的高值带和 I_c 的低值带,对应着太平洋副高的干燥带。在这一干燥带以南,是 OLR 的低值和 I_c 的高值带,对应着北半球在 5°～10°N 之间的热带辐合带(ITCZ)。类似地,在印度洋(0°～5°S,60°～100°E)也有一条 OLR 的低值和 I_c 的高值带,这与南半球的热带辐合带相对应(图略)。

"在季风后位相"图 2 上,最显著的特点是一条强水汽输送带的建立,它沿着 10°～15°N 从阿拉伯穿过印度南部、孟加拉湾和印度支那到南海,然后转向北面经过中国南部直到日本以至更东的洋面上。非常有趣的是,向北输送也在孟加拉湾北部出现,这导致了西藏高原上空水汽通量的强烈辐合。在 60°E 附近或者可能在其更西面,有着强烈的穿越赤道的水汽输送,这与有名的索马里急流相对应。

沿着这条强的水汽输送带,有着强烈的水汽通量辐合,其辐合中心分别位于阿拉伯海(15°N, 70°E)、孟加拉湾(20°N, 90°E)和南海(5°N, 105°E),这三个辐合中心位于夏季季风槽中。此外,从中国南部经过东海到日本,也有一强的水汽辐合带,这与中国和日本的梅雨相联系。值得指出的是,在海南岛和台湾省以东的上空分别有一个相当于每天蒸发量为 10 毫米的水汽辐散中心,这与 Murakami[3] 以前的发现相一致(图略)。

在"季风后位相",另一个显著的特点是存在两个互相分离的 OLR 低值区(图略,参看图

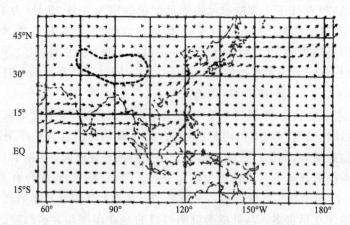

图 2 "季风后位相"(1979年6月20—30日)平均的垂直积分的水汽通量向量分布图

5)。一个是大致沿着 $10°\sim15°N$ 的纬向低值区,从阿拉伯海经过孟加拉湾,印度支那,菲律宾直到 $150°E$ 的西太平洋上空。另一个则是 SW~NE 走向的低值区,从中国南部经东海到日本。前一个低值区与南亚季风相联系,后一个则与东亚季风相联系,很有意义的是,在西藏高原的东南部亦为 OLR 的低值区,并有 I_c 的高值相配合,反映了该地区在风后"位相"强烈的对流活动,与此成为对照的是,在西藏高原的西部却为 OLR 的显著高值区。在西太平洋从台湾到日本也有一 OLR 的高值区,并有 I_c 的低值区相配合,这与太平洋副高的干燥辐散带相一致。

为了进一步清楚地看出季风建立前后水汽输送场的变化,我们计算了"季风后位相"与"季风前位相"的水汽通量的较差矢量 $(\delta\langle qu\rangle,\delta\langle qv\rangle)$,它由下式计算

$$\begin{cases}\delta\langle qu\rangle=\overline{\langle qu\rangle}(20-30/6)-\overline{\langle qu\rangle}(1-10/6)\\ \delta\langle qv\rangle=\overline{\langle qv\rangle}(20-30/6)-\overline{\langle qv\rangle}(1-10/6)\end{cases} \quad (3)$$

式中,"—"横线表示括弧中所指时段的平均,其他符号意义同前。计算结果如图 3 所示。此外,Q_2,OLR 和 I_c 的较差 $\delta(Q_2)$、$\delta(OLR)$ 和 $\delta(I_c)$ 也以同样的方法进行计算,计算结果如图 4、5 和 6 所示。

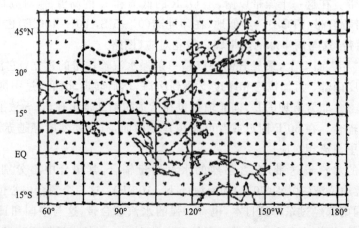

图 3 "季风后位相"与"季风前位相"的垂直积分的水汽通量向量的较差图

在图 3 中,最显著的特点是有一条非常清楚的东西向 $(\delta\langle qu\rangle,\delta\langle qv\rangle)$ 矢量轴线,这条轴线

从阿拉伯海经过孟加拉湾,印度支那,菲律宾直至 150°E 附近的西太平洋。另外沿着太平洋副高的边缘直到日本南部也有一条较差矢量轴线,这一反气旋的较差矢量型式无疑地与东亚季风相联系,并且很明显地与南亚的较差矢量轴线相区别,这两条较差矢量轴线(我们有时称它们为水汽输送的通道)在南海与华南相接并发生相互作用,但它们表现为两个相对独立的系统。与上述通道相配合,在南亚和东亚分别有一条显著的 $\delta(Q_2)$ 的负值带。此外,从菲律宾东南到 160°E 以东的赤道太平洋地区还有 $\delta(Q_2)$ 的负值带。西藏高原西部和从台湾到日本东南的西太平洋上空分别为 $\delta(Q_2)$ 的正值区(如图 4 所示)。

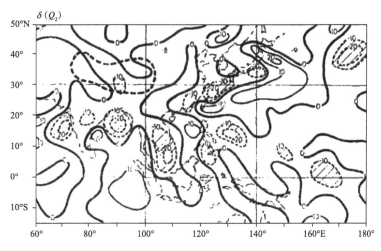

图 4 "季风后位相"与"季风前位相"Q_2 的较差图

类似的特点在图 5 上表现得特别清楚。一条很清楚的东西向的 $\delta(OLR)$ 的负值区从阿拉伯海一直延伸到菲律宾东南直至 140°E,它差不多与南亚的 $(\delta\langle qu\rangle, \delta\langle qv\rangle)$ 矢量轴线相吻合。另外,从中国南部经东海到日本,也有一 NE—SW 向的 $\delta\langle OLR\rangle$ 的负值带,其中心位于华东(30°N,120°E)和日本附近。出现在华东的中心与长江中下游梅雨建立后的对流活动加强相一致。在西太平洋(30°N,140°E)和西藏高原的西部均为显著的 $\delta(OLR)$ 的正中心。

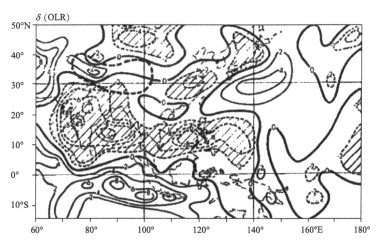

图 5 "季风后位相"与"季风前位相"OLR 的较差图

值得指出,在图 6 上,两条 $\delta(I_c)$ 的正值带仍然清楚可见。一条从孟加拉湾经过南海(12°N,115°E)到西北太平洋(12°N,135°E),另一条则从西藏高原的南部边缘到中国南部,然后伸至日本。

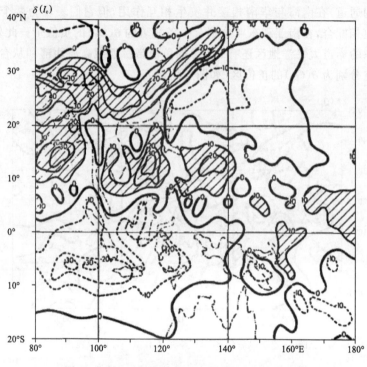

图 6 "季风后位相"与"季风前位相"I_c 的较差图

根据前面的讨论,这里的一个重要发现是东亚的季风和南亚的季风是两个相对独立的系统。这一观点已被 Asakura[10] 所提到。

总之,在印度季风建立前后,南亚和东亚地区的水汽输送场发生了显著的变化。在南亚,最重要的特点是沿着 10°～15°N 从阿拉伯海到菲律宾的强水汽输送带的建立。沿着这一强的水汽输送带有强烈的水汽通量辐合,并有 OLR 的低值带和 I_c 的高值带,这与南亚季风降水相联系。印度季风建立前位于印度的水汽通量辐散区(伴随着 OLR 的高中心和 I_c 的低中心),在季风建立后移到了高原的西部。在东亚,强的水汽输送带(伴有强的水汽辐合)早在 6 月 5 日左右(即印度季风建立前)就已在日本南部建立(图略),并导致日本南部梅雨的建立与太平洋副高相联系的水汽通量辐散区已经到达 20°N 左右。6 月中旬以后(即印度季风建立后),东亚的水汽输送和辐合进一步加强。由此可以看出,从"季风前位相"到"季风后位相"东亚的水汽输送场的向北移动不如南亚那样明显,这与大陆和海洋上的空气不同的加热速度相联系[11],然而东亚季风的建立却比南亚季风的建立要早两个星期左右。

4 水汽输送的突然变化与季风的建立

根据前面的讨论,在亚洲季风区域的各个地区水汽输送是显著不同的,为了建立水汽输送场的变化和季风的建立之间的关系,我们选取了若干地区的 Q_2、OLR 和 I_c 的时间序列,进行所谓"阶梯函数"的分析。其方法如下：

假定在时刻 t_1 和 t_2 之间，某一时间序列从一个准线而突然变化到另一个准线，而时刻之前和 t_2 时刻之后该时间序列分别围绕着不同的准线摆动，如图 7 所示。这样的阶梯式的时间变化可用下列分段线性函数 $\hat{X}(t)$ 来拟合（如图中的虚线所示）[2]。

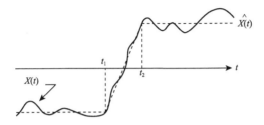

图 7 "阶梯函数"示意图。图中实线为观测序列，虚线为拟合的阶梯函数

$$\hat{X}(t) = a_1 + a_2 G(t-t_1) - a_2 G(t-t_2) \tag{4}$$

式中

$$G(t-t_k) = \begin{cases} 0 & t < t_k \\ t-t_k & t \geq t_k \end{cases}$$

(4)式与通常的线性回归方程很类似，只是增加了参变量 t_1 和 t_2。因此我们考虑所有可能的 t_1 和 t_2 的配对，对每一个 t_1 和 t_2 的配对我们采用最小二乘法求得 a_1, a_2 以及相应的拟合误差平方和。能使拟合误差平方和达到最小的 t_1 和 t_2 的配对以及相应的 a_1 和 a_2 即为方程(4)中的参变量和系数。

我们选择了孟加拉湾、南海、菲律宾、中国南部和日本等五个地区。对于 Q_2 和 OLR，地区范围为 $7.5°×7.5°$ 经纬度，对于 I_c，地区范围为 $8°×8°$ 经纬度。计算结果如图 8、9、10、11 和 12 所示。

在孟加拉湾地区，Q_2 和 OLR 的"阶梯函数"在 6 月 17—20 日呈现出突然的下降（图 8 的上、中图）。在下降之前，Q_2 呈现很小的负值，下降之后，Q_2 的负值变得很大（负值增加了近 10 个单位）。这种 Q_2 的突然变化反映了孟加拉湾地区在印度季风建立后水汽辐合的显著增强，它与 $10°\sim15°$N 的强水汽输送带的建立相吻合。I_c 的增加（图 8 底部）比 Q_2 和 OLR 的下降早若干天发生，在 6 月后半月一直维持相当的高值。Q_2 和 I_c 的上述变化反映了季风建立后强烈的对流活动，沿 $90°$E 的时间剖面图上也清楚地反映了类似的变化特点（图略）。

类似的变化也发生在南海地区和菲律宾地区（图 9、10）。图 9 中，Q_2，OLR 和 I_c 的突然变化的日期非常一致（16—19 日），变化的数值均比较大，似乎这一地区反映季风的建立比较敏感。在图 10 中，Q_2 的变化比前两个地区较为平缓，然而 OLR 和 I_c 确也显示出同样的突然变化，且变化的日期差不多与前两个地区相同。所有这些特点表明，上述三个地区季风的建立差不多是同时的。

与上述三个地区成为对照的是，在中国南部和日本（图 11、12），Q_2，OLR 和 I_c 的显著变化的日期和性质表现出与前三个地区有很大的不同，在图 11 中，Q_2 的"阶梯函数"的下降数值比前三个地区小得多。事实上 Q_2 在 15 日下降之前已是相当大的负值（图 11 的上图）。OLR 在 6 月下旬初期有一显著的变化，但 I_c 从 6 月 1—25 日一直是持续增加的（图 11 的下图）。这

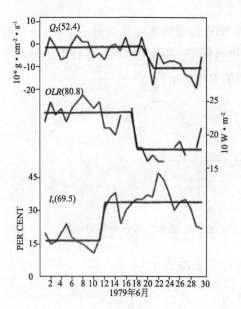

图8 上、中、下图分别为孟加拉湾地区的 Q_2、OLR 和 I_c 的时间变化曲线。上图和中图的地区范围为 (11.25°～18.75°N; 86.25°～93.75°E); 下图的地区范围为(12°～20°N; 86°～94°E)。对于 OLR 曲线若干天的记录遗缺。图中粗实线为拟合的"阶梯函数"; 左上角括弧中的数字为拟合方差百分比

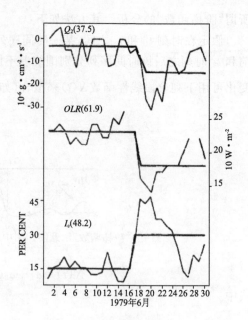

图9 南海地区。上图和中图的地区范围为 (6.25°～13.75°N; 105.0°～112.5°E); 下图的地区范围为(6°～14°N; 106°～114°E)。

其他说明同图8

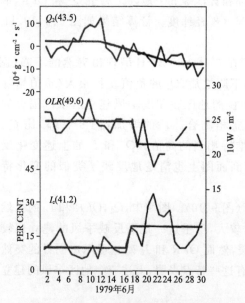

图10 菲律宾地区。上图和中图的地区范围为 (11.25°～18.75°N; 127.5°～135.0°E); 下图的地区范围为(12°～20°N; 126°～134°E)。

其他说明同图8

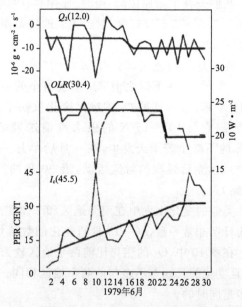

图11 南海地区。上图和中图的地区范围为 (22.5°～30.0°N; 112.5°～120.0°E); 下图的地区范围为(22°～30°N; 112°～120°E)。

其他说明同图8

表明中国南部的季风在 6 月之前就已经建立了。特别有趣的是,在日本,Q_2,OLR 和 I_c 差不多同时在 6 月 5 日左右均呈现出突然的变化(图 12)。事实上在 130°E 的时间剖面图上(图略),6 月 5 日以后,与太平洋副高相联系的水汽输送辐散区已占据了 15°～20°N 的区域,中旬以后北移到 20°～25°N。日本南部(30°N 左右)一直为水汽输送的辐合区,并有 OLR 的低值区和 I_c 的高值区相配合。时间剖面图上的这些变化特点与图 10 完全相一致。这些特征表明日本南部的梅雨确实在 6 月 5 日就开始建立了。

上述结果仅根据 1979 年 6 月一个月的资料所得,因此需要进一步地完善和验证。本研究是由美国国家科学基金会支持的。

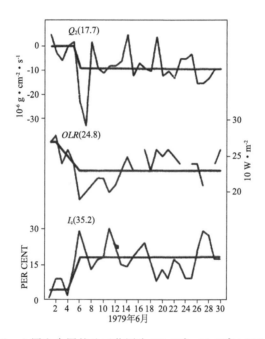

图 12　日本地区。上图和中图的地区范围为(26.25°～33.75°N;127.5°～135.0°E);下图的地区范围为(26°～34°N;126°～134°E)。其他说明同图 8

参考文献

[1] Sutcliffe R C. Water Balance and the General Circulation of the Atmosphere[J]. Quart J R Met Soc,1956,82:385-395.

[2] Flohn H. Large Scale Aspects of the "Summer-Monsoon" in South and East Asia[J]. 75th Anniversary Volume of J Met Soc,Japan,1957:180-186.

[3] Murakami T. The General Circulation and Water-Vapor Balance over the Far East during the Rainy Season[J]. Geophysical Magazine,Tokyo,1959,29(2):137-171.

[4] 徐淑英. 我国的水汽输送和水分平衡[J]. 气象学报,1958,29(1):28-42.

[5] Saito N. A Preliminary Study of the Summer Monsoon of Southern and Eastern Asia[J]. J Met Soc Japan,1966,44(1).

[6] Asakura. Transport and Source of Water Vapor in the Northern Hemisphere and Monsoon Asia[C]//

Yoshino M M. Water Balance of Monsoon Asia. University of Tokyo Press,1971.
[7] Saha K. Air and Water Vapor Transport across the Equator in Western Indian Ocean during Northern Summer[J]. Tellus,1970,22(6):681-687.
[8] Cadet, et al. Water Vapor Transport over the Indian Ocean during Summer 1975[J]. Tellus,1981,S3(5):476-487.
[9] Rao, Ramanamurty. Water Vapor Transport and Vergence Patterns over India during two Contrasting Summer Monsoons[J]. Pure and Applied Geophysics,Basel,115(3):491-502,1971.
[10] Asakura. Water Resources over Monsoon Asia[M]. ed. by Yoshino (in Japanese),1973:14.
[11] Kurashima A, Hiranuma Y. Synoptic and Climatological Study on the Upper Moist Tongue Extending from Southeast Asia to East Asia[C]//Yoshino M M. Water Balance of Monsoon Asia. University of Tokyo Press,1971.
[12] Ropelewski, Jalickee. An Objective Analysis of the Boundary Layer Thermodynamic Structure during GATE:Part Ⅱ Analysis[J]. Mon Wea Rev,1980,108:767-777.

江淮入梅的年际变化及其与北大西洋涛动和海温异常的联系*

徐海明[1] 何金海[1] 董敏[2]

(1. 南京气象学院,南京 210044;2. 国家气候中心,北京 100081)

摘要:文中首先采用简单相关和合成分析的方法研究了江淮入梅的年际变化与前期冬季环流和前期冬、春全球海温的关系。研究结果表明江淮入梅的早晚与前期冬季北半球大型环流存在显著的相关:入梅早的年份,其前期冬季北大西洋涛动强,北半球只有一个强的极涡并位于格陵兰上空,东亚大槽弱;入梅晚的年份,则其前期冬季环流表现为,北大西洋涛动弱,北半球存在两个极涡,其中一个仍然位于格陵兰上空,而另一个则位于西伯利亚上空,东亚大槽较常年强。江淮入梅的年际变化与前期冬春北大西洋海温的相关分析表明:入梅早的年份,北大西洋海温较常年偏暖;入梅晚的年份,前期冬春北大西洋海温较常年偏冷。文中还用 CCM3 模拟了冬、春季北大西洋海温增暖对后期江淮入梅和梅雨期降水的影响,并探讨了其影响的物理机制。

关键词:江淮入梅;年际变化;北大洋涛动;海温异常

1 引 言

梅雨是指每年 6 月中旬至 7 月上旬在中国江淮流域出现的连阴雨天气,雨量很大。由于梅雨期降水具有很大的年际变化,所以梅雨期降水的多寡可直接导致江淮干旱涝的形成,从而对中国国民经济建设产生重大影响,因此,对江淮梅雨的研究一直是中国气象工作者研究的重要课题[1-3]。吴仪芳②、杨广基③用多个旱涝年的月平均场资料研究江淮旱涝前期环流特征,发现江淮流域旱涝发生期及前期平均季节环流有显著差异。黄荣辉[4]、Nitta[5]通过研究西太平洋暖池海温演变来建立它与未来副热带高压(以下简称副高)位置的关系以预测江淮流域降水。陈烈庭[6]、符崇斌[7]则研究了 El Niño 现象和赤道东太平洋海温距平对北半球环流及江淮旱涝的影响。但以往大量的研究都集中在江淮梅雨期降水量的多寡及其成因上,而对江淮梅雨入梅早晚年际变化的研究则相对较少。众所周知,江淮流域梅雨的开始和结束是每年初夏东亚大气环流发生急剧变化的结果,而入梅的早晚则在一定程度上体现了每年东亚大气环流发生突变的早晚,也在一定程度上反映了东亚夏季风的强弱及其推进到江淮流域的早晚。因此,对江淮流域梅雨开始早晚的年际变化及其成因作一研究具有重要的学术意义和应用价值。

---

* 本文发表于《气象学报》,2001 年第 59 卷第 6 期,694-706。
② 吴仪芳,李麦村. 江淮旱涝形成的长期天气过程. 中国科学院大气物理研究所集刊,第 13 号,15-29。
③ 杨广基. 长江流域中下游地区夏季旱涝的前期特征. 同上,30-40。

2 江淮入梅早晚的年际变化及其与江淮 6 月降水的关系

为了研究江淮入梅的年际变化,文中采用南京气象学院气象台统计的 1957—1991 年共 35 a 的资料(表 1),气象台根据以下的标准来确定:(1)120°E 的副热带高压脊线位置北跳到 20°N 以北且稳定在 18°~25°N 之间达 4 d 以上;(2)副热带高压北跳以后,5 d 内至少有 4 d,10 d 内至少有 7 d 的阴雨日;(3)连阴雨开始以后 5 d 内有一次大到暴雨以上的雨日;(4)日平均温度入梅前后连续 4 d 以上≥22℃。这与江苏气象台确定每年入梅的标准是一样的。

从表 1 中可以清楚看到,江淮入梅日期有很大的年际变化,在这 35 a 中入梅最早的年份是 1991 年,该年早在 5 月 19 日便开始入梅,入梅最晚的年份是 1982 年,迟至 7 月 10 日才入梅,最早和最晚入梅日期相差 50 d 之久。为了更清楚地揭示入梅的年际变化,作了江淮入梅日期的年际变化曲线(如图 1 所示),从图中可以看到,这 35 a 的平均入梅日期为 6 月 17 日,与历史上 80 a 平均相比[8],入梅日期明显推迟(历史上 50% 入梅日期在 6 月 6—15 日)。以与平均入梅日期差 5 d(1 候)来定义入梅早和晚的年份,从图中可以看到,在这 35 a 中入梅早的年份有 10 a(1960,1961,1971,1974,1976,1980,1984,1989,1990,1991 年),入梅明显晚的年份有 12 a(1959,1963,1969,1978,1981,1982 和 1985 年)。从图中还可以看到入梅日期不仅有显著的年际变化,而且还有一定的年代际变化特征,20 世纪 60 年代入梅日期普遍偏晚,70 年代普遍偏早,而 80 年代以后入梅早晚则起伏较大。

表 1 南京气象学院气象台 1957—1991 年入梅日期统计表

年份	入梅日期	年份	入梅日期	年份	入梅日期	年份	入梅日期	年份	入梅日期
1957	6 月 19 日	1964	6 月 23 日	1971	6 月 9 日	1978	6 月 22 日	1987	6 月 22 日
1958	6 月 21 日	1965	6 月 30 日	1972	6 月 20 日	1979	6 月 19 日	1986	6 月 19 日
1959	6 月 27 日	1966	6 月 25 日	1973	6 月 16 日	1980	6 月 9 日	1987	6 月 18 日
1960	6 月 8 日	1967	6 月 23 日	1974	6 月 10 日	1981	6 月 22 日	1988	6 月 15 日
1961	6 月 6 日	1968	6 月 24 日	1975	6 月 16 日	1982	7 月 10 日	1989	6 月 6 日
1962	6 月 17 日	1969	6 月 23 日	1976	6 月 9 日	1983	6 月 19 日	1990	6 月 7 日
1963	6 月 23 日	1970	6 月 18 日	1977	6 月 13 日	1984	6 月 7 日	1991	5 月 19 日

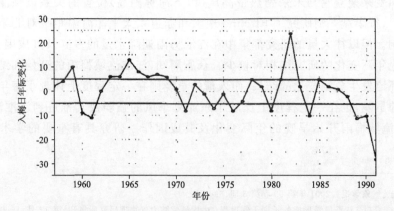

图 1 1957—1991 年江淮入梅日的年际变化曲线(零线为平均入梅日期 6 月 17 日)

为了探讨江淮入梅的早晚与江淮梅雨期降水的关系,文中计算了江淮入梅日期的年变化与同期 6 月中国降水的相关(图 2),从图中可以清楚看到,江淮入梅的早晚与江淮 6 月降水存在显著负相关,负相关系数绝对值达 0.5 以上,通过了置信度 0.01(相应相关系数为 0.42)的 T 检验,由此可见,江淮入梅早的年份江淮降水偏多,入梅晚的年份江淮降水偏少。此外,还计算了入梅早晚与 6—7 月降水的关系,发现入梅的早晚与 6—7 月的华南和江淮降水具有同样的相关关系(图略)。由此可见,江淮入梅的早晚与夏季江淮洪涝干旱也有一定的相关关系。

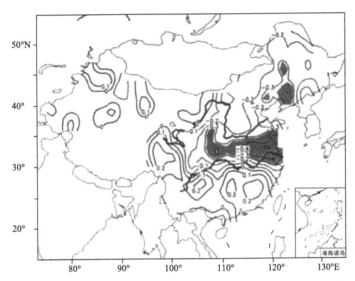

图 2 江淮入梅的年际变化与 6 月降水的相关系数分布
(等值线间隔:0.1,阴影区为通过 0.05 信度检验区)

3 江淮入梅的年际变化与前冬北半球环流的关系

如前所述,江淮梅雨与东亚大气环流的季节突变相关,而江淮入梅的早晚基本上反映了东亚环流从冬季型环流向夏季型环流转变在时间进程上的差异,已有的研究表明[3,9]北半球冬季环流的变化对后期夏季环流存在一定的影响,那么江淮入梅的早晚与北半球冬季环流又有何关系呢?下面用 500 hPa 高度场和海平面气压场来讨论江淮入梅早晚与前期冬季环流的关系。

3.1 与前期冬季 500 hPa 高度场的关系

为了探讨这个问题,首先对江淮入梅的年际变化与前期冬季 500 hPa 高度场进行了相关分析(图 3)。图 3 给出了 1957—1991 年 35 a 的入梅年际变化与前期冬季(12 月—次年 2 月)的 500 hPa 高度场的相关分布,从图中可以看到在中北大西洋上存在一个大范围的显著负相关区,而在其北方格陵兰上空则存在一个大范围的显著正相关区,正负相关区成南北向偶极子分布,其分布特征与 Wallace 等[10]所揭示的北大西洋涛动相一致,由此可见,中国江淮入梅的早晚与北大西洋涛动的强弱存在显著的相关。另外,从图中还可以看到,在黑海附近上空也存在一个大范围的负相关区,而在东亚西伯利亚东部上空也存在一片显著负相关区,表明江淮入梅的早晚与前期东亚冬季环流也存在一定的关联,采用王绍武等[11]提出的相关场显著性检验的方法,计算结果也表明整个相关场在5%的信度标准下也是显著的。

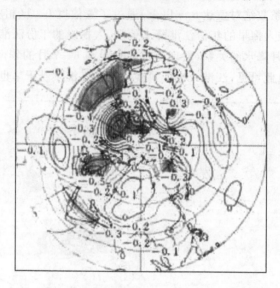

图 3　江淮入梅的年际变化与前期冬季 500 hPa 位势高度场的相关
(等值线间隔:0.1,阴影区为通过 0.05 信度检验区)

为了进一步揭示江淮入梅早晚年前期冬季环流的不同特征,根据上文所定义的早梅雨和晚梅雨年的 500 hPa 高度场距平场作了合成分析(图略),发现入梅早的年份,在中北大西洋上为一大范围的正距平区,中心距平值在 20 gpm 以上,而在格陵兰上空则为一强的负距平区,中心值在 50 gpm 以上,两者成南北偶极子分布,大大增强了南北向的位势梯度,有利于北大西洋上空强西风急流的维持,而在东西伯利亚上空则存在一大范围的正距平,有利于维持弱的东亚大槽;在入梅晚的年份,500 hPa 高度场的距平分布则与入梅早年的情形成基本相反,即中北大西洋上为一负距平区,格陵兰上空为强正距平区,东西伯利亚上空为一负距平区,有利于维持强的东亚大槽、弱的北大西洋西风急流。这种距平场分布反映在 500 hPa 平均高度场上(图 4)则表现为,入梅早的年份,北极极涡只有一个,它位于格陵兰岛附近上空,强度较强;而在入梅晚的年份,极涡分裂成两个,其中一个位于格陵兰上空,而另一个则位于西伯利亚上空。由此可见,江淮入梅的早晚与前期冬季环流密切相关。

3.2　与前期冬季海平面气压的关系

为了进一步探讨江淮入梅年际变化与前期冬季环流的关系,文中也计算了江淮入梅的年际变化与前期冬季海平面气压的相关(图 5),从图中可以看到,最大的正相关中心位于北美加拿大至格陵兰附近,最大相关系数已达 0.5 以上,通过了信度 0.01 的 T 检验(相应的相关系数为 0.4182),而在北大西洋上则为显著的负相关区,其相关分布型与 500 hPa 上高度场相关分布相一致,同样,采用王绍武等[11]提出的相关场显著性检验的方法,计算结果也表明整个相关场在 5% 的信度标准下也是显著的。这种相关分布型在海平面气压距平合成图上(图略)则表现为,入梅早的年份,冰岛及格陵兰附近为一大范围的负距平区所控制,北大西洋上为一正距平区,表明入梅早的年份,冰岛低压和北大西洋高压都增强,对应于强的北大西洋涛动;而在

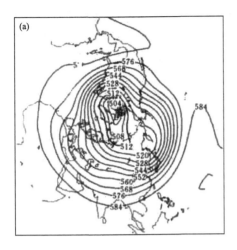 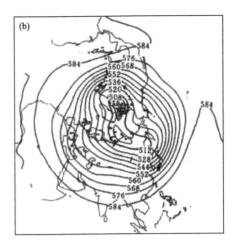

图 4　江淮入梅早年(a)和晚年(b)北半球冬季 500 hPa 位势高度合成分布(等值线间隔:40 gpm)

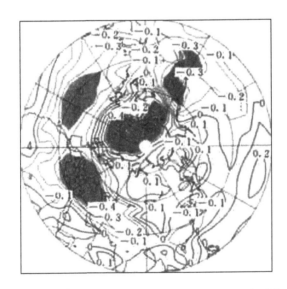

图 5　1957—1991 年 35 a 江淮入梅的年际变化与前期北半球冬季海平面气压相关
(等值线间隔:0.1,阴影区为通过 0.05 信度检验区)

入梅晚的年份,则刚好相反,冰岛低压和北大西洋高压均减弱,对应于弱的北大西洋涛动。

4　江淮入梅的年际变化与前期冬、春季海温的关系

　　从前一节的讨论,可以看到江淮入梅的早晚与前期冬季环流异常密切相关。那么前期冬季这种大尺度的环流异常又是由什么因素造成的呢? 已有的研究都表明,大范围的环流异常主要由像海温、海冰、积雪和土壤湿度等外强迫因子所造成的,其中以海温异常的影响最为重要。图 6 给出了 1957—1991 年的入梅日期与前期冬季、春季全球海温的相关分布,从图中可以清楚地看到江淮入梅日期的年际变化与北大西洋海温存在显著的负相关,该显著负相关区从冬季一直维持到前期春季,且相关系数逐渐增大,在春季其相关系数的最大值已达 0.5 以

上,通过了信度为0.01的T检验。另外,从图中还可以看到,江淮入梅早晚与赤道中东太平洋海温成正相关,与中北太平洋海温成负相关,该两块大范围的正负相关区,在前期冬季没能通过0.05的信度检验,而到了春季有小范围的海区能通过信度为0.05的T检验,但相关强度从冬季至春季是逐渐增大的。由此可见,江淮入梅的早晚与前期北大西洋海温存在显著相关,而与前期的赤道中东太平洋和北太平洋海温只存在弱的相关。这种相关分布型在前期的海温距平合成图上(图略)则表现为,在北大西洋上,入梅早年与入梅晚年相比前期冬、春海温偏暖;而在太平洋上,入梅早年前期与入梅晚年相比冬、春赤道中东太平洋海温偏冷,而中北太平洋海温偏暖。这对于预测江淮入梅的早晚具有一定指示意义。

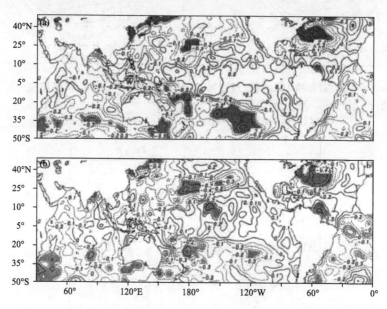

图6　江淮入梅的年际变化与前期冬季(a)和春季(b)海温的相关系数
(等值线间隔:0.1,阴影区为通过0.05信度检验区)

5　数值模拟

以上分析表明北大西洋冬、春季海温与后期江淮梅雨入梅年际变化存在显著相关,这说明前期北大西洋海温异常与否是判断后期江淮入梅早晚的一个重要指标。为了进一步研究北大西洋海温异常是通过何种途径和机制影响中国江淮梅雨入梅和降水,将采用一个全球气候模式来模拟北大西洋海温异常对江淮入梅和降水的影响。

5.1　模式和试验方案

本文所用模式为美国全国大气研究中心的第3代公共气候模式(CCM3),采用e-P混合坐标的全球谱模式,垂直分为18层,大气顶取在2.9 hPa,水平分辨率为T42(相当经纬度间距2.8°左右,全球共计128×64个格点)。模式包括了辐射、对流、垂直扩散、边界层及陆面过程等各种物理过程,模式还包括了日变化,辐射通量每隔1 h计算一次,在这之间辐射通量保持不变。时间积分步长为20 min。它与CCM2相比,在云的参数化、晴空长波辐射、深对流、边界层过程和陆面过程都作了改进,经过这些改进,模式的辐射偏差已大为减少,减弱了过强的水循环,使潜热释放及降水率均接近观测,辐射收支已接近平衡。关于模式的详细说明和改进详见文献[12,13]。

在 CCM3 模式中所用的海温为多年平均的海温资料,在模式的运行过程中保持海温的季节变化而没有年际变化。为了研究冬、春季北大西洋异常对后期江淮入梅和梅雨降水的影响,在这里设计如下两个数值试验:

第 1 个试验,采用经过多年积分以后而得到 9 月 1 日的模式场作为本试验的初始场,然后再积分 320 d,每 5 d 取一次平均作为一次结果输出,该试验称之为控制试验(CTL)。

第 2 个试验,初始场、积分时间及输出频率与试验 1 相同。与第 1 个试验不同之处在于模式积分过程中,12 月至下一年 5 月的北大西洋海温气候场上叠加了一个正海温异常场,其分布见图 7,该正海温异常分布位于西北大西洋上,中心最大值为 2.5℃,其水平分布与入梅早、晚年北大西洋海温距平之差分布相一致,该试验称之为北大西洋正海温异常试验(NAP)。

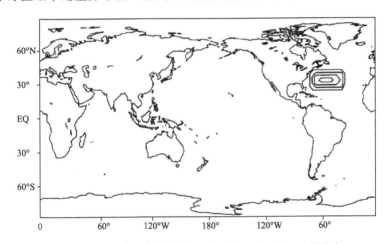

图 7 北大西洋正海温试验(NAP)中所加的海温异常分布
(等值线间隔:0.5)

5.2 试验结果分析

5.2.1 北大西洋海温异常增暖对北半球冬季大气环流的影响

图 8 分别给出了北半球冬季 CCM3 控制试验和北大西洋正海温异常试验在模式第 10 层(约 400 hPa)上的流场以及两者之差。由图 8a 可见:CCM3 很好地模拟出了北半球冬季大气环流"三槽三脊"的主要特征,3 个大槽分别位于亚洲东岸、北美东部和欧洲东部,3 个脊分别位于阿拉斯加、西欧沿岸和青藏高原北部。当 CCM3 中西北大西洋海温异常偏暖时,由图 8b 可以看到,阿拉斯加高压脊明显增强,北美大槽、东亚大槽也有不同程度的加深,而位于欧洲东部的大槽则明显变的平浅,可见西北大西洋海温异常偏暖不仅对其上空的大气环流产生影响,而且对整个北半球的大型环流也会产生一定的影响。这种大型环流的改变在差值流场上(图 8c)则表现为,在格陵兰岛以南的北大西洋北部为一个气旋性的差值环流,而在整个北大西洋的副热带地区则为反气旋性的差值环流所控制。这清楚表明西北大西洋海温异常偏暖时,WA 遥相关型增强、相应美国东部至西北大西洋上的西风急流得到增强,这与前面的诊断结果

是一致的。另外,从图8c还可以看到,西北大西洋海温异常偏暖可在阿拉斯加上空以及整个欧亚大陆上空激发出反气旋性的差值环流,而在俄罗斯远东至鄂霍次克海上空激发出气旋性差值环流,从而对整个北半球大气环流也产生一定影响。

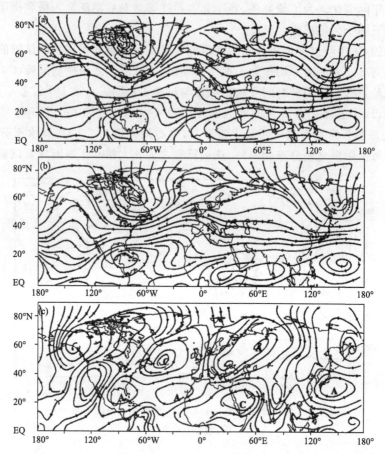

图8 北半球冬季在模式第10层(约400 hPa)上的流场和差值流场
(a. 控制试验(CTL),b. 北大西洋正海温异常试验(NAP),c. 两试验之差(NAP-CTL))

5.2.2 北大西洋海温异常增暖对江淮入梅和降水的影响

图9给出了北大西洋正海温异常试验(NAP)和控制试验(CTL)在江淮流域(28°~34°N, 105°~120°E)区域平均的降水量之差随时间的变化曲线。从图中可以看到,冬、春季的北大西洋海温的异常偏暖可以使江淮流域2—3月降水明显偏少,到了4月上旬以后,北大西洋海温异常偏暖渐渐地使江淮流域的降水增多,而到了5月中旬以后,北大西洋海温异常偏暖使江淮流域的降水迅速增加,最大偏差达4.5 mm·d^{-1},该大的异常降水偏差一直维持到6月底。由前面的分析结果表明江淮入梅的平均日期为6月中旬左右。而北大西洋海温冬、春季的异常偏暖可以使中国江淮流域的降水在5月中旬以后便迅速增加,由此可见,北大西洋海温冬、春季的异常偏暖可以使江淮梅雨入梅日期明显提前,该结果与前面的诊断分析结果是一致的。

图10分别给出了控制试验(CTL)和北大西洋正海温异常试验(NAP)5—6月的平均降水量以及两者的降水量之差。由图10a可见,在亚洲热带季风区,主要的降水位于南海北部和孟

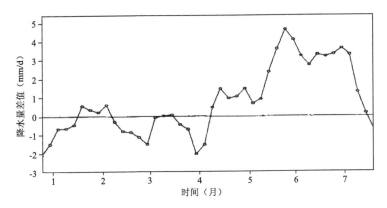

图9 北大西洋正海温异常试验(NAP)与控制试验(CTL)在江淮流域(28°~34°N,105°~125°E)区域平均降水量差随时间的变化曲线

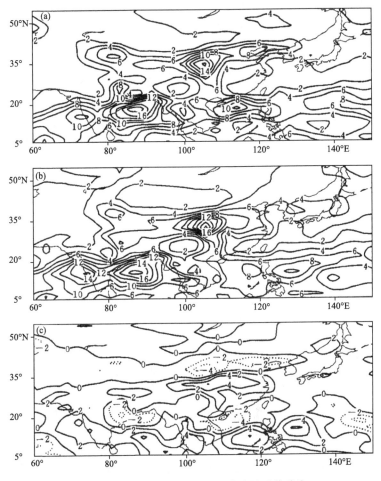

图10 5—6月平均降水量和降水量差值分布
(a. 控制试验(CTL), b. 北大西洋正海温试验(NAP), c. 两试验差(NAP-CTL); 等值线间隔: 2 mm·d^{-1})

加拉湾北部至印度半岛，分别对应于5—6月的南海夏季风和6月印度夏季风，这与实际的观测结果相一致。在中国大陆上，中国华北地区存在一条东西向的雨带，而在江淮流域没有明显的降水雨带，与实际的观测结果相比华北降水偏强而江淮流域降水偏弱，这说明CCM3模式对亚洲热带夏季风降水有较强的模拟能力，而对东亚副热带夏季风降水的模拟仍然存在一定的偏差。当在模式中考虑了北大西洋冬、春季海温异常偏暖以后，江淮流域5—6月的降水量明显增多，在NAP试验中，自青藏高原东部至中国江淮流域明显存在一条降水雨带(图10b)，华北降水雨带强度明显减弱，南海北部的降水量也明显减少。在降水差值图上(图10c)则表现为自青藏高原东部至中国江淮流域为一正的降水偏差区，而在其南北两侧的中国华北和南海北部至华南沿海为负的降水差值区。另外，从差值图中还可以看到在孟加拉湾北部和印度半岛东北部也为一负的降水差值区。由此可见，冬、春季北大西洋海温异常偏暖不仅对后期中国江淮梅雨入梅的早晚和降水存在影响，而且对初夏的南海夏季风和印度夏季风也存在一定的影响。

5.2.3 北大西洋海温异常对江淮梅雨入梅和降水影响的机制

为了研究北大西洋正海温异常对中国江淮入梅和降水的影响机制，图11给出了6月控制试验(CTL)和NAP试验在模式第10层(约400 hPa)上的流场和两者之间的差值流场。从图

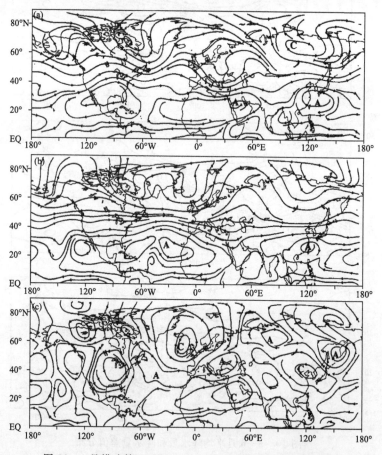

图11 6月模式第10层上(约400 hPa)的流场和差值流场
(a. 控制试验(CTL), b. 北大西洋正海温异常试验(NAP), c. 两试验之差(NAP-CTL))

11a中可以看到,在东亚沿海的日本上空和北美东海岸上空分别为两个大槽所控制,贝加尔湖以北为一个气旋性环流、乌拉尔山地区为一个反气旋性高压脊,副热带高压中心位于长江中下游至西太平洋上一带,伊朗至北非上空也为一个宽广的高压所控制,这与6月的实际观测结果相一致。图11b则给出了NAP试验6月的流场,从图中可以清楚地看到,北大西洋海温的异常偏暖可使乌拉尔山高压脊明显增强,鄂霍茨克海上空的环流也由控制试验中的气旋性低压槽转变为反气旋性高压脊所控制,副热带高压中心位置与CTL试验中的副热带高压中心位置相比明显偏南,大约位于22°N附近。在两者的差值流场(图11c)上则表现为北大西洋中纬度地区为一个反气旋性差值环流,冰岛以南的东北大西洋上为一个气旋性差值环流所控制,在贝加尔湖西北地区上空为一个反气旋性差值环流所控制,中国华北至日本海上空又为一个气旋性差值环流,这4个反气旋性、气旋性相间的差值环流中心连线类似于Wallace等[10]所揭示的欧亚波列。众所周知,江淮入梅和梅雨期的持续稳定与乌拉尔山和鄂霍茨克海高压脊或阻塞高压的建立和维持密切相关。由此可见,北大西洋海温异常偏暖通过激发欧亚波列影响欧亚上空的大气环流,使乌拉尔山和鄂霍茨克海高压脊异常发展和加强,从而使江淮提前入梅并使梅雨期降水异常增多。

在对流层低层的差值流场上(图略)则表现为,在长江中下游地区为一个明显的气旋性差值环流中心所控制,从孟加拉湾经中南半岛至中国南海北部和华南为一致的西南风差值,孟加拉湾北部也为一个气旋性差值环流中心,而在印度半岛上空则为一反气旋性差值环流所控制。可见,北大西洋海温的异常偏暖可使东亚西南季风明显增强,有利于江淮梅雨期江淮气旋的维持和发展,而使印度夏季风和季风降水明显减弱。

6 总结和讨论

文中利用1957—1991年的江淮入梅日期资料和中国160个测站的月平均降水资料、北半球500 hPa月平均高度场、海平面气压场以及全球海温月平均场研究了江淮入梅的年际变化与前期冬季环流和前期冬、春全球海温的关系,而后又用CCM3模拟了冬春季北大西洋海温异常增暖对中国初夏入梅和降水的影响,得到如下几点结论:

(1)江淮入梅日期存在明显的年际变化,而且与江淮6月降水存在显著相关,即入梅早年,江淮6月降水较常年偏多;入梅晚年,江淮降水则较常年偏少。

(2)江淮入梅的早晚与前期冬季北半球大型环流存在显著的相关。对于入梅早的年份,其前期冬季北大西洋涛动强,北半球只有一个强的极涡并位于格陵兰上空,相应东亚大槽弱;对于入梅晚的年份,则其前冬环流表现为,北大西洋涛动弱,北半球存在两个极涡,其中一个仍然位于格陵兰上空,而另一个则位于西伯利亚上空,相应东亚大槽较常年强。

(3)江淮入梅的年际变化与前期冬春北大西洋海温也存在显著的相关。即入梅早年,北大西洋海温较常年偏暖;入梅晚的年份,前期冬春北大西洋海温较常年偏冷,这对于预测江淮入梅早晚的年际变化具有很好的指示意义。

(4)数值试验结果表明,冬、春季北大西洋海温的异常偏暖可导致初夏中国江淮流域提前入梅和梅雨期降水的增多。结果进一步表明,北大西洋海温的异常偏暖可通过激发欧亚波列使乌拉尔高压脊和鄂霍茨克海上空的高压脊明显增强,从而使江淮梅雨明显提前和降水量的增多。

已有研究表明[6,7],中国东部汛期降水与赤道太平洋海温(如El Niño)之间也存在一定的关系,赤道太平洋海温异常主要通过影响热带环流和副热带高压对中国东部汛期降水产生影响,但这也只能解释了引起江淮降水异常的其中一部分原因。众所周知,江淮汛期降水的多寡既受到低纬度系统(如热带季风系统、副热带高压)的影响,同时也受中高纬大气环流(如阻塞高压)以及相应的冷空气活动的影响,而冬春中高纬欧亚大陆上空中高纬大气环流的异常显然受到其上游大气环流以及外强迫源(如SSTA)异常的影响,这也很好从Hoskins提出的大圆理论(EU波列)来得到解释,本文的数值试验结果也清楚地证明了这一点。因此,认为江淮汛期降水的异常应既受到赤道太平洋海温异常的影响,同时也受到北大西洋海温异常的影响,至于两者影响相对的重要性及如何协同作用则需进一步研究。另外,文中仅做了两个敏感性试验来探讨北大西洋海温异常对江淮入梅影响的机制,其结果还有待做更多的数值试验来进行验证。

参考文献

[1] 陶诗言,徐淑英.夏季江淮流域持续性旱涝现象的环流特征[J].气象学报,1962,32(1):1-10.

[2] 杨广基,梁佩娴.长江流域中下游持久性旱涝与低纬度热带环流型的关系[J].高原气象,1982,1(3):43-50.

[3] 陶诗言,朱文妹,赵卫.论梅雨的年际变化[J].大气科学,1988,12(特刊):2-13.

[4] 黄荣辉,李维京.夏季热带西太平洋上空的热源异常对东亚上空副热带高压的影响及其物理机制[J].大气科学,1988,12(特刊):107-116.

[5] Nitta T. Long-term variations of cloud amount in the western Pacific region[J]. J Meteor Soc Japan, 1986, 64:373-390.

[6] 陈烈庭.东太平洋赤道地区海水温度异常对热带大气环流及中国汛期降水的影响[J].大气科学,1977,1(1):1-12.

[7] 符崇斌.中国夏季的气候异常与厄尔尼诺南方涛动的关系[J].大气科学,1988,12(特刊):133-141.

[8] 徐群.八十年长江中下游的梅雨[J].气象学报,1965,35(4):507-518.

[9] 孙淑清,孙柏民.东亚冬季风环流异常与中国江淮流域夏季旱涝天气的关系[J].气象学报,1995,53(4):440-450.

[10] Wallace J M, Gulzter D S. Teleconnection in the geopotential height field during the Northern Hemisphere winter[J]. Mon Wea Rev, 1981, 109:784-812.

[11] 王绍武,胡增臻.气象要素相关场显著性检验的统计模拟方法[C]//王绍武,黄朝迎.长江黄河旱涝灾害发生规律及其经济影响的诊断研究.北京:气象出版社,1993,215-221.

[12] Kiehl J T, Hack J J, Bonan G A, et al. Description of the NCAR Community Climate Model (CCM 3) [J]. NCAR Tech. Note, 1996, NCAR/ TN- 420+STR, 152.

[13] Kiehl J T, Hack J J, Bonan G A, et al. The National center for atmospheric research community climate model: CCM3[J]. J Climate, 1998, 11:1131-1150.

Vertical Circulation Structure, Interannual Variation Features and Variation Mechanism of Western Pacific Subtropical High[*]

He Jinhai[1] Zhou Bing[2] Wen Min[3] Li Feng[3]

(1. Nanjing Institute of Meteorology, Nanjing 210044; 2. LASG, Institute of Atmospheric Physics, Chinese Academy of Sciences, Beijing 100029; 3. Nanjing Institute of Meteorology, Nanjing 210044)

Abstract: The paper investigates the vertical circulation structure of the western Pacific subtropical high (STH) and its interannual variation features in relation to East Asian subtropical summer monsoon and external thermal forcing by using the high-resolution and good-quality observations from the 1998 South China Sea Summer Monsoon Experiment (SCSMEX), the NCEP 40-year reanalysis data and relevant SST and the STH parameters. It is found that the vertical circulation structures differ greatly in features between quasi-stationary and transient components of the western Pacific STH. When rainstorms happen in the rainband of East Asian subtropical monsoon on the STH north side, the downdrafts are distinct around the ridge at a related meridian. The sinking at high (low) levels comes from the north (south) side of the STH, thereby revealing that the high is a tie between tropical and extratropical systems. The analyses of this paper suggest that the latent heat release associated with subtropical monsoon precipitation, the offshore SST and East Asian land-sea thermal contrast have a significant effect on the STH interannual anomaly. Our numerical experiment shows that the offshore SSTA-caused sensible heating may excite an anomalous anticyclonic circulation on the west side, which affects the intensity (area) and meridional position of the western Pacific STH.

Key words: western Pacific subtropical high; meridional/zonal circulation; interannual variation; offshore sea surface temperature

1 Introduction

The western Pacific STH is a major member of the East Asian summer monsoon system[1], on the north side of which is the East Asian subtropical monsoon rainband[2]. As time progresses from spring to summer, the STH jumps northward at first and then stays motion-

---

[*] 本文发表于《Advances in Atmospheric Sciences》,2001 年第 18 卷第 4 期,497-510.

less, allowing the rainband to reside for long over the middle and lower reaches of the Yangtze River, which induces persistent precipitation to produce a flood there, the good examples being the exceptional rainstorm events in the summers of 1991 and 1998. For this reason, the study of the meridional and zonal movement of the STH and its intensity change has been a heated problem in a meteorological context.

The classical theory holds that the STH is related to the subsidence of mean meridional (or Hadley) circulation from low to middle latitudes[3]. Due to the zonal asymmetry of land-sea distribution, a banded STH frequently splits into many cells so that STH genesis is often interpreted by means of a sinking leg of a local meridional circulation, which, however, has a fatal weakness in such a way that the northern STH is markedly stronger in summer than in winter as opposed to the Hadley circulation, especially in the western Pacific. Instead, an attempt was conducted to make use of the sinking leg of the zonal circulation for the purpose. Yeh et al.[4] proposed that the zonal circulation excited by Tibetan heating plays an important role in STH genesis. Chen[5] showed that the genesis and maintenance are associated with the eastward progression of the Rossby wave energy excited by the plateau heating in the framework of linear quasi-geostrophic steady wave theory. If so, significant positive correlation should be found between the plateau thermal condition and STH intensity. In recent years, through the scale analysis of a complete form vertical vorticity equation and numerical experiments, Wu et al.[6] and Liu et al.[7] pointed out that on a climatic time scale, the key factor controlling the summer STH locality and intensity in the eastern hemisphere is predominantly the latent heat released from East Asian monsoon precipitation and, to less extent, terrain and sensible heating of land surface. In the light of the theory, a high correlation of some or other form should be found between the monsoon rainfall and STH position/strength and the correlation should be confirmed by meteorological data. However, there is often little precipitation in the area where the main body of the western Pacific STH stays, then what is the forcing factor? which affects the interannual variation of the western Pacific STH. Besides, on a short time scale, how does the East Asian monsoon, especially torrential rain, affect the vertical circulation of the western Pacific STH? Those are problems we are concerned about. An attempt is made in this work to deal with relevant features of the STH vertical circulation structure and its interannual variation in order to reveal their relationships to subtropical monsoon rainfall and external forcing, whereupon is examined the mechanism of STH variation.

2 Data and method

Data used consists of daily rainfall from the 1998 SCSMEX and the NCEP daily averaged basic element reanalysis in the same period; 1958—1993 summer monthly mean rainfall from 160 stations and the western Pacific STH parameters (intensity, area index, and ridge latitude) compiled by China National Meteorological Center. Intensity is defined as the accumulated value of the part of grid point height greater than 587 dagpm in the domain from 110° to 180°E. Area index refers to the number of grid points whose height are greater than 587 dagpm in the same domain, and ridge latitude to the mean of the latitude on which the maximum value of the grid point heights greater than 587 dagpm lies in the domain from 110° to 150°E. Also, in this research we have utilized the index of East Asian summer monsoon land-sea thermal difference [8] and the moisture sink $\langle Q_2 \rangle$ and vertical velocity throughout the troposphere inversely calculated by an indirect technique (Zhou et al., see footnote Below[①]).

3 STH vertical circulation features

3.1 Features on a meridional cross section

In July 1998, the STH has its ridge in 20° to 25°N farther southward of mean and its intensity and area index higher than usual (see Fig. 1a), but on the 125°E section through the core (Fig. 1b), a robust rising leg, that originates from a low-latitude monsoon circulation, emerges in the vicinity of 20°~25°N. It is observed from Fig. 1c that the vertical velocity associated with the meridional movement of the western Pacific STH near 20°N shows a wavelike pattern, i.e. a stronger updraft in the first half of July and a downdraft in the second half of July. In particular, when the updraft near 30°N is strong (corresponding to the heavy rain in the lower reach of the Yangtze river), the downdraft in the south side is also distinct. However, as far as monthly mean is concerned, there is an updraft in the center (ridge) of the western Pacific STH with the maximum ascending motion in the middle troposphere. The longitude-time cross section of vertical velocity along 20°N (omitted) also reveals the similar feature of vertical motion. This feature is widely divergent from the traditional concept that there is a downdraft in the central area of STH. However, can we assume no downdraft to be in the neighborhood of the core and ridge? The problem will be dealt with in the following.

As shown before, Wu et al.[6] emphasized the effect of latent heat release in East Asian monsoon rainfall on the western Pacific STH. And for this reason we selected from the 1998 SCSMEX rainfall record the intervals between 19-24 July and 11-16 August (7-12 July) as the flooding (drought) periods over the Yangtze River valley to separately examine the downdraft condition around the ridge at different longitudes.

Fig. 2a is a plot of the vertical circulation structure features in July 19-24, 1998 during an exceptional torrential rainfall at Wuhan. One sees that robust updraft occurs over the city and its thereabouts around 30°N with its rising at -20 hPa $\cdot$ (6 h)$^{-1}$ together with rich moisture convergence and noticeable latent heating $\langle Q_2 \rangle$ in good correspondence to the rain-

---

① Zhou Bing, Tan Yanke and He Jinhai: 1998: Analysis of vertical structure of shearing vorticity and non-adiabatic heating in a severe rainstorm at Wuhan in 1998 (submitted to A. M. S. in 1999).

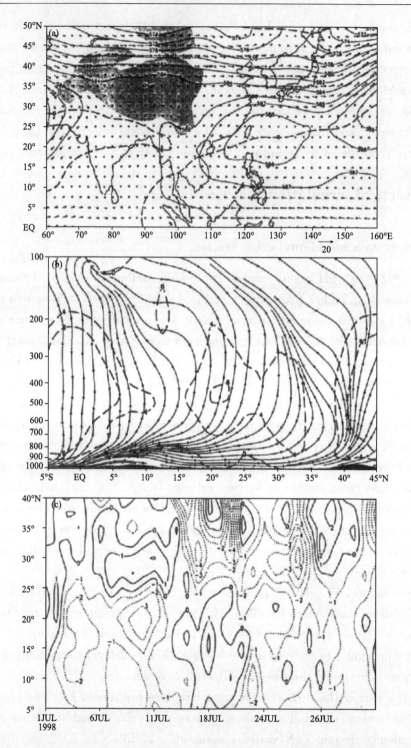

Fig. 1 500-hPa height and wind (a, dashed line denoting zero-value wind), 125°E section of a meridional circulation through the STH center (b, dashed line giving ω-isopleth in the rising region units of $-hPa \cdot (6\ h)^{-1}$) and the latitude-time cross section of ω along 125°E (c) in July 1998. The blacked regions at the bottom in (b) show actual top graphic height (the same below)

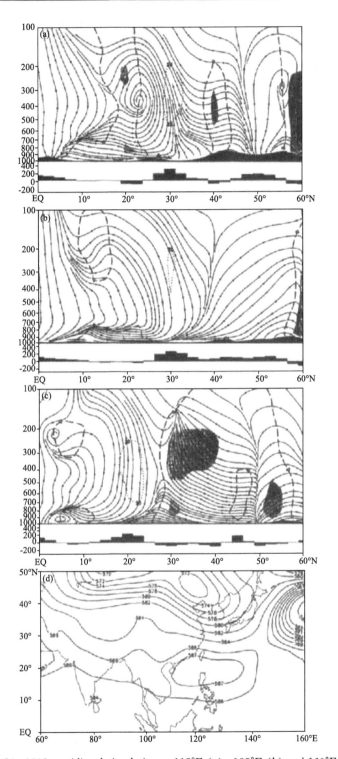

Fig. 2 July 19—24, 1998 meridional circulation at 115°E (a), 125°E (b) and 160°E (c), and related whole-extent vaper sink (bottom block) and 500-hPa circulation configuration (d). ω is in $-$hPa $\cdot$ $(6\ h)^{-1}$ and $\langle Q_2 \rangle$ in W $\cdot$ m^{-2}. Dotted line denotes an updraft center, thick dashed line the zero-value rising and the shaded sector the subsidence (-5 units of vertical velocity)

fallhit band; about 20°N, considerable downdraft emerges throughout the air column, accompanied by remarkable negative $\langle Q_2 \rangle$ just at the same latitude as the ridge in that time span (refer to Fig. 2d). And further inspection shows that the air-column downdraft at upper levels comes from updraft in the rainstorm area to the north, forming a quite strong circulation cell core between 300-hPa and 400-hPa and downdraft below 500-hPa air column has its origin in a low-latitude circulation. In contrast, the feature somewhat differs, as viewed on the 125°E section (Fig. 2b) in such a way that around 30°N updraft and vapor convergence happen except their lower intensity compared to the condition at Wuhan, and that the updraft turns southward, descending about 15°N and falling short of 400 hPa level. At 20°N in the vicinity of the ridge there arise weak downdraft of low latitude origin and $\langle Q_2 \rangle$ close to zero below 800 hPa and updraft occupies 800~400-hPa extents. The above features indicate that the robust updraft in the rainstorm belt contributes significantly to the upper-air downdraft around the ridge. On the 160°E section (Fig. 2c), the ridge is between 35°N and 40°N with low-level downdraft still coming from low latitudes but the upper-level flow feature differs considerably from what was given above, which may be attributed to downdraft behind the summer mid-ocean trough.

Fig. 3a illustrates 130°E meridional circulation on August 11-16, 1998 for another flood

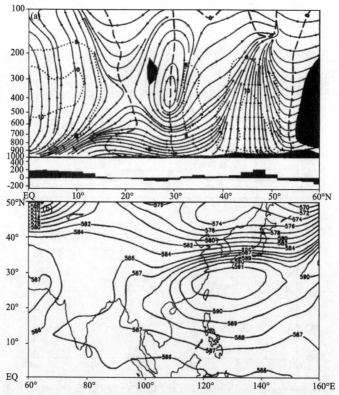

Fig. 3 Meridional circulation across the STH center and its related whole-extentmmoisture sink (a) and 500-hPa circulation pattern (b) on August 11-16, 1998

interval over the valleys. We see therefrom that a large-range updraft region is roughly between 30°N and 50°N, with its robust centers located about 33°N and 48°N, accompanied by intense vapor convergence and $\langle Q_2 \rangle$ and corresponding to a rainband in the Jiang-Huai and Nenjiang River basins, respectively; strong downdraft takes place throughout the air column between 20°N and 30°N, in company with conspicuous negative $\langle Q_2 \rangle$, meanwhile the 130°E STH core resides in the neighborhood of 28°N (see Fig. 3b). Similar to Fig. 2a, a strong circulation center is around 400-hPa and updraft from the Jiang-Huai rainband turns at the level, forming downdraft near the STH center, with low-level downdraft still coming from low latitudes.

Fig. 4 shows the characteristics analogous to what has been said before but one can see significant contribution of low-latitude flow to downdraft in the ridge belt around 30°N (see Fig. 4b) and the vertical motion is quite feeble over the extensive band to the north of 32°N and robust updraft is at 50°N. It is probably the situation that makes it unlikely to keep the STH around 30°N, leading to the STH withdrawal southward soon after in such a way that the Meiyu in the middle and lower reaches of the Yangtze River occurred once again in the year.

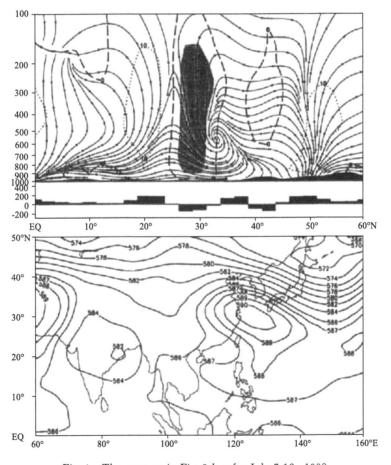

Fig. 4 The same as in Fig. 3 but for July 7-12, 1998

Based on the foregoing analysis, we are led to believe that in the mean meridional section and on a pentad scale robust downdraft and related vapor divergence arc always detectable in the vicinity of the ridge, particularly when the rain band is quite active and associated vigorous updraft occurs during the rainstorm, which make a great contribution to high-level downdraft around the ridge but low-level downdraft is dominantly of low-latitude origin, thereby suggesting that the STH is a tie between mid- and low- latitude flows.

3.2 Circulation features on a zonal cross section

As stated in Introduction, the STH genesis is interpreted in terms of a descending arm of a zonal circulation instead of the Hadley circulation for its defect. Then, what is about the western Pacific STH in 1998? For illustrative purpose we have prepared maps of zonal vertical circulations across the STH core to show the features in the flood interval for July 19-24 (Fig. 5a) and the dry period for July 7-12 (5b) in the middle and lower reaches of the Yangtze River.

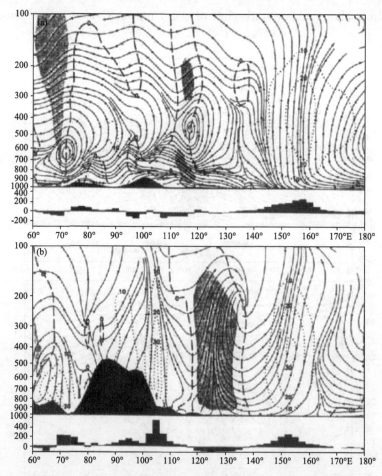

Fig. 5 Zonal circulation along 20°N on July 19-24 (a) and 30°N on July 7-12 (b) and their related whole-larger moisture sink for the ridgeline at different latitudes. Otherwise as in Fig. 2

Fig. 5a depicts substantial downdraft throughout the air column in 110°~120°E in good correspondence to the subsidence around the ridge given in Fig. 2a; mid-high levels east of 120°E are covered with updraft, and robust rising motion lies in the vicinity of 150°~160°E, accompanied by quite intense $\langle Q_2 \rangle$. It is worth noting that the high levels westward of 150°E to 60°E or so are occupied by easterly wave-form flow whilst the lower levels eastward of 60°E to 140°E by the westerly counterpart, resulting in two circulation centers, one at 450-hPa around 120°E and the other at 750-hPa about 70°E. It is apparent that flow from the western Pacific and monsoon flow from India south of the Tibetan Plateau contribute greatly to the high- and low-level downdraft of the STH, respectively, thus suggesting the interaction between the STH and Indian monsoon / the mid-western Pacific circulation at low latitudes.

Fig. 5b portrays that updraft emerges over the plateau and on both its sides, in company with vapor convergence and noticeable positive $\langle Q_2 \rangle$; robust downdraft is seen in the air column of the STH center in 30°N, 120°~130°E; vigorous updraft remains in the neighborhood of 150°~160°E over the mid-western Pacific. Is the configuration of circulations related to the fact that the STH central downdraft is induced by strong updraft on the east side of the plateau and over the mid-western Pacific or to the fact that it is the STH downdraft that intensifies updraft on both its sides? which awaits further research.

4 Related features of the STH interannual variation and the mechanism

The STH position (meridional and zonal) and intensity or area (the western Pacific STH parameters) undergo not only seasonal but also interannual variations. Especially the interannual variation bears a direct relation to the summer rainfall patterns and wetness / dryness occurrence in China[9]. For this reason, the study of the interannual variation features and the mechanism has been a great concern for some meteorologists. Huang[10] reported his systematic research on STH activities. Dong and Chou[11] indicated that the physical essence of STH genesis and variation is the result of integrated interaction between internal and external forcings on the atmosphere. We shall focus on the correlation of STH variation with precipitation over the Yangtze River valley and external forcing, followed by investigating possible mechanism of STH variation.

4.1 STH interannual variation associated with summer rainfall over the middle and lower reaches of the Yangtze River

Liu et al.[12] emphasized the key effect of latent heat released from monsoon rainfall over East Asia on the eastern hemisphere summer STH. As stated before, the active monsoon rainband at the subtropics and updraft during a torrential rain are the source of high-level downdraft around the ridge. If so, is there a relationship between the valley summer rainfall and STH interannual variation? Therefore, we constructed Table 1 that summarizes high negative (positive) correlation between the rainfall and STH latitudes (the rainfall and STH intensity / area index), meaning that the precipitation, stronger (weaker) than usual, corresponds to the STH positioned southward (northward) of mean, its higher (lower) intensity and larger (smaller) area. Evidently, these correlation features show that a close relation or a kind of interaction does exist between the valley rainfall (and related latent heat release) band, on one side, and meridional locality of the ridge and STH intensity (or the area), on the other. Such correlation or interaction is in concord with the conclusion as regards the key role of latent heat release from monsoon precipitation in the STH variation[6] and also with the self-adjustment mechanism inherent in the monsoon system[13].

Table 1 Correlation coefficients between the summer rainfall over the middle and lower reaches of the Yangtze River and STH parameters in 1958-1993

	Ridge's Latitude	STH Intensity	Area Index
Rainfall	−0.485	0.476	0.408

The thresholds passing $\alpha=0.05$ and 0.01 significance tests are 0.325 and 0.418

4.2 STH ridge latitude in relation to the land-sea thermal difference index

The western Pacific STH ridge variation on an interannual basis has been a focus in the study of floods/droughts in China[9]. Miao and Ding[14] and Liu and Tao[12] addressed STH meridional shift and seasonally abrupt change in terms of the seasonal variation in solar radiation and meridional or zonal thermal contrast, which leads us to take land-sea contrast as a basic factor of ridge latitude variation on an interannual basis.

Sun et al.[8] made calculation of the index of land-sea thermal difference (I_{LSTD}) from ground temperature (T) and sea surface temperature (TSS) in four key regions with the aid of

$$I_{LSTD} = (T_{EC} - T_{SS,STNWP}) \times \left[\frac{4}{5} + (T_{SC} - T_{SS,SCS}) \times \frac{1}{5}\right]$$

in which EC stands for East China (27°~35°N, east of 105°E), STNWP for the subtropical northwestern Pacific (15°~30°N, 120°~150°E), SC for South China (south of 27°N, east of 105°E) and SCS for the South China Sea (5°~18°N, 105°~120°E). To explore effects of land-sea thermal contrast on the STH, we investigated the correlation between I_{LSTD} and the ridge latitude at 120°E, reaching the correlation of 0.64 and gave their interannual curves in Fig. 6. We see therefrom that when the I_{LSTD} is lower (higher), the ridge is southward (northward) of mean, the valley rainfall is more (less) and the area index (intensity) is larger (smaller). The I_{LSTD} is, in fact, an index characterizing the strength of eastern Asian summer monsoon. As such, the above correlations show that the land-sea thermal contrast really represents a basic factor responsible for the interannual difference in the STH ridge latitude, East Asian summer monsoon intensity and wetness/dryness over the Yangtze River valley.

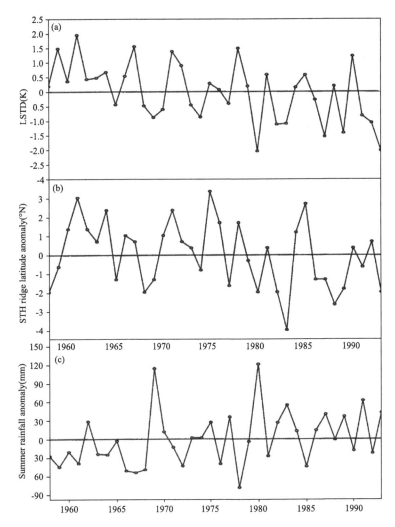

Fig. 6 Interannual variation in the related variables with East Asian summer I_{LSTD} (a). STH ridge latitude (b) and summer rainfall anomaly for the middle and lower reaches of the Yangtze River (c). The correlation coefficients $r_{ab}=0.64$, $r_{ac}=-0.73$ and $r_{bz}=-0.49$, reaching $r_{0.01}=0.42$

4.3 Numerical experiment of offshore SST influence on STH

Recent studies (e. g. , Wen et al. [15]) have stressed the impact of China offshore SST on its climate and flood / drought in summer. From the perspective of air-sea interplay, the offshore SST is thought to have more direct effect than the equatorial eastern Pacific SST on the western Pacific STH. Table 2 indicates that the offshore SST is remarkably correlated with the anomaly of STH strength and its area index, to a confidence level more than 0.01, and the South-China offshore SST (15°~25°N, 120°~140°E) bears a significantly negative correlation (−0.46) with the ridge latitude. The correlation does show an intimate relation between the offshore SST and the STH.

Table 2 Correlation coefficients between China offshore SST and STH parameters in 1958-1993

	Ridge Latitude	STH Intensity	Area Index
South-China offshore SSTA(15°~25°N, 120°~140°E)	−0.464	0.613	0.626
Offshore SSTA(15°~30°N, 130°~150°E)	−0.235	0.485	0.510

To further reveal the intrinsic mechanism for the correlation between the SST and STH interannual anomalies, Wang et al.[16] adopted the L9R15 spectral model developed by Wu et al.[17] to perform an experiment on sensitivity of the STH to the offshore sea surface temperature anomaly (SSTA). For the control run we utilized the climatic mean on January 31 as the initial field and the 1979-1998 mean SST as the model sea surface temperature, followed by integrating for 243 model days, achieving a well-simulated East Asian summer monsoon system including an STH based on the June-August output (figure not shown). The imposed SSTA pattern for the sensitivity run is given in Fig. 7a, with the center in (25°N, 140°E) and the domain covering the two offshore SSTA areas in Table 2. April-June positive and negative SSTA were introduced, separately, into the model (denoted as EX+ and EX−, respectively), followed by getting the difference in June-August simulations (EX+ minus EX−), leading to a difference-value circulation pattern at 500-hPa in summer (Fig. 7b). One can see that an anticyclonic center is around the southern Taiwan, suggesting that offshore positive SSTA will excite an anomalous anticyclonic circulation in the offshore sea. Obviously, it is the anomalous circulation that contributes to the enhancement of the STH intensity (or area) and its ridge southward of mean in the presence of positive SSTA. It is worth particularly noting that the anomalous difference-value anticyclonic circulation occurs just on the west side of the offshore SSTA.

The above offshore SSTA area is situated in the domain where the main body of the western Pacific STH resides and precipitation is small, as such, it can be considered that the effect of the SSTA, especially the effect of sensible heat, on the western Pacific STH is important. Following the theoretical analysis of Wu et al.[17], positive SSTA-produced atmospheric sensible heating is thought to excite northerly anomaly over the layer of maximum heating, leading to an anomalous anticyclonic circulation on the west side of the heating band. The high consistency of results from numerical experiments and theoretical analysis shows that the correlation between offshore SSTA and STH interannual variation is based on an intrinsic dynamic mechanism. As a new clue, the consistency enlightens us on the study of the STH anomaly on an interannual basis.

5 Conclusions

From the foregoing analysis we come to the conclusions as follows:
(1) The vertical circulation structure of the western Pacific STH shows a significant difference in features on its monthly mean and pentad or daily mean maps, the former displa-

ying updraft around the center or the ridge line and the latter noticeable downdraft in some parts close to the ridge, especially when a rainstorm takes place in the subtropical monsoon rainband on the north side of the STH, and latent heat release-produced local meridional circulation makes great contribution to downdraft in the neighborhood of the ridge at a related meridian. This well indicates that the quasi-stationary STH has the vertical circulation structure features significantly different from those of the transient STH.

(2) In the vicinity of the STH ridge, high (low) level downdraft comes from the north (south) side of the STH, suggesting that the STH is a tie of interaction between circulation systems at high and low latitudes. On a zonal section the features of zonal circulation consisting of the STH central downdraft and the updraft on both its sides change conspicuously with the latitude at which the ridge is located.

(3) High correlation exists between the interannual variation in summer STH parameters (intensity, area and ridge latitude) and rainfall over the middle and lower reaches of the Yangtze River, implying an interaction and interrelation between the subtropical summer monsoon rainband on the STH north side and the STH itself, which mirrors the intrinsic link between latent heat released from summer monsoon precipitation and STH variation on an interannual basis. The meridional and zonal land-sea thermal contrast in the East Asian monsoon belt is a key factor responsible for the interannual anomaly of the STH ridge latitude and rainfall over the middle and lower reaches of the Yangtze River.

(4) Sensitivity experiments on offshore SSTA show that with higher (lower) SSTA in the offshore sea, the associated sensible heating will excite an anomalous anticyclonic (cyclonic) circulation on the west side of SSTA. It is such a circulation that leads to the high correlation between the offshore SSTA and interannual variation of the STH.

References

[1] Tao Shiyan, Chen Longxun. A review of recent research on the East Asian summer monsoon in China [C]// Chang C-P, Krishnamurti T N. Monsoon Meteorology. Oxford University Press, 1987:60-92.

[2] Zhu Qiangen, He Jinhai, Wang Panxing. A study of circulation differences between East-Asian and Indian summer monsoons with their interactions[J]. Adv Atmos Sci,1986,3(4):466-477.

[3] Zhu Qiangen, Lin Jinrui, Shou Shaowen, et al. Principles and Methodology of Synoptic Meteorology[M]. Beijing: China Meteorologica Press,1981 (in Chinese).

[4] Yeh Tu-Cheng, Yang Guangji, Wang Dongxing. Mean vertical circulations over East Asia and the Pacific. Part I: Summer[J]. Scientia Atmosphere Sinica,1979,3(1):1-11 (in Chinese).

[5] Chen Ping. On the origin and seasonality of subtropical anticyclones and cyclones[J]. submitted to J Atmos Sci,1999.

[6] Wu Guoxiong, Liu Yimin, Liu Ping. Role of space inhomogeneous heating in the genesis and variation of subtropical highs, Part I: Scale analysis[J]. Acta Meteorologica Sinica,1999,57(3):257-263 (in Chinese).

[7] Liu Yimin, Wu Guoxiong, Liu Hui, et al. Role of space inhomogeneous heating in the genesis and variation

of subtropical highs. Part Ⅲ: Latent heating related to south Asian highs and western Pacific subtropical highs[J]. Acta Meteorologica Sinica,1999,57(5):525-538 (in Chinese).

[8] Sun Xiurong, Chen Longxun, He Jinhai. The interannual variation of East Asian land-sea thermal difference in association with circulation and rainfall[J]. Acta Meteorologica Sinica,1999 (in Chinese).

[9] Huang Ronghui. Teleconnection among East Asian circulation anomalies and the physical mechanism for summer flood / drought in China[J]. Scientia Atmosphere Sinica,1990,14(1):108-117 (in Chinese).

[10] Huang Shisong. Study on subtropical highs' activities and forecasting[J]. Scientia Atmosphere Sinica,1978,2(2):159-165 (in Chinese).

[11] Dong Buwen, Chou Jifan. Case study of the seasonal variation of the ridge latitude of western Pacific subtropical highs with numerical simulation[J]. Acta Meteorologica Sinica,1988,46(3):361-363 (in Chinese).

[12] Liu Chongjian, Tao Shiyan. On the northerly jumping of subtropical highs and cusp abrupt change[J]. Scientia Sinica (B),1983,13(5):474-480 (in Chinese).

[13] Yu Shihua, Wang Shaolong. On the circulation mechanism for the advancement / withdrawal of western Pacific subtropical highs in the middle of their lifecycle[J]. Acta Oceanographica Sinica,1989,11(3):372-377 (in Chinese).

[14] Miao Jinhai, Ding Minfang. On the abrupt and seasonal variation in atmospheric equilibrium and northward jumping of subtropical highs under thermal forcing[J]. Scientia Sinica (B),1989,19(1):87-96 (in Chinese).

[15] Wen Min, Luo Jingjia, He Jinhai. Flood/drought years and air-sea interaction[J]. J Nanjing Institute of Meteorology,1997,20(3):341-347 (in Chinese).

[16] Wang Lijuan, He Jinhai, Xu Haiming. Effects of Kuroshio SSTA on the onset time of South-China Sea summer monsoon with numerical study[J]. J Nanjing Institute of Meteorology,2000,23(2):211-217 (in Chinese).

[17] Wu Guoxiong, Liu Hui, Zhao Yucheng. A nine-layer atmospheric general circulation model and its performance[J]. Adv Atmos Sci,1996,13(1):1-18.

宁夏春季沙尘暴与北极海冰之间的遥相关关系[*]

杨建玲[1]　何金海[1]　赵光平[2]

（1. 南京气象学院大气科学系，南京 210044；2. 宁夏气象防灾减灾重点实验室，银川 750002）

摘要：根据宁夏沙尘暴发生次数资料、北极海冰密集度资料和 NCEP/NCAR 再分析 500 hPa、850 hPa 高度场、风场资料，得出了宁夏春季沙尘暴发生次数的变化规律及其与北极海冰面积之间的年代际和年际相关关系，发现宁夏春季沙尘暴发生次数与欧亚大陆北部的喀拉海、巴伦支海和格陵兰海冰面积之间存在较显著的年代际、年际相关关系。通过合成和相关分析知，宁夏春季沙尘暴偏多、偏少状况有明显不同的环流背景场，秋季格陵兰海冰异常变化通过影响其后一段时间的大气环流背景场，从而对宁夏沙尘暴产生影响。初步得出当格陵兰海秋季海冰面积增大（减小），次年春季蒙古至西伯利亚一带 500 hPa、850 hPa 高压场降低（升高），风场有明显的气旋性（反气旋性）特点，在宁夏至新疆一带西风明显偏强（偏弱），说明冷空气活动次数偏多（少），对应宁夏春季沙尘暴发生次数偏多（少）。通过海冰将全球气候变暖和宁夏（我国北方）沙尘暴总减少趋势联系起来，初次提出在环境总体恶化情况下，我国沙尘暴发生次数总体趋于减少，很可能是全球气候变暖所致。

关键词：北极海冰；宁夏；沙尘暴；遥相关

我国西北地区是沙尘暴多发区，沙尘暴发生的基本条件有三个：一是大风，二是沙源，三是局地不稳定能量，即沙尘暴是在特定的地理环境和气象条件下发生的[1]。近几十年来，我国北方地区的环境和植被覆盖总体趋于恶化[2]，而我国北方沙尘暴发生次数近几十年来总体呈减少趋势[3-4]，单从地表状况来看，似乎存在着矛盾。所以沙尘暴的减少原因必然与天气气候的异常变化有关。

近百年来全球气候逐渐变暖，其后果之一是引起了北极海冰的消融。赵玉春等[5]的研究结果表明，北极海冰面积总体呈减小趋势。黄士松等[6]认为北极海冰异常对区域气候影响方面具有与 El Nino 同等重要的作用，在某些情况下，其影响甚至可以超过后者。作为冷源的北极海冰，在北半球高纬度的海气相互作用中，有着突出的贡献，极地海冰不仅具有明显的局地效应，而且对区域大气具有显著作用。而以往有关沙尘暴的研究结果表明，几乎每一次沙尘暴的发生都与极地南下冷空气形成的锋面系统相联系，因此北极海冰的异常变化很可能影响我国西北地区沙尘暴的发生。

尚可政等[7]研究了赤道太平洋海温与河西沙尘暴发生次数之间的遥相关关系，然而与极地冷空气活动密切相关的极地海冰的异常变化对沙尘暴影响的研究迄今却还是个空白。

本文将主要讨论两个方面的问题，一是在分析宁夏沙尘暴发生次数变化规律的基础上，研

---

[*] 本文发表于《南京气象学院学报》，2003 年第 26 卷第 3 期，296-307。

究宁夏春季沙尘暴发生次数与北极各海区海冰面积之间的年代际、年际关系;二是合成分析宁夏春季沙尘暴发生次数异常的环流背景场,研究格陵兰海秋季海冰面积变化影响次年春季环流场的特征,以寻求格陵兰海冰影响宁夏春季沙尘暴的可能途径。本文进一步通过海冰将全球气候变暖和宁夏(我国北方)沙尘暴总体减少的趋势联系起来,试图解释我国沙尘暴在环境总体逐渐恶化的情况下变化趋势减少的原因。

关于海冰与其他地方沙尘暴的关系及它们之间的物理、动力机制还需进一步研究。

1 资料和方法

本文研究主要使用了三种资料:第一种是选取宁夏具有代表性的 18 个站(银川、永宁、贺兰、惠农、平罗、陶乐、吴忠、青铜峡、灵武、中宁、中卫、兴仁、盐池、韦州、同心、固原、西吉、海原)1960—2000 年 41 a 沙尘暴月发生次数资料,这里每个站的一次沙尘暴是指某一天(以北京时间 20:00 为界)只要出现沙尘暴(由于强风将地面大量沙尘吹起,使空气很混浊,水平能见度小于 1 km 的天气现象),而不管该日有几个时段发生沙尘暴。第二种资料是1968—2000 年北极(41°N 以北,纬度×经度=3°×1°)月平均海冰密集度资料(海冰密集度的值为 0～1.0,0 表示网格点范围内无冰盖,0.1 表示网格点范围内有 10%的冰盖,依次类推)。第三种资料是 NCEP/NCAR 再分析的 1960—2000 年月平均 500 hPa、850 hPa 高度场、风场资料。

采用的研究方法有:统计相关方法、lepage 突变检验法、合成分析及 t 统计检验法。这种 t 检验是用来检验两个总体均值差异性是否显著的方法,其公式为

$$t = \frac{\overline{x}-\overline{y}}{\sqrt{\frac{(n_1-1)s_1^2+(n_2-1)s_2^2}{n_1+n_2-2}}\sqrt{\frac{1}{n_1}+\frac{1}{n_2}}}$$

式中,x 和 y 的样本量为 n_1 和 n_2,样本均值为 $\overline{x}$ 和 $\overline{y}$,s_1^2 和 s_2^2 分别表示两个样本的方差。

2 宁夏沙尘暴发次数的年代际和年际变化规律

将宁夏 18 个站 1960—2000 年的沙尘暴发生次数分别按春季(3—5 月)和全年求和,为了便于比较,将资料进行标准化处理。由图 1 可以看出,宁夏近 41 a 春季和全年沙尘暴发生次数,总体都呈下降趋势,这与牛生杰等[8]的研究结果相吻合,也与全国沙尘暴总体呈减少的趋势相一致。值得注意的是它们均在 1984 年左右发生了明显的突变,1984 年之前大部分年份沙尘暴发生次数大于 41 a 的平均值,而在 1984 年以后,全部小于平均值。宁夏 18 个代表站春季沙尘暴发生次数 41 a 的平均值为 73 次,1984 年之前平均为 95 次,1984 年之后仅为 38次,前者是后者的 2.5 倍。5 a 滑动平均线更明确地表示了这种年代际变化特点。为了检验突变的存在,作了春季沙尘暴发生次数序列的 lepage 统计量的突变检验,子序列长度取为 10 a。由图 2 知春季沙尘暴在 1984 年确实发生了较明显的突变,lepage 统计量在 1984 年有一个峰值,其值为 8.69,大于置信水平为 95%的临界值 5.99。另外沙尘暴的发生还存在明显的偏多和偏少年,有明显的年际变化特征,而且具有周期性。因此可以说,宁夏春季沙尘暴发生次数的变化存在明显的年代际和年际变化特征。

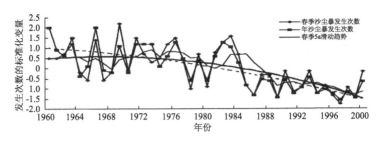

图1 1960—2000年宁夏春季、年沙尘暴发生次数(经标准化处理)的年际变化及其趋势
(粗实线和粗虚线分别是春季和年的二次多项式拟合线)

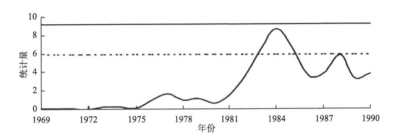

图2 宁夏春季沙尘暴发生次数的lepage检验统计量
(实直线和虚直线分别为置信水平为99%、95%的临界值)

春季和全年沙尘暴的年际变化一致,相关系数高达0.90,这是因为沙尘暴多发生于春季。春季总的沙尘暴占全年总数的57.9%,其中4月是沙尘暴发生的高峰月份。9—10月,沙尘暴发生最少(图略)。因此下面只以春季沙尘暴为例进行讨论。

3 宁夏春季沙尘暴发次数与北极海冰面积之间的相关分析

由于地理位置的不同,不同区域的海冰与大气和海洋间的相互作用有许多不同的地方。Walsh[9]指出,海冰研究应按地理位置不同细分为不同海域进行。参考武炳义等[10]、彭公炳[11]的做法,将北极海冰分成6个不同的海区,A海区:喀拉海和巴伦支海海区(20°~100°E,41°~82°N);B海区:格陵兰海区(45°W~0°~20°E,41°~84°N);C海区:巴芬湾、戴维斯海峡(45°~80°W,41°~80°N);D海区:白令海区(160°E~180°~150°W,41°~65°N);E海区:将整个北极(180°W~0°~180°E,41°~84°N)范围的海冰作为一个整体来考虑;F海区:欧亚大陆北部附近海域,即A、B海区合并起来(45°W~0°~100°E,41°~84°N)。分E、F区的目的是为了研究北极和欧亚大陆北部海域海冰整体面积变化对沙尘暴发生次数的影响。

分别将各海区1968—2000年33 a的海冰密集度资料按年、春、夏、秋、冬5个不同时段分别求和,组成30个33 a的时间序列,作为北极海冰面积指数的基本资料。再将宁夏18个站1968—2000年共33 a春季沙尘暴发生次数求和,组成1个33 a的时间序列,作为沙尘暴发生次数序列资料。

首先计算宁夏春季沙尘暴发生次数与6个不同海区同期春季以及前一年全年、春、夏、秋、冬6个时段海冰面积指数之间的相关系数。由表1可以看出:宁夏春季沙尘暴与北极海冰面积之间的相关系数几乎全部为正,但是不同海区不同时段相关系数大小之间有较大差异。从海区来看,与巴芬湾、戴维斯海峡和白令海海冰面积之间的相关系数较小,相关系数绝大部分达不到置信水平为95%的临界值0.349,而与A、B、E海区海冰面积之间的正相系数较大,绝大多数相关系数大于置信水平为99%的临界值0.448。从时间上看宁夏春季沙尘暴与同期海冰面积的相关系数并不是最大,相关系数相对较大的是与前一年秋季的海冰,为0.6732,其次是前一全年,其他则一般按时间往前推(前一年冬、夏、春),相关系数依次减小。

表1 宁夏春季沙尘暴发生次数与北极海冰面积指数之间的相关系数

海冰时段	A海区	B海区	C海区	D海区	E海区	F海区
当年春	0.3522	0.5030	0.3149	0.4198	0.5269	0.4652
前一年	0.5432	0.5797	0.2513	0.2405	0.5353	0.5874
前一年春	0.3495	0.4503	0.2259	0.1514	0.4635	0.4422
前一年夏	0.4892	0.5520	0.2832	0.3252	0.4664	0.5404
前一年秋	0.5629	0.6732	0.2568	0.1661	0.6233	0.6415
前一年冬	0.4696	0.5837	0.1733	0.3192	0.5029	0.5488

然而这种一致的正相关和特定海区持续的高相关系数并不一定能够说明沙尘暴和海冰之间的年际变化关系,因为沙尘暴在1960—2000年之间的总体变化呈明显的负趋势,在计算相关系数时包含了年代际趋势,若海冰的年代际变化也呈较明显的负趋势时,无论年际变化关系如何,如上所计算的相关系数一定是较大的正相关。事实正是如此,相关系数较大海区(如E、F海区)海冰面积总体呈减少趋势,而且与宁夏春季沙尘暴发生次数总趋势一致。因此说,宁夏春季沙尘暴与A、B、E海区海冰面积的年代际变化呈一致的减少趋势,有较好的年代际变化关系。

那么,宁夏春季沙尘暴发生次数与北极各海区海冰之间的年际变化关系究竟如何呢? 为此,在计算相关系数时应采用年际变率的方法去除趋势项,即将原序列求年际变率,得到年际变率的时间序列,再求年际变率序列的相关系数,计算内容和表1相同,得到年际变率的相关系数(表2中只列出部分有相关系数通过置信水平为95%检验的时段)。可以看到,年际变化的相关系数有正有负,而且大部分比较小,只有较少数通过了显著性检验,其中与前一年秋季格陵兰海冰的相关系数最大,为0.471。

因此,宁夏春季沙尘暴发生次数与北极海冰在年代际和年际变化趋势上都有密切关系。从以往对沙尘暴的研究发现,几乎每一次沙尘暴的发生都与极地冷空气南下形成的锋面系统相联系[12],而极地海冰作为冷源,其异常变化必然引起其上极地气团的性质及其活动变化,因此地处极地冷空气南下通道上的沙尘暴多发区之一——宁夏地区,其沙尘暴发生次数与极地海冰面积变化之间有密切关系是合理的。

表2 宁夏春季沙尘暴发生次数年际变率与北极海冰面积指数年际变率之间的相关系数

海冰时段	A海区	B海区	C海区	D海区	E海区	F海区
前一年春	−0.2551	−0.3713	0.2377	−0.1608	−0.1825	−0.3600
前一年秋	0.2782	0.4710	0.2744	−0.2472	0.3665	0.3836

4 宁夏春季沙尘暴发生次数异常的大尺度环流背景场特征

宁夏春季沙尘暴发生次数的异常变化必然有其异常的环流背景,为了突出其主要特点,应对其异常情况的环流背景进行合成分析。因为宁夏春季沙尘暴的发生同时具有明显的年代际和年际变化特点,因此合成分析时,对与年代际和年际变化相关联的大气环流形势分别进行讨论。

4.1 与沙尘暴年代际异常变化相联系的大气环流特征

以1984年为界,将宁夏春季沙尘暴突变前1960—1984年和突变后1985—2000年分别作为沙尘暴偏多和偏少阶段,研究其所对应的500 hPa、850 hPa高度场和风场特征。

由偏多阶段北半球500 hPa高度距平场(图略)可以看出,有三个正变高中心,分别位于格陵兰、东西伯利亚和阿留申群岛及其以南地区,主要负中心位于亚洲的蒙古、贝加尔湖至西西伯利亚、北美加拿大、北大西洋和地中海地区,另外在低纬度较宽广范围内为负变高区域。850 hPa与500 hPa形势相似,只是正负变高的强度和范围略有差别,位于亚洲的变高中心范围更大一些。偏少阶段与偏多阶段的形势基本相反。为了判断上述两种状态差异的显著性,对它们做了统计显著性差异t检验。在偏多阶段减偏少阶段高度场差值t检验图(图3a、b)上,500 hPa上亚洲的蒙古、贝加尔湖、北美加拿大、北大西洋、地中海和低纬度的较大范围为通过置信水平为95%检验的负差异区,850 hPa的差异显著区与500 hPa的相比,位于亚洲蒙古、贝加尔湖一带的负差异区范围更大一些,而北美加拿大地区的范围缩小了,欧洲南部的负差异显著区和低纬度的大范围负差异显著区连成一片。

对应偏多阶段500 hPa风场距平图(图略),在影响我国天气的上游关键区蒙古、贝加尔湖和西西伯利亚一带,有一明显的花生状闭合气旋性环流,在咸海附近有一范围较小的反气旋性环流,宁夏地区正好位于气旋性环流底部和反气旋性环流顶部西风异常偏强区,偏少阶段与偏多阶段的形势基本相反。偏多阶段850 hPa风场距平图上(图略),以蒙古为中心的闭合气旋性环流非常明显,宁夏位于气旋性环流底部的西风异常偏强区内,在蒙古至西西伯利亚一带,为一Ω状的气旋性环流形势,偏少阶段与偏多阶段的形势也基本是相反的。500 hPa、850 hPa风场偏多阶段减偏少阶段的差值图(图3c、d)清楚地表明了上述风场的环流特点。由偏多阶段减偏少阶段的纬向风、经向风差值t检验图(图3e-h)可以看出,超过置信水平为95%检验的区域,500 hPa上分布比较分散,范围较小,850 hPa上大片的纬向风和经向风的差异显著区域连在一起。特别值得注意的是,850 hPa纬向风差值t检验图上,在宁夏、甘肃、青海至新疆一带的我国西北地区,有通过置信水平为95%的西风异常。因为沙尘暴发生的必要条件之一就是大风,所以这种西风异常的特点是非常重要的。

因此说,宁夏春季沙尘暴偏多阶段和偏少阶段的大气环流背景场确实存在明显差异。沙尘暴偏多(少)阶段,在关键区蒙古、贝加尔湖、西西伯利亚一带,500 hPa、850 hPa高度场偏低(高),风场有明显的气旋(反气旋)环流。说明在沙尘暴偏多(少)阶段,关键区内极地南下的冷空气、气旋活动次数偏多(少),这种大气环流的异常是造成宁夏沙尘暴年代际异常的环流背景场。全林生等[13]的研究表明,中国北方春季的沙尘暴天气频次与春季850 hPa上的气旋活动次数呈非常显著的正相关关系。本文的分析结果与全林生等[13]的研究结果相吻合。

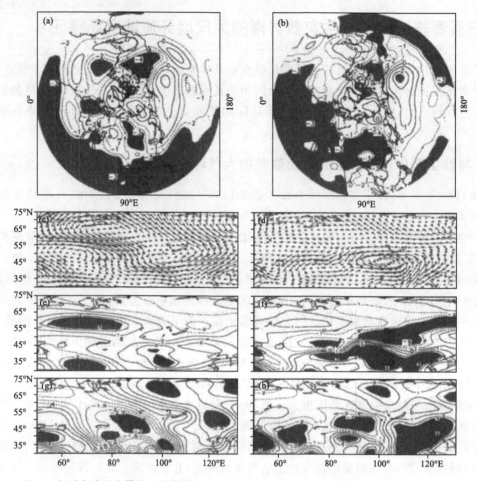

图3 宁夏春季沙尘暴发生次数偏多阶段(1960—1984年)减偏少阶段(1985—2000年)高度场和风场差值图(阴影区通过置信水平为95%的t检验)

a. 500 hPa 高度场；b. 850 hPa 高度场；c. 500 hPa 水平风场；d. 850 hPa 水平风场；
e. 500 hPa 纬向风；f. 850 hPa 纬向风；g. 500 hPa 经向风；h. 850 hPa 经向风

4.2 与沙尘暴年际异常变化相关联的大气环流特征

若在1960—2000年的41 a中按同一个标准划分沙尘暴偏多年、偏少年，就会使选出的偏多年集中在1984年以前，偏少年集中在1984年之后。这样分析时，实际并没有能够分离出年际变化并真正对年际变化进行说明，因为其中还混杂了年代际变化的特点。为了避免这种情况，在沙尘暴发生突变前1960—1984年25 a中进行分析(1984年以后的时间段太短，不便进行分析)，因为在该时间段内沙尘暴的变化趋势是平稳的。计算该时间段的平均值及其均方差，将大于和小于平均值0.5个均方差的年份分别作为沙尘暴偏多年和偏少年，得到8个偏多年是1963、1966、1969、1971、1975、1976、1982、1983年，7个偏少年是：1964、1965、1967、1968、1970、1978、1980年。分析内容同4.1。

由500 hPa、850 hPa高度场的差异性t检验图可以看出，偏多、偏少年(图4a、b)与偏多、偏少阶段(图3a、b)的形势相比有不同之处。图3中低纬度的大范围负差异显著区在图4中没有,北美加拿大的负差异中心向西移动大约 15 个经度。但在影响关键区内的形势是相似的,即在蒙古、贝加尔湖至西伯利亚一带是通过置信水平为 95% 检验的负差异区,只是负差异的区域范围和强度都有减小。850 hPa 上的差异性特点比 500 hPa 的明显一些。

对应 500 hPa、850 hPa 风场距平图(图略),在上述影响关键区内,偏多年有气旋性环流,偏少年有反气旋性环流。由偏多年减偏少年的差值图(图 4c、d)及纬向风和经向风差值 t 检验图(图略)可以看出,这种气旋性、反气旋性差异与偏多、偏少阶段相比,虽然差异性大小以及通过置信水平为 95% 检验的区域范围也有减小,但形势是相似的。

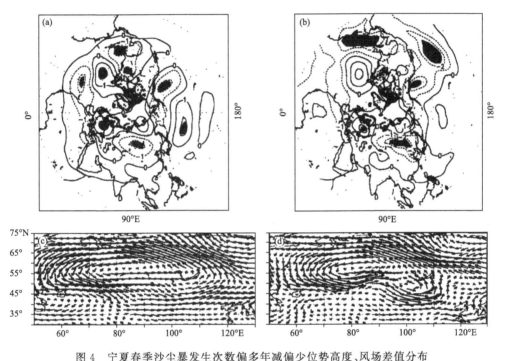

图 4　宁夏春季沙尘暴发生次数偏多年减偏少位势高度、风场差值分布
(阴影区通过置信水平为 95% 的 t 检验)
a. 500 hPa 高度场;b. 850 hPa 高度场;c. 500 hPa 水平风场;d. 850 hPa 水平风场

综上所述,在影响关键区内,与宁夏春季沙尘暴年代际变化相关联的偏多与偏少阶段,以及与年际变化相关联的偏多与偏少年,它们各自之间都存在明显不同的环流背景场,而偏多阶段与偏多年、偏少阶段与偏少年各自之间的环流背景场比较相似。因此说宁夏春季沙尘暴发生次数的异常变化,有其明显不同的环流背景场。当影响关键区的环流背景场发生异常变化时,宁夏沙尘暴发生次数就会发生相应的异常变化。

5　格陵兰海冰变化影响宁夏春季沙尘暴的可能途径

由前面的分析知,宁夏春季沙尘暴与北极部分海区海冰面积之间存在密切关系,这里选取与宁夏春季沙尘暴发生次数年代际和年际变化相关都比较显著的秋季格陵兰海冰,进一步研究它与沙尘暴的变化关系以及对沙尘暴影响的可能途径。

由图 5 可以看出,宁夏春季沙尘暴和前一年秋季格陵兰海冰面积总体变化都呈减少趋势,且总趋势非常一致,二次多项式的趋势线几乎重合,在没有去除趋势项之前,两者的相关系数高达 0.6732。在 1984 年以前和 1992 年以后这两个时段内沙尘暴和海冰都呈一致的稳定偏多和偏少状态,在 1985—1991 年海冰处于由偏多阶段向偏少阶段转变的振荡时期。在 20 世纪 70 年代中后期到 80 年代中后期(1978—1987 年)和整个 90 年代(1991—1999 年),宁夏春季沙尘暴发生次数和前一年秋季格陵兰海冰面积的距平符号完全一致。两者的年际变化趋势在大部分时段内也是一致的,32 a 中有 22 a 变化趋势一致,二者的距平符号相同率达 78.1%,32 a 中有 25 a 距平符号相同,只有 4 a 是明显相反的。年际变率的相关系数为 0.471,由图 6 可以看出,两者的年际变率在大部分时段相一致,特别是 80、90 年代相关更好。沙尘暴和海冰的年代际和年际变化关系都很明显。

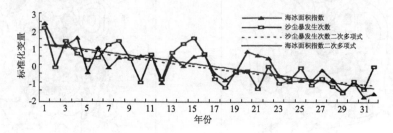

图 5　宁夏春季沙尘暴发生次数(1969—2000 年)和前一年秋季格陵兰海冰面积指数(1968—1999 年)年际变化及其趋势

方之芳[14]、蒋全荣等[15]的研究结果表明极地海冰的异常对 500 hPa 的大气环流有着重要影响,可以激发 500 hPa 遥相关型。方之芳[14]认为这种遥相关型可以看成二维 Rossby 波列,具有相当正压结构,在西风带中沿着固定的波导自高纬向低纬分布,从而对影响地区的大气环流和天气产生影响。Walsh 等[16]研究了北极海冰变化对 SLP 和温度等要素的影响,发现海冰异常对其后一段时间的气象要素场有明显的影响。陈明轩等[17]的研究也表明,海冰作为慢变介质,会稳定地影响其后一段时间的环流和天气形势。由上文分析知宁夏春季沙尘暴发生次数的异常变化年有其不同的环流背景,而北极海冰对宁夏沙尘暴的影响首先必须表现在大尺度环流的变化上,因此,选取与宁夏沙尘暴发生次数相关系数最大的格陵兰海区,从大尺度环流背景的变化去考虑海冰对沙尘暴的影响,以寻求海冰异常变化对沙尘暴影响的可能途径。

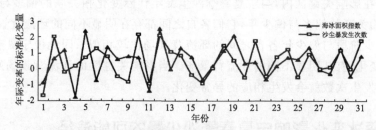

图 6　宁夏春季沙尘暴发生次数变率(1970—2000 年)和前一年秋季格陵兰海冰面积指数变率(1969—1999 年)年际变化

首先对秋季格陵兰海冰面积与次年春季北半球 500 hPa、850 hPa 高度场进行相关分析。由图 7 可以看出,在 500 hPa 上有三个区域相关系数通过置信水平为 95% 的检验,其中较大的一个区域位于我国上游的西伯利亚地区,中心区域相关系数达置信水平为 99% 的检验。从我国北方经蒙古、贝加尔湖、西伯利亚新地岛以南的较大范围内负相关系数较大。在影响关键区内 850 hPa 与 500 hPa 相关形势相似,只是 850 hPa 上相关显著的范围更大,整个亚洲几乎都是负相关,最大负相关区位于蒙古、贝加尔湖一带。将相关系数图 7a、b 与图 3a、b 和图 4a、b 相比较,发现在影响关键区内形势非常相似,尤其是与图 3a、b 较相似,这似乎说明海冰面积异常变化影响环流场异常,从而影响沙尘暴的异常,而且对年代际变化的异常影响似乎更大。

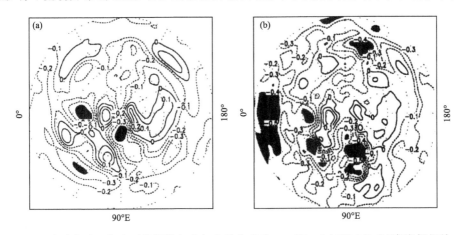

图 7 秋季格陵兰海冰面积指数与次年春季北半球 500 hPa、(a)850 hPa(b)高度场相关
(阴影区通过置信水平为 95% 的 t 检验)

由于这种相关分析包含了年代际趋势变化,因此并不一定能够说明海冰与高度场的年际相关关系(因为实际上海冰有明显的负趋势,而高度场也有正趋势变化),但是能够大致分辨出相关较显著的区域。从 500 hPa、850 hPa 上分别选出通过置信水平为 95% 检验的相关显著区 A 区(70°~110°E,55°~70°N)和 B 区(90°~110°E,40°~65°N),并将这两块区域内的春季高度场分别求和,组成两个 33 a 的时间序列 I_a、I_b 作为高度场的指数序列,分别研究 I_a、I_b 与前一年秋季格陵兰海冰面积之间的相关关系。在没有去除年代际变化之前,海冰与 I_a、I_b 的相关系数分别为 -0.476 和 -0.5552。去除年代际变化趋势后的年际变率相关系数分别为 -0.392 和 -0.3397,即年际变化也是显著的负相关。因此认为,格陵兰秋季海冰面积的变化确实与次年春季影响关键区的高度场有显著负相关关系。即当秋季格陵兰海冰面积增大(减小)时,影响我国北方天气的上游关键区蒙古、贝加尔湖、西伯利亚一带 500 hPa、850 hPa 高度场降低(升高),这正好与宁夏沙尘暴偏多(少)状况的环流背景场一致。因此可以说格陵兰海冰面积的异常变化通过影响其后一段时间的大气环流背景场而对宁夏春季的沙尘暴产生影响,海冰和沙尘暴之间的联系具有一定的环流背景。

6 宁夏(我国北方)沙尘暴发次数总体呈减少趋势的可能原因

我国北方沙尘暴发生次数总的来说呈减少趋势,宁夏沙尘暴总的变化趋势与全国一致。朱震达[2]的研究表明,20 世纪以来,随着人口压力的增大,人为活动的频繁,我国各个地带的荒漠化土地在原有的基础上进一步扩大,以我国北方地区为例,50 年代到 70 年代中期平均每年增加荒漠化土地 1560 km²,年增长率为 1.01%,70 年代中期到 80 年代中期平均每年增加 2100 km²,年增长率为 1.47%,到了 90 年代其年扩大面积达 2460 km²,荒漠化程度进一步加重。因此我国北方地区环境和植被覆盖总体上趋于恶化,单从地表状况分析,我国沙尘暴总的来说应该是增加的,但事实却相反,原因究竟是什么呢? 本文的研究表明:欧亚大陆北部海域海冰面积变化与沙尘暴发生次数之间存在着较好的年代际和年际相关关系。海冰与沙尘暴的年代际减少趋势一致,北极海冰面积的变化通过影响其后一段时间的大气环流,从而影响宁夏乃至我国北方地区沙尘暴的发生。因此,认为沙尘暴减少的原因至少部分与北极海冰面积的变化有关。而近百年来全球气候的逐渐变暖,其后果之一是引起极地海冰消融、海冰面积减少,所以可以认为宁夏(我国北方)沙尘暴发生次数总体呈减少趋势极可能是全球气候变暖所致。当然,全球气候变暖的效应是复杂的,对沙尘暴的影响也是多方面的,需要更深入细致地研究。

综上所述,设想有下面这样一种关系,但还需进一步的证明。

全球气候变暖→北极海冰面积减少→极地冷空气减弱→南下的极地冷空气活动次数减少→这样即使是环境总体恶化的情况下,沙尘暴发生次数总体仍呈减少趋势

7 结论与讨论

(1)宁夏沙尘暴发生次数和北极海冰面积有明显的年代际、年际变化特点,总体都呈减少趋势。

(2)宁夏沙尘暴的发生次数与欧亚大陆北部的喀拉海、巴伦支海和格陵兰海冰面积之间存在显著的年代际和年际相关关系,与整个北极海冰面积的相关性也比较高。

(3)格陵兰海冰面积的变化通过影响其后一段时间的大气环流,从而影响宁夏乃至我国北方地区沙尘暴的发生,海冰和沙尘暴之间的联系具有一定的环流背景。当秋季格陵兰海冰面积增大(减小)时,次年春季蒙古至西伯利亚一带 500 hPa、850 hPa 高压场降低(升高),对应风场有明显的气旋性(反气旋性)特点,在宁夏及甘肃、青海至新疆一带西风明显偏强(偏弱),说明冷空气活动次数偏多(少),对应宁夏春季沙尘暴发生次数偏多(少)。这为沙尘暴的短期气候预测提出了新的思路,并可以进一步研究我国其他地方沙尘暴与北极不同海域海冰的相关关系。

(4)我国沙尘暴发生次数在环境总体恶化的情况下反而呈下降趋势极可能是由于全球气候变暖所致。

本文只是研究它们之间的统计关系,并未深入探讨它们的物理过程和动力机制,这些工作有待进一步深入研究。

参考文献

[1] 钱正安,贺慧霞,瞿章,等.我国西北地区沙尘暴的分级标准和个例谱及其统计特征[C]//中国沙尘暴研究.北京:气象出版社,1997:1-10.

[2] 朱震达.中国沙漠、沙漠化、荒漠化及其治理的对策[M].北京:中国环境科学出版社,1999:659-660.

[3] 王式功,董光荣,杨德保,等.中国北方地区沙尘暴变化趋势初探[J].自然灾害学报,1996,5(2):86-94.

[4] 杨东贞,房秀梅,李兴生.我国北方沙尘暴变化趋势的分析[J].应用气象学报,1998,9(3):352-358.

[5] 赵玉春,孙照渤,王叶红.南、北极海冰的长期变化趋势及其与大气环流的联系[J].南京气象学院学报,

2001,24(1):119-126.

[6] 黄士松,杨修群,谢倩.北极海冰对大气环流与气候影响的观测分析和数值实验研究[J].海洋学报,1992,14(6):32-46.

[7] 尚可政,孙黎辉,王式功,等.甘肃河西走廊沙尘暴与赤道中、东太平洋海温之间的遥相关关系[J].中国沙漠,1998,18(3):239-243.

[8] 牛生杰,孙继明,桑建人.贺兰山地区沙尘暴发生次数的变化趋势[J].中国沙漠,2000,20(1):55-58.

[9] Walsh J E. The role of sea ice in climate variability [J]. Atmos Ocean, 1998, 23(3):551-577.

[10] 武炳义,高登义,黄荣辉.冬春季节北极海冰的年际和年代际变化[J].气候与环境研究,2000,5(3):249-258.

[11] 彭公炳.气候与冰雪覆盖[M].北京:气象出版社,1992:19-27.

[12] 方宗义.中国沙尘暴研究[M].北京:气象出版社,1997:1-330.

[13] 全林生,时少英,朱亚芬,等.中国沙尘天气变化的时空特征及其气候原因[J].地理学报,2001,56(4):477-485.

[14] 方之芳.夏季北极海冰激发的 500 hPa 遥相关[J].大气科学,1991,15(1):53-60.

[15] 蒋全荣,王春红,徐桂玉,等.北极Ⅰ区海冰面积变化及其与大气遥相关型的联系[J].气象学报,1996,54(2):241-246.

[16] Walsh J E, Johnson C M. Interannual atmospheric variability and associated fluctuation s in Arctic sea ice extent [J]. J Geophy Res,1979,84(c11):6915-6928.

[17] 陈明轩,徐海明,管兆勇.春季格陵兰海冰与夏季中国气温和降水的关系[J].南京气象学院学报,2001,24(4):483-490.

西北地区夏季降水异常及其水汽输送和环流特征分析[*]

何金海[1]　刘芸芸[2]　常越[1,2]

(1. 南京信息工程大学大气科学系,江苏 南京 210044；
2. 广东省广州市气象局,广东 广州 510080)

摘要：利用 NCEP/NCAR 1951—2000 年共 50 a 的再分析资料及我国西北地区内 31 个代表站的降水资料,对西北地区近 50 a 来夏季降水异常的时空特征、环流形势及水汽输送进行了研究。分析发现,西北地区夏季降水异常与东部沿海地区的降水呈反位相分布,说明西北地区和东部沿海地区的降水影响系统不同,影响时期不一致。分析结果表明,影响西北地区夏季多、少雨年的相关区域环流特征和水汽输送特征有显著差异。

关键词：西北地区；夏季降水异常；环流特征；水汽

我国西北地区包括新疆、甘肃、宁夏、青海和陕西 5 省、自治区,位于东亚内陆,其干旱主要表现为降水量极少,除了新疆北部地区以外,夏季又是西北绝大部分地区降水最为集中的季节。关于西北地区降水变化的规律已经有了不少研究[1-5]。白肇烨[3]等指出准 3 a 周期是西北降水量最重要的周期。李栋梁[5]认为,西北地区受青藏高原的影响,缺乏水汽来源,降水少且稳定性差。吴统文等[6]利用近期资料,用多年平均场和干湿年对比分析,从观测事实和模拟实验构成了西北干旱气候形成的物理图像。王秀荣[7-9]等利用近 41 a 的测站降水资料,运用旋转经验正交函数分解(REOF)以及功率谱分析等方法分析发现,西北地区夏季降水区域性较强,存在多时空尺度特征,平均具有准 3 a 和 4.8 a 的周期变化,并指出西北全区夏季在 20 世纪 60 年代中期以前和 80 年代中期以后,大气中水汽含量较多,70 年代前后较少。

过去对水汽问题的研究大多讨论的是我国东南部和热带低纬度地区[10-13],而对我国西部地区尤其是西北地区关注较少。过去的一些研究受资料和计算条件的限制,使用的资料年代短。本文运用国家气象中心整编的 1951—2000 年共 50 a 160 站的月平均降水资料,选取西北地区 31 个代表站,考虑西北地区地理差异较大,将西北地区分为 4 个降水异常区域,初步研究了西北地区夏季降水异常的时空特征,并通过对西北地区多、少雨年时期的环流形势及整层水汽输送的分析,试图解释造成西北地区夏季降水异常的原因。

1 资料和计算方法

本文利用 NCEP/NCAR 1951—2000 年共 50 a 的再分析资料(u、v、q、p_s,格距为 $2.5°\times 2.5°$),以及我国西北地区内 31 个代表站 1951—2000 年夏季(6—8月)降水资料,运用了小波分析、合成分析等诊断方法。计算整层水汽输送的计算方法[①]为：

---

[*] 本文发表于《干旱气象》,2005 年第 23 卷第 1 期,10-16。
[①] 刘芸芸,何金海,梁建茵,等.亚澳季风区整层水汽输送的季节转换特征.热带气象学报,2006,22(2):138-146.

单位边长整层大气的水汽输送通量矢量 Q 的计算公式为

$$Q = \frac{1}{g}\int_{p}^{p_s} vq\,\mathrm{d}p \tag{1}$$

式中,v 为该单位气柱内各层大气的风速矢量,q 是各层大气的比湿,p_s、p 分别是大气柱下界气压(地面气压)和上界气压(取 300 hPa),g 是重力加速度,Q 的单位为 $kg \cdot m^{-1} \cdot s^{-1}$。

月平均水汽输送通量通过逐日资料计算并算术平均得到。

2 西北地区夏季的降水异常区域

西北地区位于东亚内陆,横跨青藏高原、蒙古高原和黄土高原,境内有南疆盆地、河西走廊以及天山、秦岭等高山,地形复杂,因此对西北地区降水进行分区讨论是必要的。运用旋转正交分解(REOF)方法对西北夏季降水进行分析后发现,西北地区夏季降水可以划分为 4 个降水异常区:北疆区(代表站吐鲁番等)沙漠盆地区代表站为(若羌等)、青藏高原区(代表站为玛多等)和陕北地区(代表站为榆林等)。此划分方式与以往文献[14]的结果基本一致。

从西北地区近 50 a 夏季降水的空间分布看(图 1),降水量总的特点是东南多西北少。南疆、北疆南部和东部、甘肃河西西部的夏季月平均降水量都不足 40 mm。而降水量＞60 mm 的地区都主要集中在西北地区东南部,即青藏高原东部、河西走廊东南部及陕南、甘肃陇南,有 2 个高值中心区分别位于陕南和青藏高原东南部。

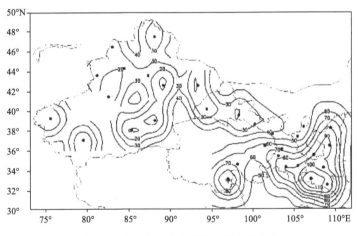

图 1 西北地区夏季月平均降水量分布
(黑点表示西北代表站的空间分布)

3 西北地区夏季降水异常的时空特征分析

3.1 西北地区夏季降水异常的小波分析

小波变换的方法把时间和频率作为独立变量,将唯一信号在时间和频率 2 个方向上展开,因此可以较好地揭示信号中各种频率的时间结构。图 2 是西北地区夏季降水的 Marr 小波变换,分析其实部特征可以看出,50 a 来,西北地区夏季降水年代际变化存在准 11 a 和准 28 a 的周期,而年际变化则有明显的准 3 a 振荡周期。另外 20 世纪 50 年代有准 5 a 的周期,60 年代到 80 年代中期还存在 7 a 左右的周期。总之,西北地区夏季降水周期比较稳定,这是我们将来进行气候预测的依据。

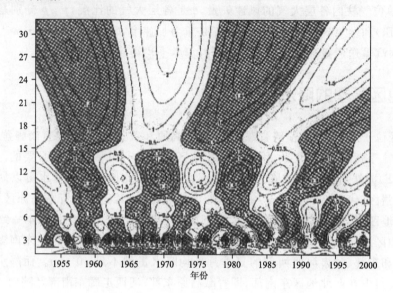

图 2　西北地区 50a 夏季降水标准化距平的小波分析图

3.2　50 a 西北地区夏季降水变化趋势的时空分布

趋势系数表示要素在变化中的升降程度,用来判断气候因子在长期变化过程中的上升或下降趋势。从西北地区 50 a 夏季降水趋势系数的空间分布(图 3)可以看出,4 个降水异常区域差别显著,北疆中部及陕北地区的降水趋势系数为负其,其中心分别位于吐鲁番和榆林,而其他地区趋势系数均为正,其系数最大值中心位于沙漠盆地的若羌,其中高原东部区的趋势系数接近零值。

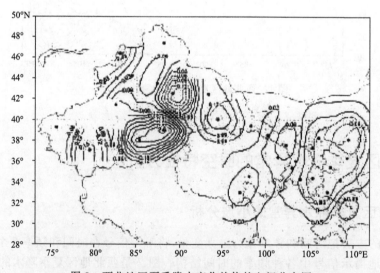

图 3　西北地区夏季降水变化趋势的空间分布图

考虑西北地区夏季降水的年代际周期为 11 a，运用二项式系数加权平均法分别对西北地区 4 个降水异常区域的夏季降水标准化距平值进行 11 a 平滑，见图 4。由图可见，4 个降水异常区域年代际变化差异明显，陕北地区在 20 世纪 70 年代以前基本是多雨的，20 世纪 70 年代以后多、少雨年基本平衡，因此降水趋势系数为负；青藏高原东部区分别有 3 次多雨期（20 世纪 50 年代、70 年代中期、90 年代前期），3 次少雨期（20 世纪 60 年代、80 年代、90 年代后期），使得总体降水趋势变化不大；沙漠盆地在 20 世纪 70 年代中期有一次明显的气候突变，之前持续少雨，1976 年以后降水迅速增加，在 1981 年达到极值，因此总体降水趋势是增加的；而北疆地区在 20 世纪 50 年代中后期 80 年代末 90 年代初距平值为正，其他时期均为负值，因此总体降水趋势也为负值，但比陕北地区趋势稍缓。值得指出的是，西北地区长期降水趋势存有显著的地区性差别，特别是在降水较多的西北东部降水趋势是负的，因此对西北地区水资源短缺的缓解不可太乐观。

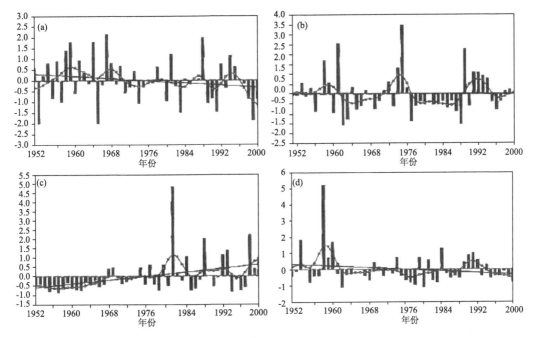

图 4　西北地区夏季降水的标准化距平、11 a 平滑曲线和趋势系数图
（a.陕北地区；b.青藏高原东部区；c.沙漠盆地区；d.北疆区）

3.3　西北地区夏季降水偏多、偏少年的选取

根据文献[8]划分旱、涝年的方法，用西北地区夏季降水资料将西北地区划分为多雨夏季、偏多雨夏季、正常夏季、偏少雨夏季和少雨夏季 5 个等级进行合成分析，对多、少雨夏季年份的各个物理量进行了对比。具体定义方法为：

多雨夏季：$R_i > (\bar{R} + 1.17\sigma)$

偏多雨夏季：$(\bar{R} + 0.33\sigma) < R_i \leqslant (\bar{R} + 1.17\sigma)$

正常夏季：$(\bar{R} - 0.33\sigma) < R_i \leqslant (\bar{R} + 0.33\sigma)$

偏少雨夏季：$(\bar{R} - 1.17\sigma) < R_i \leqslant (\bar{R} - 0.33\sigma)$

少雨夏季：$R_i < (\overline{R} - 1.17\sigma)$

式中，$\overline{R}$ 代表 6—8 月多年平均降水量；R_i 代表逐年 6—8 月降水量；σ 代表标准差。少雨夏季为 1963、1965、1969、1971、1974、1991、1997 年共 7 a；偏少雨夏季为 1955、1957、1960、1962、1972、1975、1977、1982、1986、1999 年共 10 a；多雨夏季为 1956、1958、1961、1979、1981、1988、1993 年共 7 a 和偏多雨夏季 1954、1959、1964、1970、1976、1992、1996、1998 年共 8 a。其他为正常夏季。

3.4 夏季多、少雨年全国同期降水异常的空间分布特征

为了揭示西北地区夏季降水异常与中国其他地区的关系，选取西北地区夏季多、少雨年对全国同期降水距平百分率进行合成(图 5)。由图可见，西北地区夏季多雨时(图 5a)，我国大部分地区都为多雨，只在东部地区包括华南东部、山东半岛及东北东南部等近海地区为少雨区；西北地区少雨时(图 5b)，全国内陆大部是干旱的，而在华南西南部、华东东部及东北近海地区呈多雨分布。因此，西北地区夏季降水多、少雨年都与我国沿海及近海地区呈反位相，说明西北地区夏季降水与我国东部沿海地区的主要影响系统不一样，影响时期不同。

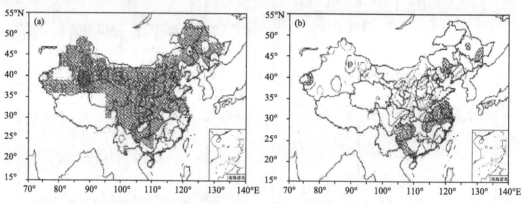

图 5 西北地区夏季多、少雨年合成的降水距平百分率图(a. 代表多余年；b. 代表少雨年)

4 西北地区夏季多、少雨年的环流特征分析

为了分析西北地区多、少雨年的环流形势，考虑青藏高原及天山等海拔高度较高，所以利用 NCEP 再分析资料中的 500 hPa 风场，进行多、少雨年的合成分析(图 6)。由图可以看出，多、少雨年的距平风场形势基本上呈反位相分布，多雨年(图 6a)，从东西伯利亚经过贝加尔湖到巴尔喀什湖有一偏东西走向的横槽，气旋环流中心位于贝加尔湖以东，我国以东地区则有一从日本延伸至我国华南地区的东北—西南向的大脊；西北地区处于 2 大系统的交界处，西北西部受气旋南侧的偏西气流控制，西北中东部有小槽存在，偏西气流和偏南气流 2 支在我国西北地区东部汇合，冷暖气流的教会、抬升作用，造成西北地区的降水偏多。而少雨年则相反(图 6b)，从东西伯利亚到巴尔喀什湖为一接近东西走向的大脊，贝加尔湖位于反气旋环流中心，我国东部为东北—西南向的大槽，这样的环流形势使得气流在西北地区辐散，不利于降水。

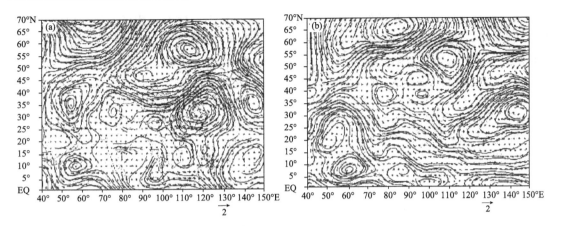

图 6　西北地区夏季多、少雨年合成的 500 hPa 距平风场图(a.代表多雨年；b.代表少雨年)

5　西北地区夏季多、少雨年的水汽输送分析

5.1　夏季多、少雨年的整层水汽输送通量距平场分析

　　水汽输送和积聚是形成降水的重要条件,西北地区深处内陆,远离水汽源地,周围被高大地形环绕,使得低层气流难以爬越。故水汽的来源、输送方式及其在降水区的积聚是该地区降水天气研究和预报的重要问题[15]。垂直积分的水汽输送综合体现深厚气层中的流场和湿度场,能清晰地显示出大尺度的环流系统(如副热带高压、ITCZ 等)。由于资料的限制,在前文选取的多雨年中挑出后 6 a 代表多雨年,少雨年年份不变。给出多、少雨年的整层水汽输送距平合成图(图 7),可以看到,水汽输送的方向在多、少雨年有很大不同,基本与距平风场一致。多雨年(图 7a),高纬地区来自北冰洋的偏北距平输送气流从新疆西北部流入西北地区,并转向东输送,而低纬的来自孟加拉湾的西南风距平水汽输送从中南半岛北部进入我国并向北输送,同时还有一支源自西太平洋经我国东南部的偏东南风距平水汽输送气流,上述 3 支水汽输送水平气流在我国西北地区东部汇合(图中方框所在区域),给西北地区带来充足的水汽；此时

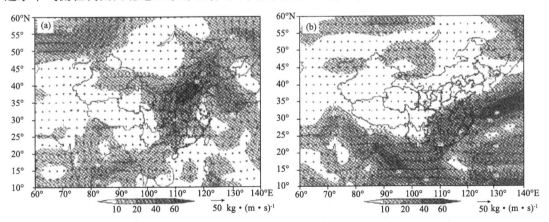

图 7　西北地区夏季多、少雨年合成的整层水汽输送距平图(a.代表多雨年；b.代表少雨年)

我国整个东部地区都为较强的偏南风水汽输送,丰富的水汽条件使得我国大部分地区都降水偏多,这与全国同期降水异常的空间分布十分吻合。少雨年则相反(图 7b),整个西北地区没有明显的水汽流入,低纬的西南水汽输送从中南半岛北部进入我国,最北只能到达陕西以南地区,然后转向东输送,主要给长江以南地区带来水汽,而全国其他大部分地区都为水汽输送低值区,不利于降水,这与全国同期降水异常的空间分布基本一致。

5.2 夏季多、少雨年的整层水汽收支距平场分析

由于西北地区降水量大值区主要集中于西北东部,且西北多、少雨年该区域的水汽输送形势有很大不同,为进一步了解该地区的水汽输送特征,对西北东部区域做了整层大气垂直积分的水汽收支分析(图 8)。由图可见,西北地区多、少雨年的水汽收支呈现相反的情况。多雨年时,西北东部地区主要由偏南气流供应水汽,来自海洋的偏南水汽输送对西北东部地区的降水起着重要影响,整个区域为净辐合,水汽盈余;而少雨年时,西北东部地区各边界的水汽通量距平都与多雨年时相反,且流入、流出量都比多雨年时减弱许多,整个区域为水汽通量距平的辐散区。因此可以说,西北东部地区降水的多(少)主要由上述偏西风输送和偏南风输送的增强(减弱)造成,尤其是偏南风输送的增强(减弱)。

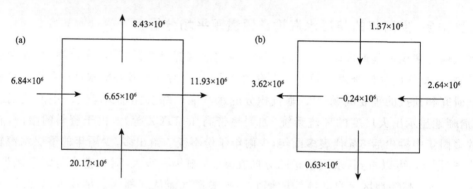

图 8 西北东部地区(32.5°~37.5°N,102.5°~110°E)夏季多、少雨年合成的垂直积分通过各边界的水汽通量距平(a.代表多雨年;b.代表少雨年,单位:kg·s^{-1})

6 结 论

(1)西北地区夏季降水年代际变化存在准 11 a 和准 28 a 的周期,而年际变化则有明显的准 3 a 振荡周期。运用 REOF 将西北地区分为 4 个降水异常区,发现其 50 a 的夏季降水趋势有明显不同,其中北疆和陕北地区降水有减少的趋势,沙漠盆地地区则降水增长,而青藏高原东部区 50 a 的降水趋势基本不变。

(2)选取西北地区多、少雨年各 7 a 和全国同期降水进行合成分析,西北地区夏季降水异常与全国大部分地区同位相,只与东部沿海地区的降水呈反位相分布,说明西北地区和东部沿海地区的降水影响系统不同,影响时期不一致。

(3)西北地区夏季降水多、少雨状况与 500 hPa 环流配置密切相关,多、少雨年的 500 hPa 距平风场基本反位相分布,多雨年,从东西伯利亚经贝加尔湖以东到巴尔喀什湖有一偏东西走向的横槽,从日本到我国华南地区则有一东北—西南走向的大脊,偏西气流和偏南气流在我国

西北地区东部汇合,有利于西北地区降水,而少雨年则相反。

(4)西北地区夏季降水多、少雨年的水汽输送距平场与降水空间分布基本一致。多雨年,来自北冰洋的偏西风距平输送气流和源于孟加拉湾及西太平洋的偏南风距平输送气流在西北东部地区汇合,表现为较强的辐合中心,给西北地区带来充足的水汽,提供有利的降水条件;而少雨年则正好相反,西北地区没有水汽的流入,尤其是西北东部地区为水汽通量距平的辐散区,因而不利于降水。水汽收支分析表明,西北东部地区降水的多(少)主要由上述偏西风输送和偏南风输送的增强(减弱)造成,尤其是偏南风输送的增强(减弱)。

参考文献

[1] 叶笃正,高由禧.青藏高原气象学[M].北京:科学出版社,1979.
[2] 徐国昌,董安祥.我国西部降水量的准三年周期[J].高原气象,1982,1(2):11-17.
[3] 白肇烨,徐国昌.中国西北天气[M].北京:气象出版社,1991:55-56.
[4] 崔玉琴.西北内陆上空水汽输送及其源地[J].水利学报,1994,9:79-87.
[5] 李栋梁.中国西部降水资料的稳定性研究[J].应用气象学报,1992,3(4):451-458.
[6] 吴统文,钱正安.夏季西北干旱区干、湿环流和高原动力影响差异的对比分析[J].高原气象,1996,15(4):387-396.
[7] 王秀荣,徐祥德,庞昕.西北地区夏季降水异常的时空特征分析[J].气象科学,2002,22(4):402-409.
[8] 王秀荣,徐祥德,姚文清.西北地区干、湿夏季的前期环流和水汽差异[J].应用气象学报,2002,13(5):550-559.
[9] 王秀荣,徐祥德,苗秋菊.西北地区夏季降水与大气水汽含量状况区域性特征[J].气候与环境研究,2003,8(1):35-42.
[10] 徐淑英.我国的水汽输送和水分平衡[J].气象学报,1958,29(1):33-43.
[11] 黄荣辉,张振渊,黄刚,等.夏季东亚季风区水汽输送特征及其与南亚季风区水汽输送的差别[J].大气科学,1998,22(4):460-469.
[12] 谢安,宋焱云,毛江玉,等.南海夏季风期间水汽输送的气候特征[J].气候与环境研究,2001,6(4):425-434.
[13] 谢安,毛江玉,宋焱云.长江中下游地区水汽输送的气候特征[J].应用气象学报,2002,13(1):67-77.
[14] 李栋梁,谢金南,王文.中国西北地区夏季降水特征及其异常研究[J].大气科学,1997,21(3):331-340.
[15] 丁一汇,胡国权.1998年中国大洪水时期的水汽收支研究[J].气象学报,2003,61(2):129-145.

华南前汛期不同降水时段的特征分析

池艳珍[1]　何金海[1,2]　吴志伟[1]

(1. 南京信息工程大学大气科学系,江苏 南京 210044;
2. 中国气象局广州热带海洋气象研究所,广东 广州 510080)

摘要:利用 1957—2001 年华南地区 74 个测站逐日降水资料和同期 NCEP/NCAR 逐日再分析格点资料,对华南前汛期(4—6 月)不同降水时段的特征进行了比较。分析发现,华南前汛期由锋面降水和夏季风降水两个时段组成。锋面降水时段主要集中在 4 月,为典型的由冬到夏过渡的环流形势,华南地区高空为平直的副热带西风急流,大气层结稳定,水汽来源主要是阿拉伯海的西风输送和西太平洋副高南侧东风的转向输送;南海夏季风爆发前,副高仍控制南海地区,华南地区水汽输送主要来自源于阿拉伯海的西风输送和西太平洋副高南侧东风的转向输送及孟加拉湾的西南输送;南海夏季风爆发后,副高东撤退出南海地区,南半球越赤道水汽输送加强并与孟加拉湾水汽输送连通,华南区域内对流发展;夏季风降水时段盛期主要集中在 6 月,此时南亚高压跃上高原,华南地区处于南亚高压东部,对流发展极其旺盛,强大的南半球越赤道水汽输送越过孟加拉湾和南海向华南地区输送。

关键词:华南前汛期;锋面降水;夏季风降水

华南地区受到热带季风和副热带季风的共同影响,是我国雨量最充沛的一个区域。与国内其他地区比较,平均年雨量最大,暴雨次数最多,而且雨季汛期也最长,早至 3 月底、4 月初开始直到 9 月甚至 10 月才结束。4—6 月为第一个多雨期(称为前汛期),占全年降水 40%～50%或更多,降水集中,雨量大,故历来受到重视。这方面的研究很多,并取得一定的成果[1-2]。已有的研究表明,华南前汛期降水有显著的准 3 a 周期,降水主要受大尺度西风带锋面系统影响[3]。此外,西太平洋副高、海温、青藏高原积雪、南极海冰和季风爆发等也可以影响华南前汛期降水[4-10]。

上述研究在讨论华南前汛期降水的时空演变特征、影响因子和影响机理时,大多把华南前汛期作为一个整体来讨论,但事实上,东亚地区从春到夏存在副热带季风雨季和热带季风雨季[11],相应地华南前汛期降水包含前期的锋面降水和后期的夏季风降水[12],因此,将其作为一个整体来研究它的影响因子、影响机理和预测方法是不恰当的。本文将华南前汛期分为"锋面降水"时段和"夏季风降水"时段,比较了这两种性质的降水在热力学、动力学和环流特征等方面的差异,试图为华南前汛期降水的预测提供新的思路。

1 资料及处理

本文所用资料包括:(1)国家气候中心整编的 740 站逐日站点降水资料,时间为 1957—

---

* 本文发表于《南京气象学院学报》,2005 年第 28 卷第 2 期,163-171.

2001年;(2)同期NCEP/NCAR的温度、水平风场、位势高度、比湿及地面气压等逐日再分析格点资料,水平分辨率为2.5°×2.5°。以上资料时段均取为3—6月。

为研究整个华南地区的前汛期降水特点,本文选取106°~121°E,20°~27.5°N作为研究区域,但剔除降水资料长度不足45 a的站点,最后选取74个测站代表华南地区,其分布见图1。

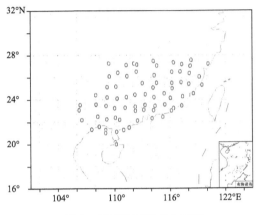

图1 华南地区站点分布图

本文计算了垂直积分的水汽通量,计算方法如下:

设在p坐标系中单位时间通过垂直于风向的底边为单位长度,高为整层大气柱的面积上的总的水汽通量(即垂直积分的水汽通量)Q(单位 kg·m^{-1}·s^{-1}),则有

纬向水汽输送通量:

$$Q_u = \frac{1}{g}\int_{p_t}^{p_s} qu\,dp$$

经向水汽输送通量:

$$Q_v = \frac{1}{g}\int_{p_t}^{p_s} qv\,dp$$

式中,p_s为地表面气压,p_t取为300 hPa,q为比湿(kg·kg^{-1}),g为重力加速度。

为了研究华南地区的大气层结状况,本文还通过1000~300 hPa各等压面层的温度和比湿计算各层的假相当位温(θ_{se})[13],并用它随高度的变化来讨论大气稳定度的变化情况。

2 华南前汛期降水特征

2.1 降水量的时间演变

图2是1957—2001年3—6月多年平均华南地区的候平均降水量演变。由图2可以看出:从春季到初夏,华南地区候平均降水量总体呈逐渐增加的趋势,但是在4月第5候至第6候降水量有所减少,至5月1候降水量出现剧增,而在5月5候的略微减少之后,5月第6候后又明显增加。华南前汛期降水存在两个较明显的峰值,第一个峰值主要集中于4月,最大降水量约为6 mm·d^{-1};第二个峰值出现在6月,量值上明显大于第一个,特别是6月中旬,候平

均降水量可达 10 mm·d^{-1} 左右。

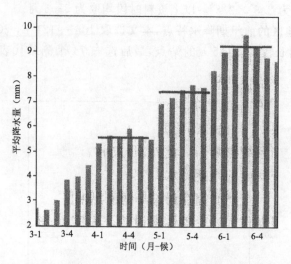

图 2　1957—2001 华南地区候平均降水量演变

2.2　水汽输送

为了解从春到夏华南地区水汽输送的演变,绘制了多年平均的垂直积分的逐候水汽通量分布图(图 3)。由图可以看出,4 月 1 候(图 3a)水汽输送仍属于较典型的冬季输送形势:中纬度为平直的西风输送,低纬度为带状的东风输送,而 100°～110°E 为西太平洋副热带高压(下称副高)西南侧的东风转向气流输送。在这种水汽输送的背景下,华南地区的水汽来源主要有两个:一是源自阿拉伯海的西风水汽输送;二是副高南侧的东风水汽输送,前者向东伸展直接进入华南地区,后者在中南半岛和南海转向进入华南地区。相比较而言,4 月 6 候(图 3b)输送形势发生了较大变化,最突出的特征是热带东风水汽输送带崩溃,70°～90°E 的热带印度洋上出现了较强的西风输送,105°E 出现了非常弱的越赤道输送,但此时索马里越赤道气流仍未形成。

5 月 1 候(图 3c)输送形势变得更加明朗,华南地区除了 4 月第 6 候的两个水汽通道仍然维持外,由于热带印度洋上的西风输送大大增强,来自孟加拉湾地区的水汽输送显著加强。此时虽然索马里越赤道气流输送不断增强,但并没有和孟加拉湾的水汽输送相连通,这种形势一直维持到 5 月 3 候(图略)。5 月 4 候(图 3d),输送形势发生了显著变化,此时副高东移撤出南海地区,其南侧的转向气流对华南地区已基本不产生影响;源于阿拉伯海的西风水汽输送显著减弱但仍然存在;而来自南半球强大的索马里越赤道水汽输送(中心值大于 400 kg·m^{-1}·s^{-1})与孟加拉湾中南部水汽输送汇合,并转向流经中南半岛和南海源源不断向华南地区输送,105°E 越赤道气流加强,整个南海地区为西南气流控制,南海夏季风全面爆发。

到了 6 月(图 3e),副高持续东撤至 130°E 菲律宾以东洋面,其南侧的东风转向输送气流对华南地区已完全没有贡献,源自阿拉伯海的西风输送逐渐消失;此时来自南半球的强大的越赤道水汽输送汇合源自阿拉伯海、孟加拉湾的水汽和 105°E 越赤道水汽,形成一条行星尺度的水汽输送带,向印度半岛、中南半岛和我国华南地区、江淮地区直至日本输送水汽。

综上所述,4—6 月影响华南地区的水汽输送形势前后期有较大的区别,4 月华南地区的水汽来源主要是西风输送和副高南侧东风转向气流输送;5月南海夏季风爆发前,副高南侧的东风输送维持,源自阿拉伯海的西风输送虽然强度逐渐减弱但仍影响华南,此时孟加拉湾的西南水汽输送逐渐形成;而南海夏季风爆发后(特别是6月),华南地区水汽主要来源于南半球越赤道水汽输送在孟加拉湾和南海的转向输送。

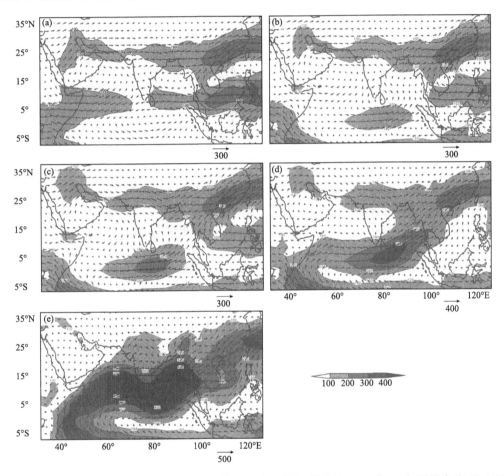

图3 1957—2001年逐候垂直积分的水汽输送通量Q分布(单位:$kg \cdot m^{-1} \cdot s^{-1}$;阴影表示$Q \geqslant 100$)
a.4月1候;b.4月6候;c.5月1候;d.5月4候;e.6月3候

2.3 层结稳定性

由于对流性不稳定的判据是用假相当位温(θ_{se})随高度的递减率来进行判断的,因而假相当位温随高度的变化就能很好地反映大气的层结稳定性,从而可以揭示对流发展状况。图4给出了时段平均的高、低层的假相当位温差值图,其中高、低层分别以600~300 hPa的平均值和1000~700 hPa的平均值代替。

由图4a可以看出,4月我国东部地区均为$\Delta\theta_{se}$正值区,黄海洋面附近中心值达到16 K。此时华南大部分地区处在正值区中,仅在西南角有一很小范围的负值区,说明此时段华南地区总的来说大气层结是稳定的,不利于对流发展,降水所需的垂直运动主要以锋面抬升为主。南海夏季风爆发前(图4b),随着$\Delta\theta_{se}$零线的向北推进(从10°N到20°N附近),华南区域负值区逐渐向东推移,但我国东部地区仍为正值区覆盖,此时南海地区已为负值控制,说明该地区不稳定性增加;南海夏季风爆发后(图4c),虽然东部地区仍为正值,但高低层差值明显减小,华南地区负值向东北推移更加明显;到6月(图4d),华南地区已全为负值所替代,大气层结不稳定性大大增加,此时段的对流发展旺盛,符合以对流为主的夏季风降水性质。

从图4c和图4d比较也可看出,南海夏季风爆发后,强对流区域由南向北迅速扩展。

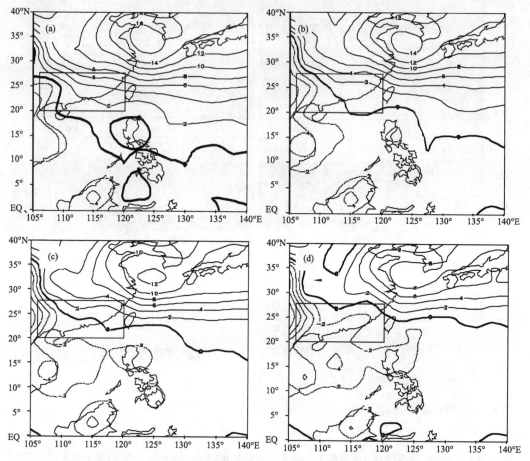

图4 时段平均的 $\Delta\theta_{se}(\theta_{se}(600\sim300\ hPa)-\theta_{se}(1000\sim700\ hPa))$ 分布(单位:K)
a. 4月1—6候;b. 5月1—3候;c. 5月4—6候;d. 6月1—4候

为了进一步分析不同降水时段大气的稳定性,给出了相应时段区域平均的层结曲线(图5),根据图1所示站点分布情况,取区域为:110°~120°E,22°~27.5°N。由图5a可知,4月不稳定层结仅在很浅薄的低层(850hPa以下),从850 hPa以上θ_{se}随高度迅速增加,总的来说大气层结是对流性稳定的;至6月(图5b),不稳定层结伸展至700 hPa,700 hPa至600 hPa之间为近中性层结,说明此时段低层对流较强,系统深厚,可到达对流层中部,对流层中上层虽为稳定层结,但θ_{se}随高度增幅很小;700 hPa以下θ_{se}随高度显著减小,差值接近−15K,说明该层以下对流性不稳定很强,对流发展旺盛。南海夏季风爆发前后(图略),大气层结状况为4月和6月的过渡形势,即不稳定层结虽向高层发展,但仍在700 hPa以下。

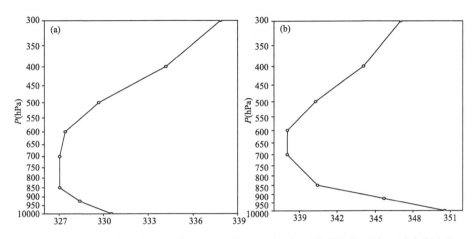

图 5　1957—2001 年 4 月 1—6 候(a)和 6 月 1—4 候(b)时段平均的区域 θ_{se} 随高度变化

综上所述，4 月华南地区大气层结稳定，对流活动不强，因而降水应是锋面降水；南海夏季风爆发后特别是 6 月该地区大气层结不稳定，对流发展旺盛，因而降水应为夏季风降水。因此，本文把 4 月称为华南前汛期锋面降水时段，而将 6 月称为华南前汛期夏季风降水盛期。

2.4　温度场分布

为了讨论华南前汛期对流层中上层温度场的演变特征，制作了多年平均的沿 $105°\sim 120°$E 平均经向温度偏差的时间—纬度剖面图(图 6)。由图可以看出，5 月 1 候前，20°N 以南为温度正偏差，20°N 以北为温度负偏差，4 月前正偏差中心位于南半球，呈明显的南暖北冷形势；5 月 1 候后，20°S 以北开始出现温度负偏差，5 月 4 候前后，赤道附近虽仍是温度正偏差，但其北部的偏差值明显大于其南侧，北暖南冷的经向温度分布形势开始建立；到了 6 月，赤

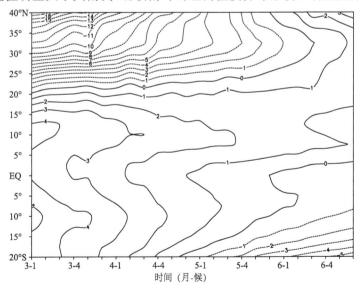

图 6　对流层中上层(500～200 hPa)沿 $105°\sim 120°$E 平均经向温度偏差的时间—纬度剖面
(单位：K，实线表示正偏差，虚线表示负偏差)

以南地区为温度负偏差,而20°N附近出现较大的温度正偏差,为显著的北高南低的经向温度分布形势。因此,华南前汛期不同时段对应着不同的温度经向分布形势,南海夏季风爆发时间与该经度南冷北暖的经向温度梯度的建立时间基本是一致的,从而表明南海夏季风爆发过程本质上是经向温度梯度转向的结果。

2.5 高空环流特征

图 7 展示了不同降水时段平均的 200 hPa 环流形势。从图中可见,锋面降水时段(图 7a)南亚高压位于菲律宾以东洋面上,处于南亚高压北侧的北半球副热带高层是一致的强大的西风,副热带急流由印度半岛北部向东经我国南部的广大地区至日本以南的西北太平洋,此时华南地区处在西风急流的控制之下,而赤道地区为热带东风区。5月,南亚高压西移北上控制中南半岛(图略),到夏季风降水盛期(图 7b),南亚高压继续北上,跃上高原,控制着东亚和南亚大范围地区,此时华南地区处于南亚高压的东北部,为偏西北气流控制。

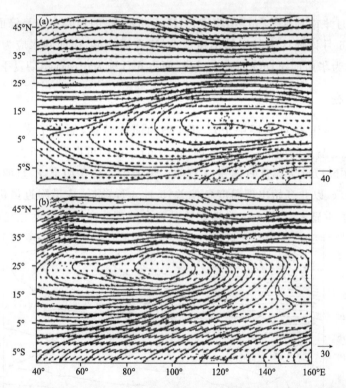

图 7 1957—2001 年 4 月 1—6 候(a)和 6 月 1—4 候(b)时段平均的 200 hPa 环流场(单位:m·s^{-1})

2.6 西风急流

以上分析表明,华南地区在锋面降水时段位于西风急流入口区,而夏季风降水时段处于西风急流出口区,下面进一步分析华南地区上空西风急流的演变特征。图 8 是多年平均的 200 hPa 纬向风沿 110°~120°E 的时间—纬度剖面图。由图可以看出,10°N 以南维持东风分量,而 4 月 6 候以后,东风分量逐渐向北扩展,这与南亚高压西移至中南半岛北部后再上高原有关。4 月 5 候前急流轴位于 30°N 附近,急流强度逐步减弱,但急流轴南北位置相对稳定。

从 4 月第 6 候开始,急流强度减弱的趋势变得缓慢,急流风速稳定,位置明显向北移动。5 月 4 候南海夏季风爆发时明显北扩,轴线到达 35°N 附近,急流强度有所增强。在 6 月 3 候前虽然急流强度逐渐减弱,但轴线稳定在 35°N 附近,此时华南地区位于高空急流入口区南侧的强辐散区,为夏季风强对流性降水提供了良好条件。6 月 4 候后急流轴线再次北移,江淮梅雨期开始。

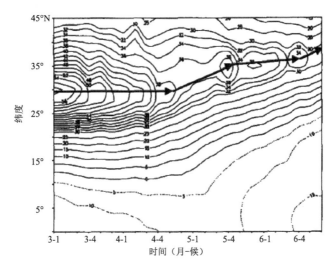

图 8 多年平均的 200 hPa 纬向风沿 110°~120°E 的时间—纬度剖面(单位:m·s^{-1})

(虚、实线分别表示东、西风)

3 小结与讨论

(1)华南前汛期(4—6 月)存在两个降水较集中阶段,即存在双峰,锋面降水集中于 4 月,夏季风降水盛期集中于 6 月。

(2)4 月华南地区水汽来源有两个,一是西风水汽输送;二是西太副高南侧的东风水汽输送,前者向东伸展至华南地区,后者在中南半岛和南海地区转向进入华南地区。南海夏季风爆发后特别是 6 月该地区水汽主要来源于南半球越赤道气流在孟加拉湾和南海的转向输送。

(3)华南前汛期锋面降水时段,华南地区上空为平直的西风气流,高空急流轴位于 30°N 附近,大气层结稳定,降水主要是北方冷空气和西太平洋副热带高压南侧转向的西南暖湿气流交绥的产物。

(4)华南前汛期夏季风降水盛期,南亚高压位于青藏高原南侧,西风急流轴北抬,华南地区位于南亚高压的东北部、西风急流轴右侧,大气层结不稳定,对流发展极其旺盛;

(5)对流层中上层,华南前汛期锋面降水时段,经向温度分布呈明显的南高北低形势;而夏季风降水盛期,温度经向分布形势逆转,呈显著的南低北高形势。

本文仅从气候平均的角度对华南前汛期不同降水时段的特征进行了比较,但其年际变化情况、前汛期锋面降水和夏季风降水的影响因子和影响机理等问题,有待进一步研究和探讨。

参考文献

[1] 鹿世瑾.华南气候[M].北京:气象出版社,1990.
[2] 郭其蕴,沙万英.华南前汛期降水变率的分析[J].应用气象学报,1998(增刊):9-15.
[3] 吴尚森,梁建茵.华南前汛期旱涝时空分布特征[J].热带气象学报,1992,8(1):87-92.
[4] 梁建茵.6月西太平洋副高脊线的年际变化及其对华南降水的影响[J].热带气象学报,1994,10(3):274-279.
[5] 覃武,孙照渤,丁宝善,等.华南前汛期雨季开始期的降水及环流特征[J].南京气象学报,1994,17(4):455-461.
[6] 邓立平,王谦谦.华南前汛期(4—6月)降水异常特征及其与我国近海海温的关系[J].热带气象学报,2002,18(1):44-55.
[7] 高波,陈乾金,任殿东.江南南部—华南北部前汛期严重旱涝诊断分析[J].应用气象学报,1999,10(2):219-226.
[8] 蔡学湛.青藏高原雪盖与东亚季风异常对华南前汛期降水的影响[J].应用气象学报,2001,12(3):358-367.
[9] 张爱华,吴恒强,覃武,等.南半球大气环流对华南前汛期降雨影响初探[J].气象,1997,23(8):9-15.
[10] 吴恒强,张爱华,蒋伯仁,等.华南前汛期降水与南极海冰变化的关系[J].南京气象学报,1998,21(2):266-273.
[11] 陈隆勋,李薇,赵平,等.东亚地区夏季风爆发过程[J].气候与环境研究,2000,5(4):345-355.
[12] 陈隆勋,朱乾根,罗汉邦,等.东亚季风[M].北京:气象出版社,1991.
[13] 丁一汇.现代天气学中的诊断分析方法[Z].北京:中国科学院大气物理研究所,1984.

北半球环状模和东北冷涡与我国东北夏季降水关系分析[*]

何金海[1,2] 吴志伟[1] 祁莉[1] 姜爱军[2]

(1. 南京信息工程大学大气科学系,南京 210044;2. 江苏省气象科学研究所,南京 210009)

摘要:利用国家气象信息中心提供的1951—2004年全国160个测站月平均降水资料和欧洲中心提供的 ERA-40再分析资料,对近50多年东北地区夏季降水、东北冷涡与前期北半球环状模和海温的关系进行了统计分析,定义了一个夏季(6—8月)东北冷涡强度指数(NECVI)。结果表明:NECVI能够较好表征东北低涡的气候效应;夏季东北冷涡强度与降水存在显著的正相关,东北冷涡强年,降水偏多,前期2月北半球环状模(NAM)偏弱;东北冷涡偏弱年,降水偏少,前期2月 NAM 偏强。此外,夏季东北冷涡与前期的中国近海海温存在显著的负相关,前期 NAM 和中国近海海温的异常可以作为夏季东北冷涡异常的一个前兆信号,进而为东北地区夏季降水异常的预测提供参考依据。

关键词:北半球环状模;东北冷涡;夏季降水;环流异常

　　东北地区的夏季旱涝异常长期以来一直受到广泛的关注并进行了大量的研究。早在20世纪30—40年代,竺可桢和涂长望等[1-2]就指出,包括东北在内的中国东部地区的夏季降水是夏季风推进的产物并与季风的进退紧密相关。朱乾根、何金海[3]明确提出东亚季风区可分为南海—西太平洋热带季风区和大陆—日本副热带季风区。陶诗言和陈隆勋的研究[4]显示在印度西南季风强的夏季,东亚夏季风往往偏弱。黄荣辉等[5]提出,东亚夏季风雨带的季节内变化还与西太平洋暖池的热状态关系密切。吴国雄等[6]和张琼等[7]的研究表明华南、江南夏季旱涝与中国近海海温(如南海、印度洋等)关系密切。He 等[8]的统计分析结果显示,夏季西太平洋副热带高压的3个参数(面积、强度和脊线纬度)与东亚副热带季风降水存在很高的相关。综观前人关于中国东部地区夏季降水的研究大多集中于热带低纬地区系统的影响,而源自中高纬地区的天气系统的作用近年来则考虑较少。

　　事实上,中高纬天气系统对我国东部地区降水同样有着重要的作用,而东北冷涡正是这样的一个天气系统,它是东亚中高纬大气环流系统中的重要组成部分,一次东北冷涡活动的生命周期一般为5~7 d,因此它是一个天气尺度的系统。孙力等[9]从天气学的角度对东北冷涡的影响和某些天气学特征进行了研究。事实上,频繁的东北冷涡活动,不仅影响到中短期天气,而且对短期气候同样有着较大影响。这些也是本文的研究出发点。开展这些研究,不仅可以了解中高纬系统对东北地区夏季降水的作用,而且可以为该地区的短期气候预测提供具有参考依据。

---

[*] 本文发表于《气象与环境学报》,2006年第22卷第1期,1-5.

1 资料和方法

本文所用 1951—2004 年全国 160 个测站夏季降水资料由国家气象信息中心提供；再分析资料取自欧洲中心 ERA-40 资料集（水平分辨率为 2.5°×2.5°），海温资料选用 Reynolds 海温（水平分辨率为 2.0°×2.0°）；1958—2002 年 NAM[10] 指数由中国科学院大气物理研究所 LASG 提供。

孙力等[9] 对 35 a 资料（1956—1990 年）的统计分析表明，6，7，8 月平均出现东北冷涡的日数为 15.1,12.5,11.3 d，其对夏季天气影响的最为显著的特征是影响地区出现持续低温连阴雨，尤以持续低温特征更为显著。东北冷涡活动越频繁，强度越强，低温事件也越频发，累积低温时间也越长，进而导致东北冷涡影响区域季节平均气温偏低；反之，当东北冷涡活动较少，强度较弱时，东北冷涡影响区域季节平均气温偏高。基于此，本文首先定义一个夏季（6—8 月）东北冷涡强度指数（NECVI：Northeast Cold Vortex Index）。参考 Ting 和 Wang[11] 采用统计手段划分降水区的方法，计算全国 160 个测站 6—8 月地面平均气温标准差，得到东北地区地面气温变率最大的测站为乌兰浩特；然后以该站为基点，计算其与 160 个测站的单点相关，得到气温空间相关图（图 1），相关系数在基点处为 1，然后向周围逐渐减小。图 1 中阴影所覆盖的显著相关区域代表了较为一致的气温变率。将阴影区内站点的夏季平均气温的相反数定义为夏季东北冷涡强度指数 NECVI（表 1）这里取相反数的目的是为了使指数更符合通常的使用习惯，即指数越高，东北冷涡越强；指数越低，东北冷涡越弱。本文将标准化 NECVI 值大于 1 和小于 -1 的年份分别定义为夏季 NECVI 指数异常高年和异常低年，如表 2。

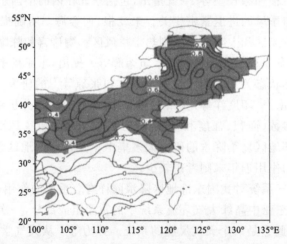

图 1 以乌兰浩特站为基点的 6—8 月地面气温单点相关分布
（阴影区为置信度检验达 95% 的区域）

2 东北地区夏季 NECVI 异常年同期 500 hPa 高度场和相对涡度场的特征

已有研究[9] 表明，东北冷涡的显著特征为 500 hPa 的位势高度的低值中心和正相对涡度中心相对应。为验证本文所定义的东北冷涡强度指数 NECVI 指数是否能描述东北冷涡的物理图像,定量刻画其强度特征,我们计算了夏季 NECVI 与同期 500 hPa 高度场和相对涡度场的相关系数(图 2)。

表 1 1951—2004 年夏季标准化东北冷涡强度指数(NECVI)

年份	NECVI	年份	NECVI	年份	NECVI	年份	NECVI	年份	NECVI
1951	−0.52	1962	0.23	1973	0.08	1984	0.56	1995	−0.32
1952	−0.13	1963	−0.26	1974	0.62	1985	0.15	1996	0.27
1953	−0.67	1964	0.95	1975	−0.43	1986	0.62	1997	−2.40
1954	1.39	1965	0.85	1976	2.34	1987	0.57	1998	−0.50
1955	−1.08	1966	−0.10	1977	0.51	1988	−0.51	1999	−1.38
1956	1.51	1967	−0.33	1978	0.52	1989	0.64	2000	−2.72
1957	1.46	1968	0.18	1979	0.90	1990	−0.16	2001	−2.25
1958	0.24	1969	1.09	1980	0.12	1991	−0.61	2002	−1.19
1959	0.11	1970	0.46	1981	−0.29	1992	1.16	2003	0.25
1960	0.06	1971	0.45	1982	−0.17	1993	1.27	2004	−0.25
1961	−0.94	1972	0.55	1983	0.90	1994	−2.15		

表 2 1951—2004 年夏季 NECVI 异常高(低)年对应的标准化指数值和标准化降水量

年份	高 NECVI	$\Delta R/\sigma$	年份	高 NECVI	$\Delta R/\sigma$
1976	2.34	0.36	2000	−2.72	−1.51
1956	1.51	1.74	1997	−2.40	−1.14
1957	1.46	1.18	2001	−2.25	−1.61
1954	1.39	1.45	1994	−2.15	−0.85
1993	1.27	0.60	1999	−1.38	−1.34
1992	1.16	−0.66	2002	−1.19	−1.38
1969	1.09	1.57	1955	−1.08	−0.67

从图 2 中可以清楚看到,从蒙古至我国东北地区存在显著的位势高度负相关区(图 2a),表明夏季 NECVI 高指数年我国东北地区有低值中心发展,而低指数年则对应着低值中心的填塞和高值中心的发展。从相对涡度相关系数来看(图 2b),东北地区为大范围的显著正相关区所覆盖,表明夏季 NECVI 高指数年东北地区有正相对涡度中心发展,而低指数年则有负涡度中心发展。从上述分析可以看到,夏季 NECVI 值基本反映了东北低涡的主要物理特征。因此,可以将 NECVI 值作为一个衡量东北冷涡强度的定量指标。

3 夏季 NECVI 与东北地区夏季降水、东亚夏季风的关系

从夏季 NECVI 与东北地区夏季降水的相关分布(图 3)来看,东北大部分地区均为正相关区所覆盖并有一片显著区。虽然在东北的东部有小范围的负相关区,但未能通过显著性检验。这些表明,当夏季 NECVI 偏高时,则东北地区夏季降水很可能偏多;当夏季 NECVI 偏低时,则东北地区夏季降水很可能偏少。如果将阴影所覆盖地区的平均降水量定义为东北地区的一

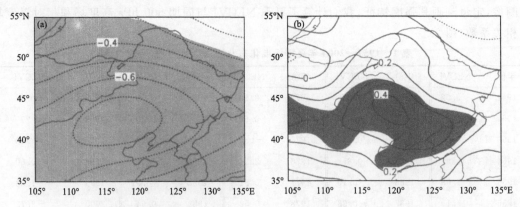

图2 夏季东北冷涡强度指数 NECVI 与 500 hPa 位势高度(a)和涡度(b)的相关系数
(阴影区为置信度检验达 95%的区域;浅色区为显著负相关区;深色为显著正相关区)

个夏季降水指数 SRI(图4),可以看到 SRI 与 NECVI 时间序列的变率趋于一致,计算 NECVI 时间序列和 SRI 时间序列的相关系数为 0.41,通过 99% 的置信度检验。这些均显示东北冷涡与东北地区夏季降水确实存在显著的正相关关系,也就是说当东北冷涡偏强时,东北地区夏季降水偏多;当东北冷涡偏弱时,东北地区夏季降水偏少。

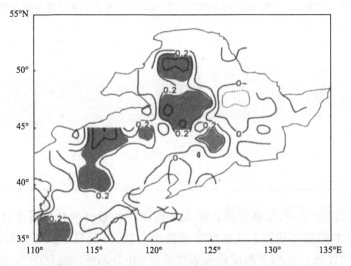

图3 夏季 NECVI 与东北地区夏季降水的相关系数
(阴影区为置信度检验达 95%的区域)

为进一步验证上述东北冷涡与东北地区夏季降水的相关关系,我们计算了夏季 NECVI 异常年东北地区的夏季标准化降水量。从表2可以看到(ΔR 为夏季降水距平,σ 为其标准差),东北冷涡异常强的 7 a 中,有 6 a 夏季降水偏多,有 4 a 偏多超过 1 倍标准差;东北冷涡异常弱的 7 a 中,有 6 a 夏季偏少,其中 5 a 甚至偏少超过 1 倍标准差。这些均与前面分析得到的结论相一致。

为了解夏季东北冷涡异常年东亚夏季风的异常特征,本文给出了夏季 NECVI 异常高年与异常低年的 850 hPa 合成风差值场(图5)。由图5可见,在东北地区有一差值低涡存在(如实线圆所示),这正好表明高(低)指数年对应低涡的加强(减弱),说明定义的低涡指数能描述低

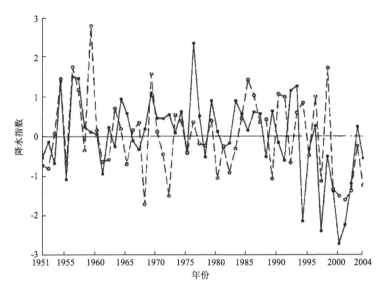

图 4 1951—2004 年 NECVI 时间序列与东北地区夏季降水指数 SRI 时间序列

涡的特征。另外,从华南经我国东部至东北东部有显著的偏西南风或偏南风差值气流,这一差值气流造成高(低)指数年水汽输送的强(弱),进而有利于东北地区高(低)指数年降水偏多(少)。

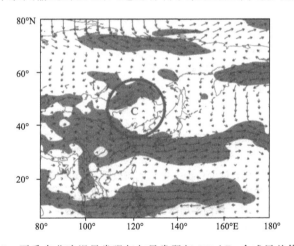

图 5 夏季东北冷涡异常强年与异常弱年 850 hPa 合成风差值场
(阴影区为置信度检验达 95% 的区域;单位为 m·s^{-1})

4 夏季东北冷涡异常年前期北半球环状模和海温的异常特征

鉴于东北冷涡是北半球中高纬地区的大气环流系统,而北半球大尺度大气环流异常往往伴随着北半球环状模(NAM)的异常。为研究夏季东北冷涡异常年前期北半球大尺度大气环流的异常特征,我们首先计算了夏季 NECVI 与其前期 1~6 个月(从前一年 12 月至当年 5 月)NAM 指数时间序列的相关系数,如表 3 所示。由表 3 可见,在前期 6 个月当中,NECVI 与 NAMI 在 2 月相关最显著达到 −0.38(通过了 99% 的置信度检验)。就就是说,当前期 2 月

NAM偏强时,夏季东北冷涡往往偏弱;当前期2月NAM偏弱时,夏季东北冷涡往往偏强。

表3 1951—2004年夏季NECVI与其前期1—6个月的NAMI的相关系数

月份	前1年12月	当年1月	当年2月	当年3月	当年4月	当年5月
与NAM的相关	−0.03	−0.04	−0.38*	−0.20	−0.04	−0.04

注:a为通过了99%置信度检验的异常显著的相关系数。

为进一步验证NECVI与NAM的这种可能的相关关系,我们还计算了夏季NECVI与其前期2月北半球纬向平均海平面气压的相关系数,如图6。由图6可见,NECVI与30°N附近的纬向平均海平面气压存在显著的负相关,而与60°N纬向平均海平面气压存在显著的正相关,这表明NECVI偏高(低)时,其前期2月中高纬地区的海平面气压显著升高(降低),而北半球中低纬地区的气压显著下降(升高),因而北半球中低纬和中高纬地区气压梯度减小(加大),造成2月NAM偏弱(偏强)[10]。至于为什么夏季的东北冷涡与前期2月北半球环状模(NAM)存在这样一种时间上的"遥相关"或者称之为"韵律"的关系,值得深入研究。

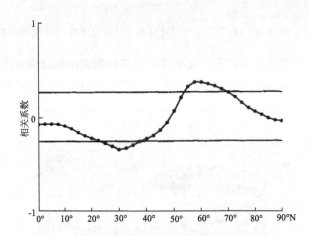

图6 夏季NECVI与前期2月北半球(0°~90°N)纬向平均海平面气压的相关系数
(直线为置信度检验达95%的阈值线;曲线为相关系数)

因此,前期2月的NAM可以作为夏季东北冷涡异常的一个前兆信号,进而为东北地区夏季降水异常的预测提供参考依据。

鉴于海温对大尺度大气环流异常起着重要作用,为研究前期海表温度对夏季东北冷涡强度的影响,计算了夏季NECVI指数与其前期1~6个月的海表温度SST的相关系数,如图7。由图7可见,由前一年12月至当年5月,可以看到从南中国海到西北太平洋、孟加拉湾、阿拉伯海以及印度洋北部均为显著负相关区,说明夏季东北冷涡异常强年前期,上述海区的海表温度持续偏低;而夏季东北冷涡异常弱年前期,上述海区的海表温度持续偏高。前期中国近海海温的异常是导致夏季东北冷涡异常的可能原因之一。

5 结 论

东北地区夏季降水异常时有发生,本文从季节平均的时间尺度对东北冷涡与东北地区夏季降水、前期大尺度海气异常的关系进行了研究,重点分析了夏季东北冷涡异常年同期和前期

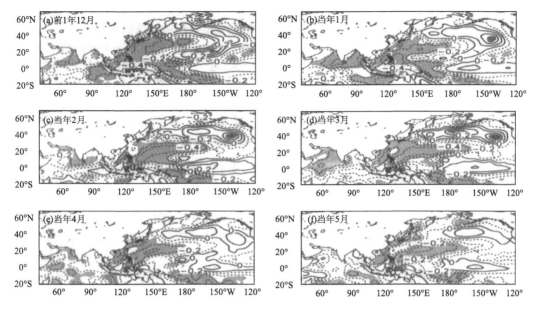

图 7 夏季 NECVI 与前期 6 个月海温场的相关系数
(阴影区为置信度检验达 95% 的区域)

大尺度大气环流和海温异常的一些基本特征,得出以下结论。

(1)东北地区关键区的平均气温能较好地表征东北冷涡的活动特征。本文所定义的东北冷涡指数能较定量地描述东北冷涡的气候效应。

(2)夏季东北冷涡与东北地区降水有着显著的正相关关系;东北冷涡偏强时,则该地区夏季降水偏多;而东北冷涡偏弱时,则该地区夏季降水偏少。

(3)夏季东北冷涡与前期 2 月 NAM 存在显著的负相关关系:当前期 NAM 偏强(弱)时,夏季东北冷涡往往偏弱(强)。此外,夏季东北冷涡与前期中国近海海温存在显著的负相关。

(4)上述结果可以为东北地区夏季降水预测提供一定的前兆信号。然而,由于东北冷涡是很深厚的系统,从低层到高空都有系统反映,如何更好地定义东北冷涡,尚需进一步探讨,而夏季东北冷涡与东北地区降水这种相关关系的内在机制还有待于进一步研究,本文的工作还是初步的。

致谢:感谢国家气象信息中心和欧洲中心提供的相关气象资料,感谢中国科学院大气物理研究所 LASG 李建平研究员提供的 NAM 指数。

参考文献

[1] 竺可桢. 东南季风与中国之雨量[J]. 地理学报,1934,1(1):1-27.
[2] 涂长望,黄士松. 中国夏季风之进退[J]. 气象学报,1944,18:1-20.
[3] 朱乾根,何金海. 亚洲季风建立及其中期振荡的高空环流特征[J]. 热带气象学报,1985,1:9-18.
[4] Tao S Y,Chen L X. The East Asian summer monsoon. Proceedings of International Conference on Monsoon in the Far East[J]. Tokyo,1985,5-8:1-11.
[5] 黄荣辉,孙凤英. 热带西太平洋暖池的热状态及其上空的对流活动对东亚夏季气候异常的影响[J]. 大气

科学,1994,18(2):141-151.

[6] 吴国雄,刘还珠. 降水对热带海表温度异常的邻域响应 I. 数值试验[J]. 大气科学,1995,19(4):422-434.

[7] 张琼,刘平,吴国雄. 印度洋和南海海温与长江中下游旱涝[J]. 大气科学,2003,27(6):992-1006.

[8] He J H, Zhou B, et al. Vertical circulation structure, interannual variation features and variation mechanism of western Pacific subtropical high[J]. Adv Atmos Sci,2001,18(4):497-510.

[9] 孙力. 东北冷涡持续活动的分析研究[J]. 大气科学,1997,21(3):297-307.

[10] Li J P, Wang J. A modified zonal index and its physical sense[J]. Geophy Res Lett,2003,30(12):1632.

[11] Ting F M, Wang H. Summertime U. S. precipitation variability and its relation to Pacific sea surface temperature[J]. J Climate,1997,10:1853-1873.

[12] 陈隆勋,朱乾根,罗会邦,等. 东亚季风[M]. 北京:气象出版社,1991:200-210.

东北冷涡的"气候效应"及其对梅雨的影响[*]

何金海[1]　吴志伟[2]　江志红[1]　苗春生[1]　韩桂荣[1,3]

(1. 南京信息工程大学气象灾害重点实验室,南京 210044; 2. 中国科学院大气物理研究所大气科学和地球流体力学数值模拟国家重点实验室(LASG),北京 100029;
3. 江苏省气象台,南京 210009)

摘要:东北冷涡是东亚中高纬地区重要的天气系统。尽管其时间尺度为天气尺度,但是频繁的东北冷涡活动具有显著的"气候效应"。这种"气候效应"不仅会影响东北地区对流层低层的月平均气温,而且对东亚地区梅雨期降水也具有影响作用。利用欧洲中期天气预报中心(ECMWF)提供的 ERA-40 再分析资料,对东北冷涡的"气候效应"及其与东亚梅雨的关系进行研究,发现:梅雨期东北冷涡和降水量存在显著的相关关系,东北冷涡越强,梅雨量很可能偏多,东北冷涡越弱,梅雨量很可能偏少;东北冷涡强年,东北冷涡引导北方"干冷"空气南侵,与低层强盛西南暖湿气流在梅雨区北缘交汇,形成"上干下湿"的不稳定层结,在上升运动的触发下,最终导致梅雨量偏多,东北冷涡弱年情况正好相反;前期北太平洋海温的异常可能是导致梅雨期东北冷涡异常的因素之一,夏季的海陆热力对比对东北冷涡起着促进作用,而冬季的热力对比对东北冷涡起着抑制作用。所有这些为东北冷涡和梅雨的短期气候预测提供了具有参考意义的结果。

关键词:东北冷涡;气候效应;梅雨

梅雨(在韩国被称为 Changma,在日本被称为 Baiu)是东亚夏季风推进的产物[1-2],它发生于北半球夏季 6—7 月,梅雨区范围从我国长江中下游地区向东延伸直达朝鲜半岛和日本。梅雨长期以来一直是我国乃至世界短期气候预测的重要内容。许多国家和地区的科学家们对其进行了大量研究[3-9]。综观这些研究,对于热带低纬地区的影响因子考虑较多,而对源自北半球中高纬地区的环流系统关注较多,详细研究较少。事实上,中高纬大气环流对包括梅雨在内的东亚夏季风降水同样有着重要的影响[10-12]。众所周知,东北冷涡是亚洲中高纬大气环流系统中的重要组成部分,一次东北冷涡活动的生命周期一般为 5~7 天,因此它是一个天气尺度的系统。然而,从空间上看东北冷涡不仅对东北地区的天气气候有着很大影响,而且可引导高纬的冷空气南下影响中低纬度地区[13];从时间上看频繁的东北冷涡活动所产生的"气候效应",不仅影响到中短期天气,而且对长期天气同样有着重大影响。所有这些也是本文的研究出发点。开展这些研究,不仅可以了解亚洲中高纬环流系统对梅雨的作用,而且可以为梅雨区的短期气候预测提供具有参考意义的结果。

---

[*] 本文发表于《科学通报》,2006 年第 51 卷第 23 期,2803-2809.其英文版发表于《Science Bull》,2007 年第 51 卷第 5 期,671-679.

1 资料和方法

本文所用再分析资料取自欧洲中期天气预报中心(ECMWF)的 ERA-40 资料集[14](水平分辨率为 2.5°×2.5°),海温资料取自美国国家海洋和大气管理局(NOAA)的扩展重建海温(ERSST)[15](水平分辨率为 2.0°×2.0°)。

孙力[16]统计分析表明,东北冷涡对夏季天气影响最为显著的特征是影响地区出现持续低温连阴雨,尤以持续低温特征更为显著。东北冷涡活动越频繁,强度越强,低温事件也越频发,累积低温时间也越长,进而导致东北冷涡影响区域季节平均气温偏低,反之当东北冷涡活动较少,强度较弱时,东北冷涡影响区域季节平均气温则偏高。基于此,我们将因东北冷涡活动异常所引起的季节平均气温和降水异常,称为东北冷涡活动的"气候效应"。为定量描述东北冷涡活动的"气候效应"对东亚梅雨的影响,这里定义了一个梅雨期(6—7月)东北冷涡强度指数 NECVI (Northeast Cold Vortex Index),具体做法如下:对东亚地区(100°～145°E,20°～160°N)1000 hPa 梅雨期平均气温进行旋转 EOF 分解[17],初始旋转的主分量取 EOF 分解的前 6 个主分量,旋转后第一模态的方差贡献占总方差的 22.9%(图1 所示),可以看到在(127°～145°E,37°～145°N)区域为负值中心所覆盖(图1 中的阴影区),该地区正是东北冷涡活跃的区域之一[16]。我们将图1 中阴影区范围内的 6—7 月 1000 hPa 平均气温的相反数,定义为梅雨期东北冷涡强度指数 NECVI(表1)。这里取相反数的目的是为了使指数更符合通常的使用习惯,即指数越高,东北冷涡越强,指数越低,东北冷涡越弱。本文将标准化 NECVI 值大于 1 和小于 −1 的分别定义为梅雨期高、低 NECVI 指数年。

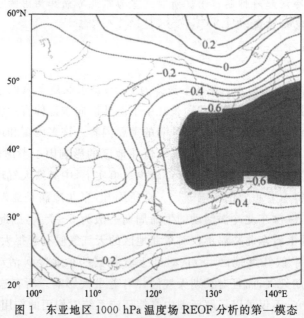

图 1 东亚地区 1000 hPa 温度场 REOF 分析的第一模态
阴影区的绝对值超过 0.6

2 梅雨期东北冷涡和降水量的关系

图 2 给出了梅雨期 NECVI 与降水量的相关系数。从长江中下游地区经东海至朝鲜半岛和日本,为大范围显著正相关区,表明当 NECVI 值越高(即东北冷涡越强)时,上述区域的降水很可能偏多,当 NECVI 值越低(即东北冷涡越弱时),上述区域的降水很可能偏少。图 2 中显著正相关区基本与东亚梅雨区重合[18],因此梅雨期东北冷涡偏强年,梅雨量很可能偏多,而东北冷涡偏弱年,梅雨量很可能偏少。由于梅雨降水分布是一种偏态分布,并非正态分布,所以在许多降水分析中,用 Z 指数[18]来描述降水量的变化。为进一步验证东北冷涡与梅雨的相关关系,本文引用文献[18]所定义的梅雨 Z 指数,具体的做法就是根据文献[18]所划定的梅雨区范围,先求出梅雨量 γ 分布概率,再计算其正态分布的逆累积分布函数,并对其进行标准化即得到 Z 指数值。Z 指数值越大,则表示梅雨降水越偏多,Z 指数值越小,则表示梅雨降水越偏少。图 3 给出了 1958—2002 年东北冷涡 NECVI 指数时间序列和梅雨 Z 指数时间序列,

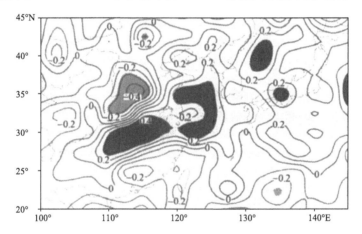

图 2 1958—2002 年梅雨期 NECVI 与东亚地区降水量的相关系数
阴影区置信度检验达 95%,深色为显著正相关区,浅色为显著负相关区

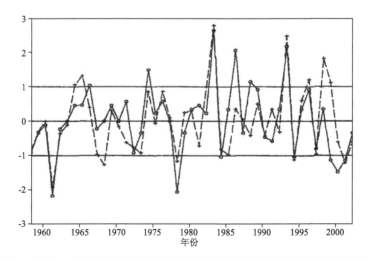

图 3 1958—2002 年梅雨期 NECVI 指数(实线)和 Z 指数(虚线)时间序列

可以看到 Z 指数时间序列和 NECVI 时间序列具有相似的变化趋势,计算两者的相关系数达到了 0.44,超过了 99% 的置信度检验,表明 NECVI 值越大,则 Z 指数值很可能也偏大,NECVI 值越小,则 Z 指数值很可能也偏小。

综上所述,梅雨期东北冷涡和梅雨存在显著的相关关系,东北冷涡越强,梅雨量很可能偏多,东北冷涡越弱,梅雨量很可能偏少。

表 1 1958—2002 年梅雨期标准化东北冷涡强度指数(NECVI)

年份	NECVI	年份	NECVI	年份	NECVI	年份	NECVI	年份	NECVI
1958	−0.92	1967	−0.23	1976	0.57	1985	0.34	1994	−1.03
1959	−0.34	1968	0.00	1977	0.00	1986	2.06	1995	0.34
1960	−0.11	1969	0.46	1978	−2.06	1987	−0.34	1996	0.92
1961	−2.18	1970	0.00	1979	−0.34	1988	1.15	1997	−0.80
1962	−0.23	1971	0.57	1980	0.34	1989	0.92	1998	0.34
1963	0.00	1972	−0.92	1981	0.46	1990	−0.46	1999	−1.15
1964	0.46	1973	−0.34	1982	0.23	1991	−0.57	2000	−1.49
1965	0.46	1974	1.49	1983	2.64	1992	0.34	2001	−1.15
1966	1.03	1975	0.23	1984	−1.03	1993	2.18	2002	−0.34

3 东北冷涡影响梅雨的可能机制分析

已有研究[16]表明东北冷涡在 500 hPa 天气图上表现为冷性的低压中心。为验证本文所定义的东北冷涡强度指数(NECVI)是否能定量描述东北冷涡的物理图像,刻画其结构特征,我们计算了前汛期东北冷涡异常强(高 NECVI)年与异常弱(低 NECVI)年的 500 hPa 合成位势高度和温度差值场(如图 3 所示)。从高度场来看(图 4a),从我国华北、东北经朝鲜半岛至日本为大范围的显著差值低压中心,说明高 NECVI 年上述地区有低压中心发展,而低指数年上述地区为低压填塞或高压发展;从温度场来看(图 4b),上述地区为显著差值低温中心所覆盖,范围比高度场的略小,表明高 NECVI 年该地区有冷中心发展,而低 NECVI 年则正好相反。从上述分析可以看到本文定义的东北冷涡强度指数(NECVI)值基本反映了东北低涡的主要特征,因此可以用它来定量描述东北冷涡的强度。

图 5 给出了梅雨期 NECVI 分别与 850,100 hPa 相对涡度场的相关系数分布。从东北冷涡活动区域高、低空相对涡度场配置来看,低层从内蒙古的东部直至日本以东洋面基本为显著正相关区所覆盖(图 5a),高层该地区同样为显著的带状正相关区控制(图 5b),表明东北冷涡活跃地区高、低空相对涡度场基本上具有相同的变化趋势,由此可以看出东北冷涡是一个深厚的系统,具有相当正压结构特征。在高 NECVI 年,上述地区正涡度发展,低 NECVI 年则有负涡度发展。我们还注意到,低层长江中下游地区存在一显著正相关区,黄河中下游以南为一显著负相关区(图 5a),高层情况正好相反(图 5b),表明这些地区高、低空相对涡度场变化趋势相反,斜压性加大,不难发现这些地区正好位于东北冷涡影响区域的底部,也是梅雨带活跃的地区。此外,我们知道正(负)涡度往往伴随着上升(下沉)运动发展。因此,东北冷涡强年,长江中下游地区往往对应着正涡度和上升运动的发展,而黄河中下游以南地区则对应着负涡度和

下沉运动发展。东北冷涡弱年则情况相反。

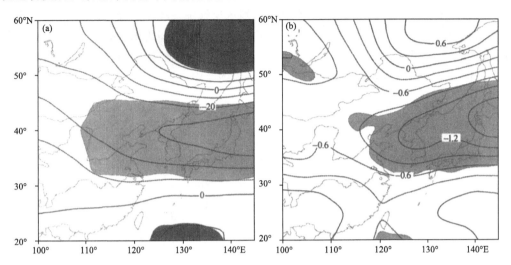

图 4　梅雨期东北冷涡异常年 500 hPa 合成位势高度(a,单位:gpm)和温度(b,单位:0.1℃)差值场
高 NECVI 年减低 NECVI 年;阴影区置信度检验达 95%,浅色为显著负值区,深色为显著正值区

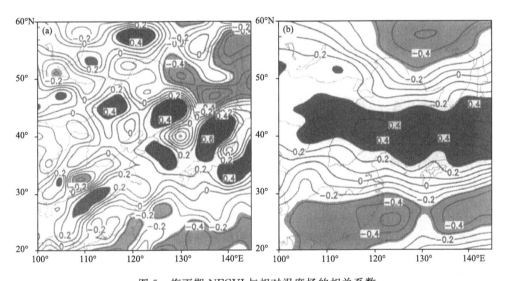

图 5　梅雨期 NECVI 与相对涡度场的相关系数
阴影区置信度检验达 95%,浅色为显著负相关区,深色为显著正相关区;(a)850 hPa;(b)100 hPa

水汽输送的强弱对梅雨量的多少有着重要影响[8]。下面我们分析东北冷涡异常年梅雨期对流层低层的水汽输送情况。图 6 为梅雨期东北冷涡异常年 850 hPa 合成水汽输送差值场,从长江中下游—朝鲜半岛—日本以东洋面为水汽输送差值辐合线(图中黑粗线),辐合线以南为显著的西—西南水汽输送差值区所覆盖,以北为显著的偏北水汽输送差值区所控制。东北冷涡偏强年,辐合线以南的西—西南水汽输送差值和以北的偏北水汽输送差值均增大,使得该区域低层水汽输送和辐合均增强,为梅雨量的增多提供了有利的水汽条件;东北冷涡偏弱年情况正好相反。究其原因,在东北冷涡偏强年,由于对流层低层长江以北直至东北的大部分地区有正涡度发展(图 5a),有利于该地区气旋性环流发展,长江以北地区正好位于气旋性环流异

常的底部,因而形成了异常的西—西北风水汽输送(图6),而在副热带西太平洋(图5a)东北冷涡强年往往伴随着异常的负涡度发展,有利于该地区反气旋性环流增强,进而导致西太平洋副热带高压的增强,梅雨区正好位于西太平洋副热带高压的西北侧边缘,因而该地区的南侧形成了异常的西—西南风水汽输送(图6)。从上面分析不难发现,东北冷涡和西太平洋副热带高压的共同作用可能是导致东亚梅雨区水汽辐合增强的原因。

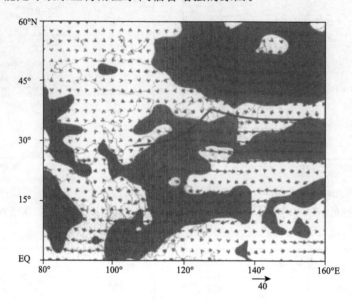

图6 梅雨期东北冷涡异常年850 hPa合成水汽输送差值场(高NECVI年减低NECVI年,单位:g·cm^{-1}·s^{-1};阴影区置信度检验达95%;黑粗线为水汽输送差值辐合线)

层结状况是梅雨降水的另一重要影响因素[19]。我们计算了梅雨期NECVI分别与1000,100 hPa相对湿度场的相关系数分布(图7)。在东北冷涡偏强年,低层梅雨区(长江中下游—朝鲜半岛—日本以东洋面)基本为正相关所覆盖(图7a),显著正相关中心分别位于长江中下游及以南地区、东北至朝鲜半岛和日本东南部洋面上,表明这些地区低层相对湿度增大,在高层(图7b),30°~150°N为显著的带状负相关区,表明该地区相对湿度减小,这就形成了"上干下湿"的高低空配置,加剧了梅雨区的不稳定层结,有利于梅雨量增多。东北冷涡偏弱年情况相反。

除此之外,上升运动也是影响梅雨降水的一个主要因子[19]。这里给出了东北冷涡异常年经向(沿120°E)和纬向(沿27.5°N)的垂直剖面差值流场(图8)。从经向剖面来看(图8a),在25°~30°N东亚地区的对流层主要为显著的差值上升气流所控制,而在30°~35°N之间对流层基本上为显著的差值下沉气流所控制。从纬向剖面来看(图8b),105°~135°E低层(850 hPa以下)为显著的差值上升气流所控制,高层(100 hPa附近)为显著的差值下沉气流所控制。所有这些表明,在东北冷涡强年,(25°~30°N,105°~135°E)区域内对流层上升运动发展,(30°~35°N,105°~135°E)区域内下沉运动发展。东北冷涡弱年,情况正好相反。

通过上述分析,可以得到东北冷涡影响梅雨的可能机制:东北冷涡强年,东北冷涡引导北方"干冷"空气南侵,与低层强盛西南暖湿气流在梅雨区北缘交汇,形成"上干下湿"的不稳定层结,在上升运动的触发下,最终导致梅雨量偏多;东北冷涡弱年情况正好相反。

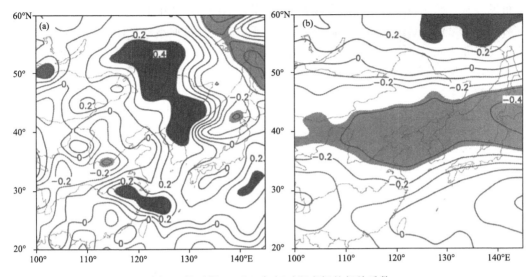

图 7　梅雨期 NECVI 与相对湿度场的相关系数

阴影区置信度检验达 95%；浅色为显著负相关区，深色为显著正相关区；(a)1000 hPa；(b)100 hPa

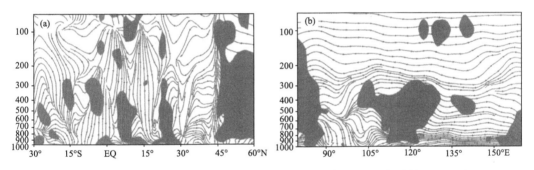

图 8　梅雨期东北冷涡异常年合成垂直剖面差值流场（高 NECVI 年减低 NECVI 年）
(a)沿 120°E 的经向剖面（阴影区置信度检验达 95%）；(b)沿 27.5°N 的纬向剖面图
（阴影区置信度检验达 99%）

4　北太平洋海温与东北冷涡异常之间的联系

鉴于大气环流异常与海温异常之间的紧密联系，我们计算了梅雨期 NECVI 与全球海温在前期 6 个月（从前一年 12 月到当年 5 月）和同期（当年 6—7 月）中的相关，发现与北太平洋的海温相关最为显著（如图 9 所示）。前一年 12 月（图 9a）北太平洋中部出现了小范围的负相关区，其后范围逐月扩大并向西北方向扩展（图 9b~g），至当年 7 月（图 9h）整个西北太平洋为大范围负相关区所覆盖，中心位于 40°~45°N，表明当梅雨期东北冷涡偏强的前期或同期时，上述海区海温往往显著偏低，这有利于上述海区异常的下沉运动发展。与此同时，随着东亚大陆的热力性质发生由冬到夏的转变（冬季东亚大陆为冷源，风由大陆吹向海洋，夏季东亚大陆转变为热源，风由海洋吹向大陆），由于东北冷涡偏强年夏季西北太平洋海温往往偏低，西北太平洋和东亚大陆之间主要为东西向的海陆热力差异，因而低层东亚大陆和西北太平洋之间会产生东风异常。同时西北太平洋的海温偏低，使得该地区与东亚中低纬地区的热力差异加大，造成高层西风急流增强。在东北冷涡出现时，东亚中高纬地区原本就存在上升运动，这样就在东亚中高纬和西北太平洋之间就形成了一个异常气流的闭合环流，有利于东北冷涡的进一步活跃和加强。反之，则东北冷涡偏弱。因此，前期北太平洋海温的异常可能是导致梅雨期东北冷涡异常的因素之一。

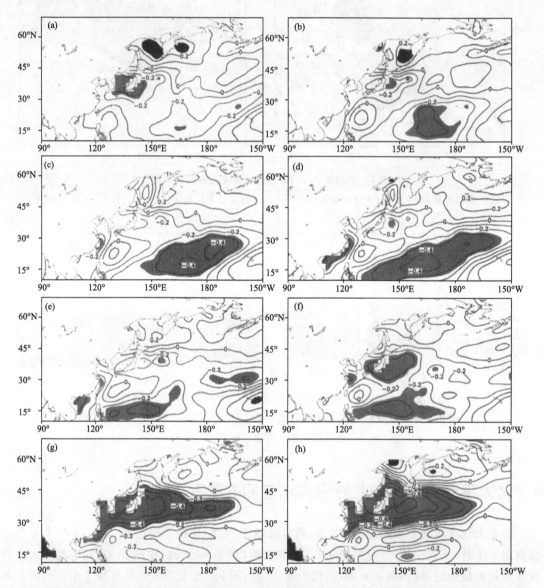

图9 梅雨期NECVI与北太平洋海温的相关阴影区置信度检验达95%，浅色为显著负相关区，深色为显著正相关区；(a)前一年12月；(b)当年1月；(c)2月；(d)3月；(e)4月；(f)5月；(g)6月；(h)7月

不难看出夏季的海陆热力对比对东北冷涡起着促进作用，而冬季的则对东北冷涡起着抑制作用。这是因为，在冬季西北太平洋是热源，而东亚大陆是冷源，因而在低层风由东亚大陆吹向西北太平洋，即为偏西风，其方向与高层一致，不利于东亚中高纬地区上升运动的发展，这

可能也是东北冷涡在夏季的发生频率远远多于其他季节的原因。

5 结 论

东北冷涡在东亚中高纬地区夏季时有发生,长期以来从天气学角度出发对其开展了一些研究,但对频繁东北冷涡活动产生的"气候效应"研究不多。本文重点分析了东北冷涡的"气候效应"对梅雨的影响,提出了可能的物理机制,从海气相互作用的角度分析了导致梅雨期东北冷涡异常的原因。根据以上的分析,我们可以得出以下一些结论:

(1)梅雨期东北冷涡和降水量存在显著的相关关系,东北冷涡越强,梅雨量很可能偏多,东北冷涡越弱,梅雨量很可能偏少。

(2)东北冷涡强年,东北冷涡引导北方"干冷"空气南侵,与低层强盛西南暖湿气流在梅雨区北缘交汇,形成"上干下湿"的不稳定层结,在上升运动的触发下,最终导致梅雨量偏多;东北冷涡弱年情况正好相反。

(3)前期北太平洋海温的异常可能是导致梅雨期东北冷涡异常的因素之一。夏季的海陆热力对比对东北冷涡起着促进作用,而冬季的对东北冷涡起着抑制作用。

致谢:感谢欧洲中心(ECMWF)和中国气象局国家气象信息中心提供相关气象资料。本工作受国家重点基础研究发展计划项目(批准号:2004CB418303)、国家自然科学基金(批准号:40605022)及上海市气象局"副热带季风研究及预报技术开发项目"联合资助。

参考文献

[1] 竺可桢.东南季风与中国之雨量[J].地理学报,1934,1(1):1-27.

[2] 涂长望,黄士松.中国夏季风之进退[J].气象学报,1944,18:1-20.

[3] Tao S Y,Chen L X. A Review of Recent Research on the East Asian Summer monsoon in China[C]// Chang C P,Krishnamurti T N. Monsoon Meteorology. Oxford:Oxford University Press,1987:60-92.

[4] 符淙斌,腾星林.ENSO与中国夏季气候的关系[J].大气科学,1988,特刊:133-141.

[5] 吴国雄,刘还珠.降水对热带海表温度异常的邻域响应Ⅰ.数值试验[J].大气科学,1995,19(4):422-434.

[6] Zhang R H. Relations of water vapor transport from Indian Monsoon with that over East Asia and the summer rainfall in China[J]. Adv Atmos Sci,2001,18(5):1005-1017.

[7] Zhang Q,Wu G X,Qian Y F. The bimodality of the 100 hPa South Asia High and its relationship to the climate anomaly over East Asia in summer[J]. J Meteor Soc Japan,2002,80(4):733-744.

[8] 徐祥德,陶诗言,王继志,等.青藏高原—季风水汽输送"大三角扇型"影响域特征与中国区域旱涝异常的关系[J].气象学报,2002,60(3):257-67.

[9] 黄荣辉,顾雷,徐予红,等.东亚夏季风爆发和北进的年际变化特征及其与热带西太平洋暖池热状态的关系[J].大气科学,2005,29(1):20-36.

[10] 张庆云,陶诗言.亚洲中高纬度环流对东亚夏季降水的影响[J].气象学报,1998,56(2):199-211.

[11] Li J P,Wang J. A modified zonal index and its physical sense[J]. Geophys Res Lett,2003,30(12):1632,doi:10.1029/2003GL017441.

[12] Ju J H,Lu J M,Cao J,et al. Possible impacts of the Arctic Oscillation on the interdecadal variation of summer monsoon rainfall in East Asia[J]. Adv Atmos Sci,2005,22(1):39-48.

[13] 苗春生,吴志伟,何金海.近50年东北冷涡异常特征及其与前汛期华南降水的关系分析[J].大气科学,2006,30(6):1249-1256.

[14] Uppala S M,Kallberg P W,Simmons A J,et al. The ERA-40 re-analysis[J]. Q J R Meteorol Soc,2005,131:2961-3012. doi:10.1256/ qj.04.176.

[15] Smith T M, Reynolds R W. Improved extended reconstruction of SST (1854—1997) [J]. J Climate, 2004,17(12):2466-2477.

[16] 孙力.东北冷涡持续活动的分析研究[J].大气科学,1997,21(3):297-307.

[17] Horel J D. A rotated principal component analysis of the interannual variability of the Northern Hemisphere 500 mb height field[J]. Mon Wea Rev,1981,109(10):2080-2092.

[18] 吴志伟,江志红,何金海.近50年华南前汛期降水、江淮梅雨和华北雨季旱涝特征对比分析[J].大气科学,2006,30(3):391-401.

[19] 陶诗言,赵煜佳,陈晓敏.中国的梅雨[C]//王明星.东亚季风和中国暴雨——庆祝陶诗言院士八十华诞文集.北京:气象出版社,1998:3-46.

北太平洋中纬度负海温异常对副热带高压影响的数值试验[*]

巩远发[1,2,3]　何金海[3]　段廷扬[1]　潘攀[1]

(1. 成都信息工程学院,四川 成都 610041;
2. 中国科学院东亚区域气候-环境重点实验室,北京 100029;
3. 南京信息工程大学,江苏 南京 210044)

摘要:利用中国科学院大气物理研究所的 $T_{42}L_9$ 全球大气环流模式,研究了 1993 年夏季北太平洋中纬度海温异常对西太平洋副热带高压的影响。数值试验结果表明:(1)中纬度地区负距平海温异常对其南侧太平洋副热带高压的形态和脊线位置都有影响,负距平海温异常试验的 500 hPa 副热带高压的形态明显比气候平均海温试验更接近于 NCEP 资料实况,对流层中下层副热带高压脊线的平均位置也比气候平均海温试验要偏南,与实况比也更接近了;(2)在 500 hPa 上,北太平洋中纬度负距平海温异常在其南侧产生的高度扰动使副热带高压向南偏移,而在其北侧产生的高度扰动则沿北太平洋中高纬度、北美中高纬度、北大西洋和欧洲南部以及地中海和北非地区的大圆路径传播,并影响和改变整个北半球大气环流的变化。1993 年夏季北太平洋中纬度异常的低海温可能是该年盛夏 6—7 月北太平洋副热带高压位置异常偏南的重要原因之一。

关键词:北太平洋中纬度地区;负距平海温异常;副热带高压;数值试验

1　引　言

　　副热带高压是位于南北两个半球副热带地区的高压带,它也是连接中、低纬度环流系统的纽带,它的活动和变异直接关系着全球大气环流系统的变化和气候异常。关于副热带高压的预报和研究涉及很多方面,吴国雄[1]、刘屹岷[2]在最近作了非常系统的总结,其中海温与副热带高压的关系研究就是人们非常关注的问题之一,如:Bjerknes[3-4]在 20 世纪 60 年代就研究了 Hardly 环流的变化对赤道异常海温的响应机制以及整个大气环流变化与赤道太平洋海温的遥相关关系;黄荣辉等[5]利用西太平洋暖池区的表层、次表层海温和高云量等观测资料分析了暖池的热状态及其对东亚气候的影响指出:当暖池增暖时,菲律宾附近对流活动加强,西太平洋副热带高压则偏北,反之则偏南;Sun 等[6]比较了强弱副热带高压年 OLR、垂直环流和海温异常的差异,发现与副热带高压异常有关的垂直环流是由热带东太平洋海温异常导致的季风环流的变化引起的;应明等[7]利用多年的观测资料,研究了冬、夏季西太平洋副高对同期及前期不同海域海温异常的响应,分析了夏季副热带高压与前冬、前春及同期各月全球海温的最佳耦合模态,进而研究了海气相互作用对副热带高压影

---

[*] 本文发表于《热带气象学报》,2006 年第 22 卷第 4 期,386-392.

响的物理过程;龙振夏等[8]用数值试验的方法研究了赤道东太平洋海温正异常对西太平洋副热带高压的影响;最近,陈海山等[9]研究了Nino C区秋季海温异常对东亚冬季大气环流的可能途径,认为秋季SST正(负)异常引起西太平洋地区出现负(正)WP遥相关型的环流异常,最终导致偏弱(强)的东亚冬季风活动;等等。这些研究都表明热带海温异常会对副热带高压产生不同程度的影响。

关于中纬度海温变化对大气环流影响的研究也逐渐为人们所重视,如:Lau和Nath[10-11]用大气环流模式研究了大气对热带外海温异常的响应;杨英等[12]用EOF方法对比分析了冬季中纬度北太平洋和赤道中、东太平洋SST异常与北半球大气环流的相互影响,揭示了中纬度海洋大气协同作用的事实和机制;李丽平等[13]在研究北太平洋区域浅层海温(ST)与海平面气压场年代际异常关系时发现,北太平洋中部前期ST持续偏高,相应的1(7)月阿留申低压(北太平洋副高)偏弱、位置偏西(变化不明显),而东亚冬(夏)季风偏弱,等等。但是,相对于热带海温影响的研究,中纬度海温异常影响的研究还是较少。

我们在的分析和研究中发现,从1993年春季的3月开始,北太平洋中纬度日界线及其东部的30°~40°N就出现了1℃以上负异常海温区,并不断加强,到盛夏,北太平洋中纬度就出现较大范围的海温负距平;而这一年夏季西太平洋副热带高压异常偏南,是一个东亚夏季风异常偏弱的年份[14],并且夏季我国的长江流域降水也异常偏多,造成该地区大范围的洪涝灾害,华北和华南降水却相对较少[15]。为了研究中纬度海温异常对副热带高压的影响,用数值试验的方法研究了1993年夏季中纬度负海温异常对副热带高压变化的影响,期望得到部分关于中纬海温异常对副热带高压和大气环流异常变化影响的认识,为我国夏季短期气候的变化和预测研究提供一些参考。

2 模式和试验方案

本文使用的模式是中国科学院大气物理研究所的$T_{42}L_9$全球大气环流谱模式,模式的动力框架和有关物理过程参见文献[16-17],最近,孙淑清等[18]还用这一模式研究海温异常对东亚夏季风及长江流域降水的影响数值试验,得到了比较好的结果。我们选择了4个模拟试验,试验1(T1)和试验2(T2)分别以1993年6月1日和7月1日的NCEP再分析资料作为初始场,6月和7月的NCEP气候平均海温作为海温的背景场;试验3(T3)和试验4(T4)的初始场也分别与T1和T2相同,但用了1993年6月和7月的NCEP的实际海温作为海温的背景场。4个试验都积分了1个月,1天输出一次结果。T1和T2作为控制试验,目的是为了检验模式在月季时间尺度上的模拟(或预报)性能;T3和T4作为敏感性试验,为了研究中纬度海温异常对西太平洋副热带高压的影响及其对北半球大气环流的作用机制。

图1是1993年6月和7月的北太平洋地区NCEP的实际海温与其气候平均海温的差。从图中可见:在1993年的6—7月,北太平洋海域中纬度的大范围地区的海温都要比气候平均的海温低很多,6月海温的负距平达-1℃以上的范围在150°E~145°W、26°~42°N,中心的值达-2℃以上;7月海温的负距平值还有一定的增加,负距平中心区域的差值接近-3℃,范围北移到30°~45°N。

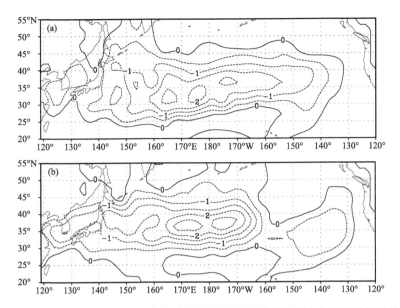

图 1　1993 年 6 月(a)和 7 月(b)北太平洋海温的距平(实际海温与气候平均海温的差)

3　控制试验结果

为了考查模式的性能,我们将 T1 和 T2 积分 30 天的平均 500 hPa 高度场以及太平洋地区 120°E～120°W 纬向平均的纬向风和气温的高度—纬度垂直剖面与对应 NCEP 资料进行比较。图 2 是 1993 年 6 月 NCEP 资料的东亚及北半球太平洋地区 500 hPa 平均高度场和 T1 积分 30 天平均模拟的 1993 年 6 月 500 hPa 高度场。从图 2a 中可以看到,在 NCEP 资料中,30°～50°N 的中纬度地区是平直、多小波动纬向环流型;30°N 以南的低纬度地区是西太平洋副热带高压,副高西部的脊线位置大约在 20°N 以南,副高东部中太平洋地区的脊线位置大约在 20°N 以北;50°N 以北的高纬度地区则是经向环流为主。将 T1 模拟的结果(图 2b)与图 2a 比较可知,模式比较好的模拟出整个流场的环流形势,环流系统的分布与实况大体一致。如中纬度的平直纬向环流,低纬度的副热带高压、高纬度 130°～150°E 的高脊和阿留申地区的低压等系统都模拟得很好,这说明了模式有较好的性能,能够较准确地模拟环流的基本状态。但是,T1 模拟的环流系统的具体范围与 NCEP 的实况资料还有一些差别:T1 中低纬副热带高压的位置明显比实况要偏东偏北一些,阿留申地区的低压位置也比实际要偏西、偏南一些。

图 3 是 NCEP 资料 1993 年 7 月东亚及北半球太平洋地区的 500 hPa 平均高度场和 T2 积分 30 天平均模拟的 1993 年 7 月 500 hPa 高度场。对比图 3a 和图 3b 也可以看到,模拟的环流系统与实况也基本一致,高纬度地区主要是经向环流,中纬度是小的纬向波动、锋区较强;但在 T2 中,副热带高压有两个中心,中太平洋上的东部副高中心位置比 NCEP 资料的中心位置明显偏北,在阿留申群岛附近的低值系统也比 NCEP 资料的低值系统要偏北偏东一些。

更进一步,我们还分别对比分析了 T1 和 T2 积分 30 天平均的、在太平洋地区 120°E～120°W 纬向平均的纬向风和温度的垂直剖面及相应的 NCEP 的实况资料。结果表明 T1 和 T2

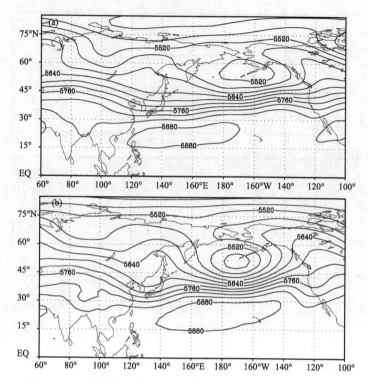

图 2　1993 年 6 月东亚和北太平洋的 500 hPa 平均高度场(a. NCEP 资料；b. T1 积分 30 天的平均)

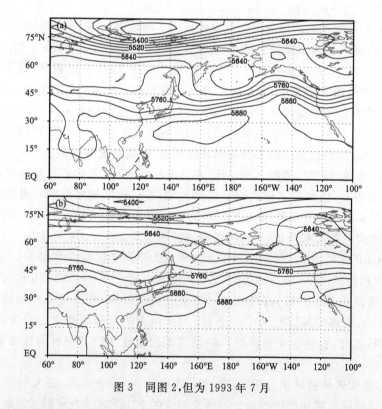

图 3　同图 2,但为 1993 年 7 月

都较好地模拟出了1993年6月和7月太平洋地区纬向平均的纬向环流和温度的分布。图4和图5是1993年7月NCEP的实况和T2的结果图,1993年6月NCEP的实况和T1的结果图略。从图4、5可以看到,在南北两个半球200 hPa上副热带西风急流轴中心的位置及其下方对流层中下层对应位置上的中纬度锋区、北半球高纬度极锋急流的位置、赤道热带地区东风区的范围等系统在两个图上都基本一致,只是急流轴中心的强度及其对应锋区在模拟试验中均较NECP资料弱一些。

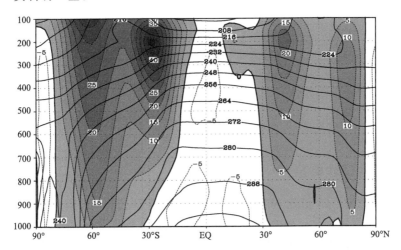

图4　1993年7月NCEP资料在120°E～120°W纬向平均的纬向风和温度的垂直剖面
虚线是等风速线,等值线间距是5 m·s^{-1}。阴影区是西风。粗实线是东西风零线。
实线的等温线,等值线间距是8K;下同

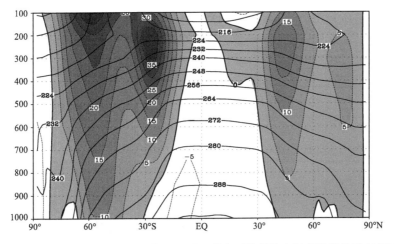

图5　T2积分30天平均在120°E～120°W纬向平均的纬向风和温度的垂直剖面
说明同图4

从热带到中纬度地区,对流层中下层东西风的交界处(纬向风风速等于零的位置),通常可以作为副热带高压脊线的位置[19]。因此,在图4和图5中还可以看到:在1993年7月的NECP资料中,北半球对流层中下层400～850 hPa的副热带高压脊线在北太平洋地区的平均位置均位于30°N以南。但在T2试验中,模拟出的对流层中下层副高脊线均较NCEP的实况位置偏北,500 hPa 上大约位于 30°N,500 hPa 以下都在 30°N 以北,这也是 T2 的存在的问题之一。总之,T1 和 T2 能比较好地模拟出 1993 年 6 月和 7 月北太平洋副热带高压和大气环流的主要特征,模式的性能是可靠的。但在副热带高压中心和脊线的具体位置的模拟方面还存在一些问题。

4 海温异常的敏感性试验

T1 和 T2 的控制试验是以多年的气候平均海温作为海洋对大气影响的因子之一的试验试验模拟的一些环流系统(如副热带高压脊线)的具体位置和实况还有差异。但在第 2 部分的介绍中,我们知道 1993 年 6—7 月在北太平洋中纬度地区海温的负距平达到了近 −3℃,为了分析和研究 1993 年 6—7 月中纬度异常低的海温对副热带高压的影响,在 T1 和 T2 的基础上,用 1993 年 6 月和 7 月的实际海温替代气候平均海温,作了海温异常对大气环流变化影响的敏感性试验,即 T3 和 T4。

4.1 海温异常的试验结果

图 6 和图 7 分别是 T4 积分 30 天的 500 hPa 平均高度场和 120°E～120°W 纬向平均的纬向风和温度的垂直剖面。比较图 6 和图 3b 可以看到,T4 的结果明显比 T2 中副热带高压的范围及其脊线位置要偏南、整个副热带高压范围也要偏西和偏南;在中高纬度阿留申群岛的低压差别不大。再将其与实况(图 3a)相比较,西太平洋副热带高压的形态和位置已经比较接近,T4 的结果显然比 T2 中的结果有较大的改进。

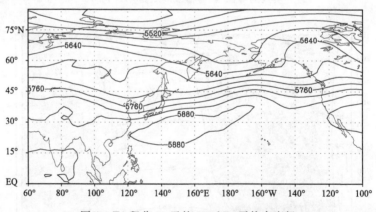

图 6 T4 积分 30 天的 500 hPa 平均高度场

比较图 7 和图 5 也可以看到,在 T4 中北半球副热带地区对流层中下层的纬向风零线的位置比 T2 中的位置偏南了 2～3 个纬距,700 hPa 以上到 400 hPa 都在 30°N 以南,同实况的 NCEP 资料(图 4)更接近;700 hPa 以下同实况比仍然有一些差异,但还是比 T2 更近于实况。因此,T4 模拟副热带高压脊线位置较接近于实际情况,比 T2 大有改进。T3 的结果同 T4 的结果一样,其模拟出的副高形态位置和脊线位置均好于 T1,限于文章的篇幅,这里就不给出详细的结果。

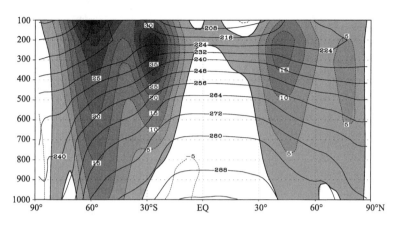

图 7　T4 积分 30 天平均在 120°E～120°W 纬向平均的纬向风和温度的垂直剖面

4.2　海温异常影响的可能机制

从 T3 和 T4 的结果分析中看到,1993 年 6—7 月北太平洋中纬度海温异常对副热带高压和大气环流可能有较大的影响,为了研究这种影响的物理机制,我们分析了 T3 和 T1 之间以及 T4 和 T2 之间积分 30 天期间的逐旬平均和整个积分的后 25 天平均的 500 hPa 高度场之差。从 T3 和 T1 之差可以看到,在 6 月的前 10 天并没有明显的差异和变化(图略),太平洋上以及亚洲和北美大陆的 500 hPa 高度场之差都不到 5 gpm,也就是说负距平海温异常对大气产生的影响还非常弱;第 11～20 天(图略),负距平海温的影响逐渐增强,500 hPa 高度场开始出现明显的差异,在北太平洋上中纬度海温负距平异常区的东部偏北地区、40°～50°N 的东太平洋上是正的高度差区域,偏南的 30°N 附近是负的高度差区域,差值虽然还较小(不超过 20 gpm),但使大洋中部副高位置南移了,亚洲和北美大陆上仅有小范围的高度差异变化区。图 8 是 T3 与 T1 积分 30 天期间 21～30 天平均(a)和整个积分的后 25 天平均(b)的北半球 500 hPa 高度场之差。从图 8a 中可以看到,第 21～30 天的 T3 与 T1 试验的差值就很明显,东太平洋上的 180°～150°W,20°N 处是负的高度差区,中心值在 40 gpm 以上,即大洋中部副高强度减弱、位置则向南向西移动;北面 35°～50°N 的东太平洋上是正高度差区,中心达 80 gpm以上,并且这些高度场负、正差值区沿北美大陆中高纬度地区的西部和东部、大西洋中高纬度地区的西部和东部、地中海和北非东部之间的中纬度地区的大圆路径传播。图 8b 与图 8a 类似,但由海温负距平产生的 500 hPa 高度扰动变化和传播的系统性更清楚。

在 T3 试验中,由北太平洋中纬度海温异常激发出的扰动由弱变强的变化和传播过程,同黄荣辉等[20]的行星波传播理论非常一致,但是 T3 是在东太平洋副热带和中纬度地区分别产生的扰动,而黄荣辉则是在西太平洋热带低纬地区对流活动异常热源产生的扰动,两者之间的联系和区别是值得进一步研究的问题。

从 T4 和 T2 之差可以看到,7 月的前 10 天(图略),中纬度负距平海温异常产生的扰动首先在日本东部和南部的西太平洋的副热带到中纬度地区出现,南侧的正变高扰动正好使得西太平洋的副热带高压向南偏移;随后,南侧的扰动逐渐加强,北侧的扰动也加强并向西扩展。图 9 是 T4 与 T2 积分 30 天期间 21～30 天平均(a)和整个积分的后 25 天平均(b)的北半球 500 hPa 高度场之差。从图 9 可以看到,负距平海温异常产生的扰动在日本东南部使副高南

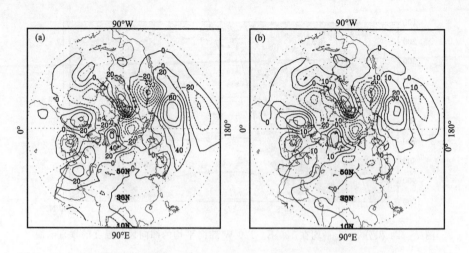

图8 T3 与 T1 积分 30 天期间 21～30 天平均(a. 等值线间距 20 gpm)和 6～30 天平均
(b. 等值线间距 10 gpm)的北半球 500 hPa 高度场之差

移,北部则沿北太平洋高纬度地区、北美大陆北部、大西洋高纬度地区、格陵兰、大西洋东部中纬度地区以及北非的路径传播,并影响北半球大气环流的变化。将图9和图8比较可知,由于负距平海温异常产生的 500 hPa 高度扰动的变化过程和传播方式基本相似,但扰动产生的地区是有一些差异的,这种差异分别使得太平洋中部和西部的副热带高压向西向南偏移,产生这些差异的原因是什么? 我们将作进一步的分析。

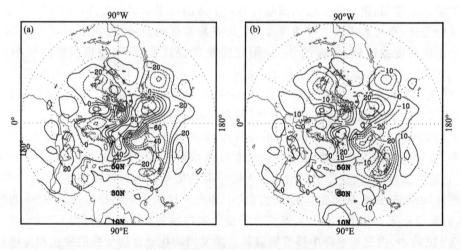

图9 同图8,但为 T4 与 T2 之差

5 小结和讨论

从上面的分析中可以看到,在 1993 年夏季的 6—7 月,北太平洋中纬负距平海温异常不仅对其南面太平洋副热带高压有影响,而且对整个北半球大气环流的变化也有较大的影响。主要表现为:

(1)北太平洋中纬度海温异常对其南面太平洋副热带高压的形态和脊线位置都有比较大

的影响。数值试验的结果表明,在加负距平海温异常试验中,北太平洋地区 500 hPa 副热带高压的形态明显比气候平均海温试验更接近于实况;对流层中下层副热带高压脊线(东西风零线)的平均位置也比气候平均海温试验要偏南 2~3 个纬距,气候平均试验中的脊线位置比实况明显偏北,负距平海温异常试验则与实况比较接近。

（2）北太平洋中纬度负距平海温异常在其南侧副热带地区激发的 500 hPa 高度扰动不仅使副热带高压向西向南偏移,而在其北侧中高纬度地区激发的 500 hPa 高度扰动则经北太平洋中高纬度地区、北美中高纬度地区、北大西洋和欧洲南部地区、地中海和北非地区的大圆路径传播,影响和改变整个北半球大气环流的变化。

上面的研究结果虽然仅是 1993 年 6 月和 7 月个例试验的结果,但从中我们可以看到,中高纬度海温异常在海气相互作用的过程中对大气的影响虽不及热带海洋敏感,但仍然是值得我们探索的一个科学问题,如上面分析中提到即使负距平海温异常的地区和性质差异不是太大,但由其作用产生的扰动的地区分布以及扰动传播路径却有一些差别,弄清其中的物理过程的差异,对研究短期气候的变化机制和短期气候预测都有重要的意义。

参考文献

[1] 吴国雄,丑纪范,刘屹岷,等.副热带高压形成和变异的动力学问题[M].北京:科学出版社,2002:314.
[2] 刘屹岷,吴国雄.副热带高压研究回顾及几个基本问题的再认识[J].气象学报,2000,58(4):500-512.
[3] Bjerknes J. A possible response of the atmospheric Hardly circulation to the equaterial anomalies of ocean temperature[J]. Tellus,1966,18(3):820-829.
[4] Bjerknes J. Atmosphere teleconnection from the equaterial Pacific[J]. Mon Wea Rev,1969,97(4):163-172.
[5] 黄荣辉,孙凤英.热带西太平洋暖池的热状态及其上空的对流活动对东亚夏季气候异常的影响[J].大气科学,1994,18(2):456-464.
[6] Sun Shuqing,Ying Ming. Subtropical high anomalies over the western Pacific and its relations to Asia monsoon and SST anomaly[J]. Adv Atmos Sci,1999,16(4):559-568.
[7] 应明,孙淑清.西太平洋副热带高压对热带海温异常响应的研究[J].大气科学,2000,24(2):193-206.
[8] 龙振夏、李崇银.赤道东太平洋海温正负异常对西太平洋副热带高压影响的数值模拟研究[J].大气科学,2000,25(2):146-159.
[9] 陈海山,孙照渤,倪东鸿.Nino C 区秋季海温异常对东亚冬季大气环流的影响[J].热带气象学报,2002,18(2):148-156.
[10] Lau N C,Nath M J. A general circulation model study of the atmospheric response to extratropical SST anomalies observed in 1950-79[J]. J Climate,1990,3(9):965-989.
[11] Lau N C,Nath M J. A modeling study of the relative roles of tropical and extratropical SST anomalies in the variability of the global atmosphere-ocean system[J]. J Climate,1994,7:1184-1207.
[12] 杨英,孙照渤.中纬度北太平洋 SST 异常与大气环流的关系[J].南京气象学院学报,1995,18(2):192-199.
[13] 李丽平,王盘兴,李泓,等.北太平洋区域浅层海温与海平面气压场年代际异常关系的分析[J].热带气象学报,2003,19(4):357-366.
[14] 祝从文,何金海.东亚季风指数及其与大尺度热力环流年际变化关系[J].气象学报,2000,58(4):391-402.

［15］陈桂英.1993年北半球环流特征及其对我国天气气候的影响［J］.气象,1994,20(4):23-26.

［16］纪立人,陈嘉滨,张道民,等.一个包含非绝热加热物理过程的全球谱模式及其初步试验结果［C］//中期数值天气预报论文集(三).北京:气象出版社,1990:27-40.

［17］Zhang Daomin,Li Jinlong,Ji Liren, et al. A global spectral model and test of its performance［J］. Adv Atmos Sci,1995,12(1):67-77.

［18］孙淑清,马淑杰.海温异常对东亚夏季风及长江流域降水影响的分析及数值试验［J］.大气科学,2003,27(1):36-51.

［19］刘平,吴国雄,李伟平,等.副热带高压带的三维结构特征［J］.大气科学,2000,24(5):577-584.

［20］黄荣辉,李维京.夏季热带西太平洋上空的热源异常对东亚上空副热带高压的影响及其物理机制［J］.大气科学,1988,12(特刊):107-117.

长江中下游夏季旱涝并存及其异常年海气特征分析*

吴志伟[1,3]　何金海[1,2]　李建平[3]　江志红[1]

(1. 南京信息工程大学气象灾害重点实验室,南京 210044;
2. 中国气象局广州热带海洋气象研究所,广州 510080; 3. 中国科学院
大气物理研究所大气科学与地球流体力学数值模拟国家重点实验室,北京 100029)

摘要: 利用国家气候中心提供的 1957—2000 年中国 720 站夏季(5—8月)逐日降水资料,对长江中下游地区夏季旱涝并存现象进行研究,并定义了一个季时间尺度的旱涝并存指数(I_{DFC}),再结合 NCAR/NCEP 的 500 hPa 高度场、850 hPa 风场等再分析资料以及 Reynolds 海温资料,对该地区旱涝并存异常年的海气背景特征进行了统计分析。结果表明:近 50 年长江中下游夏季旱涝并存异常的发生频率呈现上升趋势;夏季旱涝并存异常年,西太平洋副高空间活动范围较大,同期东亚夏季风偏弱;在其前期 6 个月中,阿拉伯海、孟加拉湾和中国南海海温显著偏高,另外赤道东太平洋海温呈现上升趋势,对应着 El Niño 的成熟阶段或发展阶段。所有这些为长江中下游夏季旱涝并存现象的预测,提供了有参考意义的前兆信号。

关键词: 长江中下游;旱涝并存;海气背景

1 引　言

在过去的 70 年里,关于长江中下游夏季降水的气候特征和环流形势方面已有许多研究成果。早在 20 世纪 30—40 年代,竺可桢[1]和涂长望等[2]就指出,长江中下游夏季降水是东亚夏季风推进的产物,其季节内变化与东亚夏季风的进退紧密相关。陶诗言等[3]对夏季江淮持久性旱涝现象的环流特征进行了研究,发现持久性旱涝在 500 hPa 高度场表现有一定的流型。吴国雄等[4]和张琼等[5]的研究表明长江中下游的旱涝与中国近海海温(如南海、印度洋等)关系密切。黄荣辉等[6]提出,东亚夏季风雨带的季节内变化还与西太平洋暖池的热状态关系密切。He 等[7]的统计分析结果显示,夏季西太平洋副高的 3 个参数(面积、强度和脊线纬度)与长江中下游夏季降水存在很高的相关。

综观前人对旱涝的研究,多集中于空间的非均衡性,即偏重于区域性的旱灾或涝灾,对旱涝的时间非均匀分布情形考虑较少,而"旱涝并存、旱涝急转"等现象正属于后者。什么是"旱涝并存",目前尚无一个明确的定义。旱涝并存现象是指一段时间特别旱,而另一段时间又特别涝,旱涝交替出现的情形,多发生于夏季,反映了旱涝极端事件在短期内的共存。通常,降水的短期气候预测只能针对未来一段时期内的总降水量进行预报,例如预测夏季总雨量正常,通常人们一定会认为"风调雨顺",但如果发生了"旱涝并存"异常,则意味这期间既发生了旱灾又

---

* 本文发表于《大气科学》,2006 年第 30 卷第 4 期,570-577.

发生了涝灾，其带来的危害是可想而知的。

到目前为止，对于"旱涝并存"的研究尚不多见。对其进行探讨，不仅可以更深入反映长江中下游夏季降水与东亚季风的联系，而且有利于该地区降水的气候预测，尤其是可为夏季防汛抗旱工作提供具有参考意义的结果，这也是本文的出发点。由于前人对长江流域重旱(涝)已有较多研究，故本文研究重点立足于夏季总降水正常或正常略偏多(少)的情形。

2 资料和方法

本文所用逐日降水资料是由中国气象局国家气象信息中心提供的1957—2000年中国720站地面观测资料，高度场、风场(水平分辨率为2.5°×2.5°)取自ERA-40再分析资料，海温场使用了NOAA扩展重建海温资料（水平分辨率为2°×2°）。

这里选取5—8月累积降水量来表征夏季降水。首先，参考Ting和Wang[8]划分水区的方法，先计算全国720站1957—2000年夏季降水标准差，得到长江中下游地区降水变率最大的测站为高邮，然后以高邮站为基点，计算其与全国720个站夏季降水的单点相关，得到空间相关图(如图1所示)，相关系数在基点处为1，然后向周围逐渐减小，阴影所覆盖的区域为显著相关的区域(通过95%水平的置信度检验)，该区域反映较为一致的降水变率，使用该区域内的测站（共42个测站）的降水平均值来表征长江中下游地区夏季降水。

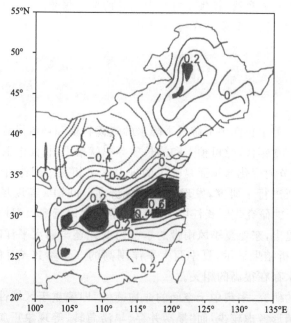

图1 全国720站夏季(5—8月)降水与高邮的单点相关图（阴影表示置信度达95%）

为更好地研究夏季"旱涝并存"现象的科学内涵和基本特征，本文首先对其作如下定义：发生于5—8月的旱涝交替出现的现象，其累积降水量在正常水平(降水距平在-0.5~0.5个标准差范围内)波动。为能定量反映"旱涝并存"的强度，我们定义了一个季时间尺度的旱涝并存指数 I_{DFC}(Dought-Flood Coexistence Index)：

$$I_{DFC} = S_{STD} \cdot 10^{(0.5-|R_{STD}|)}$$

式中，S_{STD} 为夏季标准化无雨日数，R_{STD} 为夏季标准化累积降水量。S_{STD} 反映了降水的集中程度，S_{STD} 越大，表明降雨的集中程度越大，旱和涝的强度也越大。$10^{(0.5-|R_{STD}|)}$ 为权重系数，作用是增大雨量正常夏季所占权重，减小重旱（涝）夏季所占权重。I_{DFC} 值越大，则表明雨量分布越不均匀，旱和涝的强度也越大；反之表明雨量分布越均匀，旱和涝的强度也就越小，即越"风调雨顺"。具体数值见表1。

表 1 1957—2000 年长江中下游夏季旱涝并存指数（I_{DFC}）

年份	I_{DFC}	年份	I_{DFC}	年份	I_{DFC}	年份	I_{DFC}	年份	I_{DFC}
1957	−0.85	1966	0.15	1975	−1.50	1984	−0.04	1993	−1.49
1958	−0.42	1967	0.16	1976	0.18	1985	0.16	1994	0.25
1959	0.24	1968	0.20	1977	−3.21	1986	2.27	1995	1.83
1960	0.13	1969	−0.02	1978	0.18	1987	−0.12	1996	0.04
1961	0.18	1970	−1.44	1979	0.02	1988	0.84	1997	0.94
1962	−0.45	1971	0.67	1980	0.05	1989	−0.66	1998	0.03
1963	−0.69	1972	0.41	1981	0.73	1990	1.30	1999	−0.16
1964	0.05	1973	−1.04	1982	0.07	1991	0.08	2000	2.70
1965	−1.31	1974	−0.71	1983	−0.01	1992	0.37		

3 "旱涝并存"年和"风调雨顺"年的划分及结果对比

上面所定义的 I_{DFC} 究竟能不能与本文定义的"旱涝并存"相吻合，即能否反映夏季降水总体正常但在时间分布上不均匀这一特征？

表 2 给出了 1957—2000 年 I_{DFC} 最大（小）前 10 位的 $\Delta R/\sigma_R$ 值和 $\Delta S/\sigma_S$（ΔR 和 ΔS 分别为夏季降水距平和无雨日数距平，σ_R 和 σ_S 分别为对应的标准差）。

表 2 1957—2000 年夏季最高（低）前 10 位 I_{DFC} 年的降水量和无雨日数

高 I_{DFC} 年	$\Delta R/\sigma_R$	$\Delta S/\sigma_S$	低 I_{DFC} 年	$\Delta R/\sigma_R$	$\Delta S/\sigma_S$
2000	−0.02	0.91	1977	0.05	−1.23
1986	0.08	0.88	1975	0.29	−1.03
1995	0.01	0.60	1993	0.51	−1.70
1990	−0.59	1.58	1970	0.15	−0.72
1997	−0.72	1.50	1965	0.16	−0.67
1988	−0.44	0.69	1973	−0.38	−0.90
1981	−0.93	1.84	1957	0.38	−0.75
1971	−0.28	0.37	1974	0.41	−0.68
1972	−0.59	0.42	1963	0.8	−1.63
1992	−1.11	1.21	1989	0.1	−0.31

从表 2 可以看到，夏季 I_{DFC} 值最高前 10 年中有 5 年的降水距平绝对值在 0.5σ 以内，根据前面的定义，认为这 5 年夏季总降水量属于基本正常[即排除了重旱（涝）的情形]，而对应的无

雨日数距平有4年超过0.5σ,即无雨日数偏多;另5年的降水距平虽然都小于-0.5σ(表明该5年总体偏旱),但其对应的无雨日数距平有4年在1.5个标准差以上,即无雨日数偏多异常程度大于夏季累积降水量距平偏少的程度。以上这些均说明在较少的雨日里降了更多的雨,高I_{DFC}指数前10年夏季降水的确存在时间分布上的非均匀性。I_{DFC}值越大,累积降水距平越趋于0,而降水在时间分布上的非均匀程度也越大。

在I_{DFC}值最低前10年中,有8年降水距平绝对值在0.5σ以内,表明这些年夏季总降水量基本正常,而无雨日数距平10年中有9年小于-0.5σ,即无雨日数偏少,降水日数偏多。这些说明I_{DFC}值越小,在总降水量越趋正常的同时,降水分布越趋均匀,即更为"风调雨顺"。

综上所述,旱涝并存强度指数I_{DFC}可以作为反映夏季总降水量正常而在时间分布上不均匀程度的定量指标。I_{DFC}值大,则"旱涝并存"异常强度也越大,降水分布越不均匀;I_{DFC}值越小,则降水越趋于均匀分布。

4 长江中下游夏季旱涝并存的年际变化

图2为1957—2000年长江中下游地区夏季5—8月的旱涝并存指数I_{DFC}的演变,图中粗实线为二阶多项式回归趋势线。

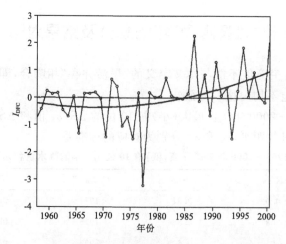

图2 1957—2000年长江中下游夏季(5—8月)I_{DFC}(标准化)的演变
(粗实线为二阶多项式回归趋势线)

由图2的回归趋势线可以清楚看到,近50年长江中下游地区"旱涝并存"异常呈现出增加的趋势。在20世纪80年代初以前,特别是60年代末至70年代末,I_{DFC}指数较小,即较为"风调雨顺",而从80年代中期开始,I_{DFC}值呈现显著增大的趋势,说明长江中下游地区夏季近20年"旱涝并存"异常的强度加大,旱涝极端事件更为频发。

从表2中也可清楚地看到这一点,在I_{DFC}最高值前10年中,有8年发生在20世纪80—90年代;而I_{DFC}最低前10年中有8年发生在1980年以前,这与图2得到的结论相一致。

5 长江中下游旱涝并存年同期大气环流的基本特征

为便于合成分析,从表1中选取I_{DFC}值最大的10年:2000、1986、1995、1990、1997、1988、1981、1971、1972和1992年来表征"旱涝并存"年,选取最小的10年:1977、1975、1993、1970、1965、1973、1957、1974、1963和1989年来表征"风调雨顺"年。

图3给出了长江中下游地区夏季"旱涝并存"高指数年与低指数年850 hPa合成风差值场,阴影区显著性检验达95%。从图3可以清楚地看到,我国大陆东部夏季5月、6月、7月和8月基本为阴影区所覆盖,都存在着显著的北风差值环流异常,这些均表明旱涝并存高指数年东亚夏季风偏弱,而低指数年东亚夏季风偏强。

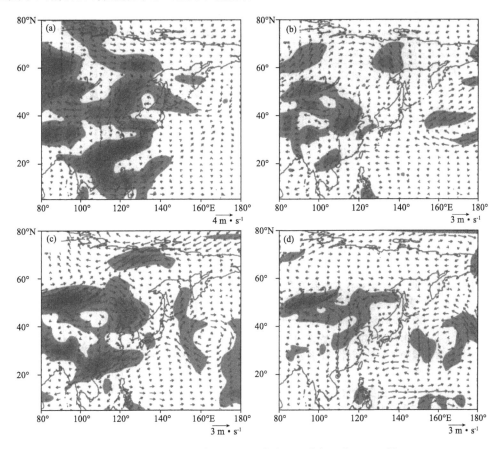

图3 高I_{DFC}年与低I_{DFC}年850 hPa合成风差值场(高I_{DFC}-低I_{DFC})
(a)5月;(b)6月;(c)7月;(d)8月。阴影:显著性检验达95%

图4为旱涝并存年(I_{DFC}高指数年)500 hPa合成高度场,阴影区为高指数年与低指数年500 hPa合成高度差值通过95%置信度显著性检验的区域。由图4可见,5月(图4a)显著的正差值区主要有两个,一个是位于从蒙古至河套的华北地区,另一个位于0°~10°N的热带地区,西太平洋副高的西部和北部边缘线不明显,这些特征表明旱涝并存异常年5月大陆冷高压势力较强,西太平洋副高位置偏南。到了6月和7月(图4b和c),我国东部大陆出现大范围显

著的位势高度正差值区,副高边缘线北抬至长江流域,表明旱涝并存年西太平洋副高显著加强西伸北抬。进入8月后(图4d),我国东部大陆的位势高度显著正差值区减弱,热带西太平洋上的显著正差值区加强,副高边缘线变得不明显,表明副高迅速东撤南退至南海及热带西太平洋上。

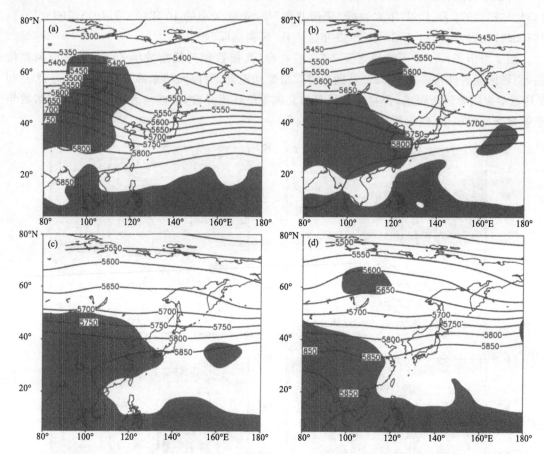

图4　旱涝并存年500 hPa合成高度场(单位:dagpm):(a)5月;(b)6月;(c)7月;(d)8月。阴影区为高I_{DFC}年与低I_{DFC}年500 hPa合成高度差值场通过95%置信度水平显著性检验的区域

从上述分析可以看出,旱涝并存异常年西太平洋副高北抬南落的空间范围非常大,活动频繁。副高的每一次北抬和南撤常常伴随着长江中下游地区的雨涝过程[7],因此,西太平洋副高季节内振荡的异常很可能是导致该地区"旱涝并存,旱涝急转"的主要因素之一。

6　长江中下游旱涝并存年前期海温场的特征

本文还计算了夏季I_{DFC}指数与其前期1~6个月全球SST的相关系数(如图5所示,阴影为置信度检验达95%的区域)。从前一年11月开始到当年4月,可以看到从南中国海到西太平洋、孟加拉湾、阿拉伯海以及印度洋北部均为显著正相关区,就是说长江中下游地区夏季旱涝并存高指数年前期,上述海区的SST也偏高,我们知道海陆热力差异是产生季风的驱动力之一[9],南海夏季风爆发前期当海温偏高时,则海陆温度差异变小,导致夏季风爆发推迟且强度偏弱;反之,夏季风偏强。因此,高 I_{DFC} 指数年(旱涝并存异常年),其东亚夏季风偏弱,这与前面 850 hPa 差值风场分析所得到的结论相吻合。可以认为,中国近海海温的偏高是导致后期长江中下游地区旱涝并存的可能原因之一。

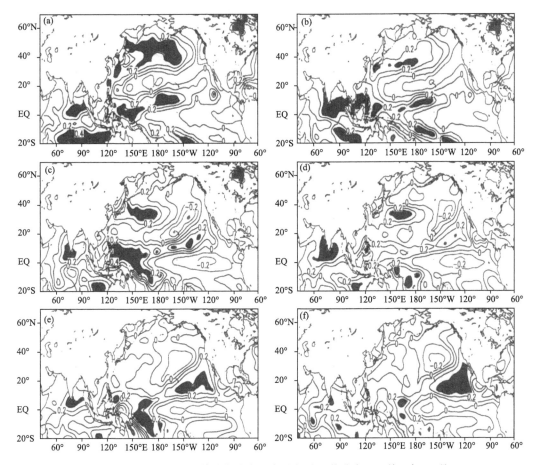

图 5 夏季 I_{DFC} 指数与其前期海表温度的相关系数分布;(a)前一年 11 月;
(b)前一年 12 月;(c)当年 1 月;(d)2 月;(e)3 月;(f)4 月。阴影为置信度检验达 95%水平的区域

另外,我们还注意到在前一年 11 月(图 5a)赤道中太平洋有显著的 SST 正相关区向东发展,并逐月向东推进(图 5b～e),至当年 4 月在北美沿岸至赤道地区出现大范围显著正相关区,表明高 I_{DFC} 指数年,赤道东太平洋海温呈现升高的趋势。事实上,在高 I_{DFC} 指数前 10 年中,有 6 年为 El Niño 年,因此,旱涝并存异常年前期对应着 El Niño 或 El Niño 的发展阶段。

7 结论

"旱涝并存"是我国夏季常见的气象灾害之一,长期以来对之研究不多。本文重点分析了长江中下游地区夏季的"旱涝并存"异常年同期大气环流和前期海温场的一些特征,根据以上分析,我们可以得出以下一些结论:

(1)近 50 年,尤其是从 20 世纪 80 年代以来,长江中下游夏季旱涝并存异常的发生频率呈现上升趋势。

(2)长江中下游地区旱涝并存与东亚夏季风的强度存在显著的相关关系,夏季旱涝并存异常年,其同期东亚夏季风偏弱。

(3)西太平洋副高季节内振荡的异常可能是导致长江中下游地区"旱涝并存"异常的主要因素之一。

(4)夏季旱涝并存异常年前期,从阿拉伯海到中国南海 SST 显著偏高,并对应着 El Niño 或 El Niño 的发展阶段。

本文主要从一个较长的时间尺度(季时间尺度)对"旱涝并存"异常的总体特征进行了分析。然而,"旱涝并存"是一个非常复杂的天气现象,如果细分,还存在短期异常(如 1 旱 1 涝 1 旱 1 涝)和中期异常(如旱旱……涝涝……),对这些情形尚需进一步分类探讨,这些工作还有待于进一步研究。

参考文献

[1] 竺可桢. 东南季风与中国之雨量[J]. 地理学报,1934,1(1):1-27.

[2] 涂长望,黄士松. 中国夏季风之进退[J]. 气象学报,1944,18:1-20.

[3] 陶诗言,徐淑英. 夏季江淮流域持久性旱涝现象的环流特征[J]. 气象学报,1962,32:1-10.

[4] 吴国雄,刘还珠. 降水对热带海表温度异常的邻域响应 I. 数值模拟[J]. 大气科学,1995,19(4):422-434.

[5] 张琼,刘平,吴国雄. 印度洋和南海海温与长江中下游旱涝[J]. 大气科学,2003,27(6):992-1006.

[6] 黄荣辉,孙凤英. 热带西太平洋暖池的热状态及其上空的对流活动对东亚夏季气候异常的影响[J]. 大气科学,1994,18(2):141-151.

[7] He Jinhai, Zhou Bing, Wen Min, et al. Vertical circulation structure, interannual variation features and variation mechanism of western Pacific subtropical high[J]. Adv Atmos Sci,2001,18(4):497-510.

[8] Ting M F, Wang H. Summer time U. S. precipitation variability and its relation to Pacific sea surface temperature[J]. J Climate,1997,10:1853-1873.

[9] 陈隆勋,朱乾根,罗会邦,等. 东亚季风[J]. 北京:气象出版社,1991:362.

江淮梅雨期降水经向非均匀分布及异常年特征分析*

竺夏英[1]　何金海[1]　吴志伟[2]

(1. 南京信息工程大学大气科学学院气象灾害重点实验室(KLME),南京 210044;
2. 中国科学院大气物理研究所大气科学与地球流体力学数值
模拟国家重点实验室(LASG),北京 100029)

摘要: 尽管江淮梅雨期降水在多数年份具有一致的空间变率,然而在有些年份却呈现出南北反相的变化特征,而此时整个梅雨区的降水量往往接近正常,这无疑增加了梅雨短期气候预测的难度。鉴于上述考虑,本文对 1951—2004 年江淮地区 15 站梅汛期(6—7 月)降水进行了 EOF 分析,发现第二特征向量主要反映了梅雨期降水的经向非均匀分布特征,据此将梅雨雨型分为南涝北旱和南旱北涝型,并利用第一套 NCEP 再分析资料和第二套扩展重建海温资料(ERSST)对梅雨降水经向分布异常年的海气背景特征进行了研究,结果发现:南涝北旱年,梅雨期低层锋区和水汽辐合中心偏南,东亚副热带夏季风偏弱,西太平洋副热带高压和 200 hPa 南亚高压位置偏南;前期 2 月北半球环状模(NAM)和南半球环状模(SAM)偏强,北半球冬、春季中国近海海温偏高。南旱北涝年,情况正好相反。此外,前期北半球冬季 ENSO 对梅雨期降水经向非均匀分布也具有一定的影响作用。

关键词: 梅雨;旱涝;空间分布;海气异常

梅雨是江淮地区夏季降水的重要组成部分,对该地区夏季旱涝异常具有重要影响。自 20 世纪 30 年代以来,前人对江淮梅雨已做了大量的研究[1-8]。纵观这些研究,对江淮地区整体旱涝异常及其相关因子的关注较多,而实际上,江淮地区除了整体偏涝或偏旱之外,也很可能出现整体正常但部分地区偏涝而其他地区偏旱,或者整体偏涝(旱)而部分地区偏旱(涝)。例如 1991 年,一般认为是涝年[9-11],但事实上,长江以南的九江、屯溪、南昌、长沙降水均为负距平,其中南昌最少,距平值达−228 mm,因此 1991 年长江以南地区应该是偏旱的,与长江以北地区偏涝正好相反。Wu 等[12] 揭示了正常夏季风年长江中下游地区"旱涝并存"的现象,即虽然长江中下游地区夏季平均降水接近正常,但在某一段时间内旱,另一段时间又涝,这是从时间上来考虑的,那么在空间上是否也存在类似的"旱涝并存",即整个江淮梅雨区降水接近正常,但长江以南和以北却同时发了旱和涝? 对于这种南北反相变化的特征,前人的研究似乎较少,但是它对江淮地区人们的生活、生产来说却同样具有重要影响,并且这种局地不一致性更增加了气象部门对梅雨短期气候预测的难度。因此本文对江淮梅雨期(6—7 月)降水量场进行了 EOF 分析,找出梅雨异常空间分布型及相应的典型年份,并着重对经向非均匀分布型及异常年海气特征进行了分析,初步提出了一些前期预报因子。

---

* 本文发表于《科学通报》,2007 年第 52 卷第 8 期,951-957。其英文版发表于《Chinese Science Bulletin》,2007 年第 52 卷第 17 期:2420-2428。

1 资料和方法

利用全国 160 站 1951—2004 年 6—7 月累积降水量,采用 Ting 和 Wang[13] 划分降水区的方法,划出江淮梅雨区空间范围,具体做法如下:先计算全国 160 站 54 年梅雨期(6—7 月)降水量标准差,得到江淮流域降水变率最大的测站为安庆,然后以安庆为基点,计算其与 160 站降水单点相关,得到空间相关图(图 1),图中阴影区即定义为江淮梅雨区,选取其中 110°E 以东分布较均匀的 15 个站作为梅雨区的特征站点。然后对这些站点 54 年的梅雨量场作 EOF 分析,得到前两个主要模态及其时间系数,其中第二特征向量的空间分布呈南北反相变化,即本文所要重点讨论的梅雨期降水经向非均匀分布特征。再利用

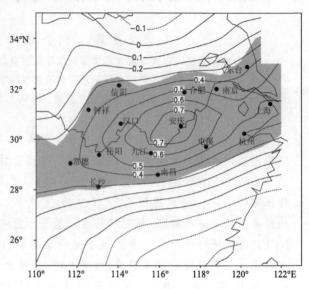

图 1 安庆为基点的 6—7 月梅汛期降水量单点相关图
阴影区相关系数超过 95% 置信度检验

NCEP 第一套月平均再分析资料集(水平分辨率 $2.5°×2.5°$)[14]和美国国家海洋和大气管理局(NOAA)提供的第二套扩展重建海温(ERSST)资料(水平分辨率 $2.0°×2.0°$)[15],分析了江淮梅雨期降水经向非均匀分布异常年的梅雨期及前期的大气环流和海温特征。

2 江淮梅雨期降水经向非均匀分布特征

对 54 年 15 站梅雨量距平场进行 EOF 分析,得到前 7 个特征向量的累积方差贡献率 V_h 达 90%(表 1),其中第一特征向量占 51%,空间分布特点是整体同位相变化,即前人主要讨论的江淮流域整体旱涝特征(图略)。第二特征向量的方差贡献率达 16%,其特点是以长江为界(图 2a),南北呈反相变化,即南涝北旱或南旱北涝,分别以东台和长沙为极值中心,说明这两个地方梅雨量年际变率最大,然后向四周扩展。结合此模态的时间系数变化曲线(图 2b),可以发现其年际变化明显,似乎有 25 年左右的振荡周期,整体变化趋势不明显。

表 1 江淮梅雨期降水距平场 EOF 分析前 7 个特征向量的方差贡献率 v_h 及累积方差贡献率 V_h

h	1	2	3	4	5	6	7
v_h(%)	51	16	8	5	4	4	3
V_h(%)	51	67	75	80	83	87	90

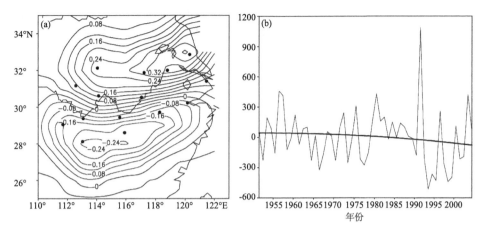

图 2　第二特征向量的空间分布(a)及对应时间系数变化曲线(b)
a 中的黑色圆点代表图 1 中的 15 个站点;b 中的粗实线为趋势线

根据前 2 个主要特征向量的时间系数的比较,选出第二模态的典型年份,具体做法是:首先分别选出前 2 个特征向量的标准化时间系数绝对值大于 0.8 的年份,对于重叠的年份则取绝对值最大的那个模态(因为当第一模态的标准化时间系数绝对值大于第二模态的标准化时间系数绝对值时,往往表现为第一模态的空间分布型),这样经过筛选之后,最后得到南涝(旱)北旱(涝)型的典型年份(见表 2)。为验证这些年份梅雨期降水经向非均匀分布特征,本文又计算了江淮地区 15 个站,长江以南 7 个站(图 2c 中负值区)和长江以北 8 个站(图 2c 中正值区)54 年标准化的梅雨量 S_{15},S_{sth7} 和 S_{nth8},并定义降水量距平绝对值小于 0.5 个标准差,即标准化降水量值在 [−0.5,0.5] 之间的为正常年份,标准化降水量值大于 0.5(小于 −0.5)的年份则为偏涝(旱)年。

表 2　江淮梅雨期南涝(旱)北旱(涝)异常年及标准化的梅雨量

南涝北旱型				南旱北涝型			
年份	S_{15}	S_{sth7}	S_{nth8}	年份	S_{15}	S_{sth7}	S_{nth8}
1964	−0.21	0.15	−0.54	1956	0.14	−0.62	0.92
1966	−0.51	−0.04	−0.88	1957	−0.27	−1.01	0.60
1973	0.18	0.57	−0.29	1960	−0.49	−0.80	−0.03
1977	0.39	0.78	−0.14	1975	−0.07	−0.68	0.61
1993	0.91	1.87	−0.37	1991	1.34	−0.53	3.02
1994	−0.39	0.26	−0.99	2003	0.22	−0.36	0.79
1995	0.42	1.19	−0.53				

* S_{15}、S_{sth7} 和 S_{nth8} 分别为江淮地区 15 站、长江以南 7 站(图 2c 中负值区)和长江以北 8 站(图 2c 中正值区)的标准化梅雨量。

由表 2 可知,梅雨期降水经向非均匀分布异常年,江淮地区整体而言,除 1966,1991 和 1993 年,大部分年份降水量距平绝对值均小于 0.5 个标准差,即整个江淮地区降水量属于正常范围内。但是长江以南和以北地区的标准化降水量值符号基本相反,最明显的是 1957,1991,1993 和 1995 年,长江以南或以北地区的标准化降水量绝对值大于 1,其他年份至少有一侧(长江以南或以北)地区标准化降水量绝对值大于 0.5,表明该地区梅雨降水异常偏多或偏少。由此可以看出,表 2 选出的江淮梅雨期降水经向非均匀分布异常年是比较合理的。

南涝北旱年,长江以南地区标准化降水量几乎都为正值(除1966年),而长江以北地区都是负值,表明长江以南地区降水偏多,以北地区降水偏少。1966年,虽然3个标准化值均小于0,但长江以南的标准化值趋于0,而事实上,常德、长沙、岳阳等地降水量明显偏多,因此仍可归为南涝北旱年。1993年,江淮地区15站总降水量明显偏多,但长江以北地区却正常偏少,而长江以南地区则异常偏多,其降水距平值将近2个标准差,故1993年也可归为南涝北旱年。为进一步验证上述结果,本文又计算了南涝北旱年降水量距平场合成图(图略),不难发现:长江以南为正距平区,最大值中心区岳阳、南昌、常德等地的降水距平超过120 mm,而长江以北为负距平区,距平最少达−136 mm左右。

南旱北涝年,长江以南地区标准化降水量大部分为负值,长江以北地区均为正值(除1960年),这刚好与南涝北旱的情况相反。对于整个江淮地区来说,6年中有5年的降水距平绝对值小于0.5个标准差,属于降水正常年份;而1991年降水距平值远远大于0.5个标准差,是偏涝的,以往的研究[9-10]均关注这一点,却很少注意到长江以南的降水量其实是偏少的,特别是南昌比正常偏少228 mm,因此1991年应该归为南旱北涝年更客观。虽然1960年3个标准化值均小于0,但长江以北的标准化值趋于0,而事实上,合肥、钟祥等地降水量明显偏多,因此仍可归为南旱北涝年。同样,由南旱北涝年降水量距平合成图(图略)易见:长江以南旱情最严重的长沙降水距平小于−160 mm,而长江以北东台的降水距平则超过230 mm。

综上所述,江淮梅雨期降水存在经向非均匀分布特征,南涝北旱的典型年份有1964,1966,1973,1977,1993,1994和1995年,南旱北涝的典型年份有1956,1957,1960,1975,1991和2003年。可以看出,80年代几乎没有出现典型的南北非均匀分布异常年,因此,这种经向非均匀分布特征可能具有年代际变化特征。对于以上这些典型年份,南旱北涝的强度似乎要比南涝北旱的强度要大得多,且大部分年份江淮地区整体梅雨量趋于正常。

3 江淮梅雨期降水经向非均匀分布异常年同期大气背景场特征

梅雨锋是冷暖空气的交汇区[16],因此梅雨雨带位置可能与冷暖空气的强弱有关。1000 hPa温度距平场合成图(图略)显示,南旱北涝年,长江以南为正距平区,长江以北为负距平区,表明冷、暖空气的势力都较强,而0距平线在长江以北,则说明冷暖空气的交汇区偏北,即锋区偏北。南涝北旱年,冷暖空气的势力仍较强,但0距平线在长江以南,表明锋区偏南。

江淮梅雨是亚洲夏季风推进到长江流域的产物[17],因此梅雨异常与亚洲夏季风异常具有密切联系。由850 hPa风场合成差值图(南涝北旱年减南旱北涝年)可见(图略),东亚副热带地区被大范围的显著东北差值气流所覆盖,表明梅雨期降水经向非均匀分布异常年东亚副热带夏季风具有很大差异,南涝北旱年西南风较弱,而南旱北涝年西南风较强,这可能有利于水汽向更北的方向输送。此外,在副热带西太平洋上空为气旋式差值环流,表明南涝北旱年西太副高很可能较南旱北涝年的强度偏弱。

850 hPa水汽输送的合成差值场(图3a)显示东亚上空为大范围的东北差值水汽输送区,说明南涝北旱年向北的水汽输送较南旱北涝年弱得多,并且这种差异主要是由偏西风的水汽输送强弱造成的,而副高南侧偏东风的水汽输送差异较小,这与卓东奇等[18]的研究结果相一致。水汽通量散度图表明虽然南北非均匀分布异常年的水汽辐合中心的强度相当,但位置差异显著,南涝北旱时水汽通量辐合中心在30°N以南(图略),而南旱北涝时水汽辐合中心则偏北,在 30°N 附近(图 3b)。

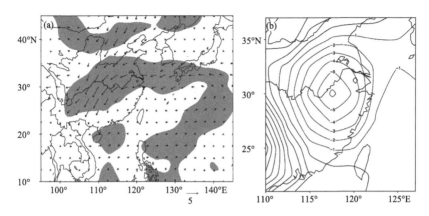

图 3 梅雨期 850 hPa 水汽通量合成差值(a)和南旱北涝年 850 hPa 水汽通量散度合成(b)
a 为南涝北旱年减南旱北涝年,单位为 $g·cm^{-1}·s^{-1}$,阴影区超过 95%
置信度检验;b 的单位为 $10^{-8} g·cm^{-2}·s^{-1}$

研究表明江淮流域夏季发生旱涝异常时 500 hPa 位势高度场有相应的异常[19]。500 hPa 位势高度距平合成图显示,南涝北旱年(图 4a),欧亚大陆东部上空从低纬到高纬为"＋－＋－"的距平波列,表明西太平洋副高偏南,高纬度鄂霍次克海阻塞形势偏强;南旱北涝年(图 4b),欧亚大陆东部上空从低纬到高纬为"－＋－＋"的距平波列,表明西太副高偏北,高纬度鄂霍次克海阻塞高压仍旧很明显。

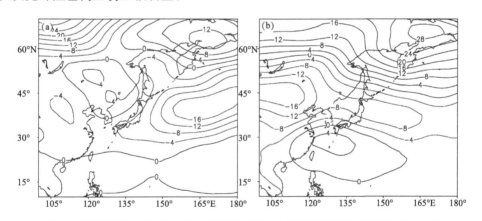

图 4 梅雨同期降水经向分布异常年 500 hPa 位势高度距平合成图(单位:gpm)
(a)南涝北旱年;(b)南旱北涝年

梅雨与南亚高压有密切联系[16]。图 5 是南涝北旱年 200 hPa 位势高度距平合成图。北半球极区为负距平区所控制,即位势高度场偏低,表明极涡偏强。欧亚大陆高纬度地区位势高度距平大部分为负值,中、低纬度地区,90°E 以西为正距平区,即位势高度偏高;90°E 以东,20°N 以北位势高度偏低,20°N 以南位势高度偏高,表明南亚高压偏西偏南。而南旱北涝年情况基本相反(图略),即极涡偏弱,南亚高压偏东偏北。

从梅雨期 6—7 月低层到高层的温度场、大气环流场及水汽输送的分析可以看出,梅雨降水南北非均匀分布异常年梅雨同期这些大气特征存在明显的差异。南涝(旱)北旱(涝)年,低层锋

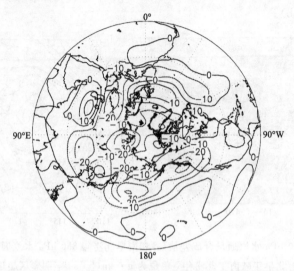

图 5　梅雨同期南涝北旱年 200 hPa 位势高度距平合成图(单位:gpm)

区和水汽辐合中心偏南(北),东亚副热带夏季风偏弱(强),西太副高和高层南亚高压偏南(北)。

4　江淮梅雨期降水经向非均匀分布异常年前期海气背景特征

由图 2 可知,第二特征向量的时间系数大于 0 表明当年梅雨带偏北,时间系数小于 0 表明梅雨带偏南,时间系数绝对值越大说明南北旱涝越严重,因此本文将标准化的第二特征向量时间系数序列作为表征梅雨量南北非均匀分布异常的参数,并称之为"南北指数"。

越来越多的研究表明,江淮梅雨与前期冬、春季的大气环流异常有密切关系[20-22],特别是 2 月[19]。本文计算了南北指数与前期的大气环流指数的相关,发现梅雨量经向非均匀分布与前期冬季 12 月—次年 2 月的北半球环状模(NAM)有较好的负相关关系,2 月相关系数最高,达 −0.29。南北指数与前期 2、4 月南半球环状模(SAM)的相关系数分别达 −0.3、−0.23,均超过 95% 的信度检验。图 6 是南北指数与前期 2 月海平面气压场的相关图,副热带及低纬度地区大部分为负相关区所控制,特别在太平洋和非洲上空相关系数达 −0.3(图 6a),在高纬度地

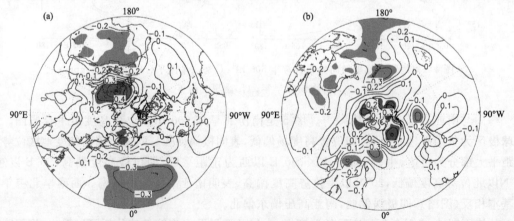

图 6　南北指数与前期 2 月海平面气压场的相关图阴影区超过 95% 置信度检验:(a)北半球;(b)南半球

区为正相关区所控制,最高相关系数超过0.4,表明在高、低纬度地区的海平面气压场在梅雨量南北非均匀分布异常年呈反位相变化特征,这验证了南北指数与南、北半球环状模的相关关系。

前人的研究指出:不仅热带地区海温异常对江淮地区夏季旱涝异常具有重要影响[23],而且中纬度中国近海(包括黑潮区)海温异常对江淮地区旱涝也有密切联系[24-27]。从南北指数与海温相关图可见,前期冬季(图7a),赤道中东太平洋为显著负相关区,说明梅雨降水南北非均匀分布的年际变化可能与ENSO有关系,即在冬季ENSO升温阶段,后期梅雨带可能偏南,长江以北少雨;在冬季ENSO降温阶段,后期梅雨带可能偏北,长江以南少雨。低纬度西太平洋为负相关区所控制,特别是南海和黑潮区为显著负相关区,表明这些海区的海温异常对后期江淮梅雨降水的经向非均匀分布可能具有一定的影响,即海温偏高时,梅雨带很可能偏南,海温偏低时,梅雨带很可能偏北。前期春季(图7b),我国邻近海域(即渤海、黄海、东海和南海北部)为显著负相关区,表明该海区有正海温异常时,后期的梅雨带很可能偏南,负海温异常时,后期的梅雨带很可能偏北。

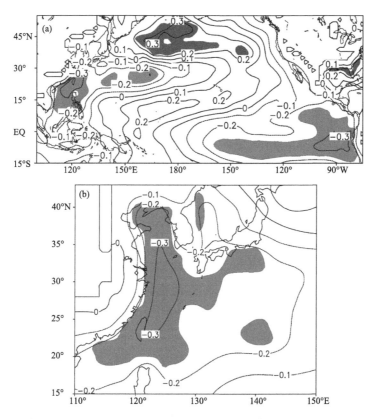

图7 南北指数与前期冬、春季海温相关图 阴影区超过95%置信度检验:(a)冬季;(b)春季

综上所述,梅雨期降水经向非均匀分布异常年前期的大气环流和海温存在明显差异,南涝(旱)北旱(涝)年,前期2月NAM和SAM偏强(弱),北半球冬、春季的中国近海(包括黑潮区)海温偏高(低),梅雨期降水经向非均匀分布异常还可能受前期北半球冬季ENSO的影响。

5 结论和讨论

江淮梅雨期旱涝异常在多数年份呈现一致的变化趋势,但在某些年份却出现南北反相的变化特征,而前人的研究似乎对此关注较少。因此,本文首先对 1951—2004 年江淮梅雨期(6—7月)降水量距平场进行了 EOF 分析,发现第二特征向量的空间分布呈南北反相的变化特征,据此将梅雨雨型分为南涝北旱和南旱北涝型并选出相应的典型年份。进一步分析发现,大多数南涝(旱)北旱(涝)年整个江淮地区的梅雨期降水量接近正常年份,这正是从空间上对 Wu 等[12]的长江中下游地区夏季"旱涝并存"的补充。然后本文运用合成分析及简单相关分析方法,分析了梅雨期降水经向非均匀分布异常年梅雨同期及前期的海气背景异常特征,结果显示:

梅雨期,南涝北旱年,低层锋区和水汽辐合中心偏南,东亚副热带夏季风偏弱,西太副高和 200 hPa 南亚高压偏南;南旱北涝年的情况相反。

前期 2 月 NAM 和 SAM 偏强(弱),北半球冬、春季的中国沿海海温偏高(低),江淮地区梅雨带偏南(北)。梅雨期降水经向非均匀分布异常还可能受前期北半球冬季 ENSO 的影响。

上述这些前期异常因子是如何影响后期的江淮梅雨期降水经向非均匀分布异常的呢? Yu 和 Ge[28]对西太平洋中纬度海温变化影响副热带高压进行了数值模拟,发现 500 hPa 环流的变化对中纬度黑潮区海洋加热场的响应存在 2～3 个月的滞后,前期春季 3—5 月中纬度黑潮区海温持续偏高,则后期 5—8 月 500 hPa 副高位置偏南,并且海温偏高越多,夏季副高南压越明显;反之亦然。Liu 等[29]则指出海洋辐射冷却在低层会激发出反气旋环流,从而对副高产生作用。就本文分析结果来看,北半球冬、春季中国近海(包括黑潮区)海温异常升高(降低),可能使得副热带地区对大气的非绝热加热异常增加(减少),在对流层中低层激发出气旋(反气旋)性环流,不利于(有利于)初夏副热带高压的北进,即 500 hPa 副热带高压偏南(北),从而使得梅雨带偏南(北)。近海海温异常究竟如何对梅雨产生影响,尤其是对江淮梅雨降水以长江为界呈南北反相的现象? 这是一个新的话题,也是一个非常有意义的问题,还有待于进一步研究。

致谢: 感谢美国国家环境预测中心(NCEP),美国国家海洋和大气管理局(NOAA)和中国气象局国家气象信息中心提供相关气象资料,感谢两位匿名审稿人的细心审阅和他们所提供的宝贵意见。本工作受国家自然科学基金项目(批准号:40605022)、国家基础研究发展规划项目(批准号:2006CB403607)和国家科技部攀登 A 项目"南海季风试验(SCSMEX)"联合资助。

参考文献

[1] 竺可桢.东南季风与中国之雨量[J].地理学报,1934,1(1):1-27.
[2] 陶诗言,赵煜佳,陈晓敏.中国的梅雨[C]//中国科学院大气物理研究所.东亚季风和中国暴雨——庆祝陶诗言院士八十华诞文集.北京:气象出版社,1998:3-46.
[3] 符淙斌.我国长江流域梅雨变动与南极冰雪状况的可能联系[J].科学通报,1981,8:484-486.
[4] 谢安,毛江玉,宋焱云,等.长江中下游地区水汽输送的气候特征[J].应用气象学报,2002,13(1):67-77.
[5] 魏凤英,谢宇.近百年长江中下游梅雨的年际及年代际振荡[J].应用气象学报,2005,16(4):492-499.
[6] 陈艺敏,钱永甫.长江中下游梅雨期降水与环流关系分析及模拟[J].热带气象学报,2006,22(1):26-33.
[7] 吴志伟,江志红,何金海.近 50 年华南前汛期降水、江淮梅雨和华北雨季旱涝特征对比分析[J].大气科学,2006,30(3):391-401.

[8] Zhou Y S, Gao S T, Shen S S P. A diagnostic study of formation and structure of the Meiyu front system over East Asia[J]. J Meteor Soc Japan, 2004, 82(6):1565-1576.

[9] 丁一汇,陆尔. 据1991年特大洪涝过程的物理分析试论江淮梅雨预测[J]. 气候与环境研究, 1997, 2(1):32-38.

[10] Wang W C, Gong W, Wei H L. A regional model simulation of the 1991 severe precipitation event over the Yangtze-Huai River Valley. Part I: Precipitation and circulation statistics[J]. J Climate, 2000, 13:74-92.

[11] 毛江玉,吴国雄. 1991年江淮梅雨与副热带高压的低频振荡[J]. 气象学报, 2005, 63(5):762-770.

[12] Wu Z W, Li J P, He J H, et al. Occurrence of droughts and floods during the normal summer monsoons in the mid- and lower reaches of the Yangtze River[J]. Geophys Res Lett, 2006, 33:L05813, doi:10.1029/2005GL024487.

[13] Ting M F, Wang H. Summer time U. S. precipitation variability and its relation to Pacific sea surface temperature[J]. J Climate, 1997, 10:1853-1873.

[14] Kalnay E, Kanamitus M, Kistler R, et al. The NCEP/NCAR 40-year reanalysis project[J]. Bull Amer Meteor Soc, 1996, 77:437-471.

[15] Smith T M, Reynolds R W. Improved extended reconstruction of SST (1854—1997)[J]. J Climate, 2004, 17:2466-2477.

[16] 朱乾根,林锦瑞,寿绍文,等. 天气学原理与方法(第三版)[M]. 北京:气象出版社, 2000.

[17] Ding YH, Chan J C L. The East Asian summer monsoon: an overview[J]. Meteor Atmos Phys, 2005, 89:117-142.

[18] 卓东奇,郑益群,李炜,等. 江淮流域夏季典型旱涝年大气中的水汽输送和收支[J]. 气象科学, 2006, 26(3):244-251.

[19] 吴仁广,陈烈庭. 长江中下游地区梅雨期降水与全球500 hPa环流的关系[J]. 大气科学, 1994, 18(6):691-700.

[20] 龚道溢,朱锦红,王绍武. 长江流域夏季降水与前期北极涛动的显著相关[J]. 科学通报, 2002, 47(7):546-549.

[21] 魏凤英,宋巧云,韩雪. 近百年北半球海平面气压分布结构及其对长江中下游梅雨异常的影响[J]. 自然科学进展, 2006, 16(2):215-222.

[22] 吴志伟,何金海,韩桂荣,等. 长江中下游梅雨与春季南半球年际模态(SAM)的关系分析[J]. 热带气象学报, 2006, 22(1):79-85.

[23] 张琼,刘平,吴国雄. 印度洋和南海海温与长江中下游旱涝[J]. 大气科学, 2003, 27(6):992-1006.

[24] 林建,何金海. 海温分布型对长江中下游旱涝的影响[J]. 应用气象学报, 2000, 11(3):339-347.

[25] 巢纪平. 大尺度海气相互作用和长期天气预报[J]. 大气科学, 1977, 3:223-233.

[26] 赵永平,McBean G A. 黑潮海域海洋异常加热对后期北半球大气环流影响的分析[J]. 海洋与湖沼, 1996, 27(3):246-250.

[27] 赵永平,陈永利,翁学传. 中纬度海气相互作用研究进展[J]. 地球科学进展, 1997, 12(1):32-36.

[28] Yu Z H, Ge X Z. Numerical simulated of seasonal movement of subtropical high ridge line Ⅱ[J]. Acta Oceanological Sinica, 1986, 5(2):183-194.

[29] Liu Y M, Wu G X, Ren R C. Relationship between the subtropical anticyclone and diabatic heating[J]. J Climate, 2004, 17:682-698.

江淮梅雨气候变化研究进展[*]

梁萍[1-3] 何金海[2]

(1. 上海市气候中心，上海 200030；2. 气象灾害省部共建教育部重点实验室，
江苏 南京 210044；3. 中国气象局 中国气象科学研究院，北京 100081)

摘要：自20世纪30年代，中国气象学家就开始了对中国梅雨的研究。随着梅雨历史数据长序列的建立，不少学者对梅雨的长期变化，特别是年代际时间尺度以上的长期变化特征进行了研究。针对江淮梅雨的气候变化研究，本文主要从梅雨特征量的变化周期、梅雨特征量的长期变化特征、梅雨特征量变化的空间非均匀性特征及梅雨长期变化的影响因子等四个方面进行了回顾和总结，并对梅雨的非典型性气候变化特征进行了解释，最后提出了值得进一步探讨的问题。

关键词：梅雨；气候变化；长江中下游

1 引 言

　　自19世纪中期以来，无论是全球尺度还是区域尺度都经历了以地面气温升高为主要特征的气候变化[1-2]。IPCC第4次评估报告指出，近年来的这种升温可能与人类活动有关[3]。Trenbert h[4]指出，地面温度的升高，一方面会使地表蒸发加剧，大气中水分可能增加，且温度升高有利于大气保持水分能力的增强，这意味着形成降水的可能性增大；另一方面，陆地表面蒸发增强，易发生局地干旱，降水分布不均匀的可能性增大。因此，以增暖为主要特征的气候变化可加速水分循环，改变降水的时空分布及强度，造成极端降水事件发生的可能性增加[5-6]。区域或局地天气气候必然会对全球增暖的气候变化有所响应。

　　作为发生在我国东部的重要天气气候现象，梅雨季降水是江淮地区汛期降水的重要组成部分。梅雨的强弱、梅雨期的长短及梅雨量的多寡等特征既反映了亚洲上空大气环流季节变化与环流调整的各种演变过程，又直接与江淮地区旱涝异常的形成与持续有关。在全球气候变暖的背景下，诸多气象工作者及公众都感觉到，与以往的典型梅雨相比，近几年梅雨的非典型程度有所增加。由此来看，以增暖为主要特征的全球气候变化反映在区域空间尺度和季节内时间尺度上，可能对江淮梅雨的时空分布变化产生一定影响。那么，自有历史记录以来，江淮梅雨是如何变化的？与气候变化的关系如何？这些问题既是气候及气候变化研究的重要课题，又是社会各界关心的热点问题。

　　近年来，关于江淮梅雨的长期变化已有不少研究成果。对江淮梅雨气候变化相关研究进行回顾和总结，有助于更深入地认识江淮梅雨及其相关东亚副热带夏季风降水的气候变化特征，同时还可为提高气象服务及应对气候变化能力提供科学依据，具有重要的学术价值和社

---

[*] 本文发表于《高原气象》，2008年第27卷增刊，8-15.

会意义。为此,本文将从梅雨特征量的变化周期、长期演变特征、梅雨变化的非均匀性以及梅雨长期变化的影响因子四个方面,对江淮梅雨的气候变化研究(涉及梅雨与全球气候变化的可能关系)进行回顾和总结,最后还提出了一些值得进一步探讨的问题。

2 研究概述

除已有关于中国全年和季节降水量的长期变化研究[7-10]外,目前关于江淮梅雨气候变化特征的研究主要包括四个方面:(1)梅雨特征量的周期变化;(2)梅雨特征量的长期变化特征;(3)梅雨特征量变化的非均匀性特征;(4)梅雨长期变化的影响因素。已有研究所采用的梅雨降水资料主要有两种来源:(1)徐群等[11]整理的长江中下游(选取上海、南京、芜湖、九江和汉口5个站点确定)1885—2000年梅雨期的梅雨参数资料;(2)在考虑梅雨的空间分布特征等问题时,通常采用多个站点6—7月的逐日降水资料。

2.1 梅雨特征量的周期变化

诸多学者利用徐群等[11]整理的长江中下游梅雨参数资料[11],对百余年(1885—2000年)梅雨参数的周期变化进行了分析。

杨义文等[12]采用非整数波计算方法,首先对梅雨各参数的周期长度进行了普查,认为普遍存在2~3年、6年、8~9年、22~23年、35~38年、72年和116年等周期变化。文献[13-15]分别采用小波方法、功率谱及最大熵谱方法对同样的梅雨资料进行了周期分析,所得到的梅雨变化周期在高频(10年内)波段与杨义文等[12]较为一致,但在低频波段的周期长度上则不一致。我国拥有丰富的历史文献资料,其中蕴涵着大量有关梅雨变化的信息。张德二等[16]根据清代晴雨录记载重建了1723—1800年长江下游地区的梅雨变化序列,在分析18世纪历史梅雨各项气候统计特征的基础上,指出梅雨变化具有9年、4~5年、2~3年的准周期性,表明现代梅雨变化的基本特征在18世纪也同样存在。最近,毛文书等[17]对江淮区域平均梅雨特征量的周期分析亦指出,江淮梅雨的5个梅雨特征量具有准3年显著周期变化特征,其中出梅日期、梅雨期长度、梅雨量还具有相同的准10年显著周期变化特征。总的来说,已有研究表明[12-17],梅雨各参数在年际变化上存在一定范围内的一致周期(如2~3年周期);但梅雨各参数在年代际变化上的具体变化周期是有差异的,不同参数的振荡周期显著不同;而且梅雨各参数的某一变化周期在百余年中的不同时段内可能具有不同的显著性。此外,梅雨的不同变化周期分别对应于不同的环流系统影响(如梅雨的3年、6年和8年的周期变化分别与低纬100 hPa高度场、热带系统以及全球陆地温度的变化有关);即使在同一变化周期中,梅雨的不同特征量可能受到不同气候因素的影响[14]。由此反映出梅雨变化周期在时间演变上的非均一性,以及变化周期影响因素的差异性,表明梅雨变化具有高度的复杂性。

基于区域性或区域平均梅雨参数,以上研究主要讨论了长江中下游或江淮梅雨的整体变化周期。那么,江淮区域内梅雨变化周期的空间分布是否一致?就此问题,王钟睿等[18]、闵屾等[19]、宗海峰等[20]研究了江淮梅雨的多尺度特征。已有研究表明,尽管在江淮梅雨区的空间分型上存在或多或少的差异,但他们的结果均表明,江淮区域内不同地区的梅雨可能具有一致的变化周期(主要是在较小的年际变化尺度上,如2~3年),但变化周期在不同时段的显著性存在不一致;而不同分区在较大的时间尺度上具有不同的变化周期;不同分区的梅雨参数存在不同的年际、年代际变化特征,对应的年际、年代际变化异常年份也各不相同。此外,江淮地区不同的降水分布型对应的北半球中高纬环流、东亚夏季风环流各子系统的强度、位置、水汽输送等也存在明显差异,其中尤以西太平洋副热带高压、梅雨槽和鄂霍次克海高压的不同配置对各雨型降水异常的形成具有决定性的作用[20]。

另外,胡波等[21]单独分析了近40年浙江省梅雨年际、年代际的变化,结果表明,22~23年和11~12年分别为梅雨总量的较稳定变化周期。前一周期与杨义文等[12]的结果一致,但后一周期则不一致。梁萍等[22]也单独对上海徐家汇站1875—2005年的梅雨参数进行了Morlet小波分析。与上述研究不同的是,上海徐家汇站的出、入梅及梅雨量都没有超过95%信度检验的变化周期,而梅雨期长度则存在64年左右的显著周期。由此来看,单个省份或单个站点的梅雨变化周期与上述长江中下游梅雨参数的变化周期存在不一致。

对比已有研究发现,对江淮区域梅雨整体而言,资料或周期分析方法的不同是导致区域梅雨变化周期各不相同的主要原因;对江淮梅雨的空间分区特征而言,由于不同研究在空间分型上所选取的站点空间分布及样本长度不尽相同,因而在江淮梅雨的分区及其变化周期特征上存在差异。我们认为,各地梅雨存在一定的一致性,也存在非均一性,而非均一性是造成江淮区域内不同局地梅雨变化周期不同的主要原因,同时也可造成不同研究在梅雨的空间分型上的不一致。

对于梅雨变化周期,各周期在梅雨变化中的相对重要性也是梅雨周期变化研究的另一内容。文献[23-24]的研究表明,年际变化、年代际变化、百余年的趋势变化在近百年的江淮梅雨变化中所占的比例依次减小,这可能是上述梅雨变化周期研究在高频波段较为一致的原因之一。

2.2 梅雨特征量的突变特征

除百余年梅雨特征量的变化周期分析外,诸多学者对长序列梅雨特征量是否发生重大转折或突变进行了讨论。章淹[25]最早对1885—1995年的梅雨变化进行了研究,并指出20世纪50年代以后江淮梅雨经历了集中降水期缩短、降水量减少等重大转变。尽管魏凤英等[14]也认为,1885—2000年梅雨强度就大尺度而言可分成强、弱两种状态,但指出梅雨的强、弱状态以1941年为界,且梅雨强度年代际振荡在20世纪50年代以后有所加强。与上述均不同的是,徐群[26]认为1885—2000年长江中下游梅雨期在70年代末发生了一次强年代际突变,即从1958—1978年的弱梅雨时段突变为1979—1999年的强梅雨时段,且两时段的出梅期由偏早转为延迟;但从2000年开始,长江中下游梅雨又转为进入新的偏少期。

魏凤英等[14]及陈艺敏等[13]对1885—2000年梅雨参数的突变情况进行了检测,他们的结果均表明,100多年来,梅雨长度和出梅日期发生了气候突变,但梅雨量并没有发生气候突变。梅雨长度和出梅日期在20世纪60年代突变之后的特征表现为出梅日期推迟、梅雨期长度变长,这与章淹[25]关于50年代以后的梅雨转变结果近于相反。除20世纪60年代的突变之外,魏凤英等[14]的研究认为,梅雨参数还存在多个突变点,并由这些突变点划分出了梅雨的6个气候阶段。

以上研究主要是采用徐群等[11]整理的百余年长江中下游梅雨参数资料进行的突变分析。徐卫国等[23]则利用我国东部180个测站的逐日降水资料,研究了1961—2000年梅雨期雨区的年代际变化,并认为雨区边界及雨区面积指数在20世纪70年代末至80年代出现了突变。

而 Wu 等[27]使用中国、日本和韩国台站雨量资料的研究亦认为,中国东南部—长江中下游—中国东北部雨型在 1977—1978 年发生转折。这些都与徐群[26]的观点较为一致。

由此来看,关于 1885—2000 年梅雨的突变特征,目前多数研究认为,20 世纪中到末期的出梅日期延迟、梅雨期变长是 1885—2000 年期间梅雨的突变特征。而梅雨量是否存在突变及其转变情况仍有争议。徐群[26]的最新研究认为,从 2000 年开始,长江中下游梅雨转为进入新的偏少期,其结果有待于事实及研究的进一步证实。

此外,针对百余年之前的梅雨变化研究目前还不多见。葛全胜等[28]采用清代雨雪分寸记载与器测降水资料重建的 1736 年以来长江中下游地区的梅雨情况,分析了 18 世纪以来东亚夏季风的强弱及与之对应的雨带位置阶段性的变化。结果表明,1736—1770 年、1821—1870 年及 1921—1970 年等时段东亚夏季风偏强,中国东部夏季风雨带多位于华北和华南,梅雨期偏短;1771—1820 年、1871—1920 年及 1971—2000 年等时段东亚夏季风偏弱,雨带多位于长江中下游地区,梅雨期偏长。

2.3 梅雨变化的空间非均匀性特征

上述梅雨的长期变化特征主要是基于区域梅雨的一致性研究。梅雨区域内各地均受到季风系统的影响,因而具有一定的共性。然而,除受季风单方面影响因素外,梅雨区内各地还可能受到本地环境因素及其周围天气系统等的影响。因此,区域梅雨表现在局地上仍可能各有差异。近年来,不少学者利用梅雨区多个站点资料针对梅汛期降水的非均匀性特征进行了研究。

关于梅汛期降水的非均匀性特征,文献[29-31]对 20 世纪 50 年代以来江淮梅汛期(6—7月)降水的 EOF 分析表明,江淮流域梅汛期降水除整体旱涝特征(同位相变化)外,还具有很大的空间差异。梅汛期降水存在约以 31°N 为界的经向非均匀性变化,即梅雨雨型可分为南涝北旱和南旱北涝型[29,31]。文献[32]分析了安徽省内江淮梅雨量的空间分布,结果表明,安徽省内江淮梅雨量的干湿变化具有较好的一致性,这与文献[29]中江淮雨型的划分也是一致的。此外,毛文书等[17]指出,江淮梅雨区除空间一致型和南北反相型外,还存在以 115°E 为界的东西反相变化型。除年际变化的非均匀性外,闵屾等[19]对江淮梅雨的分区研究还表明,江淮区域 3 个子区的梅雨在长期趋势变化上存在差异:中心区和东南区的梅雨长度和梅雨量存在上升趋势,出梅期推迟;西北区梅雨参数的变化与中心区、东南区相反。

集中降水是梅雨期间的重要特征,集中性降水易引发暴雨等强降水过程。上述研究认为梅汛期降水在空间上存在非均匀性特征,那么强降水过程在空间上是否存在非均匀性特征?根据徐群等[11]定义的梅雨期,文献[31]分析了梅汛期内暴雨降水过程的降水量、集中度和集中期的时空非均匀特征,其结果表明,长江中下游地区暴雨降水过程具有大致相同的集中时段。徐群等[11]确定梅雨期参数时重点考虑了区域内梅雨的一致性,故上述关于梅雨期暴雨降水过程一致性的结论也是基于梅雨的一致性而得到的。与此不同的是,文献[32]的研究则表明,1954—2003 年江淮南、北两区梅雨期(6—7月)暴雨特征量的候际、月际、年际和年代际变化存在显著差异:南区暴雨量呈单峰型,北区暴雨量呈双峰型;梅雨期内北区暴雨集中程度大于南区;南区梅雨期暴雨对梅雨期降水的贡献大于北区;南、北两区梅雨期暴雨量的显著主周期和次振荡周期均不同;南区梅雨期暴雨量存在显著的突变性质,江淮北区梅雨期暴雨量则不存在突变现象。尽管文献[30]将江淮地区梅雨期降水的空间特征分为东北、西北、西南和东南

4个区域,但其南部2区和北部2区的极端降水(暴雨)的差异特征与文献[31]类似。

2.4 梅雨长期变化的影响因子

关于梅雨异常的影响因子研究已有很多。下面主要总结梅雨年代际尺度以上的气候异常成因的研究成果。

诸多研究[18,34-37]认为梅雨年代际变化与海洋年代际背景状态关系密切,中东太平洋海温在梅雨的气候变化中起重要作用。中东太平洋海温异常一般通过海洋过程影响夏季北太平洋海温异常,北太平洋海温异常又强迫西太平洋副热带高压发生异常,从而造成长江中下游地区的夏季降水异常。但中东太平洋海温与降水的关系并不是稳定的[18,26,35]。Niño3区的海温与江淮流域降水的联系在一些时段是同位相的,而在另一些时段则是反位相的[18]。这在一定程度上增加了梅雨的短期气候预测难度。关于海温与长江中下游夏季降水(梅雨)关系的年代际变化,目前仍存在分歧。李峰等[35]的研究表明,中东太平洋海温与长江中下游地区夏季降水的关系在20世纪70年代中期前后出现转折:1976年前关系不明显,1976年之后两者关系密切。而关于梅雨与前期海温的关系,徐群[26]则认为,在1958—1978年(弱梅雨时段),梅雨量及出梅期与前期北太平洋海温的关系密切,而在1979—1999年(强梅雨时段)关系相对不明显。除海温之外,大气环流是梅雨长期变化的另一重要影响因素[15,18,20-21]。其中,南亚高压、东亚阻塞、印度季风、西太平洋副热带高压强度和位置等是江淮梅雨量异常的重要同期大气环流成因。关于梅雨年代际变化的同期及前期环流影响因子,徐群[26]的研究表明,在1958—1978年的弱梅雨时段和1979—1999年的强梅雨时段内,影响梅雨同期高空大气环流形势不同,主要表现为后时段东亚阻塞形势强盛,西太平洋副热带高压位置偏南,印度季风低压减弱。Wu等[28]的研究也认为,大尺度环流的年代际变化与中国东部雨型在20世纪70年代末出现的年代际变化一致。此外,前后两时段梅雨量及出梅期年际变化的前期因子也有很大差异:后时段内前期(1—3月)大气环流因子更为重要,而前期北太平洋海温的影响则锐减[26]。

除上述影响因子外,徐群[26]还指出,1958—1978年的弱梅雨时段转变为1979—1999年的强梅雨时段与地—气环境位相不同有关。在目前人类活动不断增强时期,地—气环境位相可能又发生了年代际变化,进而推测21世纪初之后的梅雨可能又将出现重大转折。

作为影响梅雨的关键环流系统,东亚副热带夏季风的强度及进程的年代际变化对江淮梅雨长期变化至关重要。根据研究[38],20世纪50—90年代期间,后期(80—90年代)东亚副热带夏季风推进为慢—快—慢过程,推进北界偏南,停滞在长江流域的时间增长,而停滞在华北的时间减少,这些与江淮梅雨在80—90年代明显增强的年代际变化一致。我们进一步对东亚副热带夏季风年代际变化的机制研究[38]表明,东亚海陆热力差异自50年代以来的年代际减小可能是东亚副热带夏季风的年代际减弱变化的一个重要原因。80年代、90年代6—7月的中高纬影响(北方冷空气)较50年代和60年代更为活跃,也不利于夏季风向北方推进。50—60年代(80—90年代),高原上空位势高度较低(高),高原东南侧较强的偏南风有(不)利于东亚副热带夏季风北推。这些因素的年代际变化是东亚副热带夏季风及江淮梅雨年代际变化的重要原因。

随着社会各界对气候变化的日趋关注,针对全球气候增暖背景下的降水变化研究已成为气候变化研究的重要方面。关于气候变暖背景下副热带地区降水的气候变化,Alexander等[39]及Frich等[40]的研究发现,降水强度大部分都存在增加的趋势。根据IPCC[41]和其他一些研究[42]。过去100年间陆地降水的器测记录表明,北半球中高纬度的多数地区降水每10年增加0.5%~1%,在总降水增加的地区大雨和极端降水事件增加更加显著。特别是在20世纪的后50年,大雨事件可能增加了2%~4%。由此表明,副热带地区的降水有可能与气候变暖存在某种联系。

针对全球平均表面气温偏冷(60—70年代)和偏暖(80—90年代)阶段,赵平等[43]研究了我国东部地区雨带的异常变动及相关联的大气环流特征。冷位相阶段的春末到夏季,我国东部强降水带表现出从华南、经过长江流域向华北移动的特征,而在暖位相时强降水主要集中在长江流域,从华南向华北移动的特征不明显。冷、暖位相阶段的雨带异常变化与东亚大气环流影响(类似于2.4节)有关,导致暖位相阶段梅雨锋加强,雨带滞留在长江流域,梅雨降水明显。丁一汇等[44]关于长江流域梅雨季降雨强度长期变化的研究发现,1885—2000年梅雨季降雨强度有明显的线性增加趋势,且猜测这种增加的长期趋势可能与1840年以来的全球变暖有某种联系存在。

2.5 关于梅雨的"非典型性"变化

在气候变暖背景下,诸多气象工作者及公众都感觉到近几年梅雨的非典型程度有所增加。那么,这种"非典型性"有何具体体现呢?最近,我们通过考察百余年上海梅雨的长期变化[45]发现,上海梅雨除入、出梅时间存在明显的推迟趋势外,还表现出梅雨期内连阴雨程度明显减弱的特征(图1)。自1875年(图1a),特别是60年代以来(图1b),上海梅雨期降水日数占梅雨期长度的百分比存在明显的减少趋势。由此表明,近年来梅雨期的降水愈来愈为集中,这与以往典型梅雨的持续性降水不同,亦可能是造成人们对当前梅雨"非典型"感觉的原因之一。但这种现象与全球或局地气候增暖的联系目前尚不清楚,有待于进一步研究。

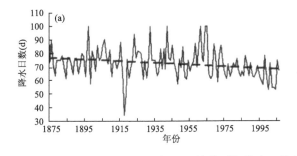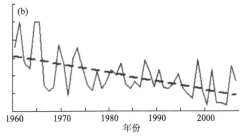

图1 上海梅雨期降水日数占梅雨期长度的百分比
(a)1875—2006年,(b)1960—2006年,虚线:线性趋势

此外,我们还考察了梅雨区降水空间分布一致性的变化特征。从1961—2006年内逐天出现降水现象(日降水量≥0.1 mm·d^{-1})站点数的变化趋势来看(图2a),华东区域(25°~35°N,110°E以东)梅雨集中期(6月中旬至7月中旬)大部分时段的降雨站数存在减少的趋势。也就是说,60年代以来华东区域梅雨季节出现一致水现象的范围存在缩小的趋势,这可能是造成近年来区域梅雨空间分布非典型的原因之一。其中,就不同量级降水的一致性出现而言,与图2a类似的一致性小—中雨(日降水量≤24.9 mm·d^{-1})出现的范围亦存在缩小的趋势;而大雨以上(日降水量≥25 mm·d^{-1})降水现象的一致性出现范围在整个6—7月存在较为明显的增大趋势(图2b)。由此表明,1961—2006年,华东地区梅雨期小到中雨降水愈来愈表现为

局地性降水;而大雨以上量级降水的全区一致性则有增强的趋势。

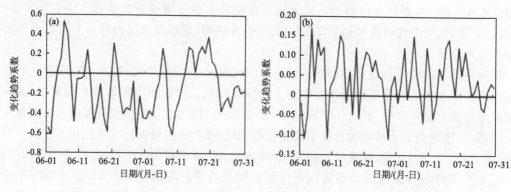

图 2　1961—2006 年 6～7 月华东区域降水站数的逐天线性变化趋势系数
(a)降水(≥0.1 mm·d^{-1})站数,(b)大雨以上(≥25 mm·d^{-1})降水站数

另外,根据 IPCC 报告[46],气候变暖在很大程度上与人类活动有关。随着经济的日益增长,城市化进程加快,相关的人类活动必然对降水变化产生影响。根据周丽英等[47]关于上海降水百年变化趋势及其城郊的差异研究,上海梅雨季节(6—7 月)存在以城区为雨岛中心的城郊差异,且近 20 年上海近郊经济发展迅速,使其降水演变由原来的趋同于远郊转为趋同于中心城区,这可能与城市化进程加快带来的热岛效应存在一定的联系。

3　讨　论

梅雨的气候变化是一个重要且复杂的问题。从上述研究来看,中国梅雨气候变化的相关研究已取得了诸多进展。但还有许多问题仍不清楚,值得进一步研究。

(1)就梅雨的长期变化而言,多数研究在高频(年际)时间尺度上得出类似的变化周期,但由于选用的资料和方法的不同,在 10 年以上的时间尺度上仍没有得到较为一致的梅雨变化周期。而关于百余年梅雨突变(重大转折性)的问题目前仍存在诸多争议。目前,梅雨特征量的确定指标因人而异,因地而异。统一区域梅雨特征量标准,建立一套公认的区域梅雨特征量的长时间序列,将对梅雨的气候变化研究发挥积极的作用。

(2)江淮流域梅雨期降水除整体旱涝的一致性特征外,还具有空间上的非均匀性特征。不同局地降水的一致性是形成区域性降水的原因。如何确定局地梅雨特征量,并对其气候变化进行分析,研究局地梅雨与区域梅雨气候变化的一致性与特殊性,是梅雨气候变化研究的另一方面。

(3)梅雨气候变化的成因分析所得到的结论多是定性的,且多是分别从海温变化、大气环流等角度来探讨对梅雨气候变化的影响。综合多个因素,考虑各因素对梅雨气候变化影响的相对重要性及其时空变化的进一步研究是有必有的。此外,目前还未对梅雨气候变化与青藏高原、大陆和海洋的三级热力差的关系进行研究,这些值得进一步探讨。

(4)目前,关于梅雨降水的数值模拟研究仍存在较大的不足,亟待加强。在此基础上,利用数值模式分析研究梅雨气候变化的事实、成因及预估问题,是梅雨气候变化的另一个重要方面。

(5)全球变暖是一个长时间尺度的过程,目前这个过程还在继续。在过去更早阶段以及未来,我国东部雨带特别是江淮梅雨呈现出何种变化特征?东亚大气环流对全球变暖不同阶段的响应机理有何差异？在更大范围上，我国其他地区降水在全球变暖的不同阶段又如何变化？这些问题都关系到如何更好地借鉴过去的经验来做未来的气候预测，因此值得继续关注和研究。

致谢：感谢上海台风研究所雷小途研究员、陈葆德研究员为本文提出的宝贵建议。

参考文献

[1] Brohan P, Kennedy J J, Haris I, et al. Uncertainty estimates in regional and global observed temperature changes: A new dataset from 1850[J]. J Geophys Res, 2006, 111: D12106.

[2] Trenberth K E, Jones P D, Ambenje P, et al. Observations: Surface and atmospheric climate change[C]// Solomon S, Qin D, Manning M, et al. Climate Change 2007: The Physical Science Basis. Cambridge, United Kingdom and New York, NY, USA: Cambridge University Press, 2007.

[3] Meehl G A, Stocker T F, Collins W D, et al. Global climate projections[C]// Solomon S, Qin D, Manning M, et al. Climate Change 2007: The Physical Science Basis. Contribution of Working Group I to the Fourth Assessment Report of the Intergovernmental Panel on Climate Change Marquis, and co-editors), Cambridge University Press, Cambridge, United Kingdom and New York, NY, USA, 2007.

[4] Trenberth K E. Uncertainty in hurricanes and global warming [J]. Science, 2005, 308: 1753-1754.

[5] Semenov V A, Bengtsson L. Secular trends in daily precipitation characteristics: Greenhouse gas simulation with a coupled AO GCM[J]. Clim Dynam, 2002, 19: 123-140.

[6] Kharin V V, Zwiers F W. Estimating extremes in transient climate change simulations[J]. J Climate, 2005, 18: 1156-1173.

[7] 陈隆勋, 邵永宁, 张清芬, 等. 近四十年我国气候变化的初步分析[J]. 应用气象学报, 1991, 2(2): 164-173.

[8] 陈隆勋, 朱文琴, 王文. 中国近45年来气候变化的研究[J]. 气象学报, 1998, 56(3): 288-296.

[9] 任国玉, 吴虹, 陈正洪. 我国降水变化趋势的空间特征[J]. 应用气象学报, 2000, 11(3): 322-330.

[10] 任国玉, 郭军, 徐铭志, 等. 近50年来中国地面气候变化基本特征[J]. 气象学报, 2005, 63(6): 942-956.

[11] 徐群, 杨义文, 杨秋明. 长江中下游116年梅雨(一)[J]. 暴雨·灾害, 2001, 1: 44-53.

[12] 杨义文, 徐群, 杨秋明. 长江中下游116年梅雨(二)[J]. 暴雨·灾害, 2001, 1: 54-61.

[13] 陈艺敏, 钱永甫. 116a长江中下游梅雨的气候特征[J]. 南京气象学院学报, 2004, 27(1): 65-72.

[14] 魏凤英, 张京江. 1885—2000年长江中下游梅雨特征量的统计分析[J]. 应用气象学报, 2004, 15(3): 313-320.

[15] 毛文书, 王谦谦, 葛旭明, 等. 近116年江淮梅雨异常及其环流特征分析[J]. 气象, 2006, 32(6): 84-90.

[16] 张德二, 王宝贯. 18世纪长江下游梅雨活动的复原研究[J]. 中国科学(B辑), 1990, 20(12): 1333-1339.

[17] 毛文书, 王谦谦, 马慧, 等. 江淮梅雨的时空变化特征[J]. 南京气象学院学报, 2008, 31(1): 116-122.

[18] 王钟睿, 钱永甫. 江淮梅雨的多尺度特征及其与厄尔尼诺和大气环流的联系[J]. 南京气象学院学报, 2004, 27(2): 317-325.

[19] 闵屾, 钱永甫. 江淮梅雨分区特征的比较研究[J]. 应用气象学报, 2008, 19(1): 19-27.

[20] 宗海峰, 张庆云, 陈烈庭. 梅雨期中国东部降水的时空变化及其与大气环流、海温的关系[J]. 大气科学, 2006, 30(6): 1189-1197.

[21] 胡波, 钟元, 潘小凡, 等. 近40年浙江省梅雨的年际与年代际演变规律[J]. 科技通报, 2001, 17(6): 57-59.

[22] 梁萍, 丁一汇. 上海百余年梅雨的气候变化特征[J]. 高原气象, 2008, 27(增刊): 76-83.

[23] 徐卫国, 江静. 我国东部梅雨雨区的年际和年代际的变化分析[J]. 南京大学学报(自然科学), 2004, 40

(3):292-301.

[24] 姚素香,张耀存.江淮流域梅雨期雨量的变化特征及其与太平洋海温的相关关系及年代际差异[J].南京大学学报,2006,42(3):298-308.

[25] 章淹.近半个世纪江淮梅雨的重大变化[J].科技导报,1997,9:58-60.

[26] 徐群.121年梅雨演变中的近期强年代际变化[J].水科学进展,2007,18(3):327-335.

[27] Wu R,Wang B. A contrast of t he east Asian summer monsoon and ENSO relationship between 1962—1977 and 1978—1993 [J]. J Climate,2002,15:3266-3279.

[28] 葛全胜,郭熙凤,郑景云,等.1736年以来长江中下游梅雨变化[J].科学通报,2007,52(23):2792-2797.

[29] 竺夏英,何金海,吴志伟.江淮梅雨期降水经向非均匀分布及异常年特征分析[J].科学通报,2007,52:951-957.

[30] 刘明丽,王谦谦.江淮梅雨期极端降水的气候特征[J].南京气象学院学报,2006,29(5):676-681.

[31] 毛文书,王谦谦,等.近50a江淮梅雨期暴雨的区域特征[J].南京气象学院学报,2006,29(1):33-40.

[32] 周后福,陈晓红.基于EOF和REOF分析江淮梅雨量的时空分布[J].安徽师范大学学报(自然科学版),2006,29:79-82.

[33] 杜银,张耀存,等.长江中下游梅汛期强降水过程非均匀特征[J].气象科学,2007,27(2):133-139.

[34] 毛天松.北太平洋海温距平对6月长江中下游夏季风影响及其机制分析[J].热带气象,1988,4(4):327-335.

[35] 李峰,何金海.北太平洋海温异常与东亚夏季风相互作用的年代际变化[J].热带气象学报,2000,16(3):260-271.

[36] 魏凤英,宋巧云.全球海表温度年代际尺度的空间分布结构及其对长江中下游梅雨的可能影响[J].气象学报,2005,63(4):477-484.

[37] 周丽,魏凤英.近百年全球海温异常变化与长江中下游梅雨[J].高原气象,2006,25(6):1111-1119.

[38] 梁萍,雷小途.副热带夏季风进程的年代际变化及机制研究[C]//2007年度国家气候中心学术年会文集.北京:2008.

[39] Alexander L V,Zhang X,Peterson T C,et al. Global observed changes in daily climate extremes of temperature and precipitation [J]. J Geophys Res,2006,111,D05109-D05109,DOI:10.1029/2005JD006290.

[40] Frich P,Alexander L V,Della-Marta P,et al. Observed coherent changes in climatic extremes during the second half of the twentieth century[J]. Clim Res,2002,19:193-212.

[41] IPCC. Climate Change 2001:The Scientific Basis[R]// Houghton J T,Ding Y,Griggs D J,et al. eds. Cambridge University Press,2001:881.

[42] Chen M,Xie P,Janowiak J E. Global land precipitation:A 50-yr monthly analysis based on gauge observation[J]. J Hydrometeorology,2002,3:249-266.

[43] 赵平,周秀骥.近40年我国东部降水持续时间和雨带移动的年代际变化[J].应用气象学报,2006,17(5):548-556.

[44] 丁一汇,柳俊杰,孙颖,等.东亚梅雨系统的天气-气候学研究[J].大气科学,2007,31(6):1082-1101.

[45] 梁萍,陈葆德,陈伯民.上海130余年汛期水资源的气候变化[J].资源科学(待发表).

[46] IPCC. Climate Change 2007:The Scientific Basis[R]//Susan Solomon,Dahe Qin,Martin Manning,et al. eds. Cambridge University Press,2007.

[47] 周丽英,杨凯.上海降水百年变化趋势及其城郊的差异[J].地理学报,2001,56(4):467-476.

江淮梅雨建立的年际变化及其前期强影响信号分析*

汪靖[1,2]　何金海[1]　刘宣飞[1]　吴彬贵[2]

（1. 南京信息工程大学大气科学学院，南京 210044；2. 天津市气象台，天津 300074）

摘要：江淮梅雨建立具有显著的年际变化特征。利用 NCEP/NCAR 再分析数据集以及 NOAA 提供的全球射出长波辐射和扩展重建海温等资料研究了梅雨建立的年际变化及其前期强影响信号，并探讨了该信号影响梅雨建立年际变化的可能机制。结果表明，前期中太平洋 ENSO（CP-ENSO）事件是影响梅雨建立年际变化的强信号，该信号具有较好的短期气候预测准确率和实用性。当前期冬季 2 月和春季呈现 CP-ENSO 暖（冷）位相时，梅雨建立最有可能偏晚（早）。CP-ENSO 主要是通过 EAP（或 JP）遥相关型影响梅雨建立，其中位于热带西北太平洋的反气旋环流起着重要作用。CP-ENSO 暖（冷）位相年，热带暖湿气流向江淮流域输送偏晚（早），赤道附近的中太平洋地区海温产生正（负）异常，通过海气相互作用，使得西太平洋副高北跳和印度西南季风建立偏晚（早），东亚上空大气环流由春到夏的季节性转换因而偏晚（早）。大气环流季节性转换和热带暖湿气流向江淮流域输送偏早（晚）是江淮梅雨建立偏早（晚）的主要原因。

关键词：江淮流域；梅雨建立；年际变化；前期强影响信号；中太平洋 ENSO

我国位于欧亚大陆东南部，濒临宽广的西太平洋和南海，具有有"世界屋脊"之称的青藏高原。受这种海陆分布和高原地形的影响，我国成为著名的亚洲季风气候区，亚洲大气环流季节性转变的特征显著。江淮流域的梅雨（Meiyu）是东亚夏季风向北推进进程中特有的雨季[1-3]，是东亚大气环流由晚春到初夏季节性转换的产物[4]，此时江淮流域常会出现连阴雨天气，在气候上表现为高湿多云，风力较小，日照时数少。梅雨在日本被称为 Baiu，在韩国被称为 Changma，梅雨也是这两个国家短期气候预测的重要内容，日本和韩国对此也开展了许多有意义的研究[5-8]。

江淮梅雨建立的早晚和梅雨量的大小密切相关。徐海明等[9]和魏凤英等[10]的研究指出，梅雨量的大小和梅雨建立的早晚存在显著的反相关关系。梅雨量的大小又和江淮流域的旱涝存在着密切的联系[11]，因此对江淮梅雨建立早晚的预测研究有助于防灾减灾工作的部署，具有重要的社会经济意义。

鉴于上述研究的重要性，中国气象工作者对江淮梅雨建立年际变化的问题进行了广泛的研究[12-18]。研究结果表明，江淮梅雨建立的年际变化与赤道中东太平洋、北大西洋、西太平洋暖池（以下简称暖池）以及马斯克林群岛附近的海温异常有一定联系。但是，前人的研究存在资料时间序列长度偏短和研究不够细致深入等问题。譬如说，赵振国[13]指出，厄尔尼诺年江

---

*　本文发表于《科学通报》，2009 年第 54 卷第 1 期，85-92. 其英文版发表于《Chinese Science Bulletin》，2009 第 54 卷第 4 期，687-695.

淮梅雨建立偏晚。但是，Niño海区可以分为Niño1+2，Niño3和Niño4区，每个海区海温异常对梅雨建立早晚的影响程度可能不一样。所以，从短期气候预测的角度考虑，我们需要在前人研究的基础上利用年代更长的气象资料来寻找梅雨建立年际变化的前期强影响信号，并对此信号进行更细致地分析研究，以提高其短期气候预测的准确率。上述这些便是本文的研究目的和动机，通过开展这些研究，为梅雨建立年际变化的短期气候预测提供有意义的参考依据。

1 资料

本文所采用的资料为(1)NCEP/NCAR再分析数据集[19]，水平分辨率为2.5°×2.5°，垂直方向分为17层；(2)美国国家海洋和大气管理局(NOAA)提供的全球射出长波辐射(OLR)资料[20]和扩展重建海温(ERSST)资料[21]，水平分辨率分别为2.5°×2.5°和2.0°×2.0°；(3)NOAA气候预测中心(CPC)提供的南方涛动指数(SOI)以及Niño1+2区、Niño3区和Niño4区海温资料，资料来源于http://www.cdc.noaa.gov/ClimateIndices/List/。

2 江淮梅雨建立的年际变化特征

周曾奎[22]的研究指出，江苏省气象台梅雨建立的划定标准充分考虑了东亚大气环流的季节性转换，是一种较合理的江淮梅雨建立划定标准。其具体的划定标准为5—6月，500 hPa印度加尔各答稳定西风结束，出现东西风相间以后，同时连续3d达到如下指标可以确定梅雨建立：(1)120°E西太平洋副热带高压(以下简称副高)脊线位置≥20°N；(2)115°、120°和125°E 588线平均位置≥25°N(或115°、120°和125°E 584线平均位置≥30°N，≤35°N)；(3)500 hPa等压面上，120°E处 −8℃等温线位置≥35°N。为此，我们根据江苏省气象台划定的梅雨建立日期[23]①计算出1954—2007梅雨平均建立日期为6月18日，并给出了梅雨建立日期距平的年际变化曲线(图1)。由图1可见，梅雨建立最早的年份是1956年(6月3日)，最晚的年份是1965年和1969年(均为6月30日)，最早年与最晚年梅雨建立日期相差近一个月。由此可见，江淮梅雨建立具有明显的年际变化特征。

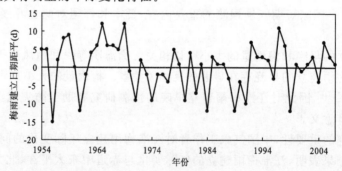

图1 1954—2007年江淮梅雨建立日期距平的年际变化曲线 水平实线表示梅雨建立的平均日期

---

① 文献[23]仅给出1954—2005年的梅雨建立日期，根据江苏省气象台的划定标准，我们划定了2006年和2007年的梅雨建立日期分别为6月21日和6月19日。

3 江淮梅雨建立年际变化的前期强影响信号分析

研究表明[24],大范围大气环流异常和海温、积雪和土壤水分含量等外界强迫因子异常有关,其中与海温异常的关系更为紧密。为此,我们从海温这个重要的外界强迫因子出发,来寻找前期(冬季和春季)梅雨建立年际变化的强影响信号。图2给出梅雨建立日期序列与前期冬季(12月到次年2月)和春季(同年3月到5月)海温的时滞相关。由图2a可见,主要显著相关区域分布在如下几个区域:低纬中太平洋地区(165°E～140°W,10°S～10°N附近)存在一个以赤道为轴的南北方向对称的大范围舌状显著正相关区;菲律宾以东的暖池存在显著负相关区;澳大利亚东面及南面的南太平洋中部存在显著负相关区;巴西东南的南大西洋中部存在一显著负相关区。上述区域相关系数绝对值的最大值均大于0.4,通过了0.01的信度检验。到了春季(图2b),上述大范围舌状显著正相关区仍然稳定维持,但暖池的负相关区缩小,相关强度减小;南大西洋中部海区相关强度减小;南美洲北部的赤道大西洋附近出现了显著正相关区域,相关系数最大值为0.3。综合图2可见,相关系数空间分布最广、持续时间最长,相关强度最强的是赤道中太平洋地区海温,可以认为它是梅雨建立年际变化的前期强影响信号:当前期赤道中太平洋地区海温偏高时,梅雨建立偏晚;反之偏早。

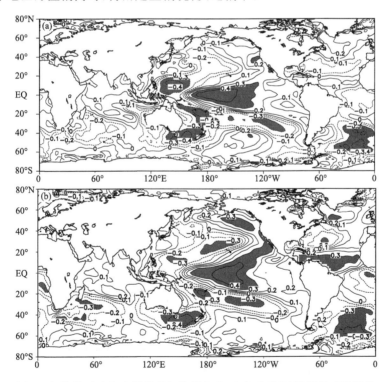

图2 江淮梅雨建立日期序列与前期冬季(a)和春季(b)海温的时滞相关
阴影表示通过0.05的统计信度

海洋与大气是耦合的,El Niño/La Niña与发生在赤道附近的南方涛动(SO)现象有密切的联系,这种海—气耦合系统被称为ENSO。图3给出了梅雨建立日期序列与前期冬、春季

SOI的标准化时间序列。可以看出,梅雨建立日期与前期冬、春季SOI呈现相反的变化趋势,计算梅雨建立日期与前期冬、春季SOI的相关系数分别为-0.44和-0.33,均通过了0.05的统计信度,表明梅雨建立的年际变化与前期SO也存在着显著的相关关系。

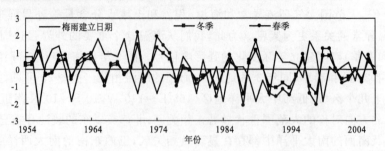

图3 1954—2007年江淮梅雨建立日期以及前期冬季和春季SOI的标准化时间序列

然而,值得指出的是,近年来关于ENSO Modoki事件(Modoki是日本用语,表示"相似而不一样")的研究表明[25-26],ENSO Modoki事件与传统ENSO事件存在着显著的差别。El Niño Modoki事件是指赤道中太平洋海温异常偏高,而其两侧的赤道东太平洋及西太平洋部分区域(主要是暖池区)海温相对较冷;La Niña Modoki事件出现时情况则相反。El Niño Modoki事件和La Niña Modoki事件统称为ENSO Modoki事件。根据ENSO Modoki事件的特征,Ashok等[27]定义了ENSO Modoki指数(EMI),具体的表达式为

$$EMI=[SSTA]_A-0.5\times[SSTA]_B-0.5\times[SSTA]_C \quad (1)$$

表达式中的3个方括号分别表示区域A(165°E~140°W,10°S~10°N)、B(110°W~70°W,15°S~5°N)和C(125°E~145°E,10°S~20°N)的海温距平值(SSTA)。当赤道中太平洋海温偏高而其两侧海温偏低时,EMI偏高;反之偏低。

由图2的分析结果可知,梅雨建立早晚与前期冬季A区域海温异常显著正相关,和B区域海温异常无明显关系,其影响可以忽略,而与赤道以北的C区域海温异常呈显著负相关,这表明梅雨建立早晚与前期冬季ENSO Modoki事件可能有较好的关系。但是,到了春季,赤道以北C区域的显著负相关区缩小,相关强度也减小,所以梅雨建立早晚与前期春季ENSO Modoki事件的相关程度可能减小。为了验证上述观点,我们根据(1)式计算了梅雨建立日期与前期冬季EMI的相关系数为0.46,通过了0.001的统计信度,而与春季的相关系数为0.27,仅通过0.05的统计信度,表明到了春季,ENSO Modoki事件对梅雨建立的影响程度降低。由此可见,和ENSO Modoki事件相比,梅雨建立与前期赤道中太平洋海温的关系最密切,前期赤道中太平洋海温异常是梅雨建立年际变化的强影响信号。

综合前面的分析可见,梅雨建立的年际变化不仅与前期SO有显著的相关关系,而且又与前期赤道中太平洋海温的关系最为密切。Kao等[28]将赤道中太平洋海温异常偏高的ENSO形态称为中太平洋ENSO形态(CP-ENSO Type),所以可以认为前期CP-ENSO事件是梅雨建立年际变化的强影响信号。

CPC将Niño海区分为Niño 1+2区(80°~90°W,10°~0°S)、Niño 3区(90°~150°W,5°S~5°N)和Niño 4区(160°E~150°W,5°S~5°N)。Niño 1+2区和Niño 3区的绝大部分位于赤道东太平洋,Niño 4区位于赤道中太平洋。为进一步验证前面的分析结果并提高梅雨建立短期气候预测的准确率,我们需要作更详细的分析。为此,文中计算了梅雨建立日期与前期冬、

春季海温的逐月相关系数(表1)。由表可见,梅雨建立日期与Niño 4区海温的相关性最好,其次是Niño 3区,最差的是Niño 1+2区。就Niño 4区海温而言,其春季海温异常对梅雨建立早晚的影响更大一些,3月相关系数可达0.47,表1的分析结果很好地验证了前面的分析结果。为提高短期气候预测的准确率,我们根据表1中相关系数的大小指出:当前期冬季2月和春季呈现CP-ENSO暖位相时,梅雨建立最有可能偏晚;反之最有可能偏早。

表1 江淮梅雨建立日期与前期冬季和春季Niño 1+2区、Niño 3区和Niño 4区海温的逐月相关系数

	前年12月	同年1月	同年2月	同年3月	同年4月	同年5月
Niño 1+2	0.13	0.14	0.16	0.2	0.23	0.28[a]
Niño 3	0.29[a]	0.29[a]	0.24	0.27[a]	0.26	0.21
Niño 4	0.35[b]	0.41[b]	0.43[c]	0.47[c]	0.42[c]	0.42[c]

a)表示通过0.05的统计信度;b)表示通过0.01的统计信度;c)表示通过0.001的统计信度

为更好且直观地表征冬、春季Niño 4区海温异常与梅雨建立早晚的密切关系,我们给出了梅雨建立日期和前期冬、春季Niño 4区海温的标准化时间序列(图4)。图4直观地表征了梅雨建立日期和前期冬、春季Niño 4区海温具有同位相变化趋势,它们之间的相关系数分别为0.41和0.43,均通过了0.01的统计信度。这进一步验证了前期CP-ENSO事件是梅雨建立年际变化的强影响信号。

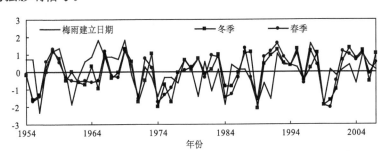

图4 1954—2007年江淮梅雨建立日期以及前期冬季和春季Niño 4区海温的标准化时间序列

为了验证上述强影响信号的准确率,文中增加了该信号在2008年梅雨建立中的应用。图5给出了2008年前期冬季和春季Niño 4区SSTA的逐月演变。由图可见,冬、春季Niño 4区海温呈现明显的负距平,负值均小于-0.5℃并持续了6个月,梅雨建立前期产生了显著的CP-ENSO冷位相事件。从2008年6月8日开始(图略),120°E副高脊线位置连续3天大于

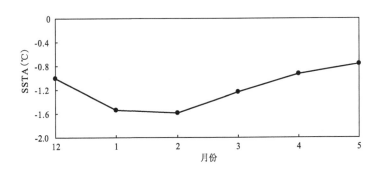

图5 2008年前期冬季和春季Niño 4区SSTA的逐月演变

20°N,115°、120°和 125°E 588 线平均位置连续 3 天大于 25°N,500 hPa 等压面上 120°E −8℃ 等温线位置连续 3 天也大于 35°N。根据第二小节中梅雨建立的划定标准可知 2008 年梅雨在 6 月 8 日建立,较平均日期提早 10 天。上述分析在一定程度上说明了该信号具有较好的短期气候预测准确率和实用性。

4　CP-ENSO 事件影响江淮梅雨建立年际变化的可能机制

上面一小节的分析揭示了前期 CP-ENSO 事件是影响梅雨建立年际变化的强信号,那么其影响梅雨建立年际变化的可能机制又是什么呢? 为探讨这种可能机制,根据图 4 先将前期冬、春季 Nino4 区海温的标准差均大于 1 的年份划定为前期 CP-ENSO 暖位相年,反之为冷位相年,然后再进行合成分析。这样,暖位相年有 1958、1969、1987、1992、1995、2003、2005;冷位相年有 1955、1956、1971、1974、1989、1999、2000。图 6 给出 CP-ENSO 冷暖位相年江淮梅雨建立同期(6 月)OLR 合成差值场和赤道地区合成垂直剖面差值流场。由图 6a 可以看出,CP-ENSO 暖位相年,赤道附近的中太平洋为大范围负差值区所控制,而亚澳"大陆桥"、菲律宾群岛以及暖池附近为大片正差值区所覆盖,表明 CP-ENSO 暖位相年赤道附近的中太平洋地区对流活跃,上升运动明显,下沉运动主要位于上述正差值区,这和图 6b 所显示的结果相一致。

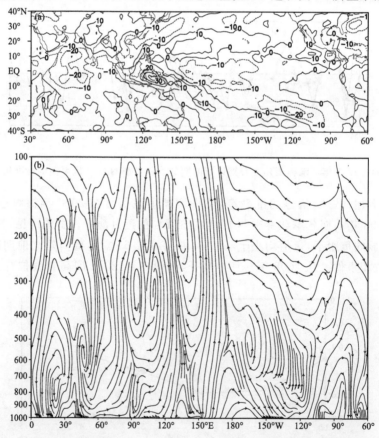

图 6　6 月 OLR 合成差值场(a,单位:W·m^{-2})及赤道地区合成垂直剖面差值流场
(b,沿 5°S～5°N 的纬向剖面图)差值场为 CP-ENSO 暖位相年减冷位相年

上述分析表明,CP-ENSO暖位相年,ITCZ偏弱,而反Walker环流偏强。此外,印度半岛大部也为正差值区所控制,对流活动受到抑制,季风低压偏弱,不利于印度西南季风建立。

图7是对流层低层850 hPa风场合成差值。由图7可见,印度半岛以南的印度洋上存在气旋性差值环流中心,孟加拉湾经印度半岛一直到阿拉伯海处有明显的东风异常,印度西南季风发展受到抑制,这和图6a所分析的结果相一致。此外,日本岛南部东海以东的副热带西太平洋地区存在明显的气旋性差值环流,菲律宾群岛附近的热带西北太平洋地区为反气旋性差值环流控制。上述三个异常环流显现出类似于EAP[29-30](东亚—太平洋型)或JP[31](日本—太平洋型)遥相关分布。Zhang等[32]和Wang等[33]的研究表明,ENSO影响东亚气候和大气环流的物理过程主要是与EAP或JP遥相关型相联系,其中位于热带西北太平洋地区的反气旋环流起着重要的作用。从图7中还可以看出,105°E苏门答腊附近越赤道气流和索马里越赤道气流偏弱,热带暖湿气流向江淮流域输送偏晚。

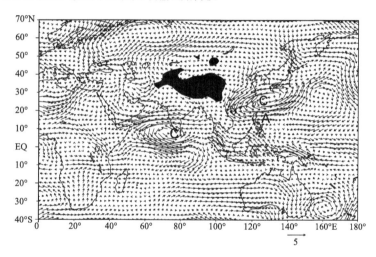

图7　同图6,但为850 hPa风场(单位:m·s^{-1})

阴影表示高于1800 m的青藏高原部分。字母A和C分别表示反气旋和气旋性差值环流中心

从500 hPa位势高度场合成差值(图8a)可以清楚看出,CP-ENSO暖位相年30°～60°N东亚沿岸为一致的负值区(东亚大槽)控制,南亚高压位置偏南(图8b)。

通过前面的分析并结合前人已有的研究[16,29-34],可以给出CP-ENSO事件影响梅雨建立年际变化的可能机制:CP-ENSO暖位相年,赤道附近的中太平洋地区海温产生正异常,通过海气相互作用,形成了前述三个异常环流,这些环流异常呈现出似EAP(或JP)遥相关分布,位于菲律宾群岛附近的热带西北太平洋异常反气旋环流和该地区的负热源异常相对应[35-37],日本岛南部气旋性环流使得赤道北侧的反Hadley环流得到发展,这种形势不利于副高北跳。东亚大槽的存在和南亚高压位置偏南也使副高不易增强北跳。副高北跳和印度西南季风建立偏晚导致东亚上空大气环流由春到夏的季节性转换偏晚,大气环流季节性转换和热带暖湿气流向江淮流域输送偏晚是梅雨建立偏晚的主要原因[23]。前述情况在CP-ENSO冷位相年相反。

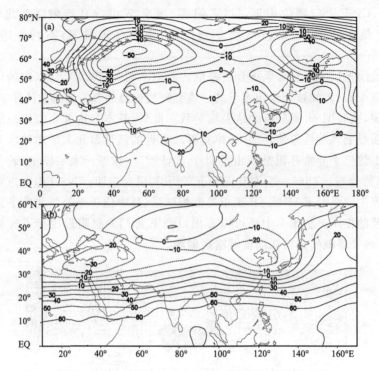

图 8　同图 6，但为 500 hPa(a)和 100 hPa(b)位势高度场（单位：gpm）

5　结论和讨论

江淮梅雨建立具有显著的年际变化特征。本文给出了江淮梅雨建立年际变化的前期强影响信号，并探讨了其影响梅雨建立年际变化的可能机制。通过上述分析，我们得到以下主要结论：

(1)前期 CP-ENSO 事件是影响梅雨建立年际变化的强信号，该信号具有较好的短期气候预测准确率和实用性。当前期冬季 2 月和春季呈现 CP-ENSO 暖位相时，梅雨建立最有可能偏晚；反之最有可能偏早。

(2)CP-ENSO 主要是通过 EAP（或 JP）遥相关型影响梅雨建立，其中位于热带西北太平洋的反气旋环流起着重要作用。CP-ENSO 暖位相年，热带暖湿气流向江淮流域输送偏晚，赤道附近的中太平洋地区海温产生正异常，通过海气相互作用，使得副高北跳和印度西南季风建立偏晚，东亚上空大气环流由春到夏的季节性转换因而偏晚。大气环流季节性转换和热带暖湿气流向江淮流域输送偏晚是梅雨建立偏晚的主要原因。CP-ENSO 冷位相年情况相反。

CP-ENSO 事件是影响江淮梅雨建立年际变化的强前期信号，但也不能排除其他因素的影响，北大西洋、暖池以及马斯克林群岛附近的海温异常对梅雨建立的年际变化也有影响，CP-ENSO 事件与上述影响信号的相对重要性仍需要数值模式和大量个例分析来加以验证。此外，文中分析结果表明，南太平洋中部以及南大西洋中部部分海区海温异常与梅雨建立早晚也有一定的联系，其影响机制又是什么？目前还不清楚，需要进一步研究。

致谢：感谢美国国家环境预测中心（NCEP）以及美国国家海洋和大气管理局（NOAA）提

供相关气象资料。感谢匿名审稿专家的细心审阅和他们所提供的宝贵意见。

参考文献

[1] Tao S Y, Chen L X. A review of recent research on the East Asian Summer Monsoon in China[C] // Chang C P, Krishnamurti T N. Monsoon Meteorology. Oxford: Oxford University Press, 1987: 60-92.

[2] Ding Y H. Summer monsoon rainfalls in China[J]. J Meteorol Soc Jpn, 1992, 70(1B): 373-396.

[3] 江志红,何金海,李建平,等. 东亚夏季风推进过程的气候特征及其年代际变化[J]. 地理学报, 2006, 61(7): 675-686.

[4] 陶诗言,赵煜佳,陈晓敏. 东亚的梅雨期与亚洲上空大气环流季节变化的关系[J]. 气象学报, 1958, 29(2): 119-134.

[5] Ninomiya K, Muraki H. Large-scale circulations over East Asia during Baiu period of 1979[J]. J Meteorol Soc Jpn, 1986, 64(3): 409-429.

[6] Lee D K. An observational study of the Northern Hemisphere summertime circulation associated with the wet summer and the dry summer in Korea[J]. J Korean Meteorol Soc, 1989, 25(4): 205-220.

[7] Kawamura R, Murakami T. Baiu near Japan and its relation to summer monsoons over Southeast Asia and the western North Pacific[J]. J Meteorol Soc Jpn, 1998, 76(4): 619-638.

[8] Krishnan R, Sugi M. Baiu rainfall variability and associated monsoon teleconnections[J]. J Meteorol Soc Jpn, 2001, 79(3): 851-860.

[9] 徐海明,何金海,姚永红. 江淮入梅的年际变化与前冬环流的联系及其可能成因[J]. 南京气象学院学报, 1999, 22(2): 246-252.

[10] 魏凤英,张京江. 1885—2000年长江中下游梅雨特征量的统计分析[J]. 应用气象学报, 2004, 15(3): 313-321.

[11] 田心如,姜爱军,高苹,等. 江苏省梅雨及其灾害影响评估分析[J]. 灾害学, 2004, 19(1): 40-45.

[12] 符淙斌. 我国长江流域梅雨变动与南极冰雪状况的可能联系[J]. 科学通报, 1981, 26(4): 484-486.

[13] 赵振国. 厄尔尼诺现象对北半球大气环流和中国降水的影响[J]. 大气科学, 1996, 20(4): 422-428.

[14] 徐海明,何金海,董敏. 江淮入梅的年际变化及其与北大西洋涛动和海温异常的联系[J]. 气象学报, 2001, 59(6): 694-705.

[15] 高辉,薛峰,王会军. 南极涛动年际变化对江淮梅雨的影响及预报意义[J]. 科学通报, 2003, 48(增刊Ⅱ): 87-92.

[16] 王钟睿,钱永甫. 海温异常对江淮流域入梅的影响[J]. 应用气象学报, 2005, 16(2): 193-204.

[17] 汪靖,刘宣飞,韩桂荣,等. 2005年江淮流域入梅偏晚的成因分析[J]. 气象, 2006, 32(12): 76-81.

[18] 竺夏英,何金海,吴志伟. 江淮梅雨期降水经向非均匀分布及异常年特征分析[J]. 科学通报, 2007, 52(8): 951-957.

[19] Kalnay E, Kanamitsu M, Kistler R, et al. The NCEP/NCAR 40-year reanalysis project[J]. Bull Amer Meteorol Soc, 1996, 77: 437-471.

[20] Liebmann B, Smith C A. Description of a complete (interpolated) outgoing longwave radiation dataset. Bull Amer Meteorol Soc, 1996, 77: 1275-1277.

[21] Smith T M, Reynolds R W. Improved extended reconstruction of SST (1854-1997)[J]. J Climate, 2004, 17: 2466-2477.

[22] 周曾奎. 江淮梅雨[M]. 北京: 气象出版社, 1996: 1-209.

[23] 周曾奎. 江淮梅雨的分析和预报[M]. 北京: 气象出版社, 2006: 1-184.

[24] Hoskins B, Karoly D. The steady linear response of a spherical atmosphere to thermal and orographic forcing[J]. J Atmos Sci,1981,38:1179-1196.

[25] Weng H Y, Ashok K, Behera S K, et al. Impacts of recent El Niño Modoki on dry/wet conditions in the Pacific rim during boreal summer[J]. Clim Dyn,2007,29:113-129,doi:10.1007/s00382-007-0234-0.

[26] Chang C W, Hsu H H, Wu C R, et al. Interannual mode of sea level in the South China Sea and the roles of El Niño and El Niño Modoki[J]. Geophys Res Lett,2008,35:L03601,doi:10.1029/2007GL032562.

[27] Ashok K, Behera S K, Rao S A, et al. El Niño Modoki and its possible teleconnection[J]. Geophys Res Lett,2007,112:C11007,doi:10.1029/2006JC003798.

[28] Kao H Y, Yu J Y. Contrasting Eastern-Pacific and Central-Pacific types of ENSO[J]. J Climate,2009,22:615-632,doi:10.1175/2008JCLI2309.1.

[29] HUANG R H, LI W J. Influence of the heat source anomaly over the tropical western Pacific on the subtropical high over East Asia [C]//Proceedings of International Conference on the General Circulation of East Asia. Chengdu, April 10-15,1987:40-51.

[30] 黄荣辉,蔡榕硕,陈际龙,等. 我国旱涝气候灾害的年代际变化及其与东亚气候系统变化的关系[J]. 大气科学,2006,30(5):730-743.

[31] Nitta T S. Convective activities in the tropical western Pacific and their impact on the Northern Hemisphere summer circulation[J]. J Meteorol Soc Jpn,1987,65(3):373-390.

[32] Zhang R H, Sumi A, Kimoto M. Impact of El Niño on the East Asian monsoon:a diagnostic study of the 86/87 and 91/92 events[J]. J Meteorol Soc Jpn,1996,74(1):49-62.

[33] Wang B, Wu R G, Fu X H. Pacific-East Asian teleconnection:how does ENSO affect East Asian climate [J]. J Climate,2000,13:1517-1536.

[34] 陶诗言,朱福康. 夏季亚洲南部100 mb流型的变化及其与西太平洋副热带高压进退的关系[J]. 气象学报,1964,34(4):385-395.

[35] 吴国雄,刘屹岷,刘平. 空间非均匀加热对副热带高压带形成和变异的影响. I. 尺度分析[J]. 气象学报,1999,57(3):257-263.

[36] 刘屹岷,吴国雄,刘辉,等. 空间非均匀加热对副热带高压带形成和变异的影响. III. 凝结潜热加热与南亚高压及西太平洋副高[J]. 气象学报,1999,57(5):525-538.

[37] Liu Y M, Wu G X, Liu H, et al. Condensation heating of the Asian summer monsoon and the subtropical anticyclone in the Eastern Hemisphere[J]. Clim Dyn,2001,17(4):327-338.

1961—2001 年青藏高原大气热源的气候特征*

钟珊珊[1]　何金海[1]　管兆勇[1]　温敏[2]

(1. 南京信息工程大学,气象灾害省部共建教育部重点实验室,南京 210044;
2. 中国气象科学研究院,灾害天气国家重点实验室,北京 100081)

摘要: 文中利用 ECMWF 逐日再分析资料,用"倒算法"计算了 1961—2001 年青藏高原上空热量源汇,并分析了高原上空大气热量分布的气候状况。结果表明:(1) 3—9 月,高原上空为热源,热源最强在 6 月;10 月—次年 2 月是热汇,热汇最强在 12 月。整个高原上空,全年大气热量状况主要表现为热源持续时间长,且热源强度较热汇要大得多。对整层热源贡献最大的因子是垂直输送项。(2)从大气加热的垂直廓线来看,热源最大值层出现的高度随季节基本没有变化,集中在 500~600 hPa,但加热的强度和厚度却随季节是变化的;而热汇最大值层和强度随季节是变化的。(3)高原整层 $<Q_1>$ 的水平分布复杂,表现出强的区域性特征:高原热源西部变化比东部迅速,4—8 月西部热源强度明显强于东部。春季,高原西部热源增强迅速,在 5 月出现 200 W·m^{-2} 中心,比东部提前一个月。7 月整个高原热源开始向南减弱,西部热源至 10 月转为热汇,比东部又提前了 1 个月。(4)自 1979 年后,各季节高原热源变化均表现出 1990 前后的气候转变信号。夏季,高原热源变率表现为南北反位相型,其他季节为高原中部—东北部与高原东南部反位相型。

关键词: 青藏高原;热量分布;加热厚度;气候转变

1　引　言

青藏高原隆起改变了亚洲气候的格局[1],而且青藏高原的气候变化又经常超前中国其他地区的气候变化[2],青藏高原对其上空及其周围地区的环流有重要影响[3],中国乃至亚洲气候变化与青藏高原气候和环境变化是分不开的。

最早研究高原大气热源的是叶笃正等[4]和巢纪平[5],他们首次计算了对流层中下层青藏高原大气视热源,得到夏季是热源而冬季是热汇的重要结果。随后,陈隆勋[6]指出夏季中国大气热源增温率最大中心位于高原东南部和云南西部,20 世纪 80 年代,陈隆勋等[7]首先提出亚洲地区夏季最强 Q_1 不在高原上空而位于孟加拉湾北部和南海—西太平洋,但孟加拉湾大气热源是在青藏高原动力和热力作用影响下形成的[8]。

高原大气热状况的异常变化与东亚大气环流关系密切[9],对中国夏季江淮地区、华南地区和华北地区的降水有一定的指示意义[10-11],当高原热源增强时,长江上游和淮河流域降水

---

* 本文发表于《气象学报》,2009 年第 67 卷第 3 期,407-416.

增加,而我国东南地区、华北地区降水偏少。青藏高原及其附近地区大气热源(汇)季节转换前后的差异是影响中国江淮地区严重干旱/洪涝的原因之一[12],但在分析高原热源异常和降水二者的关系时,必须考虑高原热力状况的空间差异[13-14]。

关于高原热力作用及其伴随的湿过程对季风的影响,国内许多研究者已经证实[15]。初夏过渡季节青藏高原非绝热加热的变化引起海—陆热力差异对比的变化,给亚洲夏季风的爆发建立了有利的背景环境,对亚洲夏季风爆发有明显的影响[16]。江宁波等[17]指出,东亚季风爆发时,高原东南部的非绝热加热作用使得高原东南部地区对流层大气爆发性增暖,而这种增暖使得高原东部以南的温度梯度改变,从而导致风场的调整—东亚季风爆发。简茂球等[18]指出5—6月高原东部对流层中高层由非绝热加热造成的显著增温对东亚夏季风的北进和维持是非常重要的。青藏高原夏季降水的时空分布不同也可导致高原及其周围地区大气热源的时空分布发生变化,这种变化使大气环流发生相应的改变,最终导致东亚夏季风爆发时间不同[19]。虽然,目前国内外季风研究学者对高原感热加热如何影响夏季风爆发的机理尚未有一致意见,但青藏高原的加热作用是夏季风爆发机制之一是毋庸置疑的。

青藏高原冬季热汇异常强时可以激发出低空异常强反气旋环流[20],形成了东亚沿海强冬季风,进而影响低纬对流异常,影响亚澳季风区向南的经向分量增强,向南异常经向冬季风到达赤道后转向为异常穿赤道西风,向东传播到东太平洋赤道后引起 SSTA 变暖即 El Niño。诊断分析和数值试验都证实了这个过程,因而高原冬季热源可以影响到全球大气环流和海洋环流[21]。青藏高原夏季大气热源感热分量可以激发气泵作用,造成高原区内上升和东侧下沉的纬向环流,调整副热带高压南北位置,甚至影响到东太平洋环流[22]。

目前,由于高原观测资料的匮乏,高原热源的正确计算是研究高原热源问题的核心也是所有研究工作的基础,高原热源结构也有待进一步的研究。本文采用 ECMWF 逐日再分析资料,计算较长时间序列的高原上空大气热源,并尝试分析高原上空大气热源的气候特征。

2 资料和方法

本文采用 1961—2001 年 ECMW 逐日再分析资料(简称 ERA)计算大气热源。大气视热源 Q_1 的计算方法采用 Yanai 等[23]给出的"倒算法"。公式简单介绍如下:

$$Q_1 = c_p \left[\frac{\partial T}{\partial t} + \mathbf{V} \cdot \nabla T + \left(\frac{p}{p_s}\right)^k \omega \frac{\partial \theta}{\partial p} \right]$$

式中,T 为温度;ω 为 p 坐标的垂直速度(采用已有资料);$k=R/c_p$,R 和 c_p 分别是干空气气体常数和比定压热容;$\mathbf{V}$ 为水平风矢量,由此可以计算各等压面层的 Q_1。从上面的公式可知,Q_1 由等式右边三项组成:第 1 项,温度局地变化项;第 2 项,温度平流项;第 3 项,温度垂直运动项。在文中我们分别简称为局地项、平流项、垂直输送项。计算对流层整层的大气热源 $\langle Q_1 \rangle$ 时,采用 Yanai 等[23]的方案,设对流顶为 100 hPa,其上 $\omega=0$。因而大气热源整层的垂直积分值 $\langle Q_1 \rangle$ 则由下式计算:

$$\langle Q_1 \rangle = \frac{1}{g} \int_{p_t}^{p_s} Q_1 \mathrm{d}p$$

式中,p_s 为地面气压;p_t 为大气顶气压(设 $p_t=100$ hPa)。我们近似地选取地面以上最近的格点所在的等压面层往上进行积分。本文所用的倒算法计算程序采用 Yanai 设计的程序包。选取夏季热源的计算结果与高原 71 个站夏季实测降水资料进行了比较,二者的强中心位置基本相似(图略);与赵平[10]的结果进行了比较,分布基本一致,只是值略偏大。

3 青藏高原上空大气热源的区域平均气候状况

3.1 青藏高原上空整层大气区域平均的热状况

为了解青藏高原上空整层大气的热状况,我们选取海拔高于 3000 m 区域对 1961—2001 年共 41 年 $\langle Q_1 \rangle$ 的气候平均状况进行分析。

1961—2001 年,青藏高原海拔≥3000 m 的范围内,各月 $\langle Q_1 \rangle$ 41 年的气候平均状况表现为(图 1 中空心圆实线):3—9 月,高原上空 $\langle Q_1 \rangle$ 值大于零,为热源,热源从 3 月开始逐渐增强,到 6 月达到极值(达 214 W·m^{-2}),然后逐渐减弱。10 月—次年 2 月 $\langle Q_1 \rangle$ 值小于 0,是热汇;最强热汇出现在 12 月,约为 -84 W·m^{-2}。可见,对整层大气热源来说,全年大气热量状况主要表现为热源持续时间长(约为 7 个月),且其强度较热汇要大得多,热源的峰值约为热汇的 2.5 倍,呈现出年循环中的显著非对称性。

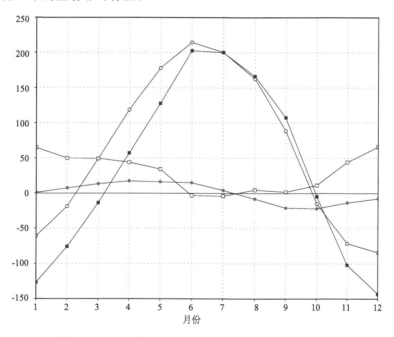

图 1 1961—2001 年青藏高原上空整层热源区域平均各月变化曲线(单位:W·m^{-2};空心圆实线为整层热源$\langle Q_1 \rangle$;实心圆实线为局地项;空心方框实线为平流项;实心方框实线为垂直项)

由上文的公式得知,$\langle Q_1 \rangle$ 分别受局地项、平流项、垂直项三项共同作用,而哪项对高原 $\langle Q_1 \rangle$ 的贡献最大?在此,我们对各分量也进行整层(到 100 hPa)积分并与 $\langle Q_1 \rangle$ 进行比较(图 1):

(1)局地项的强度总体变化不大,其值在 -21~+17 W·m^{-2} 之间变化,变化曲线呈正弦函数。1—7 月为正,峰值在 4 月;8—12 月为负值,峰值出现在 10 月。

(2)平流项的值在 -4~66 W·m^{-2} 之间变化,全年变化表现为年头年尾高,年中平的"盆地"形状曲线。平流项月平均值从 1 月下降,到 6 月强度已非常弱,接近为零,此后一直维持至 9 月;9 月之后平流项月平均值逐渐加强,至 12 月基本达到最强,为 66 W·m^{-2}。因此平流项对$\langle Q_1 \rangle$的作用,主要集中在春、冬两季,而在夏季几乎不起作用。也就是说,在春、冬两季温度平流项对$\langle Q_1 \rangle$影响较大。

(3)垂直输送项的值在 $-143 \sim 202$ W·m^{-2} 之间变化,变化趋势与整层热源$\langle Q_1 \rangle$变化趋势基本一致,最大值也出现在 6 月,最小值出现在 12 月。不同的是,1—6 月和 11—12 月垂直输送项的值都小于$\langle Q_1 \rangle$,而从 6 月开始直到 10 月,其大小与整层热源相当。可见,夏季,垂直速度的增强使得高原上空热量的垂直输送对热源的贡献最大。

综上所述,从整个月平均时间序列来看,除 3 月和 10 月外,对整层$\langle Q_1 \rangle$贡献最大的是垂直输送项。3 月,垂直输送项与局地项的作用相抵消,$\langle Q_1 \rangle$的值主要是平流输送项的贡献。10 月,局地项对$\langle Q_1 \rangle$的贡献最大,但整层$\langle Q_1 \rangle$及其各分量的值都较小,是热源向热汇的过渡月份。单位气压厚度(质量)的$\langle Q_1 \rangle$以及各项随季节的变化与上述结果一致(图略)。

3.2 青藏高原上空大气区域平均的垂直廓线分布

青藏高原由于海拔高,在对流层大气中形成被抬升的热源,那么其上空的加热廓线随高度是如何变化的? 为此,图 2 给出了青藏高原(海拔≥3000 m 区域)上空 41 年来 Q_1 值及其各项月平均值随高度变化的曲线。

从 1—12 月的 Q_1 的垂直廓线可以看出,Q_1 随高度的变化基本呈"S"形,廓线上面部分为大气热汇层,下面部分为大气热源层。中间部分为热源向热汇的过渡层。1 月热汇层的振幅和厚度远大于热源。2 月,随着热源厚度的增加,热汇层厚度变薄,二者的过渡区在 500~400 hPa,热汇的强度几乎是热源的两倍。3 月,热源层的厚度进一步增厚,热源层顶已上升到 300 hPa,在此高度以上为热汇层,而此时,热源强度和热汇强度是相当的。4 月,大气的热源层已占优势,热源层顶已上升到 200 hPa,200~50 hPa 厚度层则表现为热汇。热源最大值层的强度是热汇最大值层的 2.5 倍。5 月,热源层进一步向上扩展到 150 hPa 高度层,而热汇层变薄到 150~50 hPa。热源最大值是热汇的 4.6 倍。6 月,廓线下部振幅变至最大,即热源层的强度达到最大(4.93 K·d^{-1}),其厚度几乎贯穿整个对流层。热汇只在 100~50 hPa 的薄层有所表现。7 月,热源层的高度继续增高超过 100 hPa 层(达到最高),其上到 50 hPa 为弱热汇。8 月,其热源的垂直廓线与 7 月的类似,只是热源和热汇的强度都减小,此时热汇层厚度最薄强度最小。9 月,热汇层强度增强并开始向下层扩展,250~70 hPa 为热汇层,其中 250~150 hPa 之间,热汇随高度几乎没有变化。300 hPa 以下为热源控制。热源最大值是热汇的 3.6 倍。10 月,热汇的强度和厚度都大于热源层,其过渡层在 500~400 hPa 之间,400~70 hPa 为热汇控制。500 hPa 以下为热源。11 月,从近地面层到 100 hPa 附近几乎都为热汇控制,热源廓线呈"C"型,整层大气表现为热汇。12 月,Q_1 的廓线与 11 月相似,热汇几乎贯穿整个对流层高度,整层热源强度达到最大,热汇最大层在 200 hPa 附近。

由以上分析可见,高原上空 Q_1 随高度的变化主要表现为:热源层与热汇层强度变化趋势相反,两者变化过程与发动机"汽缸"工作过程相像,其热汇与热源之间的过渡层则相当于"活塞",把整层大气分为热汇层与热源层上下两个"汽室",随时间的推移,热源层与热汇层"体积"(厚度)不断增加或减小。当热源层"体积"(厚度)变大,"活塞"上行,热汇层厚度变小;热源层"体积"(厚度)变小,"活塞"下行,热汇层厚度变大。随着时间变化,7—8 月"活塞"高度达到最

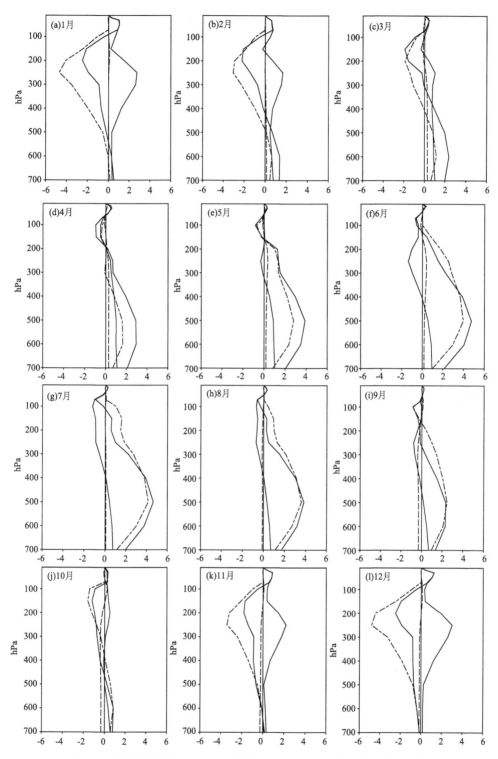

图 2 青藏高原(海拔≥3000 m 范围)上空 Q_1 及各项月平均的垂直廓线
(单位:K·d^{-1};实线为 Q_1,长虚线为局地项,点线为平流项,长短虚线为垂直项)

高,热源层的厚度最厚,10—12月"活塞"高度最低,青藏高原上空基本为热汇控制。

热源最大值层出现的高度随季节基本没有变化,集中在500~600 hPa,但其强度变化较大,6月强度最大,然后分别向年头和年尾减小,12月各层大气基本表现为热汇。而热汇最大值层和强度随季节是变化的。夏季,6—8月,热汇层薄而且弱,在11月、12月、1月热汇层的厚度最厚,最强在200 hPa层。

从热源Q_1的各项的垂直廓线可以看出(图2),热源的局地项随高度的变化较其他两项要小得多,暂时不予考虑,影响Q_1值变化的主要是平流项和垂直输送项。

冬季(12月—次年2月),平流项和垂直输送项的垂直廓线很相似,但呈反位相变化,对Q_1的贡献刚好相反。平流项(正)和垂直项(负)随高度增强,二者在250 hPa层达到极值,后者强度略大,因此整层,〈Q_1〉表现为热汇,Q_1极值在200 hPa高度。

春季(3—5月),对流层上层平流项与垂直输送项强度都逐步减小,而对流层下层则逐步增大,相应的热汇层厚度逐步减小,热源层厚度加大,整个青藏高原上空逐步由热汇向热源过渡。平流项与垂直输送项都出现上负下正的"S"型分布,但二者变化不同。平流项的强度较冬季明显减弱,绝对值不超过$1\ \mathrm{K\cdot d^{-1}}$,并且正值层厚度逐渐变小。垂直输送项的变化较大,正值层高度上升厚度加大,且强度变强,负值层厚度和强度都变小。垂直输送项在春季对〈Q_1〉影响逐步加大。

夏季(6—8月),平流项基本以400 hPa高度为界,在400 hPa以下为正值。强度为平流项全年变化的"盆底"(图1)。垂直项的强度进一步增强,从近地面到70 hPa的大气层,垂直输送项都为正值,只是在7—8月150、500 hPa高度出现两个极值。因此,在夏季Q_1的强度主要取决于垂直项的强度,其廓线与垂直项近似,在100 hPa高度以下为正值,青藏高原上空整个对流层表现为热源。

秋季(9—11月),垂直输送项的变化与春季正好相反,下层正值强度和厚度都减小,到11月整个对流层垂直项都转为负值。平流项强度在9、10月依然很弱,11月突然增强。至此,平流项与垂直输送项二者位相相反,但后者略大。青藏高原上空热源强度减弱,厚度减小;热汇强度加强,厚度加大。整个青藏高原上空由热源向热汇过渡。

由上分析可见,夏季,高原上空大气Q_1的垂直廓线与垂直项的相似,因而主要是由垂直输送项的作用造成;冬季,大气热汇是平流项和垂直项共同作用所致,二者位相相反,垂直项略大于平流项;春季与秋季分别是由热汇向热源、热源向热汇过渡季节。

4 青藏高原上空整层〈Q_1〉的水平分布

青藏高原上空整层〈Q_1〉水平分布表明(图3):1月,整个青藏高原海拔≥3000 m的范围几乎都为热汇。热汇中心在高原的东南部,该热汇中心位置正好位于东西走向的山脉(唐古拉山脉、喜马拉雅山脉东段)和南北走向山脉(横断山脉)的交汇处。中心值超过$-150\ \mathrm{W\cdot m^{-2}}$以上。2月,高原热汇中心位置不变,范围向南收缩。而此时,源自伊朗高原和四川盆地的热源逐渐增强并分别从东、西两个方向向高原扩展。3月,热汇范围进一步缩小,只剩唐古拉山脉附近及其以南的$-50\ \mathrm{W\cdot m^{-2}}$的热汇中心(30°~33°N,90°~95°E)。高原的东部和西部热源继续向高原中部扩展,并在高原中部偏北(35°~37°N,90°~95°E)即昆仑山脉东段附近连通。4月,高原全区为热源控制,热源中心位于高原西部的昆仑山脉中段和冈底斯山脉东段之间(30°~

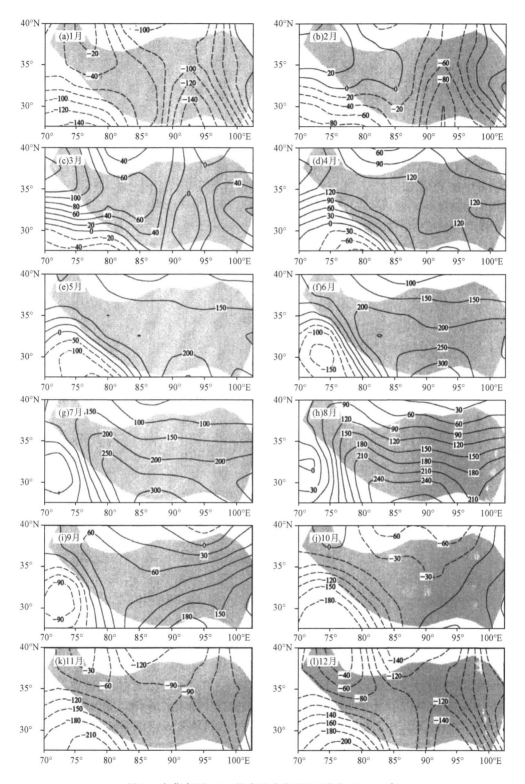

图 3 青藏高原⟨Q_1⟩的水平分布状况(单位:$W \cdot m^{-2}$)

35°N,85°E),中心强度为 150 W·m^{-2} 以上。原热汇中心虽也转为热源,但其值仍较四周的低。若以 90°E 为界,高原西部的热源要强于高原东部。5 月,高原的热源继续增强。中心依然在 90°E 以西的高原西部,而此热源 200 W·m^{-2} 中心向东南方向延伸,与孟加拉湾西北侧的热源相连通。6 月,高原东部热源突增,中心强度也达到 200 W·m^{-2} 以上,但 200 W·m^{-2} 等值线在高原西部比东部的纬度偏北,也就是说高原西部热源仍强于东部。唐古拉山脉附近热源值仍较四周的小。此时,高原东、西部热源与中心在孟加拉湾西北侧的热源连成一个整体,高原热源强度为全年最强。7 月,随着孟加拉湾北部热源的减弱,高原热源 200 W·m^{-2} 等值线南撤。与 6 月有所不同的是,此时,中南半岛北部热源达到最强,喜马拉雅山脉东段的强热源中心沿着山脉向西北方向伸展。8 月,200 W·m^{-2} 等值线南撤的同时在喜马拉雅山脉和横断山脉的交界处出现断裂。中南半岛北部和孟加拉湾北部热源分离。9 月,随中南半岛热源南缩,高原西部热源迅速减弱,100 W·m^{-2} 热源等值线明显表现为东北—西南向的倾斜,即高原东部热源强于西部。10 月,高原西部出现热汇,东部是弱的热源,热源中心向四川盆地收缩。11 月,整个高原均为热汇控制,分布与 1 月相似。12 月,热汇的强度达到最强(低于 -150 W·m^{-2}),中心在高原东南部。

从 1961—2001 年高原$\langle Q_1 \rangle$的月平均水平分布来看,4—8 月,高原西部热源强度较东部强,相同的热源等值线在西部位置较东部偏北。春季,高原西部热源增强迅速,5 月出现 200 W·m^{-2} 中心,而高原东部热源直到 5—6 月才迅速增强,6 月才出现 200 W·m^{-2} 中心,晚了一个月。7 月开始,热源向南减弱,高原西部热源减弱迅速,10 月,高原西部转为热汇,东部为 11 月,晚 1 个月。

5 青藏高原上空整层热源的时空演变

我们仅截取了 1979—2001 年的资料进行 EOF 分析(标准化),目的是为了在后面的研究中便于与其他资料进行对比分析。我们将高原整层热源分 4 个季节做 EOF 分析,并取方差贡献比较大的第一模态进行讨论(图 4)。

春季,高原整层热源的 EOF 第一模态的空间型表现为高原中部—东北部与高原东南部反位相变化,正值中心在高原中部(32°N,84°E),负值中心在高原东南侧。而且,高原中部的变率中心值是东南部的两倍。从时间序列来看,1981—1990 年振幅为正,1991—2001 年为负值,也就是说,在 1990/1991 年高原的热源异常变化的趋势发生了转变,在 1990 年以前,高原中心—东北部为热源正异常变率中心,1991 年以后转为热源负异常变率中心,而东南部的变化则相反。

夏季,高原整层热源的 EOF 第 1 模态的空间型表现为约以 32°N 为界的南北反位相型,南部为负值,北为正值。而热源变率的最大中心不在高原上空,而在高原以南有 3 个中心,分别在喜马拉雅山脉南侧的(27°~30°N,75°E),(27°N,82.5°E),以及横断山脉的西南侧。从时间序列来看也存在热源异常变化的气候转变,在 1981—1987 年为正值,1994—1998 年为负值,1988—1993 年为过渡期。也就是说,在夏季高原 32°N 以南,20 世纪 80 年代表现为热源变率的负异常,而在 80 年代末至 90 年代初转变为热源变率的正异常。

秋季和冬季的时空分布类似与春季(图略),均为高原中部—东北部和高原东南部反位相的空间型分布,时间序列依然表现为 1990/1991 年的气候转变。

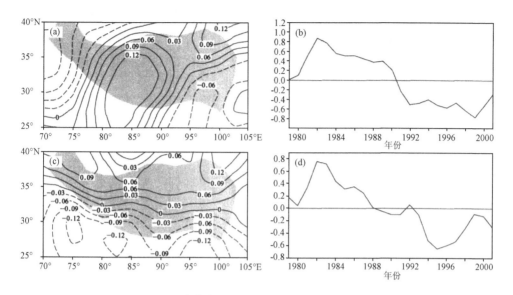

图 4 青藏高原整层大气 $\langle Q_1 \rangle$ 的 EOF 分析

a. 春季 EOF1 的空间分布，b. 春季 EOF1 的时间序列，
c. 夏季 EOF1 的空间分布，d. 夏季 EOF1 的时间序列

可见，高原整层热源 EOF 分析的第 1 模态均表现为气候转变型。所不同的是，春季、秋季和冬季，热源的变率中心都在高原上空。而夏季，热源的变率中心不在高原上，而位于高原南侧，即中南半岛和印度半岛北部。这可能是由于高原热源在夏季并非独立的热源中心，而只是属于其南部中南半岛、印度半岛及二者之间的孟加拉湾热源的一部分。夏季的空间模态主要与经圈环流有关，而春、秋、冬季的空间型分布则主要与西风气流有关。

6 结论与讨论

本论文采用 ERA 逐日资料用"倒算法"计算了 1961—2001 年高原上空热量源汇，并分析了青藏高原上空大气热量分布的气候状况。我们得到以下结论：

(1) 在青藏高原海拔 ≥ 3000 m 范围内，3—9 月，高原上空 $\langle Q_1 \rangle$ 值大于零，为热源，热源最强在 6 月（达 214 W·m^{-2}）；10 月—次年 2 月 $\langle Q_1 \rangle$ 值小于 0，是热汇，热汇最强在 12 月，约为 -84 W·m^{-2}。可见，对高原整层大气来说，全年大气热量状况主要表现为热源持续时间长（共 7 个月），且热源强度较热汇要大得多，其峰值约为热汇的 2.5 倍，呈现出年循环中的显著非对称性。而对整层热源贡献最大的因子是垂直输送项。

(2) 热源最大值层出现的高度随季节基本没有变化，集中在 500~600 hPa，但其强度变化较大，6 月强度最大，然后分别向年头和年尾减小；而热汇最大值层和强度随时间是变化的，夏季，6—8 月，热汇层薄而且弱，在 11 月—次年 1 月热汇层的厚度最大，12 月的 200 hPa 高度层强度最强。

(3) 高原整层 $\langle Q_1 \rangle$ 的水平分布复杂，表现出强的区域性特征。4—8 月，高原西部热源强度较东部强，热源最大等值线 200 W·m^{-2} 在西部位置较东部偏北。春季，高原西部热源增强迅速，而高原东部热源直到 5—6 月才迅速增强，高原西部在 5 月出现 200 W·m^{-2} 中心，而东部

6月才出现,晚了1个月。整个高原热源从7月开始向南减弱,而西部热源减弱迅速,至10月高原西部转为热汇,而东部为11月,晚1个月。

(4)高原热源各季节的变率均表现为1990前后的气候转变型。夏季,高原热源变率表现为南北反位相型,其他季节为高原的中部—东北部与高原东南部反位相型。

致谢:感谢中国气象科学研究院陈隆勋老师和张博博士的指导和帮助!

参考文献

[1] 吴国雄,刘屹岷,刘新,等.青藏高原加热如何影响亚洲夏季的气候格局[J].大气科学,2005,29(1):47-56.

[2] 张耀存,钱永甫.青藏高原隆升作用与大气临界高度的数值试验[J].气象学报,1999,57(2):157-167.

[3] Ding Y H. Effect of the Qinghai-Xizang (Tibetan) Plateau on the circulation features over the plateau and its surrounding areas[J]. Adv Atmos Sci,1992,9:112-130.

[4] 叶笃正,罗四维,朱抱真.西藏高原及其附近的流场结构和对流层大气的热量平衡[J].气象学报,1957,28(1):108-121.

[5] 巢纪平.从冬季东亚定常流型计算热源分布的初步研究[J].气象学报,1956,27(2):167-179.

[6] 陈隆勋,龚知本,陈嘉滨,等.东亚地区大气辐射能收支(三):云天大气的地气系统热量收支[J].气象学报,1965,35(1):6-17.

[7] 陈隆勋,李维亮.亚洲季风区夏季大气热量收支 M 全国热带夏季风学术会议文集(一)[M].昆明:云南人民出版社,1981:86-101.

[8] Kuo H L, Qian Y. Numerical simulation of the development of mean monsoon circulation in July[J]. Mon Wea Rev,1982,110:1879-1897.

[9] 段安民,吴国雄.7月青藏高原大气热源空间型及其与东亚大气环流和降水的相关研究[J].气象学报,2003 61(4):447-456.

[10] 赵平,陈隆勋.35年来青藏高原大气热源气候特征及其与中国降水的关系[J].中国科学(D辑),2001,31(4):327-332.

[11] 柏晶瑜,徐祥德,周玉淑.春季青藏高原感热异常对长江中下游夏季降水影响的初步研究[J].应用气象学报,2003,14(3):363-368.

[12] 吕雅琼,巩远发.2001及2003年夏季青藏高原及附近大气热源(汇)的变化特征[J].高原气象,2006,25(2):195-202.

[13] 赵勇,钱永甫.青藏高原地表热力异常与我国江淮地区夏季降水的关系[J].大气科学,2007,31(1):145-154.

[14] 段安民,刘屹岷,吴国雄.4—6月青藏高原热状况与盛夏东亚降水和大气环流的异常[J].中国科学(D辑),2003,33(10):997-1004.

[15] He J H, Li J, Zhu Q G. Sensitivity experiments on summer monsoon circulation cell in East Asia[J]. Adv Atmos Sci,1989,6:120-132.

[16] 刘新,吴国雄,刘屹岷,等.青藏高原加热与亚洲环流季节变化和夏季风爆发[J].大气科学,2002,26(6):781-793.

[17] 江宁波,罗会邦.高原加热与南海季风爆发[J].中山大学学报(自然科学版),1996,35(增刊):194-199.

[18] 简茂球,罗会邦.1998年青藏高原东部及其邻近地区大气热源与南海夏季风建立的关系[J].高原气象,2001,20(4):381-387.

[19] 巩远发,纪立人,段廷扬.青藏高原雨季的降水特征与东亚夏季风爆发[J].高原气象,2004,23(3):

313-322.

[20] Krishnamurti T N,et al. Tibetan high and upper tropospheric tropical circulation during northern summer [J]. Bull Amer Meteor Soc,1973,54(12):1234-1248.

[21] Zhao P,Chen L X. Interannual variability of atmospheric heat source/sink over the Qinghai-Xizang (Tibetan) Plateau and its relation to circulation[J]. Adv Atmos Sci,2001,18(1):106-116.

[22] Ye D Z,Wu G X. The role of the heat source of the Tibetan Plateau in the general circulation[J]. Meteor Atmos Phys,1998,67:181-198.

[23] Yanai M,Li C,Song Z S. Seasonal heating of the Tibetan Plateau and its effects on the evolution of the Asian summer monsoon[J]. J Meteor Soc Japan,1992,70(1):319-350.

一种新的 El Niño 海气耦合指数*

陈圣劼 何金海 吴志伟

(南京信息工程大学大气科学学院/气象灾害省部共建教育部重点实验室,南京 210044)

摘要:利用 1980—2010 年月平均 Hadley 中心海表温度、美国全球海洋资料同化系统(GODAS)海洋温度和 NCEP/NCAR 大气环流再分析资料,通过对 2 个海洋要素(海表温度 SST、上层热含量 HC)和 5 个大气要素(海平面气压 SLP、850 hPa 风场、200 hPa 速度势和对外长波辐射 OLR)的多变量经验正交函数展开(multivariate EOF,简称 MV-EOF)探讨了热带太平洋的主要海气耦合特征。结果表明,MV-EOF 分析的前两个耦合模态分别很好地对应了传统型 El Niño 和 El Niño Modoki 的海气耦合特征:传统型 El Niño 期间,伴随着赤道中东太平洋 SST 的异常增温,HC、SLP、200 hPa 速度势等要素总体呈东西反相的"跷跷板"变化,低层 850 hPa 赤道中太平洋出现较强西风距平,西北太平洋上空为反气旋性异常环流;El Niño Modoki 期间,SST 持续增温和 HC 正异常中心均显著西移至中太平洋,低层 SLP 和高空 200 hPa 速度势均呈现纬向三极型异常分布,低层异常强西风向西移至暖池东部,西北太平洋上空呈现气旋性异常环流。两类 El Niño 的海气耦合特征存在显著差异,较优的 El Niño 指数应不仅可以客观描述和区分 El Niño 现象本身,更要紧密联系两类事件所产生的大气响应。以往定量表征 El Niño 年际变化的指标大多立足于 SST 或 SLP,本文选取 HC 作为研究指标,定义了一组新的 El Niño 指数 HCEI 和 HCEMI。较以往基于 SST 的 El Niño 指数,HCEI 和 HCEMI 不仅能更清楚地表征和区分两类 El Niño(如 1993 年的传统型 El Niño 和 2006 年的 El Niño Modoki),而且能更好地反映和区分两类 El Niño 与大气间的海气耦合特征,为 El Niño 的监测和短期气候预测工作提供了一个新工具。

关键词:El Niño;El Niño Modoki;海气耦合特征;热含量指数;MV-EOF

1 引 言

符淙斌和弗莱彻[1]曾较早指出 El Niño 事件存在两种赤道增暖类型,一类是经典的赤道东太平洋增温型,另一类增暖中心位于日界线附近的中太平洋地区。20 世纪 90 年代以来,第二类增温型 El Niño 的频繁发生[2],再次掀起了科学界对不同类型 El Niño 现象研究的热潮,Ashok 等[3]将其命名为"El Niño Modoki",也有学者称之为"Dateline El Niño"[4]、"Central Pacific (CP) El Niño"[5]或"Warm Pool El Niño"[6]。伴随着海表温度异常(SSTA)中心位置

---

* 本文发表于《大气科学》,2013 年第 37 卷第 4 期,815-828.

的差异,两类 El Niño 所引起的环流异常及其气候效应也都表现出不同的空间分布特征[3,5-12]。正确认识和区分两类 El Niño 对全球气候的监测和预测工作有着重要的科学意义和实用价值。

近年来,不少学者相继提出了几组 El Niño 指数来定量描述和区分传统型 El Niño 和 El Niño Modoki 事件,如 Niño3 和 EMI[3]/IEMI 指数[13]、KaoEP 和 KaoCP 指数[5]、RenCT 和 RenWP 指数[14]、YuEP 和 YuCP 指数[15]等(详见表2),曹璐[16]定义了一个用 Niño 3 和 Niño 4 指数构建的指标组来监测不同类型 ENSO 事件。众多指标监测的结果有联系也有差别,如1992年被 EMI[3]确立为 El Niño Modoki 年,而根据 IEMI 和 RenWP 指数判定,该年已不满足 El Niño Modoki 标准,结合 Niño 3,确立为传统型 El Niño;2006 年,根据 IEMI 判定是正常年,根据 RenWP 指数则被列为 El Niño Modoki 年。这种争持不可避免地给实际业务工作带来决策的困难[16]。哪个指数描述这两类事件最客观,目前尚无统一定论,其原因主要是两类 El Niño 的定义和划分并不明确以及判别指数优越性的标准无法统一。Ashok 等[3]认为热带太平洋 SSTA EOF 分析的前两个模态很好地描述了传统型 El Niño 和 El Niño Modoki 的海温异常分布特征,第一模态中暖异常覆盖整个赤道中东太平洋区域,西界达到170°E以西,第二模态呈"马蹄形"纬向三极型分布,异常暖中心位于赤道中太平洋160°E~140°W区域,其东西两侧异常偏冷,如2004年的 El Niño 事件。Kao 和 Yu[5]提出另一种分型,利用联合回归(Combined regression)-EOF 方法[17]分别剔除 Niño4 和 Niño1+2 区海温的影响,得到东部(EP)和中部(CP)型 El Niño(见文献[5]的图 3),EP 型暖异常主要位于东太平洋冷舌区,西界在日界线以东160oW附近,CP 型暖异常覆盖中太平洋,范围较大,可伸展至东太平洋。可见,这两种分型的冷暖异常空间分布存在一定差异,分型的差异必然导致判别指数优劣标准的不统一。Li 等[13]选择前一种分类,比较各指数与 SSTA EOF 分析前两个主分量之间的正交性和一致性,认为 Niño3 和 IEMI 分别是描述传统型 El Niño 和 El Niño Modoki 的最优指数,而 Niño 3.4 和 Niño 4 指数均存在混淆两类事件的可能。曹璐[16]选择了后一种分型,指出 Niño 3 指数可以有效地识别 EP 型 ENSO 事件,EMI 和 Niño 4 指数对 CP 型具有较强的监测能力。两类 El Niño 的划分和指数择优工作仍是 ENSO 研究领域的重点和难点。

纵观前人对两类 El Niño 的划分及其指数的设计,均仅仅考虑了海洋 SSTA 的空间分布,然而 El Niño 现象是热带太平洋大尺度海气相互作用的产物[18],在气候研究领域里,比 El Niño 增温现象本身更受关注的是其海气耦合过程及其气候效应。研究已证实两类 El Niño 的遥相关型和气候效应均存在显著差异。Weng 等[7,8]通过分析 El Niño Modoki 对冬夏季环太平洋地区干湿气候条件的影响发现两类 El Niño 会引起北美和东南亚地区不同的降水分布型,El Niño Modoki 冬季美国西部北干南湿,我国华南和台湾、日本南部等地区降水偏少,夏季西北太平洋夏季风偏强,东亚夏季风偏弱;而传统型 El Niño 使得冬季美国西部大部偏湿,东亚南部降水偏多,夏季西北太平洋夏季风偏弱。Feng 和 Li[19]和 Zhang 等[20]分别研究了春、秋季两类 El Niño 对我国降水的影响,发现两类事件引起的华南春、秋季降水异常截然相反。因此较优的两类 El Niño 指数应不仅可以客观描述和区分 El Niño 现象本身,更要紧密联系两类事件所产生的大气响应。本文着眼于不同类型 El Niño 的大气响应特征,提取两类 El Niño 期间热带太平洋海气各要素相协同变化的成分,设计一种新的 El Niño 和 El Niño Modoki 指数,以期全面反映两类 El Niño 的海洋热力异常及其大气响应特征。

文章第三部分利用多变量经验正交函数方法(Multivariate EOF,简称 MV-EOF)[21-23]

对1980—2010年热带太平洋的海气耦合特征进行分型,以此划分两类El Niño。对于其中海洋热力要素的选取,本文加以考虑了与表层和次表层海温均有关的上层热含量(HC)这一热力特征量。观测事实表明El Niño的海洋热力异常信号不仅出现在海洋表面,在深层也有重要体现。赵永平等[24]也曾强调指出SST仅是ENSO事件在海洋表面的表现形式,温跃层深度的变化是研究ENSO现象的最佳指标。HC与次表层海温存在显著而稳定的密切联系,可近似反映温跃层深度的变化[25]。在ENSO年际循环的形成机制,如"延迟振子"、"充放电"理论中,HC的变化也都直接起着重要的动力作用[26]。而且在海气相互作用研究中,HC比SST具有更好的稳定性,对天气气候持续发展的作用也更大[27]。以往的El Niño指数均基于SST,本文选取HC作为El Niño现象的研究指标,第四、五部分基于HC设计了一种新的El Niño海气耦合指数,并将该指数与各SST指数作比较,探究其优越性。新指数从海气耦合的角度对传统型El Niño和El Niño Modoki事件作区分,强调了新指数对不同类型El Niño大气响应特征的反映,为两类El Niño的监测和预测提供了一些新想法。

2 资料和方法

本文所用资料有:(1)英国气象局Hadley中心全球月平均海表温度(SST)资料,空间分辨率为1°×1°;(2)美国全球海洋资料同化系统(GODAS)月平均海表高度(SSH)和各层海温资料,水平分辨率为1°×1/3°,海温垂直方向共40层:5、15、25、35、45、55、65、75、85、95、105、115、125、135、145、155、165、175、185、195、205、215、225、238、262、303、366、459、584、747、949、1193、1479、1807、2174、2579、3016、3483、3972、4478 m;(3)NCEP/NCAR 2.5°×2.5°月平均再分析资料,所涉及的环流要素有海平面气压(SLP)和各层 u、v 风场;(4)NOAA向外长波辐射(OLR)数据,格点分辨率为2.5°×2.5°。分析时段为1980年1月至2010年12月。冬季是指前一年12月至当年2月。

一些海气要素的计算如下:(1)参考文献[26]计算热含量的方法,本文将366 m以上(27层)的垂直平均海温作为HC的估计值。(2)速度势 ϕ 的定义为 $u_\phi = \partial\phi/\partial x$;$v_\phi = \partial\phi/\partial y$。

本文采用MV-EOF方法对热带太平洋的主要海气耦合特征进行分型。MV-EOF方法在同时表现多个要素的空间分布以及各要素之间的空间联系方面具有优势,可用来探讨复杂系统中相互作用的耦合过程[21]。

3 热带太平洋海气耦合过程的主要模态

3.1 El Niño型和El Niño Modoki型的海气耦合特征

El Niño现象是热带太平洋海洋热力异常年际变化的主要模态,在这种强异常事件的作用下,各海气要素的时空变化往往是相互联系、相互协同的[18,28,29]。Wang和Wang[22]曾利用MV-EOF方法分析了1971—1992年期间的热带太平洋海气耦合过程,将前两个主模态称为暖事件成熟期和发展期,其中第二模态中的海温异常分布恰巧与90年代后的El Niño Modoki事件相似。由于Wang和Wang[22]分析时段较早且较短,为此本文再次利用该方法详细探讨了近30多年的热带太平洋海气耦合特征。为全面反映海洋热力—大气环流—降水三方面的耦合特征,本文对经标准化处理后的 7 个要素场(SST、HC、SLP、850 hPa 纬向风 u、850 hPa 经向风 v、200 hPa ϕ 和 OLR)进行 MV-EOF 展开。各模态的方差贡献如图 1a 所示,根据 North 等[30]的法则前两个模态较为显著且相互独立,方差贡献率分别达 22.5% 和 9.0%,可重点分析这两个模态的时空特征。图 1b 给出了这两个模态分别可解释上述 7 个要素场的平均方差贡献率,第一、二模态均较好地表征了 SST、HC、SLP 和 200-hPa ϕ 的年际变化,第一模态可解释这 4 个要素场的相对方差贡献均超过 20%,SST 最大,HC 次之;第二模态均超过 5%,此时 HC 最大,SST 次之。表明 SST、HC、SLP 和 200-hPa ϕ 这四个要素场整体耦合得较为紧密,较为稳定,而 850 hPa 风场和 OLR 的变化可能受到非 El Niño 因素的影响,平均解释方差较小。

前两个模态的空间分布如图 2 所示,与 Wang 和 Wang[22]结果相一致但也存在差异。第一模态在海洋热力—大气环流—降水三方面均表现为传统型 El Niño 分布特征,亦即赤道西太平洋 SST 冷异常"V 形"包围中东太平洋暖异常;HC 异常变化的空间分布(图 2c)与 SST 有所不同,其零线位置位于中太平洋,较 SST 偏东约 30 个经度,纬向上呈东西"跷跷板"式偶极振荡,其中西太平洋负异常中心非常突出,覆盖整个暖池区,与东太平洋正异常中心具有同等量值,达到 -0.8,该模态可解释这些区域热含量距平年际变化局地方差的 64% 以上。与此相适应,海平面气压异常呈典型南方涛动(SO)分布,赤道中太平洋(160°E~150°W)西风增强,一直可延伸至南美沿岸,暖池以西印度洋上出现明显东风距平。中太平洋南北半球的环流异常整体呈气旋性切变,菲律宾附近和澳大利亚西北沿岸均出现一个闭合的反气旋性环流异常,我国东南沿的偶极型海西南风异常增强。对流层高层 200 hPa 速度势和 OLR 场也都表现出东西反相分布(图 2e),中东太平洋上空辐散增强,西太平洋辐散减弱,对应着 Walker 环流的单圈异常,中东太平洋垂直上升运动加强,对流加深,进而降水增多,西太平洋情况相反。第一模态 El Niño 分布型海洋—大气—降水三者耦合紧密,较好地展现了 Wang 等[29]提出的"太平洋—东亚"遥相关型。第二模态所反映的海气耦合特征与 El Niño 型有显著差异,各要素变化分布总体由纬向偶极型变为纬向三极型,Wang 和 Wang[22]将此模态称为 ENSO 发展期,然而分析 SST、HC 及各气象要素异常的空间特征,发现该分布与 Ashok 等[3]提出的 El Niño Modoki 事件及其大气响应较为相似,主要特征表现为当中太平洋 SST 异常增温时,东、西太平洋却异常偏冷,中太平洋暖异常"马蹄形"包围东太平洋的冷异常,其中太平洋暖异常中心偏北,东太平洋冷异常中心偏南;HC 正异常中心也位于赤道中太平洋但偏向南半球,正异常可向西伸展到暖池东部 160°E,表明 El Niño Modoki 期间中太平洋温跃层有所加深。Ashok 等[3]研究指出中太平洋温跃层的变化不仅与暖池西部经向风辐合及暖池东部低层异常强西风对暖水的向东输送紧密联系,也与赤道东太平洋冷异常带来的东风异常有关。从 MV-EOF2 低层 850 hPa 风场上(图 2f)也可以看到,赤道西风异常大值区西移至暖池东部,位置较传统型 El Niño 偏西偏北,且东太平洋出现一定的东风异常。传统型 El Niño 期间的菲律宾和澳大利亚西北沿岸反气旋异常此时西移至印度洋,同时在西北和西南太平洋均出现气旋性异常环流,暖池西部经向风辐合明显。与海洋热力场、风场相适应,低层海平面气压 SLP 呈现中间低两边高的异常分布,闭合负异常中心位于中太平洋北部地区,与南方涛动分布相区别。高层 200 hPa 速度势负异常(辐散增强)中心由 140°W 西移至日界线附近,中心偏北,即意味着 Walker 环流异常上升支位置的西移,同时东、西太平洋上高层辐合加强,对应着赤道太平洋纬向上的双圈型 Walker 环流异常。在这种异常环流形势下,深对流异常区也西移至暖池东部且偏向北半球,与赤道西风异常区有较好的重合,暖池西南部和160°W以东太平洋上对流减弱,降水减少。由此可见,热带太平洋海气各要素 MV-EOF 前两个模态所反映的异常空间型与 El Niño 和 El Niño Modoki 型海气耦合过程有很好的对应,两者存在显著差异,这种差异也会给全球天气气候带来显著不同的影响。

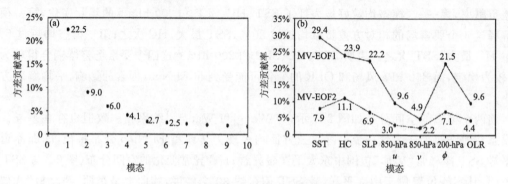

图 1　(a)MV-EOF 前 10 个模态的方差贡献率;(b)MV-EOF 第一、二模态分别解释各要素场的平均方差贡献率

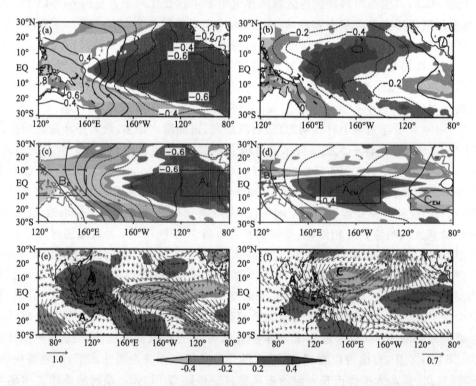

图 2　热带太平洋海气各要素 MV-EOF 展开的第一空间模态(a,c,e)和第二空间模态(b,d,f)
(a,b)SST(阴影)和 SLP(等值线);(c,d)HC(阴影,图中方框为下文热含量指数计算区)和 200 hPa 速度势(等值线);(e,f)OLR(阴影)和 850 hPa 风场(箭头)。C:气旋;A:反气旋

Wang 和 Wang[22]曾将 MV-EOF2 称为 ENSO 发展期,是由于 MV-EOF 展开的前两个模态的时间系数 PC2 与 PC1 有较高的超前相关,在超前 8 个月时,相关系数达到 0.50,超过了水平 0.01 的显著性检验,且 PC1×MV-EOF1 的量级是 PC2×MV-EOF2 的三倍。然而,

Ashok 等[3]利用 EOF、旋转 EOF、复 EOF 等方法分析了热带太平洋 SSTA 年际变化的主要模态,将 SSTA 第二模态(与本文图 2b 相似)空间型称为 El Niño Modoki 事件,计算 PC2 超前 PC1 的最大相关为 0.43,认为 SSTA 的前两个主模态不能互相解释。Li 等[31]与 Ashok 等[3]观点一致,进一步强调了 El Niño Modoki 事件与传统型 El Niño 的线性独立性,El Niño Modoki 并不能看作是传统型 El Niño 发生发展的一个位相。El Niño Modoki 事件的提出也已得到国内外众多学者的认可。因此,为进一步说明这两个海气耦合模态之间的相互联系,本文计算了 PC1 与 PC2(图 3a、b)的超前滞后相关,发现在 PC2 超前滞后 PC1 8 个月时相关系数达到最大,分别为 0.39 和 −0.49,从相关系数来看,El Niño 型前期有可能出现 El Niño Modoki 型海气耦合过程,而后期可能出现 La Niña Modoki 型海气耦合过程,后者可能性更大,与 Yu 等[15]研究指出的强中部型 La Niña 常常发生在强东部型 El Niña 之后的结论相一致。结合典型个例年[20]分析(图 3a、b),发现近 30 年 El Niño 和 El NiñoModoki 年分别与 MV-EOF 前两个主分量 PC1 和 PC2 大于 1.0 的峰值年有较好的对应关系。除 1982 年、1997/1998 年两个强 El Niño 前期存在 El Niño Modoki 型海气耦合过程,1994/1995 年 El Niño Modoki 后期表现出一定传统型 El Niño 海气耦合特征外,传统型 El Niño(El Niño Modoki)年与 PC2(PC1)大于 1.0 的峰值年基本无对应关系。而且 2000 年以来的 El Niño Modoki 之后均没有发生传统型 El Niño,如 2002/2003 年、2004/2005 年等,很难将 El Niño Modoki 看成是传统 El Niño 的发展位相。因此,尽管关于 El Niño Modoki 是否是 El Niño 的发生发展位相这一问题,国际上争论激烈且尚无统一定论,但近 30 年里热带太平洋上稳定确实存在上述两种类型的海气耦合过程,且第二模态空间型有其自身独特性,为此本文将满足上述两类海气耦合特征且能维持一段时间的事件分别称为传统型 El Niño 和 El Niño Modoki。

3.2 El Niño 型和 El Niño Modoki 型海气耦合模态的时间演变特征

利用 M-K 突变检验、Morlet 小波分析等方法简单分析了 PC1、PC2 的演变特征。从图 3 可以看出,传统 El Niño 型时间系数近 30 年来呈显著下降趋势,1998/1999 年(图 3c)发生突变,El Niño(La Niña)型发生频率明显减少(增加);而 El Niño Modoki 型时间系数近 30 年有上升趋势,尤其是 1990 年后(图 3d),El Niño Modoki 型海气耦合模态频繁出现。利用 Morlet 小波分析方法分析了 PC1、PC2 的周期变化特征(图略),El Niño 型和 El Niño Modoki 型海气耦合模态都表现出显著的年际变化特征,90 年代以前,PC1 具有显著的 4~5 a 的周期信号,之后表现为 2~6 a 的周期演变。与 PC1 相比较,PC2 早期 5~6 a 周期信号显著,90 年代后期显著周期缩短为 3 a 左右。以上分析可知,这两个模态的演变特征同样存在显著差异,90 年代后的差异更明显。

4 一种新的 El Niño 海气耦合指数

在实际工作中,PC1 和 PC2 的计算较为复杂,不便于直接作为监测两类 El Niño 海气耦合特征的指标,为方便研究和监测,可基于某一特征量(如 SST)来设计指数定量刻画 El Niño 现象,同时该指数应对 El Niño 的大气响应也有较好的反映。以往的 El Niño 指数大都基于 SST,Kao 和 Yu[5]和 Yu 等[15]曾指出两类 El Niño 期间,赤道中东太平洋 SSTA 存在一定空间重叠,有时较难直接区分。El Niño 的海洋热力异常信号不仅仅出现在海洋表面,在海洋次

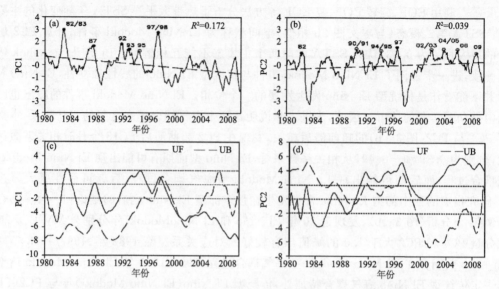

图 3 （a,b)热带太平洋海气各要素 MV-EOF 前两个模态的时间系数 PC1(a)、PC2(b)
（虚线：线性趋势线；R^2：趋势线的决定系数；年份：PC 大于 1.0 的峰值年；实心圆：传统
El Niño 年；空心圆：El Niño Modoki 年)；(c,d)PC1(c)和 PC2(d)的 M-K 突变检验

表层也有重要体现，甚至更明显[24,26,32-35]。本文选取与表层和次表层海温均有关的 HC 作为 El Niño 现象的研究指标，对两类 El Niño 的海气耦合过程进行监测和描述。在传统 El Niño 型和 El Niño Modoki 型海气耦合过程中，HC 异常同样表现出独特的空间分布（图 2c、d)。本文 3.1 节也分析指出前两个耦合模态可分别解释 HC 变化的较大方差，HC 在两类海气耦合特征分析中具有代表性，尤其是第二模态。

4.1 一种基于 HC 定义的 El Niño 新指数

为了更直接更客观地描述和区分两类 El Niño 的海气耦合特征，本文参考 Li 等[13] IEMI 指数的设计思想，基于 HC 定义两个新指数，其原则是 El Niño(El Niño Modoki)指数与 PC1 (PC2)相关最大，与 PC2(PC1)相关最小，即希望 El Niño(El Niño Modoki)热含量指数 HCEI (HCEMI)尽可能地表征 El Niño(El Niño Modoki)型海气耦合过程，而不反映 El Niño Modoki(El Niño)型的海气耦合特征，该两个指数所表征的两类事件彼此独立且区分明显。由此设计出的两个热含量指数 HCEI 和 HCEMI 的计算过程如下：

$$I_{HCE} = A_{HCA,A_E} - B_{HCA,B_E} \tag{1}$$

$$I_{HCEM} = A_{HCA,A_{EM}} - 0.2 \times B_{HCA,B_{EM}} - 0.4 \times C_{HCA,C_{EM}} \tag{2}$$

式中，I_{HCE}、I_{HCEM} 分别表示 HCEI 和 HCEMI，等号右端下标 HCA 为热含量距平，下标 A_E、B_E、A_{EM}、B_{EM}、C_{EM} 分别表示指数的计算区域（见图 2c、d，AE：10°S～10°N，120°～80°W；BE：10°～10°N，120°～160°E；AEM：15°S～5°N，170°E～140°W；BEM：0°～10°N，120°～140°E；CEM：20°～5°S，115°～80°W）；公式中各项系数通过多元线性回归分析所得。两个热含量指数之间的相关系数为 0.08，低于 KaoEP 和 KaoCP（相关为 −0.48，文献[15])、YuEP 和 YuCP(0.27，文献[15])以及 RenCT 和 RenWP(0.13，文献[14])，可以更明显地区分两类事件。同时，这两个指数与 PC1 和 PC2 的相关均分别高达 0.97，对两类事件的海气耦合特征具有较好的表征能力。

4.2 HC 指数表征的两类 El Niño 海气耦合特征

利用 HCEI 和 HCEMI 确立两类 El Niño：当标准化 HCEI(HCEMI)经 3 个月滑动平均后，达到阈值 1(0.7)倍的标准差并至少维持 5 个月时，构成一次 El Niño(El Niño Modoki)事件，其中 El Niño Modoki 的临界值较低，是由于传统型 El Niño 以强事件为主，而 El Niño Modoki 强度较弱，多为中等强度[16]。表 1 和图 4 分别给出了 1980—2010 年期间共发生的 5 次 El Niño 和 6 次 El Niño Modoki 事件。图 4 中 Year(0)表示发生当年，Year(+1)表示次年，结合图表可发现 5 次传统型 El Niño 均发生在 2000 年以前，6 次 El Niño Modoki 两次发生在 20 世纪 90 年代，其余 4 次均发生在 2000 年后。传统型 El Niño 强度明显强于 El Niño Modoki，盛期平均强度约是 El Niño Modoki 的 1.5 倍。传统型盛期平均在冬季，1982/1983 年和 1997/1998 年两次强 El Niño 事件较为典型，而 El Niño Modoki 盛期多为当年秋季。HCEI、HCEMI 所表征的 5 次 El Niño 和 6 次 El Niño Modoki 事件年份与前人的统计结果[16,20]基本一致，稍有区别的有 1993 年，Niño 3 和 Niño 3.4 指标认为是一般年份；2006 年，IEMI 指标认为是正常年，而 RenWP 和 HCEMI 均认为此次事件已达 El Niño Modoki 标准；1990 年 8 月—1992 年 7 月，曹璐[16]将此事件定义为东部型峰值、中部型传播，EMI 指标认为是 El Niño Modoki，IEMI 和 HCEMI 指数均认为该时段应分为两次独立事件：1990 年 8 月至 1991 年 11 月期间为 El Niño Modoki，1991 年 12 月—1992 年 7 月为传统型 El Niño。另外，在 1997/1998 强 El Niño 前期(1997 年 1 月—1997 年 4 月)也存在较大的 HCEMI 值，大于 0.7，但其持续时间较短，不作为 El Niño Modoki 事件列出。

表 1 El Niño 和 El Niño Modoki 事件特征

	序列	起始年月	结束年月	峰值年月(季节)
El Niño	1	1982-08	1983-08	1982-12(冬)
	2	1987-01	1987-11	1987-05(春)
	3	1991-12	1992-07	1992-03(春)
	4	1993-03	1993-07	1993-05(春)
	5	1997-05	1998-05	1997-12(冬)
El Niño Modoki	1	1990-08	1991-11	1991-10(秋)
	2	1994-07	1995-01	1994-09(秋)
	3	2002-01	2003-07	2002-10(秋)
	4	2004-02	2005-03	2004-08(夏)
	5	2006-07	2006-12	2006-10(秋)
	6	2009-08	2010-01	2009-11(秋)

为检验 HCEI/HCEMI 表征两类 El Niño 海气耦合特征的能力，同时进一步说明前两个模态所揭示出的两类海气耦合的主要特征不仅仅是 MV-EOF 数学分析的结果，我们对上述 5 次 El Niño 和 6 次 El Niño Modoki 事件各月海气耦合各特征量的距平(SST、HC、SLP、OLR、850 hPa 风场和 200 hPa 速度势及赤道太平洋垂直各层海温)进行了合成分析(图 5)。图中两类 El Niño 海气耦合各特征量的异常分布与 3.1 节分析较一致。在海洋热力要素场上，传统

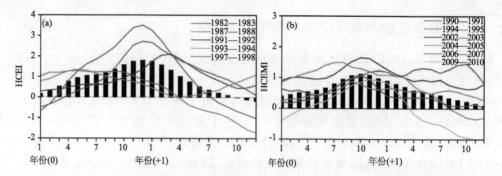

图4 各 El Niño 事件的(a)HCEI 指数值和(b)El Niño Modoki 的 HCEMI 指数值。
黑色曲线表示事件的指数平均值；横坐标的年份(0)表示发生当年,年份(+1)表示次年

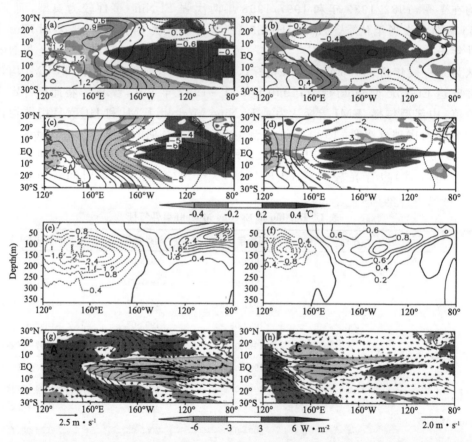

图5 El Niño(a,c,e,g)和 El Niño Modoki(b,d,f,h)海气各要素的距平合成。(a,b)SST(阴影,单位:℃)
和 SLP(等值线,单位:hPa);(c,d)赤道太平洋地区(10°S～10°N)海温距平的经度—深度剖面
(单位:℃);(e,f)HC(阴影,单位:℃)和 200 hPa 速度势(等值线,单位:5 10^5 m^2·s^{-1});
(g,h)OLR(阴影,单位:W·m^{-2})和 850 hPa 风场(箭头,单位:m·s^{-1})。C:气旋;A:反气旋

型 El Niño 期间,赤道中东太平洋 SST 持续增温明显;海洋各深度海温异常 OTA 也表现出特定的空间分布(图5c),温跃层东(西)部变深(浅),西太平洋冷异常在次表层向东部暖异常楔入,使得中太平洋海温异常上正下负,值得注意的是,海温异常极值中心均不位于海表,而是位

于次表层深度上。因此，HC异常呈东西偶极型，正负中心分别位于东西太平洋上。而El Niño Modoki期间，赤道中太平洋SST持续增温明显，西太平洋和东南太平洋出现较弱负距平；中太平洋的持续增温信号在海洋表层和次表层均有表现，最大正异常中心位于次表层 100~150 m深度上，西太平洋冷异常相比传统El Niño时面积显著减小，中心西移约20个经度，东太平洋次表层100~200 m深度上存在较弱负异常；相应的，HC暖异常主体位于中太平洋上，暖区西界可达160°E，冷异常主要分布于暖池西北部和热带东南太平洋。与SST相似，HC也呈纬向三极型，与传统型El Niño HC分布有显著差异，因此从热含量场也可以对两类事件作出较好的区分。与海洋热力异常相适应，热带太平洋上空的环流异常在两类El Niño期间同样呈现出各自所特有的空间分布。传统型El Niño事件中，SLP和200 hPa速度势均为东负西正的偶极反相分布；在中太平洋有异常强西风，强西风异常南北两侧经向风辐合明显且存在气旋性切变，西北太平洋上空此时呈现反气旋性异常环流。与传统型大气响应相区别，El Niño Modoki期间，SLP和200 hPa速度势负距平中心向西移至中太平洋地区，东、西太平洋上出现显著正距平；赤道低层西风异常大值区及其南北两侧经向风辐合区也西移至暖池上空，同时东南太平洋上出现小范围的东风异常，西北太平洋上表现为气旋性环流异常。以上分析验证了El Niño和El Niño Modoki型这两个海气耦合模态的客观存在，也表明HCEI和HCEMI指数可以较好地表征和区分两类El Niño及其大气响应。

对于几个有争议的个例年，如1993、1991、1992和2006年，利用HCEI和HCEMI指数可以作出较好的区分。如图6所示，1993年3月至当年7月和2006年7月至当年12月赤道中东太平洋SST均显著增加，暖距平中心达0.8℃以上，零散分布于整个中东太平洋，从SST场很难直接对这两年的El Niño事件进行识别。而HC异常有很大差异，1993年HC正（负）异常中心集中分布于东（西）太平洋上，为传统型El Niño；2006年，HC正异常大值区向西移至中太平洋，东太平洋虽然仍有HC正异常，但其南北两侧出现显著负异常，更倾向El Niño Modoki分布。环流场以SLP和OLR异常分布为例，也可看出这两年大气响应的差异，1993年SLP表现为典型的南方涛动，整个赤道中东太平洋对流加深，总体更接近El Niño型。而2006年，中太平洋SLP偏低，OLR负异常中心位于暖池东部，中东太平洋上对流大面积减弱，更接近El Niño Modoki型。

对于1991—1992年，图6e-h分别给出了1990年8月—1991年11月和1991年12月—1992年7月这两个时段的平均海气耦合特征，可以发现无论从SST、HC，还是SLP和OLR，都可以清楚地看到这两个时段海洋热力及大气响应空间分布的差异，所表现出的海气耦合特征与MV-EOF前两个模态较为相似。为此，应将1990—1992年区分为两次独立事件，前期为El Niño Modoki，后期为传统型El Niño。

5 各指数表征和区分两类El Niño海气耦合特征能力的比较

在气候研究领域，人们所关注的不仅仅是El Niño现象本身，往往更侧重于研究与海洋El Niño现象相协同的大气响应，也即文中第3部分所阐述的El Niño海气耦合过程，为此本文将热力异常与环流联系的紧密性作为指数的择优标准，比较选择一组既能识别El Niño和El Niño Modoki现象，又能紧密联系和准确区分两类El Niño大气响应的指数。表2给出了各El Niño指数与PC1、PC2和一些海气监测指数的相关系数，分析发现HCEI与代表El Niño

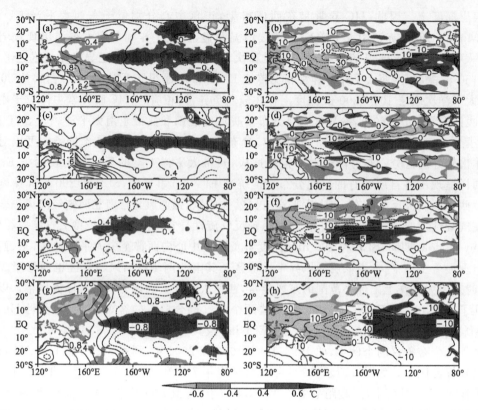

图 6 (a,b)1993 年 El Niño、(c,d)2006 年 El Niño Modoki、(e,f)1990/1991 年 El Niño Modoki 和 (g,h)1991/1992 年 El Niño 海气各要素异常分布。(a,c,e,g)SST(阴影,单位:℃)和 SLP (等值线,单位:hPa);(b,d,f,h)HC(阴影,单位:℃)和 OLR(等值线,单位:W·m^{-2})

型海气耦合特征的 PC1 相关最高,与代表 Modoki 型的 PC2 相关也达到最小,Niño 3、Niño 3.4 次之。值得注意的是,表征东部型 El Niño 的 KaoEP 指数并不是描述本文传统型海气耦合特征的较优指数,相反更倾向于反映 La Niña Modoki 的海气耦合过程,其原因主要是两类事件的划分存在一定差异,KaoEP 表征的东部型 El Niño,偏暖区主要位于赤道东太平洋冷舌区[5],西界在日界线以东 160°W,暖异常面积较传统型明显减小,有可能会与 La Niña Modoki 事件发生混淆。对于 El Niño Modoki 指数,与 PC2 一致(相关系数 0.70 以上)且与 PC1 正交(相关系数低于 0.30)较明显的指数有 IEMI 和 HCEMI,其中 HCEMI 与 Modoki 型 PC2 相关最高,与传统 El Niño 型 PC1 相关最小。另外发现 Niño 4 指数与 PC1 相关 0.79,而与 PC2 相关仅有 0.45,更侧重反映的是传统型海气耦合特征,而非 Modoki 型。可见,对比所有基于 SST 的 El Niño 指数,两个 HC 指数与两类 El Niño 海气耦合过程的联系都更紧密。

表2 各 El Niño 和 El Niño Modoki 指数的定义及其与 PC1、PC2 和一些海气监测指数的相关系数

	指数	定义	MVPC1	MVPC2	CPI	SOI	COI
El Niño	Niño 1+2	SSTA(10°S~0°,90°~80°W)	0.71	−0.31	−0.60	−0.47	−0.69
	Niño 3	SSTA(5°S~5°N,150°~90°W)	0.85	0.03#	−0.43	−0.67	−0.82
	Niño 3.4	SSTA(5°S~5°N,170°~120°W)	0.88*	0.23	−0.30##	−0.75*	−0.83*
	KaoEP[5]	联合回归 EOF(回归因子:Niño 4)的 PC1	0.46	−0.59	−0.72*	−0.26	−0.47
	RenCT[14]	RenCT=Niño 3−αNiño 4 α={2/5,Niño 3·Niño 4>0; 0,otherwise}	0.79	−0.10	−0.52	−0.61	−0.49
	YuEP[15]	OTA(100 m 以上,5°S~5°N,90°~80°W)	0.79	−0.17	−0.53	−0.61	−0.81
El Niño Modoki	HCEI	$I_{HCE}=A_{HCA,A_E}-B_{HCA,B_E}$	0.97**	−0.02##	−0.42#	−0.76**	−0.85**
	Niño 4	SSTA(5°S~5°N,160°E~150°W)	0.79	0.45	0.00	−0.70	−0.69
	TNI[36]	Niño 1+2−Niño 4	0.28	−0.61	−0.65*	−0.08##	−0.32
	EMI[3]	EMI=[SSTA]$_A$−0.5×[SSTA]$_B$−0.5×[SSTA]$_C$(A:10°S~10°N,165°E~140°W; B:15°S~5°N,110°~70°W; C:10°S~20°N,125°~145°E)	0.39	0.72	0.45	−0.46	−0.28
	IEMI[13]	IEMI=3.0×[SSTA]$_A$−2.0×[SSTA]$_B$−[SSTA]$_C$(A:10°S~10°N,165°E~140°W; B:15°S~5°N,110°~70°W; C:10°S~20°N,125°~145°E)	0.15#	0.74	0.57	−0.27	−0.06##
	KaoCP[5]	联合回归 EOF(回归因子:Niño 1+2)的 PC1	0.62	0.58	0.25	−0.60	−0.46
	RenWP[14]	RenWP=Niño 4−αNiño 3 α={2/5,Niño 3·Niño 4>0; 0,otherwise}	0.47	0.62	0.34	−0.47	−0.35
	YuCP[15]	OTA(100 m 以上,5°S~5°N,160°E~150°W)	0.49	0.76*	0.39	−0.54	−0.47
	HCEMI	$I_{HCEM}=A_{HCA,A_{EM}}-0.2\times B_{HCA,B_{EM}}-0.4\times C_{HCA,C_{EM}}$	0.08##	0.97**	0.70**	−0.23#	−0.15#

注:**、*:最大值、次大值;##、#:最小值、次小值

为进一步说明 HC 指标在表征两类 El Niño 海气耦合特征上的优越性,本文计算了各指数与中太平洋海平面指数 CPI[37]、南方涛动指数 SOI 以及太平洋对流涛动指数 COI[38] 的相关。Kug 等[6] 2009)曾指出,冷舌 El Niño 成熟期,赤道海平面异常呈东正西负的偶极"跷跷板"型,中太平洋为正负异常过渡区,而暖池 El Niño 事件中,正异常极值中心位于中太平洋上。Behera 和 Yamagata[37] 的研究也表明 21 世纪频繁发生的 El Niño Modoki 使得中太平洋海平面异常升高。计算发现,HCEMI 与 CPI 的相关最高,达 0.70,其次是 TNI 指数,而 Niño 4 指数与 CPI 几乎正交,无法反映 El Niño Modoki 与中太平洋海平面异常的关系,即无法表征海平面场上的 El Niño Modoki 现象。对于传统型 El Niño,CPI 与各指数均为负相关,相关性较弱的有 Niño 3.4、HCEI 指数,相关性最强是 KaoEP 指数,相关达−0.72,进一步表明 KaoEP 指数更侧重反映的是东太平洋增温,中太平洋增温不明显甚至降温的 El Niño,与传统

的中东太平洋增温型存在一定差异。海洋 El Niño 指数是否也可适用到大气环流场上,即该指数是否可表征大气中的"El Niño"现象,从各 El Niño 指数与 SOI 和 COI 的相关可以发现 HCEI 与这两个大气涛动现象联系最紧密,Niño3.4、Niño 3 指数次之。Niño 4 指数与 SOI 和 COI 也存在较高相关,分别为 -0.70 和 -0.69,Niño 4 指数所反映的大气响应特征更偏向传统型。比较其他 El Niño Modoki 指数,TNI[36]与 SOI 相关最小,IEMI 与 COI 相关最小,其次均为 HCEMI 指数,表明 HCEMI 也可以很好地区分出传统 El Niño 型和 El Niño Modoki 型大气响应。利用两个 HC 指数来探究 El Niño 和 El Niño Modoki 的气候效应是否也存在一定优势?由前文分析可知,两类 El Niño 对西北太平洋上空的环流形势有着不同甚至完全相反的影响,即意味着该区域气候在不同类型 El Niño 期间存在很大差异。以 El Niño 对西北太平洋夏季风(WNPSM)的影响为例分析,通过计算几种 El Niño 和 El Niño Modoki 指数与西北太平洋夏季风强度指数 WNPSMI[39]的相关(表 3),发现冬季 El Niño 使得 WNPSM 偏弱,而夏季发生 El Niño Modoki 时 WNPSM 偏强,与 Weng 等[7,8]结论吻合。冬季 HCEI 和夏季 HCEMI 与 WNPSM 存在较高的相关,均通过水平 0.01 的显著性检验,其中 HCEMI 较 IEMI 与 WNPSM 的相关更明显,且春季 HCEMI 对 WNPSM 的影响也较为显著,通过了水平 0.01 的显著性检验,表明春季 HCEMI 对西北太平洋夏季风的变化有着一定预测作用,而 IEMI 指数不能体现这一点。由此可见,利用这两个 HC 指标也可以更明显地直接揭示出两类 El Niño 所引起的气候异常。

表 3 前期冬春季和同期夏季传统型 El Niño 和 El Niño Modoki 指数与 WNPSMI 的相关系数

	冬	春	夏
Niño 3	$-0.54**$	-0.33	0.12
Niño 3.4	$-0.48**$	-0.27	0.34
HCEI	$-0.53**$	-0.32	-0.01
IEMI	0.14	0.25	$0.47**$
HCEMI	0.27	$0.46**$	$0.55**$

注:*、**:通过水平 0.05、0.01 的显著性检验

6 结论与讨论

较优的 El Niño 和 El Niño Modoki 指数应不仅可以客观描述 El Niño 现象本身,更要紧密联系两类事件所产生的大气响应。本文着眼于不同类型 El Niño 的大气响应,提取两类 El Niño 期间热带太平洋海气各要素相协同变化的特征,选取海洋上层热含量 HC 作为 El Niño 信号的研究指标,并基于 HC 设计了一种新的 El Niño 和 El Niño Modoki 指数,得到以下主要结论:

(1)利用 MV-EOF 方法对热带太平洋海气耦合过程进行分型,发现第一、二主模态中的 SST 和 HC 异常空间型与 Ashok 等[3]定义的两类 El Niño 较一致,各环流要素的空间配置较全面地反映了传统型 El Niño 和 El Niño Modoki 的海气耦合特征,且近 30 年的 El Niño 和 El Niño Modoki 年与 MV-EOF 前两个主分量 PC1、PC2 的峰值年有较好的对应关系。

(2)与传统型 El Niño 期间海气各要素东西反相的"跷跷板"式变化相区别,El Niño

Modoki事件中,SST持续增温区主要位于中太平洋,西太平洋和东南太平洋有较弱负距平,SST异常大值区可一直向下延伸到次表层深度,因此中太平洋HC的异常增加也较为明显。有趣的是,HC负异常在暖池北部较为突出,这可能与El Niño Modoki期间东太平洋HC冷异常北半球分支的西传堆积有关,有待进一步分析验证。与此热力特征相耦合,低层SLP和高空200 hPa速度势异常也均呈现出中间低两边高的异常分布,负距平中心向西移至中太平洋上空,即意味着Walker环流异常上升支的西移,赤道太平洋纬向上表现为双圈型Walker环流异常;同时赤道低层西风异常大值区及其南北两侧经向风辐合区西移至暖池上空,东南太平洋上出现小范围的东风异常,西北太平洋上空此时呈现气旋性环流异常。两类El Niño的海气耦合过程存在显著差异。

(3)以往定量表征El Niño年际变化的指标大多立足于SST或SLP,然而El Niño的海洋热力异常信号不仅出现在海洋表面,在海洋次表层深度上也有重要体现。HC的变化与次表层海温存在显著而稳定的密切联系,可近似反映温跃层深度的变化。在诸多ENSO年际循环的形成机制中,HC的变化也都直接起着重要的动力作用[26],而且HC比SST具有更好的稳定性,对天气气候持续发展的作用也更大[27]。本文直接从HC场出发,设计了两个新的El Niño指数HCEI和HCEMI。新指数全面描述了两类El Niño的海气耦合特征,其优越性表现在:对于有争议的El Niño个例年,如1991、1992、1993和2006年,新指标从热含量场异常作出了更明确的区分;HCEI(HCEMI)与PC1(PC2)的相关高达0.97,与PC2(PC1)的相关仅有-0.02(-0.09),几乎正交,该指数较好地表征了两类El Niño的海气耦合模态;HCEI与南方涛动、太平洋对流涛动等传统型El Niño大气响应的联系最为密切,HCEMI与南方涛动、太平洋对流涛动指数的相关均较小,两个HC指数对不同类型El Niño的大气响应特征也作出了较好的区分;以El Niño对西北太平洋夏季风的影响为例,分析发现利用这两个HC指标可以更明显地直接揭示出两类El Niño所产生的不同的气候影响。

新指数HCEI和HCEMI不仅从海洋热力场上更明显地区分了两类El Niño,而且更好地反映了大气对两类El Niño的响应特征,对今后不同类型El Niño的监测及其短期气候预测工作有所帮助。但该指数对于研究两类El Niño的气候效应和提高两类El Niño的可预测性有多大程度的帮助尚需深入探讨。

参考文献

[1] 符淙斌,弗莱彻 J."厄埃尔尼诺"(El Niño)时期赤道增暖的两种类型[J].科学通报,1985,8:596-599.
[2] Yeh S W, Kug J S, Dewitte B, et al. El Niño in a changing climate [J]. Nature, 2009, 461 (7263): 511-514.
[3] Ashok K, Behera S K, Rao S A, et al. El Niño Modoki and its possible teleconnection [J]. Geophys Res Lett, 2007, 112: C11007, doi: 10.1029/ 2006JC003798.
[4] Larkin N K, Harrison D E. On the definition of El Niño and associated seasonal average U. S. weather anomalies[J]. Geophys Res Lett, 2005, 32: L13705, doi: 10.1029/2005GL022738.
[5] Kao H Y, Yu J Y. Contrasting eastern-Pacific and central-Pacific types of ENSO[J]. J Climate, 2009, 22: 615-632, doi: 10.1175/ 2008 JCLI 2309.1.
[6] Kug J S, Jin F F, An S I. Two types of El Niño events: Cold tongue El Niño and warm pool El Niño [J]. J Climate, 2009, 22: 1499-1515, doi: 10.1175/2008JCLI2624.1.

[7] Weng H Y, Ashok K, Behera S K, et al. Impacts of recent El Niño Modoki on dry/wet conditions in the Pacific rim during boreal summer [J]. Climate Dyn, 2007, 29:113-129, doi:10.1007/s00382-007-0234-0.

[8] Weng H Y, Behera S K, Yamagata T. Anomalous winter climate conditions in the Pacific rim during recent El Niño Modoki and El Niño events [J]. Climate Dyn, 2009, 32 (5):663-674, doi:10.1007/S00382-008-0394-6.

[9] Ashok K, Yamagata T. Climate change: The El Niño with a difference [J]. Nature, 2009, 461:481-484, doi:10.1038/461481a.

[10] Taschetto A S, England M H. El Niño Modoki impacts on Australian rainfall [J]. J Climate, 2009, 22:3167-3174.

[11] Taschetto A S, Ummenhofer C C, Gupta A S, et al. Effect of anomalous warming in the central Pacific on the Australian monsoon [J]. Geophys Res Lett, 2009, 36:L12704, doi:10.1029/2009GL038416.

[12] Chen G H, Tam C Y. Different impacts of two kinds of Pacific Ocean warming on tropical cyclone frequency over the western North Pacific [J]. Geophys Res Lett, 2010, 37:L01803, doi:10.1029/2009GL041708.

[13] Li G, Ren B H, Yang C Y, et al. Indices of El Niño and El Niño Modoki: An improved El Niño Modoki index [J]. Adv Atmos Sci, 2010, 27(5):1210-1220.

[14] Ren H L, Jin F F. Niño indices for two types of ENSO [J]. Geophys Res Lett, 2011, 38:L04704, doi:10.1029/2010GL046031.

[15] Yu J Y, Kao H Y, Lee T, et al. Subsurface ocean temperature indices for central-Pacific and eastern-Pacific types of El Niño and La Niña events [J]. Theor Appl Climatol, 2011, 103:337-344, doi:10.1007/s00704-010-0307-6.

[16] 曹璐. 两类 ENSO 事件的监测及大气的响应 [D]. 南京:南京大学,2011:64.

[17] An S I. Conditional maximum covariance analysis and its application to the tropical Indian Ocean SST and surface wind stress anomalies [J]. J Climate, 2003, 16:2932-2938.

[18] Bjerknes J. Atmospheric teleconnections from the equatorial Pacific [J]. Mon Wea Rev, 1969, 97:163-172.

[19] Feng J, Li J P. Influence of El Niño Modoki on spring rainfall over South China [J]. J Geophys Res, 2011, 116:D13102, doi:10.1029/2010JD015160.

[20] Zhang W J, Jin F F, Li J P, et al. Contrasting impacts of two-type El Niño over the western North Pacific during boreal autumn [J]. J Meteor Soc Japan, 2011, 89 (5):563-569.

[21] Wang B. The vertical structure and development of the ENSO anomaly mode during 1979-1989 [J]. J Atmos Sci, 1992, 49 (8):698-712.

[22] Wang B, Wang Y. Development of El Niño during 1971-1992 [J]. Transactions of Oceanology and Limnology, 1994, 2 (2):26-40.

[23] Wang B, Wu Z W, Li Jianping, et al. How to measure the strength of the East Asian summer monsoon [J]. J Climate, 2008, 21 (17):4449-4463.

[24] 赵永平,陈永利,王凡,等. 热带太平洋海洋混合层水体振荡与 ENSO 循环 [J]. 中国科学 D 辑:地球科学,2007,37 (8):1120-1133.

[25] 陈锦年,王宏娜. 西太平洋暖池热状况变化特征及其东传过程 [J]. 海洋与湖沼,2009,40 (6):669-673.

[26] Hasegawa T, Hanawa K. Heat content variability related to ENSO events in the Pacific [J]. J Phys Oceanogr, 2003, 33:407-421.

[27] 吴晓芬,许建平. 海洋上层热含量的分布特征、变化模态及观测手段综述 [J]. 海洋学研究,2010,28 (1):46-54.

[28] Rasmusson E M, Carpenter T H. Variations in tropical sea surface temperature and surface wind fields associated with the Southern Oscillation/El Niño [J]. Mon Wea Rev, 1982, 110:354-384.

[29] Wang B, Wu R G, Fu X H. Pacific-East Asian teleconnection: How does ENSO affect East Asian climate [J]. J Climate, 2000, 13:1517-1537.

[30] North G R, Bell T L, Cahalan R F, et al. Sampling errors in the estimation of empirical orthogonal functions [J]. Mon Wea Rev, 1982, 110:699-706.

[31] Li G, Ren B H, Yang C Y, et al. Traditional El Niño and El Niño Modoki revisited: Is El Niño Modoki linearly independent of traditional El Niño? [J]. Atmospheric and Oceanic Science Letters, 2010, 3 (2):70-74.

[32] 李崇银, 穆明权. 厄尔尼诺的发生与赤道西太平洋暖池次表层海温异常 [J]. 大气科学, 1999, 23 (5):513-521.

[33] Chao J P, Yuan S Y, Chao Q C, et al. A data analysis study on the evolution of El Niño/La Niña cycle [J]. Adv Atmos Sci, 2002, 19 (5):837-844.

[34] Qian W H, Zhou Y F, Liang J Y. Potential contribution of maximum subsurface temperature anomalies to the climate variability [J]. Int J of Climatol, 2004, 24 (2):193-212.

[35] 黄玮, 曹杰, 黄潇, 等. 热带太平洋海温等20℃深度面的演变规律及其与ENSO循环的联系 [J]. 热带气象学报, 2011, 27 (1):82-88.

[36] Trenberth K E, Stepaniak D P. Indices of El Niño evolution [J]. J Climate, 2001, 14:1697-1701.

[37] Behera S, Yamagata T. Imprint of the El Niño Modoki on decadal sea level changes [J]. Geophys Res Lett, 2010, 37:L23702, doi:10.1029/2010GL045936.

[38] 郭艳君, 翟盘茂. 一个新的ENSO监测指标的研究[J]. 应用气象学报, 1998, 9 (2):169-177.

[39] Wang B, Wu R G, Lau K M. Interannual variability of the Asian summer monsoon: Contrasts between the Indian and the western North Pacific-East Asian monsoons [J]. J Climate, 2001, 14:4073-4090.

秋季北极海冰与欧亚冬季气温在年代际和年际尺度上的不同联系*

何金海 武丰民 祁莉
（南京信息工程大学 气象灾害教育部重点实验室，南京 210044）

摘要：北极海冰的急剧消融在近年来欧亚大陆频发的低温事件中扮演着关键角色。秋季北极海冰的偏少对应着冬季欧亚大陆的低温天气，然而二者的联系在年代际和年际两种时间尺度上存在显著区别。本文运用1979—2012年哈德莱中心第一套海冰覆盖率（HadISST1）、欧洲中心（ERA_Interim）的2 m温度、风场、海平面气压场、高度场等资料，分别研究了年代际和年际时间尺度上前期秋季北极海冰与欧亚冬季气温的联系。结果表明，欧亚和北极地区（0°~160°E,15°~90°N）的冬季气温具有显著的年代际和年际变化。在年代际尺度上，温度异常分布在21世纪初由北极冷-大陆暖转为北极暖-大陆冷。这一年代际转折与前期秋季整个北极地区的海冰年代际减少联系密切。秋季北极全区海冰年代际偏少对应冬季欧亚大陆中高纬地区的高压异常，有利于北大西洋的暖湿气流北上和北极的冷空气南侵，造成北极暖-大陆冷的温度分布；在年际时间尺度上，温度异常分布主要由第一模态的年际变化部分和第二模态组成，且第一模态包含的年际变率信号也存在显著的年代际变化。年际尺度上全区北极海冰对欧亚冬季气温的影响远不及位于北冰洋西南边缘的巴伦支海、喀拉海和拉普捷夫海西部（30°~120°E,75°~85°N）的关键区海冰影响显著。关键区内海冰的偏少会引发冬季的北大西洋涛动负位相，导致北大西洋吹往欧亚大陆的暖湿气流减弱和欧亚大陆中高纬地区的气温偏低。

关键词：秋季北极海冰；冬季欧亚低温；年代际；年际

1 引 言

21世纪以来，北极的气候系统正在发生着显著而快速的变化[1]。其中最引人注目的两个变化就是北极海冰的急剧消融[2]和北极地表的迅速升温[3-4]。2007年秋季，北极地区的海冰面积只有$4.17×10^6 km^2$，创造了有观测数据以来的最低记录。而2012年秋季这一记录又被刷新，海冰面积的最低值比2007年还要偏少18%[5]。Francis等[6]指出，1980年以来北极夏季海冰面积的减少量相当于美国面积的40%。与此同时，北极地表以超过整个北半球增温两倍的速度迅速变暖，这被称为"北极放大"（Arctic Amplification）。两者之间的正反馈作用可能是北极增暖和海冰消融的主要原因[7-8]。

---

* 本文发表于《地球物理学报》，2015年第58卷第4期，1089-1102。

在北极气候系统发生显著变化的同时,紧邻的欧亚大陆中高纬地区则频频经历严冬的肆虐。大规模寒潮频繁袭击欧亚大陆,暴雪、持续低温事件频发。2012年12月到2013年1月初,中国的平均地表气温创造了最近28年来的最低记录[9]。2012年1月下旬到2月上旬席卷欧亚大陆的极端寒潮致使一些地区的温度突破了三十年来的最低值,造成了600多人死亡[10]。此外,2008年中国南方的低温雨雪冰冻灾害和2009—2010年冬季欧亚大陆的极端低温事件也引发了日益增多的关注[11-13]。

欧亚大陆低温频发的原因可能是非常复杂的。而北极海冰的减少及与之对应的极地大气环流改变被认为是造成欧亚冬季低温的重要原因之一。这一因子由于近年来北极地区的显著气候变化正在受到越来越多的关注。许多学者以前期秋季北极海冰的异常作为切入点研究欧亚大陆的冬季低温天气。Honda等[14]指出,冬季欧亚大陆的低温天气与前期秋季北极海冰的减少有显著联系;Wu等[15]认为,前期秋季北极海冰的偏少是触发2008年中国华南雨雪冰冻灾害的重要因子;Liu[16]和Ghatak等[17]的研究则表明,北半球冬季的低温和暴雪均受到前期秋季北极海冰的影响;武炳义等[18]的研究结果显示,秋季的北极海冰异常可以影响西伯利亚高压进而造成大陆中高纬地区的低温。这些研究都表明,前期秋季的北极海冰对欧亚大陆冬季温度有显著影响。

然而,虽然有不少研究指出前期秋季的北极海冰可以作为冬季欧亚温度的一个预测因子,但对其中的年际和年代际两种时间尺度的影响却往往未加区分,这可能对研究结果产生不可忽视的影响。许多研究都显示,一个气候现象的影响因子在年际和年代际尺度上可能存在显著不同,对年际和年代际尺度的影响因子分开研究是十分必要的[19-21]。冬季的北极增温与欧亚大陆降温是否也同时存在年代际转折和年际变化?它们与北极海冰异常的联系机制在两种时间尺度上是否存在显著区别?对这一问题的研究有助于进一步理解秋季北极海冰对欧亚大陆冬季温度的影响机理,并为欧亚冬季的温度预测提供可靠线索。本文将在前人工作的基础上,分析1979—2012年34年来欧亚大陆及其北部北极地区的温度变化特征,对其中包含的年代际和年际两种时间尺度的信号进行分离,并分别研究其与秋季北极海冰之间的联系。此外,文章还通过分析大气环流异常初步探讨了其中的可能机制。

2 资料和方法

所用资料:哈德莱中心的海冰覆盖率资料,单位为百分率(%),水平分辨率 $1.0°\times1.0°$[22];欧洲中期天气预报中心(ERA_Interim)的2 m气温(SAT)、海平面气压(SLP)、850 hPa风场、500 hPa和200 hPa高度场资料,水平分辨率 $1.5°\times1.5°$[23]。本文的北大西洋涛动指数来自李建平个人网站(http://ljp.lasg.ac.cn/dct/page/1),其海平面气压资料来自NCEP。以上资料的时间都是1979—2012年共34年。文中秋季是指9—11月,冬季是指当年12月到次年2月。在计算区域平均的海冰序列时,每个点的资料都乘以了权重因子 $\cos\theta$(θ为纬度),以消除纬度对网点面积的影响。

本文虽然主要研究欧亚大陆的冬季低温,但由于欧亚大陆冬季气温不仅由区域性环流决定,且受极地的影响极大[24]。因此在对温度场进行回归分析和EOF分析时,选取的范围为 $0°\sim160°E,15°\sim90°N$ 区域,将欧亚大陆北端的北极地区也包含在内。为方便讨论,将所选区域称为欧亚和北极地区,简称EAA(Eurasia and Arctic)地区。需要指出的是,由于非均匀网点必须内插成等面积网点后再进行EOF分析才能避免产生失真的特征向量场[25-26]，故本文在对温度场进行EOF分析时，借鉴了Chung等[27]的方法，采用权重因子$\sqrt{\cos\theta}$（θ为纬度），对均匀经纬格点网上的温度序列进行了订正。该权重因子也常用于关于NAO的分析[28]。

采用高斯滤波[29]的方法，将秋季海冰标准化序列进行9年平滑，将9年平滑结果作为该序列年代际指数，将原序列与年代际指数的差定义为该序列的年际指数。文中分离EOF时间系数等序列的年代际与年际信号时也采用了相同的方法。利用相关分析分别找出了年代际和年际两种时间尺度上影响气温的海冰关键区，并对海平面气压场和中高层高度场和风场进行了回归和合成分析，以探讨秋季海冰影响冬季气温的可能机制。此外，在研究序列的周期时，采用了小波分析方法。

由于本文是研究海冰减少的影响，故在回归分析的图中显示了负的回归系数，表示海冰减少一个标准差对气象场的影响。本文采用的显著性检验为t检验，且采用了Chen[30]和施能[31]的方法考虑了有效自由度。

3 秋季全区北极海冰在两种尺度上对EAA冬季温度的不同影响

许多研究都表明，前期秋季北极海冰的偏少对应冬季欧亚大陆的低温天气[14,16,18]。由于秋季是北极的结冰季，北冰洋边缘广大地区的海冰有很大的年代际变率[32]和年际变率。故在研究秋季北极海冰对冬季欧亚低温的影响时，一般选取整个北极地区的海冰作为影响因子。

如图1a所示，将全区北极(0°～360°E,66.5°～90°N)的海冰标准化距平序列作为全区北极海冰指数，可以发现秋季北极海冰具有显著的年代际变化。在1979—2000年的22年，秋季

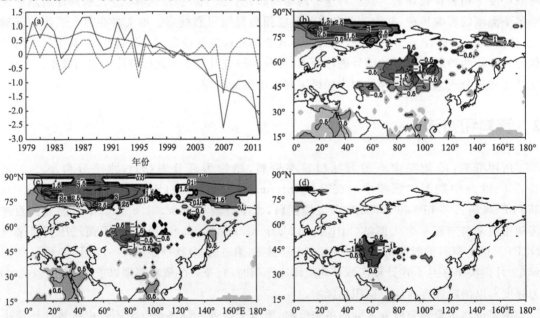

图1 (a)秋季全区北极海冰指数（实线）、全区年代际指数（点虚线）和全区年际指数（长虚线）；b,c,d分别为冬季SAT回归到秋季全区北极海冰指数、全区年代际指数和全区年际指数的负回归系数分布；阴影区域超过了0.10的显著性水平

海冰标准化序列只有三年是负距平,且绝对值都小于 0.5 个标准差;之后的 12 年则全部是负距平。同时,该指数也具有显著的年际变率。那么,在年代际和年际两种时间尺度上,全区北极海冰对欧亚大陆冬季温度的影响有何不同呢?为研究这一问题,把全区北极海冰指数分成全区年代际指数和全区年际指数(详见资料方法)。将温度场分别回归到三个指数上,如图 1 所示。可以发现,温度场对全区海冰指数与全区年代际指数的回归结果非常类似,都呈现一种北极暖—大陆冷的温度异常分布。北极的暖中心位于新地岛以西的北冰洋地区,而大陆的冷异常位于以贝加尔湖为中心的欧亚大陆腹地。然而,对海冰年际指数的回归结果则与前两者存在显著不同。只在中亚地区有显著的温度负异常分布。而在中高纬的欧亚大陆大部分地区都几乎没有显著的温度负异常。

以上的分析表明,全区秋季北极海冰对冬季欧亚温度的影响,在年代际和年际两种时间尺度上存在显著差异。在年代际尺度上,全区秋季的北极海冰偏少可以造成北极暖—大陆冷的温度分布;而在年际尺度上,全区秋季北极海冰偏少对欧亚腹地温度的影响是不显著的。那么,是不是在两种时间尺度上影响欧亚冬季温度的海冰在空间分布上存在显著区别呢?为了研究这一问题,我们将 EAA 地区的温度进行 EOF 分析,分离温度变化的年代际和年际信号。然后分别分析在两种时间尺度上影响欧亚冬季温度的秋季海冰关键区。

4 EAA 地区冬季气温与秋季北极海冰在年代际和年际尺度上的不同联系

4.1 EAA 地区冬季气温的年代际和年际变化特征

为了分析 EAA 地区冬季温度的主要变化模态及特征,我们选取 $0°\sim160°E,15°\sim90°N$ 范围,对 1979—2012 年 34 年冬季的 SAT 距平场进行了 EOF 分析。得到了 EAA 地区冬季 SAT 变化的主要模态。其中前两个模态的方差贡献分别为 41.6% 和 22.8%,两者都通过了 North 准则显著性检验,是相互独立且显著的模态。

图 2 给出了 EOF 分析的前两个模态及其时间序列。如图所示,第一模态的最显著特点是北极与中高纬的欧亚大陆温度呈反向变化,这与图 1b、c 中的温度异常分布非常类似。也就是说,当北地群岛以西的喀拉海和巴伦支海 SAT 偏高时,欧亚大陆的西伯利亚、蒙古和中国的东北及西北等广大地区呈显著温度负异常,反之亦然。从第一模态的时间系数上看,该模态存在明显的年代际转折:即在 21 世纪初时间系数由负转正。从 20 世纪 70 年代末到 21 世纪初(1979—2001)基本为负指数位相,负指数位相年占 73.9%;之后(2002—2012)则以正指数位相为主,正指数位相年占 81.8%。其中 2004—2012 年的 9 年时间系数全为正值。2004 年、2005 年和 2011 年的第一模态解释方差都达到了 80% 以上,2012 年也达到了约 60%。图 2 中 PC1 的年代际指数(图 2b,红色虚线)也清楚地表明:最近的 10 年,北极暖大陆冷的温度异常分布有显著增强的趋势。

为了进一步验证这一年代际变化,将温度距平场先用高斯滤波去除 9 年以下年际信号后再做 EOF,发现第一模态跟未做滤波的 EOF 第一模态型式非常类似(图略),且方差贡献可达 70.4%。这表明未做滤波的 EOF 第一模态主要反映了年代际信号。1979—2001 年与 2002—2012 年的冬季温度合成差值分布(图略)也与第一模态的分布型非常类似,这些都进一步说明

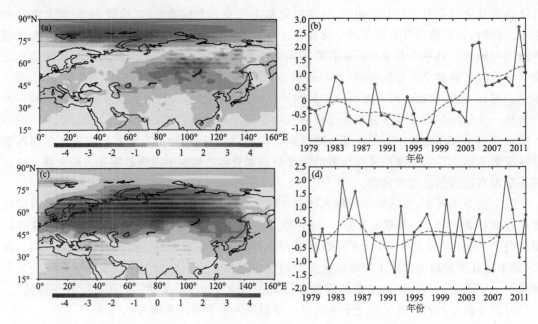

图2 上行为1979—2012年EAA冬季SAT EOF分析第一模态的空间分布型(a)及标准化时间系数(b)。图中阴影为SAT(单位:℃),实线为标准化的时间序列,虚线为时间序列的9年平滑。下行同上行,但针对第二模态

EAA地区的SAT异常具有显著的年代际变化。

第一模态的时间序列不仅具有明显的年代际信号,还具有显著的年际变化。图3a显示了对第一模态时间序列PC1进行小波分析的结果。曲线表示置信度为90%下的边界效应,即当时间因子落在曲线范围之内时,边界效应对小波变换谱的影响不超过10%。由图3a可以明显看出,PC1的周期主要为4~7年信号,这一周期信号贯穿于20世纪80年代初到21世纪初,且在90年代以后变得显著。在2004年之后,主导周期逐渐变为2~4年。这说明,第一模态包含的年际变率周期也存在年代际变化。除了这两种周期信号外,其他的周期信号均不显著。由于样本长度较短,功率谱中未能显示显著的年代际周期。

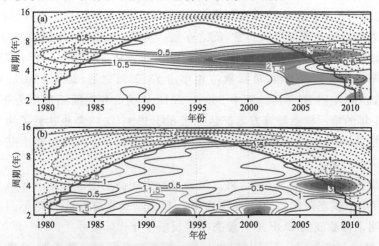

图3 PC1(a)和PC2(b)的小波功率谱,阴影区通过了0.10的显著性检验

第二模态的显著特点是,冷异常控制了 40°N 以北的欧亚大陆及大陆北侧的北冰洋边缘,冷中心位于 60°～75°N 的欧亚大陆高纬地区。欧亚大陆的中低纬地区(15°～40°N)则是暖异常分布。在第一模态中,大陆的冷中心基本位于乌拉尔山以东,纬度上则基本位于 60°N 以南的中纬地区;而第二模态的冷中心位于 45°N 以北,在东西方向上则基本覆盖了欧亚大陆全境。同时,北极地区虽然也有暖异常,但范围很小,强度也很微弱。实际上,这一模态的温度异常分布与北大西洋涛动(NAO)对 EAA 地区 SAT 的影响非常类似(图略),第二模态的时间系数 PC2 与 NAO 指数的相关系数高达－0.78,去掉 9 年以上年代际信号后,该相关高达－0.80。这说明,第二模态是由北大西洋涛动主导的。已有很多的研究表明,冬季 NAO 对欧亚大陆的温度具有重要调制作用[33-37]。从时间系数上看,第二模态并没有显著的年代际转折,是一种年际变化模态。对第二模态时间序列 PC2 的小波分析则表明(图 3b),第二模态的周期以 2～4 年信号为主,但周期随着时间显著变化,并没有一个贯穿整个时间段的显著周期。在 1994 年和 2001 年前后有显著的 2 年左右周期信号存在。2005 年后 4 年左右的周期信号变得显著。

前两个模态是 EAA 地区冬季 SAT 的最主要模态,累计方差贡献达到了 64.4%。以上的分析表明,EAA 地区的冬季温度既有显著的年代际变化,又有显著的年际变化。年代际变化主要是第一模态在 21 世纪初由北极冷—大陆暖转为北极暖—大陆冷;年际变化主要由第一模态中的年际变化部分和第二模态的年际变化组成。

4.2 秋季北极海冰与冬季欧亚温度在年代际尺度上的联系

上文的分析已表明,秋季全区北极海冰的偏少对应北极暖—大陆冷的温度分布。这正是 EOF 第一模态的分布特征。秋季全区海冰指数与第一模态时间序列 PC1 的相关系数高达－0.56,二者年代际指数的相关系数更是高达－0.92,均远高于 0.01 的显著性水平。这说明,全区北极海冰的年代际减少与 EOF 第一模态的年代际增强具有很好的相关。那么,影响第一模态年代际变化的秋季海冰异常有着怎样的空间分布呢?为此,将 PC1 和 PC1 的年代际指数分别与秋季海冰场做相关,相关系数分布如图 4 所示。由图可知,海冰相关显著区与近 30 年来秋季北极海冰线性趋势的分布基本一致,即在年代际尺度上与欧亚温度存在显著负相关的海冰区域正是在近 30 年来海冰具有显著减少趋势的区域。这说明,那些与第一模态在年代际尺度上相关不显著的海冰区域,主要是因为其在 30 年来比较稳定,减少趋势不显著。故用整个北极的秋季海冰标准化距平序列作为秋季海冰指数,可以代表与 PC1 在年代际尺度上相关显著区域的海冰变化。

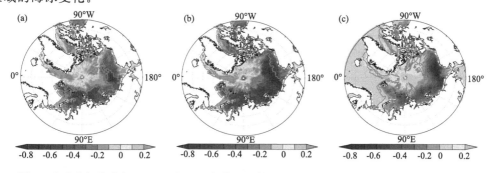

图 4 秋季北极海冰与 PC1(a)和 PC1 年代际指数(b)的相关系数分布;(c)为 1979—2012 年秋季北极海冰的线性趋势分布(单位:%/10 年)。绿色等值线内区域通过了 0.05 的信度检验

上文分析了 EOF 第一模态与北极海冰在年代际尺度上的关系。下面我们将对冬季的大气环流场进行回归和合成分析,以寻找二者联系的可能机制。合成分析中,根据全区北极海冰的年代际变化将 1979—2012 年的 34 年分为两段:一段为 1979—2000 年共 22 年,为海冰偏多阶段;一段为 2001—2012 年共 12 年,为海冰偏少阶段。图 5 分别为高度场(海平面气压场)和风场回归到海冰年代际指数的负回归系数和少冰年与多冰年的差值分布。

在冬季低层大气的气候平均状态下,欧亚大陆被强大的西伯利亚高压所控制,在大陆东侧的北太平洋是阿留申低压,而在大陆西侧则是冰岛低压。这三个大气活动中心对北半球冬季气候都有非常重要的影响。西伯利亚高压与来自高纬度地区的冷空气活动以及北大西洋涛动和北极涛动均有联系,其强度对亚洲的气候有显著的影响。一般来说,西伯利亚高压越强,欧亚大陆的冷空气活动越频繁,温度也就越低[18,38];冰岛低压和阿留申低压与北半球温度及 500 hPa 西风急流均存在密切联系:弱的阿留申低压和冰岛低压对应 500 hPa 西风急流偏弱和欧亚大陆的冷冬[39-41]。回归与合成的结果一致显示,在地表(图 5e、f),欧亚大陆的中高纬都被高压异常控制。高压异常中心位于大陆西北部的乌拉尔山一带。此外,北太平洋地区也有显著的高压异常。这种气压场异常分布使得西伯利亚高压加强,阿留申低压和冰岛低压则显著减弱。欧亚大陆中高纬地区的低层大气中有显著的东北风异常(图 5g、h),这有利于阻塞形势的形成和北极冷空气的南侵,造成欧亚大陆的低温天气。

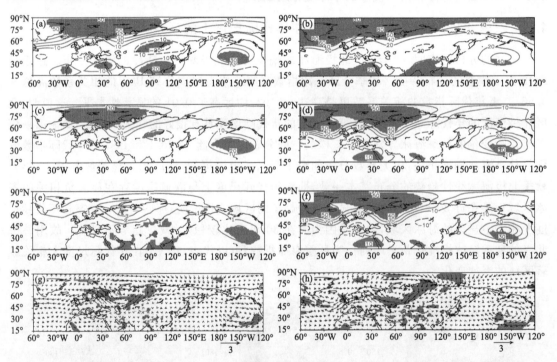

图 5 冬季 200 hPa 高度场(a、b)、500 hPa 高度场(c、d)、海平面气压场(e、f)和 850 hPa(g、h)风场异常。(a,c,e,g)为对北极海冰年代际指数的负回归系数分布;(b,d,f,h)为冰少年与冰多年的合成差值(2001—2012 减 1979—2000)。阴影区显著性水平通过了 0.05

在中高层的气候平均态中,除东亚大槽的偏北气流控制东亚地区外,欧亚大陆上空基本被西风气流控制。西风的强弱与 EAA 的温度密切相关。当西风强时,极涡加深,冷空气被捕获在极地地区,欧亚大陆中高纬地区偏暖;而西风偏弱时,急流南扩,极涡减弱,冷空气频繁暴发,欧亚大陆中高纬地区偏冷。由地转风原理,西风的强弱与极地和大陆的气压梯度成正比,位势高度差越大,西风越强,反之越弱[42]。如图5所示,在200 hPa和500 hPa的中高层,北冰洋西南侧有显著的高压异常,贝加尔湖上空则有位势高度负异常,在北太平洋有显著的位势高度正异常。即在北极-欧亚大陆-北太平洋有一个显著的"+""-""+"的波列分布。这种位势高度异常使得中高纬的位势高度差减小,西风气流偏弱,经向活动加强,这有利于北极的冷空气南下,造成欧亚大陆的冷冬。

4.3　年际尺度上影响冬季欧亚温度的海冰关键区

秋季北极海冰和冬季EAA温度第一模态在年代际尺度上具有显著联系,其中的机制是非常复杂的,可能受到全球变暖等因素的共同影响。如果把秋季北极海冰作为EAA冬季温度的预测因子,必须在年际尺度上进行研究。

由上文可知,EAA地区冬季温度的年际变率主要由第一模态变率的年际变化分量和第二模态变率组成。故为了考察秋季海冰与冬季SAT在年际尺度上的相关,将PC1的年际指数与PC2分别与秋季海冰场做相关,如图6所示。可以发现相关分布与年代际尺度的相关分布(图4)存在显著区别。只在喀拉海、巴伦支海北部以及拉普捷夫海西北部有显著的负相关分布。而在东西伯利亚海和波弗特海等海域甚至变成了微弱的正相关。这说明,年代际和年际两种时间尺度上,影响冬季EAA温度的北极海冰区域存在显著区别。波弗特海、楚科奇海和东西伯利亚海的海冰与EAA地区冬季SAT的相关只在年代际尺度上显著,而在年际尺度上不显著;而巴伦支海、喀拉海和拉普捷夫海西北部的海冰在年代际和年际两种时间尺度上都对EAA地区的SAT分布有显著影响。

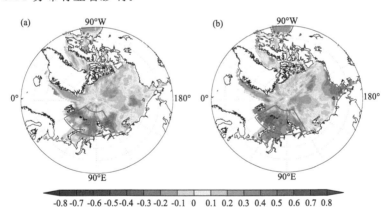

图6　秋季北极海冰分别与PC1的年际指数(a)和PC2(b)的相关系数分布,绿色等值线内区域通过了0.10的信度检验,红色实线包围的范围为关键区(30°～120°E,75°～85°N)

基于以上讨论,我们划定了30°～120°E,75°～85°N区域为秋季北极海冰在年际尺度上影响EAA地区冬季SAT的关键区(图6),将该区域的海冰标准化距平定义为关键区海冰指数,并采用上文中的滤波方法分别定义了关键区年代际指数和关键区年际指数,如图7a所示。关键区海冰指数与全区海冰指数的相关系数为0.82,二者在年代际尺度上的相关高达0.94,而在年际尺度上的相关则为0.65。这表明,在年代际尺度上,关键区海冰的变率与全区北极海冰高度一致。而在年际尺度上,二者存在显著区别,需要分开讨论。冬季 SAT 回归到关键区海冰指数的结果(图 7a)呈明显的北极暖—大陆冷的分布,且大陆的冷异常覆盖范围更大。SAT 对关键区海冰年代际指数的回归结果(图 7c)与图 1b 非常类似,这是因为关键区的海冰与全区海冰在年代际尺度上具有极高的相关。在年际尺度上,关键区海冰指数与 SAT 也有非常显著的联系,如图 7d 所示,北极的暖异常范围有所减小,而大陆的冷异常区域依然覆盖了欧亚大陆中高纬的广大地区。这进一步表明,在年际尺度上,影响欧亚大陆冬季温度的海冰主要位于 30°~120°E,75°~85°N 的关键区内而不是整个北极地区。

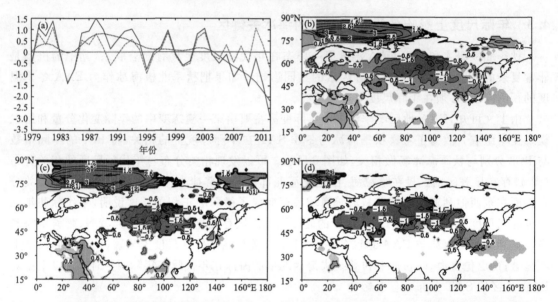

图 7 (a)为秋季关键区(30°~120°E,75°~85°N)海冰指数(实线)、关键区年代际指数(点虚线)和关键区年际指数(长虚线);(b,c,d)分别为冬季 SAT 回归到秋季北极关键区海冰指数、关键区年代际指数和关键区年际指数的负回归系数分布;阴影区域超过了 0.01 的显著性水平区域

下面用回归和合成的方法分析关键区海冰年际指数对高度场(海平面气压场)的影响。在合成分析中,依据关键区海冰年际指数,以 0.7 倍标准差为标准,选取大于 0.7 倍标准差的年份:1980、1982、1988、1998、2003、2010 年共 6 年为多冰年;选取小于 −0.7 倍标准差的年份:1984、1985、1995、2012 年共 4 年为少冰年。

与年代际结果(图 5)类似的是,在年际尺度上,高纬地区的低层也有显著的高度场(SLP)正异常(图 8e、f)。但与年代际结果相比,大陆上的高压异常中心西移到欧亚大陆西北角和北冰洋西南边缘的挪威海及格陵兰海。且在中纬度的亚索尔群岛附近有一个显著的低压异常。这是一种典型的北大西洋涛动(NAO)负位相分布。北大西洋涛动对北大西洋区域[43-45]和欧亚大陆的气候[33-34]具有重要影响,尤其是对欧亚大陆的冬季温度变化具有最大贡献[36-37,46]。当 NAO 正位相时,北大西洋吹向欧亚大陆的暖湿气流偏强,经向活动偏弱,欧亚大陆温度偏高;反之当 NAO 负位相时,北大西洋吹向欧亚大陆的西风减弱,经向活动增强,容易引发阻塞活动和欧亚大陆低温(图 8g、h)。实际上,由表 1 可以看出,关键区海冰指数与冬季 NAO 指数的相关高达 0.46,在扣除年代际信号后,二者的相关依然有 0.41,均通过了 0.05 的信度检验。相比之下,全区与非关键区的海冰指数与 NAO 的相关均不显著。此外,在年际尺度上,关键区海冰指数与 PC1 和 PC2 都有较好的相关(0.10 显著性水平),而非关键区海冰只在年代际尺度上与 PC1 有很好的相关,在年际尺度上,非关键区与 PC1、PC2 和 NAO 指数的相关均不显著。偏相关的结果也表明,在年际尺度上,关键区海冰与欧亚冬季温度和 NAO 联系明显优于全区海冰和非关键区海冰。

500 hPa 和 200 hPa 的位势高度异常依然是显著的 NAO 负位相(图 8a~d)。在欧亚大陆上空有显著的低压异常,其中心位于贝加尔湖地区。这种高度场异常使得中高层极地与大陆的气压梯度减小,西风减弱,北风加强,有利于北极的冷空气向南入侵造成欧亚大陆的低温。

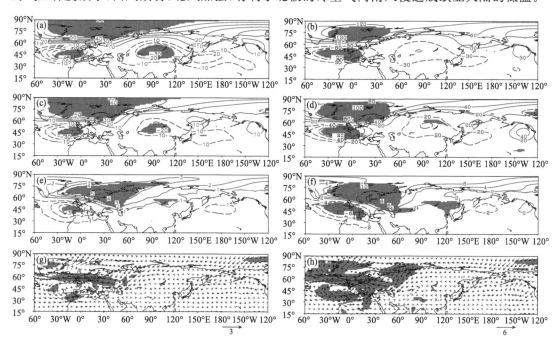

图 8　同图 5,但回归和合成均针对关键区(30°~120°E,75°~85°N)海冰的年际指数
"C"表示气旋性环流中心

表 1　秋季各北极海冰指数与 PC1、PC2 和 NAO 指数的相关和偏相关系数

	全区海冰	关键区海冰	非关键区海冰	关键区海冰*	非关键区海冰*
PC1	**−0.56**/0.02	**−0.59**/−0.32	**−0.50**/0.17	−0.39/**−0.41**	−0.17/0.31
PC2	−0.15/−0.25	**−0.32**/**−0.32**	−0.08/−0.16	**−0.36**/−0.28	0.20/−0.06
NAO	0.29/0.22	**0.46**/**0.41**	0.21/0.10	**0.44**/**0.40**	0.16/−0.05

注:* 表示偏相关,其中与关键区海冰的偏相关控制非关键区海冰不变,与非关键区海冰的偏相关则控制关键区海冰不变。斜线前为原指数的相关,斜线后为二者年际指数的相关。粗体表示通过了 0.05 的显著性水平

5　结论与讨论

本文利用 1979—2012 年哈德莱中心的海冰资料、欧洲中心的再分析资料,探讨了 EAA 地区冬季 SAT 的年代际与年际变率特征,并分析了其在年代际和年际两种尺度上与前期秋季北极海冰的联系,主要结论如下:

(1)EAA地区的冬季SAT存在显著的年代际变化和年际变化。年代际变化主要表现为第一模态在2001年左右由北极冷—大陆暖的分布型变为北极暖—大陆冷的分布型;年际信号主要是第一模态中的年际振荡部分和第二模态的年际变化。且第一模态包含的年际变率主导周期在2004年以后由4~7年缩短为2~4年。

(2)在年代际尺度上,第一模态时间序列PC1与几乎整个北极边缘地区的海冰都有显著负相关。全区北极秋季海冰的年代际偏少会引发低层欧亚大陆中高纬地区的高压异常,以及中高层北极上空的位势高度正异常和贝加尔湖上空的位势高度负异常,这种高度场分布可以导致欧亚大陆的北风加强和气温偏低。

(3)在年际尺度上,影响EAA冬季SAT的海冰异常主要分布在巴伦支海、喀拉海和拉普捷夫海西部(30°~120°E,75°~85°N)的关键区内。全区北极海冰年际指数与冬季欧亚温度的联系不显著。关键区海冰偏少导致的高度场异常呈典型的NAO负位相分布,北大西洋吹向欧亚大陆的暖湿气流显著减弱和欧亚大陆北风的加强导致欧亚大陆中高纬地区温度偏低。

由于冬季北冰洋海冰标准差最大的海域位于巴伦支海和喀拉海一带,其他海域的海冰覆盖率很高且较稳定,故在研究同期海冰对冬季北半球大气环流的影响时,许多研究者都选取了巴伦支海和喀拉海海域[47-49]。而秋季北极海冰的标准差大值区覆盖了北冰洋边缘的大部分海域,在研究前期秋季海冰对冬季大气环流的影响时,研究者们往往选取整个北极的海冰进行研究[16,50-51]。本文的研究结果则表明,研究年际尺度上前期秋季北极海冰对欧亚冬季SAT影响时,关键区海冰作为预测信号比整个北极地区的海冰具有更好的效果。

无论是在年代际尺度还是在年际尺度上,北大西洋涛动对欧亚大陆甚至整个北半球的冬季温度都有极为关键的影响[46,52]。然而,秋季的NAO与冬季的NAO基本没有相关(相关系数仅为0.17),冬季温度对秋季NAO的回归场上EAA区域也几乎均未通过0.10的信度检验。这说明秋季NAO不能用于预测冬季的温度。Seierstad等[53]指出,同期冬季的海冰异常偏少可以引发NAO负位相,但同期的显著联系对季节预测的意义不大。Liu等[16]认为秋季北极海冰偏少引发的大气环流异常与传统的AO存在显著区别。由于Liu等计算的是整个北冰洋的海冰面积,这与本文全区秋季海冰指数与冬季NAO相关不显著的结论是一致的。而本文更进一步指出,虽然全区秋季北极海冰与冬季NAO相关不显著,但前期秋季关键区的海冰变化对冬季NAO有着不可忽视的影响。这为欧亚大陆冬季SAT的季节预测提供了一个较为可靠的线索。当然,欧亚冬季低温的成因非常复杂。不仅包含了大气内部变化[54],还受到北大西洋涛动和赤道太平洋海温[55]等因素的影响,北大西洋海表温度变化也是区域气候变化的重要驱动因子[56]。故关于秋季关键区海冰影响冬季SAT的具体机制,尚需要进一步的数值模拟试验进行深入研究。

致谢:感谢美国夏威夷大学Tim Li教授和南京信息工程大学张文君教授对本研究的宝贵建议。

参考文献

[1] Zhang X D, Sorteberg A, Zhang J, et al. Recent radical shifts of atmospheric circulations and rapid changes in Arctic climate system[J]. Geophys Res Lett, 2008, 35(22), L22701, doi:10.1029/2008GL035607.

[2] Stroeve J C, Serreze M C, Holland M M, et al. The Arctic's rapidly shrinking sea ice cover: a research

synthesis[J]. Climatic Change,2012,110(3-4):1005-1027.

[3] Serreze M C,Francis J A. The Arctic amplification debate[J]. Climatic Change,2006,76(3-4):241-264.

[4] Serreze M C,Barrett A P,Stroeve J C,et al. The emergence of surface-based Arctic amplification[J]. Cryosphere,2009,3(1):11-19.

[5] Zhang J,Lindsay R,Schweiger A,et al. The impact of an intense summer cyclone on 2012 Arctic sea ice retreat[J]. Geophy Res Lett,2013,40(4):720-726.

[6] Francis J A, Vavrus S J. Evidence linking Arctic amplification to extreme weather in mid-latitudes[J]. Geophy Res Lett,2012,39(6),L06801,doi:10.1029/2012GL051000.

[7] Screen J A,Simmonds I. The central role of diminishing sea ice in recent Arctic temperature amplification [J]. Nature,2010,464(7293):1334-1337.

[8] Screen J A, Simmonds I. Increasing autumn-winter energy loss from the Arctic Ocean and its role in Arctic temperature amplification[J]. Geophyl Res Lett,2010,37(16),L16707, doi:10.1029/2010GL044136.

[9] Wu B Y, Handorf D, Dethloff K,et al. Winter weather patterns over Northern Eurasia and Arctic Sea ice loss[J]. Mon Wea Rev,2013,141(11):3786-3800,doi:10.1175/MWR-D-13-00046.1.

[10] WMO. Cold spell in Europe and Asia in late winter 2011/2012 [R]. https://www.wmo.int/pages/mediacentre/news/documents/dwd 2012 report.pdf (last access:10 January 2013).

[11] 丁一汇,王遵娅,宋亚芳,等.中国南方2008年1月罕见低温雨雪冰冻灾害发生的原因及其与气候变暖的关系[J].气象学报,2008,66(5):808-825.

[12] 李崇银,顾薇.2008年1月乌拉尔山阻塞高压异常活动的分析研究[J].大气科学,2010,34(5):865-874.

[13] Cattiaux J, Vautard R, Cassou C, et al. Winter 2010 in Europe:a cold extreme in a warming climate[J]. Geophy Res Lett,2010, 37(20),L20704,doi:10.1029/2010GL044613.

[14] Honda Meiji,Inoue Jun,Yamane Shozo. Influence of low Arctic sea-ice minima on anomalously cold Eurasian winters[J]. Geophy Res Lett,2009,36(8):L08707,doi:10.1029/2008GL037079.

[15] Wu Z W,Li J P,Jiang Z H,et al. Predictable climate dynamics of abnormal East Asian winter monsoon: once-in-a-century snowstorms in 2007/2008 winter[J]. Climate Dynamics,2010,37(7-8):1661-1669.

[16] Liu J P,Curry J A,Wang H J,et al. Impact of declining Arctic sea ice on winter snowfall[J]. Proceedings of the National Academy of Sciences,2012,109(11):4074-4079.

[17] Ghatak D,Frei A,Gong G,et al. On the emergence of an Arctic amplification signal in terrestrial Arctic snow extent[J]. J Geophys Res,2010,115(D24):D24105. doi:10.1029/2010JD014007.

[18] 武炳义,苏京志,张人禾.秋—冬季节北极海冰对冬季西伯利亚高压的影响[J].科学通报,2011,56(27):2335-2343.

[19] 陆日宇.华北汛期降水量变化中年代际和年际尺度的分离[J].大气科学,2002,26(5):611-624.

[20] 徐桂玉,杨修群,孙旭光.华北降水年代际、年际变化特征与北半球大气环流的联系[J].地球物理学报,2005,48(3):511-518.

[21] 平凡,罗哲贤,琚建华.长江流域汛期降水年代际和年际尺度变化影响因子的差异[J].科学通报,2006,51(1):104-109.

[22] Rayner N A, Parker D E. Horton E B,et al. Global analyses of sea surface temperature,sea ice, and night marine air temperature since the late nineteenth century[J]. J Geophy Res,2003,108(D14),4407, doi:10.1029/2002JD002670.

[23] Dee D P,Uppala S M,Simmons A J,et al. The ERA-Interim reanalysis:configuration and performance of the data assimilation system[J]. Q J R Meteorol Soc,2011,137(656):553-597,doi:10.1002/qj.828.

[24] 沈柏竹,廉毅,张世轩,等.北极涛动,极涡活动异常对北半球欧亚大陆冬季气温的影响[J].气候变化研究进展,2012,8(6):434-439.

[25] 丁裕国,江志红.非均匀站网 EOFs 展开的失真性及其修正[J].气象学报,1995,53(2):247-253.

[26] 罗小莉,李丽平,王盘兴,等.站网均匀化订正对中国夏季气温 EOF 分析的改进[J].大气科学,2011,35(4):620-630.

[27] Chung C, Nigam S. Weight of geophysical data in principal component analysis[J]. J of Geophys Res, 1999,104(D14):16925-16928.

[28] Zhou T J, Zhang X H, Yu R C, et al. The North Atlantic Oscillation Simulated by Version 2 and 4 of IAP/LASG GOALS Model[J]. Adv Atmos Sci,2000,17(4):601-616.

[29] 吴洪宝,吴蕾.气候变率诊断和预测方法[M].北京:气象出版社,2005.

[30] Chen W Y. Fluctuation in Norther Hemisphere 700mb height field associated with Southern Oscillation [J]. Mon Wea Rev,1982,110(7):808-823.

[31] 施能,2009.气象统计预报[M].北京,气象出版社.

[32] 武丰民,何金海,祁莉,等.海冰消融背景下北极增温的季节差异及其原因探讨[J].海洋学报,2014,36(3):39-47.

[33] Hurrell J W. Decadal trends in the North Atlantic Oscillation: Regional temperatures and precipitation [J]. Science,1995,269(5224):676-679.

[34] Yu R C, Zhou T J. Impacts of winter-NAO on March cooling trends over subtropical Eurasia continent in the recent half century[J]. Geophys Res Lett,2004,31(12),L12204,doi:10.1029/2004GL019814.

[35] Xin X G, Zhou T J, Yu R C. Increased Tibetan Plateau snow depth: An indicator of the connection between enhanced winter NAO and late-spring tropospheric cooling over East Asia[J]. Adv Atmos Sci, 2010,27(4):788-794,doi:10.1007/s00376-009-9071-x.

[36] 武炳义,黄荣辉.冬季北大西洋涛动极端异常变化与东亚冬季风[J].大气科学,1999,23(6):641-651.

[37] Wu B Y, Wang J. Possible impacts of winter Arctic Oscillation on Siberian high, the east Asian winter monsoon and sea-ice extent[J]. Adv Atmos Sci,2002,19(2):297-320.

[38] Ding Y H. Monsoons over China[M]. Kluwer Academic:Dordrecht,1994:1-419.

[39] 施能,邓自旺,谌芸.近百年北半球冬季海平面气压场与冬季气温的多时间尺度相关[J].南京气象学院学报,2000,23(4):519-527.

[40] 朱小洁,孙即霖.冬季西北太平洋阿留申低压一南北向海温差一西风急流正反馈过程分析[J].科学通报,2006,51(9):1097-1102.

[41] 洪芳玲.冬季冰岛低压一组环流指数在气候研究中的应用[D].南京:南京信息工程大学,2011.

[42] 吕美仲,侯志明,周毅.动力气象学[M].北京:气象出版社,2004.

[43] 周天军.全球海气耦合模式中热盐环流对大气强迫的响应[J].气象学报,2003,61(2):164-179.

[44] 周天军,宇如聪,郜永琪,等.北大西洋年际变率的海气耦合模式模拟Ⅰ:局地海气相互作用[J].气象学报,2006,64(1):1-17.

[45] 周天军,宇如聪,郜永琪,等.北大西洋年际变率的海气耦合模式模拟Ⅱ:热带太平洋强迫[J].气象学报,2006,64(1):18-29.

[46] Hurrell J W. Influence of variations in extratropical wintertime teleconnections on northern hemisphere temperature[J]. Geophys Res Lett,1996,23(6):665-668.

[47] 武炳义,黄荣辉,高登义.冬季北极喀拉海、巴伦支海海冰面积变化对东亚冬季风的影响[J].大气科学,1999,23(3):267-275.

[48] Outten S D, Esau I. A link between Arctic sea ice and recent cooling trends over Eurasia[J]. Climatic Change,2012,110(3/4):1069-1075.

[49] Petoukhov V, Semenov V A. A link between reduced Barents - Kara sea ice and cold winter extremes over northern continents [J]. J Geophys Res: Atmospheres (1984-2012), 2010, 115 (D21), doi:

10.1029JD013568.

[50] Overland J, Wang M. Large-scale atmospheric circulation changes are associated with the recent loss of Arctic sea ice[J]. Tellus A,2010,62(1):1-9.

[51] Francis J A, Chan W H, Leathers D J, et al. Winter Northern Hemisphere weather patterns remember summer Arctic sea-ice extent[J]. Geophy Res Lett,2009,36(7), L07503,doi:10.1029/2009GL037274.

[52] Li J P, Sun C, and Jin F F. NAO implicated as a predictor of Northern Hemisphere mean temperature multidecadal variability[J]. Geophy Res Lett,2013,40:5497-5502. doi:10.1002/2013GL057877.

[53] Seierstad I A, Bader J. Impact of projected Arctic Sea Ice reduction on extratropical storminess and NAO[J]. Climate Dynamics,2009,33(7-8):937-943.

[54] Screen J A, Simmonds I, Deser C, et al. The atmospheric response to three decades of observed Arctic Sea ice loss[J]. J Climate,2013,26(4):1230-1248.

[55] Merkel U, Latif M. A high resolution AGCM study of the El Niño impact on the North Atlantic/European sector[J]. Geophys Res Lett,2002,29(9),doi:10.1029/2001GL013726.

[56] 李建,周天军,宇如聪.利用大气环流模式模拟北大西洋海温异常强迫响应[J].大气科学,2007,31(4):561-570.

第三部分 叁

何金海论文选

季风低频变化

On the 40-50 day Oscillations During the 1979 Northern Hemisphere Summer
Part I : Phase Propagation[*]

Murakami Takio Nakazawa Tetsuo[①] He Jinhai[②]
(Department of Meteorology, University of Hawaii)

Abstract: Based on FGGE Level IIIb data, the structural features of 40-50 day oscillations over an extensive region (30°S~30°N, 30°E~150°W) during the 1979 summer are detailed. The analysis confirms earlier investigations that these low frequency modes are primarily associated with the zonal wind oscillations. These 40-50 day perturbations propagate northward and eastward, which is most clearly defined over the monsoon region north of the equator from 60° to 150°E. The monsoon region is characterized by prominent spectral peaks in the 850-hPa meridional winds with periods shorter than 10 days, probably reflecting the activities of monsoon disturbances. However, the local Hadley circulation, as defined by averaging the meridional component of the wind between 60° and 150°E, exhibits a distinct spectral peak in the period range of 40-50 days. Similarly, the square of the meridional winds, which is a measure of synoptic-scale disturbance activity, also changes with a period of 40-50 days. These features, which are similar to the group velocity phenomena, are pronounced only over the central monsoon region (10°~20°N, 60°~150°E). The low frequency modes propagate northward and become most intensified near 10°~20°N through mutual interaction between synoptic-scale disturbances, the local Hadley circulation, and the zonal mean flows over the monsoon region.

At the equator, the 40-50 day zonal wind perturbations propagate systematically eastward (500 km·d^{-1}) and upward (0.7 km·d^{-1}). In the equatorial region, the low frequency oscillations owe their existence to a lateral geopotential wave-energy flux from the monsoon region, which represents the major energy source for 40-50 day perturbations via the conversion from potential to kinetic energy.

Compared to the equator, the phase propagation of zonal wind perturbations along 15°N, although moving eastward, is not as systematic. At this latitude, zonal wind perturbations are pronounced in the lower troposphere over the monsoon region, and also in the upper troposphere over the western Pacific. As an integral part of E-W interaction between these two regimes, there occurs downward progression of westerly (or easterly) perturbations over to the Arabian Sea region. The downward phase of westerly (easterly) modes corresponds to the commencement of "active" ("break") monsoons over South and Southeast Asia.

---

[*] 本文发表于《Journal of the Meteorological Society of Japan》,1984 年第 62 卷第 3 期,440-468.
[①] On leave from the Meteorological Research Institute, Tsukuba, Japan.
[②] On leave from the Nanjing Institute of Meteorology, Nanjing, People's Republic of China.

1 Introduction

Based on spectral analysis on 10 years (1957-1967) of rawinsonde data at Canton Island, Madden and Julian[1] confirmed the existence of a significant low frequency mode oscillation with a period range of 40-50 days. They showed that these low frequency oscillations are primarily associated with changes in the zonal component of the wind, but with a character different from atmospheric Kelvin waves. Their further analysis[2] indicated that the 40-50 day oscillations are on a global scale with a zonal wavenumber 1, and an out of phase relationship existing between zonal winds in the upper and lower troposphere. They postulated that low frequency oscillations owe their existence to large-scale convective heating over the Indian Ocean and the western Pacific Ocean.

The importance of low frequency oscillations during the Northern Hemisphere summer was first emphasized by Yasunari[3-4]. His analysis of station data over the Indian Ocean and Indian subcontinent revealed an interesting northward propagation of 30-50 day perturbations, which originate in the equatorial Indian Ocean and dissipate near the foothills of the Tibetan Plateau. The meridional passage of these low frequency modes is related to the phase changes between active and break monsoons. Furthermore, these modes appear to be associated with a transient local Hadley circulation over the monsoon region and act as a modulator on the activity of synoptic-scale monsoon disturbances.

Although the long-period (~40 day) oscillations are present during any monsoon season, they were abnormally enhanced during the 1979 FGGE/MONEX summer[5]. Several investigators documented the nature of long-period oscillations during this particular summer. Krishnamurti and Subrahmanyam[6] presented a mapping of the 30-50 day filtered wind fields at 850-hPa over an extensive region between 30°S and 40°N and from 30° to 150°E during the period from 2 May to 28 July 1979. They observed the steady meridional propagation of a train of troughs and ridges (anomaly) from the equator to about 30°N, with the amplitude of the winds around $3 \sim 6 \, m \cdot s^{-1}$, the meridional phase speed of about 0.75°latitude per day, and the meridional scale of about 3000 km. Interestingly, they also noted the existence of long-period perturbations even in early May, i. e., before the onset of the summer monsoon. Utilizing Japanese satellite infrared data, M. Murakami[7] detected the systematic northward progression of highly convective cloud bands across the Indochina-Bay of Bengal region with a period range of 30 to 40 days during the 1979 summer. He further showed that enhanced deep convection occurs over regions, of strong monsoonal westerlies at 850-hPa, while the suppressed convective region coincides with an area of weaker than normal monsoon westerlies. The existence of 40-50 day oscillations throughout the 1978-79 FGGE year was confirmed by Lorenc[8]. Based on an empirical orthogonal function analysis of the velocity potential fields, he detected eastward propagating divergent modes with a zonal wavenumber 1 and a period range of 40-50 days.

Even though the synoptic studies cited above are valuable and informative, there are still more unanswered questions than answered. The role of short-period, synoptic-scale disturbances upon the phase propagation of long-period, planetary-scale perturbations is still unknown. Other unexplained phenomena include: Why do the long-period perturbations propagate northward across the monsoon region ? What mechanisms are responsible for the development of long-period oscillations over South and Southeast Asia? Are there any significant changes in the large-scale apparent heat and moisture sources in association with long-period perturbations? These questions are intimately related to a better understanding of the effect that short-period monsoon disturbances and long-period planetary oscillations have on the maintenance of the Northern Hemisphere summer monsoon. In the present study, these problems are examined by using the FGGE Level IIIb wind, geopotential height, temperature, and humidity data for the 1979 summer. This type of study is a prerequisite for the improvement of theoretical models and numerical experiments on the summer monsoon circulation, where long-period oscillations are an integral part.

2 Data and computational procedures

This study utilized daily u, v, Φ and r data at seven levels (100, 200, 300, 500, 700, 850, and 1000-hPa) for the period from 1 May to 31 September 1979. These data were extracted from the FGGE Level IIIb data set prepared by the European Center for Medium Range Weather Forecasts (ECMWF), at a reduced resolution of 3.75° longitude-latitude intervals over a region extending from 30°E to 150°W and from 30°S to 30°N.

FGGE Level IIIb data includes both objectively analyzed fields and derived parameters. The basic objectively analyzed fields are horizontal u and v wind components and geopotential height fields. These ECMWF objective analyses are based on the best overall data coverage ever obtained in our region of interest. Temperature is determined from the initialization scheme. Instead of using this initialized temperature, we first calculated layer mean virtual temperature from the geopotential fields using the relation $\Delta\Phi/\Delta\ln p = -RT_v$. We then corrected layer mean virtual temperature T_v, to layer mean temperature T, using the layer mean mixing ratio q. (Incidentally, these calculated temperature T values are nearly identical to those of initialized temperatures during the five month period from May to September 1979.) In the FGGE Level IIIb data set, relative humidity is determined from the layer mean water content and the initialized temperature. A simple successive approximation scheme (not an optimum interpolation scheme) is used for the analysis of layer mean moisture. Since the moisture data in the FGGE Level IIb observations is not significantly better than those available for the FGGE Level IIa data set, the resulting moisture analysis is sensitively biased towards the forecast model used to provide the first guess. Thus, the moisture data are of questionable accuracy over certain data void regions (for example, the Indian Ocean). As will be shown in Section 5, however, the quality of moisture data is quite satisfactory over the monsoon region covering the Arabian Sea, the Bay of Bengal, the South China Sea, and the Western Pacific between the equator and 30°N.

We used the calculated temperature T and mixing ratio q for the evaluation of daily values of large-scale apparent heat and moisture sources by the following equations:

$$Q_1 = \frac{\partial s}{\partial t} + u\frac{\partial s}{\partial x} + v\frac{\partial s}{\partial y} + \omega\frac{\partial s}{\partial p} \tag{1}$$

$$Q_2 = -L\left(\frac{\partial q}{\partial t} + u\frac{\partial q}{\partial x} + v\frac{\partial q}{\partial y} + \omega\frac{\partial q}{\partial p}\right) \tag{2}$$

where all meteorological variables are associated with large-scale atmospheric motions[9]; s is the dry static energy (i.e., $s = c_p T + \Phi$) and q the water vapor mixing ratio. (A list of symbols is given in Appendix I.)

Vorticity ζ and divergence D were computed from the u and v data such that:

$$\zeta = \frac{\partial v}{\partial x} - \frac{\partial u\cos\phi}{\cos\phi \partial y} \tag{3}$$

$$D = \frac{\partial u}{\partial x} + \frac{\partial v\cos\phi}{\cos\phi \partial y} \tag{4}$$

The vertical p-velocity ω was obtained by:

$$\omega = \omega_s + \int_p^{p_s} D \, dp \tag{5}$$

Here, ω_s denotes the vertical p-velocity at the surface and is computed from

$$\omega_s = \frac{\partial p_s}{\partial t} + u_s\frac{\partial p_s}{\partial x} + v_s\frac{\partial p_s}{\partial y} \tag{6}$$

where the subscript ( )$_s$ refers to quantities at the surface. The original profile of ω from (5) was then corrected to satisfy the upper boundary condition of $\omega = 0$ at 50-hPa, by assuming a linear correction of D with increasing height.

Introducing angle braces $\langle \rangle$ to represent the five-month mean and an asterisk $(\ )^*$ to define the departure from it, u, for example, can be expressed as

$$u = \langle u \rangle + u^* \tag{7}$$

Thus, u^* represents the transient component of u. We further separate u^* into two components as follows:

$$u^* = a + b(t-t_0) + c(t-t_0)^2 + u' \tag{8}$$

where t_0 corresponds to 16 July (middle of the five-month period). The sum of the first three terms represents the seasonal trend of u^*; the coefficients a, b, and c are determined from a least square method. Accordingly, u' is associated with intraseasonal, transient variations.

In addition to the FGGE Level IIIb data, we also utilized outgoing longwave radiation (OLR) data obtained from polar orbiting satellites. As will be shown later, changes in long-period (40-50 day) OLR and computed apparent heat sources Q_1 (or moisture sinks Q_2) agreed reasonably well. This indicates that the FGGE Level IIIb data is adequate for describing the characteristic features of 40-50 day oscillations during the Northern Hemisphere summer. Unfortunately, the planned orbits for NOAA satellites were occasionally unable to provide OLR observations over certain limited regions. These missing data were linearly interpolated using OLR observations on the preceding and succeeding days. Completely missing OLR values on 17 June and 26 July were similarly time-inter-polated.

3 40-50 day oscillations

Ahlquist[5] pointed out that the 1979 Indian summer monsoon was characterized by the extreme predominance of 40-50 day oscillations. However, its areal extent is still unknown. Also unknown is the extent of the region where short-period synoptic disturbances predominate. To answer these questions, we performed a spectral analysis using the Maximum Entropy Method (MEM) for u' and v' at 850 and 200-hPa. We followed the representation of Zangvil[10] in our depiction of the power spectral density; namely, the product of the power spectral density and frequency is plotted along the ordinate, and the log of the frequency along the abscissa. We also computed spectra for the original u and v data (not reproduced here), which are nearly identical to those shown in Figs. 1 and 2 except for longer (>60 days) periods.

Fig. 1 (top) clearly indicates the pronounced 40-50 day spectral peak for 850-hPa u' over the monsoon region (0°~15°N, 0°~120°E) reaching a maximum (40 units) at 15°N, 90°E. We also find that 40-50 day oscillations of 850-hPa u' are predominant over the western North Pacific east of 120°E along 30°N. In the same region, prominent 40-50 day oscillations of v' are observed (Fig. 1, bottom). These 40-50 day u' and v' spectra are associated with long-period fluctuations of the Pacific high.

In Fig. 1 (bottom), the 850-hPa v' spectrum near 60°E along the equator shows a substantial spectral peak at 45 days. These 40-50 day oscillations of v' over the equatorial Indian Ocean are associated with the long period fluctuations of the Somali jet and are related to the 40-50 day fluctuations of 850-hPa u' over the monsoon region. Of particular interest is the predominant spectral peak with periods shorter than ten days over the Bay of Bengal (15°N, 90°E). Undoubtedly, this reflects the activities of monsoon disturbances over this area. We also see spectral peaks with shorter periods almost everywhere. For example, short-period fluctuations of 850-hPa v' at 15°S, 60°E indicate Southern Hemispheric, mid-latitude synoptic-scale disturbances. Likewise, short-period 850-hPa v' fluctuations east of 120°E at 30°N reflect the activities of middle latitude synoptic disturbances over the Northern Pacific.

One of the central problems in this study is the investigation of the relationship between 40-50 day oscillations and short-period (<10 d) synoptic disturbances. This problem will be discussed in the next section.

In Fig. 2 (top), some moderate 40-50 day spectral peaks of 200-hPa u' are found along the equator from 60° to 120°E. Note that outside of this area, there are no significant 40-50 day u' peaks along the equator. Interestingly, along 15°S from 30°~60°E, substantial 40-50 day u' oscillations are observed. At 15°N, however, we have spectral peaks of 40-50 day u'

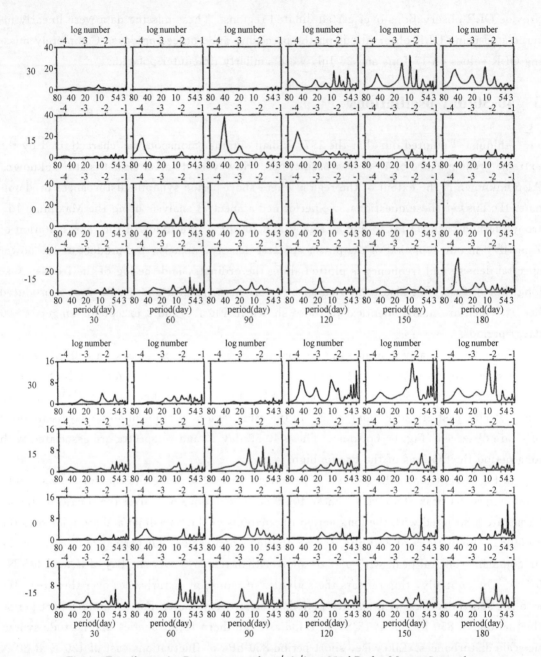

Fig. 1 Top (bottom): Power spectra for u' (v') at 850-hPa for May to September
The product of spectral power and frequency (unit: $m^2 \cdot s^{-2}$) is plotted against log-frequency

oscillations from 120°E~180° with maximum at 150°E. Thus, Fig. 2 (top) shows a SW-NE oriented band of pronounced 40-50 day oscillations of 200-hPa u' extending from 15°S, 30°E, crossing the equator at 90°E, to 30°N, 150°E. Through this channel, Southern Hemisphere upper tropospheric effects appear to penetrate deep into the Northern Hemisphere subtropics. Conversely, over the Indian subcontinent (15°N, 60°~90°E) no significant 40-50 day u' fluctuations are observed. Let's turn our attention to 200-hPa v' spectra (Fig. 2, bottom).

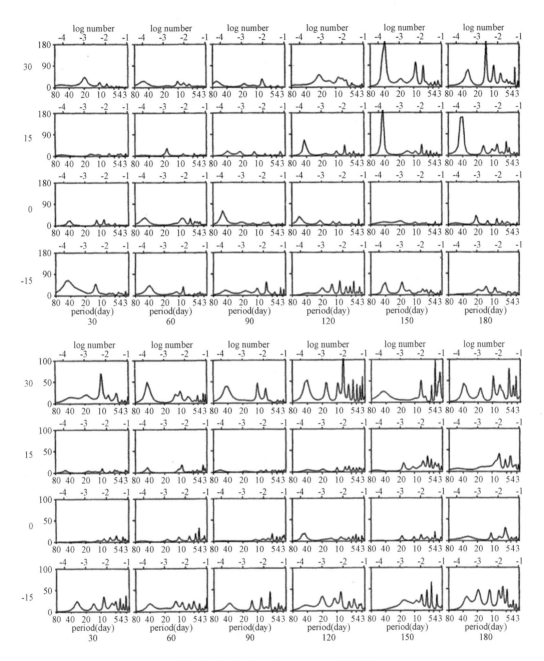

Fig. 2 As in Fig. 1, except at 200-hPa

Substantial peaks of 40-50 day v' oscillations are found at 30°N. Between 60° and 120°E, these oscillations are associated with the long-period fluctuations of the Tibetan high, whereas east of 120°E, they are closely linked with the 40-50 day fluctuations in the upper tropospheric trough (UTT) over the Northern Pacific. Also of interest is the absence of v' Spectral peaks in the 40-50 day period range along the equator except at 120°E (over the Borneo region), where the cross-equatorial northerlies are strongest during the northern summer.

Yasunari[3-4] confirmed that 40-50 day zonal wind perturbations propagate northward and eastward. To reconfirm this, we have computed the coherence and phase difference for 40 day u' oscillations by means of the MEM method[11]. In Fig. 3 (bottom) for 850-hPa u', a reference point is located at 11.25°N, 105°E; and the phase difference (degree) is shown only over regions where the coherence exceeds 0.6. The phase difference pattern clearly indicates eastward and northward propagation over the monsoon region (0°~20°N, 60°~150°E), while no systematic propagation is apparent over Africa and the western Pacific. Using a 2-level primitive equation model in which a continental cap north of 18°N is surrounded by ocean, Webster[12] showed the occurrence of systematic northward phase propagation of 15-20 day perturbations over this particular region with continent (ocean) to the north (south).

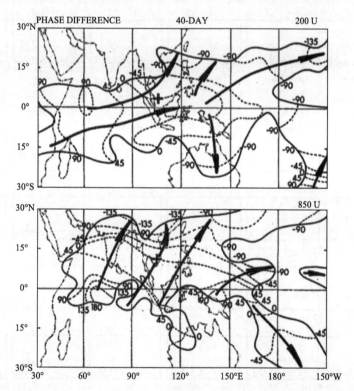

Fig. 3 Phase difference (degree) for u' perturbation at 200 (top) and 850 (bottom)-hPa. Phase (dashed line) are shown for regions where the coherence is greater than 0.6. Heavy arrows indicate the direction of the phase propagation

Fig. 3 (top) shows the phase difference at 200-hPa with a reference point at 3.75°N, 105°E. Over the Indian Ocean, phase propagation is distinctly eastnorth-eastward from the east coast of Africa (15°S, 40°E) to Borneo (0°, 120°E). 200-hPa u' perturbations propagate northeastward over the equatorial Bay of Bengal, the South China Sea, and the western North Pacific. Yet another interesting feature is the three propagation routes emitting from the equator, 130°E area. One route is toward the western Pacific east of the Philippines, another route moves toward the central subtropical Pacific (15°N, 150°W), and the third route is headed toward north eastern Australia where the wintertime jet stream at 200-hPa exhibits significant 40-50 day oscillations (refer to Fig. 2, top).

4 Relationship between short-period synoptic-scale disturbances and long-period planetary oscillations

The time-longitude section of 850-hPa v' at 15°N (Fig. 4, left) clearly demonstrates the nature of disturbance activity over South and Southeast Asia. Around 10 May, there are strong v' northerlies (anomaly; >5 m · s^{-1}) near 70°E and stronger southerlies (10 m · s^{-1}) near 90°E. This perturbation (labelled "1") is a tropical cyclone which developed in the equatorial Bay of Bengal on 7 May and reached the east coast of India (80°E, 15°N) on 11 May. A second cyclone ("2") developed over Indochina on 16 May, and a third ("3") formed near the Philippines on 20 May. Thus, between 8 and 21 May, there was a sequence of cyclone development which tracked eastward across South and Southeast Asia. The individual perturbations (of positive and negative v') tend to move westward (Fig. 4, left). However, the wave packets (regions of large v' amplitude) appear to propagate eastward. This pattern is similar to what is generally called the "group velocity" phenomena. Three more similar events with large v' amplitude occurred during the periods: 16 June-6 July, 1-21 August, and 17-28 September. These group velocity phenomena seem to be peculiar to the monsoon region, as they can not be found in Fig. 4 (right) for 150°W~30°E at 15°N. It also appears that the group velocity phenomena are prominent only near 15°N over the monsoon region (60°~150°E). This can be confirmed in the time-longitude sections of 850-hPa v' for 30°E~150°W from the equator to 30°N (not shown).

The 850-hPa (u', v') wind fields on days 52 to 58 (21 to 27 June) are shown in Fig. 5. [The reader is directed to Murakami, Nakazawa, and He's[13] paper, hereafter abbreviated as MNH, for a complete description of synoptic-scale disturbance activity during the 1979 summer, as only a brief description is given here.] Fig. 5 shows that the group velocity phenomena during this period were associated with a sequence of synoptic events that occurred over an extensive region covering the Arabian Sea, the Bay of Bengal, the South China Sea, and the western Pacific between the equator and about 25°N. On day 52, a tropical cyclone (labelled "C_1") was located off the southern tip of Saudi Arabia. The onset of the 1979 summer monsoon over central India was declared to be on day 50 (19 June), when C_1 reached the central Arabian Sea around 15°N, 65°E[14]. Krishnamurti et al.[15] investigated the mechanisms through which the onset disturbance (C_1) developed and propagated northwestward across the Arabian Sea. After day 50, C_1 moved further northwestward, while weakening considerably, and dissipated near the southern coast of Saudi Arabia sometime between days 54 and 56.

The second cyclone (signified as "C_2") formed on day 48 off the west coast of Sumatra and then moved northward along the Malaysian Peninsula, reaching southern Burma on day 50. C_2 intensified into a monsoon depression near the head of the Bay of Bengal on day 54 and dissipated by day 60 after crossing over India.

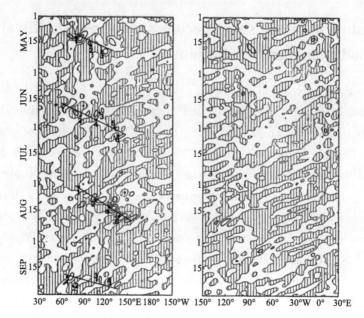

Fig. 4 Time-longitude sections of 850-hPa v' (interval: 5 m · s^{-1}) at 15°N. Shading denotes northerlies (anomaly). Heavy lines denote the group velocity phenomena determined from eqs. (9) and (10). Numerals refer to synoptic-scale disturbances. Left: For 30°E~150°W. Right: For 150°W~30°E

The WNW-ESE oriented monsoon trough (anomaly) became well-organized with three embedded cyclonic cells C_1, C_2, and C_3 by day 54. C_3 stagnated near the Vietnam coast until day 56. Two days later, C_4 formed over Burma. (Most probably, C_4 was enhanced due to the westward propagation of C_3 across Indochina.) C_4 weakened after reaching the head of the Bay of Bengal on day 62.

Between days 54 and 58, the WNW-ESE tilted monsoon trough continued to extend northward and eastward as C_5 developed over the western equatorial Pacific. To the south of this monsoon trough were widespread monsoon westerlies with a maximum speed exceeding 10 m · s^{-1} over the eastern Arabian Sea.

In Fig. 5, evidence has been presented that during the period from day 52 to 58, a series of time-clustered, partially space-overlapping disturbance development occurred over the monsoon region. Thus, the time- (space-) scale of the group velocity phenomenon is much longer (larger) than that of individual disturbances. Although individual disturbances generally moved northwestward, the monsoon trough with strong monsoon westerlies to its south clearly propagated northward and eastward. Similar northward and eastward propagation of the monsoon trough was also observed during the periods: 1-21 August and 17-28 September. Occasionally, typhoons developed when the trough eastward into the western Pacific. In

Fig. 4 (left), two disturbances (labelled "4" and "5", respectively) on 13 and 19 August intensified into major typhoons when they approached the Philippines region.

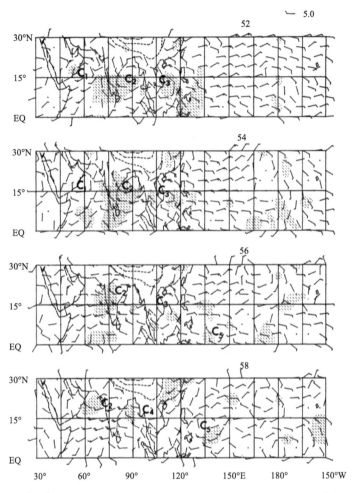

Fig. 5 850-hPa (u', v') wind fields on day 52 (21 June) to 58 (27 June). Full barb is $5 \text{ m} \cdot \text{s}^{-1}$. Major synoptic disturbances ($C_1, C_2, \cdots$) are numbered sequentially. Shading denotes regions of OLR' less than $-50 \text{ W} \cdot \text{m}^{-2}$

In order to determine the group velocity of the four events in Fig. 4 (left), we applied a similar procedure as employed by Hayashi[16]. He expressed wave packets with wavenumber band Δk as follows:

$$w_{\Delta k}(x,t) = R_e \sum_{\Delta k} w_k e^{ikx} = R_e a(x,t) e^{i\phi(x,t)} \tag{9}$$

where w_k is the complex Fourier coefficient for a wavenumber band from $k=5$ to 20, and a and ϕ are the envelope and phase functions defined by:

$$a(x,t) = |\sum_{\Delta k} w_k e^{ikx}| = (c^2 + s^2)^{1/2}$$

and

$$\phi(x,t) = \arg\{\sum_{\Delta k} w_k e^{ikx}\} = -\tan^{-1}(s/c) \tag{10}$$

where c and s are the real and imaginary parts of the complex Fourier series.

The group velocity was determined by linearly connecting the regions of large envelope (indicated by heavy lines in Fig. 4, left). The group velocities were $8.5° \cdot d^{-1}$, $5.2° \cdot d^{-1}$, $3.9° \cdot d^{-1}$, and $8.6° \cdot d^{-1}$ for the first through fourth events, respectively. The average eastward phase speed is $6.6° \cdot d$.①

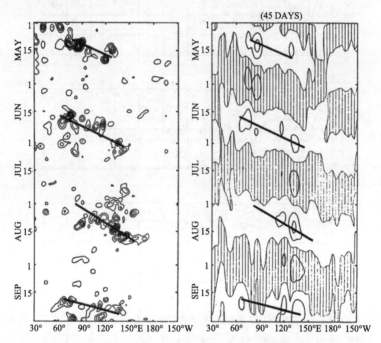

Fig. 6 Left: Time-longitude sections of 850-hPa v^{*2} (interval: 20 m² · s⁻²) along 15°N. Right: 40-50 day filtered, 850-hPa $\widetilde{v^{*2}}$ (interval: 10 m² · s⁻²) along 15°N. Shading denotes negative $\widetilde{v^{*2}}$. Refer to Fig. 4 (left) for further information

To further confirm the group velocity phenomena, we prepared Fig. 6 (left) which shows the longitude-time section of v^{*2} at 15°N. [Here, the v^{*2} data was used because of a possible contribution from seasonal trends (refer to eq. 8) to the group velocity phenomena. However, it turned out that the v^{*2} data was nearly identical to the v'^2 data.] In Fig. 6. (left), one notes a good correspondence between regions of large v^{*2} and the areas of large envelope determined from Hayashi's technique. Here, the salient feature is the occurrence of large v^{*2} with an approximate period of 45 days over the monsoon region from about 60° to 150°E. Thus, although the time-scale of v^* (or v') fluctuations over the monsoon region is generally less than 10 days [refer to v' spectra in Fig. 1 (bottom), and also to the longitude-

---

① Based on daily surface pressure data, Krishnamurti et al.[17] investigated the formation of Bay of Bengal disturbances as a result of westward propagating group velocity phenomena initiated by a typhoon or a tropical storm over the western Pacific. As shown in Fig. 4, the group velocity phenomena associated with 40-50 day oscillations clearly propagate eastward. This time-scale is much longer than the time-scale of westward moving group velocity phenomena considered by Krishnamurti et al.[17].

time section for v' in Fig. 4 (left)], the time-scale of the corresponding v^{*2} variations is definitely much longer, about 45 days. Undoubtedly, this is indicative of a strong interaction of short-period synoptic disturbances with long-period ($\sim$45 day), planetary-scale oscillations. This point will be further elaborated on in the next section.

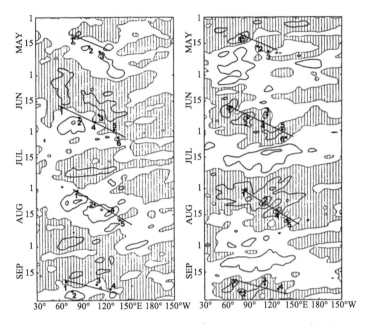

Fig. 7 Left: Time-longitude sections of 850-hPa u' (interval: 5 m · s^{-1}) along 15°N. Shading denotes easterlies (anomaly). Right: For 850-hPa ϕ' (interval: 20 m) along 15°N. Shading denotes negative ϕ'. Refer to Fig. 4 (left) for further information

Prefiltering was then applied to daily $\langle v^{*2} \rangle$ data at 15°N. Here, the same procedure as described by M. Murakami[18] was used to satisfactorily reduce oscillations with periods shorter (longer) than 40(50) days, while retaining about 99 percent of the original amplitude for 45 day oscillations. Thus, using this method, 40-50 day oscillations contribute the most to the filtered time-series data. Fig. 6 (right) depicts a longitude-time section of 40-50 day filtered $\widetilde{v^{*2}}$ data along 15°N. Here, note that positive or negative $\langle \widetilde{v^{*2}} \rangle$ anomalies represent departures from the five-month average $\langle v^{*2} \rangle$ values. It appears that the phase propagation for 40-50 day filtered $\widetilde{v^{*2}}$ is east-ward with an approximate speed of 7 m · s^{-1} over the monsoon region between about 60° and 150°E. In regions west of 60°E and east of 150°E, the phase propagation is not as systematic as that over the monsoon region.

Figs. 7 and 8 demonstrate that the group velocity phenomena over the monsoon region have strong association with changes in 850-hPa u', ϕ', q', and OLR' over the same region. In Fig. 7 (left), it appears that 850-hPa u' (not 40-50 day filtered $\widetilde{u'}$) at 15°N becomes westerly (anomaly) several days after the occurrence of the group velocity phenomena (i.e., $\widetilde{v^{*2}} > 0$) at the same latitude. In comparison, when $\widetilde{v^{*2}}$ is above (below) the seasonal mean,

850-hPa ϕ' simultaneously becomes below (above) normal, while 850-hPa q' is greater (less) than usual and OLR' is below (above) the seasonal mean. This can be confirmed by comparing Fig. 6 (right) with Figs. 7 (right), 8 (left) and 8 (right), respectively. Namely, when the group velocity phenomena occur, the monsoon trough becomes stronger than usual (i.e., $\phi'<0$) with many embedded monsoon disturbances (i.e., $v^{*2}>0$). At this time, moisture increases in and around the monsoon trough (i.e., $q'>0$) with widespread convective activity (i.e., $OLR'<0$) over the entire monsoon region. Thus, the monsoon is "active" during this period. In contrast, no group velocity phenomena occur during "break" monsoons with $\phi'>0$, $q'<0$, and $OLR'>0$.

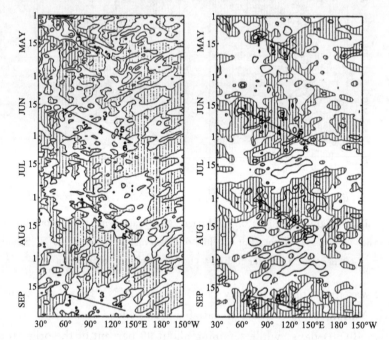

Fig. 8 Left: Time-longitude sections for 850-hPa q' (interval: 2 g·kg^{-1}) along 15°N. Right: For OLR' (interval: 40 W·m^{-2}) along 15°N. Shading denotes negative values. Refer to Fig. 4 (left) for further information

5 Northward propagation of 40-50 day planetary-scale monsoon perturbations

As shown in Fig. 3 (bottom), 850-hPa u' perturbations exhibit systematic northward progression over an extensive region encompassing the Arabian Sea (60°E), the Bay of Bengal, the South China Sea, and the western North Pacific (150°E). This region, covering 90° of longitude, probably corresponds to the extent of the Northern Hemisphere summer monsoon. In this study, changes in the zonal mean winds averaged between 60° and 150°E are used to investigate the northward propagation of 40-50 day perturbation over the monsoon region.

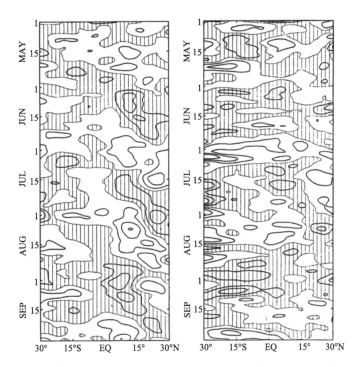

Fig. 9 Time-latitude sections of zonal mean winds at 850-hPa, averaged between 60° and 150°E. Left: For $\bar{u}'$ (interval: 2 m · s^{-1}). Right: For $\bar{v}'$ (interval: 0.5 m · s^{-1}). Shading indicates anomalous easterlies (northerlies)

Fig. 9 (left) depicts a distinct northward movement of positive and negative zones of daily 850-hPa $\bar{u}$ (not 40-50 day filtered $\tilde{\tilde{u}}$) anomalies poleward of the equator. Although 850-hPa $\bar{u}$ fluctuations are not purely periodic, they appear to consist largely of long-period variations with a period range of about 40-50 days. (The same features are noted in the time series of daily 850-hPa $\bar{u}$ along any longitude between 60° and 150°E.) Long-period 850-hPa $\bar{u}$ fluctuations become most pronounced when approaching 5°~15°N. These pronounced 850-hPa $\bar{u}$ variations appear to be correlated with the changes in 850-hPa $\bar{v}$ winds at around 5°~10°N (Fig. 9, right). For example, on day 1 to 5 July (1 to 10 August), strong $\bar{u}$ westerlies (anomaly) are associated with anomalous $\bar{u}$ southerlies between 5° and 10°N. Undoubtedly, these $\bar{v}$ fluctuations at 850-hPa reflect changes in the low-level Hadley circulation over the monsoon region. Note also that 850-hPa $\bar{v}$ fluctuations poleward of the equator appear to consist largely of long-period oscillations with an approximate period of 45 days. This stands in sharp contrast to what we observed in Fig. 1 (bottom) which shows the dominance of short-period ($<$10 d) v' spectra due to the development of regional monsoon disturbances.

A further inspection of Fig. 9 (left and right) indicates a definite phase difference between 850-hPa $\bar{u}$ and $\bar{v}$ near 5°~10°N. The lag-correlation computed at 7.5°N reveals the occurrence of maximum 850-hPa $\bar{v}$ westerlies (easterlies) three days after the maximum 850-hPa $\bar{v}$ southerlies (northerlies). A compositing technique is then employed to investigate the changes in various meteorological variables, at different levels and locations, with refer-

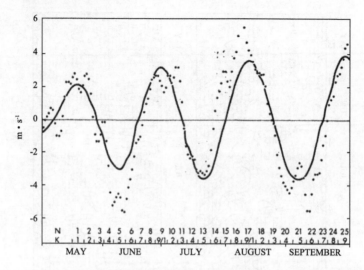

Fig. 10 Time-series of 40-50 day filtered zonal mean $\tilde{\tilde{u}}$ at 850-hPa averaged between 60°~150°E along 11.25°N. Original daily $\bar{u}$ plotted by dots. Also shown are subperiods N, and categories k. Refer to the text for more information

ence to variations of 40-50 day filtered $\tilde{\tilde{u}}$ at 850-hPa averaged from 60° to 150°E along 11.25°N. This reference was chosen because it exhibits a systematic and periodical oscillation which appears to be a key element for further investigation into the large-scale aspects of 40-50 day oscillations during the summer. The time series of 40-50 day filtered $\tilde{\tilde{u}}$ fluctuations at 11.25°N is shown in Fig. 10. Note that 40-50 day filtered $\tilde{\tilde{u}}$ is a maximum on 18 May, followed by a minimum on 9 June, and reaching another maximum on 30 June. Thus, the period from 18 May to 30 June constitutes one cycle of 40-50 day filtered $\tilde{\tilde{u}}$ fluctuations. A total of three cycles, with an average period of about 45 days, were selected from the 850-hPa $\tilde{\tilde{u}}$ time-series data for the compositing technique. These three cycles of 850-hPa are expressed as 25 subperiods; i.e., $\tilde{\tilde{u}}_n$, $n=1, 2, \cdots, 25$ (refer to Fig. 10). Here, $\tilde{\tilde{u}}_n$ represents a 5 or 6 day mean value for daily $\tilde{\tilde{u}}_n$ at 850-hPa.

The compositing technique used in this study consists of assigning an appropriate category (numbered 1 to 9) to every $\tilde{\tilde{u}}_n$ value. In this case, category 1 is assigned to the 3 subperiods of maximum 850-hPa $\tilde{\tilde{u}}$; i.e., $\tilde{\tilde{u}}_1, \tilde{\tilde{u}}_9$ and $\tilde{\tilde{u}}_{17}$, while category 5 is defined for the 3 subperiods of minimum 850-hPa $\tilde{\tilde{u}}$; i.e., $\tilde{\tilde{u}}_5, \tilde{\tilde{u}}_{13}$ and $\tilde{\tilde{u}}_{21}$. Category 9 is assigned to $\tilde{\tilde{u}}_9, \tilde{\tilde{u}}_{17}$ and $\tilde{\tilde{u}}_{25}$. The remaining categories, 2-4 and 6-8, occupy intermediate 850-hPa $\tilde{\tilde{u}}$ phases. Category values were then computed for the three select cycles of $\tilde{\tilde{u}}_n$ data:

$$\tilde{\tilde{u}}_k = (\tilde{\tilde{u}}'_k + \tilde{\tilde{u}}'_{8+k} + \tilde{\tilde{u}}'_{16+k})/3, k=1,9 \qquad (11)$$

where we assume that $\tilde{\tilde{u}}_1 = \tilde{\tilde{u}}_9 = (\tilde{\tilde{u}}_1 + \tilde{\tilde{u}}_9)/2$. Similarly, category values were also computed for the other zonal mean variables ($\bar{v}', \bar{\phi}', \bar{T}', \bar{q}', \bar{\zeta}, \bar{D}', \bar{\omega}', Q_1$ and Q_2) at every 3.75° latitude grid point for each standard pressure level over the monsoon region. Hence, composite charts for each of these meteorological variables were constructed by plotting the corresponding category values at their respective grid point and level. For any variable, composite charts for categories 5 to 8 are nearly identical to those for categories 1 to 4, respectively, except for a change of sign. We then computed the difference in composited quantities; for example,

$$\Delta \widetilde{\bar{u}}_k = (\widetilde{\bar{u}}_k - \widetilde{\bar{u}}_{k+4})/2 \qquad k=1,2,3,4 \qquad (12)$$

The statistical significance of $\Delta \widetilde{\bar{u}}_k$ was estimated following Panofsky and Brier[19]. The confidence level of these quantities generally exceeds the 90% significance level over key areas of the summer monsoon region.

If we assume that all meteorological variables oscillate in a purely periodic manner, we find,

$$\begin{aligned} A_i &= \Delta \widetilde{\bar{A}}_i & i &= 1,2,3,4 \\ A_i &= -A_{i-4} & i &= 5,6,7,8 \end{aligned} \qquad (13)$$

and

$$A_1 = A_9$$

where A denotes any of the variables, i.e., $\bar{u}'$, $\bar{v}'$, $\bar{\phi}'$, $\bar{T}'$, $\widetilde{q}'$, $\bar{\zeta}'$, $\bar{D}'$, $\bar{\omega}'$, $\bar{Q}'_1$ or $\bar{Q}'_2$.

Since categories 1 to 9 were defined by only one meteorological parameter (40-50 day filtered $\widetilde{\bar{u}}'$ at 11.25°N and 850-hPa), the phase relationship between $\widetilde{\bar{u}}'$, $\widetilde{\bar{v}}'$, $\widetilde{\bar{\phi}}'$, $\widetilde{\bar{T}}'$, $\widetilde{\bar{q}}'\widetilde{\bar{\zeta}}$, $\widetilde{\bar{D}}'$, $\bar{\omega}'$, $\bar{Q}'_1$, or $\bar{Q}'_2$ at various grid points and pressure levels can be examined by comparing the category values of these quantities.

In Fig. 11 (middle) 850-hPa $\bar{v}_i$ near 5°N reaches a maximum (southerly) at category 8 with its minimum (northerly) at category 4. In comparison, due to our category definition, 850-hPa $\bar{u}_i$ at 11.25°N (reference latitude) becomes a maximum (minimum) at category 1 (5); namely, 850-hPa $\bar{v}_i$ at 5°N leads 850-hPa $\bar{u}_i$ at 11.25°N by about one category. Therefore, it is evident that the Coriolis acceleration is, at least partially, responsible for the phase changes in 850-hPa at 11.25°N. A comparison between Fig. 11 (top and bottom) reveals an approximate quarter of a cycle phase difference between 850-hPa $\bar{u}_i$ and $\bar{\Phi}_i$ poleward of about 10°N. This can be exemplified by the association of high $\bar{\Phi}_i$ anomalies near 17°N at category 4 with easterly anomalies at 11.25°N at category 5 and westerly $\bar{u}_i$ anomalies around 23°N at category 3. Thus, an approximate geostrophic balance between $\bar{\Phi}_i$ and $\bar{u}_i$ poleward of about 10°N is implied. In contrast, the equatorial region between about 5°N and 10°S is characterized by an out of phase relationship between $\bar{\Phi}_i$ and $\bar{u}_i$; namely, high (low) $\bar{\Phi}_i$ is nearly coincident with easterly (westerly) $\bar{u}_i$. Refer to Table 2 of MNH for further information on the structural features of 40-50 day oscillations near the equator.

In Fig. 11 (middle), category 4 is associated with strong northerly $\bar{v}_i$ anomalies near 5°N while category 8 is characterized by prominent southerly $\bar{v}_i$ anomalies at the same location. This indicates the occurrence of indirect (direct) Hadley circulation anomalies at category 4

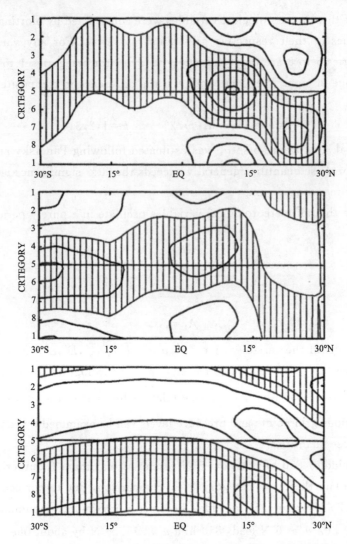

Fig. 11 Category-latitude sections of 40-50 day filtered data at 850-hPa averaged between 60° and 150°E. Top: For $\bar{u}_i$ (interval; 1 m·s^{-1}). Middle: For $\bar{v}_i$ (interval; 0.2 m·s^{-1}). Bottom: For $\bar{\Phi}_i$ (interval; 5 m). Shading denotes negative anomalies

(8). At category 8, there exists low-level convergence ($\partial \bar{v}_i/\partial y < 0$) around 15°N and low-level divergence ($\partial \bar{v}_i/\partial y > 0$) south of the equator. The reverse is true at category 4. These features reflect the phase changes in the vertical Hadley overturnings (anomaly) between categories 4 and 8. This can be further verified from an inspection of Fig. 12, which shows that 387-hPa $\overline{Q}_{1i}$ and 592-hPa $\overline{Q}_{2i}$ are both maximum (heating) at category 8 near 15°N. At the same time, $\overline{OLR_i}$ becomes minimum (above normal convection). Thus, the monsoon is "active" at category 8 near 15°N. In this vicinity, "break" monsoons occur at category 4. At this time, a large negative $\overline{Q}_{2i}$ value near 15°N is contrasted with pronounced positive $\overline{Q}_{2i}$ values to the north (~25°N) and south (0°~5°S). This agrees with our experience that, when a "break" monsoon occurs over central India, rainfall becomes above normal further to the north and south.

The sequence of events mentioned above is undoubtedly related to the northward propagation of 40-50 day perturbations across the monsoon region. In Fig. 12, regions of above normal 387-hPa $\overline{Q}_{1i}$ and 592-hPa $\overline{Q}_{2i}$, and below normal $\overline{OLR_i}$ definitely move northward with respect to category. This, in turn, indicates that regions of "active" monsoon move progressively northward with an average phase speed of one degree of latitude per day. Similarly, regions of "break" monsoon with below normal 387-hPa $\overline{Q}_{1i}$ and 592-hPa $\overline{Q}_{2i}$, and above normal $\overline{OLR_i}$, propagate northward. Thus, a substantial regional difference in the timing of "active" and "break" monsoons is apparent. In this paper, the term "active" or "break" monsoon refers to the active (break) monsoon occurring over the central South Asia region (central India, Indochina, and the South China Sea) around 15°N.

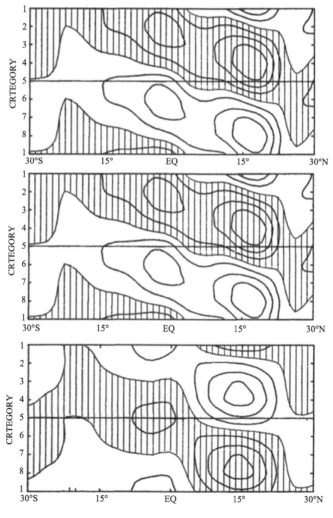

Fig. 12 As in Fig. 11 except for 387-hPa $\overline{Q}_{1i}$ 592-hPa $\overline{Q}_{2i}$ (interval: 0.2 ℃ · d^{-1}), and $\overline{OLR_i}$ (interval: 5 W · m^{-2})

Yet another feature of interest in Fig. 12 is that regions of low (high) $\overline{OLR_i}$ are in good agreement with zones of anomalous heating (cooling) in 387-hPa $\overline{Q}_{1i}$ (or 592-hPa $\overline{Q}_{2i}$). This agreement suggests that the FGGE Level IIIb wind, temperature, and moisture data, from which $\overline{Q}_{1i}$ and $\overline{Q}_{2i}$ were computed, are of sufficiently good quality over the monsoon region north of the equator.

Figs. 11 and 12 were obtained from the composite technique (eq. 13) under the assumption of pure periodic conditions. We also constructed composite charts using eq. (11). They are nearly identical to those shown in Figs. 11 and 12. However, their significance level is lower than that obtained from eq. (13). Composite charts were also constructed at numerous other levels so as to identify the vertical structures of 40-50 day oscillations. For example, an inspection of Figs. 27 and 28 of MNH reveals an approximate inverse relationship between $\overline{v}_i$ oscillations in the lower and upper troposphere for the equatorial region.

Next, an attempt is made to identify the mechanisms by which 40-50 day perturbations propagate northward. Following the same procedure as proposed by Murakami and Sumi[21], the equations of motion for category i can be written as:

$$\frac{\partial u_i}{\partial t} = AU1 + AU2 + AU3 + AU4 + F_{xi} \qquad (14)$$

$$\frac{\partial v_i}{\partial t} = AV1 + AV2 + AV3 + AV4 + F_{yi} \qquad (15)$$

where

$$AU1 = -\left(\langle u\rangle\frac{\partial u_1}{\partial x} + \langle v\rangle\frac{\partial u_1 \cos\phi}{\cos\phi \partial y} + \langle \omega\rangle\frac{\partial u_1}{\partial p}\right)$$

$$AU2 = -\left(u_i\frac{\partial \langle u\rangle}{\partial x} + v_i\frac{\partial \langle u\rangle \cos\phi}{\cos\phi \partial y} + \omega_i\frac{\partial \langle u\rangle}{\partial p}\right)$$

$$AU3 = -\left(\frac{\partial u^* u^*}{\partial x} + \frac{\partial u^* v^* \cos^2\phi}{\cos^2\phi \partial y} + \frac{\partial u^* \omega^*}{\partial p}\right)$$

$$AU4 = +\left(fv_i - \frac{\partial \phi_i}{\partial x}\right)$$

$$AV1 = -\left(\langle u\rangle\frac{\partial v_1}{\partial x} + \langle v\rangle\frac{\partial v_1}{\partial y} + \langle \omega\rangle\frac{\partial v_1}{\partial p} + \frac{\tan\phi}{a}\langle u\rangle u_i\right)$$

$$AV2 = -\left(u_i\frac{\partial \langle v\rangle}{\partial x} + v_i\frac{\partial \langle v\rangle}{\partial y} + \omega_i\frac{\partial \langle v\rangle}{\partial p} + \frac{\tan\phi}{a}-u_i\langle u\rangle\right)$$

$$AV3 = -\left(\frac{\partial u^* v^*}{\partial x} + \frac{\partial v^* v^* \cos\phi}{\cos\phi \partial y} + \frac{\partial v^* \omega^*}{\partial p} + \frac{\tan\phi}{a}u^* u^*\right)$$

$$AV4 = -\left(fu_i + \frac{\partial \phi_i}{\partial y}\right)$$

In eq. (14), AU1 and AU2 are linearly related to the composite quantities for categories 1 through 8, therefore, these terms can be evaluated by utilizing, u^*, v^*, and ω^* values. AU1 denotes the advection of u^* perturbations by the seasonal mean flows, while AU2 indicates the advection of seasonal mean winds by u^*, v^*, and ω^*. perturbations. The AU3 term, which is nonlinear with respect to transient wind components (u^*, v^*, ω^*), requires additional 40-50 day time-filtering of daily, u^*u^*, u^*v^* and $u^*\omega^*$ products, and composites for category i. The resulting AU3 term represents the rate of momentum transfer to 40-50 day perturbations at category i via nonlinear coupling between transient disturbances of all period ranges. AU4 denotes the effect due to non-geostrophic components of winds at category i. Lastly, Fx_i is the frictional effect. In order to avoid repetition, the physical meaning of each term in eq. (15) is not presented here.

The zonal mean equations of motion can be expressed as follows:

$$\overline{DU} = \frac{\Delta \overline{u}_i}{\Delta t} = \frac{\overline{u}_{i+1} - \overline{u}_i}{\Delta t} = \overline{AU1} + \overline{AU2} + \overline{AU3} + \overline{AU4} + \overline{REU} \tag{16}$$

$$\overline{DV} = \frac{\Delta \overline{v}_i}{\Delta t} = \frac{\overline{v}_{i+1} - \overline{v}_i}{\Delta t} = \overline{AV1} + \overline{AV2} + \overline{AV3} + \overline{AV4} + \overline{REV} \tag{17}$$

Here, $\Delta \overline{u}_i$ denotes the difference in the zonal mean winds for the $(i+1/2)$ category. Consequently, each term on the right side of eq. (16) was evaluated by averaging values at categories i and $i+1$. This corresponds to an integration of the zonal mean equation with a time increment of Δt (a one category difference). Thus, these computed results may exhibit significant truncation errors when compared with the observed. $\Delta \overline{u}_i$ In this study, $\overline{REU}$ was evaluated as a residual representing the remaining factors that are not explicitly included in eq. (16). Likewise, $\overline{REV}$ was estimated as a residual in eq. (17). [For convenience, all terms in eqs. (16) and (17) are multiplied by Δt and expressed in m · s^{-1}.]

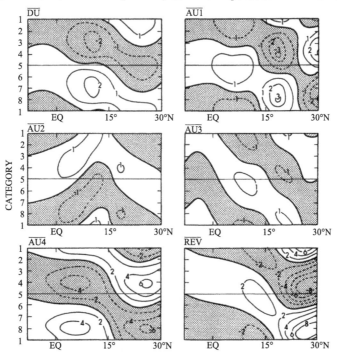

Fig. 13 Category-latitude sections of the terms in eq. (16) at 850-hPa. Shading denotes negative anomalies. Full (dashed) lines represent positive (negative) contours. Intervals are 1 m · s^{-1} for, $\overline{DU}$, $\overline{AU1}, \overline{AU2}, \overline{AU3}$, and 2 m · s^{-1} for $\overline{AU4}, \overline{REU}$

Computed results for each term in eq. (16) are shown in Fig. 13. Let us examine the northward propagation of the $\bar{u}_i$ perturbations near 7.5°N where $\overline{DU}$ variations are most pronounced at 850-hPa. In this vicinity, $\overline{AU2}$ is nearly out of phase with $\overline{DU}$. Furthermore, $\overline{AU2}$ propagates southward and, thus, contributes little to the required $\overline{DU}$ tendency. It appears that $\overline{DU}$ is the result of the combined effects of $\overline{AU1}, \overline{AU3}$ and $\overline{AU4}$ each of which exhibits a systematic northward propagation. $\overline{AU1}$ and $\overline{AU3}$ are both leading $\overline{DU}$ by a quarter of a cycle, while $\overline{AU4}$ lags behind $\overline{DU}$ by about a quarter of a cycle. The resulting net $(\overline{AU1}+\overline{AU3}+\overline{AU4})$ value is approximately in phase with $\overline{DU}$.

A comparison between Figs. 11 (top) and 13 (bottom-right) reveals an approximate inverse relationship between $\bar{u}_i$ and $\overline{REU}$ near 7.5°N. Presumably, $\overline{REU}$ represents the planetary boundary layer friction and momentum dissipation due to vertical mixing in convection. By comparison, $\overline{AU4}$ is in phase with $\bar{u}_i$, i.e., the product $\bar{u}_i(f\bar{v}_i-\overline{\partial\phi_i/\partial x})$ is positive. In this case, the $\overline{fu_iv_i}$ term contributes the most, indicating kinetic energy transfer from the meridional $\bar{v}_i$ to the zonal $\bar{u}_i$ perturbations. $\overline{AU3}$ is nearly out of phase with $\bar{u}_i$ near 7.5°N, resulting in a negative $\bar{u}_i \overline{AU3}$ product. This implies that the zonal mean $\bar{u}_i$ motions furnish kinetic energy that enhances transient disturbance activity.

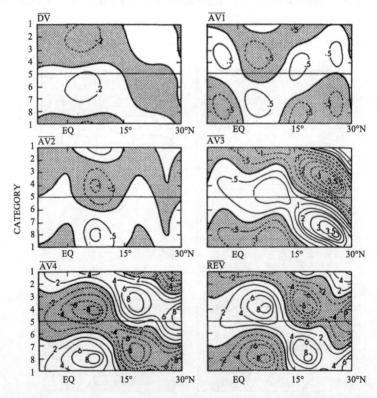

Fig. 14 As in Fig. 13 except for the terms in eq. (17). Intervals are 0.2 m·s^{-1} for $\overline{DV}$, 0.5 m·s^{-1} for $\overline{AV1}, \overline{AV2}, \overline{AV3}$, and 2 m·s^{-1} for $\overline{AV4}, \overline{REV}$

The relationship between changes in the $\bar{v}_i$ meridional circulations and transient disturbance activity can be ascertained by examining $\overline{AV3}$ near 7.5°N (Fig. 14, middle-right) which is approximately out of phase with $\overline{v}_i$, i.e., the product $\overline{v}_i \overline{AV3} < 0$. This is indicative of the increase in kinetic energy for transient eddy motions at the expense of the $\overline{v}_i$ kinetic energy.

$\overline{AV4}$ appears to be nearly in phase with $\overline{v}_i$; around 7.5°N. Therefore, when is positive (southerly), the term $-(f\overline{u}_i + \partial \overline{\Phi}_i/\partial y)$ is positive, implying that $\overline{u}_i$ is subgeostrophic. Since southerly $\overline{v}_i$ is associated with westerly $\overline{u}_i$, i.e., $-f\overline{u}_i \overline{v}_i < 0$, the correlation between $\overline{v}_i$ and $-\partial \overline{\Phi}_i/\partial y$ must be positive and substantial. Namely, southerly $\overline{v}_i$ is downgradient of $\partial \overline{\Phi}_i/\partial y$, crossing the φ_i height contours from high in the south to low in the north. A similar argument applies to northerly $\overline{v}_i$ where flows are from high φ_i in the north to low φ_i in the south. Thus, the generation of kinetic energy for $\overline{v}_i$ perturbations is implied. However, this surplus energy is not used to intensify the $\overline{v}_i$ meridional circulation, but rather the $\overline{u}_i$ zonal motions via the Coriolis effect, $-f\overline{u}_i \overline{v}_i$.

In Fig. 14 (top-left), one notes the smallness of $\overline{DV}$ fluctuations whose amplitude amounts to only 0.2 m·s^{-1} between the equator and 7.5°N. This makes it extremely difficult to investigate the phase propagation of $\overline{v}_i$ perturbations. In contrast, $\overline{AV4}$ is about forty times larger than $\overline{DV}$. This large $\overline{AV4}$ is nearly compensated for by $\overline{REV}$. Here, caution must be exercised when interpreting REV, which was evaluated as a residual in eq. (17). $\overline{REV}$ may be subject to large observational errors in estimating $\frac{\partial \varphi_i}{\partial y}$ in the tropics and also significant truncation errors with respect to time and space when evaluating the remaining terms in eq. (17). In comparison, the $\overline{AV3}$ term, which is related to the products between transient (u^*, v^*, and w^*) wind fields, may be more reliable. In. Fig. 14, $\overline{AV3}$ leads $\overline{DV}$ by about one to two categories near 7.5°N. Thus, transient disturbances generated north of 7.5°N tend to contribute, at least partially, to further propagate $\overline{v}_i$ perturbations northward.

The relationship between changes in the $\overline{u}_i$ and $\overline{v}_i$, circulations and transient disturbance activity can be seen more clearly through an inspection of Fig. 15 which depicts the latitudinal dependency of $\overline{u}_i$, $\overline{\Phi}_i$ and $\overline{AU3}_i$ in category 1 (top). and $\overline{v}_i$, $\overline{\Phi}_i$ and $\overline{AV3}_i$; in category 8 (bottom), at 850-hPa. In Fig. 15 (top), an approximate inverse relationship exists between $\overline{u}_i$, and $\overline{AV3}_i$. Here, the meridional derivative $-\partial \overline{(u^* v^*)}/\partial y$ is the largest contributor to $\overline{AV3}_i$;. Since $\overline{AV3}_i$ is zero near 16°N, $\overline{(u^* v^*)}_i$ must be a maximum in this vicinity. Thus, the northward westerly momentum transport $\overline{(u^* v^*)}_i$ is most pronounced about 5° north of the $\overline{u}_i$ westerly maximum; namely, $\overline{(u^* v^*)}_i$ is maximum in regions of strong cyclonic $\overline{u}_i$ shear. (Refer to Fig. 5 for strong disturbance activity about 5°~10° north of the monsoon westerlies.) In the vicinity of the trough ($\overline{\Phi}_i < 0$) with near zero $\overline{u}_i$, there exists westerly momentum convergence, i.e., $-\partial \overline{(u^* v^*)}_i/\partial y > 0$, resulting in an increase in $\overline{u}_i$ i.e., $-\partial \overline{u}_i/\partial t > 0$. Conversely, in the vicinity of the $\overline{u}_i$ maximum near 11°N, there exists westerly momentum divergence, and, hence, $-\partial \bar{u}_i/\partial t<0$. Accordingly, $\bar{u}_i$ tends to propagate northward due to the transient disturbances effects. Along with the northward shift of $\bar{u}_i$ westerlies, regions of northward westerly momentum transport, $\overline{(u^* v^*)}_i$ also propagate northward, since the transient disturbance activity is largely controlled by the $\bar{u}_i$ latitudinal profile. A similar argument also applies to category 5, when strong u_i easterlies approach 11°N. At this time, $\overline{(u^* v^*)}_i$ is negative (southward westerly momentum transport), reaching a minimum near 16°N. To the north (south) of 16°N, $\partial \bar{u}_i/\partial t$ is negative (positive), implying the northward progression of $\bar{u}_i$ easterlies due to transient disturbances.

In Fig. 15 (bottom), $\bar{v}_i$ is zero slightly north of 15°N in category 8. This location nearly coincides with the position of the $\bar{\Phi}_i$ rough with the downgradient $\bar{v}_i$ southerlies (northerlies) to its south (north). In and around the $\bar{\Phi}_i$ trough, transient disturbances are quite active with large $\overline{(v^* v^*)}_i$, while $-\partial(\bar{v}^* \bar{v}^*)_i/\partial y$ is near zero, or only slightly larger than zero, as is $\overline{AV3}_i$. This is evidential of the nearly standing nature of $\bar{v}_i$ perturbations at 7.5°N (refer to Fig. 11, middle). Further south, $\bar{v}_i$ is a maximum (southerly) near 5°N, where there exists meridional momentum divergence, i.e., $-\partial(\bar{v}^* \bar{v}^*)_i/\partial y<0$. Accordingly, $-\partial v_i/\partial t$ is negative due to transient disturbance effects. Conversely, around 18°N, strong northerly $\bar{v}_i$ is associated with a positive $\partial v_i/\partial t$ tendency.

As mentioned previously, the sum of the three terms $(\overline{AU1}+\overline{AU3}+\overline{AU4})$ is approximately in phase with $\overline{DU}$. Of these three terms, $\overline{AU4}$ is the most dominant. Similarly, $\overline{AV4}$ is the leading term in the $\overline{DV}$ equation (17). $\overline{AU4}$ and $\overline{AU4}$ are related to the zonal $(\bar{u}_i)$ and meridional $(\bar{v}_i)$ mean circulations averaged over the monsoon region. As will be discussed later, the meridional circulation enhances the development of transient disturbances in and around the zonally averaged monsoon trough. Transient disturbances then contribute to the intensification and northward phase propagation of the monsoon trough, the Hadley circulation, and eventually the monsoon zonal flows. Thus, via these mutual interactions with transient disturbances, $(\bar{u}_i, \bar{v}_i)$ perturbations become prominent and propagate northward over the monsoon region.

6 Northward propagation of the N-S vertical overturnings over the monsoon region

Fig. 16 shows the vertical profiles of $(\bar{v}_i, \bar{\omega}_i)$ overturnings averaged over the monsoon region for categories 1 to 4. In category 1, the vertical circulation cell (designated as "D_1") is located near 5°N with updrafts to the north around 15°N and downdrafts near the equator. After category 1, the D_1 vertical cell moves progressively northward. At category 3, D_1 is centered near 20°N with weak ascending motions to the north. Thus, north of 20°N monsoons become slightly more active than normal in category 3.

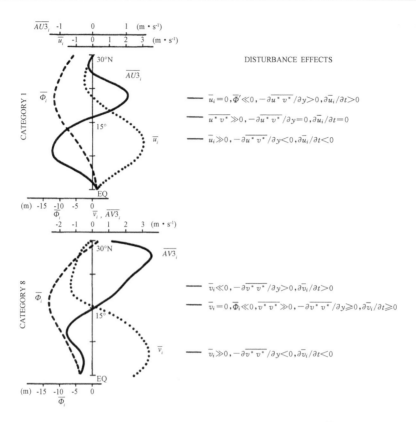

Fig. 15　Zonal mean quantities at 850-hPa between 60° and 150°E. Top: For $\bar{u}_i$ (dotted line; 1 m·s^{-1} unit), $\overline{AU3}_i$ (full line; 1 m·s^{-1} unit), and Φ_i (dashed line; 1 m unit), at category 1. Bottom: For $\bar{v}_i$ (dotted line; 0.1 m·s^{-1} unit), $\overline{AV3}_i$ (full line; 1 m·s^{-1} unit) and $\bar{\Phi}_i$ (dashed line; 1 m unit), at category 8

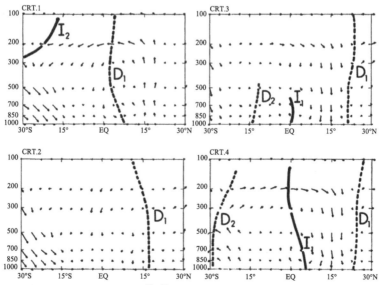

Fig. 16　Height-latitude sections of $(\bar{v}_i, \bar{\omega}_i)$ vectors for category 1 to 4 averaged between 60° and 150°E. Unit vector is 1 m·s^{-1} for $\bar{v}_i$, and 10^{-4} hPa·s^{-1} for $\bar{\omega}_i$

Looking again at category 3, we see a new vertical circulation cell (denoted as "I_1") forms in the lower troposphere near the equator. North of this I_1 cell are downdrafts with the strongest descent portion approximately coincides with near zero $\bar{u}_i$ (Fig. 17, top-right), abnormally high $\bar{\Phi}_i$ (Fig. 11, bottom), and below normal disturbance activity, i.e., $\overline{v^{*2}} < 0$, in the lower troposphere below about 700-hPa. Between categories 3 and 4, the downdraft center intensifies and tends to shift slightly northward in association with the northward displacement of the low-level portion of the I_1 cell (Fig. 16, right). The downdraft becomes most intense near 15°N at category 4. At this time, $\bar{\Phi}_i$ is well above the seasonal mean ($\bar{\Phi}_i \gg$), and disturbance activity is depressed, i.e., $\overline{v^{*2}} \ll 0$, near 15°N. Thus, category 4 corresponds to a "break" monsoon over central South Asia around 15°N. From category 5 to 7 (same as category 1 to 3, except for opposite sign), the I_1 cell moves northward, following the same pattern of movement as the D_1 cell discussed above. At category 8, the monsoon has reached its highest level of activity with an abnormally intensified ascending motion ($\bar{\omega}_i <$ 0), and strong disturbance activity ($\overline{v^{*2}} \gg 0$) in and around an unusually deepened monsoon trough ($\bar{\Phi}_i \ll 0$) over central South Asia near 15°N. The region of "active" monsoon tends to shift further northward with the progression from category 1 to 4 of the subsequent cycle.

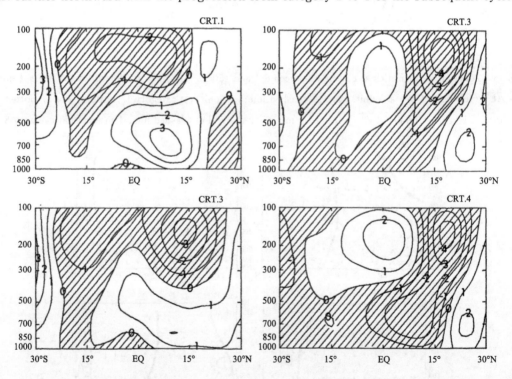

Fig. 17 Height-latitude sections of $\bar{u}_i$; (interval; 1 m · s^{-1}) for category 1 to 4 averaged between 60° and 150°E. Shading denotes anomalous easterlies

Height-latitude sections of $\bar{u}_i$ for categories 1 through 4 are shown in Fig. 17. Starting from category 1, let us trace the region of negative (easterly) $\bar{u}_i$. Here, a distinct zone of easterly $\bar{u}_i$ is located near 5°N at 200-hPa. This easterly zone shifts northward, reaching 14°N at category 2. Further northward propagation takes place between categories 2 and 3. Simultaneously, the regions of easterly $\bar{u}_i$ end to extend downward. In fact, in the lower troposphere over the equatorial latitudes (0°~15°N), westerly $\bar{u}_i$ perturbations in category 2 are replaced by easterly $\bar{u}_i$ winds in category 3. At category 4, easterly $\bar{u}_i$ perturbations are prominent not only in the upper troposphere (-4 m·s^{-1} at 150-hPa, 18°N), but also in the lower troposphere (-2 m·s^{-1} between 500 and 850-hPa, equator to 15°N). At category 5 (same pattern as category 1, except with opposite sign), easterly $\bar{u}_i$ perturbations in the lower troposphere become more pronounced (-3 m·s^{-1} at 700-hPa, near 10°N) than their counterpart in the upper troposphere (-1 m·s^{-1} at 200-hPa, near 20°N). In the vicinity of 10°~15°N, low-level easterly $\bar{u}_i$ perturbations weaken considerably and propagate further downward between categories 5 and 6. Conversely, near the equator, regions of easterly $\bar{u}_i$ tend to propagate upward, reaching the 200-hPa level by category 7 (same pattern as category 3, with a reversal of sign). Between categories 7 and 8, the zone of easterly $\bar{u}_i$ moves northward along 200-hPa, reaching 5°N at category 1 of the subsequent cycle. This completes one cycle of changes in the easterly $\bar{u}_i$ perturbations.

Near the equator, u_i perturbations propagate upward and eastward. This is clearly demonstrated in Fig. 18, which depicts the height-longitude section of u_i at the equator for categories 1 through 4. In category 1, a distinct zone of positive (westerly) u_i tilts westward with increasing height from about 700-hPa near 105°E to 300-hPa at 30°E. Between categories 1 and 3, this positive u_i zone propagates upward and eastward across the Indian Ocean and the western Pacific. Positive u_i perturbations have intensified considerably by the time they reach the upper troposphere around 150-hPa over the monsoon region (60°~150°E) in category 4. At this time, the positive u_i zone extends westward and upward from 850-hPa at 150°W to around 100-hPa at 30°E. The positive u_i zone continues to propagate eastward and upward between categories 5 and 7 (same as categories 1 to 3, except for opposite sign). By category 8, all that remain, are weak residual u_i westerlies limited to the upper troposphere between 100 and 300-hPa over the central Pacific east of 150°E. These residual u_i eventually leave the eastern boundary (150°W) of our region between categories 1 and 2 of the subsequent cycle. Simultaneously, an easterly u_i regime enters our region of interest from the west with easterly u_i winds in the lower troposphere below 700-hPa near the east coast of Africa. Between categories 2 and 5 in the subsequent cycle, this easterly u_i regime propagates eastward and upward. After category 5 of the subsequent cycle, this region of easterlies then follows the same pattern of movement as the region of westerlies discussed earlier. Thus, it takes about one cycle for the region of easterlies (or westerlies) to propagate eastward across our region which covers an area from East Africa, through the Indian Ocean, to the eastern Pacific (i.e., half way around the globe). In this equatorial region, the approximate eastward phase speed of u_i perturbations at 200-hPa is 500 km per day (refer to Fig. 40 of MNH), while their upward phase speed is about 0.7 km per day.

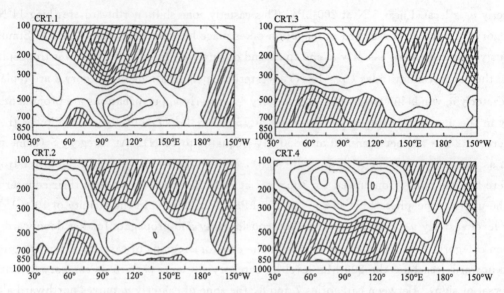

Fig. 18 Height-longitude sections of u_i perturbations at the equator for categories 1 through 4. Shading indicates regions of negative (easterly) u_i. Intervals of u_i are 1 m·s^{-1}

The vertical structural features of 40-50 day u_i perturbations at 15°N for categories 1 to 4 are shown in Fig. 19. Compared to u_i perturbations at the equator, the phase propagation of u_i along 15°N, although moving eastward, is not as systematic. Furthermore, their vertical structures are more complicated with large amplitude in the lower troposphere (below 700-hPa) over South and Southeast Asia (60°~130°E). Amplitude is largest in excess of 10 m·s^{-1} in the upper troposphere over the Pacific (130°E~150°W). The downward progression of easterly $\bar{u}_i$ between categories 3 and 4 (Fig. 17, right) occurs as an integral part of E-W interactions between the upper level u_i perturbations over the Pacific and the low-level u_i perturbations over the monsoon region. To demonstrate this, let us trace the region of easterly u_i which dominates in the upper troposphere between 100 and 300-hPa east of 90°E in category 2 (Fig. 19, bottom-left). This upper-level easterly zone shifts eastward, while intensifying, from category 2 to 3. At category 3, u_i becomes most intensified, exceeding -10 m·s^{-1} at 150-hPa near 140°E. In this vicinity, large u_i variations occur in association with changes in location and intensity of the upper tropospheric trough (UTT). Further eastward propagation of easterly u_i takes place from category 4 to 5. Looking again at category 3, we see a downward extension of easterly u_i over to the Arabian Sea near 60°E. This is congruent with the downward progression of u_i at category 3, as shown in Fig. 17 (top-right). By category 4, low-level easterly u_i becomes prominent (<-4 m·s^{-1}) below 700-hPa near 70°E. As stated earlier, category 4 corresponds to the "break" monsoon phase. Similarly, between categories 7 and 8, regions of westerly u_i extend downward over the Arabian Sea, where "active" monsoons commence at these categories. Low-level westerly u_i perturbations become most pronounced in category 1 of the subsequent cycle (Fig. 19, top-left), exceeding 5 m·s^{-1} at 850-hPa near 90°E. Between categories 2 and 4 of the subsequent cycle, the region of westerly u_i moves progressively eastward and downward. Concurrently, their intensity weakens considerably while crossing South and Southeast Asia. Over the Pacific east of about 150°E, low level u_i perturbations are confined below 700-hPa and their phase propagation becomes irregular.

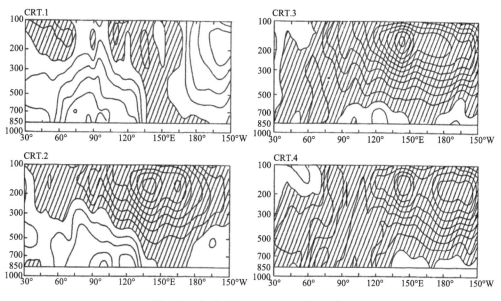

Fig. 19　As in Fig. 18, except for 15°N

In Figs. 18 and 19, we have emphasized the difference in the structural features (vertical tilt, zonal and vertical phase propagation, etc.) of u_i perturbations between the two latitudes, i. e., 15°N and the equator. Fig. 20 exhibits a large regional difference in the energetics of 40-50 day perturbations from one latitude to another. [The reader is directed to MNH for a detailed description of the energetics, as only a brief outline will be presented here.] The numerical computation of $[\overline{\omega_i T_i}]$ averaged between 30°E and 150°W from category 1 to 8 (Fig. 20, top-left), reveals negative values with large magnitude in the mid-troposphere (300~500-hPa) around 15°N. A negative sign for the product $[\overline{\omega_i T_i}]$ defines a conversion from potential to kinetic energy. In this source region near 15°N, the vertical wave-energy flux $[\overline{\omega_i \Phi_i}]$ is directed upward (downward) above (below) 400-hPa (Fig. 20, bottom-left). Downward $[\overline{\omega_i \Phi_i}]$ fluxes are required to maintain low-level 40-50 day modes against friction. The upward $[\overline{\omega_i \Phi_i}]$ flux is a maximum at 200 hPa. The wave-energy flux $[\overline{v_i \Phi_i}]$ is then directed equatorward in the upper troposphere. A positive $[\overline{\omega_i T_i}]$ product in the Southern Hemisphere tropics (0°~15°S), certainly implies that the 40-50 day perturbations in this region owe their existence to a lateral wave-energy flux from the monsoon region around 15°N. Of interest is the existence of downward wave-energy flux $[\overline{\omega_i \Phi_i}]$ between the equator and 15°S.

Fig. 20 (top-right) reveals sensible heat being transported equatorward through the upper troposphere from the monsoon region where the seasonal mean temperature is quite high.

At the equator, the $[\overline{v_i T_i}]$ sensible heat flux from the Northern to Southern Hemisphere at 200 hPa is significant, in excess of -0.2 ℃ m·s^{-1}. In Fig. 20 (bottom-right), the cross-equatorial westerly momentum transport $[\overline{u_i v_i}]$ exceeds 2 m^{2}s^{-2} around 150 hPa. Thus, upper tropospheric 40-50 day perturbations tend to reduce the seasonal mean easterly momentum associated with the easterly jet over the monsoon region.

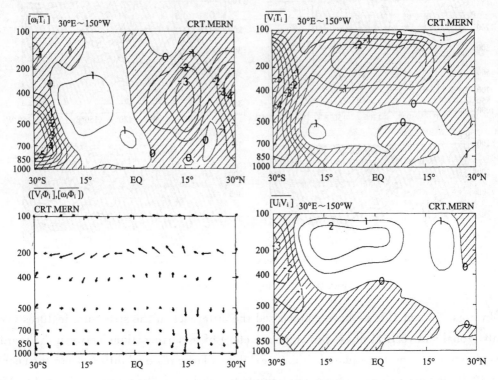

Fig. 20 Category mean $[\overline{\omega_i T_i}]$ (10^{-5} hPa·s^{-1}·℃ intervals), ($[\overline{v_i \Phi_i}]$, $[\overline{\omega_i \Phi_i}]$) vectors (unit: 5 m^2·s^{-1} for $[\overline{v_i \Phi_i}]$ and 5×10^{-4} hPa·ms^{-1} for $[\overline{\omega_i \Phi_i}]$), $[\overline{v_i T_i}]$ (0.1 m·s^{-1}·℃ intervals), and $[\overline{u_i v_i}]$ (1 m^2·s^{-2} intervals), averaged between 30°E and 150°W. Shading denotes negative values

Between 20° and 25°N at 200 hPa, $[\overline{v_i \Phi_i}]$ is directed southward. Presumably, this is indicative of the mid-latitude contribution to the maintenance of 40-50 day perturbations over the monsoon region (0°~20°N). However, this mid-latitude effect represents only a minor contributor to the 40-50 day perturbations over the monsoon region, which primarily owe their existence to large negative $[\overline{\omega_i T_i}]$ in situ.

In Fig. 20 (top-left), also note the presence of large negative $[\omega_i T_i]$ in the Southern Hemisphere extratropics poleward of about 20°S. Undoubtedly, this reflects the wintertime activity of the Southern Hemisphere midlatitude disturbances. In the lower troposphere below about 700 hPa, $[\overline{v_i \Phi_i}]$ originating near 30°S is directed equatorward. In fact, at 1000 hPa, it reaches as far north as 5°N. Associated with this is negative $[\overline{v_i T_i}]$ below about 700 hPa over the entire Southern Hemisphere region (30°S~0°). These features are consistent with low-level, southerly cold surges during the Southern Hemisphere winter.

The monsoon region between 60° and 150°E represents the major contributor to the net conversion term $[\overline{\omega_i T_i}]$ near 15°N. Kinetic energy generated in the mid-troposphere is transported downward, where it enhances the low-level 40-50 day perturbations over the monsoon region (Fig. 19, top-left). A large portion of kinetic energy is also transported upward. However, this surplus energy is not used to intensify upper-level 40-50 day perturbations over the monsoon region [refer to Fig. 19 for small 200 hPa u_i perturbations over the same region]. Rather, kinetic energy generated by the conversion process over the monsoon region is transported to the north (Tibetan Plateau-China), south (Indian Ocean), and east (subtropical North Pacific). In particular, convective activity is generally weak over the subtropical North Pacific during summer, resulting in weak conversion processes. Consequently, the strong activity of upper-level 40-50 day perturbations over the subtropical North Pacific (Fig. 19) is primarily due to the horizontal flux convergence of geopotential energy. In this vicinity, nongeostrophic upper winds are always directed down the gradient of the geopotential heights, facilitating the development of 40-50 day perturbations in and around the UTT.

7 Discussion

Section 5 was devoted to the investigation of the behavior of the 40-50 day perturbations zonally averaged over the monsoon region from 60° to 150°E north of the equator. However, the question arises as to the adequacy of this region for investigating 40-50 day oscillations. In Fig. 1 (top), note the existence of a large 45-day spectral peak for 850 hPa u' at 15°S, 180°. Significant 45-day peaks are also found over the central South Pacific east of the dateline. Nevertheless, the question of whether these low frequency modes over the central South Pacific are related to low-level 40-50 day perturbations over the monsoon region needs to be addressed. To answer this question, we have computed the coherence and phase difference for 40-50 day oscillations at 850 hPa with a reference point at 15°S, 180° (not shown). Here, phase propagation is directed southeastward from the Southern Hemisphere tropics (7°S, 170°E) through the reference point (15°S, 180°) to the Southern Hemisphere mid-latitude region (30°S, 150°W). Thus, these 40-50 day perturbations have their origin in the Southern Hemisphere tropics, and are distinctly separate from the low-level fluctuations over the monsoon region. This supports our choice of region for investigating 40-50 day perturbations over the monsoon region in this study.

The power spectra for 200 hPa u' over the South Pacific also exhibits significant peaks at a 40-50 day period range (not shown). Here again, we examined the coherence and phase difference for 40-50 day perturbations at 200 hPa. Near the reference point (15°S, 150°W), phase propagation is northeastward with its origin in the Southern Hemisphere mid-latitude region around 30°S, 170°W; thus, indicating that these prominent 40-50 day oscillations over the South Pacific are independent of the 200 hPa low frequency modes over the North Pacific.

As shown in Fig. 18, 40-50 day perturbations propagate systematically eastward along the equator. In comparison, eastward phase propagation is not systematic along 15°N (Fig. 19), where 40-50 day perturbations are most pronounced. This makes it very difficult to investigate the exact mechanisms through which these low frequency modes move eastward. Perhaps, the best approach is to apply the Fourier analysis technique to the global tropics. Recall that the low frequency modes considered here might be a part of the zonal wavenumber 1, long-period (~45 days) oscillations as discussed by Madden and Julian[2]. To examine this possibility, an extensive study is now underway at the University of Hawaii utilizing the FGGE Level IIIb global data set.

8 Concluding remarks

In this study, evidence has been presented that FGGE Level IIIb data adequately describes some of the characteristic features of 40-50 day oscillations over the Northern Hemisphere summer monsoon region. To facilitate this study, prefiltering was used to extract 40-50 day oscillations from daily u, v, T, Φ, q, ω, OLR, Q_1, and Q_2 data for the 1979 summer over an extensive region (30°S~30°N, 30°E~150°W). A composite technique was then applied to these 40-50 day filtered data with reference to changes in the zonal mean winds at 850 hPa averaged between 60° and 150°E along 11.25°N. This reference was chosen because it exhibits a systematic and periodic oscillation of the low-level monsoonal westerlies (Fig. 10). A total of three cycles, with an average period of 45 days, was selected for the compositing technique.

To summarize, the main results of this study are:

1. The spectral analysis of 850 hPa u' indicates pronounced 40-50 day peaks over the monsoon region (0°~20°N, 60°~150°E). In comparison, the 850 hPa v' spectrum over the monsoon region shows peaks with periods shorter than ten days, probably reflecting the activities of monsoon disturbances. At 200 hPa, no significant 40-50 day u' fluctuations are observed over the Indian subcontinent, while they are predominant over the subtropical North Pacific (15°~30°N, 150°E~150°W) where the upper tropospheric trough is subjected to changes in its location and intensity.

2. In Fig. 4, evidence has been presented that a series of time-clustered, partially space overlapping disturbances occurred over the monsoon region. These events are similar to the group velocity phenomena. Fig. 6 clearly shows that the time-scale of the group velocity phenomena is about 45 days, which is much longer than the time-scale of the individual disturbances. Although individual disturbances move westward, the group velocity is directed eastward. Interestingly, the local Hadley circulation, as defined by averaging the meridional component of the wind between 60° and 150°E, also exhibits a distinct spectral peak in the period range of 40-50 days.

3. The northward progression of 40-50 day perturbations is most clearly defined over the monsoon region north of the equator between 60° and 150°E (Fig. 3, bottom). Composite

charts for zonally averaged (60°~150°E) $\bar{u}_i$, $\bar{v}_i$, and $\bar{\varPhi}_i$ at 200 and 850 hPa indicate a phase speed of about 1° latitude per day. At 7.5°N, where 40-50 day perturbations are most pronounced, $\bar{u}_i$ at 850 hPa is nearly out of phase with $\bar{u}_i$ at 200 hPa (Fig. 17). Similarly, an approximate inverse relationship exists between $\bar{v}_i$ at 850 and 200 hPa, reflecting 40-50 day fluctuations of the local Hadley circulation over the monsoon region (Fig. 16).

4. When $\bar{u}_i$ westerlies (anomaly) at 850 hPa approach 10°N, the northward westerly momentum transport (anomaly) due to transient eddies, i. e., $\overline{(u^* v^*)}_i$, becomes maximum in regions of strong cyclonic $\bar{u}_i$ shear around 16°N. In the vicinity of the $\bar{\varPhi}_i$ trough (anomaly) with near zero $\bar{u}_i$, there exists westerly momentum convergence, i. e., $\partial \overline{(u^* v^*)}_i / \partial y < 0$, resulting in a positive $\partial \bar{u}_i / \partial t$ tendency. Conversely, in the vicinity of the maximum westerly $\bar{u}_i$ near 10°N, there exists westerly momentum divergence and, hence, $\partial \bar{u}_i / \partial t < 0$. Accordingly, $\bar{u}_i$ westerlies tend to propagate northward due to the transient disturbance effects. By a similar argument, $\bar{u}_i$ easterlies (anomaly) approaching 10°N vicinity tend to move northward due to transient disturbances.

5. During the active monsoon phase (category 8), the local Hadley circulation is enhanced between the equator and 15°N (Fig. 16), followed by the intensification of the monsoonal zonal flows. This induces a subsequent development of transient disturbances in and around the monsoon trough (anomaly). Transient disturbances contribute to the intensification of convective activity, Q_{1i} heat sources, Q_{2i} moisture sinks, the monsoon trough, the local Hadley circulation, and eventually the monsoonal zonal flows. Thus, via these mutual interactions with transient disturbances, 40-50 day perturbations become most prominent near 15°N. During the break monsoon phase, the activities of transient disturbances and associated convections are both depressed well below normal.

6. At the equator, 40-50 day zonal wind perturbations propagate systematically eastward and upward. Their phase speed is 500 km·d^{-1} (eastward) and 0.7 km·d^{-1} (upward). At 15°N, eastward phase propagation of low-frequency modes is not as well defined. At this latitude, zonal wind perturbations are prominent in the lower troposphere over the monsoon region, and also in the upper troposphere over the western Pacific. As a part of the E-W interaction between these two regimes, there occurs a downward progression of westerly (or easterly) perturbations over to the Arabian Sea region. The downward phase of the westerly (easterly) modes corresponds to the commencement of the "active" ("break") monsoons. The region of low-level westerly (easterly) perturbations then moves progressively eastward and downward across South and Southeast Asia along 15°N.

7. The monsoon region represents the major energy source for 40-50 day perturbations. In the equatorial region, the low-frequency oscillations owe their existence to a lateral geopotential wave-energy flux from the monsoon region. In and around the upper tropospheric trough over the subtropical North Pacific, non-geostrophic upper winds at 200 hPa are always directed down the gradient of the geopotential height, implying the generation of kinetic energy through pressure work.

Acknowledgements. The authors are indebted to Mrs. Dixie Zee for her assistance in data processing, computer programming, and final editing of the manuscript. They also thank Mr. Louis Oda for drafting the figures and Ms. S. N. H. Chock for typing the manuscript.

This research has been supported jointly by the National Science Foundation and the National Oceanic and Atmospheric Administration, Washington, D. C., under research Grant No. ATM-8206350-A01.

Appendix I: List of Symbols

λ, ϕ longitude and latitude

dx, dy $a \cos\phi d\lambda$ and $ad\phi$; a, radius of earth

p pressure

t time

u, v zonal and meridional winds

T temperature

Φ geopotential height

q water vapor mixing ratio

r relative humidity

ζ, D vorticity and divergence

f Coriolis parameter

Q_1 large-scale apparent heat source

Q_2 large-scale apparent moisture sink

ω vertical p-velocity

c_p heat capacity of air at constant pressure

L latent heat of condensation

$(^-)$ zonal average between 60° and 150°E, or from 30°E to 150°W

$\langle\ \rangle$ five-month average (May-September)

$(\ )^*$ departure from five-month average

$(\ )'$ intraseasonal transient variations defined in eq. (8)

$(\sim)$ 40-50 day filtering

$(\ )_i$ composites for category i

$[\ ]$ category mean between categories 1 to 8

References

[1] Madden R A, Julian P R. Detection of a 40-50 day oscillation in the zonal wind in the tropical Pacific[J]. J Atmos Sci, 1971, 28: 702-708.

[2] Madden R A, Julian P R. Description of global scale circulation cells in the tropics with a 40-50 day period [J]. J Atmos Sci, 1972, 29: 1109-1123.

[3] Yasunari T. A quasi-stationary appearance of 30-40 day period in the cloudiness fluctuations during the summer monsoon over India[J]. J Meteor Soc Japan, 1980, 58: 225-229.

[4] Yasunari T. Structure of an Indian summer monsoon system with a period around 40 days[J]. J Meteor Soc Japan,1981,59:336-354.

[5] Ahlquist J. Dominant empirical modes in the Indian summer monsoon flow patterns at 850-hPa[R]. Paper presented at the International Conference on Early Results of FGGE and Largescale Aspects of its Monsoon Experiments. Tallahassee,1981,12-17 January.

[6] Krishnamurti T N, Ardanuy P, Ramanathan Y, Subrahmanyam D. The 30-50 day mode at 850mb during MONEX[J]. J Atmos Sci,1982,39:2088-2095.

[7] Murakami M. Analysis of the deep convective activity over the western Pacific and Southeast Asia. Part II: Seasonal and intraseasonal variations during northern summer[J]. J Meteor Soc Japan, 1983, 62: 88-108.

[8] Lorenc A C. The evolution of planetary scale 200mb divergences during the FGGE year[J]. MET 0 20 Technical Note II/210, Met. Office, Bracknell, Berkshire, RG 12,2SZ,1983.

[9] Yanai M, Esbensen S, Chu J-H. Determination of bulk properties of tropical cloud clusters from large-scale heat and moisture budgets[J]. J Atmos Sci,1973,30:611-627.

[10] Zangvil A. Temporal and spatial behavior of large-scale disturbances in tropical cloudiness deduced from satellite brightness data[J]. Mon Wea Rev,1975,103:904-920.

[11] Hayashi Y. Space-time cross spectral analysis using the Maximum Entropy Method[J]. J Meteor Soc Japan,1981,59:620-624.

[12] Webster P. Mechanisms of low-frequency monsoon variability[J]. J Atmos Sci,1983,40: 2110-2124.

[13] Murakami T, Nakazawa T, He J. 40-50 day oscillations during the 1979 Northern Hemisphere summer [J]. UHMET 83-02. Department of Meteorology, University of Hawaii,1983.

[14] Sikka D R, Grossman R. Summer MONEX chronological weather summary[J]. International MONEX Management Center, New Delhi, India,1980.

[15] Krishnamurti T N, Ardanuy P, Ramanathan Y, Pasch R. On the onset vortex of the summer monsoon[J]. Mon Wea Rev,1981,109:344-363.

[16] Hayashi Y. Vertical-zonal propagation of a stationary planetary wavepacket[J]. J Atmos Sci,1981,38: 1197-1205.

[17] Krishnamurti T N, Molinari J, Pan H-L, Wong V. Downstream amplification and formation of monsoon disturbances[J]. Mon Wea Rev,1977,105:1281-1297.

[18] Murakami M. Large-scale aspects of deep convective activity over the GATE area[J]. Mon Wea Rev, 1979,107:994-1013.

[19] Panofsky H A, Brier G W. Some applications of statistics to meteorology[J]. The Pennsylvania State University,1958:219.

[20] Murakami T, Sumi A. Southern Hemisphere summer monsoon circulation during the 1978-1979 WMONEX. Part II: Onset, active and break monsoons[J]. J Meteor Soc Japan,1982,60: 646-671.

[21] Murakami T, Iwashima T, Nakazawa T. Heat, moisture, and vorticity budget before and after the onset of the 1978-79 Southern Hemisphere summer monsoon[J]. J Meteor Soc Japan,1983,62:69-87.

1979年夏季我国东部各纬带水汽输送周期振荡的初步分析*

何金海[1]　于新文[2]

(1. 南京气象学院,南京 210044;2. 新疆气象台,乌鲁木齐 830002)

摘要:本文根据1979年5—8月逐日850 hPa的探空和测风资料计算了我国东部各纬度带平均的经向水汽输送,并进行了功率谱分析,结果发现与我国夏季雨带季节性北移紧密联系的水汽输送随着地理位置的变化呈现明显不同振荡特性。南海到华南地区,存在准40天和4天左右的显著周期;长江流域存在80天和8天左右的显著周期;华北地区存在5天左右的显著周期。并且综合分析表明,我国华南和东部地区的降水也呈现出40~50天的振荡特性,且与来自南海地区水汽输送的变化位相一致。

1 引　言

　　水汽的输送和辐合是产生降水的基本条件。为了研究东亚和我国大陆东部夏季风降水的变化规律,国内外许多作者对有关的水汽来源和水汽输送特点问题进行了分析研究,得到了许多有意义的结果。

　　关于水汽来源问题,Murakami[1]在1959年指出,东亚水汽输送在很大程度上受二股气流的控制。即700 hPa上的印度西南季风和环绕副高的近地面东南气流,前者在东亚季风的早期阶段起主要作用,后者则在晚期阶段起重要作用。沈建柱[2]进一步指出,就气候平均而言,由南面输入我国东部地区的水汽流主要有二支,一支为西南水汽流,另一支为南和东南水汽流。水汽主要由南界(广东和广西)输入,其量值变化主要和西南季风强弱有关。沈如桂和黄更生[3]利用700 hPa轨迹追踪法指出我国西南地区和长江流域地区降水过程中的水汽主要靠来自孟加拉湾的热带季风和来自太平洋及南海的副热带季风的输送。陈世训等[4]和罗绍华[5]从计算水汽收支出发除了讨论我国大陆东部的水汽来源问题外,还讨论了旱、涝年水汽输送的不同特征。他们都指出,当涝年时,水汽辐合区位于我国东部大陆;旱年时,输入的水汽量小于输出量,大陆成为水汽的辐散区。

　　关于我国东部大陆上水汽输送的周期性至今却很少有人涉及,金祖辉[6]指出过南海地区的水汽输送有明显的周期性,其显著周期有13~14天、8~10天和3~4天。本文作者曾经在文献[7]中详细讨论过亚洲季风区域40~50天的周期振荡及其水汽输送场的变化,并指出,随着阿拉伯海南部穿越赤道的水汽输送的周期性振荡,在中国南海和西太平洋地区向北的水汽输送也发生显著的周期性振荡,最大时为1410单位,最小为688单位(如图1所示),它们之间

---

* 本文发表于《热带气象》,1986年第2卷第1期,9-16.

联系的纽带为40天振荡的向东和向北的位相传播,时间滞后为10~12天,这里我们自然要问,由中国南海北部向北的水汽输送的振荡如何向北传播影响中国大陆东部的夏季风降水？除了40天周期振荡之外,中国大陆东部的水输送还有哪些盛行周期,和造成这些周期的可能原因是什么？这些就是本文要分析的主要问题。

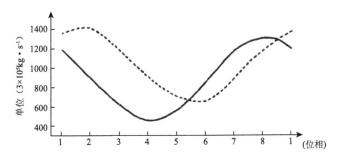

图1 水汽输送随位相的变化(图中实线为阿拉伯海地区穿越赤道的输送,虚线为穿越中国南海北界的输送,引自文献[7])

2 资料和计算方法

本文采用1979年5—8月逐日08点850 hPa的探空和测风资料,计算了15°~50°N和100°~120°E区域内40个站的水汽通量。在此基础上计算了南海地区(平均纬度18°N)和五个纬带(22.5°N,27.5°N,32.5°N,37.5°N和42.5°N)上的平均的经向水汽通量值。南海地区用西沙、崖县、海口和阳江四站平均值来代表,考虑到经向水汽输送可能对夏季风降水的作用更为重要,因此我们只对平均的经向水汽输送进行功率谱分析。计算方法采用文献[8]中的方案,样本数$n=123$,最大后延$m=41$。计算表明,当后延数为1时,各序列的自相关系数$r(1)$都普遍为较大的正值,故采用红噪声过程进行显著性检验。

比湿q是通过露点温度T_d和气压计算得到的。其计算公式如下:

$$E=\begin{cases}6.11\times 10^{\frac{7.5\times T_d}{237.3+T_d}} & \text{当 } T_d\leqslant -20\text{℃}\\ 6.11\times 10^{\frac{9.5\times T_d}{265.5+T_d}} & \text{当 } T_d> -20\text{℃}\end{cases}$$

$$q=622\times\frac{E}{P}$$

式中,q以$g\cdot kg^{-1}$为单位。

通过截面积高为1 hPa,底边长为1 cm的经向水汽通量值F_m由下式得到:

$$F_m=\frac{1}{g}qv$$

式中,g为重力加速度,v为经向风速,F_m的单位为$g\cdot(cm\cdot hPa\cdot s)^{-1}$。

为了研究某种周期的位相传播,常用的方法是计算不同地区的交叉谱,通过分析位相角的分布型式来考察其移动,本文采用另外一种比较简便的方法,即对经向水汽输送的某种显著周期,采用一种带通滤波的方法将它与其他周期分离出来,然后考察其位相传播。这种带通滤波的方法由Shanks提出并由M. Murakami做了改进,具体公式参阅文献[7]中的有关部分。

3 计算结果分析

3.1 各纬度带上经向水汽输送的盛行周期

各纬度带上平均的经向水汽输送的功率谱如图 2 所示。

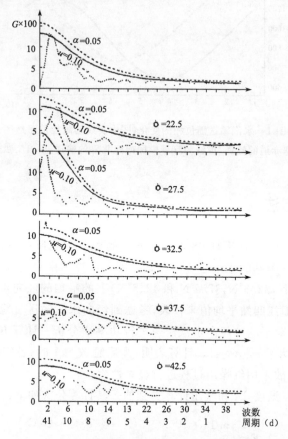

图 2　我国东部各纬带上经向水汽输送的功率谱曲线

在南海和 22.5°N 纬带上,均有一个最显著的谱峰值,其对应的周期为 40 天,均超过 0.10 的信度。我们认为这一地区的 40 天周期振荡与文献[7]中指出的水汽输送 40~50 天的振荡紧密相关。另外,还有一个周期为 4 天的谱峰值,它们接近或起过 0.10 信度线(22.5°N 上超过 0.05 信度线),这与文献[6]指出的 3~4 天周期是一致的:这种水汽输送的短周期与南海地区天气尺度的季风扰动相联系。

在 27.5°N 和 32.5°N 纬带上,均有一个 80 天周期的显著谱峰值,在 32.5°N 纬带上,这一峰值超过 0.05 的信度线,它可能与我国雨带的季节性北跳和南撤相联系。另外还有一个 7~8 天周期的谱峰值,它们接近或超过 0.10 的信度线(27.5°N 上超过 0.05 的信度线)。徐淑英和司有元等[9]指出,梅雨锋存在 8~9 天的振荡周期。陆龙骅等[10]指出,准一周振荡是 1979 年夏季青藏高原地区各气象要素及热力参数普遍存在的一种主振荡周期,并由西向东传播。

因此,这一区域的水汽输送的 7~8 天周期是与这一主振荡周期相关联的。除了 80 天和 7~8 天的周期外,还有 2~3 天的周期谱峰值,这与梅雨锋上的短期扰动有关。顺便指出,周静亚等[10]利用 1980 年夏季降水资料也发现存在 80 天和 8 天的显著周期。

在 37.5°N 和 42.5°N 纬度带上,表现最为显著的是 5 天左右周期的谱峰值,在 42.5°N 上的峰值甚至超过了 0.05 的信度。这一纬带上的温度和水汽输送的时间演变曲线(图略)表明,水汽输送的负值谷点总是在温度谷点 1~2 天之后,滞后相关系数达到 0.05 的显著性水平。这说明,这一地区水汽输送的 5 天周期是中纬度冷空气活动周期的表现。

总之,南海和华南地区存在的 40 天和 4 天周期,长江流域存在的 80 天和 7~8 天周期以及华北地区存在的 5 天周期,不仅具有统计上的显著性,且具有空间上的连续性,同时天气学意义比较清楚。

3.2 40 天周期的向北移动

为了进一步讨论低纬存在的 40 天振荡,我们对各纬带和南海地区的水汽通量序列进行了 40~50 天的带通滤波。滤波结果如图 3 所示。在南海和 22.5°N 纬带上 40~50 天的周期振荡的振幅显著,而至 27.5°N 纬带,振幅则显著减弱(图略),这与谱分析的结果一致。说明 40 天的振荡传播到 25°~30°N 之间就显著减弱了。因此可以把长江流域作为 40~50 天振荡的北界。

从图 3 还可以看出,6 月中旬以后(即印度季风爆发和长江流域梅雨建立),40~50 天振荡的振幅显著加大了,何金海和村上多喜雄[11]指出,在印度季风爆发后,最显著的特点是一条强水汽输送带的建立,它大约沿着 10°~15°N,从阿拉伯海穿过印度南部、孟加拉湾和中印半岛到南海,然后折向中国南部直到日本以至更远的洋面上。振幅加大可能与这条强水汽输送带的建立相关联。

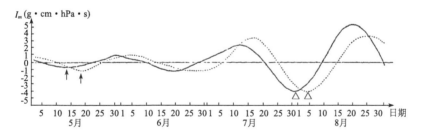

图 3　40~50 天带通滤波的水汽输送曲线(实线表示南海地区,虚线代表 22.5°N 纬带,图中符号 ↑ 和 △ 均指示水汽输送曲线典型移速的谷点)

图 3 也清楚地显示出 40 天振荡向北的位相传播,可以看出,南海地区(平均纬度 18°N)和 22.5°N 纬带上具有大约 5 天的位相差,其位相移动的速度约为 1 纬距/天,这与 Krisnamurd 等和柯史钊等指出的 40~50 天周期向北移动的速度是一致的。

这里我们还必须注意到这样一个有意义的事实:南海地区水汽输送 40~50 天振荡的每一次增强过程均对应着一次重要的大型天气过程。5 月中旬的第一次增强过程对应着南海季风的建立,6 月中下旬的第二次增强过程对应着长江流域的梅雨建立,7 月底以后的第三次增强过程对应着雨带在华北地区的稳定和维持。文献[7]指出,40~50 天周期扰动的向北和向东的位相传播紧密联系着南亚中部和东南亚季风的活跃和中断的位相转变,那么,我们是否可以从上述事实出发提出南海地区水汽输送的 40~50 天周期振荡的位相转变紧密联系着我国雨带的季节性位移和在不同地区的稳定维持。当然,由于我国大陆东部没有青藏高原的阻挡,中高纬的冷空气对我国大陆东部雨带的季节性移动和中期振荡必有重要的影响,因此雨带推进的实际过程不会如同上面所描述的那样简单。但是无论如何,南海地区作为夏季水汽输入我国大陆的关键地区,那里水汽输送的振荡对于我国雨带季节性推进应该是有重要指导意义的。

为了考察经向水汽输送的准 40 天周期振荡对我国东部降水的可能影响及它们之间的位相关系,我们对我国东部三个纬带(20°N、30°N、40°N)上每日一次的降水资料采用与文献[7]中完全类似的方法进行综合分析,其位相 1~8 的确也是根据 850 hPa 上沿 11°~25°N,60°~150°E 之间平均的 40~50 天滤波纬向风 $\vec{u'}$ 的时间序列的周期特点来进行的。这样做的目的是为了便于将中国东部降水和文献[7]中所指出的各位相水汽输送的变化联系起来加以分析,从而也可以检验文献[7]最后关于"索马里急流的变动通过 40~50 天的周期振荡的向东向北传播对中国东部降水发生重要影响的推测"是否具有一定的客观依据。另外应该指出本文在对降水资料进综合分析时没有进行 40~50 天的滤波,而只是进行 5 天滑动平均以消去短期部分。这样一点改变并不影响综合分析的结果。

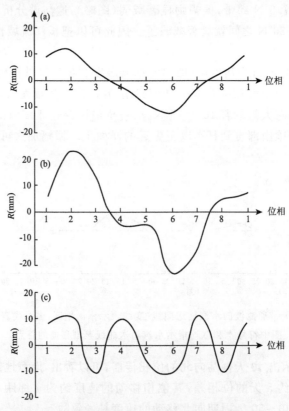

图 4　我国东部地区和各纬带降水随位相的演变曲线
a 为东部地区,b 为 20°N,c 为 30°N

由降水各位相综合演变曲线(图 4)我们可以看到,20°N 纬带上的平均降水和我国整个东部的降水(20°~40°N,105°~120°E)在第 2 位相有一个极大值,在第 6 位相有一个极小值(图 4a 和图 4b 所示)。这就是说它们在 8 个位相时段内(40~50 天)完成了一个循环,而在 30°N

纬带上的降水在8个位相时段内却大约完成了三个循环,就是说准40～45天的振荡在这一纬带上已经不很明显了,这是否因为长江流域的降水更具中低纬冷、暖空气相互作用的性质值得考虑。我们注意到,图4a和b和图1中的虚线有很一致的关系,这对文献[7]中所作的推测是一个有力的佐证,证明了索马里越赤道气流的准40天振荡引起的水汽输送场的变化及其向东向北传播确实对我国华南和整个东部地区的降水有着巨大的影响。

3 长江流域水汽输送的季节性突变

水汽输送作为表征夏季风的一个特征参数能够较好地描述我国雨带的推进和季节性突变。下面,我们试以27.5°N纬带(代表长江流域)上的5—8月的水汽输送的演变为例来说明水汽输送的季节性突变和梅雨带变动之间的关系。

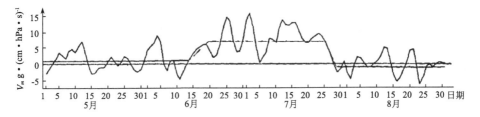

图5　5—8月27.5°N纬带上的经向水汽输送的演变曲线

如图5所示,长江流域的水汽输送在振荡过程中具有两次急剧的变化。在6月中旬以后有一次南风输送的急剧增强过程,6月下旬以后直至7月下旬以前一直维持强水汽输送。在7月下旬有一次输送的急剧下降过程,此后水汽输送值一直在较弱的水平上振荡。我们知道,1979年长江流域的梅雨是6月19日建立,出梅在7月下旬(中央台确定)。因此上述水汽输送的两次急剧变化分别对应着梅雨在长江流域的建立和结束。

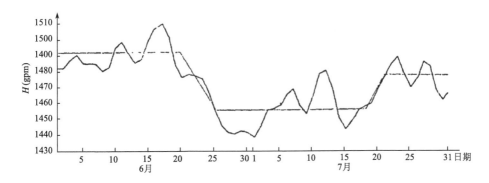

图6　5—7月长江流域850 hPa高度演变曲线

为了考察水汽输送的突变和梅雨锋带变动的关系,我们以27.5°N纬带上平均的(取56区492站,57区749站,汉口,长沙,58区的725站,南京等6个站的平均)850 hPa高度的变化表征梅雨锋带的变动。如图6所示,长江流域850 hPa的高度曲线和水汽输送变化有非常类似的特点。6月下旬的急降表明与梅雨相配合的850 hPa切变线在长江流域建立,7月下旬的急升表明与梅雨锋配合的切变线迅速北移,即长江流域梅雨结束,雨带北移。

综上所述,水汽输送的显著变化确实与雨带的季节性变动密切有关,具有鲜明的天气意义。

4 结果讨论

(1)1979年夏季我国东部大陆经向水汽输送存在80天,40~50天,7~8天和5天左右的周期振荡。

40~50天的周期振荡在低纬显著,向北传播到25°~30°N附近,振幅显著减小,传播速度大约为1纬距/天。降水也有类似的振荡特性。

长江流域存在80天和7~8天的周期振荡,高纬地区以5天左右的振荡为最显著。

(2)众所周知,1979年40~50天的周期振荡在南亚季风区域是非常显著的,并且联系着南亚季风活跃和中断的位相转变。这种振荡向东向北传播也影响到中国大陆东部夏季风的进程。正如上文所指出,南海地区经向水汽输送40天振荡的每一次增强过程都与中国大陆东部雨带的变动有某种联系。但是,由于中国大陆东部没有像青藏高原那样的山脉阻挡,加之大部分地区处于副热带,因此中国大陆东部夏季风的进程具有比南亚季风更为复杂的特点。40天振荡向北传播不能到达很高的纬度,中高纬冷空气的活动是影响我国大陆大部夏季风进程的另一重要因素。由于中低纬系统和波动的相互作用使得我国大陆东部的水汽输送从低纬到高纬呈现出明显不同的地域特点。

参考文献

[1] Murakami T. The General Circulation and Water-Vapour Balance over the Far East during the Rainy Season[J]. Geophysical Magazine,1959,29(2):137-171.
[2] 沈建柱.我国大气中的水分平衡[C]//全国热带夏季风学术会议文集.昆明:云南人民出版社,1982.
[3] 沈如桂,黄更生.1980年夏季热带季风环流与我国南方降水水汽输送的关系[C]//全国热带夏季风学术会议文集.昆明:云南人民出版社,1981.
[4] 陈世训,高绍风,杨崧,5—6月我国南方降水的水汽来源及其异常[C]//全国热带夏季风学术会议文集.昆明:云南人民出版社,1981.
[5] 罗绍华.盛夏期间我国大陆东南部低纬地区的水汽输送和水分平衡[C]//全国热带夏季风学术会议文集.昆明:云南人民出版社,1981.
[6] 金祖辉.1979年夏季南海地区水汽收支[C]//全国热带夏季风学术会议文集.昆明:云南人民出版社,1981.
[7] 何金海,Murakami T,Nakazawa T.1979年夏季亚洲季风区域40-50天周期振荡的环流及其水汽输送场的变化[J].南京气象学院学报,1984,2.
[8] 黄嘉佑,李黄.气象中的谱分析[M].北京:气象出版社,1984.
[9] 徐淑英,司有元,殷延珍.1979年季风系统振荡的初步分析[C]//青藏高原气象科学实验文集(2).北京:科学出版社,1984.
[10] 陆龙骅,朱福康,陈咸吉,等.1979年夏季青藏高原地区的中期振荡特性[C]//青藏高原气象科学实验文集(2).北京:科学出版社,1984.
[11] 何金海,村上多喜雄.1979年6月东亚和南亚上空的水汽通量[J].南京气象学院学报,1983,2.

The Southern Hemisphere Mid-Latitude Quasi-40-day Periodic Oscillation with Its Effect on the Northern Hemisphere Summer Monsoon Circulation[*]

He Jinhai　Chen Lizhen

(Nanjing Institute of Meteorology, Nanjing 210044)

Abstract: By using the method of power spectrum combined with band-pass filtering with May-September 1982 grid data from ECMWF, the spatial structure and propagation characteristics are tentatively examined of the SH (Southern Hemisphere) mid-latitude quasi-40-day (ranging from 30-60 days) periodic oscillation (QPO) together with the relation to NH (Northern Hemisphere) summer monsoon, with the result that there exists similar periodicity in the development of baroclinicity and activities of cold air at the same latitudes, and in response to this the zonal wind shows profound QPO with a nearly vertical axis of disturbance and that the air activities can act as periodic external forcing for the monsoon, which intensifies the west wind on the south side of the Mascarene or Australian high, and then the system itself, leading to the reinforcement of the SE trade wind on the north side, followed by the strengthening of cross-equatorial flow that, in turn, causes active monsoon with its northward march over the eastern part of China. The process is responsible for the low-frequency oscillation propagated in a meridional direction, which confirms the speculation of the author.

1 Introduction

Quasi-40-day oscillation has increasingly become one of the problems of great concern for meteorologists of the world. At the 1987 PRC-U. S. Kunming Monsoon Workshop an enlightened discussion was made during which many significant problems were raised that deserved further investigation.

Many researchers have studied and generalized the large-scale features of the QPO (see, for example[1-2]), but uncertainties still remain. For instance, the eastward propagation of QPO's phase has been regarded as one of its essential characteristics. Some studies, howev-

---

[*] 本文发表于《Acta Meteorologica Sinica》,1988 年第 2 卷第 3 期,331-339。其中文版发表于《南京气象学院学报》,1989 年第 12 卷第 1 期,11-18。

er, show that a westward transfer is occasionally observed (see, for example[3]). What is the caused? Besides, a considerable northward displacement of the phase occurs in the Asian summer monsoon [4], which is assumed to be associated with periodic impact due to the SH cold air activities [5]. If so, is there any possibility for QPO of the winter monsoon to show the phase transfer toward the south? With regard to the origin of the QPO there have been two hypotheses: one asserting that it is due to free waves created by the dynamic mechanism within the atmosphere and the other assuming that it is due to forcing external to the atmosphere[6]. And generally the former views the QPO as a type of Kelvin wave, and the relevant study[7], however, indicates that there exists conspicuous disparity in dynamics between the QPO and Kelvin waves and that the theory itself fails to interpret the QPO's meridional propagation, while the latter can, only with a forcing source to be found. Which seems to be more reasonable or how they are combined for use represents a problem that is worth further investigation.

The author[5] calculated the conditions of sensible heat, potential energy and momentum flux caused by the QPO in a meridional vertical cross section using the 1979 FGGE data and analyzed their properties at various phases with the result, in particular, that available for the monsoon QPO are two energy sources: one in the mid and upper troposphere over the monsoon trough (near 15°N) and the other in the mid and lower troposphere around 30°S. The former is related to the release of latent heat due to periodic variation of the monsoon circulation and the latter to the cold air outbreak and vigorous warming there, both taking turns in action with the phase difference of 10-12 days. Obviously such disposition favors the maintenance of the monsoon QPO, and thereupon the writer concludes that the QPO is the representation of the interaction between the SH/NH circulations (forming a unified monsoon circulation) by way of continued impacts on the NH monsoon circulation of the quasiperiodic outbreak of the SH cold air. The questions follow: Is the cold air activity really marked by such quasiperiodicity? Does it actually have an inherent relation to active and break monsoon? How does the northward impact affect the monsoon behavior? All these are problems the author is seeking to answer in the present paper.

2 Data and method

For the study, once-daily grid data of u, v, T are employed between May-September 1982 from the European Center of Medium-range Weather Forecasts (ECMWF), with the longitude/latitude distance 5°5°, covering 35°S~35°N and 30°~150°E.

The method [8] is used for the power spectrum analysis. But before the operation is performed of any of the sequences of $X(\tau)$, three-day running mean has to be found for the elimination of fluctuations. The sample size is 153 days and the largest lag for the lag-correlation coefficient is assumed to be $\tau=45$, leading to a maximum resolvable period of 90 days.

In order to examine the QPO over the period of May to September the band-pass filtering is used of time series of all the meteorological elements as well and chiefly 30-60 day oscillations are retained from the filtered series. To lessen the amount of computation, only mean sequences are calculated of the subregions, each spanning 30 longitudes, i. e. , 30°~60°, 60°~90°, 90°~120° and 120°~150°E between 30°~150°E, all covering 10 latitudes. Amplitudes of the filtering curve show the QPO's intensity in each subregion and the orderly-ranged peak- and valley-values on the curve indicate its propagative character and interrelation between the oscillations of these elements. For details of the filtering scheme refer to He et al.[9].

3 Structure and zonal propagation of QPO at SH mid-latitudes

To reveal the periodicity of the SH cold air behaviors, the baroclinicity index ΔT_{850} ($T_{25°S} - T_{35°S}$) and 850-hPa temperature T are chosen as parameters to perform power spectrum analyses.

The results are delineated in Fig. 1. It is apparent that the mean baroclinicity index $\Delta \overline{T}$ shows considerable spectral peak value (solid line) in the period of 30 days. Tao et al.[10] attaches much importance to the baroclinicity, indicating that the growth of the SH mid-latitude baroclinicity over 40°~160°E would bring about the outbreak of cold air on a large-scale basis and formation of the corresponding high-pressure belt, with the southeasterly flow on the north side of the belt serving as the very origin of cross-equatorial current; the establishment of the current in relevant areas results in the onset and occurrence of summer monsoon in India, Indo-China Peninsula up to the South China Sea, and the western North Pacific, respectively; after this the development of the large-scale mid-latitude baroclinicity would give rise to the northward march and intensification of the monsoon. It follows that the quasiperiodic character of the baroclinicy is of much importance to the behavior of the wind, and it has been substantiated by the analyses (see Fig. 1). This figure also depicts curves for the coastal eastern Africa (30°~60°E), the Indian (60°~90°E), the western Australia (90°~120°E), respectively. Obviously, they give larger spectral peak values for the 30-day period. Also, the power spectrum analyses illustrate that the meridional wind, particularly on the southeast side of the Mascarene high (60°~90°E), shows conspicuous 30-day periodicity (figure not shown).

Fig. 2 depicts the power spectrum curves of the 850-hPa zonal wind component u of the four subregions at 30°S (averaged over 25°~35°S), which show that the areas between 30°~150° E have spectral peak values in the domain of 30 or 45 days for u, indicating that the component does possess profound low-frequency oscillations, which is related to the periodic behavior of the baroclinicity of these subregions. In addition, it can be seen that there exists quite a close similarity of the spectral structure of the area 30°~60°E to that of 60°~90°E and of 90°~120°E to the one of 120°~150°E, yet with some disparity between the western two and eastern cases, which are associated with the variation in location and intensity of the

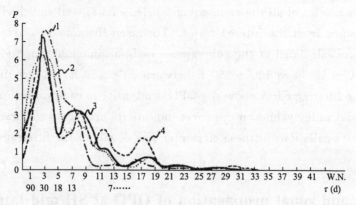

Fig. 1 850-hPa power spectrum curves of the baroclinicity parameter and temperature for 30°S. Curve 1 is for $\overline{T}(30°\sim60°E)$, 2 for $\overline{T}(60°\sim90°E)$, 3 for $\Delta\overline{T}(30°\sim150°E)$ and 4 for $\Delta\overline{T}(90°\sim120°E)$. W. N. is wave number for short

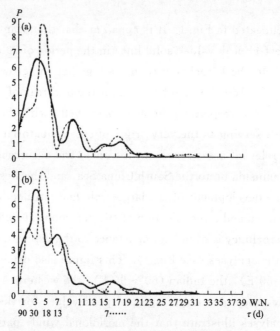

Fig. 2 Power spectrum curves of the 850-hPa zonal wind at 30°S for (a) 30°~60°E (solid line) and 60°~99°E (broken line); (b) 90°~120°E (solid line) and 120°~150°E (broken line)

Australian and Mascarene high, respectively.

Figs. 3 and 4 illustrate the filtering① curves of the 850- and 200-hPa zonal wind of the four sections between 30°~150°E for 30°S, respectively. Here we can see that the zonal wind shows evident oscillation, with a mean amplitude of 2, the maximum exceeding 3 at 850-hPa and of 3-5, the maximum in excess of 10 m·s^{-1} at 200-hPa, in good agreement with

---

① In this paper, "filtering" refers to band-pass filtering of 30-60 days, or simply quasi-40-day.

the analyses of power spectrum. Of course, these curves show their variations with time, which indicates seasonal properties of the low-frequency oscillation. Closely looking at the 850-hPa curve, we shall notice that the changes in curves (a) and (b) (denoting 30°~60° and 60°~90° E, separately) are opposite to those of curves (c) and (d) (the others), that is, small amplitudes in the western sections versus greater in the eastern for May-June and the reversal occurring after June. It remains a question to be studied whether the time-dependent amplitudes reflect the difference in their seasonal behaviors between the two highs mentioned.

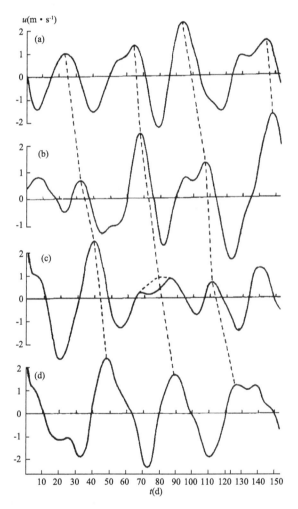

Fig. 3 Filtering curves of the 850-hPa zonal wind at 30°S for 30°~60° (a), 60°~90° (b), 90°~120° (c) and 120°~150°E (d)

We shall now deal with the properties of the propagation of the zonal wind oscillation. Fig. 3 shows that the eastward movement of the anomalous west wind has been revealed three times over 30°~150° E at 850-hPa, as indicated by broken lines. The inclination of the line characterizes the rate at which the wind moves towards the east, with an average speed of 3-4 longitudes per day, such being the case with the anomalous east wind. The eastward-moving features (Fig. 4) appear less systematic and evident at 200 than at 850-hPa, and particularly the curve for 60°～90°E is strongly distorted for the period after July, thus making the systematic propagation twisted. Nevertheless, the feature of its eastward transfer is basically existent.

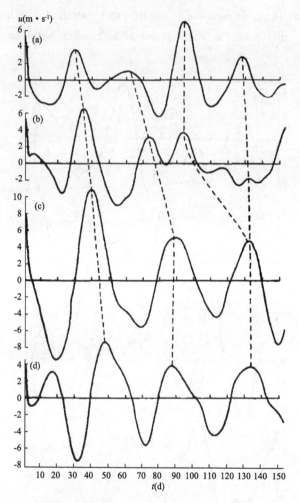

Fig. 4　Same as in Fig. 3 except for 200-hPa

　　Comparing the positions of the crests and tips in Figs. 3 and 4, we can easily see that there are 20 instances showing that the 200- and 850-hPa winds are almost in phase and only 7 anti-phase cases (i. e., the strong west wind at 200 and vigorous east wind at 850-hPa or vice versa, with 6 out-of-phase occasions). For 120°～150° E, in particular, except a slight disorder observed prior to May 20, the 200- and 850-hPa oscillations are almost in phase with the crests and tips nearly coincided, respectively, as shown by curve d in Figs. 3 and 4, implying that, with height, the axis of low-frequency disturbance of the zonal wind approaches the vertical and the amplitude is increased, a fact which is distinct from the vertical structure of the zonal wind oscillation at tropical latitudes where, as indicated by Krishnamurti[11], the upper and low oscillations are commonly in anti-phase, with strong inclination of the disturbance axis with height.

Comparison of the filtering curve of temperature to that of the zonal wind shows that both are in approximate anti-phase, i. e., the anomalous negative temperature matching abnormal west wind or vice versa. Fig. 6 illustrates curves b and c in an approximate anti-phase over 30°~150° E, i. e., the 200-hPa anomalous west (east) wind matching the 850-hPa low (high) temperature. This seems to follow the law of thermal wind, because, if the large-scale drop of temperature at mid latitudes suggests the strengthening of temperature gradients, then the reinforcement of upper west can be expected. According to this, can we make a further inference that the SH mid-latitude zonal wind QPO satisfies, at least partially, the requirements of geostrophic wind?

4 Effect on NH monsoon of quasiperiodic activities of SH cold air

The foregoing analyses have borne out the existence of a cold-air source at 30° S. If the source could actually exert significant influence on the monsoon, then the latter would reasonably show the same periodicity in response. To illustrate this, power spectrum analyses are performed of the divergence series averaged over 30°~150° E for 850 (200-hPa) and 20 (25°N), respectively, with the results shown in Fig. 5. It can be seen that the power spectrum curves of 200- and 850-hPa divergence field in the vicinity of the trough have considerable low-frequency spectral peak values, with a period of 30 and 45 days for 850 and 200-hPa, respectively. Also, the quasi-weekly and biweekly values can be discerned in the same figure. All these periods are known to be those typical of the monsoon circulation system.

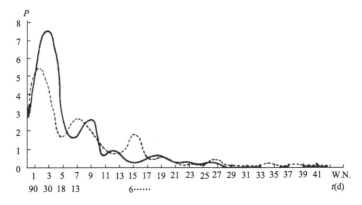

Fig. 5 30°~150°E averaged divergence power spectra in the neighborhood of the Asian monsoon trough for 850-hPa at 20°N (solid line) and 200-hPa at 25°N (broken line)

In an attempt to show the response of the monsoon to the SH cold air activity, we choose the zonal wind on the south side of the trough (15°N) as a parameter of the monsoon behavior and carry out the filtration of it, and the results are shown in Fig. 6. It can be seen that for May-September the SH large-scale activities over 30°~150°E are observed three times, each in correspondence to the intensification of the west wind on the south side of the trough, whereas the SH warming occurs as many times, each accompanied by the reduction of the wind. This is clearly denoted by the connecting line of the figure. Surprisingly enough, the lags average 10~12 days, which is the same result as the author has achieved[5], showing the same correspondence available for 1979 and 1982.

Whether the discovery is of general value remains a problem to be further investigated.

Also, Fig. 6 delineates the anti-phase oscillation between the curve of the 200-hPa zonal wind and that of the 850-hPa temperature, which was pointed out before. It is, however, of interest that the fact reveals more harmonious phase pattern between the oscillations of the SH west wind and the NH summer monsoon. This is so-called "resonance" concerning the variations in the circulations of both hemispheres, acquired by Tao et al.[10] using 1979 data. Therefore, we may as well conclude that some form of resonance does exist between the two circulations.

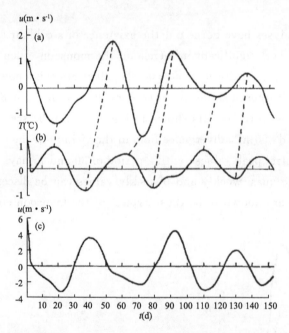

Fig. 6 Relationship among the NH monsoon, the SH cold air activity and the SH 200-hPa west wind over 30°~150°E and the filtering curve of 850-hPa $\bar{u}$(10°~20°N) (a); 850-hPa $\bar{T}$ (25°~35°S) (b); 200-hPa $\bar{u}$(25°~35°S) (c)

From the above it is certain that the SH mid-latitude cold air behavior does strongly affect the Asian summer monsoon. Then, in what manner is this realized? For each of the four subregions, calculation is made of the 850-hPa cross-equatorial flow and the results are filtered, and then the corresponding phases on the curves of temperature, cross-equatorial flow and monsoon are examined. Analyses show that the SH effect on the monsoon is conveyed chiefly by way of the subregion 30°~60°E and 90°~120°E as shown in Figs. 6 and 7, respectively, in good correspondence to the Somali and near 105°E cross-equatorial current, sepa-

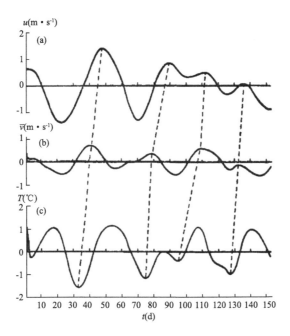

Fig. 7 Filtering curves of 850-hPa parameters at different zonal belts over 30°~60°E, with $\bar{u}$ at 15°N (a); $\bar{v}$ at the equator (b); $\bar{T}$ at 30°S (c)

rately.

By referring to Fig. 7 it is apparent that over 30°~60° E profound strengthening of the cross-equatorial flow and the west wind over the Arabian Sea always takes place in response to vigorous cold air activity, with the lag, on average, of 10 days or so and the similar event is also found in Fig. 8. Over 90°~120°E three instances of vigorous outbreak are observed, of which the first two are accompanied by significant reinforcement of the cross-equatorial flow and then the monsoon, with almost the same lag length, but on the last occasion some disorder happens and the NH west wind is little intensified. It is worth noting that for 30°~60°E the intense cold air event in the first decade of September ($t=128$) shows no effect, which may suggest that such influence is likely to involve the features of large-scale circulation of the NH, for by that time the summer monsoon is at the stage of retreat and large-scale circulation in the transition.

In addition to analyses with the filtering curves, the cross-spectra are calculated of the SH mid-latitude temperature and the NH 850-hPa zonal divergence at 20°N. Results show a peak value in a period of about 30 days, with a lag of 8~10 days for the two elements, which indicates that results from the curve analyses have statistical significance.

5 Results and discussion

(1) Quasi-40-day periodicity is really existent in the development of large-scale baroclinicity and cold air activities at the SH mid-latitudes, and it is also typical of the zonal and

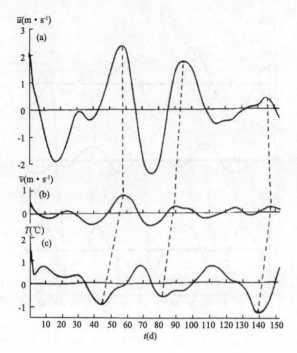

Fig. 8 as in Fig. 7 but for 90°~120°E

meridional wind. Analyses indicate that more or less difference exists in the spectral structures in close association with the position and intensity of the Mascarene and Australian highs.

(2) The SH mid-latitude zonal wind has profound quasi-40-day periodic oscillation, with low-level mean amplitude of 2 m·s^{-1} and upper of 3-5, the maximum exceeding 10. And the oscillation is eastward-propagated, the phase speed being of 3-4 longitudes per day and the disturbance axis nearly vertical, especially over South Australia. The upper zonal wind has its oscillation in anti-phase with that of temperature and is in a "resonance" state with the west wind on the south side of the monsoon trough.

(3) Diagnostic analyses show that the SH cold air activities certainly serve as a periodic external forcing for the NH summer monsoon, which first strengthens the west wind on the south side of the Mascarene or Australian high and the system itself, resulting in the intensification of the southeast trade on the north side of the high in relevant regions, followed by the reinforcement of cross-equatorial flow, which, in turn, brings about the active monsoon of the NH and its northward march over eastern China. This process is responsible for the meridional transfer of the low-frequency oscillation. This confirms the conclusion of the author [5] concerning the possible synoptic processes of the QPO.

References

[1] Murakami T, Nakazawa T, He Jinhai. 40-50 day oscillations during the 1979 northern hemisphere sum-

mer[J]. NHMET,1983,2.

[2] Chen Longxun, Xi An. Relationship between EI Nino event and 30-60 day oscillation with OLR data [C]//in collected Papers of Met Sci and Tech, No. 11. China Met Press,1987,26-35(in Chinese).

[3] Sumathipala W L, Murakami T. ,The role of low-level northerly surge upon the enhancement of 30-60 day equatorial oscillations during winter [J]. NHMET,1987,10.

[4] Yasunari T. A quasi-stationary appearance of 30-40 day period in the cloudiness fluctuations during the summer monsoon over India[J]. J Met Ser Japan,1980,58: 225-229.

[5] He Jinhai. The transfer of physical quantities in a quasi-40-day periodic oscillation and its relation to the interaction between the northern and southern hemisphere circulations[J]. Adv Atmos Sci, 1988, 5: 107- 116.

[6] Lau K M, Peng L. Origin of, low frequency (intraseasonal) oscillations in the tropical atmosphere, Part I: the basic theory[M]. J A S,1987,44: 950-972.

[7] He Jinhai. The structure and evolution of the circulation with a quasi-40-day oscillation in a zonal cross section of the Asian monsoon[J]. J Trop Met (to tie published,in Chinese),1987.

[8] Huang Jiayou, Li Huang. Spectral Analyses in Meteorological Research[M]. China Met Press,1984: 13-52 (in Chinese).

[9] He Jinhai. Circulation with 40-50 day oscillation and changes in moisture transfer over monsoon Asia in 1979 summer[J]. J of Nanjing Inst of Met,1984: 163-175 (in Chinese).

[10] Tao Shiyan, He Shixiu, Yang Zufang. An observational study on the onset of the summer monsoon over eastern Asia in 1979 MONEX[J]. Sci Atomos Sin,1983,7: 347-355 (in Chinese).

[11] Krishnamurti T N, Gadgil. On the structure of the 30-50 day mode over the globe during FGGE[J]. Tellus 37A,1985: 336-360.

Meridional Propagation of East Asian Low-Frequency Mode and Midlatitude Low-Frequency Waves[*]

He Jinhai   Yang Song

(Nanjing Institute of Meteorology, Nanjing 210044)

Abstract: Based on June to September 1981 ECMWF grid datasets analysis is done of the characteristics of the propagation and structure of low-frequency (quasi 40 day) oscillation over eastern Asia. Results show a separating (confluence) belt for the meridional propagation of low-frequency zonal (meridional) winds at higher (lower) levels over subtropical latitudes at 120°E, revealing that the oscillation of the zonal winds is quasi-geostrophic in nature and in phase in the high- and low-level. It is also found that the eastward propagation of the high-level zonal winds around 35°N in East Asia is the result of eastward march of midlatitude low-frequency waves with 60—90 longitude wavelength and speed of 1.5—2.0 longitudes per day. In addition, such low-frequency vortices, when moving over the coastwise region, tend to develop, accompanied by sharp oscillation in the westerly jetstream over eastern Asia.

1 Introduction

30—50 day periodic oscillation (hereafter designated LFO) is marked by propagation as one of its basic features. Any theories of its essence and origin must be able to interpret the propagation. Madden and Julian[1] and Murakami et al.[2] indicated that as a rule, LFO travels eastward at equatorial latitudes and northward in summer monsoon area. Lau and Peng[3] documented that the eastward propagation of near-equatorial LFO is accounted for in terms of the response of Kelvin waves on the east side of the disturbing heat source after the introduction into their study the mobile wave CISK mechanism. They, however, failed to explain its northward propagation and the travelling of extratropical LFO either.

Obviously, the meridional propagation of LFO should be viewed as an equally important feature, and however, study in this respect is relatively insufficient. Some researchers (e.g., [4-6]) reported the meridional march of LFO in South Asia together with its relation

---

[*] 本文发表于《Acta Meteorologica Sinica》,1990 年第 4 卷第 5 期,536-544. 其中文版发表于《气象学报》,1992 年第 50 卷第 2 期,190-198.

to Indian summer monsoon. In contrast, still fewer papers were published concerning the meridional propagation over eastern Asia. In their numerical study of the effect of the Southern Hemisphere midlatitude low-frequency cold air activity on Asian summer monsoon using 1982 data, He et al.[7] indicated that low-level meridional wind anomalies due to the impact of the cold air on the south side of Australian high have influence upon eastern Asian summer monsoon and China's Meiyu during its northward match from the NW side (around 10°S) of the high via 100°E. On the other hand, is there any limit to this northward march? Is there any difference in travelling between higher and low levels, and what are the characteristics of the propagation for other years in view of interannual variability?

LFO is of global nature, i.e., it occurs both in tropics and in extratropics. It is safe to say that tropical LFO has been much examined compared to its counterpart. In his study of global LFO using FGGE datasets, Krishnamurti et al.[8] proposed the concept of "low-frequency storms" indicating that many of the "storms" either cyclonic or anticyclonic, are stationary, with the exclusion of part of them characterized by meridional movement, which, as anomalous systems, all tend to move eastward in the neighborhood of annually-averaged vigorous wester-lies. However, the wave characteristics of these "storms" remain obscure.

Chen and Xie[9] analysed fully the characteristic propagation of LFO in terms of day-to-day OLR data of 1975-1983, indicating that statistically LFO travels dominantly westward over 15°~30° N and eastward around 37.5° N and at 120°E in tropics/subtropics (extratropics) LFO propagates northward (southward), resulting in the meeting about the subtropical belt. Obviously, the following questions are of interest. Do such propagation characteristics continue to be in wind field? What is the cause of the eastward travel of midlatitude (around 37.5° N) LFO?

East Asia is a place of the co-existence and vigorous display of tropical and subtropical monsoons[10] and a region where LFO shows evident meridional propagation as well. Apparently, investigation of propagation features, particularly the meridional behaviors, of eastern Asian LFO is of much consequence to the appreciation of the properties of LFO at tropical and midlatitudes, to the exploration of the interaction between LFO of mid and low latitudes, and to the understanding of the nature of midlatitude LFO. This paper aims at the examination of LFO travelling meridionally in this area, and relevant features of midlatitude low-frequency waves together with its relation to eastern Asian LFO.

For this study, the ECMWF grid datasets of June-September 1981 were used. The filtered wind and height field data were obtained by means of the band-pass filtering (in the range of 35-45 days) technique used in the raw wind and height data from which the seasonal trend had been removed. For the details see He et al.[11].

2 Meridional pattern and propagation of LFO

East-Asia (120 °E) is a well-known monsoon area marked by the co-existence of tropical and subtropical monsoons. Therefore, LFO has the characteristics of its pronounced regionality and hence complexity in its distribution and propagation.

Fig. 1 is the 120°E cross section of 200-hPa time-varying zonal wind for June through September. It can be seen that between 60°N and 40°S there are 5—6 LFO centers of the wind, with the maximum amplitude of 10 m·s^{-1}, especially notable being the centers over 30°~40°N, 5°~10°N and 30°~40°S. One can also find that these centers are arranged almost in a crosswise fashion, showing the pronounced component of standing wave oscillation, a characteristic pattern associated with the particular meridional distribution of eastern Asian summer large-scale circulation systems. The first center in the midlatitude westerly wind on the north side of the South-Asian high (SAH) is related to the changing intensity and position of eastern Asian westerly jetstream, and it is also the result of the jet oscillation in association with the development of midlatitude low-frequency vortices that move eastward to this area, as will be discussed in Section Ⅲ. The second center in tropical easterly jetstream on the south side of the SAH reflects quasi-40-day variation of the jet in this area and the periodic change in the southward-spreading flow over the monsoon trough in the South-China Sea. The third center is none other than the manifestation of low-frequency variation in the Australian westerly jet-stream. In addition, a weaker oscillation center is seen both at 50°N and at the Southern Hemisphere tropical-subtropical latitudes. Noticeable is the fact that after June 20, the centers on both sides of the SAH become farther apart (which is caused mainly by the northward shift of the oscillation center over China's mainland), followed by another weaker center between 20°~30°N around July 20 that is located in easterly flow on the south side of the SAH, with propagation in various directions, indicating remarkable features of standing waves.

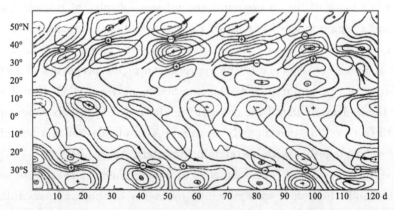

Fig. 1 120°E time cross-section of 200-hPa zonal wind LFO. Signs + and − denote the positive and negative center in the height field, respectively. Arrows indicate the direction the LFO moves. Heavy sold line is the zero valued with isopleth interval of 2 m·s^{-1}

Another feature exhibited by Fig. 1 is that north of 50°N LFO propagates northward (after the first decade of September the course is reversed), and the easterly jet oscillation moves systematically equatorward to 20°~30°S, at times even to midlatitudes, with the propagation separating belt in 20°~30°N. In association with this is a meeting belt of meridional propagation of meridional wind LFO between 20°~30°N at 850-hPa level. Inspection of Fig. 2 shows that this LFO around 10°S can travel poleward to South China where the equatorward meridional wind LFO of the NH extratropics can reach, too. The discovery of the meeting and separating belt is a fact of consequence, which bears relation to the findings of Chen and Xie[9] and He et al.[12]. Perhaps the existence of low-level confluence and high-level separating belts is somewhat associated with the orientation of basic flow about 120°E. We may just as well take the 20°~30°N belt as the division of LFO between tropical and extratropical latitudes. It can therefore be assumed that the LFO characteristics should be considerably different on both sides of this belt.

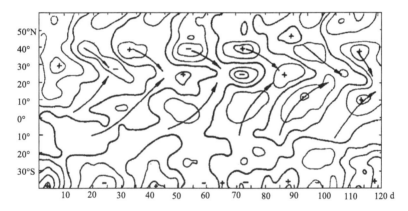

Fig. 2 120°E time cross-section of 850-hPa meridional wind LFO. Arrows and heavy solid lines denote the same as in Fig. 1, but with the isopleth interval of 1 m • s^{-1}

Fig. 3 displays the time cross-section of u component at 850-hPa, showing 5 weaker LFO centers zonally, with maximum amplitude of 4 m • s^{-1}. Three out of the five centers are notable, two situated around 30°~35°N and 20°N, separately, which move farther equatorward after 70 days, i. e., the first decade of August, but with no other center coming into view as in the case of 200-hPa. The third center is between 30°~40°S. indicating less regular propagation than that at 200-hPa but the Southern Hemisphere midlatitude LFO still has quite significant equatorward propagation. As for the NH, the extratropical LFO travels equatorward to around 30°N prior to the second decade of August, followed by chaotic propagation. Overall, the meridional propagation of LFO if opposite in direction at high- and low-levels of extratropical latitudes for both the hemispheres.

Comparison of positions of high- and low-level zonal wind LFO centers on the cross section (Fig. 4) indicates that for the SH midlatitude high- and low-level, the centers always remain unchanged in relative position, and for the NH north of 30°N, they are roughly unaltered. both with more or less meridional deviation in September; south of 30°N, strong LFO centers at both levels show great difference in phase, even in opposite phase with each other.

One can see that the LFOs at both levels for extratropics are essentially in phase with each other, a result in agreement with the conclusion of Krishnamurti et al.[5].

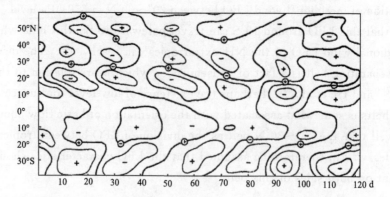

Fig. 3　120°E time cross-section of 850-hPa zonal wind LFO. Otherwise as in Fig. 1

Further inspection of Figs. 1 and 3 indicates that the zonal wind LFO centers are characteristic of appearance in pairs zonally, quite distinct at high and low levels. To elucidate this, we have to examine features of the height oscillation.

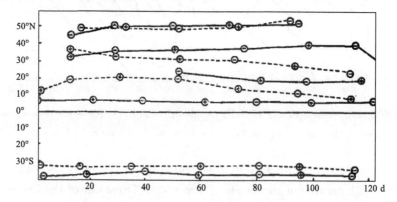

Fig. 4　Positions of zonal wind LFO centers at high and low levels on the cross-section.
Solid (broken) line connects centers at high (low) levels

Figs. 5 and 6 illustrate time cross-sections of LFO height at 200 and 850-hPa, respectively. Fig. 5 shows a row of noticeable height oscillation centers at midlatitudes for both hemispheres. Interestingly, after July 20, i. e. 50-day integration, a row of height oscillation centers shows up along 30°N, which is a match for zonal wind LFO centers (refer to Fig. 1). The propagation of height oscillation, albeit less regular than that of zonal wind, is evident in its northward march north of 30°N with equatorward travel south of 30°N. That is, a separating belt is available for height oscillation meridional propagation as well. Another fact of interest is that north of 30°N the axis lines of maxima and minima of the height oscillation coincide with the E-W zero-value line of zonal wind oscillation with the negative (positive) center of the wind oscillation located on the south (north) side of the positive (negative) center of the height oscillation (see Figs. 1 and 3), an allocation that reflects that the LFO in this area is quasi-geostrophic in character. So is the SH midlatitude oscillation. No such a pattern is, however, seen around the equator. The 850-hPa zonal wind and height fields show a similar pattern. Of particular interest is the fact that between 20°~30°N there is a row of centers of height oscillation with zonal wind oscillation centers appearing in pairs on both sides of the height oscillation centers, in agreement with the quasi-geostrophic relation. It is worth noting that the row of centers had important effect on the low-frequency variation in China's mainland weather for the summer of 1981. Preliminary analysis indicates that it bears intimate relation to the southward advance of the LFO cold air in the land and variation in the position of the subtropical high.

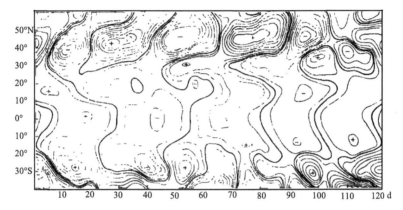

Fig. 5 120°E time cross-section of 200-hPa height LFO. Heavy solid line is the zero-value line
Isopleth interval is 80 gpm

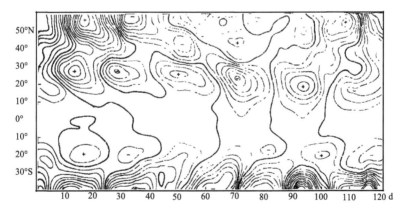

Fig. 6 As in Fig. 5 but for 850-hPa and the interval is 40 gpm
phase with each other, and the LFO around the equator loses its quasi-geostrophy

To summarize, around 120°E of East Asia there is a separating and a confluent belt for LFO meridional propagation at high and low levels, respectively, the division being at 20°~30°N, with marked different vertical structure and properties of the oscillation on either side. On the north side the zonal wind LFO's at high and low levels are in phase and quasi-geostrophic in nature, while on the south side the LFOs are remarkably out of phase for both the levels, even in opposite with each other, and the LFO around the equator loses its quasi-geostrophy.

3 Eastward propagation of zonal wind LFO and midlatitude low-frequency waves

Chen and Xie[5] indicated the eastward travel of LFO north of 37.5°N. However, they did not make further discussion of the propagation features of midlatitude, oscillation, perhaps in view of the applicability of the OLR datasets to extratropics. In fact, the mid-latitude LFO propagation is an essential problem concerning the dynamics and origin of the oscillation. Unfortunately, little is studied of the issue.

Fig. 7 depicts temporal variation in the 35°N zonal wind LFO. One can see in the Eastern Hemisphere two noticeable LFO centers, generally more than 6 m·s^{-1}, the maximum reaching over 10 m·s^{-1}, and they move westward over a great distance as the time goes on. For instance, they are about 30°~40°E and 130°E, respectively, in the second decade of June and move to around 10°E and 80°E for the same -decade of September, accompanied at the time by another stronger oscillation center between 130°~140°E, whose intensity tends to grow and which travels westward.

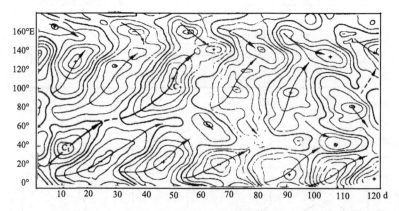

Fig. 7 35°N time cross section of 200-hPa low-frequency zonal wind. Arrows denote the direction of propagation and heavy-lined arrows the process under discussion. C_1 and C_2, are strong LFO centers

Inspection of Fig. 7 shows that the oscillation travels basically eastward to the west of 130°~140°E with a break (around 50°E) in the eastward march between the centers. And the oscillation east of 130°~140°E is featured by westward propagation. Such a transition or break band in zonal propagation is also found in other zonal belts. This transition band is noteworthy because it shows that LFO does not travel east- or westward all the way around the globe but has striking regionality. Causes may be many. Preliminary analysis indicates that the transition band in zonal propagation is associated with the meridional movement of low-frequency waves. We can possibly say that the zonal transition band serves as the passage for LFO meridional propagation or interaction between mid- and low-latitude LFO.

"Regionality" implies not merely the association of the direction the LFO progresses in a particular area, but that the existence of an oscillation center in a certain zone as the standing wave component of LFO should be related to the existence of the quasi-stationary circulation system or the center of atmospheric activity. Within a certain geographic locality, the regular propagation of oscillation is merely the manifestation of its progressive wave component, whose speed must be dependent on the period of local change of LFO. Only in this way can we explain the quasi-40-day period of LFO. Then the following questions arise: What is the process by which such coordination occurs? What are the system and dynamic mechanism responsible for the change?

Careful inspection of all eastward propagating processes in Fig. 7 indicates that they have many in common. Therefore we select one eastward propagating process from the Mediterranean to East Asian coast as marked by heavy-lined arrows to discuss the problems mentioned above (refer to Fig. 7).

Fig. 8 illustrates the horizontal pattern of the LFO wind field at 5 intervals of time for the process. Since the filtered wind field is slow-changing, the horizontal wind distributions at 5 different times are used to represent the respective fields at the five intervals.

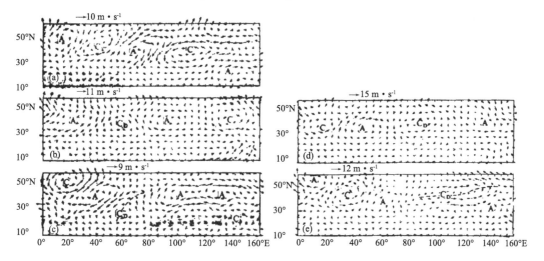

Fig. 8 200-hPa horizontal distribution of the LFO wind at five different times. C denotes the cyclone, A the anticyclone and the cyclone under consideration. (a) on day 10; (b) on day 20; (c) on day 30; (d) on day 40; (e) on day 50

It is evident from Fig. 8 that there lies a nearly E-W train of cyclones and anticyclones at 30°~50°N of the Eastern Hemisphere, with mean wavelength about 60~90 degrees of longitudes. By closely following the track, one can notice that they all move eastward fairly uniformly, with speed of 1.5~2.0 longitudes per day, suggesting that the period of oscillation in the local wind field caused by the eastward wave system is roughly 40 days. This characteristic time scale is naturally in agreement with the result of the wind field treated by bandpass filtering. One important discovery in our analysis is what circulation system gives rise to the quasi-40-day oscillation in the local wind field.

To examine the way the zonal wind travels eastward and quasi-40-day oscillation of the local zonal wind due to the movement of low-frequency vortices we consider the eastward shift of a vortex C_D and the west wind on the south side.

It can be seen that on day 10 (i. e. , June 10, and the other day can be inferred from June 10 for day 10) C_D is at around 40°N, 40°E, with a strong west wind in 30°~40°E on its south side (35°N), which is the locality of the first center C_1 in Fig. 7. Subsequently, as the train keeps on moving eastward on day 20, C_D is west of 60°E, accompanied by the west wind going in the same direction, but with decreased intensity. On day 30, the center of C_D moves southward to around 25°N, with a considerable trough stretching out in the NE direction from it. but the midlatitudes are still covered by the pronounced wave train. Because of the southward advance of C, the westerly wind is suddenly veered to the easterly, resulting in the break its the eastward travel of the west wind shown in the cross section. Later, the trough in the north gets apart from the center in the south and continues to move eastward and starts to develop. On day 50 it develops into a vigorous low-frequency vortex between 90°~140°E, and the strong west wind on its south side corresponds to the second center C on the cross section and also to the vigorous center formed on day 50 in 30°~40°N (Fig. 1). The development of the low-frequency vortex is related to the wet season of North China. The eastward travel of the west wind as seen in Fig. 7 happens to be at the speed of 1.5~2.0 longitudes per day in agreement with the speed of the low-frequency vortex, suggesting that the eastward travel of the west wind is indeed caused by the eastward advance of the vortex.

Above analyses show that in a certain area at midlatitudes, the propagation properties of the oscillation are associated with the movement of low-frequency waves and the standing wave nature of the oscillation, i. e. , the existence of the oscillation center with development of the low-frequency waves. The growth of low-frequency vortices indicated above is probably related to the supply of barotropic instability energy in the jetstream area. Xu (1989)[*] showed that the barotropic instability in the neighborhood of jetstream is really a possible mechanism for providing low-frequency disturbance with energy. From the above we can arrive at the conclusion that midlatitude low-frequency waves are quasi-geostrophic high- and low-level waves roughly in phase. Study of these waves is of consequence to the understanding of the essence of midlatitude LFO.

4 Results and discussions

(1) For 120° E of East Asia a confluent (separating) belt is available for the meridional travel of low-frequency meridional (zonal) wind perturbance at low (high) levels. The belt is

---

[*] Xu Jianjun(1989). Observation and analysis of 30—50 day periodic oscillation in atmospheric motions and diagnostic study of the effect upon disturbances of the variation in the kinetic energy and basic flow. M. S. thesis, Naniing University.

found in 20°~30°N inside which the propagation is more changeable in direction, and the LFO systems on both sides of the belt greatly differ in properties. Therefore we take the belt as the division between midlatitude and tropical LFOs.

(2) The eastward travel of zonal wind LFO is evident along 35°N in East Asia, with a pronounced oscillation center in the Mediterranean and eastern Asia, respectively. This characteristic demonstrates that the zonal wind LFO at this belt is featured by the components of both progressive and standing waves. Successive analyses of the horizontal wind fields indicate that the characteristic oscillation propagation depends on the eastward advance of low-frequency waves and that the properties of standing waves are the result of the low-frequency vortices developing over the belt, which may be associated with the supply of barotropical instability energy in the central area of the jetstream.

(3) Midlatitude low-frequency waves are Ca. 60~90 longitudes in wavelength, with the eastward speed of 1.5~2.0 longitudes a day. The two features are good enough to account for the quasi-40-day periodicity of the local wind fields. The midlatitude waves are quasi- geostrophic in dynamics and approximately in phase for high and low levels. Numerical study show's ([13], to appear) that nonlinear solitary Rossby wave travels westward and the linear Rossby mode including basic flow goes eastward, with the composite pattern kept quasi-stationary, travelling eastward at 1.3~1.5 longitudes per day and having wavelength of roughly 60 longitudes. This agrees well with the diagnostic result.

The important discovery in the study of midlatitude low-frequency waves helps to appreciate their essence and origin. Evidently, the diagnostic result needs to be verified in terms of even more comprehensive datasets.

References

[1] Madden R A, Julian P R. Description of global scale circulation cells in the tropics with a 40-50 day period [J]. J Atmos Sci,1972,29:1109-1123.

[2] Murakami T, Nakazawa T, He Jinhai. On the 40-50 day oscillations during the 1979 northern hemisphere summer, Part Ⅰ:Phase propagation[J]. J Meteor Soc Japan,1984,62:440-468.

[3] Lan K M, Peng L. Origin of low frequency (intraseasonal) oscillations in the tropical atmosphere, Part Ⅰ: The basic theory[J]. J Atmos Sci,1987,44:950-972.

[4] Yasunari T. A quasi-stationary appearance of 30-40 day period in the cloudiness fluctuations during the summer monsoon over India[J]. J Meteor Soc Japan,1980,58:225-229.

[5] Krishnamurti T N, Subrahmanyam D. The 30-50 day mode at 850 mb during MONEX[J]. J Atmos Sci, 1982,39:2088-2095.

[6] Webster P J. Mechanism of monsoon on low frequency variability:Surface hydrological effects[J]. J Atmos Sci,1983,40:2110-2124.

[7] He Jinhai, Chen Lizhen. The SH midlatitude Quasi-40-day Periodic Oscillation with its effect on the NH summer monsoon circulation[J]. Acta Meteorologica Sinica,1988,2:331-339.

[8] Krishnamurti T N, Gadgil S. On the structure of the 30-50 day mode over the globe during FGGE[J].

Tellus,1985,37A:336-360.

[9] Chen Longxun, Xie An. Westward propagation low-frequency oscillation and its teleconnections in the Eastern Hemisphere[J]. Acta Meteorologica Sinica,1988,2:300-312.

[10] Zhu Qiangen, He Jinhai, Wang Panxing. A study of circulation differences between East-Asian and Indian summer monsoon with their interaction[J]. Adv Atmos Sci,1986,3:466-477.

[11] He Jinhai, et al. Circulation with 40-50 day oscillation and changes in moisture transfer over monsoon Asia in the summer of 1979[J]. J Nanjing Inst of Meteor,1984:163-175 (in Chinese).

[12] He Jinhai, Li Jun, Li Yongping. Numerical experiment with processes for effect of Australian cold air activity on East-Asian summer monsoon[J]. Acta Meteorologica Sinica,1989,4:51-59.

[13] Luo Zhexian, He Jinhai. A numerical study of low-frequency oscillation of midlatitude flow pattern[J]. to appear.

Numerical Experiment with Processes for Effect of Australian Cold Air Activity on East-Asian Summer Monsoon[*]

He Jinhai[1]　Li Jun[1]　Li Yongping[2]

(1. Nanjing Institute of Meteorology, Nanjing;
2. Lianyungang Weather Station of Salt Industry, Lianyungang, Jiangsu Province Received)

Abstract: Based on diagnostic results, a numerical study is made of the processes of Australian cold air activity affecting East Asian summer monsoon by using Kuo-Qian P-σ incorporated coordinate five-layer primitive equation spherical band model. Analysis is done of the response to the Southern Hemisphere circulation with and without cold air activity in Australia of the flow, rainfall and diabatic heating fields in the monsoon area of Asia, especially, East Asia, with special attention to the intensification and northward march of the monsoon due to the activity. It is found that the processes for the effect transmission are very analogous to the meridional propagation of quasi-40-day oscillation, together with the meridional wind disturbance showing south-north travel and the flow/rainfall fields exhibiting corresponding movement in this direction, only with a 12-day lag.

1 Introduction

The relationship between the SH (Southern Hemisphere) midlatitude cold-air activity (SMCA) and Asian summer monsoon and the interaction between the two hemispheric circulations have already been subjects of much interest for Chinese meteorologists[1-3]. As far back as the early 1960s, Tao et al. indicated that during the prevailing East-Asian meridional circulation, the SH midlatitude meridional circulation is robust, providing more possibilities for the near-equatorial air to travel from the SH to NH(Northern Hemisphere). Obviously, the strengthened south wind is linked to the vigorous SMCA. Later Tao et al.[4] found that the development of large-scale baroclinity at the SH midlatitudes is responsible for the establishment of cross-equatorial flow in some places, leading to the building up and development of summer monsoon in relevant Asian areas, and after the wind establishment, the baroclinity development would result in the reinforcement and northward march of the monsoon.

---

* 本文发表于《Acta Meteorologica Sinica》,1990 年第 4 卷第 1 期,51-59. 其中文版发表于《气象学报》,1991 年第 49 卷第 2 期,162-169.

Wang and Li[5] showed that the cross-equatorial flow originating in Australia bears a close relation to the monsoon circulation and rainfall pattern over China, with emphasis on the consequence of the Australian high to China's summer monsoon.

Chen et al.[6] and He et al.[7] made an approach to the relationship between the interaction of both hemispheric circulations and meridional propagation of the quasi-40-oscillation in the monsoon area. Chen et al. indicated that the quasi-40-oscillation represents the manifestation of the oscillation of the SMCA traveling to the Northern Hemisphere ITCZ, whereby the interhemispheric effect occurs. He et al. emphasized the important role of the quasi-40-day SMCA in forcing the tropical atmosphere and subsequently the Asian summer monsoon by indicating that the cold-air outbreak on the south side of Mascarene and Australian highs causes the highs to be enforced, then resulting in enhancing cross-equatorial flow and subsequently in the strengthening and northward march of the monsoon in India and East Asia. The aim of this paper is to, along the line, illustrate the relationship by further evidence through numerical experimentation regarding the specific processes and their association with the meridional propagation of the quasi-40-day oscillation.

2 Model, data and scheme

As mentioned in Abstract, the Kuo-Qian model was employed which had five layers, the top two in P and the remaining in σ coordinates, with an essential coupling layer at ground or below msl making allowance for earth-atmosphere interaction. The spherical band covered the whole belt over 35°S~45°N, including primarily the area of the SMCA and the NH summer monsoon.

In this model involved are a variety of diabatic processes, e.g., solar shortwave radiation, longwave radiation of earth-atmosphere-system, large-scale condensation, cumulus convection, and underlying-surface sensible/latent heating, with the actual topography for the run. For details of the model, refer to Kuo and Qian[8].

With a view to verifying the model capacity to reproduce the summer monsoon circulation and system, the mean June flow field was used as the initial one for integration for 15 days. And then the 8~12 day mean field was compared with the average July field from observations, thus indicating that the model is quite effective in simulating the evolution of the NH summer monsoon (figure not shown).

The scheme for the run was designed in the following fashion. Following the diagnostic results[7], the quasi-40-day SMCA has considerable effect on the Asian monsoon, and therefore the wean June 1982 field was taken as the basic initial field, upon which the strong cold air and cold-air-absent anomalies in Australian (blocked) area were superimposed, respectively. They were used as the modified initial fields. Diagnostic analyses show that around 18 June stronger west wind in southern Australia is related to a more intense SMCA event. Therefore, for height Z_1 and moisture Q_1 of the box, the mean June fields were replaced by the 16-20th day average ones. For the case with no cold air activity, the fields of $Z_0 - (Z_1 - Z_0)$ and $Q_0 - (Q_1 - Q_0)$ were taken for run in place of the mean June fields of Z_0 and Q_0. Both experiments are hereafter referred to as MAS and MAW, respectively. Initial data included 5-level height (100, 300, 500, 700 and 1000 hPa) and 4-level moisture (300,500,700, …) was taken in geostrophic equilibrium for mid and high latitudes, and obtained in terms of the simplified balance equations for the area around the equator. The initial field of temperature was derived from the height field with the aid of the hydrostatic equilibrium relation. Then the model levels for the initialized fields were interpolated by the data of corresponding standard isobaric surfaces. Based on these, the model was run out with the period length of 20 days each.

3 Analyses of numerical results

3.1 Difference in results between MAS and MAW

Figs. 1a, b show the initial field at the 4th layer[①] for MAS and MAW, separately. As illustrated in the box of Fig. 1a, initially a remarkable anticyclonic circulation is shown up in Australian region, with a strong W-SW flow in the south. The temperature-pressure fields at upper and lower levels indicate a vigorous south-north directed temperature gradient in this region, with the 700-hPa difference ΔT ($T_{15°S} - T_{35°S}$) of 14℃ along 110°E. A large-amplitude trough orientated south-north is seen at 500-hPa along 140°E, with the surface cold high centred just in the rear of the trough extending equatorward. Such a circulation background

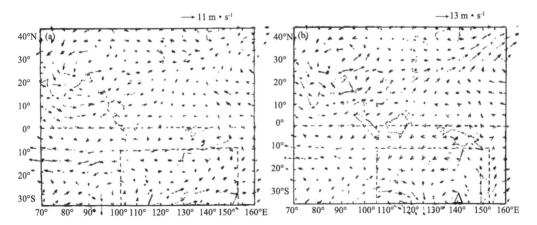

Fig. 1 The initial flow field for the 4th layer (around 810-hPa) of the model for the experiment MAS (a) and MAW (b). Vigorous cold-air activities are seen in the blocked region of Australia for MAS but none for MAW. Outside the block, June mean fields are used as the initial ones

---

① This layer is the σ surface whose height is 812.5-hPa if the surface pressure is 1000-hPa.

demonstrates that a robust SMCA is available with significant development of the troposphere baroclinity. In contrast, there is not any anticyclonic circulation in the region only with a stronger northerly in the south, especially between 110°~140°E, as depicted in Fig. 1b. The upper- and lower-level temperature-pressure fields show that the S-N temperature gradient is small throughout the troposphere. Such a background indicates a weaker baroclinity of the region with warming or no cold air activity.

Comparison of Figs. 1 a, b shows that the difference of the initial field between MAS and MAW exists only in the Australian region and the vicinity of its boundaries[①], with the area outside the Australian taking the mean June 1982 field as the initial one for run. The initial field shows the basic circulation background including the SH near-equator easterlies, Asian summer monsoon around 15°N, Asian continental cyclonic circulation and the western North Pacific subtropical anticyclonic circulation.

As indicated in He et al.[7], a roughly 10~12 day lag is obtained between the Australian cold air outbreak and the reinforcement of the Asian monsoon. If the diagnostic analyses do reveal the fact, then appreciable difference should occur in the monsoon trough or eastern Asian flow pattern after a 10-day integration for the two experiments. For this reason, a detailed comparison is done of the difference in the flow fields averaged for the 10—14th day between MAS and MAW.

Fig. 2 delineates the 4th-layer 5-day (10-14th) mean flow difference (MAS minus MAW), with the difference indicated by arrows showing relative speed (by the shaft length) and direction. Evidently, three belts of strong flow difference are shown in the monsoon area. The first is the belt of vigorous southerly (SW-SE) difference stretching from the Bay of Bengal to East China via the Indo-China Peninsula, which suggests that when a strong SMCA event takes place, the Asian summer monsoon mainly as the southerly component gets considerably intensified. The second belt is of the conspicuous difference in anticyclonic circulation, which suggests that the western North Pacific subtropical high is more intense in MAS than in MAW, particularly in the southeasterly flow in the west. The two belts of strong flow difference merge into one between the Peninsula and South China Sea and move toward the north, reaching eastern China through the Yunnan-Guizhou Plateau with the northernmost limit near 30° N, thus exerting a considerable influence on rainfall all the way, which will be dealt with later on. Additionally, noteworthy is the fact that the difference in northerly current is observed over the most part of North India and the adjacent Peninsula, which suggests that the Indian monsoon simulated in MAS is somewhat weaker, and in other words, the SMCA has opposite effect on the East-Asian and Indian monsoon. This confirms again the viewpoint that the monsoon is in anti-phase for the two regions.

With the flow difference associated are regions of difference in rainfall, vertical motion

---

① When the anomaly is superimposed for the Australian region, smoothing is needed for the vicinity, which results in some discrepancy.

and heating. Fig. 3 is the rainfall difference (MAS minus MAW), showing a positive belt extending from equatorial Indonesia to the Changjiang-Huaihe Basins by way of the Indo-China Peninsula and the Yunnan-Guizhou Plateau, with three appreciable difference centers: one of 3.0 mm · d^{-1} in equatorial Indonesia around 105 °E; the next between the northern Peninsula and Plateau around 100 °E (which corresponds with the locality of a vigorous low in MAS); The last of smaller difference over the Basins in eastern China about 30°N, which is related to the convergence there due to the northward travel of the intense difference in flow, which, as we will see, is the result of the northeastward extension of the robust monsoon precipitation belt. It can also be seen that located on either side of the aforementioned positive rainfall difference belt is a negative belt, one in the western North Pacific and one in India, in agreement with the flow difference fields mentioned above. The eastern one is associated with the reinforced subsiding caused by the westward advance of the enhanced western North Pacific subtropical high ridge in MAS and the western one is the manifestation of weakened Indian summer monsoon in good concern with the flow difference field.

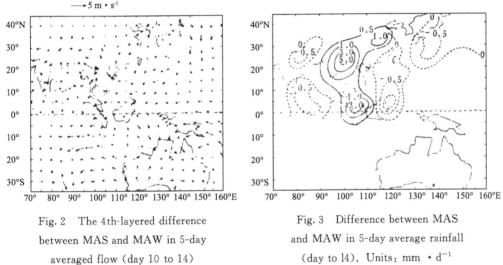

Fig. 2 The 4th-layered difference between MAS and MAW in 5-day averaged flow (day 10 to 14)

Fig. 3 Difference between MAS and MAW in 5-day average rainfall (day to 14). Units: mm · d^{-1}

The distribution of 300-hPa heating difference is very similar to the rainfall difference in both runs (see Fig. 4). Front Indonesia via the Peninsula to South and East China is a relatively positively heating zone, in rough agreement with the relatively positive rainfall belt as given in Fig. 3, except for its position moved slightly eastward. It can be reasoned out that the difference in diabatic heating obtained in MAS and MAW is probably due to difference in latent heating through condensation in the course of large-scale monsoon rainfall. What is the same thing, relative to the two negative zones of rainfall difference are those of heating difference in the upper troposphere. The conformity suggests that the latent heating in monsoon rainfall plays an important role in the diabatic heating.

3.2 Analysis of the propagative processes

The previous analyses can lead us to believe that the fields of flow, vertical motion,

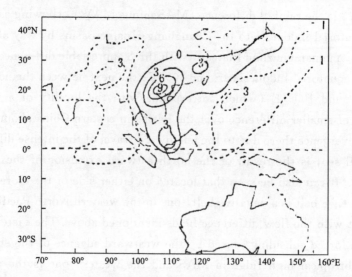

Fig. 4 Difference between MAS and MAW in day average diabatic heating (day 10 to 14) for 300-hPa. Units: 10^{-2} cal·cm^{-2}·min^{-1}

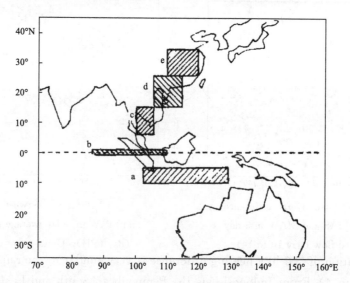

Fig. 5 The schematic view of the key regions during 5-propagation. (a) 5°~10°S, 100°~130°E (NW Australia). (b) 85°~110°E equator (equatorial Indonesia). (c) 5°~15°N, 100°~105°E (northern Indo-China Peninsula). (d) 15°~25°N, 105°~115°E (South China). (e) 25°~35°N, 11°~120°E (China's Chang jiang-Huaihe basins)

rainfall and diabatic heating over the NH monsoon area are sensitive to the SMCA, i.e., the latter has significant effect on the Asian, in particular eastern-Asian summer monsoon.

Then we should like to question: "What is the process by which the effect is transmitted?" "Is this transmission more or less similar to the meridional travel of the quasi-40-day oscillation in this monsoon area?" In order to intuitively examine the SMCA effect on the Asian monsoon, a few key regions were selected for the northward propagation of the meridional wind disturbance caused by the SMCA, as specified in Fig. 5.

Fig. 6 shows the time-latitude cross section of the meridional wind anomaly (i. c. , the wind via the key regions during the run period minus the average in MAS). The northward travel of the wind is extremely evident, suggesting that with the initial field marked by a SMCA, the south wind gets amplified on the north side of the Australian high after a 2-day integration, and on day 6 the region corresponding to the key zone marked by b (around 100°E) displays a highly-enhanced cross-equatorial flow, followed by the enforced south wind between the Peninsula and South China. On day 14 or so the strongest south wind disturbance reaches the Basins, with the maximum anomaly occurring for day10-14. It is the wind anomaly that is responsible for the strengthened monsoon (SW-SE) with its rainfall from the Indo-China Peninsula to eastern China (see Figs. 2 and 3). The propagation owes its origin to the SMCA or the development of baroclinity in the troposphere. On the contrary, the weakened monsoon takes place if warming or no SMCA is observed. It should be noted here that the northward propagation of the SMCA effect discussed above is very analogous to the synoptic processes by which the quasi-40-day oscillation travels northward, as indicated in He et al.[7]. Such a close similarity may serve as a guide for further investigation of the mechanism of the quasi-40-day oscillation meridional travel.

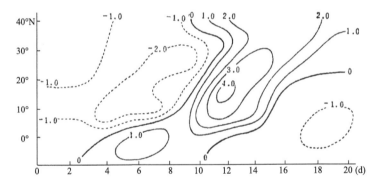

Fig. 6 The time-latitude cross-section of anomalies in the meridional wind (MAS integration minus average integration for the same period) for the five regions as shown in Fig. 5

In accordance with the monsoon advance, the rainfall is spread northward in a regular manner. Figs. 7a, b, c display the rainfall distribution for three intervals of 5—6th, 10—11th and 13—14th day. For the first interval located in the belt covering equatorial Indonesia via the northern Indo-China Peninsula to the Yunnan-Guizhou Plateau are two rainfall centers (denoted as Centers EIn and NIC, respectively, see Fig. 7a); for the second, Center EIn is moved slightly westward with little change in strength, but Center NIC becomes significantly enforced and extends northeastward to the Basins, in good agreement with the northward advance of the enhanced monsoon; and for the third, i. e. , day 13-14, Center NIC gets weakened and the rainfall is increased in intensity and coverage for the Basins area. The rainfall variations for the three intervals indicate that as the cross-equatorial flow and subsequently East-Asian monsoon strengthen and march northward, so does the relative monsoon rainfall. Comparison of Figs. 3 and 7c illustrates a similar pattern for their development, suggesting that the rainfall difference(MAS minus MAW)for day 10-14 is due to the reinforcement and northward advance of East-Asian summer monsoon rainfall in the run of MAS, meaning that the SMCA effect is involved.

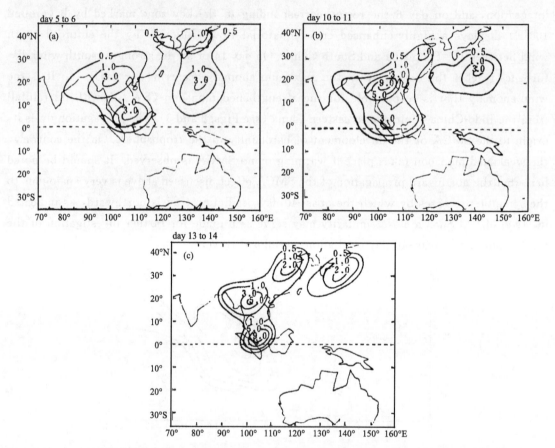

Fig. 7 Distribution of rainfall (mm · d^{-1}) at 3 intervals of time for MAS

To further investigate the processes whereby the SMCA effect is transmitted across the equator, a numerical experiment was designed with actual field involving a SMCA as the initial one (referred to as RAS), which differs from MAS in that the area outside the block (see Fig. 1) represents the actual rather than the mean June field. Figs. 8a, b, c display the 4th-layer flow fields on day 5,10,15, respectively. In comparison to the initial field (figure omitted), the flow field after a 5-day integration shows considerably enforced cross-equatorial current between 75°~115° E, with more amplified SE trade wind in 5°~15° S. In association with the 105°E cross-equatorial flow, the SW wind is stronger over the Bay of Bengal and South-China Sea, with the northernmost boundary reaching the northern part of the Sea (broken line). The day-10 field illustrates a strong S-SW wind belt covering the Bay of Bengal, Indo-China Peninsula and eastern-China Changjiang-Huaihe basins, with the northern-

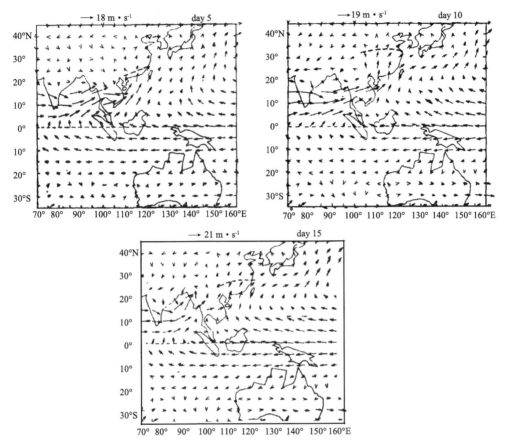

Fig. 8 The 4th-layered flow field on day 5, 10 and 15 in the experiment RAS with the actual field involving vigorous cold-air activity over Australia as the initial field. The broken line delineates the northern boundary of the strong south wind and the arrow at the top the maximum speed of the wind

most limit advanced 5-10 latitudes. The advance in conjunction with the seasonal trend would be responsible for such major events as the onset and ending of Meiyu in eastern China (refer to [4]). Fig. 8c is the field on day 15, where is shown the weakened SW monsoon with the northernmost limit located around 25°N and the negligible cross-equatorial flow near-by 100°E. All these show that the East-China summer monsoon reaches its maximum at a point during day 10-15 with the position advanced to the northernmost limit. Such intensification and northward march bear a close resemblance to those exhibited in MAS. All these suggest that the SMCA effect on the monsoon is by meridional propagation with a time lag of about 12 days.

4 Results and discussions

Based on the foregoing analyses, we obtain the following.

(1) With the initial fields involving a or no SMCA the contrast runs show that the summer monsoon in Asia, especially along the belt from the Bay of Bengal via the Indo-China Peninsula to southeastern China and the Basins is sensitive to the occurrence of the SMCA and change in the intensity of the Australian high, and, put another way, the SMCA has appreciable effect on Asian summer monsoon. As illustrated in the run, the eastern Asian monsoon exhibits a process of significant reinforcement and northward movement 10~14 days after a SMCA happens. More specifically, the southerly wind is enhanced; monsoon rainfall increases in intensity and spreads northward; the western North Pacific subtropical high ridge strengthens and extends westward with weakened ascending motion and thus decreased rainfall for the region under its control. In contrast to enforced East-Asia monsoon, the Indian SW monsoon gets weakened. The reversal happens for no SMCA.

(2) The SMCA effect on the NH monsoon is realized by the propagative travel of meridional wind disturbance. More specifically, the SMCA intensifies the SE trade wind north (northwest) of Australia and then brings about the enhanced cross-equatorial flow in Indonesia around and west of 100°E, thus causing monsoon (SW-SE) and its rainfall lying from the Indo-China Peninsula to the South China Sea, southern China and the Changjiang-Huaihe basins to be active and enforced and to move northward. The time lag is found to be 10~15 days.

(3) The northward propagation of the meridional wind disturbance and related Asian summer monsoon variation as revealed in the numerical study are attributed to the initial forcing by the SH midlatitude cold air outbreak. As noted before, He et al. [7] showed the SMCA to be of quasi-40-day periodicity and detailed the synoptic processes by which the quasi-40-day oscillation travels northward, the processes very analogous to those for the meridional wind disturbance going northward found in the present study. Whether or not the quasi-periodic SMCA is viewed as the external forcing on the tropical atmosphere and a possible mechanism for the quasi-40-oscillation northward propagation remains an open question. It should be noted here that the zone for cold air activity close to the lateral boundary under the influence of given conditions, and hence the results presented above require further verification in terms of a global model.

References

[1] Yeh Tucheng, et al. The averaged vertical circulation over the East-Asia and the Pacific area(I) in summer[J]. Sci Atmos Sin,1979,3:1-11(in Chinese).

[2] Tao Shiyan, et al. The characteristic of the zonal and meridional circulation over tropical and sub- tropical regions in Eastern Asia in Summer[J]. Acta Met Sinica,1962,32:91-102(in Chinese).

[3] Gun Qiyun, Ye Weiming. The circulation in Southern and Northern Hemispheres and the monsoons over East Asia[J]. Acta Met Sinica,1979,37:86-95(in Chinese).

[4] Tao Shiyan, et al. An observational study on the onset of the summer monsoon over eastern Asia in 1979 MONEX[J]. Sci Atmos Sin,1983,7:347-355(in Chinese).

[5] Wang Jizhi, Li Maicun. Cross-equatorial flow from Australia and monsoon over China[J]. Sci Atmos Sin, 1982,6:1-10(in Chinese).

[6] Chen Longxun, Jin Zuhui. On the interaction of circulations between the two hemispheres in the East Asia monsoon circulation system during summer[M]//Proceedings of the Symposium on Summer Monsoon in Southeast Asia. Kunming: Yunnan People's Press, 1982:218-231(in Chinese).

[7] He Jinhai, Chen Lizhen. The Southern Hemisphere midlatitude quasi-40-day periodic oscillation with its effect on the Northern Hemisphere summer monsoon circulation[J]. Acta Met Sinica, 1988,2:331-339.

[8] Kuo H L, Qian Y F. Numerical simulation of the development of mean monsoon circulation in July[J]. Mon Wea Rev, 1982,110:1879-1897.

Impacts of SST and SST Anomalies on Low-Frequency Oscillation in the Tropical Atmosphere*

He Jinhai[1,2]　　Yu Jingjing[1,3]　　Shen Xinyong[1]

(1. Jiangsu Key Laboratory of Meteorological Disaster (KLME),
Nanjing University of Information Science and Technology, Nanjing 210044;
2. Institute of Tropical and Marine Meteorology, China Meteorological Administration,
Guangzhou 510080; 3. State Key Laboratory of Numerical Modeling for Atmospheric
Sciences and Geophysical Fluid Dynamics, Institute of Atmospheric Physics,
Chinese Academic of Sciences, Beijing 100029)

Abstract: Considering the multiscale character of LFO (low-frequency oscillation) in the tropical atmosphere, the effects of SST on LFO in the tropical atmosphere are discussed by using an absolute ageostrophic, baroclinic model. Here, SST effects include sea surface heating and forcing of SST anomalies (SSTAs). Studies of the influences of sea surface heating on LFO frequency and stability show that sea surface heating can slow the speed of waves and lower their frequency when SST is comparatively low; while higher SST leads to unstable waves and less periods of LFO. Since the impact of a SSTA on ultra-long waves is more evident than that on kilometer-scale waves, long-wave approximation is used when we continue to study the effect of SSTAs. Results indicate that SSTAs can lead to a longer period of LFO, and make waves unstable. In other words, positive (negative) SSTAs can make waves decay (grow).

Key words: low-frequency oscillation(LFO); sea surface heating; sea surface temperature anomaly(SSTA); ultra-long wave; kilometer-scale precipitation

1　Introduction

Atmospheric intraseasonal oscillation of 30-60 days in the tropics, which Madden and Julian [1-2] first discovered, is one of the basic motions of the tropical atmosphere. It bears a close relation not only with long-range weather change, but also with short-range climate anomalies. Therefore, a series of studies have explored its basic features [3-7].

Based on studies of observational features of Low-Frequency Oscillation (LFO) in the tropical atmosphere, many dynamical and thermal mechanisms [8-11] have been proposed to

---

* 本文发表于《Advances in Atmospheric Sciences》,2007 年第 24 卷第 3 期,377-382。

explain LFO. In theoretical studies of LFO in the tropical atmosphere, Xu and Gao[12] summarized a series of factors affecting LFO, among which SST was one of the most important. A series of observational analyses [13-14] have shown that tropical LFO in the western and eastern Pacific are both very active, while the characteristics of SST in these two regions are obviously very different: SST in the Western Pacific is higher, with little variation; while SST in the Eastern Pacific is lower, but with a greater degree of variation. Therefore, the two effects of SST (heating and anomalies) on tropical LFO are important and require separate study.

Xu et al. [15] have demonstrated the process of sea surface heating of the atmosphere: convergence (divergence) of a stream field due to higher SST can produce large-scale vertical motion, then the release of latent heat due to condensation warms up air. The effect of this SST anomalies (SSTAs) can be described as the influence on the upper atmosphere circulation (including LFO) due to the change of ocean circulation. Gao et al. [16] point out that low-frequency variation in the atmosphere over the East Pacific is related to SSTAs in this region.

Li and Li [17] studied CISK containing the effect of SSTAs by using a baroclinic semi-geostrophic model and analyzed the influence of SSTAs on LFO in the tropics. Fu et al. [18] established a semi-geostrophic model to discuss the influence of SST, including evaporation wind feedback, and pointed out that the forcing of SST on LFO is important. Liu [19] showed that a semi-geostrophic approximation can only filter the high-frequency Rossby waves that satisfied the condition (wave number is close to 0). Further investigation showed that LFO has a multiscale characteristic in the tropics: that is, the horizontal scale of tropical LFO changes from the planetary scale ($L \sim 10^7$ m) to a smaller scale (e.g. precipitation cloud with a kilometer scale ($L \sim 10^6$ m)). Therefore, a semi-geostrophic model has a limitation in describing all scales of LFO motion in the tropics, however it is simple to discuss the equation resolution by using this model. In this paper, an absolute ageostrophicbaroclinic model is established to discuss the two effects of SST: sea surface heating and SSTAs. For a detailed discussion of equation resolution, we take appropriate means and approximations according to different scales of tropical LFO.

2 Basic equations and analysis

The linear equations of a baroclinic model on the equator β plane, which utilizes the two effects of SST to describe LFO motion in the tropics, can be written as (not considering the mean flow):

$$\frac{\partial u}{\partial t} - \beta y v = -\frac{\partial \varphi'}{\partial x} \tag{1}$$

$$\delta \frac{\partial v}{\partial t} + \beta y u = -\frac{\partial \varphi'}{\partial y} \quad (\delta = 0 \text{ or } \delta = 1) \tag{2}$$

$$\frac{\partial u}{\partial x}+\frac{\partial v}{\partial y}+\frac{\partial w}{\partial z}=0 \qquad (3)$$

$$\frac{\partial}{\partial t}\left(\frac{\partial \varphi'}{\partial z}\right)+N^2 w = Q+F \qquad (4)$$

where N is the Brunt-Vaisala frequency, $\varphi = p'/\rho_0$ (p' is the pressure perturbation against static pressure, and ρ_0 is the density of static air). In Eq. (4), Q and F show the effect of sea surface heating and the forcing by a SSTA, respectively. $\delta = 1$ represents an ageostrophic model, while $\delta = 0$ is used to describe the terminal adaptation state means zonal momentum equation, which satisfies semi-geotrophic balance.

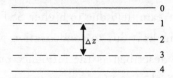

Fig. 1 Schematic diagram of the two-layer model

Using a two-layer model is one of the easiest methods to show the main characteristics of the baroclinic atmosphere. So, using a simple two-layer model (Fig. 1), we put dynamical equations and a mass continuity equation on the first and third levels. It should be noted that the boundary conditions $W_0 = W_4 = 0$ have been used in the continuity equation. By setting,

$$\begin{cases} \hat{u} = \frac{1}{2}(u_1 - u_3) \\ \hat{v} = \frac{1}{2}(v_1 - v_3) \\ \hat{\varphi} = \frac{1}{2}(\varphi_1 - \varphi_3) \end{cases}$$

Eqs. (1-4) can be rewritten as:

$$\begin{cases} \dfrac{\partial \hat{u}}{\partial t} - \beta y \hat{v} = -\dfrac{\partial \hat{\varphi}}{\partial x} \\ \delta \dfrac{\partial \hat{v}}{\partial t} + \beta y \hat{u} = -\dfrac{\partial \hat{\varphi}}{\partial y} \\ w_2 = \Delta z \left(\dfrac{\partial \hat{u}}{\partial x} + \dfrac{\partial \hat{v}}{\partial y} \right) \end{cases}$$

In this simple atmospheric model, there are no oceanic variables and we need to show the effects of SST by parameterization of atmosphere variables.

In the context of this paper, Q is related to convergence or divergence of the stream field, and thus we suppose

$$Q \sim r_T \left(\frac{\partial u_1}{\partial x} + \frac{\partial v_1}{\partial y} \right) \sim -r_T \left(\frac{\partial u_3}{\partial x} + \frac{\partial v_3}{\partial y} \right)$$

where r_T is related to SST (the higher the SST, the bigger the values of r_T and Q). Simple parameterization of SST is proposed by

$$Q \sim 2r_T \left(\frac{\partial \hat{u}}{\partial x} + \frac{\partial \hat{v}}{\partial y} \right)$$

Ran et al. [20] pointed out that the effect of divergent winds excited by external thermal forcing on low-frequency waves is important.

By including the thermodynamic equation on the second layer, and substituting the vertical differential by vertical difference, we have:

$$\frac{2}{\Delta z} \frac{\partial \hat{\varphi}}{\partial t} + N^2 w_2 = 2r_T \left(\frac{\partial \hat{u}}{\partial x} + \frac{\partial \hat{v}}{\partial y} \right) + F_2 \tag{5}$$

where the term F_2 represents the height difference between 200-hPa and 850-hPa, as in Li and Li [17]. That is, the influence on atmospheric circulation embodies the change of height difference due to SSTAs and ocean circulation. When the SSTA is positive (negative), the height difference between 200-hPa and 850-hPa increases (decreases). Therefore, the parameterization of the SSTA is proposed as $F_2 : -\alpha_s \hat{\varphi}$. When $\alpha_s > 0$, SSTA is positive; and when $\alpha_s < 0$, SSTA is negative. Equations (1-4) may be further written as:

$$\begin{cases} \dfrac{\partial \hat{u}}{\partial t} - \beta y \hat{v} = -\dfrac{\partial \hat{\varphi}}{\partial x} \\ \delta \dfrac{\partial \hat{v}}{\partial t} + \beta y \hat{u} = -\dfrac{\partial \hat{\varphi}}{\partial y} \quad (\delta = 0 \text{ or } \delta = 1) \\ w_2 = \Delta z \left(\dfrac{\partial \hat{u}}{\partial x} + \dfrac{\partial \hat{v}}{\partial y} \right) \\ \dfrac{\partial}{\Delta z} \left(\dfrac{\partial \hat{\varphi}}{\partial t} \right) + \dfrac{N^2 w_2}{2} = -\alpha_s \hat{\varphi} + r_T \left(\dfrac{\partial \hat{u}}{\partial x} + \dfrac{\partial \hat{v}}{\partial y} \right) \end{cases} \tag{6}$$

Considering the multiscale character of tropical LFO, we take the ageostrophic model ($\delta = 1$):

$$\frac{\partial \hat{u}}{\partial t} - \beta y \hat{v} = -\frac{\partial \hat{\varphi}}{\partial x} \tag{7}$$

$$\delta \frac{\partial \hat{v}}{\partial t} + \beta y \hat{u} = -\frac{\partial \hat{\varphi}}{\partial y} \tag{8}$$

$$\frac{\partial \hat{\varphi}}{\partial t} + C_2^2 \left(\frac{\partial \hat{u}}{\partial x} + \frac{\partial \hat{v}}{\partial y} \right) = -\alpha_s \Delta z \hat{\varphi} \tag{9}$$

where $C_2^2 = \dfrac{N^2 \Delta z^2}{2} - r_T \Delta z$.

Then, by eliminating $\hat{u}$ and $\hat{\varphi}$ from Eqs. (7-9), we have:

$$\left[C_2^2 \frac{\partial}{\partial t} \left(\frac{\partial^2}{\partial x^2} + \frac{\partial^2}{\partial y^2} \right) - \frac{\partial}{\partial t} \left(\frac{\partial^2}{\partial t^2} + \alpha_s \Delta z \frac{\partial}{\partial t} + \beta^2 y^2 \right) + C_2^2 \beta \frac{\partial}{\partial x} - \beta^2 y^2 \alpha_s \Delta z \right] \hat{v} = 0 \tag{10}$$

The boundary condition of Eq. (10) is: $\hat{v}|_{y \to \pm \infty} = 0 \tag{11}$

By invoking $\hat{v} = V \exp(ikx - i\sigma t)$, where k is the wave number in the x direction, and σ is angular frequency, Eqs. (10) and (11) can be converted into:

$$\begin{cases} \dfrac{d^2 V}{dy^2} + \left[-\dfrac{k\beta}{\sigma} + \dfrac{\sigma^2}{C_2^2} - k^2 + \dfrac{i\alpha_s \Delta z \sigma}{C_2^2} - \left(\dfrac{\beta^2}{C_2^2} + \dfrac{i\alpha_s \Delta z \beta^2}{C_2^2 \sigma} \right) y^2 \right] V = 0 \\ V|_{y \to \pm \infty} = 0 \end{cases} \tag{12}$$

This is a Weber-type equation, and its eigenvalues are:

$$\left(-\frac{k\beta}{\sigma}+\frac{\sigma^2}{C_2^2}-k^2+\frac{ia_s\Delta z\sigma}{C_2^2}\right)\bigg/\sqrt{\frac{\beta^2}{C_2^2}+\frac{ia_s\Delta z\beta^2}{C_2^2\sigma}}=2m+1 \quad (m=0,1,2\cdots\cdots) \tag{13}$$

Considering the low-frequency characteristic, we omit $1/\sigma$ and retain σ or σ^2, giving us:

$$\sigma=\frac{k\beta}{k^2-\dfrac{i(2m+1)^2 a_s\Delta z\beta}{2C_2^2 k}+\dfrac{(2m+1)\beta}{C_2^2}\sqrt{C_2^2-\dfrac{(2m+1)^2 a_s^2\Delta z^2}{4k^2}-\dfrac{ika_s\Delta zC_2^2}{\beta}}} \tag{14}$$

which shows that there is obvious change in the frequency due to the effects of SST heating ($r_T\neq 0$) and SSTA forcing ($a_s\neq 0$). In addition, a low-frequency wave can easily be unstable (the imaginary part of the wave solution is mainly related to a_s). Detailed discussion follows.

First, we take the effect of SST heating when the change in SST is smaller (we suppose $a_s=0$). Thus, the frequency expression from Eq. (14) can be written as:

$$\sigma=-\frac{k\beta}{k^2+(2m+1)\dfrac{\beta}{C_2}} \quad (m=0,1,2\cdots) \tag{15}$$

Note that the frequency expression is

$$\sigma=-\frac{k\beta}{(2m+1)\dfrac{\beta}{C_2}}$$

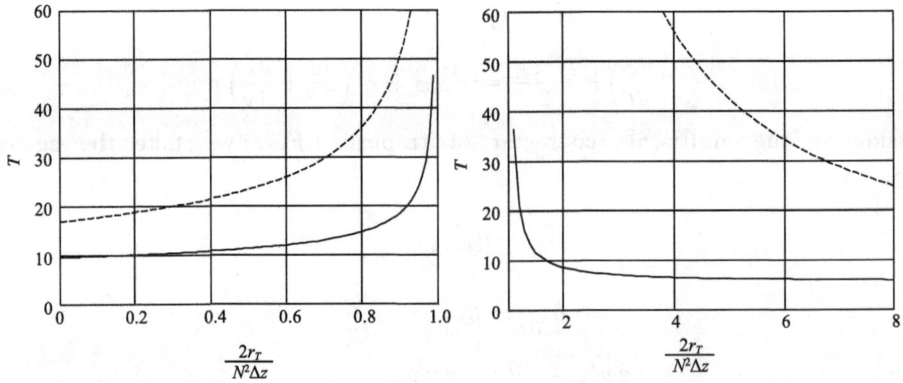

Fig. 2 Variations of period with $2r_T/(N^2\Delta z)$ for ultra-long Rossby waves ($m=1, k=1.5\times 10^{-6}$), in weaker convection $[2r_T/(N^2\Delta z)<1]$ and stronger convection $[2r_T/(N^2\Delta z)>1]$ respectively (real line is $k=1.5\times 10^{-6}$ m^{-1}; dashed line is $k=4\times 10^{-7}$ m^{-1})

and the frequency becomes larger if we use the semi-geostrophic model, especially for smaller scales (e. g. , precipitation cloud with a kilometer horizontal scale, where, $L_s\sim 10^6$ m, $k\sim 10^{-6}$ m^{-1}, $\beta/C_2\sim 10^{-12}$, k^2 cannot be omitted. Or the period of this smaller scale wave will reduce and cannot satisfy the ranges of LFO periods).

We may obtain some results from Eq. (15) as follows:
(1) If

$$r_T<\frac{N^2\Delta z}{2}$$

(low SST), angular frequency is a real number, and the wave is stable. Wave velocity is

$$C_x = \frac{\sigma}{k} = -\frac{\beta}{(2m+1)\beta/C_2 + k^2} = -\frac{\beta}{(2m+1)\beta/\sqrt{\frac{N^2 \Delta z^2}{2} - r_T \Delta z} + k^2}$$

that is, low-frequency Rossby waves are westward, and the frequency (σ) and wave velocity (C_x) becomes small because of the existence of r_T.

(2) If

$$r_T > \frac{N^2 \Delta z}{2}$$

(higher SST), angular frequency is an imaginary number, and the wave is unstable ($\sigma_i \neq 0$).

$$C_2 = \pm i \sqrt{r_T \Delta z - \frac{N^2 \Delta z}{2}}$$

$$\sigma = \frac{\beta k \left[-k^2 \mp \frac{(2m+1)\beta}{\sqrt{r_T \Delta z - \frac{N^2 \Delta z^2}{2}}} \right]}{\left[\frac{(2m+1)\beta_0}{\sqrt{r_T \Delta z - \frac{N^2 \Delta z^2}{2}}} \right]^2 + k^4} = \sigma_r + i\sigma_i$$

And we obtain

$$\sigma_r = \frac{-\beta_0 k^3}{\left[\frac{(2m+1)\beta_0}{\sqrt{r_T \Delta z - \frac{N^2 \Delta z^2}{2}}} \right]^2 + k^4}$$

$$\sigma_i = \mp \frac{\beta_0^2 k \frac{(2m+1)}{\sqrt{r_T \Delta z - \frac{N^2 \Delta z^2}{2}}}}{\left[\frac{(2m+1)\beta_0}{\sqrt{r_T \Delta z - \frac{N^2 \Delta z^2}{2}}} \right]^2 + k^4}$$

That is, when SST is higher, the propogation of Rossby waves is also westward and unsteady (high SST can make waves decay or grow).

For Kelvin waves, setting $\hat{v} = 0$ in Eqs. (7-9), we get $\sigma = C_2 k$ by using the same means. The results show that Kelvin waves are eastward when SST is lower [$r_T < (N^2 \Delta z)/2$], while Kelvin waves are unstable (angular frequency is an imaginary number) when SST is higher.

For the tropics, by taking $\beta = 2 \times 10^{-11}$ s^{-1} · m^{-1}, $N = 1 \times 10^{-2}$ s^{-1}, and $\Delta z = 5$ km in the two-layer model, we draw Fig. 2, from which we can find some differences between the figure with $k = 1.5 \times 10^{-6}$ m^{-1} (smaller scale) and that with $k = 4 \times 10^{-7}$ m^{-1} (ultra scale) under lower (higher) SST.

Fig. 2 shows the influences of SST are as follows: when SST is lower [$2r_T/(N^2 \Delta z) < 1$], period becomes longer with an increase in SST; when SST is higher [$2r_T/(N^2 \Delta z) > 1$], period becomes shorter with an increase in SST. Furthermore, periods of ultra-long waves are longer than that of long waves of kilometer-scale size under conditions of the same SST.

Disregarding two important horizontal scales of LFO in the tropics—planetary scale ($L \sim 10^7$ m) motion and precipitation cloud with a smaller scale ($L \sim 10^6$ m)—we compared the responses of the two scales to SST heating and its anomaly. Using the scale analysis, the results shows the following:

(1) For planetary scale motion in the x direction, which is familiar in the tropics, the characteristic space scales are 10^7 m in the x direction and 10^6 m in the y direction; and the corresponding characteristics of wind velocity in the x and y directions are 10 m·s^{-1} and 10^0 m·s^{-1} respectively. The characteristic value of divergence is 10^{-6} s^{-1}. We take, $\beta_0 = 10^{-11}$ m^{-1}·s^{-1}, $N = 1 \times 10^{-2}$ s^{-1} and $\Delta z = 10^3$ m. By scale analysis, we know that the characteristic value of $\hat{\varphi}$ and w_2 are 10^2 m^2·s^{-2} and 10^{-3} m·s^{-1}. Because the value of $\partial \hat{\varphi}/\partial t$ is smaller, we take long-wave approximation, i.e. the semi-geostrophic model ($\delta = 0$). In the thermal equation, if the value of $(N^2 w_2)/2$ is equivalent to the value of $\alpha_s \hat{\varphi}$, the value of parameter α_s, which can show the SSTA, is taken as $\alpha_s \sim 10^{-9}$ m^{-1}·s^{-1}; if the value of $(N^2 w_2)/2$ is equivalent to the value of

$$r_T \left(\frac{\partial \hat{u}}{\partial x} + \frac{\partial \hat{v}}{\partial y} \right)$$

the value of parameter r_T, which can show SST, is taken as $r_T \sim 10^{-1}$ m^2·s^{-1}.

(2) For large scale motion with $L_x \sim L_y \sim 10^6$, the characteristic wind velocity in the x direction and in the y direction are both 10 m·s^{-1}, and the characteristic value of divergence is 10^{-5} s^{-1}. By scale analysis, we can obtain that the characteristic value of $\hat{\varphi}$ to be 10^1 m^2·s^{-2}, and the characteristic value of w_2 to be 10^{-2} m·s^{-1}. Therefore, we must take $\delta = 1$ in Eq. (6). In the thermal equation, if the value of $(N^2 w_2)/2$ is equivalent to the value of $\alpha_s \hat{\varphi}$, and the value of $(N^2 w_2)/2$ is equivalent to the value of

$$r_T \left(\frac{\partial \hat{u}}{\partial x} + \frac{\partial \hat{v}}{\partial y} \right)$$

then the value of α_s is 10^{-7} m^{-1}·s^{-1} and the value of r_T is 10^{-1} m^2·s^{-1}.

Based on the above analysis, we know the impacts of SSTAs on ultra-long waves are more evident than on the large scale, while the impacts of sea surface heating on ultra-long waves and large-scale waves are equivalent. In addition, we also realize that the motion of large-scale waves can not be omitted for considering the effects of sea surface heating.

When the effect of the SSTA is evident, we must take $\alpha_s \neq 0$. Based on the results of scale analysis, we know the impacts of the SSTA on ultra waves are more evident than those on a large scale. Therefore, we take long-wave approximation ($\delta = 0$), that is, the term k^2 in Eq. (15) can be omitted, and the planetary scale is emphasized.

The frequency expression in the low-frequency do-main can be attained:

$$\sigma = \frac{1}{2} \left[-i\alpha_s \Delta z \pm \sqrt{\frac{4k^2(N^2 \Delta z^2/2 - r_T \Delta z)}{(1+2m)^2} - \alpha_s^2 \Delta z^2} \right] = \sigma_r + i\sigma_i \quad (m = 0, 1, 2 \cdots) \quad (16)$$

For Kelvin waves,

$$\sigma = \frac{1}{2}\left[-i\alpha_s \Delta z + \sqrt{4C_2^2 k^2 - \alpha_s^2 \Delta z^2}\right]$$

Therefore, the propagation of Kelvin waves is eastward due to the effects of SST.

For Rossby waves,

$$\sigma = \frac{1}{2}\left[-i\alpha_s \Delta z - \sqrt{\frac{4k^2 C_2^2}{(1+2m)^2} - \alpha_s^2 \Delta z^2}\right] \quad (m=0,1,2\cdots)$$

and Rossby waves are eastward due to the effects of SST.

From Eq. (16), frequency σr is decreasing with the increase of αs, that is, the periods of LFO in the tropics is becoming larger with the increasing of SSTAs. Furthermore, the impacts of SSTAs ($\alpha_s \neq 0$) can make waves unstable ($\sigma_i \neq 0$). Positive (negative) SSTAs can make waves decay (grow). These results are in agreement with the conclusions of Li and Li[17], however the results of the present study show the effects of SSTAs on the larger scale is more evident.

3 Discussion and conclusions

SST is an important factor affecting tropical LFO. The influences of SST include not only sea surface heating, but also the forcing of SST anomalies. Tropical LFO ranges from the planetary scale to a kilometer-sized scale. The response to SST anomalies on the planetary scale is much more severe than that on the kilometer-sized scale, while the response to SST heating on the planetary scale is equivalent to that on the kilometer-sized scale. Therefore, LFO on the synoptic scale cannot be neglected when sea surface heating is considered. Using the absolute ageostrophic model, conclusions can be drawn based on the discussion of the frequency equation as follows:

(1) When SST is comparatively low, sea surface heating can slow down the speed of waves and make the frequency low; and when SST is comparatively high, waves become unstable, and periods of LFO become shorter with the increase in SST.

(2) The period of LFO waves augments with the abnormal increasing of SST, and the wave is unsteady under the effect of the SST anomaly. The wave weakens with abnormally high SST, while the wave rises with abnormally low SST.

The influences of sea surface heating and SST anomalies on the frequency and stability of tropical LFO are mainly discussed in this paper. However, we only considered the effects of SST on the atmosphere. In virtue of the complicated interaction between the ocean and atmosphere, this relatively simple model has limitations in that the influences of SST are mainly manifested by parameters of the atmosphere. Therefore, it is necessary to further investigate the impacts of SST on tropical LFO waves, either theoretically or by numerical modeling.

Acknowledgements. This work was jointly supported by the National Basic Research

Program of China under No. 2006CB403607, State Key Project (Grant No. 40633018), National Natural Science Foundation of China (Grant No. 90211011) and the Key National Project "SCSMES".

References

[1] Madden R D, Julian P. Detection of a 40-50 day oscillation in the zonal wind in the tropical Pacific[J]. J Atoms Sci, 1971, 28:702-708.

[2] Madden R D, Julian P. Description of global scale circulation cells in the tropics with 40-50 day period [J]. J Atoms Sci, 1972, 29:1109-1123.

[3] He Jinhai. The structure and evolution of the circulation with quasi-40-day periodic oscillation in zonal cross-sections over Asian monsoon region[J]. Journal of Tropical Meteorology, 1988, 4(2):116-125. (in Chinese)

[4] He Jinhai. Discussion of Meridional Propagation Mechanism of quasi-40-Day Oscillation[J]. Adv Atmos Sci, 1990, 7(1):78-86.

[5] Krishnamurti T N, Subrahmanyan D. The 30-50 day mode at 850 mb during MONEX[J]. J Atmos Sci, 1982, 39:2088-2095.

[6] Knutson T R, Weic km ann K U. 30-60 day atmospheric oscillation: composite life cycle of convection and circulation anomalies[J]. Mon Wea Rev, 1987, 115:1407-1436.

[7] Li Chongyin, Wu Peili. A further inquiry on 30-60 day oscillation in the tropical atmosphere[J]. Acta Meteorologica Sinica, 1990, 4:525-535.

[8] Chao Jiping, Lin Yonghui, Wang Bin. The influence of sea surface temperature anomaly on Madden-Julian [J]. Acta Meteorologica Sinica, 1996, 54(3):257-271. (in Chinese)

[9] Krishnamurti T N, Oosterhof D K, Mehta A V. Air-sea interaction on the time scale of 30 to 50 days[J]. J Atmos Sci, 1988, 45:1304-1322.

[10] Lau K M, Chan P H. Intraseasonal and interannual variations of tropical convection: Possible link between 40-50 day oscillation and ENSO[J]. J Atmos Sci, 1988, 45:506-521.

[11] Webster P J. Mechanisms determining the atmospheric response to sea surface temperature anomalies[J]. J Atmos Sci, 1981, 38:554-571.

[12] Xu Xiangde, Gao Shouting. Forcing and Dynamics of Wave-Mean Flow Interaction[M]. Beijing: Ocean Press, 2002:7-13. (in Chinese)

[13] Li Chongyin. A dynamical study on the 30-50 day oscillation in the tropical atmosphere outside the equator[J]. J Atmos Sci, 1990, 14(1):101-112.

[14] Dong Ming, Zhang Xinqiang, He Jinhai. A diagnostic study on the temporal and spatial characteristics of the tropical intraseasonaloscillation[J]. Acta Meterorologica Sinica, 2004, 62(6):821-830. (in Chinese).

[15] Xu Xiangde, He Jinhai, Zhu Qiangen. A dynamical analysis of basic factors of low-frequency oscillation in the tropical atmosphere[J]. Acta Meteorologica Sinica, 1990, 4:157-166.

[16] Gao Shouting, Zhu Wenmei, Dong Ming. On the wave-low interaction in the low frequency atmospheric variation blocking pattern[J]. Acta Meteorologica Sinica, 1998, 56:665-680. (in Chinese)

[17] Li Chongyin, Li Guilong. A dynamical study of influence of El Niño on intraseasonal oscillation in tropic atmosphere[J]. Scientia Atmospherica Sinica, 1996, 20:159-168. (in Chinese)

[18] Fu Zuntao, Zhao Qiang, QiaoFangli, et al. Response of atmospheric low-frequency wave to oceanic forcing

in the tropics[J]. Adv Atmos Sci,2000,17:569-574.
[19] Liu Shikuo. An investigation of filtered model for the tropical atmosphere[J]. Journal of Tropical Meteorology,1990,6:106-118. (in Chinese)
[20] Ran Lingkun, Gao Shouting, Li Chongyin. Propagation of low-frequency Rossby wave with divergent wind[J]. Acta Meteorologica Sinica,2005,63(1):13-20. (in Chinese)

Another Look at Influences of the Madden-Julian Oscillation on the Wintertime East Asian Weather[*]

He Jinhai[1]　　Lin Hai[2]　　Wu Zhiwei[2]

(1. Department of Atmospheric Sciences, Nanjing University of Information Sciences and Technology, Nanjing, Jiangsu, China; 2. Meteorological Research Division, Environment Canada, Dorval, Québec, Canada)

Abstract: As one of the major predictability sources on an intraseasonal time scale, the Madden-Julian Oscillation (MJO) exerts a profound influence on the subseasonal forecast of the East Asian (EA) surface air temperature (SAT) and precipitation. In this study we examine the direct link of the MJO with the EA SAT and precipitation and its dynamical mechanism. To represent the MJO, we perform an empirical orthogonal function (EOF) analysis on pentad outgoing longwave radiation (OLR) over (20°S~20°N, 60°E~150°W). The EOF1 mode is mostly characterized by a single convection center near the maritime continent (90°~150°E), whereas the EOF2 mode has an east-west dipole structure with enhanced convection over the eastern Indian Ocean and suppressed convective activities over the tropical western Pacific. At the same pentad of a positive EOF1 phase, large areas of cold anomalies with reduced precipitation emerge in the EA region north of 20°N and persist through two pentads later. At the same pentad of a positive EOF2 phase, SAT and precipitation exhibit a zonal dipole pattern with cold and dry anomalies covering the EA region west of 120°E and warm and wet anomalies to the east. These dipole anomalies in SAT and precipitation systematically move eastward in the next two pentads. A linearized global primitive equation model is utilized to assess the cause of the intraseasonal variability in SAT and precipitation over East Asia associated with the tropical heating of the MJO. The model responses to heating sources that mimic the EOF1 and EOF2 OLR patterns match well the general features of the observed circulation anomalies. Under forcing of a positive phase of the EOF1 pattern, strengthened local Hadley cell or monsoon circulations within (90°~150°E) are reproduced, with anomalous northerly winds and dry anomalies prevailing in the lower troposphere over East Asia. The processes that influence East Asia are mainly associated with intraseasonal changes in local Hadley circulation and the Northern Hemisphere branch of the equatorially trapped Rossby wave gyres forced by the MJO heating.

---

[*] 本文发表于《Journal of Geophysical Research: Earth Surface》, 2011 年第 116 卷, D03109, doi: 10. 1029/2010JD014787.

1 Introduction

Subseasonal variations of wintertime precipitation and near surface air temperature (SAT) have a significant impact on societal and economical activities in the highly populated region of East Asia. For instance, an event with persistent heavy snow, ice, and cold temperature can possibly cause extensive damage and transportation disruption for thousands of travelers [1-4]. A useful forecast of such a persistent anomaly is thus of great value.

There is indication that the Madden-Julian Oscillation (MJO), with its intraseasonal time scale, can provide an important signal for subseasonal weather predictions[5]. The MJO is a tropical large-scale oscillation that is dominated by periods of 30~60 days and zonal wave number one propagating eastward [6]. The MJO has a direct impact on the weather in the tropical region, as it organizes convection and precipitation. The impact of the MJO on the extratropical circulation and weather has been of considerable interest. For example, Higgins et al. [7] and Mo and Higgins [8] investigated the relationships between tropical convection associated with the MJO and U.S. West Coast precipitation. Vecchi and Bond [9] found that the phase of the MJO has a substantial systematic and spatially coherent effect on subseasonal variability in wintertime SAT in the Arctic region. Donald et al. [10] reported a near-global impact of the MJO on rainfall. Lin and Brunet [11] and Lin et al. [12] revealed that the MJO has a significant influence on wintertime SAT and precipitation variability in Canada. Wheeler et al. [13] documented the MJO impact on Australian rainfall and circulation.

There have been many studies on the impact of the MJO on the South and East Asian (EA) region. Most of them, however, are for the boreal summer season[14-16]. In summer, the MJO-related intraseasonal disturbances tend to propagate north-eastward that significantly influence the "active" and "break" monsoon rainfall fluctuations[17-19]. The disturbances associated with the MJO directly modulate the rainfall over the Asian continent through its influence on the genesis of higher-frequency monsoon lows and depressions. As revealed by Goswami et al. [20], a majority of such monsoon lows and depressions develop during the wet phase of the MJO. A recent study of Zhang et al. [21] reported a significant impact of the MJO on summer rainfall in southeast China. The impact of the MJO on wintertime weather in East Asia, especially in its midlatitude region, is less well documented. Jeong et al. [22-23] studied the simultaneous relationship between the MJO and wintertime SAT and precipitation in East Asia from an observational perspective. Using a global primitive-equation model, Matthews et al. [24] investigated the possible mechanisms on how the MJO affects the extratropical circulations from a global perspective. However, East Asia has many distinct features due to unique orographic forcing: huge thermal contrasts between the world's largest continent, Eurasia, and the largest ocean basin, the Pacific, and is strongly influenced by the world's highest land feature, the Tibetan Plateau[25-29]. The global strongest Hadley cell in boreal winter is located over East Asia [30]. On the other hand, comparing to and teleconnection [31], East Asia is much closer to the MJO activity. Most previous studies about the impact of the MJO were based on simultaneous composites with respect to MJO phases [22-23, 26-29, 32]. However, as the MJO is not a perfect periodic oscillation, one phase is not followed by another with the exact time interval. Also it takes time for a tropical forcing to influence the middle latitude atmosphere. An evolution of such simultaneous middle latitude composites may not be directly associated with tropical convection anomalies of the MJO[12]. Therefore, it is necessary to reexamine the linkage between the MJO and the wintertime SAT and precipitation variability in East Asia and the possible physical mechanisms in particular.

In this study we analyze the component of the EA SAT and precipitation that is associated with the MJO, as this component is likely the most predictable on the intraseasonal time scale. We do this by using pentad data, as it can effectively filter out high-frequency components and keep the intraseasonal time scale signal. SAT and precipitation anomalies are extracted according to the main patterns of convection anomalies associated with the MJO. We consider the time lagged association of SAT and precipitation anomalies with respect to the tropical MJO convection activity. These anomalies are interpreted by the associated changes in circulation. A numerical model is used to understand the dynamics of the MJO-induced circulation response that controls the SAT and precipitation anomalies in East Asia.

Section 2 describes the data used and the analysis procedure. In section 3 lagged associations of the EA SAT and precipitation with respect to the variability of the leading modes of MJO convections are presented. The extratropical circulation anomaly associated with the MJO is discussed in section 4. Section 5 presents results of numerical experiments that are designed to simulate the atmospheric response in East Asia to the tropical MJO convection anomalies. Conclusions and discussions follow in section 6.

2 Data and analysis procedure

Grid point data are used to describe SAT and precipitation over East Asia. They are the NCEP/NCAR daily reanalysis of 2 m air temperature[33] and the pentad data of the Climate Prediction Center (CPC) Merged Analysis of Precipitation (CMAP) [34].

To represent the atmospheric circulation fields, variables used here include geopotential height at 500, 700, and 850-hPa, horizontal winds and vertical velocity at all available pressure levels, and specific humidity at 850-hPa. Stream function fields at 200 and 850-hPa are calculated from the horizontal winds. As a proxy for tropical convection, the daily averaged outgoing longwave radiation (OLR) data from the National Oceanic and Atmospheric Administration (NOAA) polar-orbiting series of satellites [35] are used. These data are provided by the NOAA/OAR/ESRL PSD (Boulder, Colorado, USA) from their Website at http://www.cdc.noaa.gov/.

The horizontal resolution for the NCEP/NCAR reanalysis, the OLR, and the CMAP precipitation is 2.5°×2.5°. The daily values of the NCEP/NCAR reanalysis and the OLR data are averaged for consecutive 5 days to construct pentad data. The analysis is conducted for extended winters from November to March. The period of our analysis starts from 1979, as the OLR data are continuously available starting from that year. For the NCEP/NCAR reanalysis, the OLR, and the CMAP precipitation, 30 extended winter seasons from 1979/1980 to 2008/2009 are used.

The seasonal cycle that is the time mean of the 30 year pentad climatology is first removed for each grid point. The mean for each winter is then removed in order to eliminate interannual variability. The resulting pentad perturbation represents an intraseasonal variability. To describe the relative magnitude of changes associated with the MJO compared to the standard deviation of each variable, the perturbation of each variable is normalized by its own standard deviation during the analysis period.

3 Connection between the MJO and East Asian SAT and precipitation

Shown in Figs. 1a and 1b (figure not shown) are the extended winter climatology and standard deviation of the pentad-to-pentad variability of SAT. A strong intraseasonal variability of SAT occurs in Siberia and Mongolia that extends to north China. Strong variability can also be found in south China. The SAT variability in north and south China seems to be connected, with a large variance along the east coast (Fig. 1b). This may be associated with low-frequency expansions of winter cold air from the north (cold surges). The climatology and standard deviation of precipitation are illustrated in Fig. 2 (figure not shown). Large values of average precipitation and variability occur in southeast China that seem to be part of the strong precipitation and variability over the ocean south of Japan, which is likely connected to the subtropical East Asian jet and storm track.

As in the work by Lin et al.[12], the two leading modes of the empirical orthogonal function (EOF) analysis of the pentad OLR anomaly for the extended winter from 1979/1980 to 2008/2009 are used to represent the tropical convection patterns of the MJO. The area used is (20°S~20°N, 60°E~150°W) where the tropical convection and its variability are strong. The two leading modes explain 11% and 10% of the total variance, respectively, that are dominant among all the EOF modes, since EOF3 or a higher-order EOF accounts for a much smaller fraction of variance. The EOF1 mode shows a large area of above normal convection anomalies centered near the maritime continent (90°~150°E), whereas a dipole structure of convection anomalies, enhanced convection in the eastern Indian Ocean west of 120°E and reduced convective activity in the western Pacific east of 120°E, characterizes the EOF2 mode (Fig. 3). The properties of these two EOFs were discussed in detail by Lin et al.[12]. In brief, their principal components (PCs) have a significant power spectrum peak at the MJO time period (around 10 pentads). A sequence of the appearance of +EOF2, +EOF1, −EOF2, and −EOF1 separated by about two pentads can be implied from the lag correlation between PC1 and PC2. The above features indicate that the two leading modes together represent the eastward propagating MJO disturbance, in agreement with previous studies that performed similar EOF analysis of OLR data[32,36]. The longitudinal distributions of these two leading modes are very similar to the OLR components of the combined EOF analysis of Wheeler and Hendon[37]. PC1 and PC2 are highly correlated with their real-time multivariate MJO (RMMs). The correlation between PC1 and RMM1 is 0.62, and that between PC2 and RMM2 is −0.76. An analysis using the OLR PCs should reach similar conclusions comparing to the one that is based on RMMs. Here we choose PC1 and PC2 to represent the MJO variability instead of RMM1 and RMM2 because we would like to construct a direct link of the major patterns of tropical convective activity related to the MJO with the EA SAT and precipitation as well as circulations. As will be seen in section 5, such tropical convection patterns contribute significantly in forcing the midlatitude circulation anomalies and intraseasonal weather changes in East Asia.

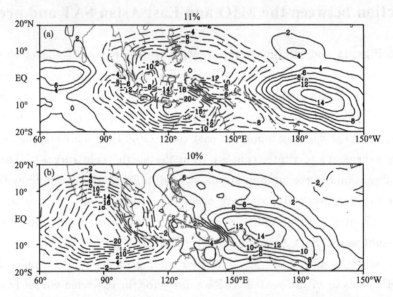

Fig. 3 (a) EOF1 and (b) EOF2 of outgoing longwave radiation (OLR) represented as regressions of pentad OLR onto the respective principal component (PC). The magnitude corresponds to one standard deviation of the PC. The contour interval is 2 W·m^{-2}. Contours with negative values are dashed. The zero contour is not plotted

In an attempt to see the relationship between the tropical MJO convection and the EA winter SAT and precipitation anomalies, lagged regressions between the two leading OLR PCs and SAT and precipitation anomalies are calculated. As discussed in section 2, PC2 leads PC1 and PC1 leads a negative phase of PC2 by about two pentads. Therefore an evolution of global atmospheric anomalies in relation to the MJO could be visualized just by analyzing the simultaneous regressions or composites with respect to PC1 and PC2, as did in many previous studies[32,38]. However, as the MJO is not a perfect periodic oscillation which is evident from the fact that the lagged correlation between PC1 and PC2 does not reach 1, the atmospheric anomalies associated with the two leading modes have differences as well as linkages. Therefore, we present here lagged regressions between atmospheric anomalies in East Asia and PC1 and PC2, separately. In this way, we would have a correct timing of the appearance of middle latitude anomalies with respect to the tropical convection anomaly patterns of the MJO. This will also help to facilitate a comparison with the numerical modeling of midlatitude response to the MJO convective anomalies that takes time to develop as will be presented in section 5.

3.1 SAT

Shown in Fig. 4 (figure not shown) are the lagged regressions of the SAT variability with respect to PC1 and PC2, where lag = n means that the SAT anomaly lags the OLR PC by n pentads. The magnitude represents an SAT anomaly corresponding to one standard deviation of the given PC. Shaded areas are those where the regression is statistically significant at the 99% confidence level according to a Student t test. As can be seen, A positive PC1 phase has significantly cold anomalies with the SAT variability over the midlatitude East Asia north of 20°N at the simultaneous period and persists through two pentads later (Figs. 4a-4c). On the other hand, when a positive phase of PC2 occurs, pronounced warm anomalies are found over a large area of the EA region east of 120°E, basically over Japan at the simultaneous pentad (Fig. 4d) and systematically move eastward from the lag-1 to lag-2 pentad (Fig. 4e and Fig. 4f). The western and central China is controlled by prominent cold anomalies, gradually. At the lag-1 pentad (Fig. 4e), the magnitude of the negative SAT anomalies increases and the area expands to the southeast coast. The maximum decrease of SAT reaches about 0.2 unit corresponding to one standard deviation of PC2. Recall that the SAT anomaly has been normalized by its pentad standard deviation. Therefore, a decrease of SAT by 20% of its standard deviation in southeast and northwest China can be expected when a positive PC2 with a one standard deviation magnitude occurs. At the lag-2 pentad (Fig. 4f), the area of negative SAT anomalies propagate eastward, with some left along the southeast coast and central China area. The lagged relationship between PC1 and PC2 is reflected in Fig. 4, as Fig. 4f resembles Fig. 4a and Fig. 4c is similar to an opposite phase of Fig. 4d. The negative simultaneous correlations between PC2 and the SAT over the East Asian land area as shown in Fig. 4d are consistent with those of Jeong et al.[22], who analyzed the wintertime SAT anomaly at East Asian stations in different phases of the MJO that were defined according to the RMMs of Wheeler and Hendon[37]. Our lagged regressions (Fig. 4), however, provide a clear picture on the evolution of SAT in the whole EA region associated with the tropical convection of the MJO.

3.2 Precipitation

Because of the skewness of the probability distribution, a square root transformation is used for the precipitation pentad data, following Richman and Lamb[39] and Mo and Higgins[8]. The transformed precipitation anomaly at each station or grid point is normalized by its pentad standard deviation.

The lagged regressions of precipitation anomalies with respect to PC1 and PC2 are presented in Fig. 5(figure not shown). Both PC1 and PC2 have significant connections with precipitation anomalies in the midlatitude East Asia. A positive PC1 phase, which corresponds to enhanced equatorial convection near the maritime continent (90°~150°E), is associated with significantly reduced precipitation in a large area of midlatitude China and Japan (Fig. 5a). The largest anomaly can be found one pentad after a positive PC1 phase, with a reduction of precipitation of about 15% of its pentad standard deviation from central China to south Japan (Fig. 5b).

For a positive PC2 phase, which corresponds to a dipole equatorial convection pattern with enhanced convection over the Indian Ocean west of 120°E and suppressed convective activities over the tropical western Pacific east of 120°E, its influence on the midlatitude precipitation is mainly over the northwest Pacific Ocean east of China, where a band of significant positive precipitation anomalies can be found at the simultaneous and one pentad after a positive PC2 phase (Fig. 5d and Fig. 5e). The west end of this band is over southeast China, which generates substantial increase of precipitation there, with a maximum positive precipitation anomaly about 15%~20% of the pentad standard deviation. This anomalous precipitation band seems to be collocated with the climatological maximum precipitation and variability near the wintertime East Asian storm track (Fig. 2). The anomalous precipitation band moves southeastward and becomes weakened, and two pentads later the positive anomaly signal in southeast China is almost gone and the band disappears in the northwestern Pacific (Fig. 5f). The simultaneous regression of the East Asian precipitation with PC2 (Fig. 5d) seems to confirm the result of Jeong et al. [23], who found significant increase of precipitation in east China, Japan, and Korea when the MJO is in its phases 2 and 3. With the lagged regressions as shown in Fig. 5, the evolution of precipitation anomaly in the whole EA region associated with the tropical convection of the MJO are clearly demonstrated.

4 Large-scale circulation anomalies

In order to understand the SAT and precipitation signal in East Asia related to the MJO as observed above and have a better confidence of this connection, in this section we examine the large-scale circulation anomalies corresponding to the tropical MJO convection. As demonstrated in many previous studies, the MJO with its tropical diabatic heating anomaly can excite not only equatorial waves but also extratropical Rossby wave trains and influence the global circulation [11,24,31,40]. The winter SAT and precipitation in Canada, for example, are significantly influenced by the MJO activity through the extratropical Rossby wave train mechanism [11-12]. East Asia is a special region, as it is not that remote to the MJO convection activity comparing to other extratropical regions. Therefore, the extratropical Rossby wave train mechanism may not apply. On the other hand, the MJO disturbance itself is limited to the tropical region and thus the impact on the midlatitude East Asia can hardly be explained as a direct extension of the MJO convection. How, then, do the MJO induced circulation anomalies influence East Asia?

We start by looking at the upper zonal wind near the jet level at 200-hPa (U200) over East Asia. Fig. 6a(figure not shown) shows the U200 climatology for the extended winter during 1979/1980 to 2008/2009. The subtropical westerly jet is seen along about 30°N with its maximum wind of over 65 m • s^{-1} near 140°E south of Japan. The pentad-to-pentad variability of U200 is large along the jet, as is clear from the standard deviation distribution (Fig. 6b). The variability near the jet core of 140°E is relatively weak. Upstream of the jet core the area of strong U200 variability is quite narrow and confined to the jet axis implying that the jet varies mainly in strength, whereas to the east of the jet core over the North Pacific the region with strong variability has a broad meridional extent indicating that here the jet undergoes variability in both strength and north-south position. To see the changes of U200 associated with the tropical MJO convection, regressions are calculated with respect to PC1 and PC2, as shown in Fig. 6c and Fig. 6d, respectively. Corresponding to a positive phase of PC1, which has above normal convective activities over the equatorial maritime continent (90°~150°E) (Fig. 3a), the westerly jet is significantly strengthened over the midlatitude East Asia. With a one standard deviation of PC1 change, the subtropical jet strength increases by more than 30% of its pentad-to-pentad standard deviation. The regression of U200 with respect to PC2, on the other hand, shows an increase of westerly wind over west China north of India and an easterly wind anomaly in the North Pacific east of the jet core (Fig. 6d).

The above changes in upper zonal wind are closely related to variability in local Hadley circulation induced by tropical MJO convections. To measure the variability in the intensity of local Hadley circulation, the meridional and vertical velocities for each pentad are regressed against PC1 and PC2. A longitudinal band of 20° is chosen to represent the local Hadley circulation corresponding to the tropical convection center. Shown in Fig. 7a is the regression of the medional-vertical flows averaged for the longitude section of 110°~130°E with respect to PC1. As is seen, a positive PC1 with enhanced tropical convection near 120°E is correlated with an intensified local Hadley cell in the Northern Hemisphere subtropics. The upper northward outflow of the tropical convection north of the equator is then influenced by the Coriolis force that intensifies the subtropical westerly jet (Fig. 6a). At the same time, this upper outflow converges near the jet latitude that leads to an increased downward motion in the troposphere (which will be shown in Fig. 9c) and reduced precipitation over a large part of midlatitude East Asia as observed in Fig. 5a. The upper zonal wind variability near the subtropical jet associated with PC2 can also be explained by changes in local Hadley cells. This time, the tropical convection anomaly has a dipole structure with enhanced convection in the Indian Ocean west of 120°E centered at 90°E and reduced convection in the western Pacific east of 120°E centered at 165°E (Fig. 3b). The reduced tropical western Pacific convection decreases the strength of the local Hadley cell near that longitude, as is illustrated in Fig. 7b of the regression for meridional-vertical flow averaged for the longitude section of 155°~175°E with respect to PC2 and weakens the midlatitude westerlies (Fig. 6d) over the northwestern Pacific. On the other hand, the enhanced convection of the tropical Indian Ocean intensifies the local Hadley cell (not shown) and midlatitude westerlies over the region north of India (Fig. 6d).

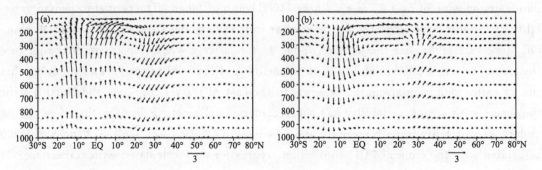

Fig. 7 Simultaneous regression of meridional-vertical wind cross section of (a) 110°~130°E onto PC1; of (b) 155°~175°E onto PC2. Note that the vertical velocity is multiplied by 10 for displaying purpose. The vector unit is m·s^{-1}

To illustrate atmospheric circulation as a response to tropical MJO heating anomaly, shown in Fig. 8 are regressions to PC1 and PC2 for stream function anomalies at 200-hPa and 850-hPa. In general, the MJO convection patterns of EOF1 and EOF2 are both associated with circulation anomalies that are symmetric to the equator, consistent with the theoretical result of Gill [41]. The anomalies have a baroclinic vertical structure within the tropical region of 30°S~30°N, with opposite signs at the upper and lower troposphere. Two Rossby wave gyres straddling the equator are located slightly west of strong tropical convection regions. To the east of convections, a structure similar to an equatorial Kelvin wave can be found. A positive phase of PC1 is associated with a pair of cyclonic circulations at 850-hPa and anticyclonic flows at 200-hPa as the equatorially trapped Rossby wave response. Their centers are slightly to the west of but very close to the enhanced tropical convection center near 120°E, except for the Northern Hemisphere 200-hPa anticyclone which is at the same longitude as the tropical heating. For PC2, the cyclonic Rossby wave gyres at 850-hPa are located near 60°E, to the west of the tropical heating, while the Northern Hemisphere 200-hPa anticyclone is located near 90°E, the same longitude as the tropical heating. Lagged regressions are calculated for 200-hPa and 850-hPa stream function with respect to PC1 and PC2, indicating systematic eastward propagation of the circulation anomalies (not shown), consistent with the SAT and precipitation anomalies.

Regressions of 850-hPa geopotential height (Z850) anomalies to PC1 and PC2 are illustrated in Figs. 9a and 9b(figure not shown), respectively. They are in general consistent with the stream function anomalies at 850-hPa (Figs. 8c and 8d), showing significant nega-

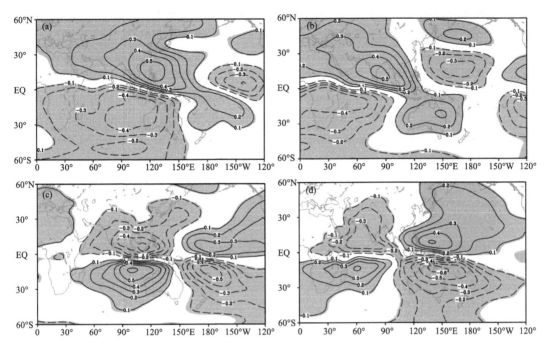

Fig. 8 Simultaneous regression of normalized 200-hPa stream function onto (a) PC1 and (b) PC2. Simultaneous regression of normalized 850-hPa stream function onto (c) PC1 and (d) PC2. The magnitude corresponds to one standard deviation of the respective PC. Contour interval is 0.1 units. The zero contour is not plotted, and contours with negative values are dashed. The shaded areas represent those where the correlation is statistically significant at a 99% confidence level

tive geopotential anomalies influencing the EA region corresponding to the Northern Hemisphere branch of the equatorial Rossby wave response. For PC1 (Fig. 9a), a negative 850-hPa geopotential anomaly is observed extending from the tropics to the northwestern Pacific. Over the ocean between the low-pressure system and the east coast of China, strong anomalous low-level wind from the northeast is produced along the isobars, leading to cold advection that may be responsible for the cold anomaly over the northwestern Pacific (Fig. 4a). The regression of Z850 to PC2 (Fig. 9b) has a large negative anomaly covering almost the whole EA region. The low-pressure system shows fine local structures that may be related to topography and land-sea distributions. Over the northwestern Pacific east of Taiwan, southwesterly wind anomaly with warm advection could explain the warm anomaly there (Fig. 4d). The northerly wind in central China is likely responsible for the cold anomaly there (Figs. 4d and 4e). The low-level trough extending from south China north-eastward to Japan corresponds well with the increased precipitation band observed in Fig. 5d.

To better understand the significant precipitation anomalies over East Asia with respect to the MJO, we present regressions of pressure velocity anomaly to PC1 and PC2 in Figs. 9c and 9d, respectively. A good correspondence can be found between the vertical motion and precipitation anomalies by comparing with Figs. 5a and 5d. A positive PC1 phase is linked to a significant downward motion over midlatitude East Asia (Fig. 9c). As discussed earlier, this could be associated with the enhanced local Hadley circulation caused by an increased convection near the maritime continent (Fig. 7a). For PC2, the intensified upward motion extending from south China to Japan could be associated with both the reduced local Hadley cell in the western Pacific (Fig. 7b) and the low-level trough of Fig. 9b.

Changes in synoptic-scale transient activity and moisture transport associated with the MJO may also contribute to precipitation anomalies. Daily geopotential height data at 700-hPa during each extended winter are band-pass filtered to retain fluctuations with periods between 2 and 6 days. The standard deviation of the filtered data is calculated for each pentad to represent transient activity. Shown in Figs. 10a and 10b(figure not shown) are regressions of transient activity to PC1 and PC2, respectively. Note that these regressions are for unnormalized perturbations. As can be seen, associated with a positive PC1, reduced storm track activity is observed over most of the EA land region. However, a positive PC2 is associated with enhanced transient eddy activity from east China to south Japan. Regressions of moisture flux at 850-hPa to PC1 and PC2 are depicted in Figs. 10c and 10d. The shaded areas represent those where the moisture flux convergence is stronger than 1.0×10^{-6} g $\cdot$ kg^{-1} $\cdot$ s^{-1}. A positive phase of PC1 is accompanied by northeasterly moisture flux and divergence over the middle latitude EA, whereas a positive PC2 is associated with southwesterly moisture flux and convergence over southeastern China and south Japan.

5 Numerical model experiments

We have observed that the intraseasonal variability of SAT and precipitation in East Asia is significantly correlated with the tropical MJO convection and the local change in the Hadley cell. To test the hypothesis that the atmospheric circulation anomalies over East Asia and the related changes in SAT and precipitation result from a direct response to the MJO convective heating, a series of numerical model experiments are performed. We use the simple general circulation model (SGCM) as described by Hall et al. [42]. In brief, it is a global primitive equation spectral model with no moisture representation. The resolution used in this study is T31 with 10 vertical levels. This model is able to reproduce the observed climatology and variability reasonably well and has been used in studies with different applications[42-44].

Here linear integrations are performed using the SGCM, with the approach as described by Lin [45]. The basic state is chosen to be the model climate of a perpetual winter (DJF) integration (control run) of 3600 days under a time-independent forcing that is calculated empirically from observed daily data as a residual for each time tendency equation of the model. In the linear experiments, a new constant forcing, which is calculated as a residual in the model equations with only time mean quantities of the model climate, is applied to maintain this basic state.

Four linear experiments are conducted with different perturbation thermal forcings at the equator. They are (1) H120 run, a heating at 120°E; (2) H90C165 run, a dipole thermal forcing with heating at 90°E and cooling at 165°E; (3) H90 run, a heating at 90°E; (4) C165 run, a cooling at 165°E. The first two experiments are designed to study the model response to a tropical thermal forcing that mimics the EOF1 and EOF2 MJO diabatic heating (Figs. 3a and 3b, respectively). Experiments 3 and 4 are conducted to further investigate the separate contribution of the west heating and east cooling of the dipole convection pattern of EOF2. The tropical heating anomaly that is added to the temperature equation is switched on at $t=0$ and persists during the integration. No forcing anomaly is applied to the vorticity, divergence, or mass equations. The heating perturbation represents deep convection in the tropics and has an elliptical form in the horizontal. All the perturbation heating and cooling sources have a semimajor axis of 40 degrees of longitude and a semiminor axis of 11 degrees of latitude. The magnitude of the heating is proportional to the squared cosine of the distance from the center. The heating anomaly has a vertical profile of $\sqrt{1-\sigma}\sin[\pi(1-\sigma)]$, which peaks at $\sigma=0.45$ with a vertically averaged heating rate at the center of $3.5℃ \cdot d^{-1}$. This rate is equivalent to a latent heating associated with a precipitation of about $1.5 cm \cdot d^{-1}$. The vertically averaged heating anomalies for the four runs are illustrated in Fig. 11(figure not shown). For each experiment, the linear model is integrated for 5 days starting from the SGCM's control run climate. To ensure a small amplitude linear response, the forcing amplitude is scaled by 10^{-2} times the anomalous forcing and the solutions are scaled back for display purposes.

We first look at the model response of large-scale circulations in the East Asian region. The model responses of 200-hPa zonal wind (U200) anomaly at day 5 in the four experiments are shown in Fig. 12(figure not shown). By comparing Figs. 12a and 12b with the observed U200 anomalies associated with PC1 and PC2 (Figs. 6c and 6d), it is clear that the linear model simulates well the changes of the subtropical westerly jet. This confirms that the upper zonal wind variability of the jet in the midlatitude East Asia is a direct response to the tropical convection anomaly of the MJO. As in the observations, in the H120 run, a band of westerly wind anomaly is generated across midlatitude China and Japan, which is associated with an increase of local Hadley circulation caused by the enhanced tropical convection near 120°E. In the H90C165 run, a strong westerly wind anomaly over west China north of India and an easterly anomaly in the North Pacific are produced (Fig. 12b), again in good agreement with the observations (Fig. 6d). Experiments with separate tropical forcings (H90 and C165 runs) confirm that the different changes of U200 along the jet to the west and east of 120°E are caused separately by the heating over the tropical Indian Ocean and the cooling over the tropical western Pacific (Figs. 12c and 12d). These changes in midlatitude U200 are related to variability in local Hadley cells induced by tropical convection centers.

The day 5 responses of 200-hPa and 850-hPa stream function anomalies are presented in Figs. 13 and 14, respectively. Comparing with Fig. 8, the linear model is found to be able to reasonably well produce all the features of the observed circulation anomalies in both the upper and lower troposphere. The longitudinal positions of all the equator-symmetric pairs of circulation flows are correctly simulated by the model. From the result of the H120 run (Figs. 13a and 14a), it is seen that the Northern Hemisphere gyre of the equatorially trapped Rossby wave cause by the tropical heating at 120°E is located directly over the East Asian region, which could lead to anomalous weather conditions there. In addition, the local Hadley cell is intensified within the region (90°~150°E). For the H90C165 run, west-east dipole circulation anomalies with an anticyclone (cyclone) to the west and a cyclone (anticyclone) to the east in the upper (lower) troposphere are generated along the mid-latitude East Asia by the tropical forcing. The relative contributions of the heating at 90°E and cooling at 165°E can be assessed from Figs. 13c and 13d or Figs. 14c and 14d. The heating at 90°E produces a strong Rossby wave response that covers a large part of the Asian continent where the cooling of 165°E has no influence. Over the northwest Pacific, both H90 and C165 have contributions, where the Rossby wave response of C165 is in phase with the Kelvin wave circulation anomaly of H90 to form a strong cyclonic (anticyclonic) circulation at 200-hPa (850-hPa). The local Hadley cell is enhanced within the EA region west of 120°E and weakened within the EA region east of 120°E.

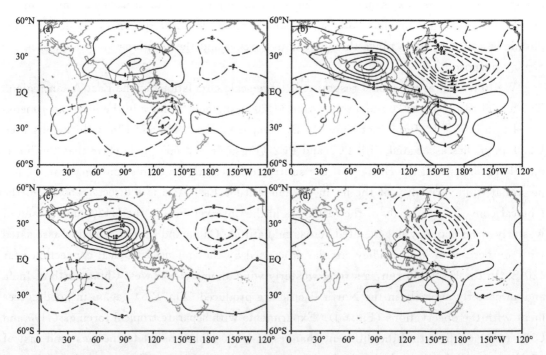

Fig. 13 The 200-hPa stream function response at day 5 for (a) H120 run, (b) H90C165 run, (c) H90 run, and (d) C165 run. The contour interval is 2×10^6 m$^2 \cdot$ s^{-1}. The zero contour is not plotted, and contours with negative values are dashed

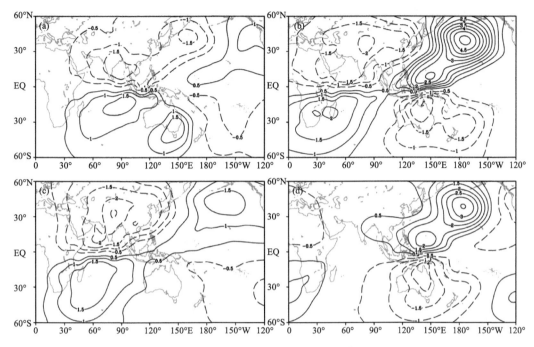

Fig. 14 Same as Fig. 13 but for 850-hPa stream function response.
The contour interval is 0.5×10^6 m² · s⁻¹

Shown in Fig. 15 (figure not shown) is the day 5 response of air temperature at 950-hPa (T950) which is the lowest model level. In the H120 run, the tropical heating at 120°E produces negative temperature anomalies over the northwestern Pacific, Indochina, and north India, with a general feature similar to the observed SAT anomalies associated with PC1 (Fig. 4a). The H90C165 run generates a negative T950 anomaly over the midlatitude continent and a northeast-southwest oriented band of warm T950 over the north-western Pacific, a distribution very similar to the observed SAT anomalies associated with PC2, except that the simulated cold anomaly over the main land is a little too far to the north. Examination of the separate response of the H90 and C165 runs (Figs. 15c and 15d) indicates that the cold anomaly over the main land is forced by the tropical heating at 90°E, whereas both the H90 and C165 forcing contribute to the strong warm anomaly over the northwestern Pacific.

Finally we examine the day 5 response of pressure velocity at 550-hPa for the four integrations (Fig. 16, figure not shown). Because the model used here has no moisture and does not produce precipitation, this forced vertical motion response can serve as a good indicator for precipitation, as an upward motion in the real atmosphere generally produces convection and precipitation while a downward motion is normally associated with dry condition. The H120 run produces a band of downward motion anomaly in the midlatitude East Asia which is in good agreement with the observations (Fig. 9c). For the H90C165 run, an intensified upward motion in the northwestern Pacific extending from south China to Japan is generated, which resembles the observed vertical motion anomaly associated PC2 (Fig. 9d). By looking at the separate response from H90 and C165 (Figs. 16c and 16d), we conclude that both the heating at 90°E and cooling at 165°E contribute to the upward motion anomaly over the northwestern Pacific.

The model response for the second pentad in the East Asian region has a similar distribution but stronger amplitude compared to that in the first pentad. As was shown by Lin et al.[12] [Figs. 13c and 13d] where the same H90C165 experiment was integrated for 15 days, the extratropical Rossby wave develops and propagates in the North Pacific and North American region after 5 days of integration.

6 Conclusions and Discussions

The MJO plays an important role in subseasonal forecast of EA SAT and precipitation and has received lots of research interests[3,22-23]. The direct link of the tropical convective activity with the EA SAT and precipitation is examined in this study. The MJO convections are represented by the leading two modes of an EOF analysis of the pentad OLR over (20°S~20°N, 60°E~150°W). The EOF1 mode is characterized by a single convection center near the maritime continent (90°~150°E), whereas the EOF2 mode has an east-west dipole structure with enhanced convection over the Indian Ocean west of 120°E and suppressed convective activities over the tropical western Pacific east of 120°E. At the same pentad of a positive EOF1 phase, large areas of cold anomalies with reduced precipitation emerge in the EA region north of 20°N and persist through two pentads later. At the same pentad of a positive EOF2 phase, SAT and precipitation exhibit a zonal dipole pattern with cold and dry anomalies covering the EA region west of 120°E and warm and wet anomalies occupying the EA region east of 120°E. These dipole anomalies in SAT and precipitation systematically moves eastward from one pentad to two pentads later. The direct link of the MJO with the EA SAT and precipitation is through the local Hadley cell and the circulation anomaly of the Northern Hemisphere branch of the equatorially trapped Rossby wave as a response to the tropical MJO diabatic heating.

Matthews et al.[24] suggested that the extra-tropical responses to the MJO are consistent with theories of Rossby wave forcing and dispersion on the climatological flow. Compared with other extratropical regions, East Asia is much closer to the MJO activities. This study suggests that on the intraseasonal time scales, the MJO impacts on the EA winter weather through changing the local Hadley cell, and the influence of the northward extension of the direct equatorial Rossby wave response. Linear numerical experiments using a primitive equation model with anomalous thermal forcings that resemble the leading modes of tropical convection are able to reproduce the observed circulation anomalies associated with the two modes, suggesting that the observed circulation anomalies result from the thermal forcing of the MJO. The relevant physical processes can be summarized as following: For the EOF1 mode, anomalous heating (cooling) near the maritime continent (90°~150°E) strengthens (weakens) the local Hadley cell or the winter monsoon circulation and brings about cold and dry (warm and wet) weather over the midlatitude East Asia[30]. For a strong (weak) EOF2 mode, anomalous heating (cooling) over the Indian Ocean west of 120°E and cooling (heating) over the tropical western Pacific east of 120°E enhances (weakens) the local Hadley cell over the western EA region and weakens (enhances) the local Hadley cell over the eastern EA region, which consequently leads to occurrence of cold and dry weather in the western (eastern) East Asia and warm and wet weather in the eastern (western) East Asia. The northern branch of the tropical Rossby wave gyre associated with the MJO convections also influences the EA weather through such processes as modified storm track activity and moisture flux.

The same analysis as discussed above for SAT and precipitation was repeated using station data over China, and very similar results were obtained. This indicates that the NCEP/NCAR reanalysis SAT and CMAP precipitation data capture well the intraseasonal variability in the East Asian region.

The MJO has predictability significantly longer than synoptic scale weather systems[46]. For example, it was shown by Reichler and Roads [47] with a perfect model approach that a useful forecast skill of the tropical intraseasonal variability is about 4 weeks. Forecasts of the MJO with statistical methods are also quite encouraging[38,48-49]. The useful predictive skill of the MJO from these empirical models can usually reach about 15-20 days lead time. Therefore, the connection of the SAT and precipitation anomalies to the tropical MJO convection as observed in this study provides useful information for extended range and long-range forecasting in the midlatitude EA region. However, it is important to keep in mind that there are many factors that can influence the low-frequency variability of the EA SAT and precipitation in winter, i. e., cold waves from the mid to high latitudes[4,28,50] and the thermal state of the Tibetan Plateau. The result of the regression analysis just represents an averaged influence of many MJO events in a statistical sense. For an individual case, the signal coming from the tropics could be overwhelmed by other influences such as those from the high-latitude weather. On the other hand, previous studies have shown that the atmospheric response to tropical forcing is sensitive to the extratropical basic flow [51]. Because the MJO response in East Asia is on a time scale of about a week, this sensitivity may be much weaker than those on seasonal time scale but still may not be negligible. In addition, since much of the MJO predictability arises from MJO events which have already formed, the low skill in predicting the MJO onset may shorten practical forecast time.

It is also important for numerical models to correctly represent and simulate the tropical-extratropical connections as observed in this study. The presented result provides useful information for validating the numerical models.

Acknowledgments. This research was supported by the Natural Science Foundation of China (NSFC 40633018), by the Canadian Foundation for Climate and Atmospheric Sciences, by the Natural Science and Engineering Research Council of Canada (NSERC), and by the Special Research Program for Public Welfare (Meteorology) of China under grant GYHY200906016. We thank three anonymous reviewers whose comments and suggestions helped to improve our paper.

References

[1] Ding Y H, Wang Z Y, Song Y F, et al. Causes of the unprecedented freezing disaster in January 2008 and its possible association with the global warming[J]. Acta Meteorol Sin, 2008, 665, 809-825.

[2] Wen M, Yang S, Kumar A, et al. An analysis of the large-scale climate anomalies associated with the snowstorms affecting China in January 2008[J]. Mon Wea Rev, 2009, 137, 1111-1131.

[3] Hong C-C, Li T. The extreme cold anomaly over southeast Asia in February 2008: roles of ISO and ENSO[J]. J Climate, 2009, 22, 3786-3801.

[4] Wu Z, Li J, Jiang Z, et al. Predictable climate dynamics of abnormal East Asian winter monsoon: Once-in-a-century snowstorms in 2007/2008 winter[M]. Clim Dyn, in press, 2010.

[5] Waliser D E, Lau K M, Stern W, et al. Potential predict- ability of the Madden-Julian oscillation[J]. Bull Am Meteorol Soc, 2003, 84: 33-50.

[6] Madden R A, Julian P R. Description of a 40-50 day oscillation in the zonal wind in the tropical Pacific [J]. J Atmos Sci, 1971, 28: 702-708.

[7] Higgins R W, Schemm J-K E, Shi W, et al. Extreme precipitation events in the western United States related to tropical forcing[J]. J Climate, 2000, 13: 793-820.

[8] Mo K C, Higgins R W. Tropical convection and precipitation regimes in the western United States[J]. J Climate, 1998, 11: 2404-2423.

[9] Vecchi G A, Bond N A. The Madden-Julian Oscillation (MJO) and northern high latitude wintertime surface air temperatures[J]. Geophys Res Lett, 2004, 31, L04104.

[10] Donald A, Meinke H, Power B, et al. Near-global impact of the Madden-Julian Oscillation on rainfall[J]. Geophys Res Lett, 2006, 33, L09704, doi: 10.1029/2005GL025155.

[11] Lin H, Brunet G. The influence of the Madden-Julian Oscillation on Canadian wintertime surface air temperature[J]. Mon Wea Rev, 2009, 137: 2250-2262.

[12] Lin H, Brunet G, Mo R. Impact of the Madden-Julian Oscillation on wintertime precipitation in Canada [J]. Mon Wea Rev, 2010, 138: 3822-3839.

[13] Wheeler M, Hendon H H, Cleland S, et al. Impacts of the Madden-Julian oscillation on Australian rainfall and circulation[J]. J Climate, 2009, 22: 1482-1498.

[14] Zhu C, Tetsuo N, Li J. Modulation of twin tropical cyclo-genesis by the MJO westerly wind burst during the onset period of 1997/98 ENSO[J]. Adv Atmos Sci, 2003, 20(6): 882-898.

[15] Zhu C, Tetsuo N, Li J, et al. The 30-60 day intraseasonal oscillation over the western North Pacific Ocean and its impacts on summer flooding in China during 1998[J]. Geophys Res Lett, 30(18), 2003, doi: 10.1029/2003GL017817.

[16] Zhan R, Li J, Gettelman A. Intraseasonal variations of upper tropospheric water vapor in Asian monsoon region[J]. Atmos Chem Phys Discuss, 2006, 6: 8069-8095, doi: 10.5194/acpd-6- 8069-2006.

[17] Yasunari T. Cloudiness fluctuations associated with the Northern Hemisphere summer monsoon[J]. J Meteorol Soc Jpn, 1979, 57: 227-242.

[18] Murakami T, Nakazawa T, He J. On the 40-50 day oscillations during the 1979 Northern Hemisphere

summer. Part I: Phase propagation[J]. J Meteorol Soc Jpn,1984,62: 440-467.

[19] Wang B,Webster P,Kikuchi K,et al. Boreal summer quasimonthly oscillation in the global tropics[J]. Clim Dyn,2006,27:661-675.

[20] Goswami B N,Ajayamohan R S,Xavier P K,et al. Clustering of low pressure systems during the Indian summer monsoon by intraseasonal oscillations[J]. Geophys Res Lett, 2003, 30(8): 1431, doi: 10.1029/2002GL016734.

[21] Zhang L,Wang B,Zeng Q. Impact of the Madden-Julian Oscillation on summer rainfall in southeast China [J]. J Climate,2009,22:201-216.

[22] Jeong J-H,Ho C-H,Kim B-M,et al. Influence of the Madden-Julian oscillation on wintertime surface air temperature and cold surges in East Asia [J]. J Geophys Res, 2005, 110, D11104, doi: 10.1029/2004JD005408.

[23] Jeong J-H,Kim B-M,Ho C-H,et al. Systematic variation in wintertime precipitation in East Asia by MJO-induced extratropical vertical motion[J]. J Climate,2008,21:788-801.

[24] Matthews A J,Hoskins B J,Masutani M. The global response to tropical heating in the Madden-Julian Oscillatio during Northern winter[J]. Q J R Meteorol Soc,2004,130: 1991-2011.

[25] Ding Y H. Summer monsoon rainfalls in China[J]. J Meteorol Soc Jpn,1992,70:397-421.

[26] Chang C-P. Preface,in East Asian Monsoon,pp. v-vi,World Sci,Singapore,2004.

[27] Wang B,Wu Z,Li J,et al. How to measure the strength of the East Asian summer monsoon[J]. J Climate,2008,21: 4449-4463.

[28] Wang B,Wu Z,Liu J,et al. Another look at climate variations of the East Asian winter monsoon: Northern and southern modes[J]. J Climate,2010,23:1495-1512.

[29] Wu Z,Wang B,Li J,Jin F-F. An empirical seasonal prediction model of the East Asian summer monsoon using ENSO and NAO[J]. J Geophys Res,2009,114,D18120.

[30] Wu Z,Li J,Wang B,et al. Can the Southern Hemisphere annular mode affect Chinese winter monsoon [J]. J Geophys Res,2009,114,D11107,doi:10. 1029/2008JD011501.

[31] Lin H,Brunet G,Derome J. An observed connection between the North Atlantic Oscillation and the Madden-Julian Oscillation[J]. J Climate,2009,22:364-380.

[32] Ferranti L,Palmer T N,Molteni F,et al. Tropical-extratropical interaction associated with the 30-60 day oscillation and its impact on medium and extended range prediction[J]. J Atmos Sci,1990,47:2177-2199.

[33] Kalnay E,et al. The NCEP/NCAR 40-year reanalysis project[J]. Bull Am Meteorol Soc, 1996, 77: 437-471.

[34] Xie P,Arkin P A. Global precipitation: A 17-year monthly analysis based on gauge observations,satellite estimates,and numerical model outputs[J]. Bull Am Meteorol Soc,1997,78: 2539-2558.

[35] Liebmann B,Smith C A. Description of a complete (interpolated) outgoing longwave radiation dataset [J]. Bull Am Meteorol Soc,1996,77:1275-1277.

[36] Lau K-M,Chan P H. Aspects of the 40-50 day oscillation during the northern winter as inferred from outgoing longwave radiation[J]. Mon Wea Rev,1985,113:1889-1909.

[37] Wheeler M,Hendon H H. An all-season real-time multivariate MJO index: Development of an index for monitoring and prediction[J]. Mon Wea Rev,2004,132:1917-1932.

[38] Jones C,Waliser D E,Lau K M,et al. Global occurrences of extreme precipitation and the Madden-Julian oscillation: Observations and predictability[J]. J Climate,2004,17:4575-4589.

[39] Richman M B,Lamb P J. Climate pattern analysis of three and seven day summer rainfall in the central United States: Some meteorological considerations and a regionalization[J]. J Appl Meteorol,1985,24:

1325-1343.

[40] Lau K-M, Phillips T J. Coherent fluctuations of extratropical geopotential height and tropical convection in intraseasonal time scales[J]. J Atmos Sci,1986,43:1164-1181.

[41] Gill A E. Some simple solutions for heat-induced tropical circulation[J]. Q J R Meteorol Soc,1980,106: 447-462.

[42] Hall N M J, Derome J. Transients, nonlinearity, and eddy feedback in the remote response to El Niño[J]. J Atmos Sci,2000,57:3992-4007.

[43] Derome J, Lin H, Brunet G. Seasonal forecasting with a simple General Circulation Model[J]. J Climate, 2005,18:597-609.

[44] Lin H, Brunet G, Derome J. Intraseasonal variability in a dry atmospheric model[J]. J Atmos Sci,2007, 64:2422-2441.

[45] Lin H. Global extratropical response to diabatic heating variability of the Asian summer monsoon[J]. J Atmos Sci,2009,66:2693-2713.

[46] Ding R, Li J, Seo K. Predictability of the Madden-Julian oscillation estimated using observation data[J]. Mon Wea Rev,2010,138:1004-1013,doi:10.1175/2009MWR3082.1.

[47] Reichler T, Roads J O. Long-range predictability in the tropics. Part II: 30-60 days variability[J]. J Climate,2005,18: 634-650.

[48] Waliser D E, Jones C, Schemm J K, et al. A statistical extended range tropical forecast model based on the slow evolution of the Madden-Julian oscillation[J]. J Climate,1999,12:1918-1939.

[49] Jiang X, Waliser D E, Wheeler M C, et al. Assessing the skill of an all-season statistical forecast model for the Madden-Julian oscillation[J]. Mon Wea Rev,2008,136:1940-1956.

[50] Chang C-P, Lau K M W. Northeasterly cold surges and near-equatorial disturbances over the winter MONEX area during December 1974. Part 2: Planetary-scale aspects[J]. Mon Wea Rev,1980,108: 298-312.

[51] Ting M, Sardeshmukh P D. Factors determining the extratropical response to equatorial diabatic heating anomalies[J]. J Atmos Sci,1993,50: 907-918.

MJO 对华南前汛期降水的影响及其可能机制[*]

李文铠[1]　何金海[1]　祁莉[1]　陈伯民[2]

(1. 南京信息工程大学气象灾害教育部重点实验室，江苏 南京 210044；
2. 上海市气候中心，上海 200030)

摘要：利用站点降水资料、实时多变量 MJO（Madden-Julian oscillation）指数、向外长波辐射（OLR）资料和 NCEP/NCAR 再分析资料，采用合成分析方法研究了 MJO 对华南前汛期降水的影响，并讨论其可能机制。结果表明，（1）在 MJO 不同位相，华南前汛期降水异常有明显的差异，并且这种差异随滞后时长而发生变化。第 2～3 位相和第 6～7 位相分别是影响华南前汛期降水的典型"湿位相"和"干位相"。（2）华南前汛期降水对 MJO 的响应需要一定时间，滞后时长约为 1～2 候。（3）MJO 活跃（受抑）对流可通过激发 Rossby 波影响华南前汛期降水。当 MJO 活跃（受抑）对流中心位于赤道印度洋附近时，非绝热加热作用激发的 Rossby 波到达并影响华南地区，华南地区出现水汽供应的增强（减弱），从而促进（抑制）华南前汛期降水。

关键词：气候学；降水异常；合成分析；MJO；华南前汛期

1　引　言

中国降水具有显著的季节内变化特征[1-3]，即在季节内尺度存在一段多雨期和少雨期的现象。这可导致短时间内的旱涝极端天气[4]，对社会和经济活动产生重要影响。然而，当今常规天气预报主要着眼于 10 d 内的预报，短期气候预测主要解决月时间尺度以上的预测，介于两者时间尺度之间的 10～30 d 预报（即延伸期预报）就成为"预报缝隙"，使预报降水的季节内变化成为难题，是目前尚未解决而又急需解决的问题之一。

寻找到有效的预报因子是做好延伸期预报的前提。研究表明，MJO（Madden-Julian oscillation）可以为延伸期预报提供预报信号[5-6]，MJO 的活动规律是制作延伸期预报的有效方法之一。MJO 最早由 Madden 等[7]发现，是热带地区季节内变化最主要的成分，它沿赤道自印度洋向太平洋东移，平均速度为 5 m·s^{-1}，振荡周期为 30～90 d，伴随对流、降水的活跃或受抑[8]。MJO 的重要性不仅在其本身，还在于它可以影响包括我国在内的接近全球范围的天气[9-10]，例如，MJO 影响我国春季降水[11-12]、夏季降水[13-17]、冬季降水和近地层气温（SAT）[18-22]的季节内变化。因此，研究 MJO 对降水的影响有利于提高延伸期预报水平，具有理论和现实意义。

华南地区是我国雨量最充沛的一个区域，易受到热带大气变化的影响，4—6 月为华南前

---

[*] 本文发表于《热带气象学报》，2014 年第 30 卷第 5 期，983-989.

汛期,降水丰富、洪涝灾害多发。研究MJO对华南前汛期降水的影响,不仅有利于理解华南前汛期降水的季节内变化,还有助于制作好华南前汛期降水的延伸期预报。章丽娜等[14]研究了MJO对华南前汛期降水的影响,指出MJO的活跃中心从印度洋进入西太平洋,华南地区的降水由偏多转为偏少,最显著的降水正负异常分别位于第4位相和第7位相,且相应的大尺度背景场也发生了季节内变化。但仍有一些问题值得继续讨论,例如,MJO活动中心以外地区的天气对MJO的响应需要一定时间,那么华南前汛期降水对MJO的响应过程是怎样的?MJO影响华南前汛期降水的可能机制又是什么?本文将围绕上述问题,进一步研究MJO对华南前汛期降水的影响,为华南前汛期降水的季节内变化寻找预报信号和科学依据。

2 资料和方法

本文采用的资料包括:(1)中国753个地面气象观测站的逐日降水资料;(2)实时多变量MJO指数(Real-Time Multivariate MJO Index,RMM指数)[23];(3)美国NOAA提供的逐日向外长波辐射(Outgoing Longwave Radiation,OLR)资料[24],水平分辨率为2.5°×2.5°;(4)美国NCEP/NCAR提供的逐日全球再分析资料[25],水平分辨率为2.5°×2.5°,垂直方向为17层。以上资料的时间范围为1975—2011年,但由于OLR资料在1978年有缺测,为统一起见均不包含1978年,共计36年。

为了滤除高频变化成分,本文使用逐候资料,处理方法是对上述资料依次求得4月1日—6月29日期间各候平均值(即4月1—5日、……、5月31日—6月4日、……、6月25—29日分别求平均)。再对逐候降水资料、OLR资料和再分析资料进行如下处理:(1)减去对应候的36年气候均值,以移除季节循环;(2)减去对应年期间华南前汛期均值,以移除年际、年代际变化。以下分析均采用经过处理后的资料。

本文用RMM指数描述MJO活动,有关该指数的定义和意义可参见文献[15]或文献[23]。RMM指数以$\sqrt{RMM1^2+RMM2^2}>1$表示强MJO,否则为弱MJO。弱MJO更接近MJO生成时的随机变化,章丽娜等[14]研究表明,弱MJO对华南前汛期降水没有显著的影响,因而本文的研究只考虑强MJO事件,即$\sqrt{RMM1^2+RMM2^2}>1$的情况。

本文采用合成分析方法,包括同期和滞后合成。具体做法:根据RMM指数(时段范围为4月1日—6月19日)确定的MJO的8个位相,将各候归类为第8~1位相、第2~3位相、第4~5位相、第6~7位相,共4类,36年间各类别对应的总候数见表1,同类别候及其滞后1、2候的降水或其他气象要素分别求平均。之所以分析MJO与降水或环流的滞后关系,是因为MJO活动中心以外地区的大气环流及天气变化对MJO的响应需要一定时间,例如Lin等[26]指出当MJO对流活动进入印度洋或西太平洋后5~15 d,北大西洋涛动的强度发生显著改变,He等[21]发现东亚地区冬季降水、SAT对MJO的响应具有1周左右的滞后。同期合成得到的结果实际上是依赖于MJO周期性的间接联系,但考虑到MJO并非严格的周期振荡,每个位相持续时间并不固定,甚至可以从任意位相开始或结束,因此有必要研究MJO与天气变化的超前、滞后关系,建立MJO与天气变化的直接联系,更充分理解MJO影响天气变化的机制。此外基于对MJO的实时监测,这种超前、滞后关系还可以直接作为预报信号。

表 1 逐候 RMM 资料中 MJO 各位相的候数

位相	8～1	2～3	4～5	6～7
候数(个)	98	83	80	82

本文对合成结果进行显著性检验以确定其统计学意义。降水距平资料不满足正态分布,对其检验时采用非参数估计的蒙特卡洛方法,每次检验均进行 10000 次随机试验。其他气象要素近似满足正态分布则使用 t 检验。

3 华南前汛期降水与 MJO 的关系

在分析华南前汛期降水与 MJO 的关系之前,先选取华南前汛期降水代表站。参考 Ting 等[27]划分降水区的方法,计算得到华南地区 36 年间 4 月 1 日—6 月 29 日逐候降水量标准差最大的站点为阳江站,再求出阳江站与所有站点在华南前汛期期间逐候降水量的相关系数,选取相关系数超过 0.05 的显著性水平检验、空间分布连续、具有相对一致的降水变率的站点作为华南前汛期降水的代表站,共 57 站(图 1),与文献[28-29]研究华南前汛期降水时选取的代表站的空间分布相似。

对 57 个代表站的逐候降水距平按上述介绍的方法分别进行合成,再对合成结果做如下分析:(1)计算得到区域平均降水距平(指平均每站的降水距平,图 2);(2)统计合成结果中超过 0.1 显著性水平检验的站点个数(表 2)。

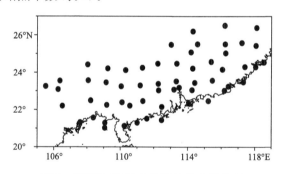

图 1 华南地区 57 个代表站的空间分布

图 2 所示的区域平均降水距平反映了不同 MJO 位相期间或滞后 1、2 候时华南地区降水的总体异常情况。在滞后时长为 0 候(Lag=0)时,MJO 的第 2～3 位相、第 4～5 位相对应降水正异常,而第 6～7 位相、第 8～1 位相对应降水负异常,这与文献[14,16-17]的研究结论相符。在滞后 1 候(Lag=1)时,MJO 的第 2～3 位相、第 4～5 位相仍然对应降水正异常,且第 2～3 位相对应的降水正异常值比 Lag=0 时增大,而第 6～7 位相、第 8～1 位相仍对应降水负异常,且第 6～7 位相对应的降水负异常值比 Lag=0 时减小。在滞后 2 候(Lag=2)时,MJO 的第 8～1 位相、第 2～3 位相对应降水正异常,而第 4～5 位相、第 6～7 位相对应降水负异常,且第 6～7 位相对应的降水负异常值比 Lag=1 时略减小。可见,在 MJO 不同位相,华南前汛期降水异常有明显差异,并且这种差异随滞后时长而发生变化。

第 2～3 位相和第 6～7 位相分别有降水正、负异常随时间滞后的增强过程,且对应降水正、负异常的峰值、谷值,而第 8～1 位相、第 4～5 位相分别是降水负异常向正异常、正异常向负异常的过渡期,表明每次完整的 MJO 周期分别有一次促进和一次抑制华南前汛期降水的过程,两次过程主要发生在第 2～3 位相和第 6～7 位相。因此第 2～3 位相和第 6～7 位相可被认为是 MJO 影响华南前汛期降水的典型"湿位相"和"干位相",作为 MJO 影响华南前汛期降水的关键位相。

区域平均降水距平合成的峰值和谷值分别对应 MJO 第 2～3 位相滞后 1 候(异常值为 1.5 mm·d^{-1})和 MJO 第 6～7 位相滞后 2 候(异常值为 -1.1 mm·d^{-1}),而非关于 MJO 位相的同期合成,这反映出华南前汛期降水对"湿位相"和"干位相"的响应需要一定时间,滞后时长为 1～2 候。

表 2 是合成结果中超过 0.1 显著性水平检验的站点个数,站点个数越多表示受到 MJO 影响且显著的站点越多。该结果揭示的 MJO 影响华南前汛期降水的关键位相、影响结果与上述分析基本一致。

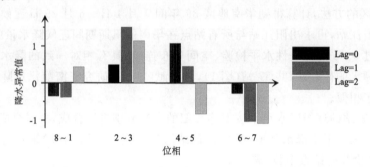

图 2　区区域平均降水距平在不同 MJO 位相的合成

(横坐标表示 MJO 位相,纵坐标表示降水异常值(单位:mm·d^{-1}),Lag 为滞后候数)

表 2　对 57 个代表站的逐候降水距平按不同 MJO 位相合成结果中超过 0.1 显著性水平检验的站点个数

位相	8～1	2～3	4～5	6～7
滞后 0 候	2(13)	10(1)	23(0)	1(3)
滞后 1 候	1(11)	35(0)	7(2)	0(29)
滞后 2 候	11(1)	26(0)	0(18)	0(27)

注:括号外和括号内的数字分别表示合成值为正、负且超过 0.1 显著性水平检验的站点个数。

4　可能机制

为了进一步理解上述预报信号的物理本质,下面将分析与 MJO 第 2～3 位相和第 6～7 位相(典型"湿位相"和"干位相")活动相联系的大尺度环流变化,以讨论 MJO 影响华南前汛期降水的可能机制。

MJO 对流中心释放凝结潜热对大气是一个外强迫热源,可引起全球大气变化[30-31],本文将从这个角度出发,解释 MJO 影响华南前汛期降水的可能机制。图 3 是 OLR 异常场在 MJO 第 2～3 位相、第 6～7 位相的合成。在 MJO 第 2～3 位相,MJO 对流中心位于印度洋,而西太平洋、东太平洋有受抑对流,但分布分散,且强度较弱。在 MJO 第 6～7 位相,印度洋地区为受抑对流,而西太平洋、东太平洋有异常对流,但分散且强度弱。总之,不同 MJO 位相反映 MJO

活跃或受抑对流活动的差异,从第 2～3 位相到第 6～7 位相的活跃或受抑对流的区域发生了显著改变。那么,与 MJO 对流活动区域发生变化的同时,大气环流场发生了怎样的变化? 又如何影响华南前汛期降水? 下文将就此进行分析讨论。

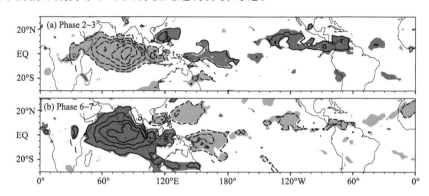

图 3　OLR 异常场在 MJO 第 2～3 位相(a)和第 6～7 位相(b)的合成(滞后 0 候)
(等值线间隔为 5 W·m^{-2},未画出零线,虚线表示负值,
深色、浅色阴影分别表示合成值为正、负且超过 0.05 显著性水平检验的区域)

4.1　与第 2～3 位相 MJO 相联系的大尺度环流

MJO 第 2～3 位相时,对流中心位于印度洋(图 3a),对流中心释放凝结潜热,相对大气来说是一个异常非绝热热源。图 4 是 700 hPa 和 200 hPa 位势高度异常场及异常风场在 MJO 第 2～3 位相的合成(包括滞后 0 候和滞后 1 候)。在 700hPa 上(图 4a),热源西北侧和西南侧有两个位势高度负异常中心,伴随气旋式异常风场,且在滞后 1 候时(图 4b)更明显,这些异常是非绝热加热引起的赤道 Rossby 波响应,符合 Gill[32] 提出的响应模型。热源东侧的海洋大陆地区,位势高度异常场表现为低压舌的东伸,异常风场表现为东风异常,是赤道 Kelvin 波响应。西北太平洋地区(150°E,30°N 附近)、澳大利亚东侧地区(170°E,35°N 附近)有位势高度正异常及反气旋式异常风场,这些异常是非绝热加热激发的 Rossby 波响应,符合 Matthews 等[30]和白旭旭等[12]通过数值试验证明的响应模型。该 Rossby 波向东北方向传播,表现为波列形式,这在 200 hPa(图 4c)上表现得更为明显,西太平洋地区为正 P-J 型(太平洋-日本型) Rossby 波列,通过北太平洋向北美传播。Lin 等[26,33-34]研究表明,MJO 可通过此波列影响北美大气环流及天气。华南地区显然在此波列的影响范围内。滞后 1 候时,整个异常系统向东移动(图 4b、4d)。

上述分析表明,Rossby 波到达华南地区后,西北太平洋地区低空出现了位势高度正异常和反气旋式异常风场,这些异常将增强西北太平洋副热带高压(简称"西太副高")的强度,从而使华南地区来自西太副高南侧转向的西南风输送水汽增多,最终促进华南地区降水。

4.2　与第 6～7 位相 MJO 相联系的大尺度环流

第 6～7 位相时,印度洋地区为受抑的对流中心(图 3b),受抑对流相对大气是一个异常非绝热冷源,若仅考虑影响过程中线性部分,4.1 节中讨论的环流异常在此时应是正负号相反的情形。

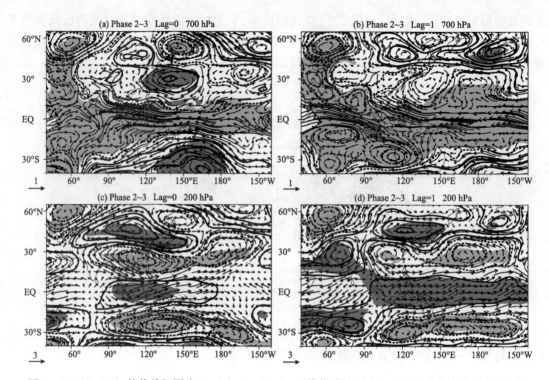

图 4 700 hPa(a、b,等值线间隔为 5 m)和 200 hPa(c、d,等值线间隔为 10 m)的位势高度异常场(等值线)和异常风场(矢量箭头,单位:m·s^{-1})在 MJO 第 2~3 相的滞后 0 候(a、c)和滞后 1 候(b、d)的合成(等值线未画出零线,虚线表示负值,深色、浅色阴影分别表示合成值为正、负且超过 0.1 的显著性水平检验的区域)

图 5 是 700 hPa 和 200 hPa 位势高度异常场及异常风场在 MJO 第 6~7 位相的合成(包括滞后 0 候和滞后 1 候)。在 700 hPa 上(图 5a),冷源东侧的海洋大陆地区表现为位势高度异常场上的高压舌及西风异常,为赤道 Kelvin 波响应。北太平洋地区(145°E,35°N 附近)、澳大利亚东侧地区(160°E,35°N 附近)分别有位势高度负异常及气旋式异常风场,为 Rossby 波响应。该 Rossby 波向东北方向传播,表现为波列形式,在 200 hPa(图 5c)上表现得更明显,西太平洋地区为负 P-J 型 Rossby 波列。但波列较第 2~3 位相时东移了,即便如此,200 hPa 上华南地区还是被位势高度正异常和反气旋式异常风场控制,表明 Rossby 波再次传播到华南地区。滞后 1 候时,整个异常系统向东移动(图 5b、5d)。

上述分析表明,Rossby 波到达华南地区后,西北太平洋地区低空出现了位势高度负异常和气旋式异常风场,这些异常将减弱西太副高的强度,从而使华南地区来自西太副高南侧转向的西南风输送水汽减少,最终抑制华南地区降水。

5 结论与讨论

本文利用站点降水资料、RMM 指数、OLR 资料和 NCEP/NCAR 再分析资料,采用合成分析方法研究了 MJO 对华南前汛期降水的影响,并讨论其可能机制。

(1)在 MJO 不同位相,华南前汛期降水异常有明显差异,并且这种差异随滞后时长而发

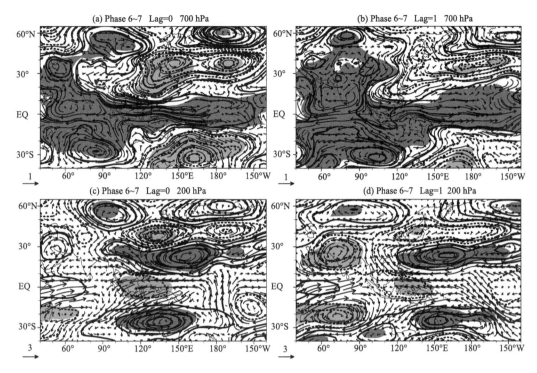

图 5　同图 4,但为 MJO 第 6～7 位相

生变化。每次完整的 MJO 周期分别有一次促进和一次抑制华南前汛期降水的过程,主要发生在第 2～3 位相和第 6～7 位相,它们可被认为是 MJO 影响华南前汛期降水的典型"湿位相"和"干位相"。

(2)华南前汛期区域平均降水距平合成的峰值和谷值分别对应"湿位相"滞后 1 候和"干位相"滞后 2 候,反映出华南前汛期降水对"湿位相"和"干位相"的响应需要一定时间,滞后时长为 1～2 候。

(3)MJO 活跃(受抑)对流区相对大气来说是一个异常热源(冷源),可通过激发的 Rossby 波影响华南前汛期降水。当 MJO 活跃(受抑)对流中心位于赤道印度洋附近时,非绝热加热作用激发的 Rossby 波向东北方向传播。Rossby 波到达华南地区后,西北太平洋地区低空出现了位势高度正(负)异常和反气旋(气旋)式异常风场,这些异常将增强(减弱)西太副高的强度,从而使华南地区来自西太副高南侧转向的西南风输送水汽增多(减少),最终促进(抑制)华南前汛期降水。

限于合成分析的局限性,本文讨论的仅是 MJO 与华南前汛期降水的总体关系,实际中还应考虑不同降水过程的具体情况,如中高纬度西风槽的配合[17]、华南前汛期不同降水时段的特征[35]等。

参考文献

[1] Lau K M,Yang G J,Shen S H. Seasonal and intraseasonal climatology of summer monsoon rainfall over Eeat Asia[J]. Mon Wea Rev,1988,116(1):18-37.

[2] 王遵娅,丁一汇.中国雨季的气候学特征[J].大气科学,2008,32(1):1-13.

[3] 梁萍,丁一汇.东亚梅雨季节内振荡的气候特征[J].气象学报,2012,70(3):418-435.

[4] 吴志伟,何金海,李建平,等.长江中下游夏季旱涝并存及其异常年海气特征分析[J].大气科学,2006,30(4):570-577.

[5] Waliser D E, Lau K M, Stern W, et al. Potential predictability of the Madden-Julian oscillation[J]. Bull Amer Meteor Soc, 2003, 84(1):33-50.

[6] 丁一汇,梁萍.基于MJO的延伸预报[J].气象,2012,36(7):111-122.

[7] Madden R A, Julian P R. Detection of a 40-50 day oscillation in the zonal wind in the tropical Pacific[J]. J Atmos Sci, 1971, 28(5):702-708.

[8] Zhang C. Madden-Julian oscillation[J]. Rev Geophys, 2005, 43(2):2003.

[9] Jones C, Waliser D E, Lau K M, et al. Global occurrences of extreme precipitation and the Madden-Julian oscillation: Observations and predictability[J]. J Climate, 2004, 17(23):4575-4589.

[10] Donald A, Meinke H, Power B, et al. Near-global impact of the Madden-Julian oscillation on rainfall[J]. Geophys Res Lett, 2006, 33(9):704.

[11] 白旭旭,李崇银,谭言科,等.MJO对我国东部春季降水影响的分析[J].热带气象学报,2011,27(6):814-822.

[12] 白旭旭,李崇银,李琳.MJO对中国春季降水影响的数值模拟研究[J].气象学报,2012,70(5):986-1003.

[13] Zhang L, Wang B, Zeng Q. Impact of the Madden-Julian oscillation on summer rainfall in Southeast China[J]. J Climate, 2009, 22(2):201-216.

[14] 章丽娜,林鹏飞,熊喆,等.热带大气季节内振荡对华南前汛期降水的影响[J].大气科学,2011,35(3):560-570.

[15] 李汀,严欣,琚建华.MJO活动对云南5月降水的影响[J].大气科学,2012,36(6):1101-1111.

[16] 林爱兰,李春晖,谷德军,等.热带季节内振荡对广东6月降水的影响[J].热带气象学报,2013,29(3):353-363.

[17] 林爱兰,李春晖,郑彬,等.6月MJO对广东降水调制与直接影响系统的联系[J].应用气象学报,2013,24(4):397-406.

[18] Jeong J H, Ho C H, Kim B M, et al. Influence of the Madden-Julian oscillation on wintertime surface air temperature and cold surges in east Asia[J]. J Geophys Res, 2005, 110(D11):104.

[19] Jeong J H, Kim B M, Ho C H, et al. Systematic variation in wintertime precipitation in East Asia by MJO-induced extratropical vertical motion[J]. J Climate, 2008, 21(4):788-801.

[20] Jia X, Chen L, Ren F, et al. Impacts of the MJO on winter rainfall and circulation in China[J]. Adv Atmos Sci, 2011, 28(3):521-533.

[21] He J, Lin H, Wu Z. Another look at influences of the Madden-Julian oscillation on the wintertime East Asian weather[J]. J Geophys Res, 2011, 116(D03):109.

[22] 贾小龙,梁潇云.热带MJO对2009年11月我国东部大范围雨雪天气的可能影响[J].热带气象学报,2011,(5):639-648.

[23] Wheeler M C, Hendon H H. An all-season real-time multivariate MJO index: Development of an index for monitoring and prediction[J]. Mon Wea Rev, 2004, 132(8):1917-1932.

[24] Liebmann B. Description of a complete (interpolated) outgoing longwave radiation dataset[J]. Bull Amer Meteor Soc, 1996, 77:1275-1277.

[25] Kalnay E, Kanamitsu M, Kistler R, et al. The NCEP/NCAR 40-year reanalysis project[J]. Bull Amer Meteor Soc, 1996, 77(3):437-471.

[26] Lin H, Brunet G, Derome J. An observed connection between the North Atlantic Oscillation and the Mad-

den-Julian oscillation[J]. J Climate,2009,22(2):364-380.
[27] Ting M,Wang H. Summertime US precipitation variability and its relation to Pacific sea surface temperature[J]. J Climate,1997,10(8):1853-1873.
[28] 吴志伟,江志红,何金海. 近50年华南前汛期降水、江淮梅雨和华北雨季旱涝特征对比分析[J]. 大气科学,2006,30(3):391-401.
[29] 强学民,杨修群. 华南前汛期开始和结束日期的划分[J]. 地球物理学报,2008,51(5):1333-1345.
[30] Matthews A J,Hoskins B J,Masutani M. The global response to tropical heating in the Madden-Julian oscillation during the northern winter[J]. Q J R Meteorol Soc,2004,130(601):1991-2011.
[31] Zhang X,Lin H,Jiang J. Global response to tropical diabatic heating variability in boreal winter[J]. Adv Atmos Sci,2012,29(2):369-380.
[32] Gill A E. Some simple solutions for heat-induced tropical circulation[J]. Q J R Meteorol Soc,1980,106:447-462.
[33] Lin H,Brunet G. The influence of the Madden-Julian oscillation on Canadian wintertime surface air temperature[J]. Mon Wea Rev,2009,137(7):2250-2262.
[34] Lin H,Brunet G,Mo R. Impact of the Madden-Julian oscillation on wintertime precipitation in Canada[J]. Mon Wea Rev,2010,138(10):3822-3839.
[35] 池艳珍,何金海,吴志伟. 华南前汛期不同降水时段的特征分析[J]. 南京气象学院学报,2005,28(2):163-171.

Meridional Propagation of the 30- to 60-day Variability of Precipitation in the East Asian Subtropical Summer Monsoon Region: Monitoring and Prediction[*]

He Jinhai[1] Chang Luyu[1,2] Chen Hua[1]

(1. Key Laboratory of Meteorological Disaster of Ministry of Education,
Nanjing University of Information Science and Technology, Nanjing 210044, China;
2. Shanghai Meteorological Service, Shanghai 200030, China)

Abstract: The meridional propagation of the 30- to 60-day intraseasonal variability (ISV) of precipitation in the East Asian subtropical summer monsoon (EASSM) region and its monitoring and prediction are investigated in the current study. Based on a multivariate empirical orthogonal function (MV-EOF) analysis of precipitation and relative vorticity at 700 hPa in East Asia, a bivariate index referred to as the EASSM-ISV index is designed using the two leading MV-EOF modes, with the objective of real-time monitoring of the 30- to 60-day variability of precipitation in the EASSM region. It is found that this index, with its eight phases, can explain the meridional propagation of the 30- to 60-day ISV in precipitation and circulation in the EASSM region. Based on a singular value decomposition technique, a statistical forecast model is developed in which the EASSM-ISV indices from the preceding five pentads are used to predict the indices in five pentads in the future. Meanwhile, the indices are used to predict the meridional propagation of the 30- to 60-day precipitation anomaly in the EASSM region. This model thus provides a useful tool for intraseasonal prediction of precipitation during the rainy season in China.

Key words: East Asian subtropical summer monsoon region; 30- to 60-day intraseasonal oscillation; meridional propagation; real-time monitoring; prediction

1 Introduction

The weather and climate in China are influenced by the East Asian monsoon, which is composed of the South China Sea (SCS)-West Pacific tropical monsoon and the East China-Japan subtropical monsoon[1], with the latter being the major factor affecting droughts and floods in China. The anomalous weather and climate in China, especially in East China, are

---

[*] 本文发表于《Atmosphere—Ocean》,2015 年第 53 卷第 2 期,251-263.

primarily influenced jointly by the tropical monsoon and mid-latitude weather and climate systems. However, the intraseasonal variability (ISV) of the subtropical monsoon system and its influence on weather and climate are still not well understood. It is also unclear how to apply the ISV to extended-range and monthly weather predictions. Thus, it is of great importance to develop techniques for monitoring and predicting the ISV in the subtropical monsoon region; this would help improve the forecast of droughts and floods in China.

The ISV is one of the most important characteristics of the Asian monsoon. Zhu and Yang[2] found significant low-frequency oscillations during the northward progression of the subtropical monsoon in China with different frequencies including quasi-40 day, two week, and two day, with the quasi-40-day period being the most prominent. Murakami, Chen, and Xie[3] used outgoing longwave radiation (OLR) data to study the ISV and its relationship with the monsoon onset and retreat. The zonal and meridional propagation of the low-frequency oscillation in the Asian monsoon region has also been examined in many studies[4-7]. Chen, Zhu, Wang, and Zhang[8] pointed out that there was an obvious 30- to 60-day oscillation from central and eastern China through the SCS and the West Pacific Ocean within $100°\sim150°E$ for the East Asian monsoon circulation system. He and Yu[9] found that water vapour was transported northward at all latitudes in East China and that water vapour and precipitation were associated with low-frequency variability at multiple time scales. Therefore, it is critical to focus on the ISV, especially the quasi-40-day oscillation, in real-time monitoring of the intraseasonal-to-seasonal progression of precipitation in the East Asian subtropical summer monsoon (EASSM) region.

Techniques for real-time monitoring of the ISV at different time scales and for different regions has been elucidated in several recent studies. Based on a Multivariate-Empirical Orthogonal Function (MV-EOF), Wheeler and Hendon[10] designed a bivariate Madden-Julian Oscillation (MJO) index to monitor the intensity and zonal propagation of the MJO. The Wheeler-Hendon index has been implemented operationally at the Chinese National Climate Center for real-time monitoring and prediction of the MJO. Lin[11] designed an index to monitor and predict the meridional propagation of the summer ISV in the East Asian and Western North Pacific (EAWNP) summer monsoon region. Lee et al.[12] used OLR and the zonal wind at 850 hPa to define two real-time indices that described most of the ISV in the Asian summer monsoon region and the characteristics of northward and northwestward propagation of the ISV.

These ISV indices were designed mainly for tropical low-frequency variability (i.e., the MJO and boreal summer ISV in the south Asian monsoon region). A method for monitoring and predicting the ISV in the subtropical monsoon region has still to be determined. Therefore, several related scientific questions are addressed in this paper. These include the following:

• What would be the design of a real-time monitoring index for the meridional propagation of 30- to 60-day precipitation in the EASSM region?

- What is the physical meaning of the proposed index?
- What is the relationship between the index and the intraseasonal movement of rainfall in EASSM?
- Is the index predictable?
- How should the forecast skill of the index be assessed?

2 Data and methodology

Two datasets were used in this study because two different resolutions were needed. One was the National Centers for Environmental Prediction (NCEP)/National Center for Atmospheric Research (NCAR) reanalysis[13], which includes daily wind, specific humidity, and vertical velocity on a 2.5°×2.5° grid for the 1979-2008 period (used only in Section 4 for the composite analysis of precipitation and circulation). The other dataset used was the ERA-Interim, a global atmospheric reanalysis produced by the European Centre for Medium-range Weather Forecasts[14], which includes the daily relative vorticity and conventional surface fields on a 1°×1° grid for the 1979-2012 period (used in the entire paper except for the analysis in Section 4).

In comparison with regular EOF analysis, MV-EOF is more suited to capturing the spatial distributions of several selected variables and the co-variability relationship among these variables[7, 15]. In this study, MV-EOF is used to obtain the dominant modes of the intraseasonal propagation in the EASSM region. Other diagnostic methods used include spectrum analysis, composite analysis, correlation analysis, and the Spatial-Temporal Projection Method (STPM) for the extended-range prediction[16-17].

3 A bivariate index for EASSM-ISV

3.1 Monitoring Period, Region and Variables

In the current study, the raining seasons in the southern Yangtze River (SYR) region, South China, the Yangtze-Huai River Valley (YHRV; where a persistent rain belt is referred to as Meiyu), and the region of North and Northeast China, constitute the EASSM with an entire cycle of rain belt movement[18-26]. To analyze this warm-season cycle of rain belt movement, the monitoring period chosen for this study was late March to late September (i.e., from 22 March to 27 September). When a year was divided into 73 pentads (5-day periods), our monitoring period was from the 17th pentad to the 54th pentad.

It is known that the result of an EOF analysis is sensitive to the choice of region analyzed. According to the definition of the EAWNP summer monsoon of Ding, Li, and Liu[27] and Lin[28], Lin[11] developed an ISV index to monitor and predict the EAWNP ISV for 10°S~40°N, 90°~150°E. However, Lin[11] focused on the ISV in the tropics rather than the subtropical monsoon region. Hence, the key region of the current study is 15°~45°N, 110°~130°E which includes the subtropical region of EASSM.

In previous studies[11-12] ISV indices were developed using OLR and upper and lower troposphere zonal winds. OLR is a good proxy for convection in the tropics. In the extratropical region, however, OLR is no longer suitable for representing convection. He et al.[24] found that the relative vorticity at a lower level corresponded well with precipitation amount and that the evolution of the positive vorticity was consistent with the intraseasonal-to-seasonal progression of the rain belt. Meanwhile, the shear line at 700 hPa is empirically indicative of the Meiyu precipitation. Therefore, relative vorticity at 700 hPa and precipitation are chosen as the two variables to design the ISV index for the EASSM region.

3.2 MV-EOF analysis

In order to isolate the dominant structures of the meridionally propagating ISV in the EASSM region, an MV-EOF analysis is performed on the pentad anomalies of zonally averaged precipitation and 700 hPa relative vorticity. The zonal average is taken between the longitudes 110° and 130°E for the two variables from 15° to 45°N, so that the data are reduced to a single dimension in latitude. Figure 1 summarizes the procedure for obtaining the first two spatial modes. The Lanczos band-pass filter[29] is applied to the pentad data before performing the MV-EOF analysis of the zonally averaged 700 hPa relative vorticity and precipitation anomalies to extract the 30- to 60-day signal. The MV-EOF analysis is then performed on the two variables for the late March to late September period (i.e., the 17th pentad to the 54th pentad of each year).

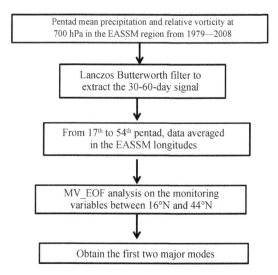

Fig. 1 Flow chart of obtaining the first two spatial major modes

The percentage of variance explained by each of the first eight modes is illustrated in Fig. 2. As can be seen, the first and second modes explain 24.9% and 20.5% of the total variance, respectively. They are well separated from the other high order modes according to the statistical test of North, Bell, and Calahan[30].

The spatial (meridional) patterns of the first two modes, denoted as EOF-1 and EOF-2, are shown in Fig. 3. For both modes, the meridional distribution of the anomalous 700 hPa relative vorticity is almost in phase with that of precipitation, indicating that enhanced (reduced) precipitation is co-located with the low level cyclonic (anticyclonic) wind shear. EOF-1 is characterized by a band of negative anomalies in the 700 hPa relative vorticity and precipitation amount in the middle from 25° to 35°N and positive values to the north and to the south. On the other hand, in EOF-2 both the 700 hPa relative vorticity and precipitation anomalies are positive near 25°N, and negative around 35°N. It appears that EOF-2 is in quadrature with EOF-1, with the former shifted northward with respect to the latter.

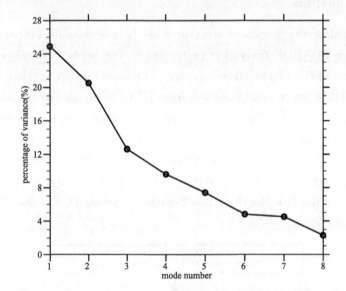

Fig. 2 Percentage of variance of the first eight modes from MV-EOF

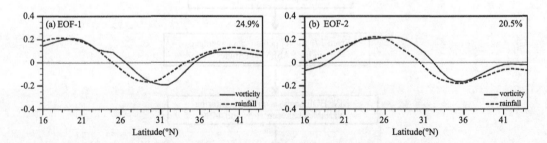

Fig. 3 Spatial patterns of the first (a) and second (b) modes from MV-EOF of the relative vorticity at 700 hPa (solid lines) and precipitation (dashed lines)

The EOF-1 and EOF-2 time series are denoted as PC1 and PC2, respectively. Figure 4 shows the autocorrelation of the PC1 time series (solid curve) and the cross-correlation between PC1 and PC2 (dashed curve). It is seen that the autocorrelation coefficient for PC1 reaches a minimum at both a lead and lag of four pentads, indicative of the existence of ISV with a quasi-40-day period. PC2 also has a period of 40 days (not shown), in agreement with Lin (2013). The cross-correlation between PC1 and PC2 reaches the maximum (minimum) when PC1 leads (lags) PC2 by two pentads, with the maximum (minimum) value exceeding the 0.01 confidence level based on the Student's t-test. It suggests that there is a quarter-cycle phase difference between PC1 and PC2. Therefore, both the spatial patterns and time series of the first two modes from MV-EOF confirm that they are in quadrature to each other. Thus, it is possible to use these two modes to characterize the meridional propagation of the 30- to 60-day oscillation of the ISV in the EASSM region.

3.3 A Bivariate Index for the EASSM ISV

In the above discussion the Lanczos band-pass filter[29] was applied to the data before performing the MV-EOF analysis of the zonally averaged 700 hPa relative vorticity and precipitation anomalies. In a real-time monitoring application, however, such time filtering is impossible because future observational data are not available. A non-filtering method is then proposed to obtain the anomalies of subseasonal variability as follows:

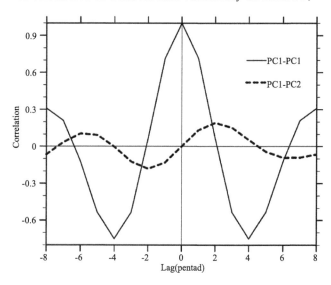

Fig. 4 The solid line shows the autocorrelation for PC1 and PC2, with negative values indicating that PC2 leads PC1 (and vice verse)

(1) Determine 5-day averages of the daily fields (700 hPa relative vorticity and precipitation) to obtain a dataset of 73 pentads per year.

(2) Remove the climatological annual cycle, which is the 30-year pentad climatology (1979-2008) smoothed with a 16-pentad (90 day) low-pass filter.

(3) Remove high-frequency signals shorter than 30 days by three pentad running mean.

(4) Remove signals longer than 60 days by subtracting the preceding 30-day average at each time step.

After these steps, it is likely that the anomalous fields mainly retain the 30- to 60-day variability. We redo the MV-EOF analysis with these pentad 700 hPa relative vorticity and precipitation anomalies. Figure 5 shows the spatial patterns and power spectra of PC1 and PC2. It is striking that the spatial patterns of the first two modes are almost the same as those obtained from the filtered data as shown in Fig. 3. From the power spectrum analysis, it is clear that there is a dominant peak at the 30- to 60-day period, exceeding the 0.05 confidence level according to the red noise significance test. Therefore, the non-filtering method is effective in extracting the 30- to 60-day intraseasonal signal.

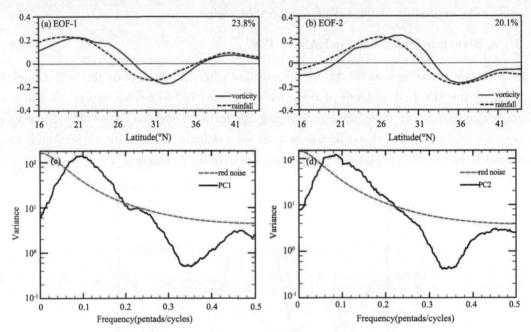

Fig. 5 Spatial patterns of the first (a) and second (b) modes from MV-EOF of the relative vorticity at 700 hPa (solid lines) and precipitation (dashed lines) using non-filtering method. Spectrum of the time series of the first (c) and second (d) modes. In (c) and (d), the red line indicates the 5% confidence level using the red noise significance test

Now we can define a bivariate ISV index to characterize the amplitude and meridional propagation of 30- to 60-day precipitation variability in the EASSM region. The 700 hPa relative vorticity and precipitation anomalies from the non-filtering method are projected onto the spatial patterns of the first two MV-EOF modes of the filtered analysis (as shown in Fig. 3) to obtain two time series (ISV_1 and ISV_2) that are used to construct the bivariate EASSM-ISV index. Similar to Lin[11], eight phases can be defined that correspond to different latitudinal positions of the precipitation anomaly. Figure 6 shows the bivariate phase distribution of ISV_1 and ISV_2 from late March to late September 1979. It is convenient to diagnose the state of SM-ISV as a point in the bivariate phase space defined by ISV_1 and ISV_2. The distance of a point from the origin in the phase space represents the amplitude of ISV. When the point is near the origin the ISV is weak. It should be noted that we use the spatial patterns of the filtered MV-EOF (Fig. 3) instead of the non-filtered patterns of Fig. 5 to define the EASSM-ISV index because these patterns are well-known from historical data, which does not prevent them being used for real-time applications. Nevertheless, as can be expected from the similarity between the spatial patterns of Fig. 3 and Fig. 5, similar results can be obtained if we use the non-filtered patterns.

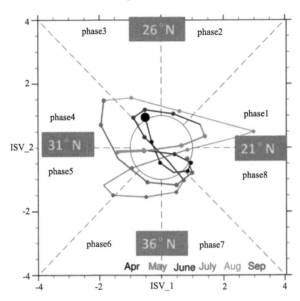

Fig. 6 Evolution of phase distribution determined by ISV_1 and ISV_2 for 1979. The x-axis is ISV_1, and y-axis is ISV_2. The space is divided into eight phases

To demonstrate that the ISV index defined above is representative of the ISV as well as its associated precipitation variability in the EASSM region, the precipitation anomaly represented by a linear regression of ISV_1 and ISV_2 is compared with the observed 30- to 60-day precipitation anomaly obtained through the non-filtered method, using 1979 as an example. Figure 7 shows the latitude-time cross-section of the low-frequency precipitation averaged over 110°~130°E in East Asia in 1979. The precipitation anomalies related to the ISV index are displayed as contours, whereas the observed precipitation anomalies are shaded. As can be seen, the agreement is evident. In general, the ISV signal propagates northward. This indicates that the ISV index defined above is useful in practice to represent the observed 30- to 60-day ISV of precipitation.

4 Phase composites of precipitation and circulation anomalies

In order to further investigate whether the EASSM-ISV index defined above can capture the main features of precipitation and atmospheric circulation variability, composites of precipitation and circulation anomalies for the eight phases with strong ISV (($ISV\_1^2 + ISV\_2^2)^{1/2} \geq 1$) are calculated for the period 1979 to 2008. The sample numbers for the eight pha-

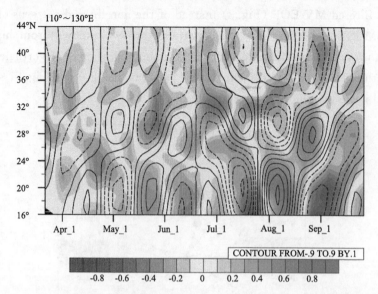

Fig. 7 Evolution of the 30-60-day precipitation averaged over 110°~130°E in East Asia in 1979. Regressions of precipitation onto the first two modes from MV-EOF are displayed by contours, and the real-time 30-60-day precipitation using non-filtering method is displayed as shading (Unit: mm · d^{-1})

ses are 65, 73, 70, 70, 80, 61, 62, and 70, respectively. Phase 5 has the most sample numbers, corresponding to above-normal precipitation near the SCS as shown in Fig. 6.

4.1 Precipitation Anomalies

Figure 8 shows the composite precipitation anomalies in the EASSM region for the eight phases of the EASSM-ISV index. A strong zonally oriented band of positive precipitation anomalies is located in the SCS and western Pacific region centred at approximately 18°N during phase 1 then moves northward to 21°N during phase 2. It then moves further north to 26°N during phase 3, to 29°N during phase 4, to 31°N during phase 5, to 35°N during phase 6, and to 38°N during phase 7. During phase 8, the positive anomaly reaches the northernmost latitude in Northeast China while, at the same time, another positive precipitation anomaly centre develops in the SCS and Philippines. Thus, the EASSM-ISV index with its different phases captures the main feature of the meridional propagation of the 30- to 60-day precipitation anomaly in the EASSM region.

4.2 Atmospheric Circulation Anomalies

Figure 9 shows the composite horizontal wind and relative vorticity anomalies at 850 hPa in the EASSM region during eight phases of the EASSM-ISV index. As can be seen, the circulation anomalies correspond well with the evolution of the precipitation anomalies shown in Fig. 8. During phase 1, there is anomalous cyclonic wind circulation (green) over the SCS where the maximum positive centre of relative vorticity also resides, indicating convergence in this area. The anomalous cyclone and positive vorticity centre move northward to South

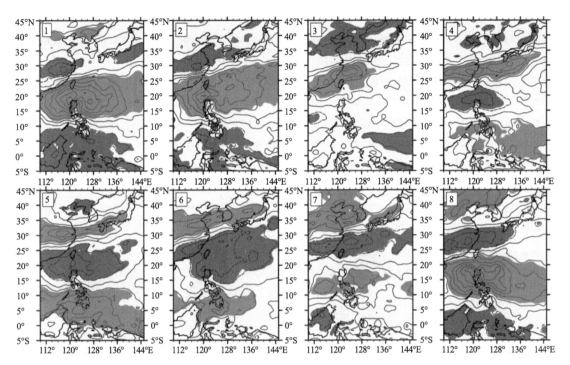

Fig. 8 Composite precipitation anomalies (mm) in EASSM region at eight phases of the ISV indices. The positive (negative) values are plotted with solid (dashed) lines (contour interval: 0.1 mm). Values significant at 5% confidence level are shaded (red and green colors represent negative and positive values respectively)

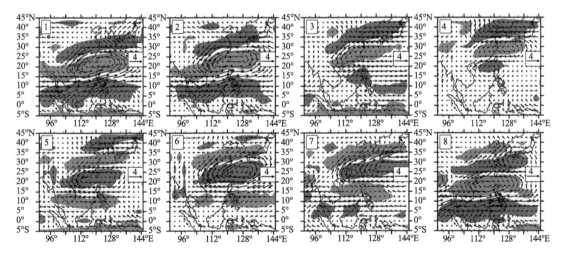

Fig. 9 Composite horizontal wind (arrows, m · s^{-1}) and relative vorticity (green and red shading indic-ating positive and negative values significant at 5% confidence level, respectively) (s^{-1}) at 700 hPa in EASSM region in eight phases of the ISV indices

China during phase 2, to South China-SYR during phase 3 with a weaker intensity, and to YHRV during phase 4. During phases 5, 6, and 7, the anomalous cyclone is too weak to form a closed circulation. But the significant positive centre of vorticity moves continuously northward to Northeast China during phase 8, when there is an evident cyclonic circulation and positive vorticity developing over the SCS and the Philippines.

The composite vertical velocity anomalies averaged over 110°~130°E in the EASSM region during eight phases of the EASSM-ISV index are given in Fig. 10. During phase 1, there is a significant negative centre in the SCS latitudes throughout the entire troposphere, suggestive of strong upward motion that favours the occurrence and strengthening of precipitation in this area. From phase 2 to phase 7, such upward motion maintains and moves northward from SCS to South China, SYR, YHRV, and North China. During phase 8, it reaches the northernmost latitude in northeast China. At the same time, another obvious upward motion develops over the SCS and Philippines, which is consistent with Fig. 9.

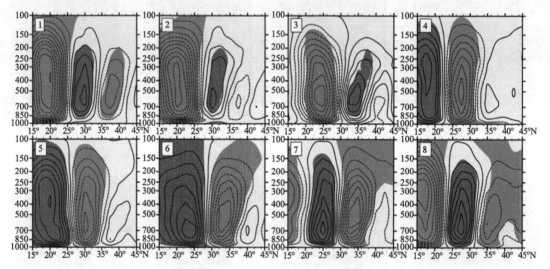

Fig. 10 Composites of vertical velocity anomalies (cm · s^{-1}) averaged over 110°~130°E in EASSM region in eight phases of the ISV indices. The positive (negative) values are plotted with solid (dashed) lines (contour interval 1 cm · s^{-1}). Values that are significant at the 5% confidence level are shaded. Negative (positive) values indicate upward (downward) motion

Figure 11 displays the composite water vapour transport and divergence anomalies in the EASSM region during the eight phases of the EASSM-ISV index. During phase 1, an anomalous anticyclone of water vapour transport exists near the equator and an anomalous cyclone exists over SCS. Convergence of water vapour occurs in the region between the anticyclone and cyclone, indicating that the northeastward water vapour transport is strong and that the convergence results in upward motion of warm and moist air over the SCS and Philippines. The centre of the water vapour convergence roughly corresponds to the positive centre of the precipitation anomaly, with a slight shift to its south. During phase 2, the anomalous system moves northward, consistent with the northward propagation of the rain belt. During phases 3, 4, and 5, the anomalous system keeps moving northward but weakens compared with earlier phases. During phase 6, the water vapour transport by the anomalous cyclone disappears, but the transport by the anomalous anticyclone increases significantly. The northeastward water vapour transport on the northern side of the anticyclone is strong, giving rise to the convergence of warm and moist air from the south in the YHRV region and increasing precipitation in this region. During phase 7, the water vapour transport by the anomalous anticyclone, and the convergence zone to its north, move further north and reach northern China, in agreement with the precipitation anomaly. The anomalous system reaches Northeast China during phase 8, when a new cyclone of water vapour transport intensifies above the SCS.

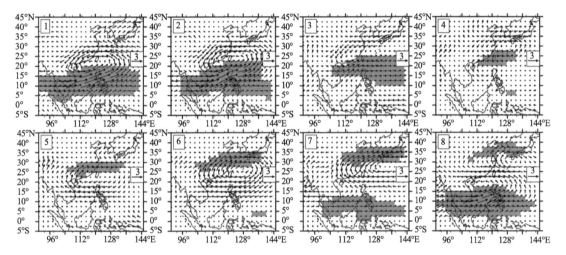

Fig. 11 Composite vertically integrated (surface to 700 hPa) water vapour transport (arrows: $kg \cdot s^{-1} \cdot m^{-1} \cdot hPa^{-1}$) and water vapour divergence. (The water vapour convergence value at 5% confidence lev-el is shaded ($kg \cdot s^{-1} \cdot m^{-2} \cdot hPa^{-1}$) in the EASSM region in eight phases of ISV indices

5 Prediction of the EASSM-ISV

The STPM has been widely used in seasonal and interannual, as well as in extended-range predictions[16-17]. In the current study, an STPM based on a Singular Value Decomposition (SVD) analysis is employed as an extended-range prediction model. The procedure is briefly described in this section (see Fig. 12). The details of the STPM can be found in Zhu et al.[17].

Several points need to be clarified. First, as introduced in Section 2, the EASSM-ISV index starts on the 17th pentad and ends on the 54th pentad each year. However, using the STPM, the prediction starting time starts on the 22nd pentad and ends on the 49th pentad, for a total of 28 pentad predictions each year. For instance, if the prediction starting time is the 22nd pentad, the predictors (left field) are the indices for the 17th, 18th, 19th, 20th, and 21st pentads and the predictands (right field) are the indices for the 23rd, 24th, 25th, 26th, and 27th pentads. This ensures that there is enough information for the 22nd pentad from the preceding five pentads to produce a meaningful forecast. Second, the training period is 1979 to 2000 (616 pentads in total), which is used to build the STPM extended-range prediction model. The prediction period is 2001 to 2012 (336 pentads in total). Third, because ISV_1 and ISV_2 in the EASSM-ISV indices have a significant autocorrelation and lead-lag correlation, in the following we discuss the forecast of the index using the index itself as a predictor.

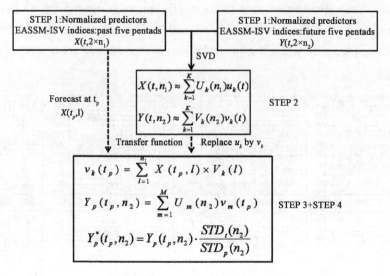

Fig. 12 The flowchart of the STPM, where X and Y are normalized predictor and predictand fields, respectively; t is the temporal grid during a training period, and n_1 and n_2 are the numbers of preceding and succeeding pentads with respect to time t; K is total number of SVD modes. U_k and V_k denote the singular vectors of predictor and predictand, respectively; m denotes the mth SVD mode that passes the 99% confidence level during the cross-validation; M is the total number of modes that pass the confidence level; u_k and v_k indicate the time series of the expansion coefficient of the kth SVD mode for the predictor and predictand. Solid arrows denote the training procedures during model development. Forecast procedures at time t_p are denoted by dashed arrows

Figure 12 shows the flowchart of the forecast model used. The EASSM-ISV indices are predicted by the STPM model using past information. The EASSM-ISV indices from the preceding five pentads are taken as predictors and the indices for the future five pentads as predictands. The forecast procedure is as follows: (i) normalize the predictors and predictands; (ii) extract the coupled patterns from the temporal-varying patterns of each predictor and predictand; (iii) perform the cross-validation to retain well-coupled modes containing persistent and useful predictability sources; and (iv) perform the hindcast procedure[11].

Following Lin, Brunet, and Derome[31] and Lin[11,28], two statistical measures are utilized to evaluate the quality of the prediction (i. e., the correlation skill (COR) and the root mean square error (RMSE)). The relationship between the predicted variable and the observation is assessed by COR, whereas the RMSE assesses the deviation of the predicted variable from the observation (Fig. 13). For the ISV_1 index, the COR values are 0.5, 0.49, 0.48, 0.38, and 0.19 for predictions with a lead time of one pentad, two pentads, three pentads, four pentads, and five pentads, respectively, with the first four passing the statistical significance level of 0.05. Note that the degrees of freedom (DOF) are re-calculated in the significance test[32] because the indices are auto-correlated. The RMSE of the predicted ISV_1 index is small, indicating that the predictions have a small error compared with the observation. Thus, based on the COR and RMSE scores, the prediction of the ISV_1 index is considered skillful when the lead time is shorter than four pentads. For the ISV_2 index, the COR values are 0.61, 0.55, 0.54, 0.47, and 0.39, for the five leads of one to five pentads, respectively, all of which are statistically significant at the 0.05 confidence level. Similar to the ISV_1 index, the RMSE for ISV_2 is small, suggestive of a small forecast error as well.

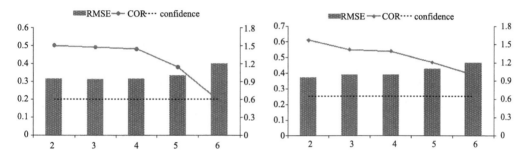

Fig. 13 Correlation (COR, red curve) and root mean square error (RMSE, blue bar) between the predicted indices (ISV_1 on the left and ISV_2 on the right) and observations. The black dashed line shows critical values at 5% confidence level of COR

Figures 14 and 15 display the time evolution of the observed and forecast ISV_1 and ISV_2 indices. For brevity, we only show the results from six years randomly chosen from the twelve years. It can be seen that the STPM model is able to predict, to a good degree, the variability, the maximum and minimum, and the amplitude of the monitoring indices.

In the following, the ability of the predicted EASSM-ISV index to capture the meridional propagation of the 30- to 60-day ISV signal in the EASSM region is discussed. The time-latitude distribution of the 30- to 60-day precipitation anomaly from the non-filtering method is compared with that obtained by reconstructing the precipitation anomaly with the two leading spatial patterns of the MV-EOF modes and the predicted ISV_1 and ISV_2 indices. Shown in Fig. 16 is an example for 2001. When the forecast lead times are 2, 3, 4, and 5 pentads, a positive centre of the predicted precipitation anomaly (contours in Fig. 16) is located at 16°-20°N in mid-May, which moves northward till early June. Another positive centre appears at low latitudes in mid-June, which also moves northward till late July. A third positive centre moves northward from 16°N in early August to 30°N in late August. Such processes are quite consistent with the meridional movement of the real-time 30- to 60-day precipitation anomaly (green and red shading in Fig. 16), indicating that the predicted indices are capable of representing the ISV characteristics of the 30- to 60-day precipitation in the EASSM region. When the forecast lead time is 6 pentads, the prediction is poor, yet the positive centres are still predicted.

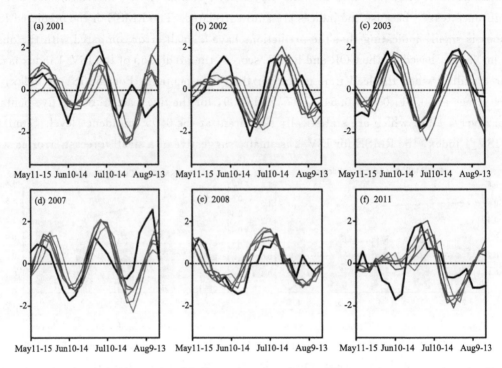

Fig. 14 The observed ISV_1 index (black curve) and prediction (blue, green, red, purple and yellow lines for forecast lead times of two, three, four, five and six pentads, respectively)

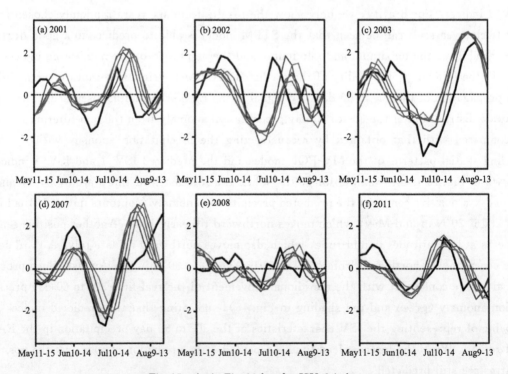

Fig. 15 As in Fig. 13 but for ISV_2 index

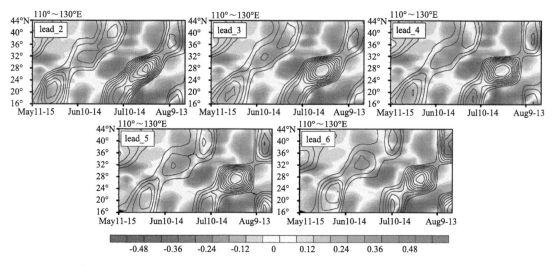

Fig. 16 The meridional distributions of the real-time 30-60-day precipitation (mm) in 2001 from the non-filtering method (shaded) and from the regression of the spatial patterns of the first two modes from MV-EOF onto the predicted ISV indices (contours, only positive precipitation anomalies are displayed, with a contour interval of 0.06 mm)

Because the analysis above is based on results from a single year, 2001, its statistical significance is not obvious. The correlation between the predicted and observed 30- to 60-day precipitation anomaly for the 2001-2012 period is thus calculated for each latitude (Table 1). When the forecast lead times are 2, 3, 4, and 5 pentads, the correlations are significant at the 0.05 confidence level, indicating that predictions for the meridional activity of 30- to 60-day precipitation are relatively skillful. When the forecast lead time is 6 pentads, the prediction is skillful except for the regions 28°~29°N and 41°~44°N. In summary, using the technique proposed in the current study it is possible to skillfully predict, 1-5 pentads ahead of time, the amplitude and meridional propagation of the 30- to 60-day precipitation anomaly in the EASSM region, and this may provide a useful tool for seamless operational predictions.

Table 1 Correlation coefficients between the predicted and the observed 30- to 60-day precipitation for the 2001-2012 period. Values exceeding the 5% (10%) confidence level are in bold (italic) font. The forecast lead times are listed in the first row. The latitude is listed in the first column

Lat(°N)	2	3	4	5	6
16	**0.351**	**0.390**	**0.377**	**0.304**	0.146
17	**0.407**	**0.438**	**0.433**	**0.355**	0.193
18	**0.439**	**0.462**	**0.463**	**0.392**	0.238
19	**0.444**	**0.460**	**0.464**	**0.406**	0.280
20	**0.437**	**0.446**	**0.453**	**0.402**	0.303
21	**0.408**	**0.402**	**0.413**	**0.371**	0.305
22	**0.388**	**0.364**	**0.370**	**0.333**	0.295

Lat(°N)	2	3	4	5	6
23	0.396	0.355	0.349	0.308	0.274
24	0.419	0.366	0.347	0.290	0.226
25	0.443	0.391	0.372	0.311	0.222
26	0.445	0.403	0.385	0.318	0.206
27	0.429	0.392	0.370	0.291	0.155
28	0.417	0.387	0.357	0.260	0.095
29	0.379	0.358	0.331	0.234	0.059
30	0.362	0.337	0.3319	0.252	0.110
31	0.413	0.397	0.385	0.334	0.221
32	0.488	0.497	0.495	0.448	0.353
33	0.574	0.579	0.584	0.542	0.457
34	0.583	0.569	0.572	0.527	0.437
35	0.569	0.526	0.519	0.475	0.398
36	0.430	0.376	0.355	0.296	0.224
37	0.254	0.212	0.187	0.112	0.034
38	0.155	0.123	0.096	0.012	−0.079
39	0.132	0.097	0.075	−0.018	−0.144
40	0.173	0.135	0.121	0.027	−0.126
41	0.199	0.164	0.159	0.074	−0.069
42	0.183	0.155	0.153	0.075	−0.046
43	0.195	0.180	0.181	0.109	0.000
44	0.231	0.225	0.232	0.165	0.049

6 Summary and discussion

The meridional propagation of the 30- to 60-day variability of precipitation in the EASSM region (15°~45°N, 110°~130°E) and its real-time monitoring and prediction were investigated in the current study. The monitoring period ranges from late March to late September every year, and the variables used are precipitation and relative vorticity at 700 hPa. The main conclusions are summarized below:

(i) Based on an MV-EOF analysis of the zonally averaged precipitation and 700 hPa relative vorticity anomalies, a bivariate index (EASSM-ISV index) was developed using the principal components (PCs) of the two leading modes. The objective is to monitor the meridional propagation of the 30- to 60-day precipitation variability in the EASSM region. The index in the phase space of PC1 and PC2 corresponds to the latitudinal location of the 30- to 60-day precipitation anomaly at a specific time, while the distance from the base point indicates the intensity of the precipitation anomaly.

(ii) From the evolution of precipitation and circulation anomalies at the eight phases of the index, the meridional propagation of the 30- to 60-day precipitation and circulation anomalies in the EASSM region is evident, which is closely related to the intraseasonal movement of the rain belt.

(iii) On the basis of the SVD analysis, an extended-range forecast model was built in which the EASSM-ISV index from the preceding five pentads is used to predict the index of the future five pentads. Meanwhile, the EASSM-ISV index was also used to predict the meridional propagation of the 30- to 60-day precipitation variability in the EASSM region. The forecast was found to be skillful. This provides a useful tool for intraseasonal prediction of precipitation during the rainy season in China.

The statistical STPM extended-range forecast model is based on the autocorrelation and lagged correlation of the EASSM-ISV index. It is known that many factors influence the meridional propagation of the 30- to 60-day precipitation in the EASSM region. To further improve the forecast, other factors should also be considered. We plan to incorporate various circulation fields as predictors in our future work.

Acknowledgements. This study was supported by the National Basic Research Program of China (Grant 2013CB430202), the Special Research Program for Public Welfare (Meteorology) of China (Grant GYHY201406018 and Grant GYHY201306030), the program for Changjiang Scholars and Innovative Research Teams in University (PCSIRT), and a project funded by the Priority Academic Program Development of Jiangsu Higher Education Institutions (PAPD). The authors would like to thank two anonymous reviewers whose comments and suggestions helped to improve the paper.

References

[1] Zhu Q, He J. On features of the upper circulation in the establishment of Asian Monsoon in 1979 and its medium-range oscillation [J]. Journal of Tropical Meteorology,1985,2(1):9-18.

[2] Zhu Q, Yang S. The northward advance and oscillation of the East-Asian Summer Monsoon [J]. Journal of Nanjing Institute of Meteorology,1989,12(3):249-258.

[3] Murakami T, Chen L, Xie A. Relationship among seasonal cycle, low-frequency oscillation and transient disturbances as revealed from outgoing long wave radiation data [J]. Mon Wea Rev, 1989, 114: 1456-1465.

[4] Murakami T, Nakazawa T, He J. On the 40-50 day oscillation during the 1979 northern hemisphere summer, Part I: phase propagation [J]. J Meteorol Soc Jpn,1984,62:440-468.

[5] Yasunari T. A quasi-stationary appearance of 30-40 day period in the cloudiness fluctuations during the summer monsoon over India [J]. J Meteoro Soc Jpn,1985,65:67-102.

[6] He J, Yang S. Meridional propagation of East-Asian low frequency oscillation and mid-latitude low-frequency waves [J]. Acta Meteorologica Sinica,1990,4:51-59.

[7] Wang B, Ding Y. An overview of the Madden-Julian Oscillation and its relation to monsoon and mid-lati-

tude circulation [J]. Adv Atmos Sci,1992,9:93-111.

[8] Chen L,Zhu C,Wang W,Zhang P. Analysis of the characteristics of 30-60 day low-frequency oscillation over Asia during 1998 SCSMEX [J]. Adv Atmos Sci,2001,8(4):623-638.

[9] He J,Yu X. A preliminary analysis of the periodic oscillation in water vapour transportation crossing different latitudes over the east of China [J]. Journal of Tropical Meteorology,1986,2(1):9-15.

[10] Wheeler M,Hendon H H. An all-season real time multivariate MJO index:Development of an index for monitoring and prediction [J]. Mon Wear Rev,2004,132:1917-1932.

[11] Lin H. Monitoring and predicting the intraseasonal variability of the East Asian-Western North Pacific summer monsoon [J]. Mon Wea Rev,2013,141:1124-1138.

[12] Lee J-Y,Wang B,Wheeler M C,Fu X,Waliser D E,Kang I-S. Real-time multivariate indices for the boreal summer intraseasonal oscillation over the Asian summer monsoon region [J]. Clim Dynam,2013,40: 493-509.

[13] Kalnay E,Kanamitsu M,Kistler R,Collins W,Deaven D,et al.. The NCEP/NCAR 40-year reanalysis project [J]. B Am Meteorol Soc,1996,77(3):437-471.

[14] Berrisford P,Dee D,Fielding K,et al. The ERA-Interim archive. ECMWF,ERA Report Series 1,2009.

[15] Wang B,Wang Y. Development of El Niño during 1971-1992 [J]. Transactions of Oceanology and Limnology,1994,2(2):26-38.

[16] Hsu P,Li T,Yu L,You L,Gao J,Ren H. A spatial-temporal projection model for 10-30-day rainfall forecast in South China [J]. Clim Dynam,2014,44:1227-1244.

[17] Zhu Z,Li T,Hsu P,He J. A spatial-temporal projection model for extended-range forecast in the tropics [J]. Clim Dynam,2014. doi:10.1007/ s00382-014-2353-8.

[18] Lau K M,Yang S. Climatology and inter-annual variability of the Southeast Asian summer monsoon [J]. Adv Atmos Sci,1997,14:141-162.

[19] Webster P J,Magana V O,Palmer T N,et al.. Monsoon:Processes,predictability,and the prospects for prediction [J]. J Geophys Res,1998,103:14451-14510.

[20] Wang B,Lin H. Rainy season of the Asian-Pacific summer monsoon [J]. J Climate,2002,15:386-398.

[21] Chen G T J. Research on the phenomena of Meiyu during the past quarter century:An overview [M]. In C. P. Chang (Ed.),East Asian Monsoon (p. 564). Singapore:World Scientific,2004.

[22] Wang Z. Climate variability of summer rainfalls in China and the possible mechanism [J] (in Chinese). Beijing:Chinese Academy of Meteorological Sciences,2007.

[23] Qi L,He J,Zhang Z,Song J. Seasonal cycle of the zonal land- sea thermal contrast and East Asian subtropical monsoon circulation [J]. Chinese Sci Bull,2007,52(24):2896-2899.

[24] He J,Zhao P,Zhu C,Zhang R,Tang X,Chen L,Zhou X. Discussions on the East Asian subtropical monsoon [J]. Acta Meteorologica Sinica,2008,66(5):683-696.

[25] He J,Qi L,Liu D,Zhu Z. Transition of East Asian general circulation from winter to summer and its features [J]. Scientia Meteorologica Sinica,2010,30(5),591-596.

[26] Zhu Z,He J. Seasonal transition of East Asian Subtropical Monsoon and its possible mechanism [J]. Journal of Tropical Meteorology,2013,29(2):245-254.

[27] Ding Y,Li C,Liu Y. Overview of the South China Sea Monsoon Experiment (SCSMEX) [J]. Adv Atmos Sci,2004,21:1-18.

[28] Lin H. Global extratropical response to diabatic heating variability of the Asian summer monsoon [J]. J Atmos Sci,2009,66:2693-2713.

[29] Duchon C E. Lanczos filtering in one and two dimensions [J]. Journal of Applied Meteorlogy,1979,18

(8):1016-1022.

[30] North G R,Bell T L,Calahan R F. Sampling errors in the estimation of empirical orthogonal functions [J]. Mon Wea Rev,1982,110:699-706.

[31] Lin H,Brunet G,Derome J. Forecast skill of the Madden-Julian Oscillation in two Canadian atmospheric models [J]. Mon Wea Rev,2008,136:4130-4149.

[32] Quenouille M H. Associated measurements. Academic Press Inc. New York. 1952.

附录1
何金海教授主要著(译)作和论文

一、科学著作(1991—2014年)

陈隆勋,朱乾根,罗会邦,何金海,董敏,冯志强. 东亚季风. 北京:气象出版社,1991.

何金海,丁一汇,陈隆勋. 亚洲季风研究的新进展. 北京:气象出版社,1996.

何金海,丁一汇,高辉,徐海明. 南海夏季风建立日期的确定与季风指数. 北京:气象出版社,2001.

吴国雄,丑纪范,刘屹岷,何金海,等. 副热带高压形成和变异的动力学问题. 北京:科学出版社,2002.

约翰. M. 华莱士,彼得. V. 霍布斯. 大气科学(第二版). 何金海,王振会,银燕,朱彬,等,译. 北京:科学出版社,2008.

Fu Congbin, Jiang Zhihong, Guan Zhaoyong, He Jinhai, et al. Regional Climate Studies of China. Springer, 2008.

何金海,祁莉,张韧,等. 西太平洋副热带高压研究的新进展及其应用. 北京:气象出版社,2010.

何金海,徐海明,钟珊珊,等. 青藏高原大气热源特征及其影响和可能机制. 北京:气象出版社,2011.

何金海,郭品文,银燕,申双和. 大气科学概论. 北京:气象出版社,2012.

孙国武,何金海,李震坤,信飞,梁萍,李耀辉,冯建英. 低频天气图延伸期过程预报方法. 中国环境出版社,2014.

二、主要科学论文(1982—2019年)

何金海. 100毫巴准常定扰动对角动量输送的长期变化的初步分析. 南京气象学院学报,1982(1):63-72.

何金海,T. 村上多喜雄. 1979年6月东亚和南亚上空的水汽通量. 南京气象学院学报,1983(2):158-172.

Zhu Qiangen, He Jinhai, Wang Panxing. A study of circulation differences between East-Asian and Indian summer monsoons with their interaction. Adv Atmos Sci, 1986, 3(4): 466-477.

Takio Murakami, Tetsuo Nakazawa, Jinhai He. On the 40-50 day oscillations during the 1979 Northern Hemisphere Summer Part Ⅰ: phase propagation. J Meteorol Soc Jpn, 1984, 62(3): 440-468.

Takio Murakami, Tetsuo Nakazawa, Jinhai He. On the 40-50 day Oscillations during the 1979

Northern Hemisphere Summer Part Ⅱ: Heat and Moisture Budget. J Meteorol Soc Jpn, 1984, 62(3): 469-484.

何金海, Murakami T, Nakazawa T. 1979 年夏季 40—50 天周期振荡的空间结构及其位相传播. 南京气象学院学报, 1985(1): 56-66.

何金海. 热带环流和季风研究动向分析——夏威夷国际会议一瞥. 南京气象学院学报, 1986(1): 97-102.

何金海, 于新文. 1979 年夏季我国东部各纬带水汽输送周期振荡的初步分析. 热带气象, 1986, 2(1): 9-16.

He Jinhai, Murakami T, Nakazawa T. Energy balance in 40-50 day periodic oscillation over the Asian summer monsoon region during the 1979 summer. Adv Atmos Sci, 1987, 4(1): 66-73.

何金海, 王盘兴. 准 40 天振荡的基本特征及其研究进展. 气象科学, 1988(3): 54.

何金海. 热带气象研究的进展和动向——1988 年澳大利亚热带气象国际会议成果简介. 南京气象学院学报, 1988, 11(4): 493-497.

何金海. 亚洲季风区纬圈剖面内准 40 天周期振荡的环流结构及其演变. 热带气象, 1988, 4(2): 116-125.

He Jinhai. The transfer of physical quantities in QDPO and its relation to the interaction between the NH and SH circulations. Adv Atmos Sci, 1988, 5(1): 97-106.

He Jinhai, Chen Lizhen. The Southern Hemisphere mid-latitude quasi-40-day periodic oscillation with its effect on the Northern Hemisphere summer monsoon circulation. Acta Meteorologica Sinica, 1988, 2(3): 331-339.

He Jinhai. Discussion of meridional propagation mechanism of quasi-40-day oscillation. Adv Atmos Sci, 1990, 7(1): 78-86.

He Jinhai, Li Jun, Li Yongping. Numerical experiment with processes for effect of Australian cold air activity on East-Asian summer monsoon. Acta Meteorologica Sinica, 1990, 4(1): 51-59.

He Jinhai, Yang Song. Meridional propagation of East Asian low-frequency mode and midlatitude low-frequency waves. Acta Meteorologica Sinica, 1990, 4(5): 536-544.

Xu Xiangde, He Jinhai, Zhu Qiangen. A dynamical analysis of basic factors of low-frequency oscillation in the tropical atmosphere. Acta Meteorologica Sinica, 1990, 4(2): 157-167.

何金海, 王坚红, 苗春生. 冬季亚澳季风环流的低频耦合过程. 热带气象学报, 1993, 9(3): 193-201.

何金海, 徐建军, 张永新, 王英宇. 北半球低频扰动能量结构特征及其与西风急流的关系. 热带气象学报, 1993, 9(1): 28-36.

He Jinhai, Xu Jianjun, Zhang Yongxin. Asia-Pacific winter LFO structure and low frequency vortex activity features. Acta Meteorologica Sinica, 1994(4): 431-439.

王坚红, 何金海, 苗春生. 冬季 SSW 低频特性及平流层低层和对流层中层环流异常. 南京气象学院学报, 1994, 17(1): 38-43.

姚永红, 何金海, 徐海明. 夏季南海—菲律宾上空低频热源对东亚环流影响的数值试验. 南京气象学院学报,1994,17(4):411-417.

何金海,智协飞.中国东部地区降水季内变化的季节锁相.热带气象学报,1995,11(4):370-374.

He Jinhai, Zhou Xueming, Ye Rongsheng. Numerical study of ural blocking high's effect upon Asian summer monsoon circulation and East China flood and drought. Adv Atmos Sci, 1995,12(3):361-370.

何金海,朱乾根,Murakami M. TBB资料揭示的亚澳季风区季节转换及亚洲夏季风建立的特征.热带气象学报,1996,12(1):34-42.

智协飞,何金海.北半球中高纬度大气低频变化的若干基本特征.南京气象学院学报,1996,19(1):76-82.

何金海,王盘兴.太平洋地区多时间尺度海气相互作用和ENSO形成关系研究.地球科学进展,1997(5):98-99.

罗京佳,何金海.亚洲夏季风建立格局和南海季风爆发特征及其成因初探.南京气象学院学报,1997(3):26-33.

周黎明,何金海,朱永褆.调整模式大气环境场的一种可行性方案及其应用.南京气象学院学报,1997,20(4):460-467.

谭言科,何金海.准四年周期的北半球夏季遥相关与东亚夏季风环流的相互关系.南京气象学院学报,1998(1):70-79.

谭言科,何金海,祝从文.欧亚冬季雪盖对北半球夏季大气环流的影响及其与东亚太平洋型遥相关的可能联系.大气科学,1999,23(2):152-160.

王劲松,丁治英,何金海,陈久康.用Zwack-Okossi方程对一次爆发性气旋的诊断分析.南京气象学院学报,1999,22(2):180-188.

Zhang Fengqi, He Jinhai, Zhang Aihua, Cheng YanJie. Air-sea coupling features from the warm phase to the cold phase of ENSO cycles in the seventies to the eighties. Acta Meteorologica Sinica,1999,13(2):164-174.

何金海,徐海明,周兵,王黎娟.关于南海夏季风建立的大尺度特征及其机制的讨论.气候与环境研究,2000,5(4):333-344.

He Jinhai, Wang Lijuan, Xu Haiming. Abrupt change in elements around 1998 SCS summer monsoon establishment with analysis of its onset. Acta Meteorologica Sinica,2000,14(4):426-432.

He Jinhai, Wang Panxing, Zhang Suping. Peculiarities of long term variation of sea surface temperature in tropical western pacific ocean. Journal of Tropical Meteorology,2000,6(2):172-178.

李峰,何金海.北太平洋海温异常与东亚夏季风相互作用的年代际变化.热带气象学报,2000,16(3):260-271.

林建,何金海.海温分布型对长江中下游旱涝的影响.应用气象学报,2000,11(3):339-347.

孙秀荣,何金海,陈隆勋.东亚海陆热力差指数与中国夏季降水的关系.南京气象学院学报,2000,23(3):378-384.

王黎娟,何金海,徐海明.黑潮地区海温影响南海夏季风爆发日期的数值试验.南京气象学院学报,2000,(23)2:211-217.

温敏,何金海.夏季季风降水凝结潜热释放效应对西太平洋副高形成和变异的影响.南京气象学院学报,2000,23(4):536-541.

祝从文,何金海,吴国雄.东亚季风指数及其与大尺度热力环流年际变化关系.气象学报,2000,58(4):391-402.

高辉,何金海,谭言科,柳俊杰.40a南海夏季风建立日期的确定.南京气象学院学报,2001,24(3):379-383.

He Jinhai, Zhou Bing, Wen Min, Li Feng. Vertical circulation structure, interannual variation features and variation mechanism of western Pacific subtropical high. Adv Atmos Sci, 2001,18(4):497-510.

徐海明,何金海,董敏.江淮入梅的年际变化及其与北大西洋涛动和海温异常的联系.气象学报,2001,59(6):694-706.

Xu Haiming, He Jinhai, Dong Min. Effect of the Indian peninsula on the course of the Asian tropical summer monsoon. Acta Meteorologica Sinica, 2001,15(3):310-322.

巢纪平,陈鲜艳,何金海.风应力对热带斜压海洋的强迫.大气科学,2002,26(5):577-594.

何金海,温敏,施晓晖,赵巧华.南海夏季风建立期间副高带断裂和东撤及其可能机制.南京大学学报(自然科学),2002,38(3):318-330.

温敏,何金海.夏季西太平洋副高脊线的活动特征及其可能的机制.南京气象学院学报,2002,25(3):289-297.

徐海明,何金海,温敏,董敏.中南半岛影响南海夏季风建立和维持的数值研究.大气科学,2002,26(3):330-342.

张秀丽,郑祚芳,何金海.近百年武汉市主汛期降水特征分析.气象科学,2002,22(4):379-386.

何春,何金海.冬季北极涛动和华北冬季气温变化关系研究.南京气象学院学报,2003,26(1):1-7.

李晓莉,何金海,毕宝贵,李泽椿,王欣.MM5模式中城市冠层参数化方案的设计及其数值试验.气象学报,2003,61(5):526-539.

柳俊杰,丁一汇,何金海.一次典型梅雨锋锋面结构分析.气象学报,2003,31(3):291-302.

谭言科,张人禾,何金海.热带印度洋海温的年际异常及其海气耦合特征.大气科学,2003,27(1):53-66.

谢付莹,何金海.华北夏季降水与哈得孙湾海冰的相关分析.南京气象学院学报,2003,26(3):308-316.

杨建玲,何金海,赵光平.宁夏春季沙尘暴与北极海冰之间的遥相关关系.南京气象学院学报,2003,26(3):296-307.

赵巧华,何金海.基于资源卫星图像对NDVI进行大气修正的一种简单方法.南京气象学院学报,2003,26(2):236-242.

周兵,何金海,吴国雄,韩桂荣.东亚副热带季风特征及其指数的建立.大气科学,2003,27(1):123-135.

丁一汇,李崇银,何金海,陈隆勋,甘子钧,钱永甫,阎俊岳,王东晓,施平,方文东,许建平,李立.南海季风试验与东亚夏季风.气象学报,2004,62(5):561-586.

韩荣青,李维京,何金海,胡国权.水平涡度与夏季风环流变化.南京气象学院学报,2004,27(3):381-390.

何金海,宇婧婧,沈新勇,高辉.有关东亚季风的形成及其变率的研究.热带气象学报,2004,20(5):449-459.

蒋国荣,何金海,王东晓,阎俊岳,姚华栋.南海夏季风期间海-气通量整体输送系数分析.气象学报,2004,62(3):338-345.

李崇银,何金海,朱锦红. A review of decadal/interdecadal climate variation studies in China. Adv Atmos Sci,2004,21(3):425-436.

李向红,徐海明,何金海.对亚洲两支越赤道气流与华南暴雨的关系探讨.大气科学,2004,24(2):161-167.

王宝鉴,黄玉霞,何金海,王黎娟.东亚夏季风期间水汽输送与西北干旱的关系.高原气象,2004,23(6):912-918.

王遵娅,丁一汇,何金海,虞俊.近50年来中国气候变化特征的再分析.气象学报,2004,62(2):228-236.

温敏,何金海,肖子牛.中南半岛对流对南海夏季风建立过程的影响.大气科学,2004,28(6):864-875.

占瑞芬,李建平,何金海.西太平洋副热带高压双脊线及其对1998年夏季长江流域"二度梅"的影响.气象学报,2004,62(3):294-307.

周波涛,何金海,陈隆勋,赵平.GISS海气耦合模式的基本气候态评估.应用气象学报,2004,15(4):500-505.

祝从文,何金海,谭言科.春夏季节转换中亚洲季风区副热带高压断裂特征及其可能机制分析.热带气象学报,2004,20(3):237-248.

池艳珍,何金海,吴志伟.华南前汛期不同降水时段的特征分析.南京气象学院学报,2005,28(2):163-171.

何金海,刘芸芸,常越.西北地区夏季降水异常及其水汽输送和环流特征分析.干旱气象,2005,23(1):10-16.

黄昌兴,李维京,何金海.正、斜压涡度拟能相互作用对乌拉尔阻塞过程的影响.高原气象,2005,24(2):178-186.

刘毅,何金海,王黎娟.近40a重庆地区夏季降水的气候特征.气象科学,2005,25(5):490-498.

罗玲,何金海,谭言科.西太平洋副热带高压西伸过程的合成特征及其可能机理.气象科学,2005,25(5):465-472.

占瑞芬,李建平,何金海.北半球副热带高压双脊线的统计特征.科学通报,2005,50(18):2022-2026.

周长艳,何金海,李薇,陈隆勋.夏季东亚地区水汽输送的气候特征.南京气象学院学报,2005,28(1):18-27.

常越,何金海,刘芸芸,梁萍.华南旱、涝年前汛期水汽输送特征的对比分析.2006,25(6):1064-1070.

巩远发,何金海,段廷扬,潘攀.北太平洋中纬度负海温异常对副热带高压影响的数值试验.热带气象学报,2006,22(4):386-392.

Gong Yuanfa, He Jinhai, Chen Longxun. Relationship between the eastern Tibetan Plateau rainfall and subtropical high shift in summer. Acta Meteorologica Sinica, 2006, 20(4): 437-446.

何金海, 吴志伟, 江志红, 苗春生, 韩桂荣. 东北冷涡的"气候效应"及其对梅雨的影响. 科学通报, 2006, 51(23): 2803-2809.

何金海, 吴志伟, 祁莉, 姜爱军. 北半球环状模和东北冷涡与我国东北夏季降水关系分析. 气象与环境学报, 2006, 22(1): 1-5.

He Jinhai, Wen Min, Ding Yihui, Zhang Renhe. Possible mechanism of the effect of convection over Asian-Australian "land bridge" on the East Asian summer monsoon onset. Sci China Ser D, 2006, 49(11): 1223-1232.

He Jinhai, Wen Min, Wang Lijuan, Xu Haiming. Characteristics of the onset of the Asian Summer Monsoon and the importance of Asian-Australian "land bridge". Adv Atmos Sci, 2006, 23(6): 951-963.

Jin Qihua, He Jinhai, Chen Longxun. Impact of ocean-continent distribution over southern Asia on the formation of summer monsoon. Acta Meteoro Sinica, 2006, 20(1): 95-108.

李飞, 胡鹏, 何金海. 南海夏季风建立特征及模拟. 台湾海峡, 2006, 25(2): 160-165.

李勇, 陆日宇, 何金海. 太平洋西部遥相关型与赤道中东太平洋海温的关联性和独立性. 自然科学进展, 2006, 16(8): 1051-1055.

刘飞, 何金海, 姜爱军. 亚洲夏季西风指数与中国夏季降水的关系. 南京气象学院学报, 2006, 29(4): 517-525.

刘芸芸, 俞永强, 何金海, 张振国. 全球变暖背景下热带大气季节内振荡的变化特征及数值模拟. 气象学报, 2006, 64(6): 723-733.

祁莉, 何金海, 占瑞芬, 张祖强. 1962年西太平洋副热带高压双脊线演变过程的特征分析. 大气科学, 2006, 30(4): 682-692.

王黎娟, 何金海, 管兆勇. 东亚副热带夏季风槽的气候特征及其与南海夏季风槽的比较. 气象学报, 2006, 64(5): 583-593.

吴伟杰, 何金海, Chung Hyo-Sang, Cho Chun-Ho, 陆日宇. 夏季东亚高空急流与天气尺度波动的气候特征之间的联系. 气候与环境研究, 2006, 11(4): 525-534.

吴志伟, 何金海, 韩桂荣, 刘芸芸. 长江中下游梅雨与春季南半球年际模态(SAM)的关系分析. 热带气象学报, 2006, 22(1): 79-85.

吴志伟, 何金海, 李建平, 江志红. 长江中下游夏季旱涝并存及其异常年海气特征分析. 大气科学, 2006, 30(4): 570-577.

张立峰, 许建平, 何金海. 西太平洋暖池研究的新进展. 海洋科学进展, 2006, 24(1): 108-116.

邹燕, 何金海, 邱章如, 杨月文. 副高及其南侧偏东气流输送在福建前汛期降水中的作用. 热带气象学报, 2006, 22(5): 515-520.

陈桦, 丁一汇, 何金海. 夏季热带东风急流的结构、变化及其与亚非季风降水的关系. 大气科学, 2007, 31(5): 926-936.

高媛媛, 何金海, 王自发. 城市化进程对北京区域气象场的影响模拟. 气象与环境学报, 2007, 23(3): 58-64.

何金海,祁莉,韦晋,池艳珍. 关于东亚副热带季风和热带季风的再认识. 大气科学,2007,31(6):1257-1265.

He Jinhai, Ju Jianhua, Wen Zhiping, Lu Junmei, Jin Qihua. A review of recent advances in research on Asian monsoon in China. Adv Atmos Sci,2007,24(6):972-992.

He Jinhai, Sun Chenghu, Liu Yunyun, Jun Matsumoto, Li Weijing. Seasonal transition features of large-scale moisture transport in the Asian-Australian monsoon region. Adv Atmos Sci,2007,24(1):1-14.

He Jinhai, Yu Jingjing, Shen Xinyong. Impacts of SST and SST anomalies on low-frequency oscillation in the tropical atmosphere. Adv Atmos Sci,2007,24(3):377-382.

胡亮,何金海,高守亭. 华南持续性暴雨的大尺度降水条件分析. 南京气象学院学报,2007,30(3):345-351.

蒋国荣,俞永强,何金海. 季节内振荡的数值模拟 I. 模拟的自然变率. 大气科学,2007,31(3):536-546.

李勇,何金海,姜爱军,周兵. 冬季西太平洋遥相关型的环流结构特征及其与我国冬季气温和降水的关系. 气象科学,2007,27(2):119-125.

梁萍,何金海,陈隆勋,李薇. 华北夏季强降水的水汽来源. 高原气象,2007,26(3):460-465.

王鹏祥,何金海,郑有飞,张强. 夏季北极涛动与西北夏季干湿特征的年代际关系. 中国沙漠,2007,27(5):883-889.

徐海明,何金海,谢尚平. 卫星资料揭示的中尺度地形对南海夏季气候的影响. 大气科学,2007,31(5):1021-1031.

朱玮,刘芸芸,何金海. 我国江淮地区平均场水汽输送与扰动场水汽输送的不同特征. 气象科学,2007,27(2):155-161.

竺夏英,何金海,吴志伟. 江淮梅雨期降水经向非均匀分布及异常年特征分析. 科学通报,2007,52(8):951-957.

何金海,胡亮,杨松. 大气中的波流相互作用研究进展. 热带气象学报,2008,24(1):1-10.

He Jinhai, Zhao Ping, Zhu Congwen, Zhang Renhe, Tang Xu, Chen Longxun, Zhou Xiuji. Discussion of some problems as to the East Asian subtropical monsoon. Acta Meteorologica Sinica,2008,(22)4:419-434.

何溪澄,丁一汇,何金海. 东亚冬季风对 ENSO 事件的响应特征. 大气科学,2008,32(2):335-344.

梁萍,何金海. 江淮梅雨气候变化研究进展. 高原气象,2008,27(增刊):8-15.

Qi Li, He Jinhai, Zhang Zuqiang, Song Jinnuan. Seasonal cycle of the zonal land-sea thermal contrast and East Asian subtropical monsoon circulation. Chinese Sci Bull,2008,53(1):131-136.

韦晋,何金海,钟珊珊,温敏. 高原南侧地形槽与孟湾槽的形成演变特征及其与南海夏季风建立的关系. 高原气象,2008,27(4):764-771.

刘伯奇,何金海,王黎娟. 4—5月南亚高压在中南半岛上空建立过程特征及其可能机制. 大气科学,2009,33(6):1319-1332.

刘煜,何金海,李维亮,陈隆勋,李薇,张博. MM5 对中全新世时期中国地区气候的模拟研究.

气象学报,2009,67(1):35-49.

汪靖,何金海,刘宣飞,吴彬贵.江淮梅雨建立的年际变化及其前期强影响信号分析.科学通报,2009,54(1):85-92.

向纯怡,何金海,任荣彩.2007/2008年冬季平流层环流异常及平流层—对流层耦合特征.地球科学进展,2009,24(3):338-348.

钟珊珊,何金海,管兆勇,温敏.1961—2001年青藏高原大气热源的气候特征.气象学报,2009,67(3):407-416.

Liu Yu, Li Weiliang, He Jinhai, Chen Longxun. Simulations of hydrological cycle changes between the LGM and the present day over China. Acta Meteorologica Sinica, 2010, 24(5):641-656.

Qi Li, He Jinhai, Zhang Zuqiang, Guan Zhaoyong. West Pacific subtropical high double ridges and intraseasonal variability of the South China Sea summer monsoon. Theor Appl Climatol, 2010, 100:385-396.

申乐琳,何金海,周秀骥,陈隆勋,祝从文.近50年来中国夏季降水及水汽输送特征研究.气象学报,2010,68(6):918-931.

王丽娟,何金海,司东,温敏,钟珊珊.东北冷涡过程对江淮梅雨期降水影响机制的分析.大气科学学报,2010,1:89-97.

许建明,何金海,阎凤霞.1961—2007年西北地区地面太阳辐射长期变化特征研究.气候与环境研究,2010,15(1):89-96.

高郁东,万齐林,何金海.三维变分同化雷达视风速的改进方案及其数值试验.气象学报,2011,69(4):631-643.

He Jinhai, Lin Hai, Wu Zhiwei. Another look at influences of the Madden-Julian Oscillation on the wintertime East Asian weather. J Geophys Res, 2011, 116, D03109, doi:10.1029/2010JD014787.

金小霞,何金海,占瑞芬,万齐林.大气季节内振荡对热带气旋活动影响的研究进展.热带气象学报,2011,27(1):133-138.

刘丹妮,何金海,姚永红,祁莉.江淮流域梅雨环流结构特征及其演变分析.热带气象学报,2011,27(4):465-474.

舒斯,何金海,刘毅,王永,蔡兆男.夏季青藏高原O_3低值与南亚高压东西振荡的关系.气候与环境研究,2011,16(1):39-46.

Wu Jin, Zhang Zuqiang, He Jinhai, et al. The analysis of the relationship between the subsystems of the Asian summer monsoon. Journal of Tropical Meteorology, 2014, 20(4):342-348.

谢潇,祁莉,何金海.1980—2009年中国东部上空温度变化特征.气候变化研究进展,2013,9(2):102-109.

Xu Kang, Zhu Congwen, He Jinhai. Two types of El Nino-related Southern Oscillation and their different impacts on global land precipitation. Adv Atmos Sci, 2013, 30:1743-1757.

杨玮,何金海,孙国武,孔春燕.低频环流系统的一种统计预报方法.气象与环境学报,2011,27(3):1-5.

杨晓君,何金海,吕江津,王颖,何群英.海面风场订正对风暴潮数值模拟的影响.气象,2011,37(3):270-275.

Zhang Y,Sun S,Olsen S C,Dubey M K,Dean S H,He J. CCSM3 simulated regional effects of anthropogenic aerosols for two contrasting scenarios:Rising Asian emissions and global reduction of aerosols. Int J Climatol,2011,31(1):95-114.

卞洁,何金海,李双林.近50年来长江中下游汛期暴雨变化特征.气候与环境研究,2012,17(1):68-80.

郭玲,何金海,祝从文.影响长江中下游夏季降水的前期潜在预报因子评估.大气科学,2012,36(2):337-349.

黄露,何金海,卢楚翰.关于西太平洋副热带高压研究的回顾与展望.干旱气象,2012,30(2):255-260.

刘慧斌,温敏,何金海,张人禾.东北冷涡活动的季节内振荡特征及其影响.大气科学,2012,36(5):959-973.

陶梦初,何金海,刘毅.平流层准零风层统计特征及准两年周期振荡对其影响分析.气候与环境研究,2012,17(1):92-102.

万齐林,何金海.海洋气象观测系统在热带气旋资料同化中的应用.中国工程科学,2012,14(10):33-42.

张高杰,何金海,周自江,曹丽娟.RHtest方法对我国降水资料的均一性检验试验.气象科技,2012,40(6):914-921.

Zhu Zhiwei,He Jinhai,Qi Li. Seasonal transition of East Asian subtropical monsoon and its possible mechanism. Journal of Tropical Meteorology,2012,18(3):305-313.

常炉予,何金海,祁莉,温敏.东亚与北美东部降水和环流季节演变差异及其可能机理分析.气象学报,2013,71(6):1074-1088.

陈圣劼,何金海,吴志伟.一种新的El Nino海气耦合指数.大气科学,2013,37(4):815-828.

程叙耕,何金海,车慧正,赵天良,郑小波.1980—2010年中国区域地面风速对能见度影响的地理分布特征.中国沙漠,2013,33(6):1832-1839.

何金海,梁萍,孙国武.延伸期预报的思考及其应用研究进展.气象科技进展:英文版,2013(1):11-17.

梁萍,何金海,穆海振.MJO在延伸期预报中的应用进展.气象科技进展:英文版,2013(1):31-38.

Liu B Q,Wu G X,Mao J Y,He J H. Genesis of the South Asian High and Its Impact on the Asian Summer Monsoon Onset. J Climate,2013,26(9):2976-2991.

尚可,何金海,朱志伟,詹丰兴.西太平洋暖池区热含量和海表温度与江南春雨的相关性对比研究.地理科学,2013,33(8):986-992.

王晓芳,何金海,廉毅.前期西太平洋暖池热含量异常对中国东北地区夏季降水的影响.气象学报,2013,71(2):305-317.

许姿,何金海,高守亭,林青.集合动力因子对登陆台风"莫拉克"(0908)暴雨落区的诊断与预报研究.大气科学,2013,37(1):23-35.

袁良,何金海.两类ENSO对我国华南地区冬季降水的不同影响.干旱气象,2013,31(1):

24-31.

朱志伟,何金海,钟珊珊,尚可.春夏东亚大气环流年代际转折的影响及其可能机理.气象学报,2013,71(3):440-451.

朱志伟,何金海.孟加拉湾低涡与南海季风爆发关系及其可能机理.热带气象学报,2013,29(6):915-923.

何金海,刘伯奇,吴国雄.春末夏初南亚高压的形成过程及其与ENSO事件的联系.大气科学,2014,38(4):670-684.

李文铠,何金海,祁莉,陈伯民.MJO对华南前汛期降水的影响及其可能机制.热带气象学报,2014,30(5):983-989.

Peng Yuping, He Jinhai, Chen Longxun, Zhang Bo. A study on the characteristics and effect of the low-frequency oscillation of the atmospheric heat source over the eastern Tibetan Plateau. Journal of Tropical Meteorology, 2014, 20(1):17-25.

祁莉,何金海,王玉清.青藏高原-东亚平原-西北太平洋三级阶梯状海陆热力差异及其对东亚气候季节转换的影响.科学通报,2014,59(19):1904.

祁莉,王晓芳,何金海,张文君,吴捷.前期西太平洋暖池热含量异常影响长江中下游夏季降水的可能途经.地球物理学报,2014,57(6):1769-1781.

武丰民,何金海,祁莉,李文铠.海冰消融背景下北极增温的季节差异及其原因探讨.海洋学报,2014,36(3):39-47.

陈立波,何金海,谭癸,王国杰.中国近20a卫星遥感土壤湿度的季节变化及其验证.气象科学,2015,35(6):744-750.

何金海,武丰民,祁莉.秋季北极海冰与欧亚冬季气温在年代际和年际尺度上的不同联系.地球物理学报,2015,58(4):1089-1102.

何金海,袁良,祁莉.冬季西太平洋暖池与华南降水关系的年代际变化.大气科学学报,2015,38(6):721-730.

He Jinhai, Chang Luyu, Chen Hua. Meridional propagation of the 30- to 60-day variability of precipitation in the East Asian subtropical summer monsoon region: monitoring and prediction. Atmos Ocean, 2015, 53(2):251-263.

刘伯奇,何金海.亚洲夏季风动力学研究综述.热带气象学报,2015,31(6):869-880.

刘嘉慧敏,张文君,何金海,祁莉.前期春季西北太平洋潜热通量与中国南方秋季降水的联系及其可能的物理机制.气象学报,2015,73(2):305-318.

罗艳艳,何金海,邹燕,孙蔡亮.华南前汛期雨涝强弱年的确定及其环流特征对比.气象科学,2015,35(2):160-166.

王迪,何金海,祁莉,栾健,蔡波.全球变暖减缓背景下欧亚秋冬温度变化特征和原因.气象科学,2015,35(5):534-542.

魏晓雯,梁萍,何金海,穆海振.汛期强降水过程与月内低频降水的联系及其可能机制.高原气象,2015,34(3):722-730.

詹丰兴,何金海,章毅之,朱志伟.江南雨季地理区域及起止时间的客观确定.海洋学报,2015,37(6):1-11.

Dong Di, He Jinhai, Li Jianping. Linkage between Indian Ocean Dipole and two types of El

Nino and its possible mechanisms. Journal of Tropical Meteorology,2016,22(2):172-181.

何金海,詹丰兴,祁莉,王迪.全球变暖减缓期陆地地表气温变化特征和CMIP5多模式的未来情景预估.大气科学,2016,40(1):33-45.

He Jinhai,Liu Boqi. The east Asian subtropical summer monsoon:recent progress. Journal of Meteorological Research,2016,30(2):135-155.

黄海燕,王亚非,何金海,陈圣劼,秦坚肇.天气尺度波列对长江中下游6月梅雨的影响.大气科学学报,2016,39(1):28-36.

Huang Jiaowen,He Jinhai,Xu Haiming,Jin Qihua. Relationship between the seasonal transition of East Asian monsoon circulation and Asian-Pacific thermal field and possible mechanisms. Journal of Tropical Meteorology,2016,22(4):466-478.

金蕊,祁莉,何金海.春季青藏高原感热通量对不同海区海温强迫的响应及其对我国东部降水的影响.海洋学报,2016,38(5):83-95.

李海燕,张文君,何金海.ENSO及其组合模态对中国东部各季节降水的影响.气象学报,2016,74(3):322-334.

Sullivan A,Luo J-J,Hirst A C,Bi D,Cai W,He J. Robust contribution of decadal anomalies to the frequency of central-Pacific El Nino. Scientific Reports,10.1038/srep38540,6,1,2016.

廖廓,何金海.福建秋冬之交降水强度变化及其环流背景分析.气象科技,2017,45(5):864-869.

Ren Q,Zhu Z W,Hao L P,He J H. The enhanced relationship between Southern China winter rainfall and warm pool ocean heat content. Int J Climatol,2017,37(1):409-419.

任倩,何金海,祁莉,张文君.中国南方冬季降水与前期暖池热含量异常的关系及可能机制.高原气象,2014,33(6):1568-1578.

王静,何金海,祁莉,吴志伟,施晓晖.青藏高原土壤湿度的变化特征及其对中国东部降水影响的研究进展.大气科学学报,2018,41(1):1-11.

于浩慧,祁莉,何金海.非ENSO事件次年大西洋海温异常对夏季青藏高原大气热源准双周低频活跃度的影响.高原气象,2018,37(3):602-613.

Jin Qihua,Liu Boqi,He Jinhai. Possible causes for the asymmetric evolution between the aerosol optical depth over East Asia and eastern United States during boreal spring. Int J Climatol,2019,39(4):2474-2483.

附录 2
何金海教授重要年表

1941年11月,出生在江苏省丹徒县(现镇江市)一个贫苦农民家庭。
1950年9月,进入当地小学学习。
1954年6月,考入当地儒里中学,成为村里第一个初中生。
1957年9月,考入江苏省常州高级中学。
1960年9月,进入南京大学气象学院(1963年独立为南京气象学院)大气物理系学习。
1965年8月,南京气象学院气象学系毕业,并被录取为中央气象局气象科学研究所(现中国气象科学研究院)天气动力学专业研究生,师从叶笃正先生。
1968年8月,中央气象局气象科学研究所研究生毕业,同年8月分配到中央气象局(现中国气象局)工作,做气象员。
1970年4月,再分配至江苏省扬中县气象站,做气象员。
1975年8月,调入南京气象学院工作。
1982年1月,以访问学者身份赴美国夏威夷大学气象系进修访问,在著名气象学家Murakami Takio教授指导下开展合作研究。
1984年1月,从夏威夷大学学成回国。
1985年2月,与朱乾根、王盘兴合作撰写的《A study of circulation differences between East-Asian and Indian summer monsoons with their interactions》论文引起国内外广泛关注。
1986年5月,晋升为南京气象学院副教授。
1989年8月,赴美国宾夕法尼亚大学参加中美季风合作成果交流会并作口头报告。
1990年5月,晋升为南京气象学院教授,任天气动力学系副主任。
1990年11月,获国家气象局气象科学奖二等奖,获奖项目名称为"东亚季风研究"。
1992年4月,任南京气象学院天气动力学系主任。
1992年10月,开始享受国务院颁发的政府特殊津贴。
1994年,以第一完成人身份获中国气象局气象科学奖二等奖,获奖项目名称为"热带准40天振荡的现象、成因及其与热带外环流的遥相关"。
1995年,获国家自然科学奖二等奖,获奖项目名称为"东亚季风研究"(第三完成人)。
1996年,主编(第一作者)出版《亚洲季风研究的新进展》。
1996年12月至1997年1月,与朱乾根教授赴日本气象研究所开展合作研究。
1997年12月,被评为江苏省高校优秀学科带头人。
1998年3月,参加南海季风试验国际合作项目,担任项目专家组成员及第一课题组组长。
1998年4月,荣获第三届江苏省优秀科技工作者称号。
1998年8月,受邀赴德国慕尼黑大学访问。
2000年8月,被评为江苏省师德模范;受聘为国家自然科学基金委员会评委/项目咨询评估专家。

2001年9月,被评为江苏省优秀教育工作者。

2002年10月,受聘为中国气象学会理事。

2004年,获教育部提名国家自然科学奖一等奖,项目名称为"中南半岛地区热力特征对南海季风爆发的影响和机理研究"(第三完成人)。

2005年11月,受聘为国家气候研究委员会委员。

2006年6月,受聘为上海气象中心科学指导委员会委员。

2006年,以第一作者身份在《Advances in Atmospheric Science》上发表论文《Characteristics of the onset of the Asian Summer Monsoon and the importance of Asian-Australian "land bridge"》。

2008年5月,受东亚副热带季风委员会委托,在东亚副热带季风学术研讨会上作题为《东亚副热带季风研究回顾、相关事实及科学问题》的主题报告。

2010年11月,受聘中国科学院大气物理研究所大气科学和地球流体力学数值模拟国家重点实验室(LASG)学术顾问委员会委员。

2010年12月,主持完成《东亚季风多尺度变率与我国旱涝机理研究》,获江苏省科学技术奖一等奖(排名第一)。

2011年2月,以第一作者身份在《Journal of Geophysical Research》上发表《Another look at influences of the Madden-Julian Oscillation on the wintertime East Asian weather》。

2014年7月,受邀前往美国檀香山参加"纪念Murakami Takio热带气象与季风研讨会"并作题为《Meridional propagation of the 30- to 60-day variability of precipitation in the East Asian subtropical summer monsoon region:monitoring and prediction》的口头报告。

2015年5月,受邀参加加拿大第39届气象与海洋学会年会,并作特邀报告。

2017年6月,受邀在吉林"弘扬陶诗言学术精神研讨会"上作题为《东亚季风季节转变与亚太热力场的联系及其可能机理》的报告。

2018年10月,受邀参加台湾大学举办的"两岸季节内振荡与台闽区持续性天气事件延伸期预报研讨会",促进两岸气象交流,分享两岸最新科研成果。

2019年8月,参加北京"天气气候变化动力学及数值模拟国际研讨会"并作题为《拉尼娜衰减的多样性及其对中国东部春夏季降水异常的影响》的报告。